D1170747

Mobile Broadband Multimedia Networks

History has shown that forecasting the future has become a science and perhaps even an art.

In: P. van der Dui and R. Kok, *Mind the gap – linking forecasting with decision making.* Telektronikk, December 2004.

Mobile Broadband Multimedia Networks

Techniques, Models and Tools for 4G

Edited by

Luís M. Correia

AMSTERDAM • BOSTON • HEIDELBERG • LONDON • NEW YORK • OXFORD
PARIS • SAN DIEGO • SAN FRANCISCO • SINGAPORE • SYDNEY • TOKYO
Academic Press is an imprint of Elsevier

Academic Press is an imprint of Elsevier
Linacre House, Jordan Hill, Oxford OX2 8DP
84 Theobald's Road, London WC1X 8RR, UK
30 Corporate Drive, Suite 400, Burlington, MA 01803, USA
525 B Street, Suite 1900, San Diego, California 92101-4495, USA

First edition 2006

British Library Cataloguing in Publication Data
A catalogue record for this book is available from the British Library

Library of Congress Catalog Number: 2006922835

ISBN–13: 978-0-12-369422-5
ISBN–10: 0-12-369422-1

For information on all Academic Press publications
visit our web site at http://books.elsevier.com

Typeset by Cepha Imaging, India
Printed and bound in Great Britain

06 07 08 09 10 10 9 8 7 6 5 4 3 2 1

Contents

Preface

This books presents the research work of COST 273 *Towards Mobile Broadband Multimedia Networks*, hence, it reports on the work performed and on the results achieved within the project by its participants. The material presented here corresponds to the results obtained in four years of collaborative work by more than 350 researchers from 137 institutions (universities, operators, manufacturers, regulators, independent laboratories and others – a full list is provided in Appendix B) belonging to 29 countries (mainly European, but also from Asia and North America) in the area of mobile radio. The objective of publishing these results as a book is essentially to make them available to an audience wider than the project. In fact, it just follows a 'tradition' of previous COST Actions in this area of telecommunications, i.e. COST 207, 231 and 259.

The main objective of COST 273 was to increase the knowledge on the radio aspects of mobile and wireless broadband multimedia networks, by exploring and developing new methods, models, techniques, strategies and tools towards the implementation of 4th generation mobile and wireless communication systems. As a secondary objective, it was intended that it should continue to play a supporting role similar to the one played by previous Actions in this area, that is, beside giving inputs to the development of systems beyond the 3rd generation, it was also expected that it would contribute to the deployment of systems that are more or less standardised, like UMTS and WLANs.

The structure of the book is based on the structure of the project itself, i.e. chapters are roughly mapped onto part of the work produced in a Working Group. There were a total of three Working Groups: WG1 – Radio System Aspects; WG2 – Propagation and Antennas; WG3 – Radio Network Aspects. A brief description of the project is given in Appendix A, so that readers who are not familiar with the COST framework in general, and with COST 273 in particular, can have a glimpse of it.

Chapters and sections were edited by some of the project's participants, being based on the contributions that all participants have presented to the project. These contributions were given in the form of Temporary Documents (TDs), to which many references appear in the text. Since they were internal documents, their distribution cannot be carried out without the authors' permission, which could pose problems for the reader in gaining access to them. To overcome this situation, references to TDs were replaced as much as possible by the corresponding ones in the open literature (e.g. papers in journals and from conferences); nevertheless, any reader interested can have access to the list of TDs and their authors via the project's web site: http://www.lx.it.pt/cost273.

As already mentioned, this book is the result of the work of many colleagues. An acknowledgement is due to all COST 273 participants, and to its management committee, for their effort and contributions; without their continued interest and participation, it would not have been possible to reach the goal of publishing this book. A special thanks is also due to section editors, as well as to all other direct contributors, for volunteering to perform the task of putting together all the information in a uniform and coherent style. A word of recognition is due as well to the colleagues who helped me in the direct management of the project: the first Vice-Chairman, Mr Per Lehne; the current Vice-Chairman, Prof. Narcis Cardona, and the Working Group Chairmen, Prof. Alister Burr, Prof. Ernst Bonek and Prof. Roberto Verdone, who served also as chapter editors, together with Prof. Pierre Degauque; the Sub-Working Group Chairpersons, Prof. Andreas Molisch, Prof. Gert Pedersen, Prof. Pertti Vainikainen and Prof. Silvia Ruiz. And definitely last but not the least, the valuable work of Drs Tricia Willink, Jan Sykora and Tomaz Javornik in establishing the guidelines for the writing of all contributions in LaTeX, and of Lúcio Ferreira and Martijn Kuipers in going through the whole manuscript, taking care of all the LaTeX edition tasks, is duly recognised.

Luís M. Correia
(COST 273 Chairman)
Lisbon

List of contributors and editors

Hamid Aghvami
King's College London, UK

Dragana Bajic
University of Novi Sad, Serbia and Montenegro

Mark Beach
University of Bristol, UK

Jean-Charles Bolomey
University of Trier, Germany

Ernst Bonek
Vienna University of Technology, Austria

Silvia Ruiz Boqué
Polytechnical University of Catalunya, Spain

Robert J. C. Bultitude
Communications Research Centre, Canada

Alister Burr
University of York, UK

Narcis Cardona
Polytechnical University of Valencia, Spain

Filipe Cardoso
Instituto Superior Técnico/Technical University of Lisbon and Escola Superior de Tecnologia/ Polytechnic Institute of Setúbal, Portugal

Batu Chalise
University of Duisburg-Essen, Germany

Laurent Clavier
École Nouvelle de Ingénieurs en Communication, France

Olivier Colas
France Télécom, France

Luís M. Correia
Instituto Superior Técnico/Technical University of Lisbon, Portugal

Andreas Czylwik
University of Duisburg-Essen, Germany

Pierre Degauque
University of Lille, France

Vittorio Degli-Esposti
University of Bologna, Italy

Eduardo Rodrigues De Lima
Polytechnical University of Valencia, Spain

Benoit Derat
Sagem, France

José Diaz
Polytechnical University of Valencia, Spain

Lyubomír Doboš
Technical University in Košice, Slovak Republic

Mischa Dohler
France Télécom R&D, France

Lúcio Studer Ferreira
Instituto Superior Técnico/Technical University of Lisbon, Portugal

Alexander Gerdenitsch
Motorola, Austria

João Gil
Instituto Superior Técnico/Technical University of Lisbon and Escola Superior de Tecnologia e Gestão/Leiria Polytechnical Institute, Portugal

Andrés Alayon Glazunov
TeliaSonera, Sweden

Paolo Grazioso
Fondazione Ugo Bordoni, Italy

Helmut Hofstetter
Forschungszentrum Telekommunikation Wien, Austria

Mythri Hunukumbure
University of Bristol, UK

Clemens Icheln
Technical University of Hamburg-Harburg, Germany

Ichirou Ida
Tokyo Institute of Technology, Japan

Emmanuel Jaffrot
École Nationale Supérieure de Techniques Avancées, France

Tuomas Jääskö
Filtronic, Finland

Stefan Jakl
Vienna Institute of Technology, Austria

Tomaz Javornik
Institut Jozef Stefan, Slovenia

Kimmo Kansanen
University of Oulu, Finland

Outi Kivekäs
Helsinki University of Technology, Finland

Jarmo Kivinen
Helsinki University of Technology, Finland

Wim A. Th. Kotterman
Ilmenau Technical University, Germany

Joonas Krogerus
Helsinki University of Technology, Finland

Christiane Kuhnert
University of Karlsruhe, Germany

Thomas Kürner
Technical University of Braunschweig, Germany

Pekka Kyösti
Elektrobit, Finland

Mario García Lozano
Polytechnical University of Catalunya, Spain

Tadashi Matsumoto
University of Oulu, Finland

Olivier Merckel
Supélec – CNRS, France

Andreas Molisch
Lund Technical University, Sweden

José F. Monserrat
Polytechnical University of Valencia, Spain

Claude Oestges
Université Catholique de Louvain, Belgium

Geir E. Øien
Norwegian University of Science and Technology, Norway

John Orriss
University of Manchester, UK

Jörg Pamp
IMST, Germany

Stephane Pannetrat
Sagem, France

Gianni Pasolini
University of Bologna, Italy

Riccardo Patelli
Wireless Future, Italy

Christian Pietsch
University of Ulm, Germany

Markus Radimirsch
University of Hannover, Germany

Alain Sibille
École Nationale Supérieure de Techniques Avancées, France

Gerhard Steinböck
ARC Seibersdorf research, Austria

Pasi Suvikunnas
Helsinki University of Technology, Finland

Jan Sykora
Czech Technical University of Prague, Czech Republic

Jun-Ichi Takada
Tokyo Institute of Technology, Japan

Werner Teich
University of Ulm, Germany

Martin Toeltsch
SYMENA, Austria

Velio Tralli
University of Ferrara, Italy

Pertti Vainikainen
Helsinki University of Technology, Finland

Emmanuel Van Lil
Katholieke Universiteit Leuven, Belgium

Fernando Velez
University of Beira Interior, Portugal

Roberto Verdone
University of Bologna, Italy

Tobias Weber
University of Kaiserslauten, Germany

Tricia Willink
Communications Research Centre, Canada

Alberto Zanella
University of Bologna, Italy

Dirk Zimmermann
University of Stuttgart, Germany

List of acronyms

1G	1st Generation
2D	2 Dimensions
2G	2nd Generation
3G	3rd Generation
3GPP	3rd Generation Partnership Project
4G	4th Generation
AAU	Aalborg University
ABC	Always Best Connected
ABL	Adaptive Bit Loading
AC	Adaptive Coding
ACK	Acknowledgement
ACLR	Adjacent Channel Leakage Power Ratio
ACM	Adaptive Coding and Modulation
ACS	Adjacent Channel Selectivity
ADI	Alternating Direction Implicit
ADPS	Azimuth Delay Power Spectrum
ADSL	Asynchronous Digital Subscriber Line
AM	Amplitude Modulation
AMC	Adaptive Modulation and Coding
AMR	Adaptive Multi-Rate
AODV	Ad-hoc On Demand Distance Vector
AP	Access Point
APL	Adaptive Power Loading
APP	A Priori Probability
APS	Angular Power Spectrum
AR	Auto-Regressive
ARQ	Automatic Retransmission Request
ARROWS	Advanced Radio Resource Management for Wireless Services – IST Project
AS	Angular Spread
ASE	Average Spectral Efficiency
AWGN	Additive White Gaussian Noise

AbBC	Absorbing Boundary Condition
AcR	Autocorrelation Receiver
AfP	Affine Projection
AnM	Antenna Module
AoA	Angle of Arrival
AoD	Angle of Departure
B3G	Beyond 3G
BCC	Business City Centre
BCCH	Broadcast Control Channel
BCJR	Bahl-Cocke-Jelinek-Raviv
BEP	Bit Error Probability
BER	Bit Error Rate
BGT	Berrou, Glavieux and Thitimajshima
BHCA	Busy Hour Call Attempt
BICM	Bit-Interleaved Coded Modulation
BL	Beam Launching
BLAST	Bell Labs Layered Space-Time
BLER	BLock Error Rate
BMIR	Bit Mapping Incremental Redundancy
BPSK	Binary Phase Shift Keying
BRAM	Block RAM
BS	Base Station
BSC	Binary Symmetric Channel
BT	Bluetooth
BTD-SAGE	Broadband Time Domain-SAGE
BTG	Bracketing Transduction Grammar
BTS	Base Transceiver Station
BoL	Body Loss
CAC	Call Admission Control
CAD	Computer Aided Design
CAI	Co-Antenna Interference
CBR	Constant Bit Rate
CC	Convolutional Codes
CD	Compact Disc
CDF	Cumulative Distribution Function
CDM	Code Division Multiplex
CDMA	Code Division Multiple Access
CFIE	Combined Field Integral Equation
CFL	Courant-Friedrichs-Lewy
CFLN	Courant-Friedrichs-Lewy Number

CG	Conjugate Gradient
CI	Cell Identification
CIR	Carrier-to-Interference Ratio
CMD	Correlation Matrix Distance
COFDM	Coded-OFDM
CORDIC	COrdinate Rotation DIgital Computer
COST	European Cooperation in the Field of Scientific and Technical Research
COST 259	COST Action 259 – Wireless Flexible Personal Communications
COST 273	COST Action 273 – Towards Mobile Broadband Multimedia Networks
CP	Channel Process
CPICH	Common Pilot Channel
CPM	Continuous Phase Modulation
CPU	Central Processing Unit
CQF	Channel Quality Feedback
CRC	Cyclic Redundancy Check
CRLB	Cramer-Rao Lower Bound
CRRM	Common Radio Resource Management
CRa	Clipping Ratio
CS	Circuit Switch
CSE	Channel State Estimation
CSI	Channel State Information
CSMA	Carrier Sense Multiple Access
CSMA/CA	Carrier Sense Multiple Access with Collision Avoidance
CSNR	Channel Signal-to-Noise Ratio
CTPC	Conventional Transmit Power Control
CTS	Clear to Send
CW	Continuous Wave
ChIR	Channel Impulse Response
CoDiT	Code Division Testbed – RACE Project
CoM	Convolution Module
CrR	Critical Region
DAB	Digital Audio Broadcast
DAR	Decision Aided Reconstruction
DC	Direct Current
DCIR	Directional Channel Impulse Response
DCS	Digital Cellular System
DDDPS	Double Directional Delay Power Spectra
DDIR	Double Directional Impulse Response
DE	Density Evolution
DECT	Digital Enhanced Cordless Telecommunications

DFE	Decision Feedback Equaliser
DFT	Discrete Fourier Transform
DGPS	Differential Global Positioning System
DL	Downlink
DMC	Diffuse Multipath Components
DMI	Direct Matrix Inversion
DPCCH	Dedicated Physical Control Channel
DPDCH	Dedicated Physical Data Channel
DS	Delay Spread
DS-CDMA	Direct Sequence-CDMA
DSCH	Downlink Shared Channel
DSL	Digital Subscriber Line
DSSS	Direct Sequence Spread Spectrum
DUT	Device Under Test
DVB	Digital Video Broadcasting
DVB-H	Digital Video Broadcast-Handheld
DVB-T	Digital Video Broadcast-Terrestrial
DWG	AutoCAD's native file format for CAD models
DXF	Data Exchange Format
DoA	Direction of Arrival
DoD	Direction of Departure
DxPSK	Differential xPSK
E-UMTS	Enhanced-UMTS
EAC	Extended Alamouti Code
EADF	Effective Aperture Distribution Function
ECC	Error Correcting Code
EDGE	Enhanced Data rate for GSM Evolution
EGC	Equal Gain Combining
EIRP	Effective Isotropic Radiated Power
EM	Electro-Magnetic
EMA	Expectation Maximisation Algorithm
EMC	Electro-Magnetic Compatibility
EMTS	Eigenmode Transmission Scheme
ER	Effective Roughness
ES	Evolution Strategies
ESPRIT	Estimation of Signal Parameters via Rotational Invariance Techniques
ETSI	European Telecommunications Standards Institute
ETSI/BRAN	ETSI/Broadband Radio Access Network
EU	European Union
EVS	Electromagnetic Vector Sensor

EXIT	Extrinsic Information Transfer
FAM	Fuzzy Associative Memory
FB	Filter Based
FD	Frequency Dependent
FDD	Frequency Division Duplex
FDMA	Frequency Division Multiple Access
FDTD	Finite Difference Time Domain
FEC	Forward Error Correction
FEM	Finite Element Method
FEQ	Frequency-domain EQualiser
FER	Frame Error Rate
FF	Frame Fixed Length
FFT	Fast Fourier Transform
FH	Frequency Hopping
FIM	Fisher Information Matrix
FIR	Finite Impulse Response
FLOPS	FLOating Point operations per Second
FOMA	Freedom of Mobile Multimedia Access
FPGA	Field Programmable Gate Array
FSMC	Finite-State Markov Chain
FSR	Fisheye State Routing
GA	Genetic Algorithm
GB	Giga Byte
GBAR	Gamma Beta Auto-Regressive
GCOD	Generalised Complex Orthogonal Design
GERAN	GSM/EDGE Radio Access Network
GF	Galois Field
GIS	Geographic Information System
GMSK	Gaussian MSK
GO	Geometrical Optics
GOF	Groups Of Frequencies
GOP	Group Of Pictures
GPRS	General Packet Radio Service
GPS	Global Positioning System
GRASP	Greedy Randomised Adaptive Search Procedure
GSCM	Geometry-based Stochastic Channel Model
GSM	Global System for Mobile Communications
GTD	Geometric Theory of Diffraction
GoS	Grade of Service
HARQ	Hybrid Automatic Repeat reQuest

HD	High Definition
HDD	Hard-Decision Demapping
HHO	Horizontal Handover
HIPERLAN	High Performance Radio Local Area Network
HIPERMAN	High Performance Radio Metropolitan Area Network
HPA	High Power Amplifier
HSDPA	High Speed Downlink Packet Access
HTTP	Hyper Text Transfer Protocol
HW	HardWare
IAI	Inter Antenna Interference
IC	Interference Cancellation
IDFT	Inverse DFT
IEEE	Institute of Electrical & Electronic Engineers
IF	Intermediate Frequency
IFFT	Inverse FFT
iid	independent identically distributed
IISI	Intra- and Inter-Symbol-Interference
IMT-2000	International Mobile Telecommunications-2000
INR	Interference-to-Noise Ratio
IO	Interacting Object
IP	Internet Protocol
IR	Impulse Response
IRE	Impulse Response Estimates
IS-95	Interim Standard-95
ISI	Inter-Symbol Interference
ISIS	Initialisation and Search Improved SAGE
ISM	Industrial, Scientific and Medical
IST	Information Society Technologies
ITTS	Iterative Tree/Trellis Search
ITU	International Telecommunications Union
JCE	Joint Channel Estimation
JD	Joint Detection
JD-CDMA	Joint Detection-Code Division Multiple Access
JT	Joint Transmission
KP	Kronecker Product
L2CAP	Link layer Control and Adaptation Protocol
LA	Location Area
LAN	Local Area Network
LBS	Location-Based Services
LDC	Linear Dispersion Code

LDD	Linear Decorrelating Detector
LDPC	Low Density Parity Check
LLR	Log Likelihood Ratio
LMDS	Local Multipoint Distribution System
LMS	Least Mean Squares algorithm
LoS	Line-of-Sight
LSD	List Sphere Decoding
LUT	Look-Up Table
MAC	Medium Access Control
MAI	Multiple Access Interference
MAP	Maximum A Posteriori
MB-OFDM	Multi-Band OFDM
MBS	Mobile Broadband Systems – RACE Project
MC	Monte Carlo
MCC	Multipath Component Cumulative power
MC-CDMA	Multi Carrier-CDMA
MCL	Microwave Consultants Ltd
MCS	Modulation and Coding Schemes
MDDCM	Multiuser Double-Directional Channel Model
MEA	Multi-Element Antenna
MEG	Mean Effective Gain
MELG	Mean Effective Link Gain
MERP	Mean Effective Radiated Power
MERS	Mean Effective Radiated Sensitivity
MF	Matched Filters
MFIE	Magnetic Field Integral Equation
MHA	Mast Head Amplifier
MIMO	Multiple-Input Multiple-Output
ML	Maximum Likelihood
MLBMIR	Modulation Limited Bit Mapping Incremental Redundancy
MLD	Maximum Likelihood Detector
MLSE	Maximum Likelihood Sequence Estimator
MLk	MIMO Link
MM	MultiMedia
MMS	Multimedia Messaging Service
MMSE	Minimum Mean Square Error
MoM	Method of Moments
MOMENTUM	Models and Simulations for Network Planning and Control – IST Project
MORANS	Mobile Radio Access Network Reference Scenarios
MPC	Multipath Component

MPCC	Multiple Parallel Concatenated Codes
MPEG	Moving Picture Experts Group
MPG	Multipath Group
MPIC	MAP decoder aided PIC
MRC	Maximal Ratio Combining
MRT	Maximum Ratio Transmission
MRTo	Magnetic Resonance Tomography
MSE	Mean Square Error
MSINR	Maximum Signal to Interference plus Noise Ratio
MSK	Minimum Shift Keying
MSRC	Mode Stirred Reverberation Chamber
MSSTC	Multi-Stratum Space-Time Code
MT	Mobile Terminal
MUD	Multi User Detection
MUSIC	Multiple Signal Classification
MVDR	Minimum Variance Distortion Response
MVM	Minimum Variance Method
NACK	Negative Acknowledgement
NB	NarrowBand
NEC2	Numerical Electromagnetics Code Version 2
NLMS	Normalised Least Mean Squares algorithm
NLoS	Non-LoS
NMS	Network Management System
NPCC	Normalised Parallel Channel Capacity
NPCG	Normalised Parallel Channel Gain
OFCDM	Orthogonal Frequency Code Division Multiplexing
OFDM	Orthogonal Frequency Division Multiplexing
OFDMA	Orthogonal Frequency Division Multiple Access
OLoS	Obstructed Line-of-Sight
OMC	Operation and Maintenance Centre
ORT	Over Roof Top
OSI	Open System Interconnection
OSIC	Ordered Successive Interference Cancellation
OVSF	Orthogonal Variable Spreading Factor
PA	Paging Area
PAM	Pulse Amplitude Modulation
PAN	Personal Area Network
PAPR	Peak to Average Power Ratio
PARSAR	Parametric Reconstruction of SAR
PAS	Power-Azimuth Spectrum

PC	Power Control
PCB	Printed Circuit Board
PCS	Personal Communications Service
PDA	Personal Digital Assistant
PDF	Probability Density Function
PDP	Power Delay Profile
PE	Parabolic Equation
PEC	Perfectly Electric Conducting
PEP	Pairwise Error Probability
PER	Packet Error Rate
PIC	Parallel Interference Cancellation
PIFA	Planar Inverted-F Antenna
PIM	Pattern Integration Method
PL	Path Loss
PLk	Physical Link
PM	Phase Modulation
PMA	Power and Modulation Adaptation
PN	Pseudo-Noise
PO	Physical Optics
PSAM	Pilot Symbol Assisted Modulation
PSK	Phase Shift Keying
PWF	Pre-Whitening Filter
PhS	Physically Stationary
PrM	Propagation Module
Q-OSTBC	Quasi Orthogonal STBC
QAM	Quadrature Amplitude Modulation
QPSK	Quaternary PSK
QR-RLS	Q-R decomposition-based Recursive Least Squares algorithm
QoS	Quality of Service
RAB	Radio Access Bearer
RACE	Research and Technology Development in Advanced Communication Technologies
RAM	Random-Access Memory
RAN	Radio Access Network
RC	Resistance Capacitor
RC2	Rivest's Cipher 2
RCS	Radar Cross Section
RDF	Resource Description Framework
RELAX	RELAXation spectral estimator
RF	Radio Frequency

RF SIM	Radio Frequency Subscriber Identity Module
RLB	Radio Link Budget
RLS	Recursive Least Squares algorithm
RMC	Reference Measurement Channel
RMS	Root Mean Square
RNC	Radio Network Controller
RNN	Recurrent Neural Network
RNP	Radio Network Planning
RR	Round-Robin
RRM	Radio Resource Management
RS	Reed Solomon
RSCP	Received Signal Code Power
RSSI	Receiver Signal Strength Indicator
RSSUS	Reference System Scenario for UMTS Simulations
RT	Ray Tracing
RTS	Request to Send
RTT	Round Trip Times
RX	Receiver
RtS	Roof-to-Street
S-V	Saleh-Valenzuela
SAGE	Space-Alternating Generalised Expectation
SAM	Specific Anthropomorphic Mannequin
SAR	Specific Absorption Rate
SAS	Smart Antenna Systems
SB	Single-Block
SC	Selection Combining
SCDMA	Synchronous Code Division Multiple Access
SDD	Soft-Decision Directed
SDM	Space Division Multiplexing
SDMA	Space Division Multiple Access
SDR	Software Defined Radio
SE	Spectral Efficiency
SEACORN	Simulation of Enhanced UMTS Access and Core Networks – IST Project
SER	Symbol Error Rate
SfC	Soft Canceller
SfISfO	Soft In Soft Out
SFMG	Scattered-Field-Measurement Gain
SFTF	Space-Frequency Transmit Filtering
SHO	Soft Handover
SI	Self-Interference

SIC	Successive Interference Cancellation
SIMO	Single Input Multiple Output
SINR	Signal-to-Interference-plus-Noise-Ratio
SIR	Signal-to-Interference-Ratio
SISO	Single Input Single Output
SL	Simple switched Lobe
SM	Spatial Multiplexing
SMP AAL	Siemens Mobile Phones in Aalborg
SNR	Signal-to-Noise-Ratio
SOVA	Soft Output Viterbi Algorithm
SP	Sum-Product
SPEAG	Schmid & Partner Engineering AG
SPIC	Selective Parallel Interference Cancellation
SPUCPA	Stacked Polarimetric Uniform Circular Patch Array
SSDT	Site Selection Diversity Transmit Power Control
SSPA	Solid State Power Amplifier
ST	Space-Time
STBC	Space-Time Block Code
STBICM	Space-Time Bit-Interleaved Coded Modulation
STF	Space-Time-Frequency
STFT	Short-Time Fourier Transform
STOBC	Space-Time Orthogonal Block Codes
StP	Stochastic Process
STTC	Space-Time Trellis Codes
STTD	Space-Time-Transmit-Diversity
SVD	Singular Value Decomposition
TA	Timing Advance
TCE	Time and Channel Estimation
TCM	Trellis Coded Modulation
TCP	Transmission Control Protocol
TD	Temporary Document
TD-CDMA	Time Division-Code Division Multiple Access
TD-SCDMA	Time Division-Synchronous Code Division Multiple Access
TDD	Time Division Duplex
TDM	Time Division Multiplexing
TDMA	Time Division Multiple Access
TDoA	Time Difference of Arrival
ToA	Time of Arrival
TPC	Transmit Power Control
TR	Transmitted-Reference

TRP	Total Radiated Power
TRS	Total Radiated Sensitivity
TRX	Transceiver
TS	Taboo Search
TSD	Transmission Selection Diversity
TSG	Technical Specification Group
TTI	Transmission Time Interval
TX	Transmitter
UCA	Uniform Circular Array
UE	User Equipment
UHF	Ultra High Frequency
UHS	Ultra High Sites
UL	Uplink
ULA	Uniform Linear Array
UMMSE	Unbiased MMSE
UMTS	Universal Mobile Telecommunications System
UPL	Uniform Power Loading
URL	Uniform Resource Locator
US	Uncorrelated Scattering
UTD	Uniform Theory of Diffraction
UTRA	UMTS Terrestrial Radio Access
UTRAN	UTRA Network
UWB	Ultra Wide Band
V-BLAST	Vertical-Bell Labs Layered-Space Time
VAA	Virtual Antenna Array
VBR	Variable Bit Rate
VD	Viterbi Decoder
VDSL	Very-High-Rate Digital Subscriber Line
VF	Variable Frame Length
VHF	Very High Frequency
VNA	Vector Network Analyser
VOWAL	Voice communication Over Wireless Local Area Network
VPL	Vertical Plane Launch
VoIP	Voice over IP
WB	Wideband
WCDMA	Wideband CDMA
WCM	Wheeler Cap Method
WDCM	Wideband Directional Channel Model
WLAN	Wireless LAN
WLL	Wireless Local Loop

WP Workshop Paper
WPAN Wireless Personal Area Network
WSS Wide-Sense-Stationary
WSSUS Wide-Sense-Stationary Uncorrelated Scattering
WWW World Wide Web
XML eXtensible Markup Language
XPD Cross Polarisation Discrimination
XPR Cross-polarisation Power Ratio
ZF Zero Forcing

1 Introduction

Luís M. Correia

This chapter contains a discussion on the evolution of mobile and wireless communications, as well as a brief description of the contents of the book. For the former, a perspective on the evolution of mobile and wireless communications is given. The latter intends to present the scope of the book, its structure, and the way the various chapters are related to each other.

1.1 Trends and evolution[1]

Mobile and wireless communications have known a success that is beyond the most optimistic initial expectations. In some countries, namely European ones, the penetration of mobile communications is around 100%, which means that virtually everyone capable of using a mobile phone has (at least) one. How did we get here? What were the reasons for this huge success? These questions have been discussed over and over, and are not addressed here. Nevertheless, one thing is certain: in the past, someone, somewhere, had a vision that we all could use a single communications device under the well-known motto 'anytime, anywhere'.

In the first years of the 1980s, Europe was at the very beginning of the commercial exploration of mobile communications services (exclusively voice at the time), almost each country presenting its own system (the so-called 1st Generation (1G)). Things taken for granted nowadays, like roaming, were not possible, in most cases not even between countries using the same system. The panorama was very much different from the one we live in today.

Then, as a consequence of the previously mentioned vision, a pan-European mobile communications system was developed, enabling European citizens to move around the common European space, communicating using their mobile phones. Hence, the Global System for Mobile Communications (GSM) was born (defining the commonly accepted concept of 2nd Generation (2G)), the rest of the success story is well known, and will not be told here.

Universal Mobile Telecommunications System (UMTS), the successor to GSM in Europe (usually designated by 3rd Generation (3G)), found itself being different from its predecessor by offering users a set of services that were not (initially) foreseen for GSM, i.e. multimedia in general (in particular, ranging from video-telephony to Internet access). At the time UMTS was being outlined, the vision was to give mobile users, already used to making calls with their mobile phones, the possibility to use services on their phone that would contribute to an increase in their means of communication.

Since then the technical evolution of mobile and wireless communication systems has witnessed the appearance of a panoply of systems, encompassing Digital Video Broadcasting (DVB),

[1] Ideas for this section came from many discussions that were held with a number of colleagues in COST 273 and other fora (namely the eMobility platform within the EU IST framework).

High Performance Radio Local Area Network (HIPERLAN) and Bluetooth, to mention only a few within European initiatives (some of which were not actually implemented into commercial products). Outside Europe, many other systems were developed, and made a path parallel to the European ones.

For the future, mobile and wireless applications and services are likely to become pervasive, with a widespread use of devices. Computation and communication capabilities are being integrated in a large variety of devices, from simple sensors and interactive appliances (cards, rings, eyeglasses, etc.) via pocket and lap-sized devices to wall or table screen working areas. The technology will undergo a transformation, from an expensive, highly visible, 'hi-tech' technology as in early mobile phones, over the current situation where (almost) everyone owns a mobile phone, to a 'disappearing technology' that is present everywhere and taken for granted. Since the current cellular mobile approach, with its excellent mobility management and coverage properties, does not scale everywhere into large bandwidths, the result will be heterogeneous infrastructures with moderate bandwidth wide area coverage, and a local high bandwidth wireless coverage. Moreover, there is clearly the need to invent new access techniques and network architectures, so that future use of mobile and wireless communications can be made in an effective way. Such a vision challenges many of the current paradigms in mobile and wireless communications.

Currently, there is lack of a unique concept for the incoming 4th Generation (4G). A new vision should be set; as in the past, one can almost say that the probability of increased success can be measured by the way this vision creates a disruption with the existing networks and systems. On the one hand, one cannot aim at far-fetched objectives, which are very unlikely to be reached within an acceptable timeframe. On the other, having as an objective something similar to the Olympic Games motto ('Citius, Altius, Fortius', meaning 'Faster, Higher, Stronger'), applied to current systems and techniques (i.e. more of the same, but just better), is too short-term.

The old vision of 'anywhere, anytime' could easily be replaced by 'any network, any device, any content'. Basically, this vision carries many implications, some of which are discussed in what follows.

It cannot be expected that users make a decision on the network to be used on their common use of the system; hence, the adaptability of networks to the type of users and information they are communicating will be an essential feature. Users will be capable of communicating how, where and whenever they want to, but it also implies that the complexity of networks and systems should be concealed from them, even if they are using more than one system and/or network simultaneously to carry the information.

The use of the system will tend to be independent of the device. One can imagine that users will just carry a kind of Radio Frequency Subscriber Identity Module (RF SIM), i.e. simply a small card that carries all their information and that communicates via RF with all the devices available in its range. This will enable users to take advantage of many types of devices, using those that are more appropriate for a given service or location. The more important role of Personal Area Networks (PANs) and ad hoc networking will set new borders. Sensor networks will become increasingly important, and the number of devices people carry with them (knowingly or unknowingly) will increase. Machine-to-machine communications will definitely need to be taken into account, because they help in increasing system intelligence and in concealing technology from users.

Users will be provided with and have access to content and information they want in a useful way, namely transmission speed. In the future, much of the user information may indeed be local, as opposed to information that does not depend on location. Moreover, peer-to-peer communications will play a key role. The provision of and access to the 'right content' at the 'right time' is perceptual, and should be provided when a user is ready to receive it, in a format that considers user privacy and present context by using any available means and network. Sometimes the user requests the information, i.e. ‘user access to information’, while other times it is the ‘information that accesses the user’, based on the user’s personal or family profile. Access to the vast range of information and content has to consider a user’s ‘techno-ability’, and must be simple and intuitive to use. The success of such vision depends very much on the simplicity of access and use of services and operation of devices.

Content clarifies that the game is no longer played only with voice, but rather it extends far beyond that, not only enabling the myriad of services and applications that everyone talks about these days, but also certainly including others that have not yet been foreseen. A future vision cannot be complete without the definition of the future application scenarios and users. A good starting point from which one should draw trends is to look at users as our children/grandchildren who will be the active population in 15–20 years’ time. Moreover, content needs to be meaningful, since, on the one hand, non-desired information (e.g. advertisement, spam, virus, etc.) and privacy are proving problematic in today’s communications, as is evident in computing (e.g. e-mail, and intrusive and destructive Internet access); on the other, it also means that filtering of information is very important, so that users get what they really want.

But there is the need to stretch the science of mobile and wireless communications beyond radio and computer science into new areas of knowledge. Taking the main priorities that are identified these days for research and development, one can identify links with them: biology (e.g. use of biometrics for user identification), medicine (e.g. measurement of the health state and transmission of an alarm in case of a problem), psychology (e.g. sensing the mood and state of mind of the user), sociology (e.g. interaction with other ‘compatible’ users), nano-technologies (incorporation of circuitry/terminals in common use objects, such as spectacles), materials (e.g. cooperation with the clothing industry for the use of jackets for virtual reality), transport (e.g. placing terminals in cars), environment (e.g. decreasing ‘electromagnetic pollution’), energy (e.g. expanding the lifetime of the batteries), among others. Finally, the information should be ‘multisensory’ making use of all five human basic senses. Clearly, the realisation of the future vision of mobile and wireless communications demands multidisciplinary research and development, crossing the boundaries of the above sciences and different industries.

Many steps are required to make this vision a reality, and clearly from the establishment of this macroscopic vision, a number of tasks and challenges need to be identified and solved at the microscopic scale. The work performed in Towards Mobile Broadband Multimedia Networks (COST 273) intended to give small contributions at the latter scale.

1.2 Scope of the book

This book is structured into eight more chapters.

Chapter 2 covers transmission techniques employed in the physical layer of the Open System Interconnection (OSI) model, addressing specific wireless standards and air interfaces. There are two important classes in the mobile and wireless systems currently actively under development, this being the reason for the division into sections of this chapter: those based on Orthogonal Frequency Division Multiplexing (OFDM), including Wireless LANs (WLANs) and proposals for future high speed mobile systems, and those based on Code Division Multiple Access (CDMA), especially for 3G mobile standards.

Chapter 3 deals with processing algorithms required to transmit and detect signals, which may be used in various systems. Again, the focus is on link level aspects, including multi-user systems and multi-user detection, covering modulation and coding techniques used for data transmission, equalisation techniques used at the receiver to overcome signal dispersion, parameter estimation, multi-user detection, adaptive air-interfaces, and link adaptation algorithms. Two other topics are addressed in this chapter: techniques for Multiple-Input Multiple-Output (MIMO) systems, although a large part of this appears in Chapter 7, with another perspective; 'turbo' techniques, which cannot be separated out, because the principle is so widely acceptable.

Chapter 4 deals with propagation modelling and channel characterisation, from various perspectives. Electromagnetic modelling covers integral methods and differential equation ones. Deterministic modelling discusses ray models and hybrid approaches. Then, channel measurement techniques and parameter estimation are described in detail, including channel sounder architectures. Finally, channel characterisation is approached from the path-loss and fading (short- and long-term) perspectives, models being presented for several scenarios, as well as the results from measurement campaigns to obtain statistical information on temporal and angular dispersion.

Chapter 5 provides new insights into antennas, including diversity schemes and Ultra Wide Band (UWB) applications. Handset antennas are addressed, by introducing figures of merit as the mean effective gain and the mean effective radiated power, taking the multipath structure of the communication channel into account, together with an insight into standardised measurement techniques. Antenna diversity is also analysed, not only from more usual perspectives, but also concerning its possible application to UWB systems and MIMO techniques. Lastly, UWB systems are considered, by describing UWB antenna characteristics and radio channel measurements and models.

Chapter 6 deals with MIMO channel modelling. Initially, canonical or reference scenarios are discussed. Along with a general discussion of modelling, details of new physically inspired modelling concepts developed in the project are then addressed. Specification of antenna arrays at both link ends, by setting the number of antenna elements, their geometrical configuration and their polarisations, is dealt with. A discussion of old and new analytical MIMO models is presented, after which the novel COST 273 MIMO Channel Model is introduced, mainly based on the numerous measurement campaigns within COST 273, which are described as well. Since the discrepancy between efforts to develop new MIMO models and validating them is striking, model validation is also considered.

Chapter 7 addresses MIMO systems. It starts by considering the various aspects of these systems from an information theory viewpoint. Multi-element array antenna-based systems are then considered, from the two usual perspectives: beamforming and array signal processing, commonly known as 'smart antennas', and the actual MIMO one. For the former, transmit diversity schemes, including space-time codes, have been included, while for the latter, transmission techniques and receiver architectures are considered. Finally, the implications of MIMO techniques for complete wireless systems are addressed, namely those concerning how they can increase the capacity of complete networks.

Chapter 8 analyses the radio network aspects that have been considered, namely those related to radio network planning, radio resource management, and the design and evaluation of techniques for radio network optimisation, which are not UMTS oriented. Within the topics that are not specifically related to particular air interfaces, the Mobile Radio Access Network Reference Scenarios (MORANS) initiative is presented, which is oriented to provide a set of scenarios, parameters and models to be used as a common simulation platform. Within the other topics, various subjects are included, ranging from packet scheduling for cellular systems to the description of services and traffic models. Next, specific networks are addressed, WLANs, Wireless Personal Area Networks (WPANs), and Wireless Ad Hoc Networks. Finally, terminal location identification is dealt with.

Finally, Chapter 9 focuses on UMTS radio network planning and optimisation, including aspects of improvement of the performance of Radio Resource Management (RRM) procedures, parameter setting to control the capacity-coverage-cost trade-off, automatic planning algorithms, real-time tuning, and results on UTRA Network (UTRAN) performance evaluation by means of simulation tools. Among other aspects, this chapter also addresses the comparison of static and dynamic system simulators and link level impact on system performance. The current evolution and additional technologies that are being incorporated into the last releases of UMTS standards are considered at the end.

2 Transmission techniques

Chapter editor: *Alister Burr*
Contributors: *Emmanuel Jaffrot, Eduardo Rodrigues de Lima, and Werner Teich*

2.1 Introduction

In this chapter and the next, we consider transmission techniques (this chapter) and signal processing (Chapter 3) for wireless mobile communication systems: that is for the most part techniques employed on the physical layer of the OSI model, both at transmitter and receiver, and including both the air interface itself and the signal processing algorithms required to transmit and detect them. Thus, it covers link level aspects (not to be confused with the link layer of the OSI model), including the design and evaluation of techniques used on individual links of a wireless network, although multi-user systems and multi-user detection are also covered. These techniques depend on the characteristics of the radio channel, as described in Chapter 4, antennas and diversity (Chapter 5), and both support and are influenced by the system level evaluation of complete wireless networks, as described in Chapters 8 and 9.

Chapter 7 'MIMO Systems' of course includes much work on transmission techniques and signal processing specifically for MIMO and space-time systems. Nevertheless, this chapter also includes work on signal processing for MIMO systems, since increasingly MIMO is being seen as an essential element of wireless systems. However, the emphasis here is on the systems and transmission techniques, rather than the MIMO aspects: Chapter 7 places the work on MIMO transmission in the context of space-time channels, which are discussed in Chapter 6.

The two chapters between them describe primarily work carried out in COST 273 WG1 (Working Group on Radio System Aspects). However, there has also been a good deal of interaction with other Working Groups: with WG2 (Working Group on Propagation and Antennas) in respect of the channel models used to evaluate the systems, and with WG3 (Working Group on Radio Network Aspects) in respect of the interaction of the link and the system levels. Note that the latter works in both directions: link level performance of course affects the performance of complete wireless systems, but also system level techniques, such as radio resource allocation, affect the link level, resulting in an increased degree of 'cross-layer' work. COST 273 has been well placed to contribute to such work, precisely because of the expertise in all these three aspects represented in the Working Groups and because of the interaction among them.

The division between the two chapters is as follows. This chapter covers specific wireless standards and air interfaces, while the next discusses the more generic signal processing techniques that may be used in various systems. For wireless systems currently actively under development there are two important classes: those based on OFDM, including WLAN and standards proposed for future high speed mobile systems, and those based on CDMA, especially 3rd generation mobile standards. For this reason the chapter is divided into two sections: Section 2.2 covers OFDM-based systems, while Section 2.3 covers CDMA. Multi Carrier-CDMA (MC-CDMA) systems, which can be regarded as a combination of the two, are discussed in Subsection 2.3.4.

2.2 OFDM systems

2.2.1 Introduction

During the past 15 years, Orthogonal Frequency Division Multiplexing (OFDM) has been gaining year after year a well-deserved reputation, demonstrating its high data rate and robustness to wireless environments capabilities. In the multipath environment, broadband communications will suffer from frequency selective fading. In such a situation, single-carrier technologies do not work optimally and a transport scheme better suited for the environment is needed. For this reason, OFDM became very popular recently [Enge02].

OFDM is an attractive modulation scheme used in broadband wireless systems that encounter large delay spreads. The complexity of Maximum Likelihood (ML) detection or even sub-optimal equalisation schemes needed for single carrier modulation grows exponentially with the bandwidth-delay spread product. OFDM avoids temporal equalisation altogether, using a cyclic prefix technique with a small penalty in channel capacity [PaNG03].

Where Line-of-Sight (LoS) cannot be achieved, there is likely to be significant multipath dispersion, which could limit the maximum data rate. Technologies like OFDM are probably best placed to overcome these, allowing nearly arbitrary data rates on dispersive channels. OFDM, in particular for broadband systems in dispersive environments, is a technology that could have a place in the 4G concept.

Although the principle of OFDM communication has been around for several decades, it was only in the last decade that it started to be used in commercial systems. The most important wireless applications that make use of OFDM are Digital Audio Broadcast (DAB), DVB, WLAN and more recently Wireless Local Loop (WLL) [Enge02].

The DAB system is seen as the future of radio as it makes more efficient use of crowded airwaves and provides CD quality sound that is noticeably better than an FM analogue broadcast. DAB makes use of an OFDM transmission scheme with differential Quaternary PSK (QPSK) modulation. The DVB system is very similar to the DAB standard, but is intended for broadcasting of digital television signals. Due to the high data rates, the DVB system uses an 8 MHz bandwidth. The subcarriers in the OFDM signal are modulated with a higher order Quadrature Amplitude Modulation (QAM) constellation, with up to 64 points.

The third generation of WLAN systems is intended to offer high data rates in the 5 GHz frequency band and more recently in the 2.5 GHz band. The communication is based on OFDM in the 20 MHz bandwidth. Per subcarrier, the modulation schemes are Binary Phase Shift Keying (BPSK), QPSK, 16-QAM and 64-QAM. Together with a variable error coding rate, this allows the data rates to be adapted from 6 Mbit/s to 54 Mbit/s depending on the propagation environment. Future WLAN standards are being studied to overcome that range.

Wireless Local Loops provide high speed Internet access and multimedia services to fixed users. WLL is a competitive technology to Very-High-Rate Digital Subscriber Line (VDSL) and cable modems. OFDM is one of the supported transport schemes for WLL technologies [Enge02].

The simplification of equalisation and the high bandwidth efficiency and flexibility are the main motivations for using OFDM, which is always a preferred alternative if a high data rate is to be transmitted over a multipath channel with large maximum delay [Corr01].

Channel coding plays an important role in OFDM systems. Due to the narrowband subcarriers and the appropriate cyclic prefix, OFDM systems suffer from flat fading. In this situation, efficient channel coding leads to a very high coding gain, especially if soft decision decoding is applied. For this reason OFDM systems will always have to make use of channel coding [AlLa87]. Furthermore, OFDM allows multiple access techniques to certain time and/or frequency regions of the channel in a very simple way [RoGr97].

Finally, OFDM signals have a large peak-to-mean power ratio due to the superposition of all subcarrier signals, therefore in each Transceiver (TRX), the power amplifier will limit the OFDM signal by its maximal output power. This also disturbs the orthogonality between subcarriers, leading to both intercarrier and out-of-band interference, which is unacceptable [Corr01].

An OFDM signal consists of N subcarriers spaced by the frequency distance Δf, thus, the total system bandwidth B is divided into N equidistant subchannels. On each subcarrier, the symbol duration $T_s = 1/\Delta f$ is N times as large as in the case of a single carrier transmission system covering the same bandwidth. Additionally, each subcarrier signal is extended by a guard interval – called the cyclic prefix – with the length T_g. All subcarriers are mutually orthogonal within the symbol duration T_s. The kth subcarrier signal is described analytically by the function $g_k(t)$, $k = 0, \ldots, N-1$. For each subcarrier a rectangular pulse shaping is applied.

$$g_k(t) = \begin{cases} e^{j2\pi k\Delta f t} & \forall t \in [-T_g, T_s] \\ 0 & \forall t \notin [-T_g, T_s] \end{cases} \tag{2.1}$$

The guard interval is added to the subcarrier signal in order to avoid Inter-Symbol Interference (ISI), which occurs in multipath channels. At each Receiver (RX), the cyclic prefix is removed and only the time interval $[0, T_s]$ is evaluated. The total OFDM block duration is $T = T_s + T_g$. Each subcarrier can be modulated independently with the complex modulation symbol $s_{n,k}$, where the subscript n refers to time and k to the subcarrier index inside the considered OFDM block. Thus, within the symbol duration T the following signal of the nth OFDM block is formed:

$$s_n(t) = \frac{1}{\sqrt{N}} \sum_{k=0}^{N-1} s_{n,k} g_k(t - nT) \tag{2.2}$$

Due to the rectangular pulse shaping of the signal, the spectra of the subcarriers are *sinc* functions, Figure 2.1. The spectra of the subcarriers overlap, but the subcarrier signals are mutually orthogonal, and the modulation symbol $s_{n,k}$ can be recovered by a simple correlation:

$$s_{n,k} = \frac{\sqrt{N}}{T_s} \left\langle s_n(t), \overline{g_k(t - nT)} \right\rangle \tag{2.3}$$

In a practical application, the OFDM signal $s_n(t)$ is generated in a first step as a discrete-time signal in the digital signal processing part of the TRX. As the bandwidth of an OFDM system is $B = N\Delta f$, the signal must be sampled with sampling time $t = 1/B = 1/N\Delta f$. The samples of the

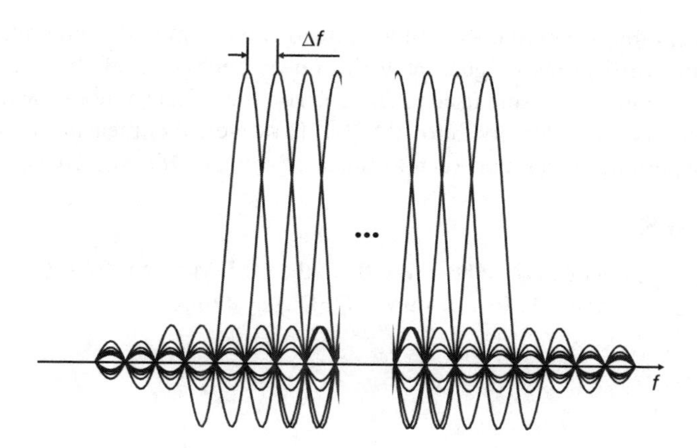

Figure 2.1 *Representation of the OFDM subcarriers.*

signal are written as $s_{n,i}$, $i = 0, 1, \ldots, N-1$, and can be calculated by an Inverse DFT (IDFT) which is typically implemented as an Inverse FFT (IFFT).

$$S_{n,i} = \frac{1}{\sqrt{N}} \sum_{k=0}^{N-1} s_{n,k} e^{j\frac{2\pi ik}{N}} \tag{2.4}$$

The subcarrier orthogonality is not affected at the output of a frequency selective radio channel; therefore, the received signal $r_n(t)$ can be separated into the orthogonal subcarrier signals by a correlation technique according to (2.3). Alternatively, the correlation at RX can be done as a Discrete Fourier Transform (DFT) or a Fast Fourier Transform (FFT) respectively:

$$R_{n,k} = \frac{1}{\sqrt{N}} \sum_{i=0}^{N-1} r_{n,i} e^{-j\frac{2\pi ik}{N}} \tag{2.5}$$

where $r_{n,i}(t)$ is the ith sample of the received signal $r_n(t)$, and $R_{n,k}$ is the recovered complex symbol of the kth subcarrier. If the subcarrier spacing Δf is chosen to be much smaller than the coherence bandwidth, and the symbol duration T much smaller than the coherence time of the channel, then the transfer function of the radio channel $H(f, t)$ can be considered constant within the bandwidth Δf of each subcarrier and the duration of each modulation symbol $S_{n,k}$. In this case, the effect of the radio channel is only a multiplication of each subcarrier signal $g_k(t)$ by a gain factor $H_{n,k} = H(k\Delta f, nT)$. As a result, the received complex symbol $R_{n,k}$ after the FFT is

$$R_{n,k} = H_{n,k} \cdot S_{n,k} + N_{n,k} \tag{2.6}$$

with $N_{n,k}$ being Additive White Gaussian Noise (AWGN).

2.2.2 OFDM: a very dynamic signal

OFDM is an attractive modulation for frequency selective channels since it allows a very simple mitigation of ISI using a guard interval. However, a well-known drawback of OFDM is that transmitted signals exhibit a Gaussian-like time domain waveform with some relatively high peaks. This results in many difficulties to guarantee the linear behaviour of the system over its large dynamic range. The common measure used to characterise the effects of non-linearities in the OFDM signal is the Peak to Average Power Ratio (PAPR). However, this measure is not necessarily the most representative parameter to report on non-linear effects on OFDM signals.

Mitigating the PAPR

Many studies have been conducted on how to reduce the OFDM signal PAPR. Figure 2.2 shows the shape of the probability of the PAPR in common OFDM systems.

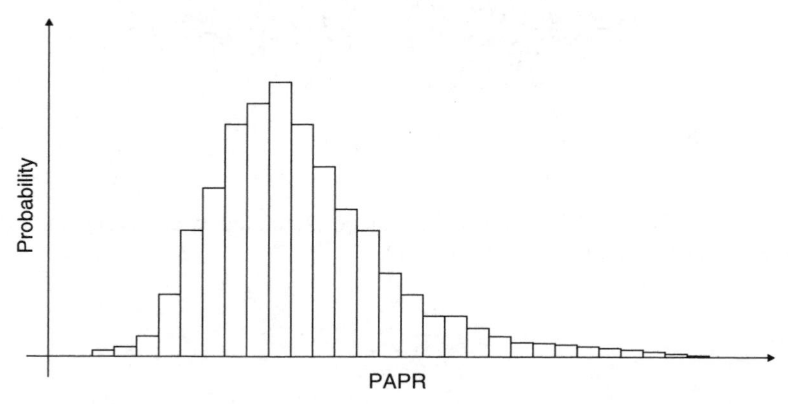

Figure 2.2 *General shape of the PAPR probability.*

The most classical way is to generate an OFDM signal with low PAPR. For this, coding solutions [WiJo95], [Nee96] and phase shifting [Frie97], [BaFH96], [MuHu97b], [MuHu97a], [Tell98] have been extensively researched.

Another solution is to clip intentionally the amplitude of the transmitted signal leading to non-linear distortion [LiCi98], [NeWi98], [O'Lo94], [DiWu98], [WuGo99], which cannot be efficiently corrected with a classical linear receiver, even using an Error Correcting Code (ECC) [TeHC03]. Kim and Stuber [KiSt99] propose an iterative non-linear decoder that corrects the clipping effect on OFDM transmissions, called Decision Aided Reconstruction (DAR). An algorithm inspired from this previous study and taking advantage of the error correcting code present in high rate Coded-OFDM (COFDM) transmissions in a turbo fashion was investigated [GeCD04].

As described on Figure 2.3, a convolutionally binary sequence is mapped onto QAM symbols $\mathbf{X}$ and modulated using an IFFT in an N-point output sequence, such as:

$$x_m = \frac{1}{\sqrt{N}} \sum_{k=0}^{N-1} X_k e^{\frac{2j\pi mk}{N}}, \qquad 0 \leq m \leq N-1 \tag{2.7}$$

where $\mathbf{X} = \{X_k\}_{k=0}^{N-1}$ is the transmitted symbol sequence, and N is the OFDM block size. A guard interval is added to the $\mathbf{X}$ sequence as:

$$x_k^G = x_{k+N-G}, \qquad 0 \leq k \leq N+G-1 \tag{2.8}$$

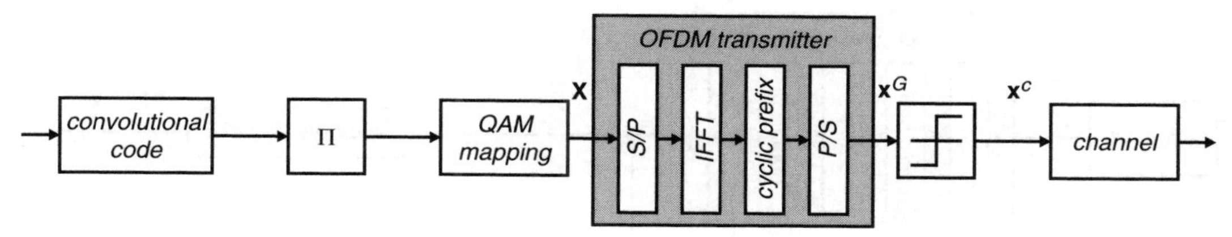

Figure 2.3 *Turbo-DAR transmission system.*

where G is the length of the cyclic prefix, and $(k)_N$ is the residue of k modulo N. Finally the clipping operation is performed on the time-sequence x_G as:

$$x_k^c = \begin{cases} x_k^G & |x_k^G| \leq A \\ Ae^{(arg x_k^G)} & |x_k^G| > A \end{cases}, \quad 0 \leq k \leq N + G - 1 \tag{2.9}$$

where $\mathbf{x}^c$ is the clipped output sequence and A is the clipping amplitude. The Clipping Ratio (CRa) is defined as:

$$CRa[dB] = 20 \log \frac{A}{\sigma_x} \, dB \tag{2.10}$$

where σ_x is the standard deviation of the x_k. After transmission through the channel and removal of the cyclic prefix, the signal can be written as:

$$y_k = \sum_{m=0}^{M} h_m x_{k-m}^c + n_k, \quad 0 \leq k \leq N - 1 \tag{2.11}$$

where h_m is the channel coefficient at the lag m and n_k is a zero-mean AWGN with variance N_0. By processing the FFT on y_k, we obtain:

$$Y_k = H_k \cdot X_k^c + n_k, \quad 0 \leq k \leq N - 1 \tag{2.12}$$

$$= H_k \cdot X_k + Q_k \tag{2.13}$$

where Q_k is the sum of the AWGN and the clipping noise and H_k represents the complex channel gain on the kth subcarrier. Equalisation is performed in the frequency domain using zero-forcing (ZF):

$$Z_k = \alpha_k Y_k, \ \alpha_k = \frac{H_k^*}{|H_k|^2} \tag{2.14}$$

These equalised symbols are used as the inputs of the Turbo-DAR algorithm. This algorithm, whose principle is given in Figure 2.4, implements an iterative technique summarised as follows:

1. First, the equalised signal $\mathbf{Z}$ is stored in memory $\tilde{\mathbf{X}}^{(0)} = \mathbf{Z}$.
2. The noisy symbols $\tilde{\mathbf{X}}^{(i)}$ are coded using the hard symbol estimator depicted in Figure 2.5: $\tilde{\mathbf{X}}^{(i)}$ is soft-demapped using the relation:

$$P(b_k = a) = \sum_{s \in \mathcal{S}} P(b_k = a | s_m P(S = s_m)) = \sum_{s_m \in \mathcal{S}'} P(S = s_m)$$

$$a = \{0, 1\}, k = 0, \ldots, \log_2 M - 1 \tag{2.15}$$

where $\mathcal{S}$ denotes the set of symbols of the M-QAM constellation and $\mathcal{S}'$ the subset of $\mathcal{S}$ such that $s_m \rightleftharpoons \{b_0, \ldots, b_{\log_2 M-1}\}$ and $b_k = a$. These bit likelihoods are deinterleaved to

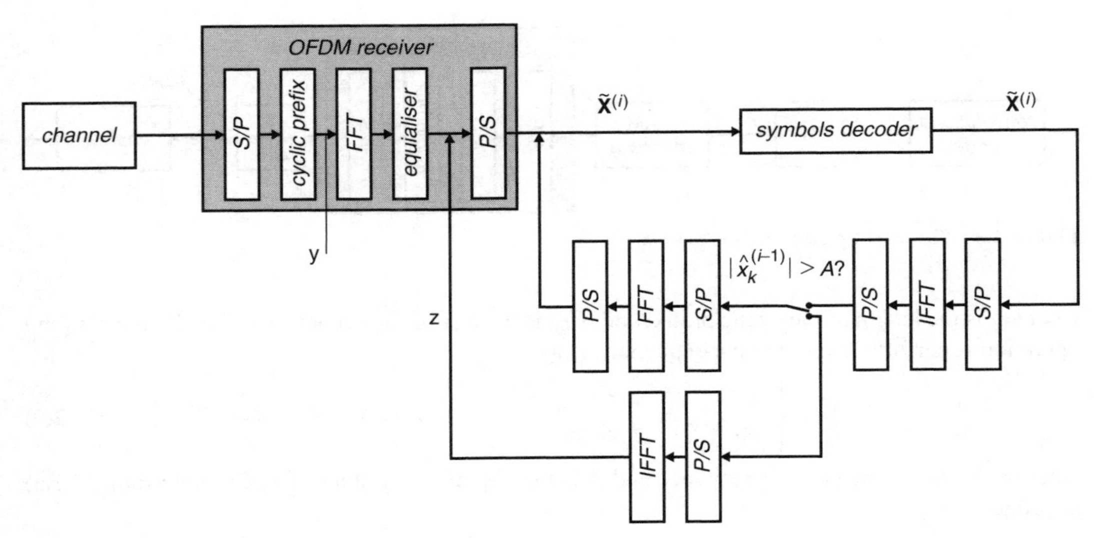

Figure 2.4 *Turbo-DAR: principle.*

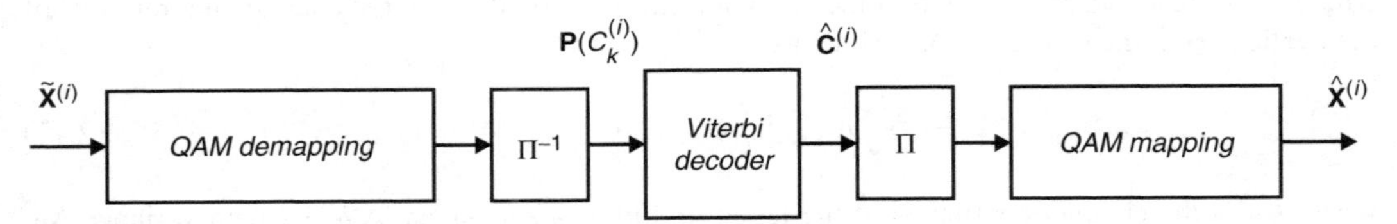

Figure 2.5 *Turbo-DAR hard symbol decoder principle.*

provide the likelihoods of the noisy codeword bits $\mathbf{P}(C_k(i))$ and are used at the Viterbi decoder input to compute the maximum likelihood decoded sequence $\hat{\mathbf{C}}^{(i)}$ and an estimation of the information bits.

3. $\hat{\mathbf{C}}$ is interleaved, mapped onto symbols and converted back to the time domain using an IFFT leading to $\hat{\mathbf{X}}^{(i)}$.
4. The detection of the clipped samples is performed in the time domain by comparing $|\hat{\mathbf{X}}^{(i)}|$ to A:

$$\hat{x}_k^{i+1} = \begin{cases} \hat{x}_k^{(0)} & |\hat{x}_k^{(i)}| \leq A \\ \hat{x}_k^{(i)} & |\hat{x}_k^{(i)}| > A \end{cases} \tag{2.16}$$

5. The estimated symbols are converted to the frequency domain by an FFT. Go back to step 2 and increment the index number $i+1$. Iterate I times.

The final decision is taken using $\tilde{\mathbf{X}}^{(I)}$ to obtain the estimated information bits by Viterbi decoding, Figure 2.5. A soft decision Turbo-DAR can be defined the same way, replacing the Viterbi decoder by a Bahl-Cocke-Jelinek-Raviv (BCJR) decoder with soft outputs.

Figure 2.6 illustrates the performance of the proposed techniques for $CRa = 1$ dB using QAM-16 on a Time Invariant Frequency Selective (TIFS) channel. Those results show a high improvement of

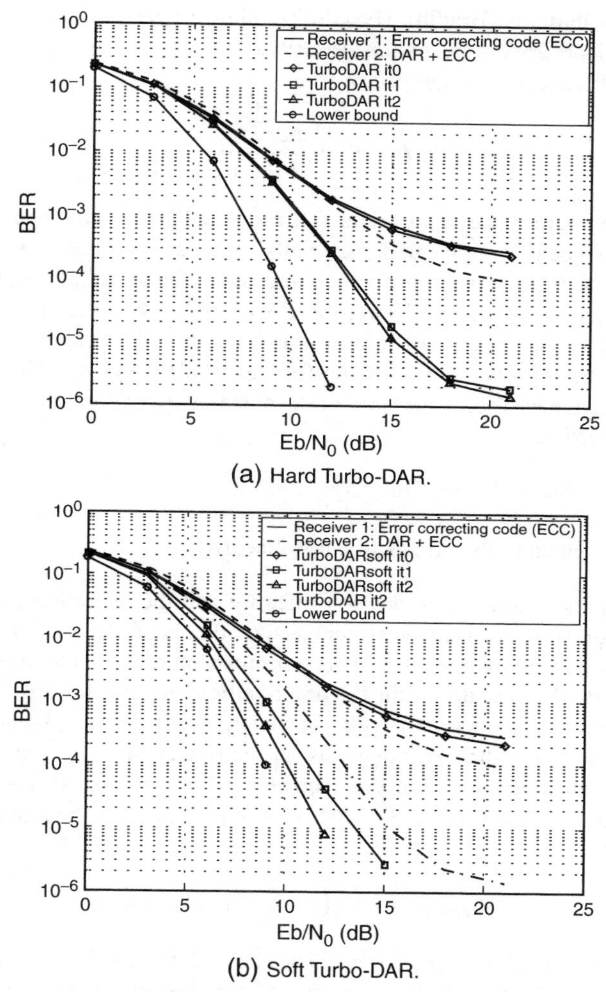

Figure 2.6 *BER comparison on TIFS channel of hard and soft Turbo-DAR with 16-QAM, CRa = 1 dB, N = 128.*

the Bit Error Rate (BER) by combining the effect of the error correcting code and a decision taken in the frequency domain in a turbo fashion.

Although current results are given for static channels, Turbo-DAR methods can also be used on time-varying channels such as in high rate wireless applications with adaptive channel estimator, e.g. [CiBi94], [MoMe01].

Exceeding power: an alternative to PAPR

It seems it is not entirely clear how well the PAPR measure is related to the real effects of non-linearities [BeEr02]. Does a reduction of the PAPR always lead to a decrease in the effects of the non-linearity?

In some recent contributions [Brai00], [BSGS02], other measures have also been mentioned for the sensitivity of multicarrier systems to non-linearity.

Classically, the PAPR measure is defined by:

$$PAPR = \frac{\max |x_n|^2}{E[|x_n|^2]} \tag{2.17}$$

Theoretically, the maximum value of a sample x_n from an OFDM symbol can get as high as N, but the probability of such a peak is very small, especially for a large number of subcarriers. In practice, in order to decrease the intercarrier interference and out-of-band radiation, it is required that the amplifier operates in its linear region. Therefore it is desirable to limit the maximum envelope of the multicarrier signal, namely the PAPR of the signal. On the other hand, reducing the PAPR does not necessarily mean an improvement in the performance of the system. However, there are other possibilities in defining a measure. A measure of the signal degradation in non-linearity should have a set of desirable properties:

- it should be independent on the non-linearity
- it should be easy to compute
- it should be highly related to the effects of a non-linearity.

The first property is desirable, since the non-linearity and the operating point of the system are typically unknown at the time of baseband system design; the second one is important in the sense that the measure must be easily attainable when using a limited set of data; the third one is important for obvious reasons. The PAPR measure fulfils the first requirement; considering the PAPR of a discrete signal, it fulfils the second property as well; however, according to [TaJa00], the performance of an OFDM system with non-linearity depends on the power of distortion, while the peak value does not show the power of the signal or the distortion. This may discard PAPR measure for some applications. Here, *Excess Power* of the OFDM symbol is defined as the amount of the power over a certain power level:

$$P_{Excess} = E[(|s_n| - |\bar{s_n}|)^2(|s_n| > |\bar{s_n}|)] \tag{2.18}$$

where $|\bar{s_n}|$ is the average envelope of the signal. The above definition is one example of a large set of measures that can be used for amplitude variations. A more general form of (2.18) can be written as:

$$P_{Excess}^{(m,L)} = \sqrt[\frac{m}{2}]{E\left[(|s_n| - L)\,|(|s_n| >> L)\right]} \tag{2.19}$$

where L is the reference level for the distortion. It is straightforward to show that $P_{Excess}^{(m,L)}$ gives a peak of signal when $m = \infty$. Hence,

$$PAPR \propto \lim_{m \to \infty} P_{Excess}^{(m,L)}, \quad \forall L < \max |s_n| \tag{2.20}$$

On the other hand, $P_{Excess}^{(m,L)}$ can be a measure of power when $m = 2$ as defined in (2.7). Both PAPR and $P_{Excess}^{(2)}$ fulfil the two first properties above. Figure 2.7 pictures the correlation coefficients between the distortion power and PAPR, and distortion power and excess power are plotted for different values of back-off. This figure shows that when the non-linearity is severe, excess power can show the distortion very well, while for high back-offs PAPR is a better measure.

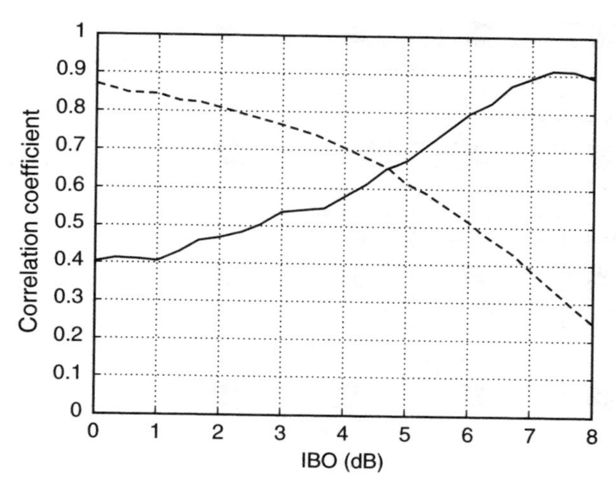

Figure 2.7 *Correlation coefficients between PAPR and P_d (solid) and P_{Excess} and P_d (dashed) vs. Input Back-Off.*

2.2.3 Enhancing OFDM systems

After encouraging applications of the OFDM technique to Digital Subscriber Line (DSL), Broadcasting (DAB, DVB-T) and WLANs (HIPERLAN/2, IEEE802.11a), OFDM is proposed for higher bit rates and wider mobile systems. Thus, the focus of the current research is to enhance the OFDM technique capacity and robustness to its imperfections, and to the propagation environments.

Many studies have been conducted in this area of interest during the past 10 years. With the increasing need of high bit rates and mobility, authors show more and more interest in highly selective channels in time and frequency. Moreover, to increase OFDM systems' capacity, the use of antenna arrays within the MIMO systems needs particular attention for channel estimation purposes.

Since the efficiency and performance of a channel estimator is highly correlated to the channel model precision and accuracy, the choice and definition of a channel model is fundamental.

Turbo channel estimation

A channel model based on Karhunen-Loève (KL) decomposition of the channel correlation matrix was proposed in [JaSi00]. This model considers time-frequency blocks, Figure 2.8, of the OFDM signal, characterising the channel in time and frequency. The theoretical channel correlation matrix is derived from the spaced-time spaced-frequency correlation function, which is given for a classical Doppler and exponential multipath intensity profile with average power $\phi(0, 0)$ as:

$$\phi(\Delta f, \Delta t) = \phi(0, 0)\frac{J_0(\pi B_d \Delta t)}{1 + j2\pi T_m \Delta f} \tag{2.21}$$

where B_d and T_m are respectively the Doppler and multipath spread of the channel and $J_0(.)$ is the 0th-order Bessel function of the first kind.

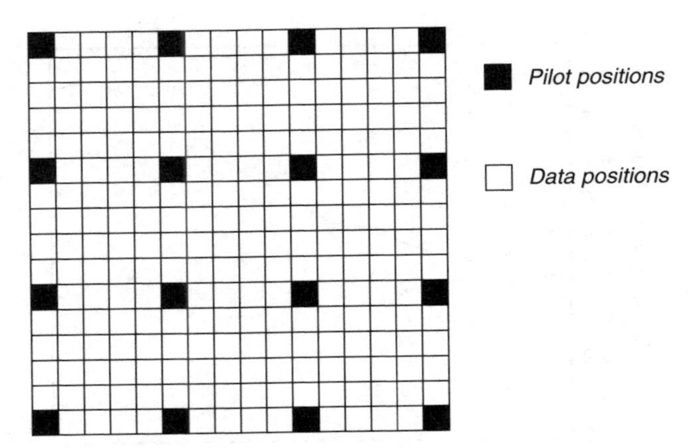

Figure 2.8 *Time-frequency block.*

The channel received on the lth diversity branch of the receiver can be expressed as:

$$\mathbf{H}^l = \sum_{k=0}^{N-1} G_k^l \mathbf{B}_k \tag{2.22}$$

where $\{\mathbf{B}_k\}_{k=0}^{N-1}$ are the orthonormal eigenvectors of the equivalent discrete channel covariance $N \times N$ matrix $\mathbf{F} = E[\mathbf{C}^l \mathbf{C}^{l^H}]$ and $\left\{G_k^l\right\}_{k=0}^{N-1}$ are independent complex zero-mean Gaussian random variables with variance equal to the eigenvalues $\{\Gamma_k\}_{k=0}^{N-1}$ of the Hermitian matrix $\mathbf{F}$. The (p, q)th entry of this matrix is given by:

$$F_{pq} = \phi\left((m(p) - m(q))\Delta f, (n(p) - n(q))\Delta t\right) \tag{2.23}$$

As depicted in Figure 2.9, the multicarrier receiver is composed of L diversity branches provided by spatially decorrelated receiving antennas. We assume that the lth diversity branch output signal associated to the symbol a_{mn} can be written as:

$$R_{mn}^l = c_{mn}^l a_{mn}^l + N_{mn}^l \tag{2.24}$$

where c_{mn}^l is the discrete channel gain factor of the lth branch seen by the symbol a_{mn} and N_{mn}^l is a complex AWGN with variance N_0.

From this representation, a channel estimator, based on the Maximum A Posteriori (MAP) criterion is derived. The MAP estimate $\left\{\hat{\mathbf{G}}^l\right\}_{l=0}^{l=L-1}$ of $\left\{\mathbf{G}^l\right\}_{l=0}^{l=L-1}$ is defined as:

$$\left\{\hat{\mathbf{G}}^l\right\}_{l=0}^{l=L-1} = \arg \max_{\{\mathbf{G}^l\}_{l=0}^{l=L-1}} p\left(\left\{\mathbf{G}^l\right\}_{l=0}^{l=L-1} \middle| \left\{\mathbf{R}^l\right\}_{l=0}^{l=L-1}\right) \tag{2.25}$$

the $\left\{\mathbf{R}^l\right\}_{l=0}^{l=L-1}$ being the L received vectors on the L diversity branches.

This equation is iteratively solved by the means of the Expectation Maximisation Algorithm (EMA). The pth component of the lth branch re-estimate $\mathbf{G}^{l(d+1)}$ is given by:

$$G_p^{l(d+1)} = \omega_p \sum_{k=0}^{N-1} R_{\delta(k)}^l B_{p\delta(k)}^* \left(\sum_{A \in \Omega} AP\left(A_{\delta(k)} = A \middle| \{\mathbf{R}\}_{l=0}^{L-1}, \{\mathbf{G}^{l(d)}\}_{l=0}^{L-1}\right)\right)^* \tag{2.26}$$

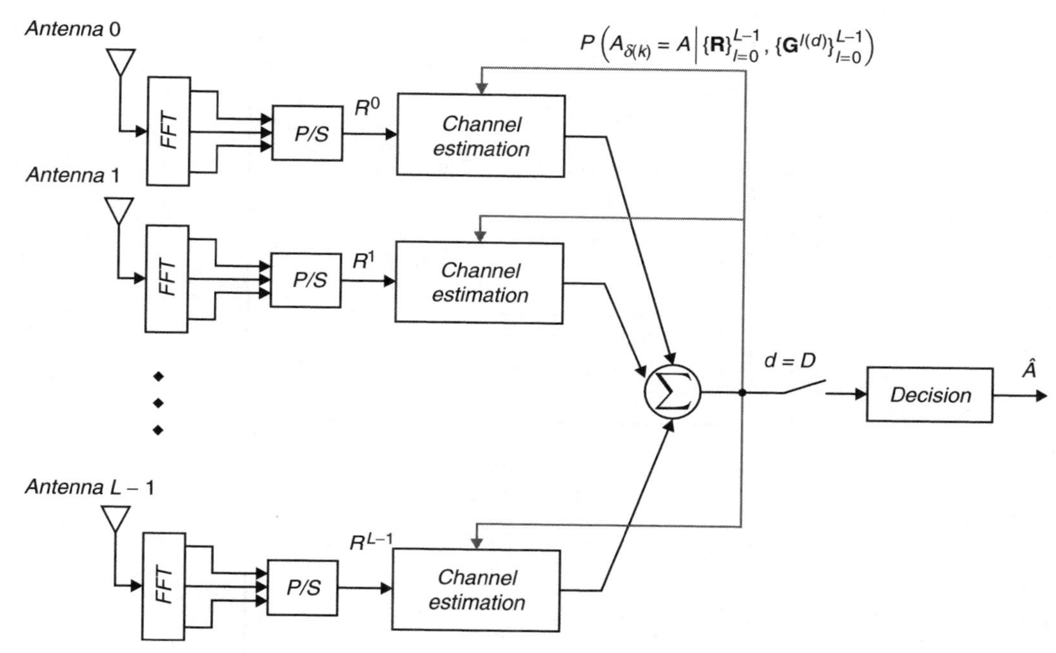

Figure 2.9 *Block diagram of the receiver.*

where $B_{p\delta(k)}$ is the kth component of $\mathbf{B}_p$ and

$$\omega_p = \frac{1}{1 + N_0/\Gamma_p} \tag{2.27}$$

We use

$$G_p^{l(0)} = \omega_p \sum_{\delta(k)\in S_P} R^l_{\delta(k)} D^*_{\delta(k)} B^*_{p\delta(k)} \tag{2.28}$$

as pth component of the initial guess $\{\mathbf{G}\}^{l(0)}$, where $D_{\delta(k)}$ is the value taken by the symbol pilot $A_{\delta(k)}$, $\delta(k) \in S_p$, S_p being the pilot positions set within the time-frequency block.

This method shows very good performance in the presence of extremely selective channels, Figure 2.10. Furthermore, very few iterations are needed for reaching the result of the MAP equation. Further complexity reduction can be achieved by representing the channel from the most powerful eigenvectors of the covariance matrix of the channel $\mathbf{F}$.

Joint channel estimation

Other techniques, taking into account more directly the multi-user case such as Joint Channel Estimation (JCE) [MWSL02] were investigated. Optimisations were made to this technique proposed by Maniatis *et al.* in [MWSL02] in terms of taking into account the shape of the receive antenna array, Figure 2.11.

Each Access Point (AP) of this system is equipped with an array antenna. The characteristics of the array will be taken into account deriving the equation of ML and Minimum Mean Square

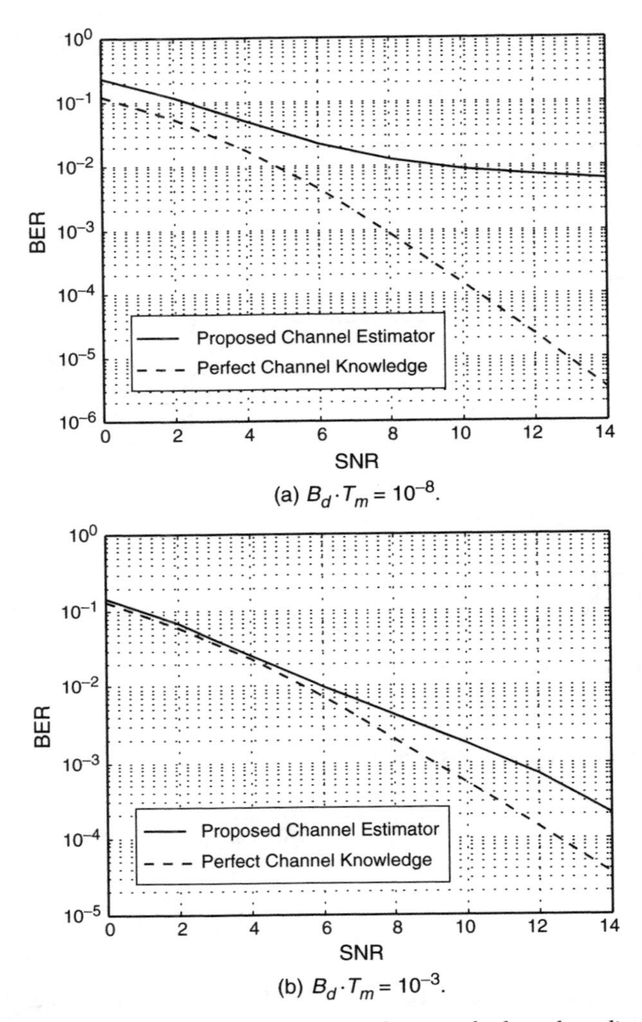

Figure 2.10 *Proposed channel estimation performance for very high and medium time and frequency selectivity.*

Error (MMSE) JCE. The ML estimates are given by:

$$\hat{\tilde{\mathbf{h}}} = \tilde{\mathbf{A}}\mathcal{F}_{W,tot}\left(\tilde{\mathcal{G}}_d^H\mathbf{R}_n^{-1}\tilde{\mathcal{G}}_d\right)^{-1}\tilde{\mathcal{G}}_d^H\mathbf{R}_n^{-1}\tilde{\mathbf{e}} \tag{2.29}$$

The MMSE estimate is given by:

$$\hat{\tilde{\mathbf{h}}} = \tilde{\mathbf{A}}\mathcal{F}_{W,tot}\mathbf{R}_{\mathbf{h}d}\tilde{\mathcal{G}}_d^H\left(\mathbf{R}_{\tilde{\mathbf{n}}} + \tilde{\mathcal{G}}_d\mathbf{R}_{\mathbf{h_d}}{}^{-1}\tilde{\mathcal{G}}_d^H\right)^{-1}\tilde{\mathbf{e}} \tag{2.30}$$

with $\tilde{\mathbf{A}}$, $\mathcal{F}_{W,tot}$, $\mathbf{R}_{\mathbf{h}d}$, $\tilde{\mathcal{G}}_d$, $\mathbf{R}_{\tilde{\mathbf{n}}}$.

The application of array antennas at the receiver of JOINT allows the inclusion of directional information in the estimation process concerning the Mobile Terminal (MT) and noise signals.

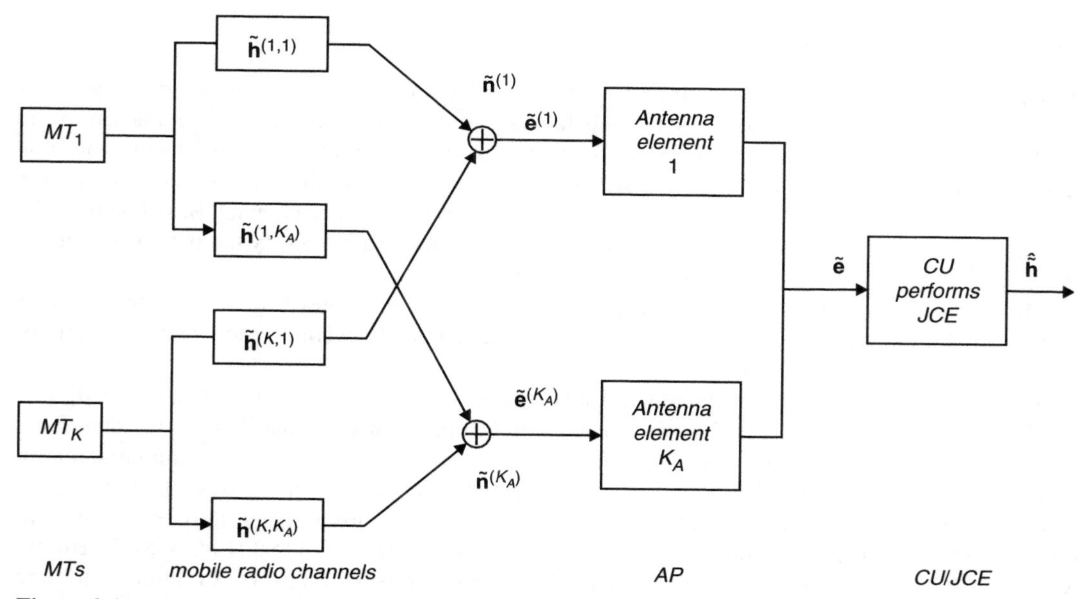

Figure 2.11 *System model considered for uplink channel estimation in JOINT.*

Together with the exploitation of frequency correlation, and of the known array geometry, the performance of ML-JCE is improved compared to the single RX antenna case. The performance of JCE is further enhanced by the application of the MMSE estimation principle, Figure 2.12. MMSE-JCE outperforms ML-JCE due to the additional channel state information included in the estimation process. Unlike ML-JCE, MMSE-JCE does not require a reduction of the number of unknown values to be estimated. On the other hand, MMSE-JCE does require a priori information about the considered point-to-point radio channels.

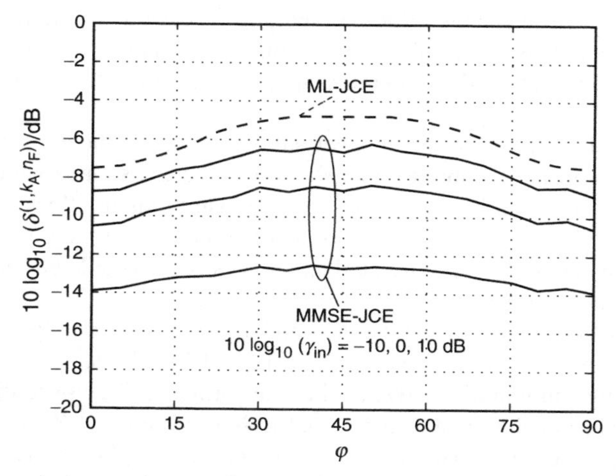

Figure 2.12 *SINR degradation vs. ϕ; omnidirectional noise scenario, $K = 4$ MTs, $K_A = 4$ RX antenna elements.*

Bit and power loading

Conventional OFDM systems use fixed constellation size and power level allocation of all subchannels. Due to multipath fading, some subchannels could experience severe degradation in the SNR resulting in high overall bit error rates, coding being a common technique to mitigate this effect. If the channel is static (e.g. in DSL) or slowly time varying, the receiver can provide the transmitter with Channel State Information (CSI) using a robust feedback channel. Based on the CSI, an adaptive transmission technique has the possibility to dynamically modify the parameters of the modulator in order to improve performance [KeHa00].

Multicarrier modulations, and OFDM in particular, have the advantage of allowing the transmitting power and bit rate of each subchannel to be changed dynamically according to channel selectivity variations (adaptive bit loading).

The first applications of bit loading algorithms appeared in DSL systems [Kale89], [Bing00]. It is well known that the theoretical channel capacity can be approached by distributing the total transmitted energy according to the *waterfilling* principle. In the realistic case, where a finite granularity in constellation size is required, the rounded bit distribution obtained starting from the *waterfilling* solution could still not be the optimum. Some sub-optimum algorithms to reduce the complexity have been proposed in the Asynchronous Digital Subscriber Line (ADSL) context [ChCB95], [FiHu96]. Campello [Camp98] gives the theoretically sufficient conditions for a discrete bit allocation to be optimal.

Recently, some studies regarding the application of adaptive bit loading algorithms to wireless channels appeared [Czyl96], [PTVG98], [BaFu01], [DMMC02]. In this case, particular attention must be paid to channel estimation and CSI update rate effects on the performance [SuSc01], [SoPi01], [YeBC02]. However, *waterfilling*-based techniques require a large overhead for CSI feedback, making them suitable only for the static or very slow time varying channels. Moreover, the modem must be able to continually change the modulation format and power on a subcarrier basis (high complexity if high data rates are requested).

It was proposed in [Dard04] to select only the strongest K subchannels (i.e. the subchannels characterised by a higher gain value) and to use higher constellation sizes by keeping the total bit rate and transmitted power unchanged. Ideally, the power loss due to the use of higher constellation size is partially compensated by the distribution of the total power over fewer subchannels. However, it is expected that the higher reliability of the strongest K subcarriers leads to an improved performance.

The receiver's task is to estimate the channel gains, H_n, select the K strongest (most reliable) subchannels and, through the feedback channel, inform the transmitter which to use in the next packet transmission. This could be accomplished by transmitting an allocation mask vector every T_{csi} seconds, Figure 2.13, where a bit in the nth position means that the nth subchannel belongs to the usable set of subchannels. A Cyclic Redundancy Check (CRC) could be added to each allocation mask vector for error detection.

Figure 2.14 shows simulations of the 4-QAM and 16-QAM systems for different loading configurations. The corresponding useful data rates are 12 and 24 Mbit/s, respectively. The $K = N/2$ and $K = 2/3N$ loading configurations give the best performance. Thanks to the diversity due to the ordering process, a considerable gain in the Signal-to-Noise-Ratio (SNR) is achieved. Figure 2.15 shows the average packet error probability $\overline{P}_{ep}$ as a function of E_b/N_0 for the 12 Mbit/s system over correlated fading with coding. Again, the best loading condition is for $K = N/2$. For $\overline{P}_{ep} = 10^{-2}$, the gain obtained to the reference scheme is around 6–7 dB.

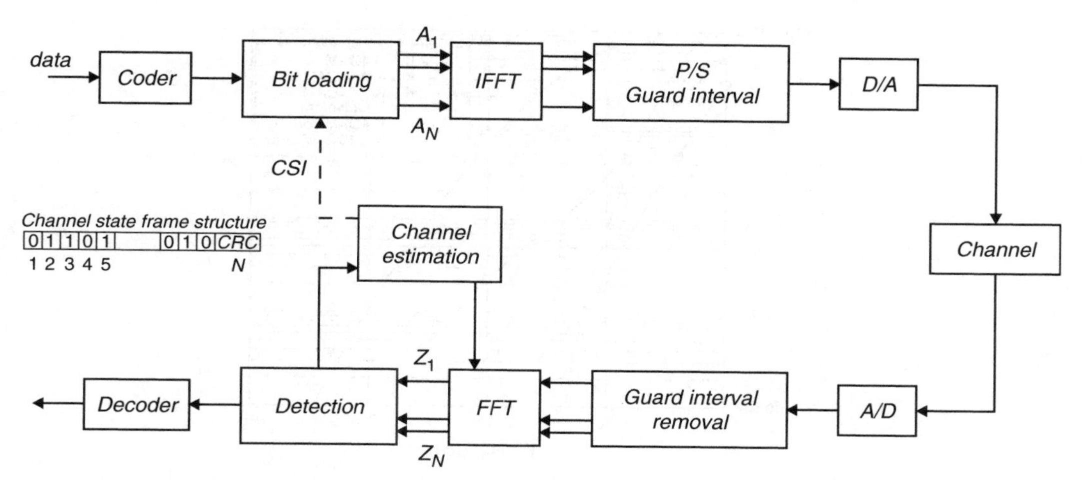

Figure 2.13 *Transmission system block diagram.*

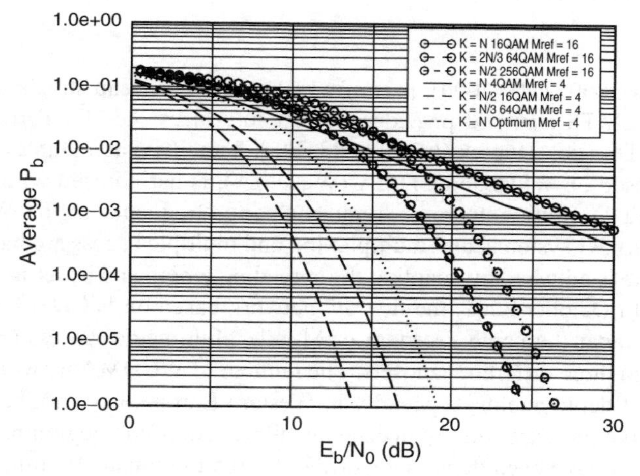

Figure 2.14 *Average bit error probability as a function of E_b/N_0 in the uncoded and independent Rayleigh fading simulations.*

Those results are obtained with a considerable lower hardware complexity compared to that required by the optimal solution due to the same constellation size and power level constraint.

2.3 CDMA systems

2.3.1 Introduction

Code Division Multiple Access (CDMA) is the most important multiple access scheme for 3G mobile communication systems. These systems follow the worldwide standard International Mobile

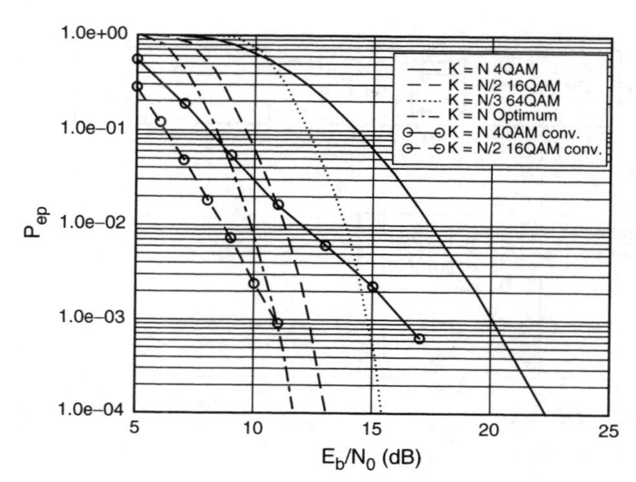

Figure 2.15 *Average packet error probability as a function of E_b/N_0. Different loading conditions: $K = N$, $K = N/2$ and $K = N/3$. 12 Mbit/s ($M_{ref} = 4$) reference system. Circled curves: simulations with convolutional coding (soft-decision).*

Telecommunications-2000 (IMT-2000), part of which is the European standard UMTS, adopted in January 1998. The UMTS standard provides two air interfaces: UMTS Terrestrial Radio Access (UTRA)-Frequency Division Duplex (FDD) and UTRA-Time Division Duplex (TDD).

UTRA-FDD is based on Wideband CDMA (WCDMA), is harmonised with other 3G WCDMA solutions, and is an FDD solution, therefore requiring paired frequency bands. WCDMA is the dominating air interface of 3G systems, and multiplexing and multiple access are based only on CDMA. The required data rate is adjusted by adapting the spreading length and by assigning up to six unique codes to each user. In October 2001, the first 3G network based on WCDMA started in the Tokyo metropolitan areas, under the name Freedom of Mobile Multimedia Access (FOMA). Subscribers could choose between three different handsets. The number of WCDMA networks started to grow in 2004. However, most deployments, especially in Western Europe, have not been driven by market demand or competitive pressure, but by licence deadlines. As of the beginning of 2005, more than 60 WCDMA networks have been deployed worldwide, and more than 16 million subscribers could choose between around 100 different WCDMA devices (handsets and PC card products).

UTRA-TDD is based on Time Division-Code Division Multiple Access (TD-CDMA) and uses TDD, therefore being adopted for the unpaired UMTS frequency bands. In this standard, multiplexing and multiple access are based on time and code division multiplexing, which leads to a narrowband CDMA technology. Starting in 2002, this technology is being deployed in various countries worldwide to offer mobile broadband Internet services competing with wired DSL solutions.

Time Division-Synchronous Code Division Multiple Access (TD-SCDMA), an enhanced version of TD-CDMA, is the Chinese contribution to IMT-2000 and will be commercially ready in 2005 according to the TD-SCDMA Forum [TDSC05].

MC-CDMA is a combination of OFDM and CDMA. It was proposed independently by several groups in 1993, and is being considered a key technology for 4G mobile communication systems. An introduction to MC-CDMA can be found in [FaKa03], [Lind99]. The high data rate mode of CDMA-2000, a family of standards based on the narrowband CDMA standard

Interim Standard-95 (IS-95), uses three 1.25 MHz wide carriers. Therefore it can be considered as a variant of MC-CDMA. It is part of IMT-2000, being deployed in several countries, especially in Asia and North America.

This section covers the system specific aspects of CDMA. Details on the various algorithms used in CDMA systems can be found in Sections 3.3 (equalisation), 3.5 (multi-user systems and multi-user detection), 7.3 (array processing and beamforming), and 7.4 (MIMO transmission techniques). The remainder of this section is divided into three parts, each one devoted to a specific CDMA system. Subsection 2.3.2 deals with topics related to WCDMA systems. Quite a few papers treat the combination of WCDMA and multiple receive and/or multiple transmit antennas (see also Sections 7.3 and 7.4) with interference cancellation techniques being another important topic (see also Section 3.5). In Subsection 2.3.3 topics related to TD-CDMA and TD-SCDMA are covered, a few papers dealing with the signal processing scheme 'joint transmission' (see also Section 3.3). Finally, Subsection 2.3.4 deals with all aspects related to systems based on MC-CDMA.

2.3.2 *Wideband CDMA*

Direct Sequence-CDMA (DS-CDMA) and FDD are transmission techniques that are currently part of the UMTS standard, and are being considered for use in future communication systems as well, e.g. [SaCl03]. All users within a DS-CDMA system are separated by a distinct code which is assigned to each user. In contrast to MC-CDMA and Orthogonal Frequency Division Multiple Access (OFDMA), codes are applied in the time domain and each user's resulting signal occupies the whole frequency band. A series of papers, Temporary Documents (TDs) [Ener01], [DMFM01], [Baum03], [MRFB03], [Pomm02], [FoBe02], [OsNA05], [LSPB04], [GaCl03], [BoPM04], [SaCl03], [MDFC05] and Workshop Papers (WPs) [MDFB03], [DPLC03] discuss various physical layer issues related to WCDMA and other systems employing DS-CDMA as a multiple access scheme. In the following, a few of these issues are addressed here. However, it should be noted that not all aspects, like, for example, Vector Detection techniques, are solely a matter of DS-CDMA systems, but have to be considered in various other systems as well. Therefore, these aspects are treated in a more general way in a dedicated section.

The cardinality of the set of spreading codes limits the number of users which may be supported by a cell. Beamforming and MIMO techniques, see also Sections 7.3 and 7.4, have been applied to DS-CDMA transmission schemes, [Ener01], [DMFM01], [FoBe02], [BoPM04], in order to increase the data rate, the number of supported users, and/or the reliability of a communication link. A particular issue that is restricted to FDD schemes is the use of different frequency bands in the up- and downlinks. The up- and downlink frequency bands are usually spaced further apart than the coherence bandwidth of the physical channel, which prevents the use of the channel impulse response from the downlink in the uplink and vice versa. In [FoBe02], it is analysed how much information may be extracted for the weighting vector of the downlink from the uplink. The analysis is based on measurement results with an eight element uniform linear antenna array in a typical urban environment. Their conclusion is that the Direction of Arrival (DoA) for channels that are separated by 200 MHz are virtually identical, despite the fact that the channels fade independently. Furthermore, they conclude that the additional computational complexity of super resolution techniques is not justified by the small performance gain that one obtains with respect to the conventional Fourier method for the estimation of the DoA. Different blind beamformers for DS-CDMA systems are compared [BoPM04]. This paper also proposes a new technique that is robust against mismatches due to estimation errors or time varying channels and at the same time outperforms the other presented techniques, e.g. the direct matrix inversion solution of the classical modified code filtering approach, if strong interferers are present. The convergence of a semi-blind algorithm, also for the uplink, is discussed [Pomm02]. The analysis is based on the least square algorithm and assumes the particular data structure of a UTRA-FDD signal. The main result of this work is an alternative simulation method with reduced computational complexity. Furthermore, it is shown that only a few iterations are necessary for the convergence of the algorithm due to a high tolerance against data estimation errors. A spatial temporal channel model for evaluating the performance of adaptive antennas in CDMA systems was proposed [Ener01]. The model includes statistical data obtained from measurement campaigns.

Orthogonal code sequences guarantee perfect orthogonality among users under ideal conditions only. Unfortunately, due to the impairment of the physical channel [LSPB04] and possible asynchronity among users especially in the uplink, orthogonality gets destroyed and one has to cope with interference among users. Therefore, CDMA has been a driving force for the development of vector detection techniques that allow for a good extraction of a user's signal from the overall received signal. As optimum detectors, which consider all users jointly and minimise the overall BER, are usually far too complex, powerful sub-optimum solutions have to be applied. Iterative interference cancellers seem to be the most promising solution that may actually reach the single user performance at times. Various TDs [DMFM01], [MRFB03], [OsNA05], [MDFC05], [MDFB03] present a detailed analysis on different interference cancellation techniques using DS-CDMA as a transmission scheme. The use of DS-CDMA in conjunction with MIMO or space-time codes, e.g. [OsNA05], puts an even higher burden on detectors at the receivers, due to the additional interference caused by the signals that each user transmits simultaneously.

Other aspects concerning UTRA-FDD have been discussed. In [Baum03], the optimum power ratio between the pilot bits and the information bits is considered. Figure 2.16 shows the results for the International Telecommunications Union (ITU) Vehicular A radio channel with mobile speeds

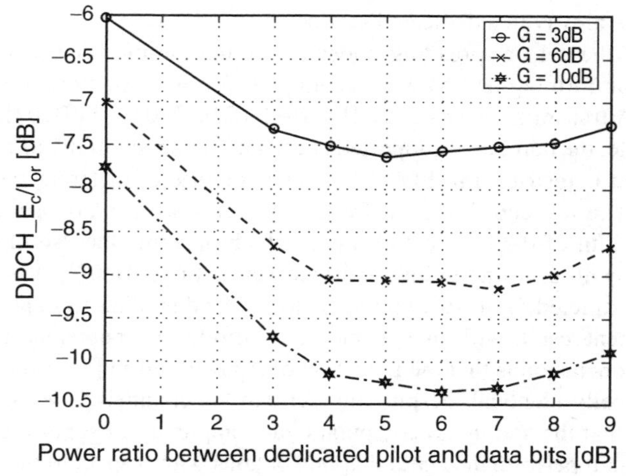

Figure 2.16 *Required DPCH$_{E_c}/I_{or}$ for BLER = 1% for the 144 kbit/s reference channel using different power ratios between dedicated pilot bits and information bits if the dedicated pilot bits are used for channel estimation.*

of 3 km/h and different geometry factors G. Their conclusion is that the performance may be improved if the power is adapted separately for the pilot bits and the information bits. Additionally, they state that the optimum solution is feasible within the framework of the current UMTS specifications. Multisensor synchronisation and demodulation algorithms for the downlink of the UMTS FDD mode are considered in [DPLC03].

In [GaCl03], the sensitivity of DS-CDMA and MC-CDMA to phase noise is discussed. Both are considered as candidates for future wireless communication systems in the 60 GHz band. Both schemes show a similar performance degradation with increasing phase noise, with a slight advantage in favour of DS-CDMA using orthogonal Walsh codes for small system load. All techniques require a βT_s smaller than 0.02. For all loads, OFDM-Time Division Multiple Access (TDMA) is the most sensitive technique because of the loss of subcarrier orthogonality.

2.3.3 Time division CDMA

Joint Detection (JD) is an advantageous detection scheme for time slotted CDMA systems, which leads to air interface concepts known as Joint Detection-Code Division Multiple Access (JD-CDMA) [Klei96], [LuBa00]. Also TD-CDMA, the air interface for the TDD bands of UMTS and IMT-2000 [HKKO00], utilises JD as an option, which allows the total elimination of intracell multiple access interference and intersymbol interference. It is expected that in many applications of TD-CDMA, for instance web browsing, the downlink has to support much higher data rates than the uplink. Therefore, in TD-CDMA the downlink tends to be the bottleneck, and it would be desirable to enhance the performance especially of this link. In what follows, we will show how such an enhancement can be achieved by a non-obvious and inexpensive combination of conventional JD-CDMA with the scheme Joint Transmission (JT) recently proposed in [MBWL00], [MBLP00], [BMWT00], [MTWB01]. As a prerequisite of this combination, the channel impulse responses valid in the downlink have to be known at the Base Station (BS). In the case of TDD systems, this knowledge is at least approximately available from the uplink channel estimation. The combination of JT with multiple transmit antennas at the BS is especially favourable. With JD it is easily possible to make optimum use of multiple receive antennas [StBl96], [BKNS94]. JT offers a solution dual to the one exploited in JD to make optimum use of multiple transmit antennas. In order to keep the presentation as concise as possible, the reader is assumed to be familiar with the contents of [Klei96], [LuBa00], [HKKO00], [MBWL00], [MBLP00], [BMWT00] and with the notations introduced in those publications.

Let us consider a conventional JD-CDMA downlink where the BS supports K MT $k = 1 \ldots K$. The K partial data vectors $\underline{\boldsymbol{d}}^{(k)}$, $k = 1 \ldots K$, to be transmitted by the BS to the individual MTs $k = 1 \ldots K$ can be stacked to form the total data vector:

$$\underline{\boldsymbol{d}} = \left(\underline{\boldsymbol{d}}^{(1)^{\mathrm{T}}} \ldots \underline{\boldsymbol{d}}^{(K)^{\mathrm{T}}}\right)^{\mathrm{T}} \tag{2.31}$$

Generally, both the BS and the MTs dispose of more than one antenna. By means of a modulator matrix $\underline{\boldsymbol{M}}$ that is a priori constituted by the utilised CDMA codes and – if multiple transmit antennas are employed at the BS – also by the antenna weights [LuBa00], the total transmit signal

$$\underline{\boldsymbol{s}} = \underline{\boldsymbol{M}}\,\underline{\boldsymbol{d}} \tag{2.32}$$

is generated. The channel impulse responses between the individual antennas of the BS and MT k, can be represented by the partial channel matrix $\underline{\boldsymbol{H}}^{(k)}$ [LuBa00]. Then, the desired signal impinging at MT k is given by

$$\underline{e}^{(k)} = \underline{H}^{(k)}\underline{s} = \underline{H}^{(k)}\underline{M}\,\underline{d} \tag{2.33}$$

where $\underline{s}$ in the middle term of (2.33) is substituted by (2.32) in order to obtain the right-hand term of (2.33). At MT k, with $\underline{e}^{(k)}$ of (2.33) and following the zero forcing algorithm to perform JD, the estimate

$$\hat{\underline{d}} = \left[\left(\left(\underline{H}^{(k)}\underline{M}\right)^{*\mathrm{T}}\underline{H}^{(k)}\underline{M}\right)^{-1}\left(\underline{H}^{(k)}\underline{M}\right)^{*\mathrm{T}}\right]\underline{e}^{(k)} \tag{2.34}$$

of $\underline{d}$ of (2.31) can be obtained [Klei96], [LuBa00], where the matrix in brackets is a posteriori determined at MT k. Of course, at MT k only the estimate

$$\hat{\underline{d}}^{(k)} = \underbrace{\left[\left(\left(\underline{H}^{(k)}\underline{M}\right)^{*\mathrm{T}}\underline{H}^{(k)}\underline{M}\right)^{-1}\left(\underline{H}^{(k)}\underline{M}\right)^{*\mathrm{T}}\right]_{kN}^{(k-1)N+1}}_{\underline{D}^{(k)}}\underline{e}^{(k)} \tag{2.35}$$

of the partial data vector $\underline{d}^{(k)}$ pertaining to MT k is of interest, where $[\cdot]_j^i$ stands for a matrix $\underline{D}^{(k)}$ consisting of lines i through j of the matrix in brackets. Nevertheless, the matrix $\left(\left(\underline{H}^{(k)}\underline{M}\right)^{*\mathrm{T}}\underline{H}^{(k)}\underline{M}\right)^{-1}\cdot\left(\underline{H}^{(k)}\underline{M}\right)^{*\mathrm{T}}$ utilised in conventional JD allows data that is not useful to a particular MT to be detected there, which is unnecessary. This feature of the conventional JD-CDMA can be considered as an unnecessary restriction when generating the transmit signal $\underline{s}$. Generating $\underline{s}$ according to (2.32) would have the effect that also irrelevant data could be detected by JD at MT k. Therefore, $\underline{s}$ should rather be generated under the sole consideration of the data of interest to this MT, being $\underline{d}^{(k)}$. This rationale would offer additional degrees of freedom when designing $\underline{s}$, which for instance could be exploited to minimise the required transmission energy and, consequently, the interference to MTs not belonging to the ensemble of the K MTs $k = 1\ldots K$.

As shown in [MBWL00], [MBLP00], [BMWT00], in the JT scheme the demodulators at the MTs $k = 1\ldots K$ are a priori determined in the form of MT specific demodulator matrices. These matrices, together with the partial channel matrices $\underline{H}^{(k)}$ already explained above, serve as the inputs for a posteriori determining a modulator matrix $\underline{M}$, which, in general, differs from the modulator matrix employed in conventional JD-CDMA.

Now, as the crux of our approach, the matrices $\underline{D}^{(k)}$ of (2.35) can be considered as the partial demodulator matrices of a transmission system using JT. As shown in [MBWL00], [MBLP00], [BMWT00] the partial demodulator matrices of a JT system can be stacked in a block-diagonal matrix termed the total demodulator matrix. With the matrices $\underline{D}^{(k)}$ of (2.35) as the partial demodulator matrices, the total demodulator matrix

$$\underline{D} = \mathrm{blockdiag}\left(\underline{D}^{(1)}\ldots\underline{D}^{(K)}\right) \tag{2.36}$$

can be formed. The partial channel matrices $\underline{H}^{(k)}$ introduced above can be put together to form the total channel matrix [MBWL00], [MBLP00], [BMWT00]

$$\underline{H} = \left(\underline{H}^{(1)^{\mathrm{T}}}\ldots\underline{H}^{(K)^{\mathrm{T}}}\right)^{\mathrm{T}} \tag{2.37}$$

According to the scheme JT, $\underline{\boldsymbol{D}}$ of (2.36) and $\underline{\boldsymbol{H}}$ of (2.37) are used to generate the JT modulation matrix [MBWL00], [MBLP00], [BMWT00]

$$\underline{\boldsymbol{M}} = \left(\underline{\boldsymbol{D}}\,\underline{\boldsymbol{H}}\right)^{*\mathrm{T}}\left(\underline{\boldsymbol{D}}\,\underline{\boldsymbol{H}}\left(\underline{\boldsymbol{D}}\,\underline{\boldsymbol{H}}\right)^{*\mathrm{T}}\right)^{-1} \tag{2.38}$$

If the total transmit signal $\underline{\boldsymbol{s}}$ of (2.32) is generated by the a posteriori determined modulator matrix $\underline{\boldsymbol{M}}$ of (2.38) instead of the a priori determined modulator matrix of conventional JD-CDMA, then the total transmitted energy $||\underline{\boldsymbol{s}}||^2/2$ is minimised [MBWL00], [MBLP00], [BMWT00]. Figure 2.17 shows the system model of a MIMO JT scheme.

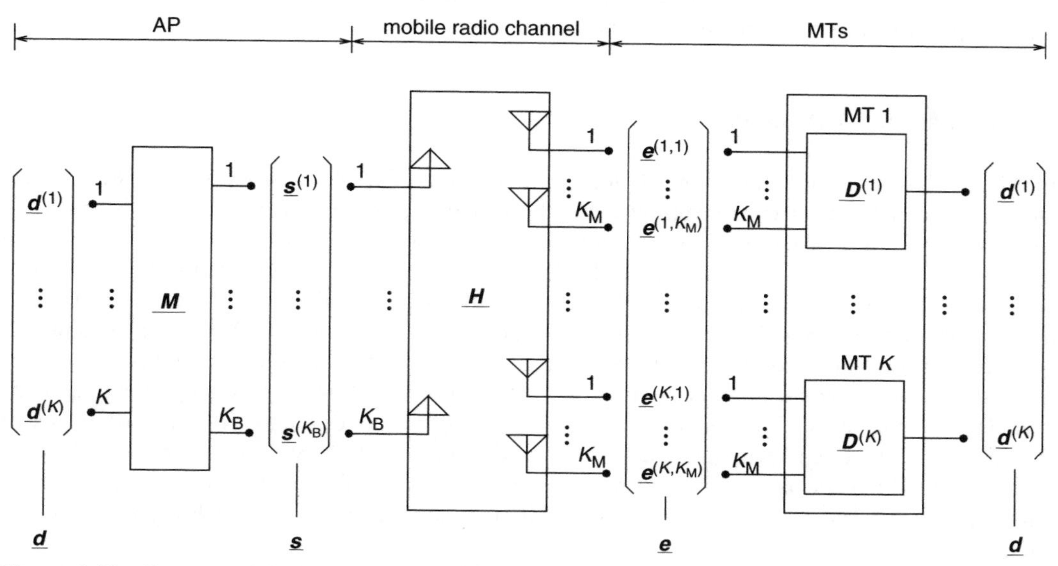

Figure 2.17 *System model of a MIMO JT scheme.*

In [MoGF03], it is also considered a downlink scenario and utilises the TDD feature of UTRA-TDD for a space-time pre-equalisation on the transmit side in order to simplify the MT receiver. An equaliser synthesis method is presented that explores both frequency redundancy of DS-CDMA and the spatial diversity achieved through the use of an array of antennas at the BS. Filter coefficients are obtained using the zero-forcing criterion. More details on the algorithm can be found in Section 3.5. Simulations have been performed with an UTRA-TDD compliant simulator using the ITU Pedestrian A channel model. Using several antennas at the BS the BER performance is identical to the one achieved in the single user case, even in a fully loaded system.

In [GiCo02], the performance of the conjugate gradient beamforming algorithm applied to a UTRA-TDD macro-cell scenario is presented. A wideband directional channel model based on the circular single bounce model is used. Simulation results mainly cover the final beamforming processing gain, focusing on its relation to scenario characteristics. The beamforming gain is not greatly affected by the MT distance to the BS. The radius of the scattering circle around the MT and the total number of scatterers have a role, the type of MT grouping having the major impact on the beamforming performance. Generally, parameter changes leading to lower angular freedom to place lobes and nulls towards the desired signal, respectively the non-desired interference lead to lower beamforming gains.

In [TeRe01], the TD-SCDMA system is described. TD-SCDMA is part of the 3G mobile communication standards IMT-2000 and UMTS. It is an improved narrowband variant of TD-CDMA. The uplink time synchronisation has been improved, such that the main paths of each user arrive at the same time at the BS. Since orthogonal codes are used, a large portion of the interuser interference can be avoided in this way. Together with the use of multiple receive antennas, the system capacity can be increased. For the same spreading length of 16, in TD-SCDMA up to 16 codes can be used, in contrast to only eight for TD-CDMA. For a system with full load, the BER performance of different non-linear VD algorithms based on recurrent neural networks (see Section 3.5) is investigated. Both for time-invariant and for time-variant multipath channels, the non-linear VD algorithms show superior performance compared to the standard linear detector.

2.3.4 Multicarrier CDMA

In [Manz02], several code allocation policies for the downlink of an MC-CDMA system are discussed. Quasi-orthogonal sequences are employed and the variable spreading factor technique is applied to support different user data rates. For the mixed allocation strategy, in turn a Walsh code followed by a quasi-orthogonal code are assigned. In this case, the performance of each user is the same, regardless of the code sequence assigned to him. For the separate allocation strategy, on the other hand, Walsh codes are assigned first, one after the other. Only when the Walsh set is depleted, quasi-orthogonal sequences are introduced in the system. Here, Walsh users experience a gradually increasing interference as more users enter the system. The very first active quasi-orthogonal code user, on the other hand, immediately perceives the maximum interference from the whole set of Walsh codes. For systems with users with different data rates, a strategy is used where Walsh codes are assigned to high rate users, while low rate users employ quasi-orthogonal sequences. The different interference experienced by different classes of users is also reflected in the BER performance of the different classes of users.

In [SäMo03], Space Division Multiple Access (SDMA) is applied to the downlink of a wireless cellular system based on MC-CDMA. Specifically, joint transmit filtering in space and frequency dimension is used. The techniques are based on maximum ratio transmission and maximisation of the signal to interference plus noise ratio. Both techniques lead to a considerable reduction of the multiple access interference, thus enabling the use of very low complex detection techniques at the MT. Furthermore, system capacity can be significantly increased by this combination of SDMA and MC-CDMA. A detailed description of the algorithm is given in Section 3.5.

TD [MoCa03] considers channel pre-equalisation for an uplink Orthogonal Frequency Code Division Multiplexing (OFCDM) scheme. TDD offers channel reciprocity between up- and downlinks. A low-complexity channel pre-equalisation avoids explicit channel estimation at the BS and, therefore, increases the data rate in the uplink, since there is no need for pilot symbols. The BER performance is very good for MTs with low velocities. For higher velocities of the MTs, a data-aided adaptive detection technique is used to compensate the increasing mismatch between channel pre-equalisation and the actual channel impulse response. The proposed algorithm showed good BER performance for moderate MT velocities up to 60 km/h. In case of large Doppler, the system load can be maximised by using small spreading factors.

In [MoSä04], the spreading sequence assignments in the downlink of an OFCDM system using multiple transmit antennas is discussed. The algorithm applies transmit beamforming in order to reduce the multiple access interference. It involves a selection of the needed subset of spreading sequences and a distribution of these selected sequences according to the spatial signature of MTs. Simulation results over realistic transmission scenarios emphasise the necessity to optimise the assignment of spreading sequences, as well as the efficiency of the proposed solution.

In [DYMT03], the performance of a MIMO-MC-Code Division Multiplex (CDM) system is compared with a MIMO-OFDM system without spreading. Applying iterative detection and decoding (turbo detection) at the receiver, it is shown that MIMO-MC-CDM outperforms MIMO-OFDM for nearly all detection schemes considered. Details on the used detection algorithm can be found in Section 3.5.

TD [HTBS05] proposes a frequency-averaged MMSE channel estimator. Results for MC-CDMA systems show that this estimator can effectively reduce estimation errors and keeps a better performance than conventional robust estimators. The estimator is also used for OFDM. For more details on OFDM, see Section 2.2, and on the estimation algorithms, see Section 3.4.

References

[AlLa87] M. Alard and R. Lassale. Principles of modulation and channel coding for digital broadcasting for mobile receivers. *EBU Technical Review*, 224:168–190, Aug. 1987.

[BaFH96] R. W. Bauml, R. F. H. Fischer, and J. B. Huber. Reducing the peak-to-average power ratio of multicarrier modulation by selected mapping. *Elect. Lett.*, 32(22):2056–2057, Oct. 1996.

[BaFu01] A. N. Barreto and S. Furrer. Adaptive bit loading for wireless OFDM systems. In *Proc. PIMRC 2001 - IEEE 12th Int. Symp. on Pers., Indoor and Mobile Radio Commun.*, San Diego, CA, USA, Oct. 2001.

[Baum03] T. Baumgartner. Optimum power ratio between dedicated pilot bits and information bits if user specific beamforming is used in UMTS FDD. TD(03)019, COST 273, Barcelona, Spain, Jan. 2003.

[BeEr02] A. Berhavan and T. Erriksson. PAPR and other measures for OFDM systems with nonlinearity. In *Proc. WPMC 2002 - Wireless Pers. Multimedia Commun.*, Honolulu, HI, USA, Oct. 2002. [Also available as TD(03)050].

[Bing00] J. A. C. Bingham, editor. *ADSL, VDSL and multicarrier modulation*. John Wiley & Sons Ltd., New York, NY, USA, 2000.

[BKNS94] J. J. Blanz, A. Klein, M. M. Naßhan, and A. Steil. Performance of a cellular hybrid C/TDMA mobile radio system applying joint detection and coherent receiver antenna diversity. *IEEE J. Select. Areas Commun.*, 12(4):568–579, May 1994.

[BMWT00] P. W. Baier, M. Meurer, T. Weber, and H. Tröger. Joint transmission (JT), an alternative rationale for the downlink of time division CDMA using multi-element transmit antennas. In *Proc. ISSSTA 2000 - IEEE 6th Int. Symp. on Spread Spectrum Techniques and Applications*, Parsippany, NJ, USA, Sep. 2000.

[BoPM04] M. Borgo, S. Pupolin, and B. Matteo. A novel robust beamforming based on convex optimization for CDMA space-time receivers. TD(04)122, COST 273, Gothenburg, Sweden, June 2004.

[Brai00] R. N. Braithwaite. Using walsh code selection to reduce the power variance of band limited forward-link CDMA waveforms. *IEEE J. Select. Areas Commun.*, 18(11):2260–2269, Nov. 2000.

[BSGS02] A. R. S. Bahai, M. Singh, A. J. Goldsmith, and B. R. Saltzberg. A new approach for evaluating clipping distortion in multicarrier systems. *IEEE J. Select. Areas Commun.*, 20(5):462–472, May 2002.

[Camp98] J. Campello. Optimal discrete bit loading for multicarrier modulation systems. In *Proc. ISIT 1998 - IEEE Int. Symp. on Information Theory*, Cambridge, ME, USA, Aug. 1998.

[ChCB95] P. S. Chow, J. M. Cioffi, and J. A. C. Bingham. A practical discrete multitone transceiver loading algorithm for data transmission over spectrally shaped channels. *IEEE Trans. Commun.*, 43(2/3/4):773–775, Feb./Mar./Apr. 1995.

[CiBi94] J. M. Cioffi and A. C. Bingham. A data-driven multitone echo canceller. *IEEE Trans. Commun.*, 42(10):2853–2869, Oct. 1994.

[Corr01] L. M. Correia, editor. *Wireless Flexible Personalised Communications: COST 259, European Co-operation in Mobile Radio Research*. John Wiley & Sons Ltd., New York, NY, USA, 2001.

[Czyl96] A. Czylwik. Adaptive OFDM for wideband radio channels. In *Proc. Globecom 1996 - IEEE Global Telecommunications Conf.*, London, UK, Nov. 1996.

[Dard04] D. Dardari. A uniform power and constellation size bit-loading scheme for OFDM based WLAN systems. In *Proc. PIMRC 2004 - IEEE 15th Int. Symp. on Pers., Indoor and Mobile Radio Commun.*, Barcelona, Spain, Sep. 2004. [Also available as TD(03)155].

[DiWu98] N. Dinur and D. Wulich. Peak-to-average power ratio in amplitude clipped high order OFDM. In *Proc. MILCOM 1998 - IEEE Military Communications Conference*, Boston, MA, USA, Oct. 1998.

[DMFM01] E. Del Re, S. Morosi, R. Fantacci, and D. Marabissi. Low complexity selective interference cancellator for a WCDMA communication system with antenna array. TD(01)051, COST 273, Bologna, Italy, Oct. 2001.

[DMMC02] D. Dardari, M. G. Martini, M. Milantoni, and M. Chiani. MPEG-4 video transmission in the 5GHz band through an adaptive OFDM wireless scheme. In *Proc. PIMRC 2002 - IEEE 13th Int. Symp. on Pers., Indoor and Mobile Radio Commun.*, Lisbon, Portugal, Sep. 2002.

[DPLC03] D. Depierre, F. Pipon, P. Loubaton, and J.-M. Chaufray. Multi-sensor synchronization and demodulation algorithms in the downlink of the UMTS FDD mode. WP(03)009, COST 273, Paris, France, May 2003.

[DYMT03] M. A. Dangl, D. Yacoub, U. Marxmeier, W. G. Teich, and J. Lindner. Performance of joint detection techniques for coded MIMO-OFDM and MIMO-MC-CDM. WP(03)017, COST 273, Paris, France, May 2003.

[Ener01] P. Eneroth. A stochastic spatial temporal channel model for evaluating the performance of adaptive antennas in CDMA-systems. TD(01)011, COST 273, Brussels, Belgium, May 2001.

[Enge02] M. Engels, editor. *Wireless OFDM Systems: How to make them work?* Kluwer Academic Publishers, London, UK, 2002.

[FaKa03] K. Fazel and S. Kaiser. *Multi-carrier and spread spectrum systems*. John Wiley & Sons Ltd., New York, NY, USA, 2003.

[FiHu96] R. F. H. Fisher and J. B. Huber. A new loading algorithm for discrete multitone transmission. In *Proc. Globecom 1996 - IEEE Global Telecommunications Conf.*, London, UK, Nov. 1996.

[FoBe02] S. E. Foo and M. Beach. Uplink based downlink beamforming in UTRA FDD. TD(02)104, COST 273, Lisbon, Portugal, Sep. 2002.

[Frie97] M. Friese. OFDM signal with low-crest factor. In *Proc. Globecom 1997 - IEEE Global Telecommunications Conf.*, Phoenix, AZ, USA, Nov. 1997.

[GaCl03] C. Garnier and L. Clavier. Sensitivity of various multiple access techniques to phase noise. TD(03)111, COST 273, Paris, France, May 2003.

[GeCD04] G. Gelle, M. Colas, and D. Declercq. Turbo decision aided reconstruction of clipping noise in coded OFDM. In *Proc. SPAWC 2004 - Sig. Proc. Advances in Wireless Commun.*, Lisbon, Portugal, July 2004. [Also available as TD(04)013].

[GiCo02] J. M. Gil and L. M. Correia. Dependence of adaptive beamforming performance on directional channel macro-cell scenarios for UMTS. TD(02)050, COST 273, Espoo, Finland, May 2002.

[HKKO00] M. Haardt, A. Klein, R. Koehn, S. Oestreich, M. Purat, V. Sommer, and T. Ulrich. The TD-CDMA based UTRA TDD mode. *IEEE J. Select. Areas Commun.*, 18(8): 1375–1385, Aug. 2000.

[HTBS05] Y. Hara, A. Taira, L. Brunel, and T. Sälzer. Frequency-averaged MMSE channel estimator for multicarrier transmission schemes. TD(05)002, COST 273, Bologna, Italy, Jan. 2005.

[JaSi00] E. Jaffrot and M. Siala. Turbo channel estimation for OFDM systems on highly time and frequency selective channels. In *Proc. ICASSP 2000 - IEEE Int. Conf. Acoust. Speech and Signal Processing*, Istanbul, Turkey, June 2000. [Also available as TD(02)025].

[Kale89] I. Kalet. The multitone channel. *IEEE Trans. Commun.*, 37(2):119–124, Feb. 1989.

[KeHa00] T. Keller and L. Hanzo. Adaptive multicarrier modulation : a convenient framework for time-frequency processing in wireless communications. *IEEE Proc. of the IEEE*, 88(5):611–640, May 2000.

[KiSt99] D. Kim and S. Stuber. Clipping noise mitigation for OFDM by decision aided reconstruction. *IEEE Commun. Lett.*, 3(1):4–6, Jan. 1999.

[Klei96] A. Klein. *Multi-user detection of CDMA signals – algorithms and their application to cellular mobile radio*. Number 423 in Fortschrittberichte VDI, Reihe 10. VDI-Verlag, Düsseldorf, Germany, 1996.

[LiCi98] X. Li and L. Cimini. Effects of clipping and filtering on the performance of OFDM. *Elect. Lett.*, 2(5):131–133, May 1998.

[Lind99] J. Lindner. MC-CDMA in the context of general multisubchannel-multiuser transmission methods. *European Transactions on Telecommunications*, 10(4):351–367, Aug. 1999.

[LSPB04] S. Lüder, C. Schneider, M. Pettersen, and L.-E. Braten. Downlink orthogonality degradation in UMTS FDD. TD(04)091, COST 273, Gothenburg, Sweden, June 2004.

[LuBa00] Y. Lu and P. W. Baier. Performance of adaptive antennas for the TD-CDMA downlink under special consideration of multi-directional channels and CDMA code pooling. *AEÜ, International Journal of Electronics and Communications*, 54:249–258, 2000.

[Manz02] U. Manzoli. Performance of a multicarrier CDMA system with multiple classes of users employing walsh and quasi-orthogonal codes. TD(02)133, COST 273, Lisbon, Portugal, Sep. 2002.

[MBLP00] M. Meurer, P. W. Baier, Y. Lu, A. Papathanassiou, and T. Weber. TD-CDMA downlink: Optimum transmit signal design reduces receiver complexity and enhances system performance. In *Proc. ICT 2000 - 7th Int. Conf. on Telecommunications*, Acapulco, Mexico, May 2000.

[MBWL00] M. Meurer, P. W. Baier, T. Weber, Y. Lu, and A. Papathanassiou. Joint transmission: advantageous downlink concept for CDMA mobile radio systems using time division duplexing. *Elect. Lett.*, 36(10):900–901, May 2000.

[MDFB03] S. Morosi, E. Del Re, R. Fantacci, and A. Bernacchioni. Improved iterative parallel interference cancellation for wireless DS-CDMA communication systems. WP(03)007, COST 273, Paris, France, May 2003.

[MDFC05] S. Morosi, E. Del Re, R. Fantacci, and A. Chiassai. Design of turbo-MUD receivers with density evolution in overloaded CDMA systems. TD(05)068, COST 273, Bologna, Italy, Jan. 2005.

[MoCa03] D. Mottier and D. Castelain. Channel pre- and post-equalization in uplink OFCDM systems with mobility. In *Proc. PIMRC 2003 - IEEE 14th Int. Symp. on Pers., Indoor and Mobile Radio Commun.*, Lisbon, Portugal, Sep. 2003. [Also available as TD(03)151].

[MoGF03] A. J. Morgado, A. M. Gameiro, and J. J. Fernandes. Constrained mean-square-error space-time pre-equalizer for the downlink channel of UMTS-TDD. In *Proc. VTC 2003 Spring - IEEE 57th Vehicular Technology Conf.*, Jeju, Korea, Apr. 2003. [Also available as TD(03)137].

[MoMe01] M. Morelli and U. Mengali. A comparison of pilot-aided channel estimation methods for OFDM systems. *IEEE Trans. Signal Processing*, 49(12):3065–3073, 2001.

[MoSä04] D. Mottier and T. Sälzer. Spreading squence assignment in the downlink of OFCDM systems using multiple transmit antennas. In *Proc. VTC 2004 Spring - IEEE 59th Vehicular Technology Conf.*, Milano, Italy, May 2004. [Also available as TD(04)163].

[MRFB03] S. Morosi, E. Del Re, R. Fantacci, and A. Bernacchioni. Improved iterative parallel interference cancellation for DS-CDMA 3G systems. TD(03)025, COST 273, Barcelona, Spain, Jan. 2003.

[MTWB01] M. Meurer, H. Tröger, T. Weber, and P. W. Baier. Synthesis of joint detection (JD) and joint transmission (JT) in CDMA downlinks. *Elect. Lett.*, 37(14):919–920, July 2001. [Also available as TD(01)003].

[MuHu97a] S. H. Muller and J. B. Huber. A comparison of peak power reduction schemes for OFDM. In *Proc. Globecom 1997 - IEEE Global Telecommunications Conf.*, Phoenix, AZ, USA, Nov. 1997.

[MuHu97b] S. H. Muller and J. B. Huber. OFDM with reduced peak-to-average power ratio by optimum combination of partial transmit sequences. *Elect. Lett.*, 33(5):368–369, Feb. 1997.

[MWSL02] I. Maniatis, T. Weber, A. Sklavos, and Y. Liu. Pilots for joint channel estimation in multi-user OFDM mobile radio systems. In *Proc. ISSSTA 2002 - IEEE 7th Int. Symp. on Spread Spectrum Techniques and Applications*, Prague, Austria, Sep. 2002. [Also available as TD(04)009].

[Nee96] R. D. J. Van Nee. OFDM codes for peak-to-average power reduction and error correction. In *Proc. Globecom 1996 - IEEE Global Telecommunications Conf.*, London, UK, Nov. 1996.

[NeWi98] R. Van Nee and A. Wild. Reducing the peak-to-average power ratio of OFDM. In *Proc. VTC 1998 - IEEE 48th Vehicular Technology Conf.*, Ottawa, Canada, May 1998.

[O'Lo94] R. O'Neill and L. B. Lopes. Performance of amplitude limited multitone signals. In *Proc. VTC 1994 - IEEE 44th Vehicular Technology Conf.*, Stockholm, Sweden, June 1994.

[OsNA05] F. S. Ostuni, M. R. Nakhai, and A. H. Aghvami. Iterative MMSE receivers for space-time trellis coded CDMA systems in multi-path fading channels. *IEEE Trans. Veh. Technol.*, 54(1):163–176, Jan. 2005. [Also available as TD(04)025].

[PaNG03] A. Paulraj, R. Nabar, and D. Gore, editors. *Introduction to Space-Time Wireless Communications*. Cambridge University Press, Cambridge, UK, 2003.

[Pomm02] P. Pommer. Convergence of a semi-blind least-squares-algorithm for the UMTS FDD uplink with adaptive antennas. TD(02)072, COST 273, Espoo, Finland, May 2002.

[PTVG98] L. Van der Perre, S. Thoen, P. Vandenameele, B. Gyselinckx, and M. Engels. Adaptive loading strategy for a high speed OFDM-based WLAN. In *Proc. Globecom 1998 - IEEE Global Telecommunications Conf.*, Sydney, Australia, Nov. 1998.

[RoGr97] H. Rohling and R. Grunheid. OFDM transmission technique with flexible subcarrier allocation. In *Proc. VTC 1997 - IEEE 47th Vehicular Technology Conf.*, Phoenix, AZ, USA, May 1997.

[SaCl03] W. Sawaya and L. Clavier. Simulation of DS-CDMA on the LOS multipath 60 GHz channel and performance with RAKE receiver. In *Proc. PIMRC 2003 - IEEE 14th Int. Symp. on Pers., Indoor and Mobile Radio Commun.*, Beijing, China, Sep. 2003. [Also available as TD(04)189].

[SäMo03] T. Sälzer and D. Mottier. Downlink strategies using antenna arrays for interference mitigation in multi-carrier CDMA. In *Proc. MC-SS 2003 - 4th Workshop on Multi-Carrier Spread Spectrum and Related Topics*, Oberpfaffenhofen, Germany, Sep. 2003. [Also available as TD(02)137].

[SoPi01] M. R. Souryal and R. L. Pickholtz. Adaptive modulation with imperfect channel information in OFDM. In *Proc. ICC 2001 - IEEE Int. Conf. Commun.*, Helsinki, Finland, June 2001.

[StBl96] A. Steil and J. J. Blanz. Spectral efficiency of JD-CDMA mobile radio systems applying coherent receiver antenna diversity with directional antennas. In *Proc. ISSSTA 1996 - IEEE 4th Int. Symp. on Spread Spectrum Techniques and Applications*, Mainz, Germany, Sep. 1996.

[SuSc01] Q. Su and S. Schwartz. Effects of imperfect channel information on adaptive loading gain of OFDM. In *Proc. VTC 2001 Fall - IEEE 54th Vehicular Technology Conf.*, Atlantic City, NJ, USA, Oct. 2001.

[TaJa00] V. Tarokh and H. Jafarkhani. On the computation and reduction of the peak-to-average power ratio in multicarrier communications. *IEEE Trans. on Commun.*, 48(1):37–44, 2000.

[TeHC03] J. Tellado, L. M. C. Hoo, and J. M. Cioffi. Maximum-likelihood detection of nonlinearly distorted symbols by iterative decoding. *IEEE Trans. Commun.*, 51(2):218–228, Feb. 2003.

[Tell98] C. Tellambura. Phase optimization criterion for reducing peak-to-average power ratio in OFDM. *Elect. Lett.*, 34(2):169–170, Jan. 1998.

[TeRe01] W. G. Teich and M. Reinhardt. Multiuser/multisubchannel detection based on a recurrent neural network structure for the mobile communication system TD-SCDMA. In *Proc. ISCTA 2001 - 6th Int. Symp. on Communication Theory and Applications*, Ambleside, UK, July 2001. [Also available as TD(02)069].

[WiJo95] T. A. Wilkinson and A. E. Jones. Minimization of peak-to-mean power ratio of multicarrier transmission schemes by block coding. In *Proc. VTC 1995 - IEEE 45th Vehicular Technology Conf.*, Rosemont, IL, USA, July 1995.

[WuGo99] D. Wulich and L. Goldfeld. Reduction of peak factor in orthogonal multicarrier modulation by amplitude limiting and coding. *IEEE Trans. Commun.*, 47(1):18–21, Jan. 1999.

[YeBC02] S. Ye, R. S. Blum, and L. J. Cimini. Adaptive modulation for variable rate OFDM systems with imperfect channel information. In *Proc. VTC 2002 Spring - IEEE 55th Vehicular Technology Conf.*, Birmingham, AL, USA, May 2002.

3 Signal processing

Chapter editor: *Alister Burr*
Contributors: *Tomaz Javornik, Tadashi Matsumoto, Jan Sykora, Laurent Clavier, and Geir E. Øien*

3.1 Introduction

While the previous chapter focused on specific wireless standards and air interfaces, this chapter now moves to more generic signal processing techniques that may be used in various systems. Again, the focus is link level aspects, including also multi-user systems and multi-user detection, and again it is primarily the work of WG1 that is reported.

Thus Section 3.2 covers modulation and coding techniques used for data transmission; Section 3.3 covers equalisation techniques used at the receiver to overcome signal dispersion, as well as some more general issues of signal detection; Section 3.4 covers issues of parameter estimation, which includes carrier phase and symbol timing recovery as well as channel estimation more generally. Section 3.5 covers multi-user detection, which of course overlaps with CDMA systems, as described in Section 2.3, but here the emphasis is on the detection techniques rather than the systems, and their application may in fact be broader than CDMA. Finally Section 3.6 considers adaptive air interfaces, and especially the link adaptation algorithms employed.

Two topics have been recurrent features of the work of WG1, described in this chapter. One has been techniques for MIMO systems, and although a large part of this appears in Chapter 7, signal processing techniques for MIMO remain an important part of this chapter, since increasingly MIMO is being seen as an essential element of wireless systems. However, in this chapter the focus is on the signal processing techniques, rather than the MIMO aspects.

The second, however, has been iterative, or 'turbo' techniques. Because the principle is so widely applicable 'turbo' techniques will be found in all the remaining sections of the chapter, and cannot be separated out. Therefore, in the remainder of this section we include a general introduction to iterative techniques, or 'turbo processing'.

'Turbo processing' takes its name from the iterative decoder used in turbo codes, which in turn is derived from the turbo-charged internal combustion engine, in which some of the output power is fed back to the input to enhance the power of the engine (in that case by pre-compression of the fuel/air mixture) [BeGT93]. In this case, it is extrinsic information from the output of the decoder concerning the data. For the turbo decoder, this is fed back from the output of one of the two constituent decoders to the input of the other. The information fed back usually takes the form of a log-likelihood ratio, which is a form of soft information, because it indicates the reliability of the information as well as the most likely symbol. It is extrinsic in the sense that it contains information about the data obtained from the second decoder only, which is therefore extrinsic to the first decoder. Information obtained from the first decoder, or directly from transmitted data, is subtracted from it. Note that this information should be as far as possible independent of the information intrinsic to the first decoder.

For this reason, in most cases there is an interleaver between the two decoders, so that the information tends to come from different parts of the code stream. The process is illustrated in Figure 3.1.

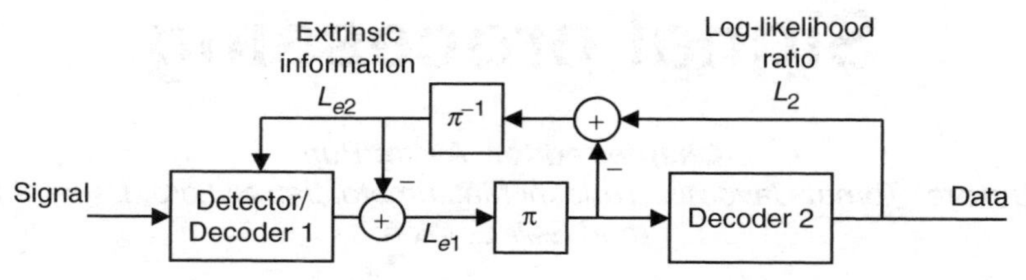

Figure 3.1 *Iterative or 'turbo' processing. (© 2004 IEEE, reproduced with permission)*

This extrinsic information is then used in the first decoder as a priori information, which allows that decoder to improve its estimate of the data in a second iteration of decoding. The first decoder in its turn passes extrinsic information to the second, resulting in an improved estimate of both the data itself and the extrinsic information. This iterative process is then repeated until it converges (hopefully) to the correct data.

The decoders used must both accept and generate soft information, and hence are commonly referred to as soft in, soft out or SISO decoders. (Since this acronym clearly conflicts with single input, single output, in the work described in this chapter it has been amended to Soft In Soft Out (SfISfO)). The decoding algorithm most often used is the MAP, or BCJR algorithm [BCJR74], which uses and provides a log-likelihood algorithm. Two simplifications of it, the MAP and the Soft Output Viterbi Algorithm (SOVA) [RoVH95] may also be used, but the soft information is not as accurate.

The application to functions other than decoding alone arises from the recognition that the extrinsic information can be used to assist in many other functions carried out in a receiver. For example, channel estimation is enhanced if the data is known, and in equalisation some knowledge of data allows intersymbol interference to be estimated and removed. The feedback of soft information also allows the reliability of the data decisions to be taken into account. The first decoder is then replaced in general by some form of detector that exploits the extrinsic information fed back to perform the function required.

Note that so far these applications nearly always involve decoding jointly with one or more other functions (though usually just one other function). This is partly because the approach lends itself particularly well to this, but also because the application of Forward Error Correction (FEC) coding, and especially turbo coding, often causes particular difficulties for other functions. In particular, coding will tend to reduce the signal to noise ratio at the front end of a detector, first because of the coding gain which is the purpose of coding, but also because the reduced rate of the signalling means that the signal to noise ratio (as opposed to the bit energy to noise density ratio) is further reduced. Since the performance of such functions as carrier phase recovery depends on signal to noise ratio, this often means that the overall performance of the coded system depends on that of the carrier recovery. Joint decoding and carrier recovery overcomes this problem, and turbo processing provides a means of implementing it much more simply than a full joint estimation.

Note also that the codes do not have to be turbo codes. As mentioned, the name 'turbo' comes from the use of extrinsic information from an outer decoder fed back to a front-end decoder/detector, and hence applies whatever this outer coding uses. Outer convolutional decoding is very popular in these applications, although it requires the convolutional decoder to be replaced with a (more complex) soft output decoder. A joint turbo decoder and detector would in fact include two turbo processing loops, one inside the other. Note, however, that turbo processing systems based on turbo codes often perform particularly well, due to the power of the code, and also that the additional complexity is often very small since the turbo decoder already provides soft output.

This detector can in general take one of two forms. One is the equivalent of the MAP algorithm in that it generates accurate estimates of the log-likelihood ratio of the data, along with likelihood functions of any parameters to be estimated, from the signal and fed-back extrinsic information. This is in principle optimum, but may be very complex in some situations. For example, for multi-user detection the detector then needs to consider all possible combinations of user data, and therefore complexity tends to be exponential on the number of users. Similarly a MAP detector for equalisation will have complexity exponential on the channel impulse response length. An alternative is an approach based on expectation, in which the expectation of the data is calculated from the extrinsic information being fed back, and then used as an estimate of the data for use in a data-aided detector/estimator. A particularly popular example of this is expectation-maximisation as used in carrier phase recovery and channel estimation [SyBu04]. One advantage of this approach is that it can be shown that it will converge under a range of commonly encountered conditions. Another is the use of soft information in parallel interference cancellation multi-user detection. This avoids the exponential dependence on the number of users by calculating the expected value of other-user interference and subtracting it. Some of the work described in this chapter has considered and compared particular examples of these approaches [SyBu04].

As mentioned above, turbo-processing is now so important to the field that it features in all the remaining sections of this chapter. Its applications can essentially be divided between estimation and detection, as in the following list, although in some cases there is a blurring of the distinction:

Estimation:

- carrier phase recovery;
- symbol timing recovery;
- channel estimation;
- interference Probability Density Function (PDF) estimation.

Detection:

- equalisation;
- MIMO detection;
- multi-user detection;
- unknown interference cancellation.

3.2 Modulation and coding

The research activities described in the section include advanced coding and modulation issues, the topics about UWB communications, methods for theoretical system performance evaluation, and implementation issues. The majority of contributions dealing with coding and modulation related to space-time coding are described in the part about MIMO systems. In advanced coding and modulation issues, the design criteria for trellis coded Continuous Phase Modulation (CPM) space-time modulation and linear diversity precoding for block fading the delay limited MIMO channel, the puncturing for parallel concatenated codes, and finally a new code specially suitable for network layer, based on integer codes, are described. The frequency domain detectors, the effects of spreading bandwidth on the performance of Rake receivers and the combining of the transmit reference signal with an autocorrelation receiver were studied in the UWB subsection. In the theoretical systems performance evaluation, the system capacity in frequency selective Rayleigh fading channel and a new approach for approximation of error curves are evaluated. The implementation issues in part focus on the impact of non-linear distortion and receiver simplification on the system performance degradation. The last part also deals with the design and implementation of the communications system on Field Programmable Gate Array (FPGA).

3.2.1 Advanced coding and modulation issues

The design rules and criteria to create efficient coding and modulation schemes for various propagation environments are included in the following subsection. The design principle for trellis coded CPM modulation in a flat fading Rayleigh channel is described first, while the remaining three contributions deal mainly with coding. In the first one a design criterion for an efficient two stage coding process with low complexity in the block fading delay limited MIMO channel is proposed. A complete new idea of code design based on the integer codes is explained next. The description of the method for optimal puncturing of multiple parallel concatenated codes finishes the subsection.

A CPM type constant envelope Space-Time (ST) modulation is an attractive option especially for very high frequency and mobile applications due to its resistance to a non-linear distortion. It is shown [Syko01] by analysing the mean squared distance of the trellis coded CPM ST modulated signal in a Rayleigh slowly flat fading spatial diversity channel with independent coefficients, that the distance evaluation depends on both the modulator trellis and the distance evaluation trellis. The latter has special properties regarding its free paths; these properties, a critical path and the distance increments behaviour, are the basis for the ST trellis code design.

The trellis code design principle maximising the mean squared distance identified in [Syko01] can be summarised in the following steps: (1) search for terminating sequences, (2) classify them according to the length and determine the critical path in distance evaluation trellis, (3) exclude the critical sequences from the stock, (4) use the remaining ones for ST trellis code.

The first two steps related to the finding of critical paths are essential parts of the design procedure. They are found with utilisation of the distance evaluation trellis. It can be shown that the distance evaluation is a procedure with memory described by distance evaluation trellis

$$\rho'^2 = \sum_{m=-\infty}^{\infty} \Delta\rho'^2(m, \Delta\mathbf{q}_m, \mathbf{\Theta}(m)) \tag{3.1}$$

where its input $\Delta\mathbf{q}_m = \mathbf{q}_m^{(1)} - \mathbf{q}_m^{(2)}$ is the channel symbol difference, and the distance evaluation state is $\mathbf{\Theta}(m)$. The final terminal state identifies the terminating sequences. We can show that squared distance increment is zero $\Delta\rho'^2(m) = 0$ if and only if $\mathbf{\Theta}(m) = \mathbf{0}$ and $\Delta\mathbf{q}_m = \mathbf{0}$. When the final state in the distance evaluation trellis is reached paths with the final state $\mathbf{\Theta}(m) = \mathbf{0}$ cease to increase the accumulated squared distance. On the other hand, the paths with the final state $\mathbf{\Theta}(m) \neq \mathbf{0}$ continue to increase the accumulated squared distance.

The overall procedure is general and is not exclusively related to the spatial channel. It can be applied to the design of a code which is applied to any modulation possessing a memory described by the finite state machine model.

A novel approach utilising the virtual multiple-access-like capacity region with symmetry condition on rates is used to develop the design criterion of linear precoding (inner code) for the block fading delay limited MIMO channel. The overall coding process is split into two parts – outer and inner code. The outer coders are supposed to work block-wise independently and the only joint processing is assumed inherently in the inner code. The motivation is to develop a two-stage coding process having the same outage capacity performance, but significantly lower complexity of the design compared with the information-theoretically optimal joint coding book.

We assume a MIMO channel model with time-block stacked matrices $\tilde{\mathbf{x}} = \tilde{\mathbf{G}}\tilde{\mathbf{q}} + \tilde{\mathbf{w}}$. We want to replace the *joint (one-stage)* coding process $\tilde{\mathbf{d}} \mapsto \tilde{\mathbf{q}}$, where $\tilde{\mathbf{d}} = [\mathbf{d}_1^T, \ldots, \mathbf{d}_M^T]^T$ is the raw information data vector and $\tilde{\mathbf{q}} = [\mathbf{q}_1^T, \ldots, \mathbf{q}_M^T]^T$ is the vector of channel coded symbols, by the *two-stage* coding. In the two-stage approach, the first stage – *block-wise coding* $\mathbf{d}_n \mapsto \mathbf{c}_n$, $\forall n \in \{1, \ldots, M\}$ (M is the frame length) – is performed by *outer coders* sharing a common codebook and the second stage – *inner code* $\tilde{\mathbf{c}} \mapsto \tilde{\mathbf{q}}$, $\tilde{\mathbf{c}} = [\mathbf{c}_1^T, \ldots, \mathbf{c}_M^T]^T$. The vector $\tilde{\mathbf{c}}$ is the vector of intermediate encoded symbols forming the input to the virtual multiple-access-like channel. The inner code is assumed to be a *linear precoding*, $\tilde{\mathbf{q}} = \tilde{\mathbf{F}}\tilde{\mathbf{c}}$, where $\tilde{\mathbf{F}}$ is the precoder unitary matrix.

The joint (one-stage) coding is the optimal one. The one-stage code effectively dissolves multiple random instances of the channel in the frame by using a long codeword spanning the whole frame. In the two-stage approach, the linear precoding creates a new virtual codebook viewed at the level of $\tilde{\mathbf{q}}$ symbols from the symbols $\mathbf{c}_n$ coded by common codebook from $\mathbf{d}_n$.

The design goal for the precoder is to provide the uniform achievable rate per one block (i.e. simultaneously for each individual block in the frame) in the two-stage case that should be as close as possible (in probabilistic sense) to the one obtained if direct one-stage coding spanning the whole frame were used

$$\frac{1}{M}\tilde{C}_M^{\tilde{\mathbf{d}} \mapsto \tilde{\mathbf{q}}} = \min_{k \in \mathcal{M}, \mathbf{s} \in \mathcal{S}(k)} \frac{1}{k}\tilde{C}_{\mathbf{s}} \tag{3.2}$$

The one-stage encoding capacity is

$$\tilde{C}_M^{\tilde{\mathbf{d}} \mapsto \tilde{\mathbf{q}}} = \log_2 \det\left(\mathbf{I}_{MN_T} + \frac{\Gamma}{N_T}\tilde{\mathbf{F}}^H\tilde{\mathbf{G}}^H\tilde{\mathbf{G}}\tilde{\mathbf{F}}\right)$$

The capacities of capacity region $\tilde{C}_{\mathbf{s}}$ are the capacities of all possible k-tuples of subchannels from the set $\{1, \ldots, M\}$,

$$\tilde{C}_{\mathbf{s}} = \log_2 \det\left(\mathbf{I}_{kN_T} + \frac{\Gamma}{N_T}\sum_{i=1}^{k}\tilde{\mathbf{F}}_{[s_i]}^H\tilde{\mathbf{G}}^H\tilde{\mathbf{G}}\tilde{\mathbf{F}}_{[s_i]}\right)$$

where Γ is signal to noise ratio. The phrase 'as close as possible' is understood in a probabilistic sense, since there is no channel state information at the transmitter and the channel observation is non-ergodic. Selected precoders were evaluated in terms of their capability to satisfy this criterion. It was shown by simulation that the precoders performing temporal-only precoding achieve a similar performance to the full spatial-temporal precoding. For details, see [SyKn04].

Algebraic coding theory has been developed mostly considering the efficiency of coder and decoder hardware realisation. The finite field theory approach led to an extremely simple hardware applied for polynomial multiplication and division. Attempts to find codes more suitable to byte-oriented computer applications led to the class of codes known as single byte error correcting double byte error detecting codes. The codes based upon integer, instead of finite fields, arithmetic are rare, one of them being described in [BaBu04].

A complete novel code is designed to correct a single error that occurs within a byte of arbitrary length [BaBu04]. The code is based upon the integer ring, and the main idea is that each single bit error (positive or negative) is a cyclic shift of another single bit error within a byte. The coding and syndrome forming procedures are simple. Error correcting is performed if necessary, without the long iterations and huge look-up tables. The code is sub-optimal, as its length is slightly shorter than that of classical codes of the same correcting capability. A simple extension within the correcting procedure makes a correction of two adjacent errors within a byte possible, with additional shortening of the code.

The proposed code has the following advantages:

- The coding and syndrome forming procedures are simple, only two modulo M additions per information byte are required.
- The codeword length is arbitrary up to the maximal limit similar to the Hamming code with the parity check bit.
- If no error correction is required the code is equivalent to the Fletcher checksum procedure for error detection.
- Error correction procedure starts only if necessary. It consists of comparison and cyclic shifts of auxiliary registers.
- The code has two versions: the first one corrects single bit error within a byte, while the other one corrects pairs of errors within a byte. Switching can be implemented within a single algorithm.
- The proposed code is well suited for FEC coding at the network layer, especially with coded route diversity.

Since the invention of turbo codes (two parallel concatenated codes), the turbo principle has been extended to symmetric and asymmetric Multiple Parallel Concatenated Codes (MPCCs). In the original turbo code, all uncoded (systematic) bits are transmitted, while the coded bits of both constituents are punctured to raise the code rate. Each of the constituent codes in Multiple Parallel Concatenated Codes (MPCC) can have their own puncturing ratio that can be chosen freely, while keeping a fixed code rate. Parallel concatenated codes with two constituent codes can be analysed using two-dimensional Extrinsic Information Transfer (EXIT) charts, while the EXIT chart for N constituents is N-dimensional; thus, EXIT analysis becomes complicated, and the technique of projecting the N-dimensional EXIT chart to the two-dimensional chart was used to find the optimal puncturing ratios, chosen to minimise the convergence threshold. The EXIT chart is used to optimise the choice of puncturing ratios over a range of code rates in [BrRG04]. For a given code rate R, the minimum SNR, γ_{min}, together with optimal puncturing Δ_{opt} can be found by

$$\begin{aligned}\left[\gamma_{min}, \Delta_{opt}\right] = \arg \min_{[\gamma,\Delta]} g(\gamma, \Delta), \quad \text{subject to} \\ \gamma > 0 \\ \Delta \in [0, 1]^{N+1}\end{aligned} \tag{3.3}$$

where the objective function $g(\gamma, \Delta)$ is defined as

$$g(\gamma, \Delta) = \begin{cases} \gamma & f_I(\gamma, \Delta) \geq J\left(2Q^{-1}(P_b)\right) \\ 0 & \text{otherwise} \end{cases}$$

where f_I is the function determining the convergence point. The optimisation can be solved for a specific R by initialising γ to a large value. For detailed procedure refer to [Brän04]. The results give an SNR-rate region, Figure 3.2, for Berrou, Glavieux and Thitimajshima (BGT) code and for three other combinations of convolutional codes.

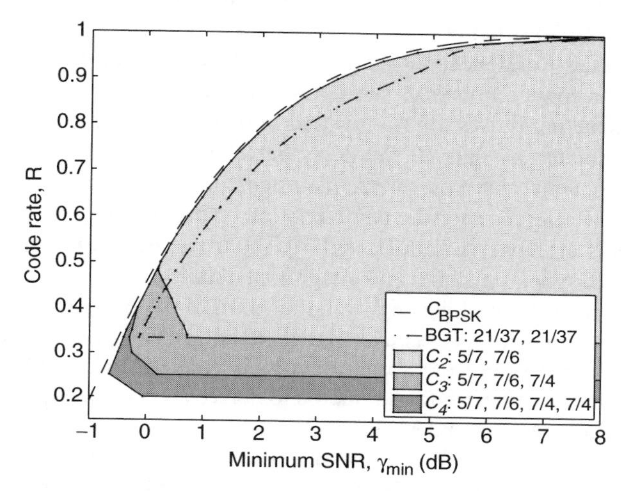

Figure 3.2 *SNR-rate regions for BTG and convolutional codes for convergence point > 0.9999.*

3.2.2 *Ultra wide bandwidth*

UWB spread spectrum multiple access techniques reach attention for future commercial and military communication systems. Short duration pulses with a low duty cycle carry information in the UWB, thus, the signal energy is spread over a wide range of frequencies, which results in desirable capabilities of communication systems, including accurate position location and ranging, lack of significant fading, multiple access and covert communications. The effects of spreading bandwidth on the performance of UWB Rake receivers, and combining the transmit reference signalling with an autocorrelation and frequency domain detector in a frequency selective multi-user environment were investigated in COST 273.

The Rake receiver architecture provides the ability to resolve multipath components and transform dense multipath channels, and creates a high degree of path diversity. For an ideal Rake receiver, with unlimited number of taps and correlators and instant adaptivity, the spreading bandwidth should be as wide as possible. However, for a finite complexity receiver with limited number of taps and correlators a limited bandwidth is optimal. In [CWVM03], the effects of spreading bandwidth on the performance of UWB Rake receivers with reduced complexity are analysed. The influence of spreading bandwidth of two receivers, selective Rake receiver (SRake), which selects the L best paths and combines them with maximum ratio combining and partial Rake receiver (PRake), which combines the first L arriving paths, was analysed. For a fixed number of Rake fingers and a fixed transmit number, the study shows that an optimal bandwidth exists. The optimal bandwidth increases with the number of Rake fingers, and it is higher for an SRake than for a PRake.

The effect of two fading statistics, namely Rayleigh and Nakagami, on the optimal spreading bandwidth was also analysed. Optimal spreading bandwidth is approximately the same for both statistics. For a low ratio L_b/L_r, where L_b denotes the number of the best paths and L_r denotes the number of available multipath components, the performance of an SRake in a Rayleigh fading channel is better than in a Nakagami fading channel, while for high ratio L_b/L_r, the receiver performance in the Nakagami fading channel is better.

Combining Transmitted-Reference (TR) signalling with an Autocorrelation Receiver (AcR) is an elegant way to obtain a low-complexity, sub-optimal UWB communication system. In the basic TR UWB signalling scheme, pulses are transmitted in pairs. The first pulse remains unmodulated and the second is modulated by data. If the delay between both pulses is small compared to the coherence time of the channel, both pulses are distorted equally by the channel, such that the AcR can use the first pulse as reference for the demodulation of the second one. Currently, many issues regarding these systems are unsolved. In [RoWi04], the usage of an alternative AcR is proposed, which enables simplified synchronisation and integration duration optimisation at the cost of a higher clock rate. The proposed AcR combines the weighted sum of the oversampled correlator output to improve the system performance. A statistical characterisation of noise and Intra- and Inter-Symbol-Interference (IISI) at the oversampled correlator output is derived, along with a method to compute the bit error probability. Two criteria are considered to derive the weighted sum coefficients, namely Maximal Ratio Combining (MRC) and MMSE.

The performance of several UWB TR systems operating on measured indoor multipath radio channels, including the effect of small-scale fading, is presented and discussed taking implementation issues into account. All these systems use the proposed AcR, but differ with respect to delay, bandwidth, sampling rate and bit rate. In the absence of IISI, TR systems with a 200 MHz bandwidth are found to outperform 400 and 800 MHz ones, Figure 3.3. The presented statistical characterisation

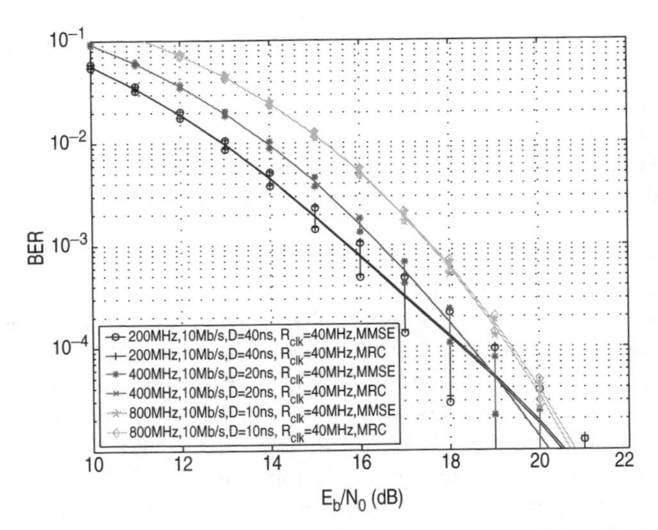

Figure 3.3 *The average bit error probability of six 10 Mb/s TR systems within an office environment including small-scale fading.*

reveals the presence of a non-Gaussian noise term within the demodulator output of which the variance grows approximately linearly with the bandwidth. This term is responsible for the performance decrease with increasing bandwidth. Furthermore, MRC is found to result in almost the same performance as MMSE weighting. On the other hand, large bandwidth systems are inherently less sensitive to IISI than their narrowband counterparts. Hence, the optimal choice of bandwidth depends on the bit rate, delay and environment.

A conventional anti-multipath approach for a single-carrier transmission is performed in the time domain, which is unsuitable for UWB transmission, where the transformation is tens of Msymbols/s and more than 30-symbol ISIs have to be taken into account. An alternative is a frequency domain detector, which was proposed for indoor downlink communications for frequency selective environments [MoBi04]. The proposed detector relies on the introduction of the cyclic prefix at the transmitter and the use of a frequency domain detector at the receiver. It can effectively cope with the orthogonality loss and the rise of Self-Interference (SI) and Multiple Access Interference (MAI). Two different detection strategies based on either the Zero Forcing (ZF) or the MMSE criteria have been investigated and compared with the classical Rake receiver. The results for four users are shown in Figure 3.4. The Rake receiver is not able to cope with MAI whose effects are increased by the long

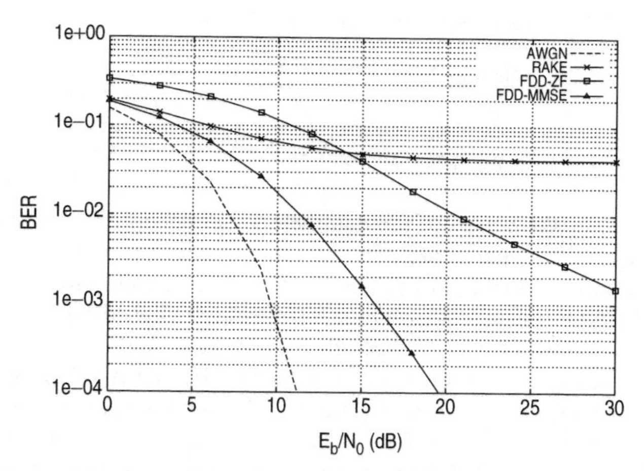

Figure 3.4 *Performance comparison of frequency domain UWB detector for four users.*

multipath spread: as a result, performance is greatly impaired and the error floor is evident also for medium to low E_b/N_0 values. On the contrary, both strategies are able to restore the orthogonality between users since they perfectly compensate the effects of the channel. While ZF performance is limited by noise enhancement, MMSE permits a great performance gain. The performance of a correlation receiver for a single-user UWB-Impulse Response (IR) system in an AWGN channel with no multipath is reported for comparison.

3.2.3 System performance evaluation

The physical layer, network layer aspects, and the efficiency of higher layers contribute essentially to the system spectral efficiency. Its evaluation requires the analysis of the system at the network layer and above. Compared to the link layer, where theoretical analyses achieve excellent results, often no mathematical model can be found for system analyses and, hence, the method of choice for performance evaluation is computer simulations.

However, detailed simulation of the entire communication system from the physical layer to the network layer would require enormous computation time. Therefore, it is necessary to reduce the simulation complexity by subsystems' models with coarse but still realistic behaviour. Two methods have been presented during COST 273:

1. A method to approximate error rate curves from literature by double exponential functions [FeRa02].
2. A model of the BER curves of trellis coded modulation by union bound technique.

A common physical layer model for system evaluation is to use error rate curves that have been derived earlier by link layer investigations, either by theoretical analysis or by computer simulation. The idea behind the first point is based on curve fitting methods. Curve fitting requires that the general approximation function with a number of adjustable parameters is known in advance. The least square method, then, is used to determine the parameters such that the squared distance between the approximation function and the sample points becomes minimal.

Two functions have been proposed for the approximation:

$$g_1(\gamma) = \left(1 - e^{ae^{b(\gamma-\delta)}}\right)^d * (\alpha - \epsilon) + \epsilon \tag{3.4}$$

$$g_2(\gamma) = \left(1 - e^{ae^{b(\gamma-\delta)}+s}\right)^d * (\alpha - \epsilon') + \epsilon', \tag{3.5}$$

where $\mathbf{k} = (a, b, d, s)^T$ is the vector of the parameters that need to be determined with the curve fitting method. The terms α, δ and ϵ' are constants that are derived from the qualitative course of the approximation functions. Most error rate curves can easily be approximated by the first proposed function g_1. Some curves with sharp edges require the use of g_2, where the derivation of the parameters for g_2 is more complicated than for g_1. Hence, if possible, g_1 should be used.

The approach has been applied to error rate curves for HIPERLAN/2 from literature [ETSI01], [KMST00] using the simpler function g_1, Figure 3.5 (the symbols are the simulation results from literature, the curves are the fitting results for g_1). Please refer to [FeRa02] for other more complicated examples from the literature with distinctive error floors that have been approximated by g_2.

The second point above is another interesting field discussed during COST 273. It deals with the evaluation of the exact union bound of coded modulation schemes in the additive white Gaussian noise channel, which requires to consider error events for all pairs of encoded and decoded sequence transmitted over the communication channel. The union bound method, which takes into account only the minimum Euclidean distance, does not give us satisfactory results; therefore, the distance spectrum, which describes all error events and the probability of their occurrence, is used instead. While the topic is well analysed for the trellis coded and non-linear modulation schemes using optimal sequence estimation, there is a lack of research for systems that use different trellis pruning techniques. The analytical studies of the union upper bound on error performance for the state space partitioning trellis detector and parametric channel is shown in [VcSy04].

The main idea of the state space partitioning is to form a group of regular states into a super state. Only one survivor in the super state is kept in the Viterbi algorithm. The complexity of the receiver is reduced due to reduced number of traced branches. The main contribution to the performance loss is truncation of the error event due to the state partitioning. An error event occurs when two states

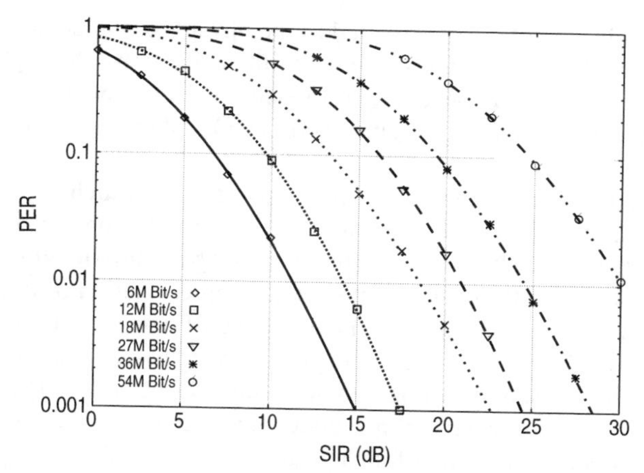

Figure 3.5 *PER samples (symbols) from [KMST00] and fitted curves for HIPERLAN/2 from [FeRa02]. (© 2002 IEEE, reproduced with permission)*

merge. The proposed method is based on the product distance matrix based on executed modification in the trellis of the decoder and gives an instrument for analytical error performance analysis of the detector with state partitioning. The method is extended to the non-ideal coherent detection in the presence of random phase error. Simulation results fit the analytical for the whole observed SNR range for optimal and reduced state space diagram.

3.2.4 Implementation issues

The demand on spectral efficiency requires the replacement of the robust one level, usually a constant envelope modulation scheme, with a multilevel one in future communication systems. The higher number of levels of the modulation schemes increases the complexity of data detection and their sensitivity on the distortions introduced by non-ideal components applied for the implementation of the communication system. The analysis of the system degradation due to non-ideal system components is a two-stage process: the first stage includes modelling of the non-ideal component, while in the second stage the impact of the component on the system performance is studied. During COST 273, the main focus was given to the modelling of the high power amplifiers and their impact on the system performance. The optimal receiver implementation for non-linear modulation schemes with multiple levels and CPM signals in MIMO systems are complex, for that reason a simplified receiver was suggested [JaKa02a] and [Syko05]. The implementation of IEEE 802.11 is described in [CVAV04].

Modelling of high power amplifier and its impact on the system performance

Two approaches exist for modelling Radio Frequency (RF) devices: the first one is based on the electrical equivalent scheme, and the second one is behavioural modelling. Usually, the electrical equivalent scheme models apply the small signal S-parameter measurements of the RF devices, such as diodes, transistors, etc., to obtain the large-signal models, which causes error propagation in complicated designs. An alternative approach is using large-signal component measurements to construct the large-signal model. As this approach is novel, its use is still rather limited in complex designs.

Also, behavioural models are not directly related to the physics of the device, but their characteristics are determined from the component responses to the excitations. A single tone excitation was applied in early approaches for the behavioural modelling. However, in a digital communication system the single tone excitation does not give useful results, therefore, a multitone excitation was applied [ScRe03].

The behavioural model is represented by the state equations, which describe dependency relations among input, output and device state variables. The modelling consists of determining the state variables, by the false nearest neighbour approach and by approximating the state functions by artificial neural networks. The steady state time domain representation of multitone excitation will generate huge amounts of data. For that reason the amount of data is reduced by the following procedure proposed in [ScRe03]. The data reduction procedure consists of sampling at the Intermediate Frequency (IF) and at each IF sampling point, the data samples of one RF period are processed. The collection of RF samples forms the data set for model building. Even when oversampling is applied, the procedure would reduce the number of data points to be handled in data processing drastically. The artificial neural networks method was applied to approximate the state functions. The artificial neural network considers one hidden layer with eight neurons and it is trained using the back propagation algorithm. An off-the-shelf RF amplifier is used to show the validity of the modelling process. The results are plotted in Figure 3.6 for five excitation tones. The five fundamental tones and four intermodulation products at the output are well predicted, while at lower intermodulation products a small disagreement between the behavioural (black arrow) and the reference simulation result using the manufacturer's model (grey arrows) occurs. A higher IF sampling rate will further improve the reliability of the proposed model.

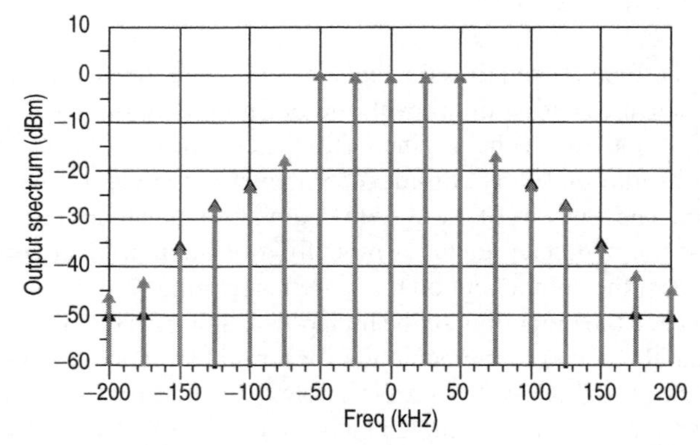

Figure 3.6 *Comparison between behavioural modelling results and results of reference simulation using the manufacturer's model.*

The High Power Amplifier (HPA) models based on multitone excitation are appropriate for spectral analysis of the HPA non-linearity on the communication systems. Nevertheless, this method is too complicated to be used in the system simulation, where the HPA block is only a small part of a complex communication system. For that reason, a simple Amplitude Modulation (AM)/AM and AM/Phase Modulation (PM) model of a Solid State Power Amplifier (SSPA) for Ka-band was derived in [WhBJ03]. The AM/AM model is based on the exponential model $(1-e^r)$ [Honk96] and the added correction factor:

$$A(r) = a\left(1 - e^{-br}\right) + cre^{-dr^2} \tag{3.6}$$

where the r is the input signal magnitude and parameters a, b, c, and d are obtained by the least squares curve fitting method. Parameter a determines the saturation level, b specifies the amplification of the HPA in the linear range, and c and d correct the AM/AM curve close to the saturation level. The AM-PM characteristic is modelled by an exponential model:

$$\Phi(r) = \begin{cases} f\left(1 - e^{-g(r-h)}\right), & \Phi(r) \geq 0 \\ 0, & \Phi(r) > 0 \end{cases} \tag{3.7}$$

where f is the magnification, g is the steepness of the curve, and h is the shift of the curve along the r-axis.

The sensitivity of square M-QAM, Differential xPSK (DxPSK) and N-Minimum Shift Keying (MSK) modulation scheme on the HPA non-linearity is studied in terms of spreading of the power spectrum in neighbouring frequency bands and increase in BER over the AWGN channel [JaKa02b]. The link between the peak to average power ratio and the spectral regrowth is demonstrated for M-QAM signals. After filtering, the 16, 64 and 256-QAM signals exhibit the similar spectral regrowth over the SSPA model. BER performance for M-QAM signals with high M is shown to degrade more quickly as the operating point is moved closer to saturation than for M-QAM with low M. DxPSK signals are less sensitive to non-linear distortion of the SSPA model. Spectral spreading into adjacent channels is observed, but differential phase detection results in no degradation in BER performance over the SSPA compared to a linear amplifier. The SSPA non-linearity causes the smaller spectrum to spread to 2-MSK rather than to M-QAM and DxPSK signals, while the influence of the non-linearity on BER performance is smaller for DxPSK than for 2-MSK signals.

The predistortion can be used for the reduction in the spectral regrowth of the square M-QAM, DxPSK and 2-MSK signals. The simulation results show that the significant reduction in the spectral spreading of the 256-QAM signal up to operating points around the -3 dB relative to the 1 dB compression point of the SSPA can be achieved. Predistortion of D8PSK signals is shown to produce low spectral regrowth at high operating points (around 0 dB relative to 1 dB compression point).

Spectral regrowth and BER performance are evaluated from a set of alternative M-QAM signals over the SSPA model, namely star, rounded and hexagonal QAM constellations. It is shown that the star 16-QAM constellation, while inherently less power efficient than square 16-QAM, degrades more slowly with increasing operating point when used with the predistorted SSPA. Ultimately, the square and star 16-QAM provide similar power efficiency for the predistorted SSPA. An input back-off of 3 dB for square 16-QAM and 2 dB for star 16-QAM from the 1 dB compression point of the predistorted SSPA is sufficient to provide low spectral regrowth and good BER performance. Alternative star and hexagonal 64-QAM modulation schemes degrade more slowly with increasing operating point than conventional square 64-QAM signal. An input back-off of 4 dB is necessary to achieve acceptable spectral and BER results over the predistorted SSPA using 64-QAM. It is slightly more power efficient to use either a rounded or hexagonal constellation rather than the square pattern. Power savings of 0.75 dB for rounded and 1.25 dB for hexagonal 64-QAM constellations are achieved at BER $= 10^{-6}$. The BER and spectral regrowth for the 64-QAM constellation with implemented predistortion is shown in Figure 3.7.

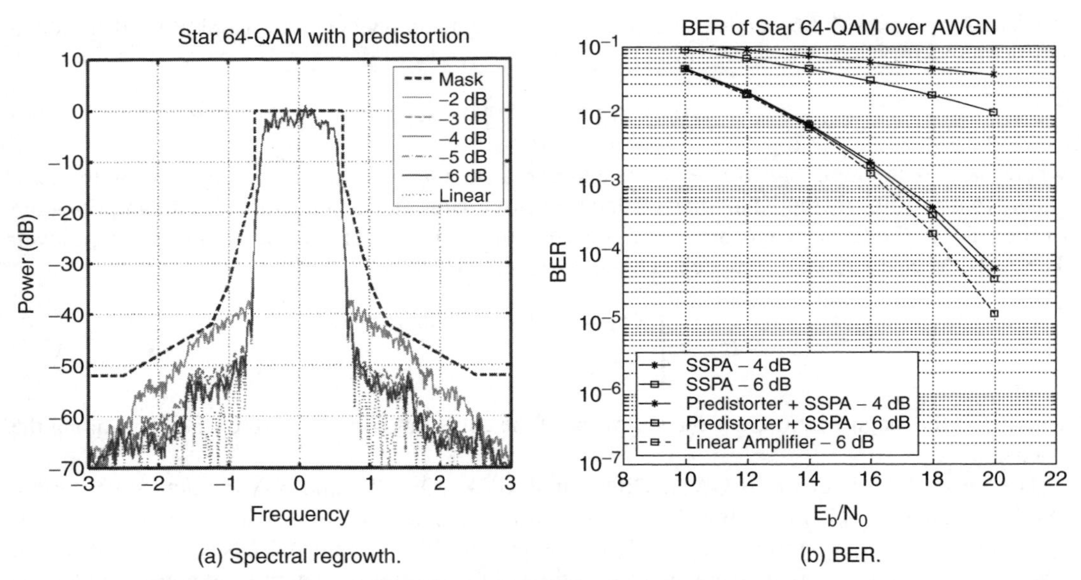

(a) Spectral regrowth. (b) BER.

Figure 3.7 *Spectral regrowth and BER for star 64-QAM with predistortion.*

Simplified receivers for CPM signals

Generally, multilevel modulation schemes are not suitable for mobile and satellite communication systems, due to their sensitivity to the distortion caused by non-linear amplifiers. The 2MSK modulation is rather insensitive to non-linear distortions, but the maximum likelihood detection, which is the optimal detection for signals corrupted by additive white Gaussian noise, is rather complex, and for that reason unsuitable for implementation. In [JaKa02a], a differential 2MSK receiver based on the regeneration of the larger MSK signal component is proposed. The block diagram of the receiver is shown in Figure 3.8.

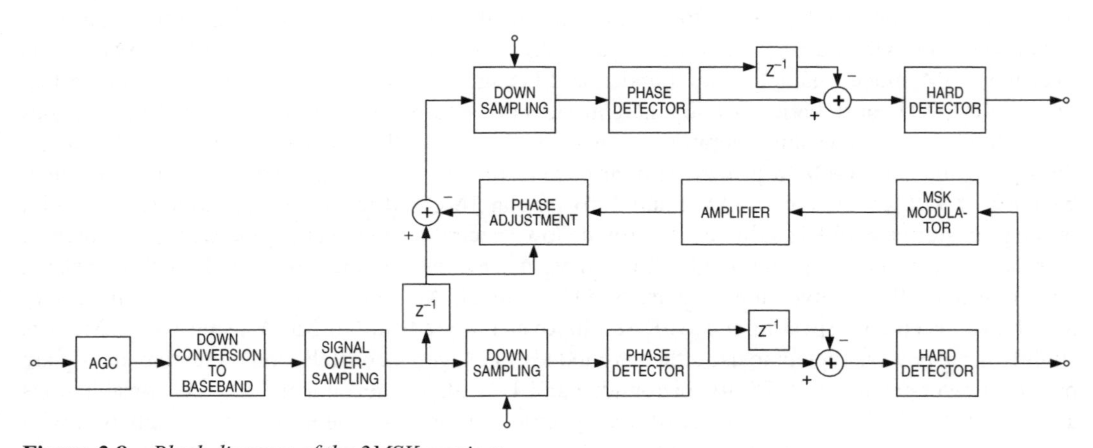

Figure 3.8 *Block diagram of the 2MSK receiver.*

The 2MSK differential receiver is based on the following two properties of the 2MSK signals:

1. The 2MSK signal is obtained by the superposition of two MSK signals with different amplitudes.
2. The phase shift between the received and one symbol interval delayed 2MSK signal depends only on the data sequence transmitted by the MSK component with the larger amplitude.

According to the described 2MSK signal properties the data transmitted by the MSK component with larger amplitude is differentially detected from the phase of the 2MSK signal. The data obtained is the input of the MSK re-modulator, which output is subtracted from the received 2MSK signal. The resulting MSK signal, which depends only on data transmitted in the smaller MSK component, is differentially detected.

The proposed receiver is simpler than the optimum maximum likelihood sequence estimator and it has only 1 dB lower power efficiency. The receiver shows almost no system performance degradation when the sampling instance is displaced by 1/32 of the symbol interval, while an additional 2.5 dB of signal power is required for a sampling error of 1/16 of the symbol interval. Larger differences from optimum sampling instances cause serious degradation of the system performance. When the remodulated MSK signal phase is shifted by less than 1/16 radian no serious degradation is observed, while higher phase shifts degrade the system performance significantly.

A novel non-linear preprocessing method for the reception of ST coded CPM signals in the MIMO system is shown in [Syko05]. The preprocessing reduces the dimensionality of the algorithm to 1 dimension per antenna in the MIMO flat block fading channel. The procedure is built on the fact that the useful received signal can be described by the information waveform space manifold. It was shown [Syko05] that there exists an isomorphism between received useful signal and one-dimensional Euclidean space of sum-phase of the CPM signals – a geometric approach for obtaining an approximate isomorphism for the received signal with additive white Gaussian noise.

FPGA implementation of the communication systems

An example of the communication system implementation was also reported in COST 273. The FPGA design and implementation of the OFDM-based transceiver for WLAN at intermediate frequency is described in [CVAV04]. The design can be applied either for HIPERLAN/2 or Institute of Electrical & Electronic Engineers (IEEE) 802.11.a,g. The receiver structure is plotted in Figure 3.9. The transceiver comprises of the three main blocks: autocorrelator, COrdinate Rotation DIgital Computer (CORDIC) and FFT processor. Each block is used in different time intervals to perform necessary operations. The autocorrelator is active during broadcast preamble for frame detection in section A, time synchronisation and coarse carrier frequency offset estimation in section B. The coarse carrier frequency offset is compensated by CORDIC, while the fine frequency offset and its compensation is the task of the CORDIC in preamble section C. Section C of the preamble is used for channel estimation in the frequency domain using the FFT/IFFT block and the CORDIC calculates the necessary divisions required in the channel estimation stage.

The whole IF transceiver is implemented on a Spartan-3 Xilinx FPGA. The resources required for each subsystem are summarised in Table 3.1. The whole IF transceiver is implemented by 1809 slices, 4 Block RAM (BRAM) and 15 embedded multipliers. The operating frequency for the autocorrelator and the CORDIC for circular mode is 20 MHz, while when the CORDIC performs the division in linear mode the required frequency is 40 MHz. The FFT and the transmitter mixer run at 120 MHz, while 60 MHz is sufficient for the receiver mixer. An XC3S400-4 Spartan-3 device with 4775 slices, 16 BRAM and 16 embedded multipliers is sufficient for the whole transceiver implementation.

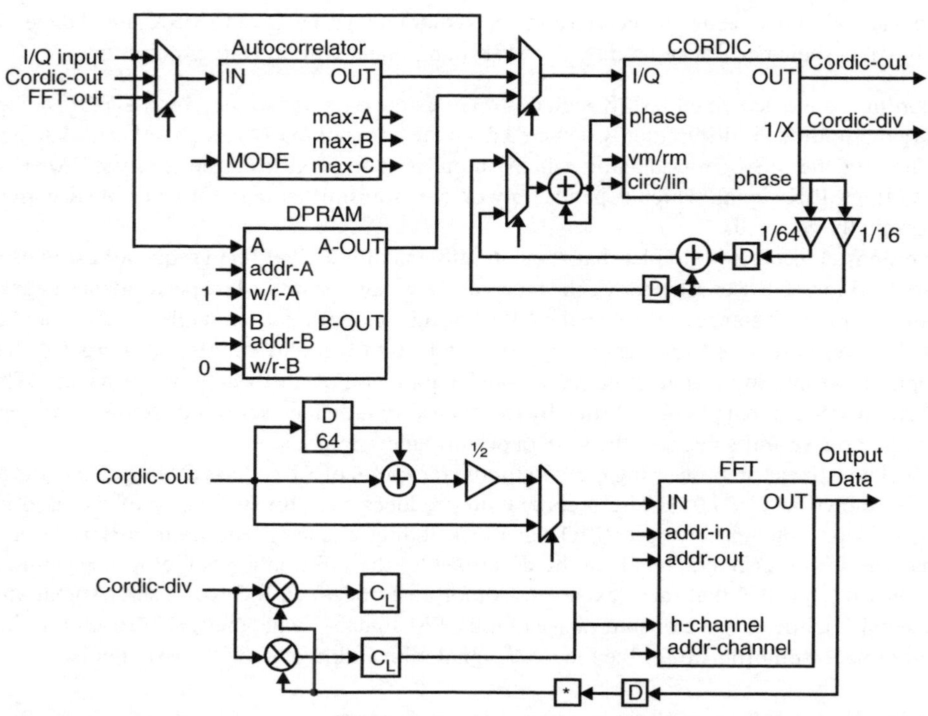

Figure 3.9 *Receiver implementation in FPGA.*

Table 3.1 *Hardware resources.*

	Slices	BRAM	Mults
Upconverter	273	0	0
Downconverter	140	0	0
Mapping	20	0	0
Demapping	62	0	0
FFT/IFFT and channel compensation	340	3	3
Autocorrelator	431	0	9
CORDIC	363	0	2
Channel estimation	85	0	0
Control	95	1	0

3.3 Equalisation

3.3.1 Introduction

Equalisation is a research topic with a long history. It has been well known that among those techniques for the compensation of ISI, maximum likelihood sequence estimation using the Viterbi algorithm can achieve the best performance. The origin of the Viterbi equaliser is found in Forney's landmark paper in 1972, and the research on its applications to time-varying ISI channels, such as frequency selective multipath fading channels, in cooperation with adaptive channel estimation, dates back to Ungerboeck's paper in 1974. Since then, a lot of techniques and algorithms have been proposed and analysed. The 'equalisation' idea has been extended in various ways: 'equalisation' of co-channel interference is known as interference cancellation, and the techniques for simultaneous detection of other users' signals are known as 'multi-user detection'.

However, as the requirement for broadband communications increases, for which ISI cancellation is mandatory, Maximum Likelihood Sequence Estimator (MLSE) techniques finally reached a deadlock due to their prohibitively high complexity. The size of the trellis diagram needed for the Viterbi algorithm increases exponentially with the channel memory lengths as well as the number of simultaneous users. This imposes a limit of its practicality.

To solve the complexity problem, various sub-optimal techniques have been researched, the main stream of which can be classified into linear and non-linear techniques: the linear class includes zero forcing and MMSE equalisation techniques, and the non-linear class includes Decision Feedback Equaliser (DFE) techniques. Some hybrid approaches between MLSE and linear and/or non-linear equalisation, such as delayed decision feedback sequence estimation, as well as MLSE with channel shortening filtering techniques, have also been proposed to solve the complexity problem.

However, obviously those sub-optimum techniques cannot achieve the best performance, and complexity and performance are always a matter of trade-off. This was noticed at least in the late 1980s, and since then the majority of the research activities have been directed towards the creation of the technological bases that can achieve good performance without explicitly having to use equalisers; spread-spectrum and orthogonal signalling techniques were intensively researched for their applications to the 3rd generation mobile communications systems and wireless local area networks, respectively. Such tendency has lasted until the advent of the turbo signal detection concept.

The discovery of the turbo codes has made a drastic change, not only in the coding techniques but also in its related signal processing techniques and algorithms. The equalisation technique based on the turbo concept is, in general, called turbo equalisation. The latest version of the turbo equaliser can asymptotically achieve the optimal MLSE performance without requiring prohibitively large computational complexity. The turbo concept can be used in various applications, among which the most attractive and interesting one is signal detection for MIMO systems.

Because of the general view of the technological evolution experienced since the advent of the turbo concept described above, the TDs describing equalisation algorithms for the receiver side are classified into two categories, iterative and non-iterative ones.

Another important technique that also falls into the 'equalisation' technique is pre-equalisation. The term 'pre-equalisation' can be understood in various ways, and some of the temporary documents have primarily contributed to MIMO 'precoding' techniques. Since especially for the multi-user MIMO scenario 'precoding' includes a lot of technological aspects such as adaptive loading and scheduling that are mostly out of the scope of 'equalisation'. Therefore, this section only focuses on 'pre-processing techniques of signals prior to transmission' and their impacts on performances.

Now, given the overall view of the TDs described above, this section has three major subsections, iterative equalisation, non-iterative equalisation, and pre-equalisation, all for single-user and multi-user MIMO cases. Before entering the equalisation topics in this section, we briefly visit MIMO transmission system's modelling assumptions. Some of the TDs deal with the model assumptions that provide us with new aspects and insights, and by using those new models, different equalisation concepts have arisen or may arise in the future.

3.3.2 Transmission model

We consider a discrete-time complex baseband model for the uplink of a $(K + K_I)$-user MIMO radio system, where K and K_I denote the number of desired users and interfering users, respectively, Figure 3.10.

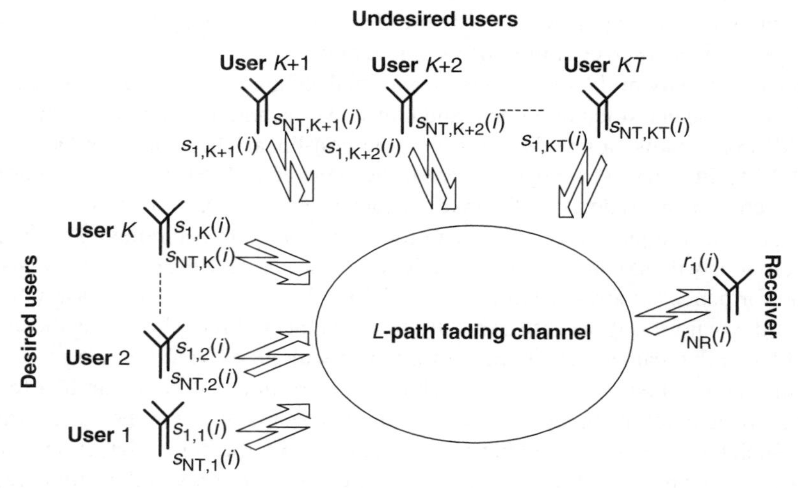

Figure 3.10 *Generic MIMO uplink transmission model.*

The term 'desired users' refers to users whose signals need to be detected; all other users will be considered as interfering users. The total number of users in the system is denoted by $K_T = K + K_I$. Without loss of generality, we assumed that every user will have N_T number of transmit antennas and that these $K_T N_T$ signals are transmitted over an L-path Rayleigh quasi-static fading channel and received by N_R number of antennas at the receiver. We also assumed that the received signals are sampled at the symbol rate $1/T$. Specifically, the transmitted symbol from the nth antenna corresponding to the kth user at discrete-time i, $1 \leq n \leq N$, is represented by $s_{n,k}(i)$, where N denotes the frame length. In general, these symbols can be BPSK modulated or they can be modulated using higher modulation formats. Furthermore, these symbols can represent the uncoded information data, or they can be the outputs of a channel encoder or a pre-equaliser, as we shall see in our further discourse. We will denote the sequence of L symbols transmitted by the nth antenna corresponding to the kth user as $\mathbf{s}_{n,k}(i) = [s_{n,k}(i) \ldots s_{n,k}(i - L + 1)]^T$.

The channel impulse response of the (n, m) antenna pair, $n = 1, \ldots, K_T N_T$ and $m = 1, \ldots, N_R$, which spans over L symbol periods, is given by $\mathbf{H}_{n,m} = [h_{n,m}(0) \ldots h_{n,m}(L-1)]$. Since we only consider quasi-static fading, the channel will be constant over a burst of N symbols and changes independently from burst to burst. These impulse responses capture the ISI effect between different transmit and receive antenna pairs caused by multipath propagation.

The received signal at time i can be modelled by an $N_R \times 1$ vector, which is given by

$$\begin{bmatrix} r_1(i) \\ \vdots \\ r_{N_R}(i) \end{bmatrix} = \begin{bmatrix} \mathbf{H}_{1,1} & \cdots & \mathbf{H}_{1,K_T N_T} \\ \vdots & \vdots & \vdots \\ \mathbf{H}_{N_R,1} & \cdots & \mathbf{H}_{N_R,K_T N_T} \end{bmatrix} \begin{bmatrix} \mathbf{s}_{1,1}(i) \\ \vdots \\ \mathbf{s}_{K_T,N_T}(i) \end{bmatrix} + \begin{bmatrix} n_1(i) \\ \vdots \\ n_{N_R}(i) \end{bmatrix}$$

$$\mathbf{r}(i) = \mathbf{H}\mathbf{s}(i) + \mathbf{n} \quad i = 1, \ldots, N \tag{3.8}$$

where $\mathbf{n}$ is a matrix of complex Gaussian random variables with zero mean and independent real and imaginary parts having the same variance $N_0/2$. While the expression given in (3.8) has been modelled for the uplink scenario, it can also be generalised for downlink transmissions. In this case, the channel experienced by all the users' signals transmitted from the same antenna will be identical.

We can separate the K desired signal contributions from the K_I interfering signals and noise by modifying (3.8) such that

$$\mathbf{r}(i) = \mathbf{H}_D \mathbf{s}_D(i) + \mathbf{H}_I \mathbf{s}_I(i) + \mathbf{n} \quad i = 1, \ldots, N \tag{3.9}$$

where $\mathbf{H}_D$ is now an $N_R \times KLN_T$ matrix representing the channel impulse responses of the desired signals while $\mathbf{H}_I$ is an $N_R \times K_I LN_T$ channel impulse response matrix of the interfering users. Hence, the first term in (3.9) denotes the desired signal and the second and third terms denote the interference and noise, respectively. Based on (3.9), it is easy to devise efficient receiver techniques that have the capability of suppressing the interferences. For instance, the signal contributions from the interference and noise can be modelled by a correlation matrix [PNTL02], [PSTL03a]. Then based on this correlation matrix, a pre-whitening filter can be derived at the receiver so that the Signal-to-Interference-plus-Noise-Ratio (SINR) is maximised [PNTL02], [PSTL03a].

Traditionally, the MIMO system is modelled assuming that the signals from all the transmit antennas associated to a particular user corresponding to the time k are arrived synchronously at each receive antenna. However, in practice, these signals will have different propagation delays, and hence they will arrive at the receive antenna at different times. These mutually unequal T-fractional delays will result in the so-called spatial interbranch interference [Syko03], [WiJo03], which can be treated similar to ISI. Hence, in order to mitigate the effects of interbranch interference, a linear MMSE spatial-temporal equaliser is proposed in [Syko03]. Alternatively, the maximum likelihood sequence detection or its equivalent sub-optimal solutions can be used to mitigate the interference [WiJo03].

3.3.3 Pre-equalisation

The basic idea of pre-equalisation is to design the transmit signals in such a way that interferences in the received data symbols are a priori avoided. In the following, it is assumed that the receivers consist of a linear filter followed by a quantiser. Both the filter impulse responses and the quantisation schemes are assumed to be a priori known at the transmitter side. Furthermore, perfect channel knowledge is assumed to be available at the transmitter side.

In the past, investigations focused on linear signal processing algorithms for pre-equalisation in combination with conventional, i.e. simply connected, quantisation schemes. The three most important linear signal processing algorithms for pre-equalisation are:

- transmit matched filtering,
- transmit zero forcing [KoMa00], [MBWL00], and
- partial transmit zero forcing.

It was only recently that non-linear pre-equalisation algorithms, which exploit the discrete nature of the modulation alphabet, attained attraction. First investigations focused on non-linear transmitter side signal processing in combination with conventional simply connected quantisation schemes on the receiver side [WeMe03], [IHRF03]. However, non-linear transmitter side signal processing alone cannot combat the problem of large required transmit energies for interference compensation. This problem can only be avoided by going to unconventional multiply connected quantisation schemes. The novel transmit non-linear zero forcing [WeMZ04] exploits multiply connected quantisation schemes for transmit energy reduction. Similar transmission schemes were published in [Toml71], [HaMi72], [Fisc02].

In the following, a scenario where each MT receives a single data symbol is considered, i.e. pre-equalisation is used for multiple access interference reduction. The data symbol to be transmitted for MT k is denoted by $\underline{d}^{(k)}$. For simplicity, it is assumed that the data symbols are QPSK modulated in the following:

$$\underline{d}^{(k)} \in \{\pm 1 \pm \mathrm{j}\}, \quad k = 1 \ldots K \tag{3.10}$$

All K data symbols of the different MTs are compiled in the data vector

$$\underline{\mathbf{d}} = \left(\underline{d}^{(1)} \ldots \underline{d}^{(K)}\right)^{\mathrm{T}} \tag{3.11}$$

On the transmitter side a transmitted signal described by the transmitted vector

$$\underline{\mathbf{s}} = \left(\underline{s}^{(1)} \ldots \underline{s}^{(K_{\mathrm{s}})}\right)^{\mathrm{T}} \tag{3.12}$$

is formed. The transmitted vector contains samples in time and also in the space domain if multiple transmit antennas are employed. With the channel matrix describing the linear channel and the noise vector $\underline{\mathbf{n}}$ follows

$$\underline{\mathbf{e}} = \underline{\mathbf{H}} \cdot \underline{\mathbf{s}} + \underline{\mathbf{n}} \tag{3.13}$$

for the received vector, Figure 3.11. For simplicity, it is assumed in the following that the elements of the noise vector $\underline{\mathbf{n}}$ are uncorrelated, Gaussian distributed with variance σ^2 of real and imaginary

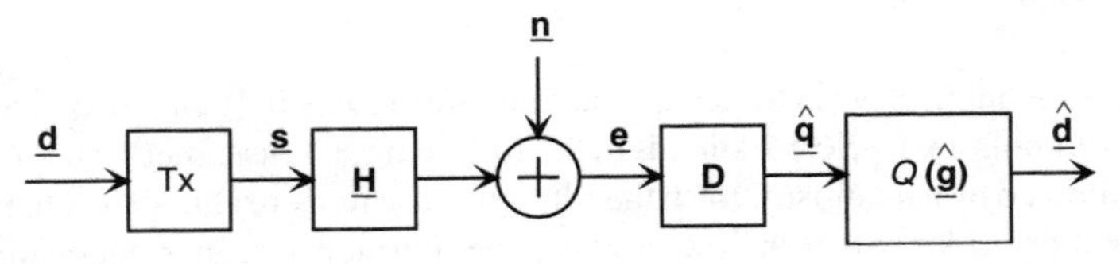

Figure 3.11 *System model.*

part [Proa95]. On the receiver side, discrete valued estimates $\hat{\underline{d}}^{(k)}$ of the transmitted data symbols $\underline{d}^{(k)}$ are obtained in a two step procedure. First of all, continuous valued estimates $\hat{\underline{q}}^{(k)}$ compiled in the vector

$$\hat{\underline{\mathbf{q}}} = \left(\hat{\underline{q}}^{(1)} \dots \hat{\underline{q}}^{(K)}\right) \tag{3.14}$$

are obtained as a linear function of the received vector $\underline{\mathbf{e}}$ characterised by the demodulator matrix $\underline{\mathbf{D}}$:

$$\hat{\underline{\mathbf{q}}} = \underline{\mathbf{D}} \cdot \underline{\mathbf{e}} \tag{3.15}$$

In a second step, the discrete valued estimates $\hat{\underline{d}}^{(k)}$ are obtained by the quantisation of the continuous valued estimates $\hat{\underline{q}}^{(k)}$, $k = 1 \dots K$.

Concerning the quantiser, in general multiple representatives $\underline{q}_{m,p^{(m)}}$, $p^{(m)} = 1 \dots P^{(m)}$, exist for each possible data symbol value $\underline{d}_m$. The Voronoi regions $\mathbb{Q}_{m,p^{(m)}}$ of the representatives $\underline{q}_{m,p^{(m)}}$, $m = 1 \dots M$, $p^{(m)} = 1 \dots P^{(m)}$, are termed partial decision regions in the following. In general, each decision region $\mathbb{Q}_m$, $m = 1 \dots M$, consists of multiple partial decision regions $\mathbb{Q}_{m,p^{(m)}}$:

$$\mathbb{Q}_m = \bigcup_{p^{(m)}=1}^{P^{(m)}} \mathbb{Q}_{m,p^{(m)}} \tag{3.16}$$

We term a quantisation scheme conventional, in which

- the decision regions $\mathbb{Q}_m$, $m = 1 \dots M$, are simply connected, that is $P^{(m)} = 1$, $m = 1 \dots M$, and
- the representative $\underline{q}_{m,1}$ is equal to the corresponding data symbol value $\underline{d}_m$.

Otherwise, we designate the quantisation scheme as unconventional. If the continuous valued estimate $\hat{\underline{q}}^{(k)}$ falls into the decision region $\mathbb{Q}_m$, then the quantiser delivers at its output the value

$$\hat{\underline{d}}^{(k)} = \underline{d}_m \tag{3.17}$$

This operation of the quantiser can be characterised by the quantisation function

$$\underline{Q}\left(\hat{\underline{q}}^{(k)}\right) = \hat{\underline{d}}^{(k)} \in \mathbb{D} \tag{3.18}$$

The notation $\underline{Q}\left(\hat{\underline{\mathbf{q}}}\right)$ in Figure 3.11 means that the quantisation function $\underline{Q}\left(\hat{\underline{q}}^{(k)}\right)$ of (3.18) is applied to each of the K components of the vector $\hat{\underline{\mathbf{q}}}$. A common example of an unconventional quantiser for QPSK modulation is the so-called modulo quantiser.

In the case of linear transmitters for the transmitted vector

$$\underline{\mathbf{s}} = \underline{\mathbf{M}} \cdot \underline{\mathbf{d}} \tag{3.19}$$

holds, i.e. in the case of linear data transmission the transmitted vector $\underline{\mathbf{s}}$ is the superposition of partial transmitted vectors

$$\underline{\mathbf{s}}^{(k)} = \left[\underline{\mathbf{M}}\right]_{K_{\mathrm{A}},k}^{1,k} \underline{d}^{(k)}, \quad k = 1 \dots K \tag{3.20}$$

each containing the information on a specific data symbol $\underline{d}^{(k)}$. The column vector $\left[\underline{\mathbf{M}}\right]_{K_{\mathrm{A}},k}^{1,k}$ is termed transmitted signature for the data symbol $\underline{d}^{(k)}$ in the following.

The idea behind transmit matched filtering is the minimisation of the average partial transmitted energy required for the transmission of a certain data symbol $\underline{d}^{(k)}$ if interferences are neglected, i.e. each transmitted signature is optimised separately. From this rationale the modulator matrix

$$\underline{\mathbf{M}}_{\mathrm{MF}} = \underline{\mathbf{H}}^{*\mathrm{T}}\underline{\mathbf{D}}^{*\mathrm{T}} \cdot \left(\mathrm{diag}\left(\underline{\mathbf{D}}\,\underline{\mathbf{H}}\,\underline{\mathbf{H}}^{*\mathrm{T}}\underline{\mathbf{D}}^{*\mathrm{T}}\right)\right)^{-1} \tag{3.21}$$

follows. If multiple data symbols are transmitted in parallel, the partial transmitted vectors $\underline{\mathbf{s}}^{(k)}$ of (3.20) obtained by transmit matched filtering may cause quite strong interferences to other data symbols.

Transmit zero forcing totally eliminates interferences, i.e.

$$\underline{\mathbf{D}} \cdot \underline{\mathbf{H}} \cdot \underline{\mathbf{M}}_{\mathrm{ZF}} = \mathbf{I} \tag{3.22}$$

holds. Remaining degrees of freedom in the choice of $\underline{\mathbf{M}}_{\mathrm{ZF}}$ are exploited for the minimisation of the required average transmitted energy. From this rationale, the modulator matrix

$$\underline{\mathbf{M}}_{\mathrm{ZF}} = \underline{\mathbf{H}}^{*\mathrm{T}}\underline{\mathbf{D}}^{*\mathrm{T}}\left(\underline{\mathbf{D}}\,\underline{\mathbf{H}}\,\underline{\mathbf{H}}^{*\mathrm{T}}\underline{\mathbf{D}}^{*\mathrm{T}}\right)^{-1} \tag{3.23}$$

is obtained. The price to be paid for the elimination of interferences is an increase of the average transmitted energy as compared to transmit matched filtering.

In transmit partial zero forcing, the elimination of interference to certain data symbols is sacrificed for the reduction of the required transmitted energy. Without loss of generality, the partial transmitted vector $\underline{\mathbf{s}}^{(k)}$ for data symbol $\underline{d}^{(k)}$ is considered in the following. Interferences to preceding data symbols $\underline{d}^{(k')}$, $k' = 1 \ldots k-1$, should be eliminated whereas interferences to subsequent data symbols $\underline{d}^{(k')}$, $k' = k+1 \ldots K$, are allowed. Exploiting the remaining degrees of freedom for energy minimisation, the transmitted signature

$$\left[\underline{\mathbf{M}}_{\mathrm{PZF}}\right]_{K_{\mathrm{A}},k}^{1,k} = \left[\underline{\mathbf{H}}^{*\mathrm{T}}\left[\underline{\mathbf{D}}\right]_{k,K}^{1,1\;*\mathrm{T}}\left(\left[\underline{\mathbf{D}}\right]_{k,K}^{1,1}\underline{\mathbf{H}}\,\underline{\mathbf{H}}^{*\mathrm{T}}\left[\underline{\mathbf{D}}\right]_{k,K}^{1,1\;*\mathrm{T}}\right)^{-1}\right]_{K_{\mathrm{A}},k}^{1,k} \tag{3.24}$$

for data symbol $\underline{d}^{(k)}$ is obtained. The computational complexity required for computing the modulator matrix $\underline{\mathbf{M}}_{\mathrm{PZF}}$ could be reduced using the QR decomposition of $\underline{\mathbf{D}}\,\underline{\mathbf{H}}$ [PTVF92], [HoJo85].

The crux of our new approach, transmit non-linear zero forcing, consists in using an unconventional quantiser, which requires only slight modifications of the receivers. Thanks to this unconventional quantiser, it is not necessary to totally eliminate interferences in order to render them harmless any more. The goal is to design the transmitter in such a way that the continuous valued estimates $\hat{\underline{q}}^{(k)}$, $k = 1 \ldots K$, are as close as possible to one of the representatives of the data symbol values to be transmitted, i.e. there exists the additional degree of freedom to choose one of the representatives. Preferably, the representatives are chosen in such a way that the average transmitted energy is minimised [PeHS03]. In general, this optimisation is very complex. In the following a sub-optimum, low complexity algorithm for the selection of the representatives, which leads to a good overall performance, is presented.

The partial transmitted vectors $\underline{\mathbf{s}}^{(k)}$ for the data symbols $\underline{d}^{(k)}$, $k = 1 \ldots K$, are designed one after the other in transmit non-linear zero forcing, starting with the partial transmitted vector $\underline{\mathbf{s}}^{(1)}$ for the first data symbol $\underline{d}^{(1)}$. This is no restriction, since MTs can be numbered in any desired order. In the following the design of the partial transmitted vector $\underline{\mathbf{s}}^{(k)}$ for one specific data symbol $\underline{d}^{(k)}$, taking the value $\underline{d}_m$, is described in detail. First of all, the interference $\underline{i}^{(k)}$ resulting from the partial transmitted vectors $\underline{\mathbf{s}}^{(k')}$ of the preceding data symbols $\underline{d}^{(k')}$, $k' = 1 \ldots k-1$, is calculated

$$\underline{i}^{(k)} = \left[\underline{\mathbf{D}} \cdot \underline{\mathbf{H}} \cdot \sum_{k'=1}^{k-1} \underline{\mathbf{s}}^{(k')} \right]_k \tag{3.25}$$

If an unconventional quantiser is used, it is only necessary to transmit a vector $\underline{\mathbf{s}}^{(k)}$ which, together with the interference, results in a continuous valued estimate $\hat{\underline{q}}^{(k)}$ that is equal to one of the representatives $\underline{q}_{m,p^{(m)}}$, $p^{(m)} = 1 \ldots P^{(m)}$. We introduce the difference signal

$$\underline{\Delta}^{(k)} = \underline{q}_{m,p^{(m)}} - \underline{i}^{(k)} \tag{3.26}$$

Preferably the representative $\underline{q}_{m,p^{(m)}}$ for which the difference signal energy $\frac{1}{2}\left\|\underline{\Delta}^{(k)}\right\|^2$ is minimum is chosen. Finally, following the idea of transmit partial zero forcing, the partial transmitted vector $\underline{\mathbf{s}}^{(k)}$ is designed in such a way that no new interferences to the previous data symbols $\underline{d}^{(k')}$, $k' = 1 \ldots k-1$, are introduced. In contrast to transmit zero forcing, interferences to subsequent data symbols $\underline{d}^{(k')}$, $k' = k+1 \ldots K$, are allowed. With (3.24) and (3.26) follows

$$\underline{\mathbf{s}}^{(k)} = \left[\underline{\mathbf{M}}_{\mathrm{PZF}}\right]_{K_{\mathrm{A}},k}^{1,k} \cdot \underline{\Delta}^{(k)} \tag{3.27}$$

Transmit non-linear zero forcing is particularly beneficial if the MTs are numbered in the order of increasing channel gain. Then the partial transmitted vectors $\underline{\mathbf{s}}^{(k)}$ of the MTs with the lowest channel gains are designed first. This way especially the average partial transmit energies for MTs with low channel gains are reduced as compared to transmit zero forcing. As the partial average transmitted energies for MTs with low channel gains have dominant influence on the average transmitted energy, also the average transmitted energy is reduced considerably.

3.3.4 SISO equalisation

Non-iterative equalisation

In this subsection, sub-optimal reduced complexity sequence detectors based on the delayed decision feedback sequence detector are introduced for reception of signals transmitted over frequency selective channels. The computational complexity of the optimal MLSE is governed by the constellation size and the memory order of the channel, and the trellis size becomes infeasibly large in channels with a high memory order. Thus, there is a considerable interest in the reduction of the computational complexity by the means of sub-optimal techniques. The considered technique reduces the number of trellis states in the Viterbi algorithm. In the original algorithm, the receiver includes prefilter and a reduced state Viterbi algorithm. The purpose of the prefilter is to process the received signal with a whitened matched filter. Similarly, the reduced state Viterbi algorithm performs sequence estimation. The performance loss due to memory truncation in the Viterbi algorithm is mitigated by per-survivor processing. The metric of each survivor is computed by using DFE with a reduced number of taps. The problem of this original algorithm is that there is no guarantee of the existence of a feedforward filter. However, by introducing the mean-square whitened matched filter as a prefilter, this problem can be solved.

The performance of the original algorithm with mean-square whitened matched filter can be improved by adopting the generalised Viterbi algorithm at the Viterbi stage. The enhanced version of the algorithm is denoted as the mean square generalised delayed decision feedback sequence detector, where several survivors are allowed to remain as candidates at each state. As a result, there are several transitions diverging from and merging into each state. At each trellis state, survivors that have merged in each state are sorted in ascending order, and the sequence associated to the number of predefined numbers having lower metrics is selected as survivors.

Numerical results shows that the mean-square whitened matched filter outperforms the original algorithm when a severely frequency selective channel is considered. Moreover, numerical results of the version applying the generalised Viterbi algorithm suggest that through a suitable choice of truncation and the number of survivors allows further degrees of freedom between complexity and performance.

Iterative equalisation

A generic approach to solving complex problems is trying to approach the optimal solutions by iterative means. For the iterative method to be useful, applying multiple iterations should result in a lower computational complexity than solving the optimal solution directly. In the case of detection of error correction coded data transmitted over a frequency selective channel, the optimal solution has the direct computational complexity depending exponentially on the combined memory of the channel and the error correcting code. In the iterative turbo approach, the channel equaliser operates on the channel memory and the channel decoder operates on the error correcting code memory. These two exchange probabilistic information on the transmitted data to iteratively improve the received data estimates.

While the complexity reduction from the optimal detector is much reduced by using the turbo principle, the resulting complexity of the channel equaliser and the decoder may still be too high for realistic applications. For instance, the trellis-based channel equalisers (MAP, etc.) using the full channel memory have complexity exponentially dependent on the channel memory order. On the other hand, many results are available on constructing high-performance channel codes with reasonable decoding complexity. Therefore, many sub-optimal turbo equalisers are designed to minimise the equaliser complexity. Much of the work reported here has been devoted to complexity reduction approaches.

The original work on turbo equalisation utilised the soft-output Viterbi algorithm for channel equalisation. The work reported in [MRST03] considers the utilisation of MMSE-DFE in place of noise-whitening filter as the prefilter for the equaliser. Due to using a bi-directional soft-input-soft-output algorithm for equalisation, the equaliser filters must account for the time-reversal. The block diagram is shown in Figure 3.12.

Another sub-optimal turbo equalisation approach that has gained wide interest is the structure combining a soft interference canceller with an MMSE filter. Reference [Dejo02] presents a unified view on MMSE turbo equalisation for single-input-single-output channels. The algorithm computes the expected value $\vec{\bar{s}}$ of the transmitted symbol $\vec{s}$ and performs interference cancellation from the received signal as

$$\vec{\tilde{r}} = \vec{r} - \vec{H}\bar{s} \tag{3.28}$$

A time-varying MMSE filter is then computed to minimise the mean-squared error

$$\arg\min_{\vec{w}} \mathrm{E}\left\{s - \vec{w}\tilde{r}\right\} \tag{3.29}$$

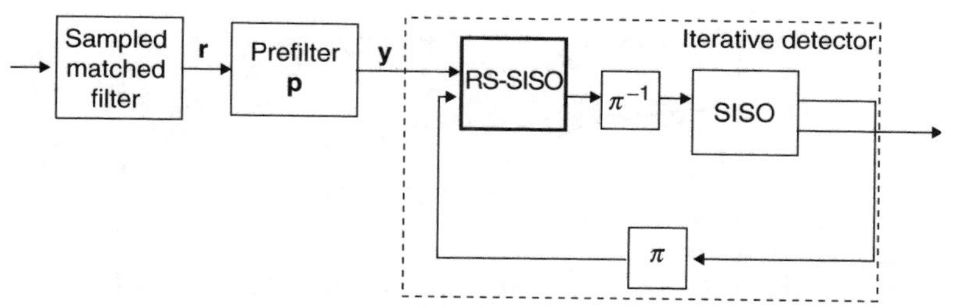

Figure 3.12 *The feedforward-feedback structure [MRST03].*

Asymptotic and convergence properties for bit-interleaved coded modulation are analysed also in [Dejo02]. The utilised symbol mapping is found to be important for the convergence and asymptotic performance of the equaliser. Gray mapping is found to offer convergence at a lower SNR than set-partitioned mapping, at the expense of poorer asymptotic performance when all interference has been suppressed. Further work related to the problem of transmission methods for turbo equalisation is reported in [KaMa03], where a multilevel coded transmission is utilised to enhance the convergence properties of the turbo equaliser. A hierarchical mapping is used with the multilevel codes to enable a simple MMSE turbo equaliser algorithm without the need for symbol-to-bit demapping as in [Dejo02], with the unequal error protection property for transmitted data. Further results on the robustness of the approach for layered transmission in MIMO systems is reported in Section 3.3.5.

In the above, it has been demonstrated that the performance of turbo equalisation depends on both the convergence and asymptotic properties of the used channel code. An approach for designing codes explicitly for the turbo equaliser is reported in [WKTM05]. The convergence characteristic of a frequency-domain MMSE turbo equaliser, as given by its extrinsic information transfer function, is computed semi-analytically by evoking the Gaussian equivalent channel approximation for the equaliser output and computing the corresponding mutual information with transmitted symbols analytically. The method is then used to generate a set of equaliser extrinsic information transfer function based on random block-static channel realisations, which are used to find an 'outage extrinsic information transfer function'. Any channel realisations with the equaliser extrinsic information transfer function being below or crossing the outage function are assumed not to converge. A low density parity check code is then constructed to fit the outage extrinsic information transfer function. The new code outperforms the corresponding unoptimised code, as shown in Figure 3.13. Reference [MKSS04] presents a pipelined method to efficiently compute the MMSE filter coefficients required in the symbol-wise MMSE filter utilised in the MMSE turbo equaliser. Another approach, which utilises an approximate fixed MMSE filter over the received frame, is evaluated with model-based as well as measured channels. It is found that the performence of the approximate version is close to the exact version.

3.3.5 MIMO equalisation

Non-iterative MIMO equalisation

This section introduces a linear frequency domain MIMO equaliser for single carrier point to point spatially multiplexed for turbo coded MIMO communication. In the past years, linear time-domain

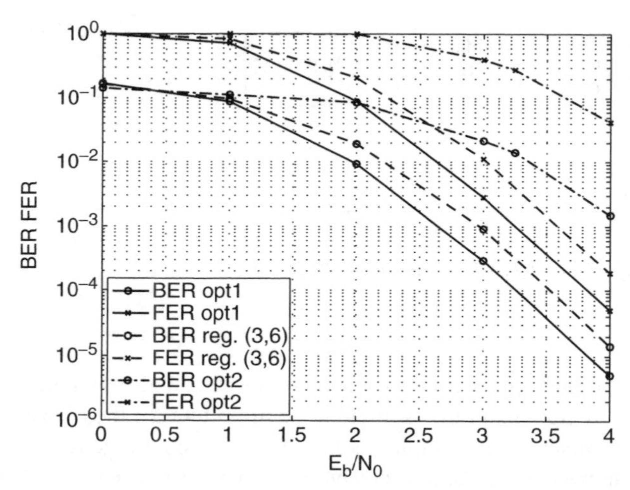

Figure 3.13 *Outage-optimised codes with MMSE turbo equalisers (opt1: rate 1/2, opt2: rate 2/3) [WKTM05]. (© 2005 IEEE, reproduced with permission)*

MIMO equalisers have been intensively studied in single carrier systems for spatially multiplexed and space-time coded MIMO schemes. However, the time-domain equalisation is not often practical, since the number of operations per symbol increases polynomially with the length of ISI. Thus, the computational complexity can be enormous in broadband time-dispersive MIMO channels. Recently, frequency-domain equalisation techniques have been recognised as one of the most promising methods to reduce the computational complexity of the equaliser. Computational complexity analysis results show that the computational complexity of the frequency-domain equaliser grows logarithmically with the block length of the FFT.

The complexity reduction achieved by the frequency-domain MIMO equaliser is based on discrete convolution theorem, which enables the utilisation of an efficient FFT algorithm. However, in order to exploit the discrete convolution theorem, the linear convolution of transmitted signal and channel must be converted into a circular one. The most common way to satisfy this requirement is to use cyclic transmission, in which an additional guard period, either cyclic prefix or training sequence, is inserted at the transmitter and discarded at the receiver. As a result, linear time domain convolution is converted to circular and time domain circular convolution can be replaced by simple element-wise multiplication in the frequency domain. Moreover, block circulant channel matrices are diagonalised with the well-known block diagonalisation process to enable efficient computation of equaliser coefficients in the frequency domain.

In the receiver signal processing is performed on a block by block basis. Thus, at each receiver antenna the guard period is first discarded and the received signal transformed into the frequency domain. In the following step, the frequency-domain equaliser's space-frequency filter coefficients are calculated. The filter coefficients are determined based on MMSE criterion to minimise the effect of ISI, co-antenna interference and thermal noise. Therefore, the linear frequency-domain MIMO equaliser aims to detect transmitted antennas, antenna by antenna, by using linear space-frequency domain filtering. Finally, after element-wise multiplication with received signal and equaliser coefficients, detected transmitted layers are transformed back to the time domain by performing an IFFT over each detected transmitted layer.

It is well known that the distribution of residual interference-plus-noise at the output of an MMSE filter can be well approximated by a Gaussian distribution. Therefore, it is reasonable that a soft-input-soft-output channel decoder assumes that the soft output of the MMSE filter represents the output of an equivalent AWGN channel. As a result, the equivalent AWGN channel parameters are computed based on the equaliser coefficients. By using the output of the MMSE filter and the calculated equivalent AWGN parameters, the log-likelihood ratios of coded bits can be computed for the turbo decoder.

The proposed algorithm is applicable to any types of channel coding of which the decoder requires soft input. Correspondingly, numerical results illustrate that frequency-domain MIMO equalisation utilises effectively available multipath diversity in frequency selective Rayleigh fading channels. However, a significant performance degradation is encountered in the presence of spatial correlation.

Iterative MIMO equalisation

This section focuses on iterative receiver design for bandwidth efficient communications based on MIMO techniques. Therefore, most of the receiver concepts presented here can be seen as a natural extension of the techniques described in Subsection 3.3.4, by means of appropriate space-time mapping. The principal transmitter and receiver block diagrams are illustrated in Figure 3.14 and they closely resemble those of Subsection 3.3.4. The non-iterative techniques of Subsection 3.3.5 can be seen as the first iteration of the corresponding iterative techniques.

It has been pointed out earlier that the receiver complexity is the main bottleneck for the practical implementation of the broadband systems based on single carrier communications. The complexity of the optimal receiver in MIMO communications is further increased by the fact that multiple transmit antennas' signals have to be detected simultaneously. This has been the main motivation for the development of low complexity iterative receiver structures. Thanks to the decoupled detection (equalisation) and decoding blocks of the iterative receiver, the complexity reduction can be performed independently at these two stages. The majority of the techniques reviewed in this section consider complexity reduction at the detection or equalisation stage only.

An interesting example is presented in [WiJo03] where the authors propose iterative search techniques to reduce the complexity of the optimal MAP MIMO detection stage at the receiver. The core of the technique is the M-algorithm, by which only M branches are retained for each active node of the trellis. Thereby, the trade-off between complexity and performance of the proposed receiver can be controlled by a factor M. Moreover, the receiver complexity is *linear* in the number of transmit antennas and can be further reduced by the use of multilevel mapping, which makes it approximately independent on the constellation size. An example of the receiver performance for different list sizes M is presented in Figure 3.15.

The approach that certainly attracted most attention recently is a low complexity MMSE equalisation enhanced by soft cancellation. The authors of [Vand02] and [Guég03] have considered such a receiver for MIMO transmission assuming a vertically encoded Bell Layered Space-Time (BLAST) structure. A list sphere decoding approach has been used in [Guég03] as an approximation of the optimal A Priori Probability (APP) receiver, for performance comparison. Although having an order of magnitude lower computational complexity than list sphere decoding, the MMSE turbo equaliser receiver was shown to have the same performance as the list sphere detection-based receiver. It should be pointed out that the performance of the list sphere detector itself is already very close to the optimal receiver performance [Guég03]. Additional complexity reduction with only a minor performance degradation can be achieved by considering the time-averaged version of the original MMSE

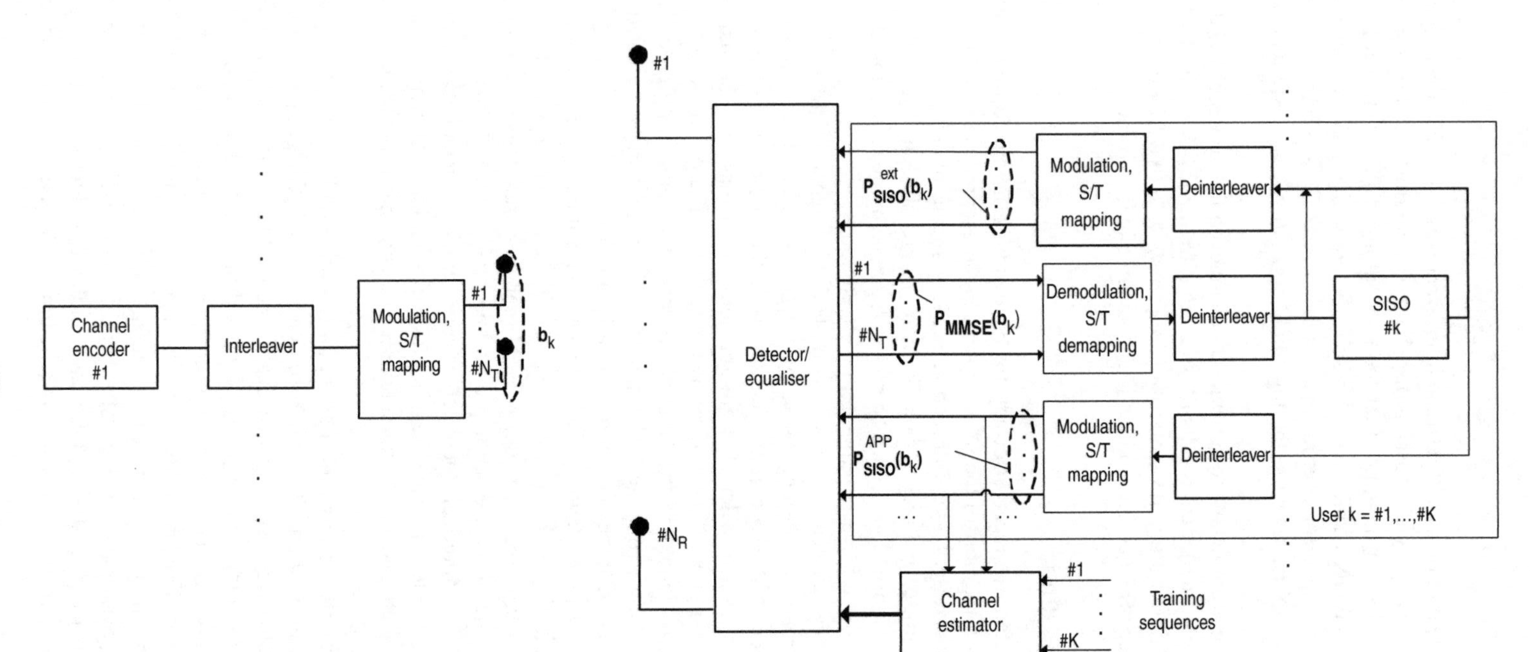

Figure 3.14 *Transmitter and receiver block diagrams for iterative MIMO equalisation.*

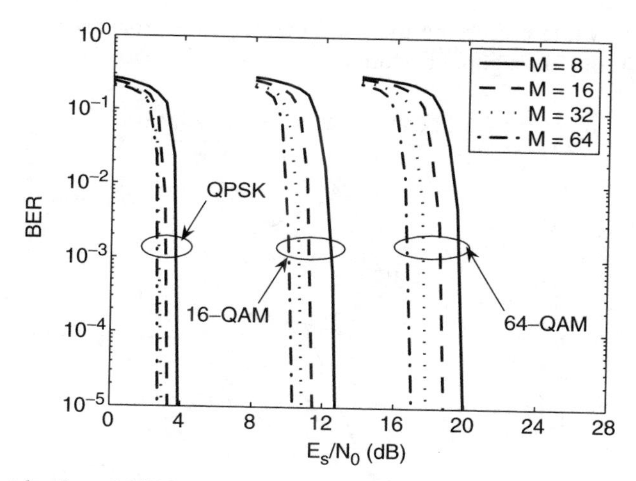

Figure 3.15 *M-algorithm-based MIMO (8 × 8) equalisation, M-branches used [WiJo03]. (© 2003 IEEE, reproduced with permission)*

turbo equaliser algorithm proposed in [Vand02]. In severely frequency selective channels with very slow fading, as found in broadband wireless systems, a frequency-domain implementation of the equaliser [KaMa04] becomes very attractive in terms of complexity and performance. Moreover, it is very effective in exploiting multipath and multiple receive antennas' diversity as well as time diversity provided by channel code, which can be seen in Figure 3.16.

The performance of the MMSE turbo equaliser receiver in the real fields is further evaluated in [STTM03] and [KSMTar]. The importance of channel multipath richness to reach high performance in MIMO scenarios is highlighted in both evaluations. In the latter, the transmission presented in [KaMa03] is combined with a smart Automatic Retransmission Request (ARQ) algorithm to reach

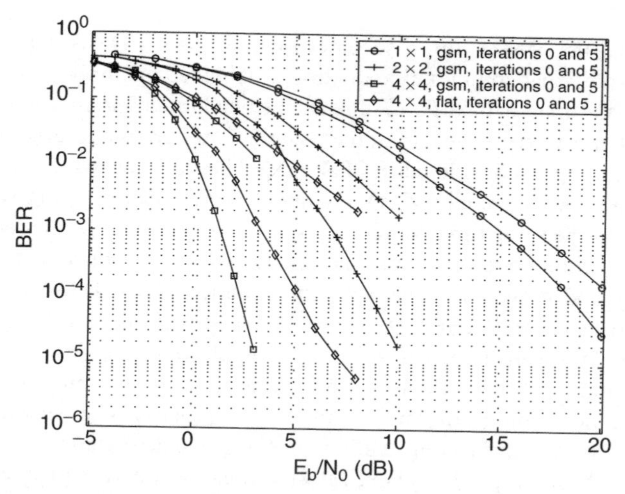

Figure 3.16 *MMSE turbo equalisation in GSM MIMO channels [Vand02].*

a high link throughput while preserving the good convergence properties of the multilevel coded transmission. The resulting throughput performance is shown in Figure 3.17.

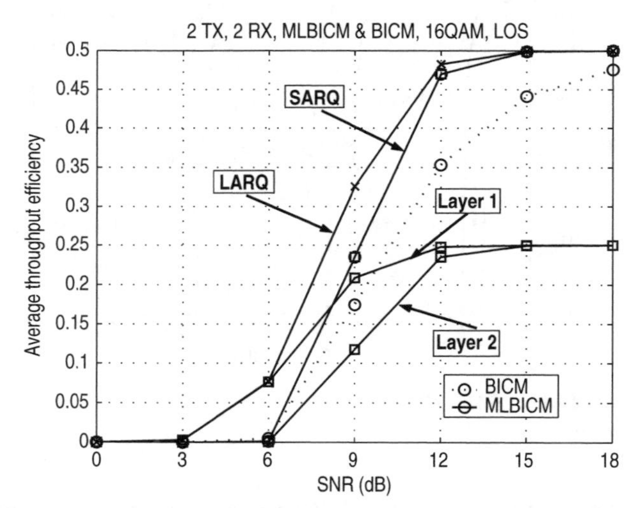

Figure 3.17 *Multilevel coding for MMSE turbo equalisation combined with layered ARQ [KSMTar]. (© 2005 IEEE, reproduced with permission)*

The results reported in [Guég03] and [Vand02] are obtained with the assumption of perfect channel state information at the receiver. On the contrary, in [VeMa03] the authors have considered an iterative MIMO MMSE turbo equaliser receiver with least squares and MMSE channel estimation algorithms. Thereby, the iterative processing is performed jointly over equalisation, decoding and channel estimation blocks. Comparison of the cases with least-squares estimated and perfect channel state information is illustrated in Figure 3.18. It can be seen that the proposed iterative equalisation, decoding and channel estimation scheme shows only about 1 dB performance loss due to the channel estimation error.

Further extension of the MMSE turbo scheme for multi-user CDMA systems can be found in [BuSh02], where an additional complexity reduction is achieved by applying matched filtering after the second iteration. This approach partly eliminates the need for covariance matrix inversion, which is computationally the most expensive part of the receiver. An iterative receiver for CDMA that does not employ channel coding is considered in [TeRe01]. The non-linear recurrent neural network is proposed there to combat co-channel and intersymbol interference, showing superior performance to linear approaches. At the same time the receivers' complexity remains moderate.

In the case of space-time coded systems, the receiver design slightly differs from that of horizontally or vertically coded BLAST schemes. An example of the MMSE turbo receiver suitable for this scenario is presented in [VeSM04], where the composite signal of multiple transmit antennas is treated as a single higher-order constellation symbol point. In addition to this, a complexity reduction method is proposed, which is based on taking into account only a significant portion of the Carrier-to-Interference Ratio (CIR) in the equalisation process. The performance evaluation by using field measurement data reveals only a slight performance degradation. A hybrid MMSE-MAP turbo receiver has also been proposed in [VeSM04] to preserve degrees of freedom of the receiver, which was shown to be beneficial in the presence of unknown interference and spatial correlation.

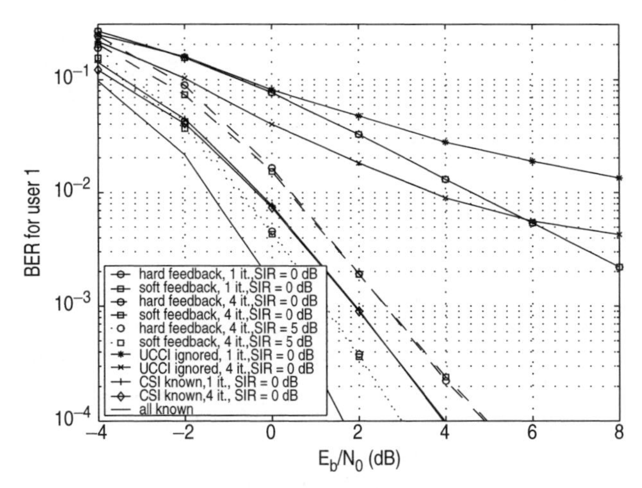

Figure 3.18 *Iterative channel estimation with MMSE turbo equalisation [VeMa03]. (© 2005 IEEE, reproduced with permission)*

Other methods to suppress unknown interference based on pdf and covariance estimation are reported in [MaVJ04].

An interesting example of complexity reduction at the decoding stage of the iterative receiver has been considered in [KaVM05]. Space-time weighted non-binary repeat-accumulate codes have been used there together with the MMSE turbo receiver. Decoding of these codes is possible using a sum-product algorithm whose complexity is about six times lower than the complexity of the MAP algorithm. That, in turn, makes them very attractive alternatives to the trellis-based codes that require complex decoding algorithms. Furthermore, it was shown that the iterative receiver of [KaVM05] is capable of achieving *full* diversity order even in heavily loaded multi-user scenarios, Figure 3.19.

In [TrMa04], it is shown that, for the choice of BPSK modulation or any other modulation that can be approximated as a pulse amplitude modulation with real-valued symbol alphabet (multi-level Pulse Amplitude Modulation (PAM), MSK, Gaussian MSK (GMSK)), the approximation of the Log Likelihood Ratios (LLRs) should explicitly consider this fact. Furthermore, the MMSE filter computation is optimised, and the numerical complexity is slightly reduced. Simulations for a multipath Rayleigh fading channel model show an improved performance especially for the first iteration. Simulations using channel sounder measurements from a microcellular environment reveal a slight improvement in performance critical situations.

3.4 Synchronisation and channel state estimation

This section focuses on selected aspects of the synchronisation and Channel State Estimation (CSE) technique for digital communications, namely in four major areas: (1) an iterative technique is the promising approach that allows to operate modern receivers at very low signal-to-noise ratios dictated by excellent bit error rate performance of modern turbo codes; (2) a range of problems associated with channel state estimation and prediction for time-varying channel is discussed next; (3) we also

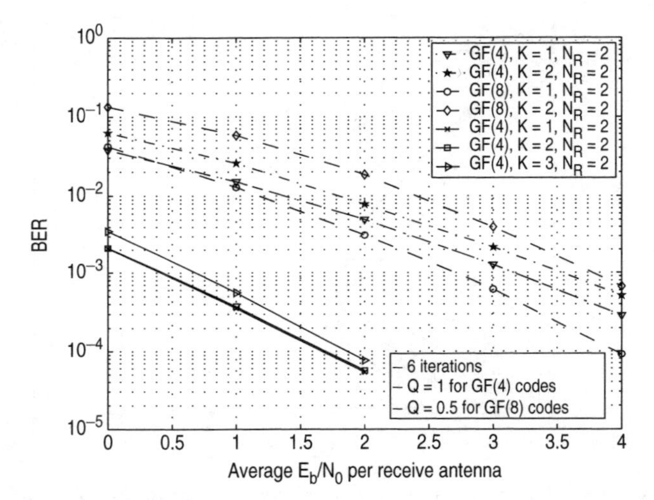

Figure 3.19 *Weighted non-binary repeat-accumulate space-time codes with MMSE turbo equalisation [KaVM05].*

analyse some aspects of the fundamental performance limits dictated by the data transmission over the parametric channel; (4) a specific topic of the sequence synchronisation technique is covered at the end.

3.4.1 Iterative technique

APP aided carrier recovery in an iterative receiver

We consider the problem of synchronisation (i.e. carrier phase and symbol timing recovery) and decoding in coded systems, especially turbo coded systems. We introduce a new concept, 'APP aided estimation', where the log-likelihood ratio obtained from a MAP decoder (as used in turbo decoding) is used to aid estimation in an iterative synchroniser/decoder. We illustrate this in detail with reference to carrier recovery in a turbo coded BPSK system, and demonstrate a substantial performance advantage over conventional separate synchronisation and decoding. We also give some outline results for turbo coded QPSK, and for timing recovery in turbo coded BPSK.

We assume an AWGN channel with received signal $r_i = A_i \exp(j\phi_i)$, A_i being the magnitude and ϕ_i the phase, where ϕ_i vary due to carrier phase error and AWGN. The log-likelihood function of the ith received signal is

$$L_i(\phi) = \ln \frac{\cosh\left(\frac{L_e(i)}{2} + \frac{A_i \cos(\varphi - \phi_i)}{\sigma^2}\right)}{\cosh \frac{L_e(i)}{2}} \tag{3.30}$$

where σ^2 is defined as the noise power, and $L_e(i) = \ln(p_i(1)/p_i(0))$ is the log-likelihood ratio of the ith data bit obtained from the turbo decoder (the a priori or extrinsic likelihood ratio from the turbo decoder). The likelihood function over a block is $\bar{L}(\phi) = \sum_{i=1}^{N} L_i(\phi)$.

In order to find the value maximising the log-likelihood function the following simplification is used. We can conveniently approximate the function by its Fourier series, and since it is even it has only cosine terms

$$L_i(\phi) = A_0(i) + \sum_{r=1}^{\infty} A_r(i) \cos\left(r(\phi - \phi_i)\right) \tag{3.31}$$

In practice, we find that harmonics above the second are negligible in amplitude, and thus,

$$L_i(\phi) \approx A_0(i) + A_1(i)\cos(\phi - \phi_i) + A_2(i)\cos\left(2(\phi - \phi_i)\right) \tag{3.32}$$

This approximation allows for a direct calculation of ϕ maximising the log-likelihood function. A similar approach can be applied to the symbol timing recovery. See [ZhBu01] for details.

Turbo channel estimation for OFDM on highly time and frequency selective channels

The general principle of the iterative CSE technique aided by soft information from the iterative detector can be applied also to the case of OFDM on highly time and frequency selective channels. However, the OFDM case brings a rather specific additional feature in the availability of properly and frequently time-frequency positioned pilot symbols that can be used as additional a priori information for the CSE. High time and frequency selectiveness of the channel requires proper two-dimensional discretisation of the channel transfer time-frequency transfer function. The OFDM transmission allows easy accomplishing of this. The CSE estimator itself is based on an iterative EMA implementation of the ML estimator. For further details, see Section 2.2 and also [Jaff01], [Jaff02].

EMA interpretation of iterative joint synchronisation and detection

It is shown how the iterative EMA may provide a maximum likelihood estimation of channel parameters (phase, delay, amplitude). A structure in which EMA iterations are combined with those of a turbo receiver is proposed. More particularly, phase-aided timing recovery and joint timing and phase recovery are considered and compared. In each case, performance of the proposed synchroniser is illustrated through simulation results: the estimator mean and mean squared error, as well as the bit error rate reached by the synchronised system, are reported.

In the EMA approach, data is considered as an unavailable observation [DeLR77]. Applying the EMA principle to the case of joint estimation of phase ϕ and symbol timing τ, we get the iterators at the nth iteration step in the form

$$\hat{\tau}_n = \arg\max_{\check{\tau}} \left| \sum_{k=0}^{K-1} \eta_k^*(r, \hat{\phi}_{n-1}, \hat{\tau}_{n-1}) y(kT + \check{\tau}) \right| \tag{3.33}$$

$$\hat{\phi}_n = \arg \sum_{k=0}^{K-1} \eta_k^*(r, \hat{\phi}_{n-1}, \hat{\tau}_{n-1}) y(kT + \check{\tau}) \tag{3.34}$$

where r is the received signal in signal space representation, $y(kT + \check{\tau})$ is the matched filter output sampled at $kT + \check{\tau}$, and K is the length of observation. The value $\eta_k(r, \hat{\phi}_{n-1}, \hat{\tau}_{n-1})$ is obtained with the use of soft information from the turbo detector

$$\eta_k(r, \hat{\phi}_{n-1}, \hat{\tau}_{n-1}) = \sum_{a \in \mathcal{A}} ap(a_k = a | r, \hat{\phi}_{n-1}, \hat{\tau}_{n-1}) \tag{3.35}$$

$\mathcal{A}$ is the symbol alphabet. Simulation results show the performance of the joint timing and phase estimator, Figure 3.20. Detailed treatment can be found in [NHDL03].

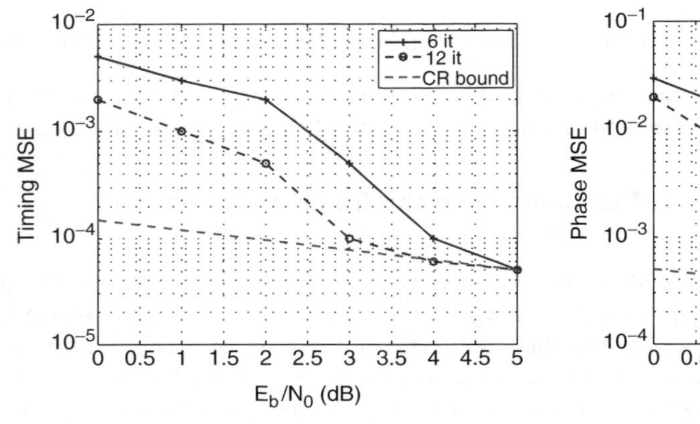

(a) Timing estimator MSE for normalised offset $\tau/T = 0.25$ after 6, 12 iterations.

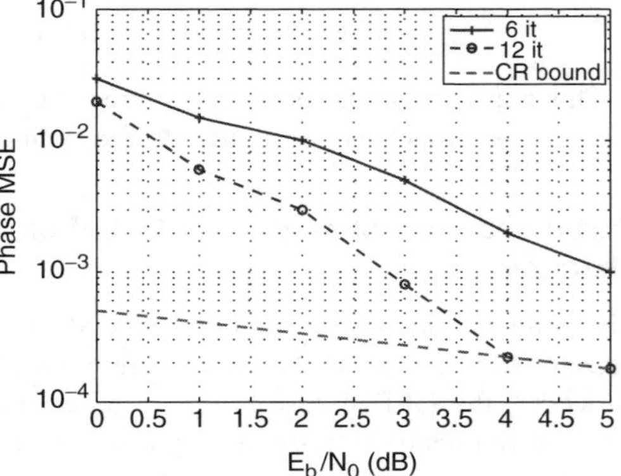

(b) Phase estimator MSE for $\tau/T = 0.25$ and $\phi = \pi/12$ after 6, 12 iterations.

Figure 3.20 *Timing and phase estimator for MSE.*

Iterative Soft-Decision Directed (SDD) linear timing estimator

Here we address the issue of iterative timing estimator in an alternative form to the EMA approach. We propose a linear timing estimator that takes benefit from the soft information delivered by the turbo system at each iteration to compute the timing estimate. This estimator is compared to a synchroniser, based on the previously described EMA approach. Performance of the proposed synchroniser is illustrated by simulation results and proves better than that of the EMA-based counterpart regarding both the convergence speed and range.

The implementation of the maximisation step in the EMA symbol timing approach can be done using a steepest-descent method. This can be shown to reduce a linear operation on a subset of the available observations. A natural step is to implement an efficient and unbiased *linear* estimator for computation of the next estimate update. The estimator update correction at the nth step is shown to be

$$\begin{aligned}\hat{\epsilon}_n = {} & \beta \sum_k \frac{\mathcal{E}\{h_I(k)\}}{\sigma^2_{w_I(k)} + \sigma^2_{e_I(k)}} \mathcal{R}\left\{ e^{-j\arg(\eta_{k,n})} \left(\dot{y}(kT + \hat{\tau}_n) - \sum_{k'} \eta_{k',n} \dot{x}_{k-k'} \right) \right\} \\ & + \beta \sum_k \frac{\mathcal{E}\{h_Q(k)\}}{\sigma^2_{w_Q(k)} + \sigma^2_{e_Q(k)}} \mathcal{I}\left\{ e^{-j\arg(\eta_{k,n})} \left(\dot{y}(kT + \hat{\tau}_n) - \sum_{k'} \eta_{k',n} \dot{x}_{k-k'} \right) \right\}\end{aligned} \tag{3.36}$$

where $\sigma^2_{w_I(k)}$, $\sigma^2_{e_I(k)}$, $\sigma^2_{w_Q(k)}$, $\sigma^2_{e_Q(k)}$ are variants of in-phase and quadrature components of self-noise and background noise respectively, $\dot{y}$ is the time-derivative of the matched filter output, $\eta_{k,n}$ is the information symbol a posteriori average value obtained using the soft information from the iterative detector, $x(t)$ is the raised-cosine filter and

$$\beta = \left(\sum_k \frac{\mathcal{E}\{h_I(k)\}}{\sigma^2_{w_I(k)} + \sigma^2_{e_I(k)}} + \sum_k \frac{\mathcal{E}\{h_Q(k)\}}{\sigma^2_{w_Q(k)} + \sigma^2_{e_Q(k)}} \right)^{-1} \tag{3.37}$$

Performance results, Figure 3.21, show that both estimators – the EMA with steepest-descent maximisation and the new proposed linear one – have the same mean after a large number of iterations. The linear one exhibits a convergence toward the actual timing offset faster than the EMA method. As expected, this difference between the two estimators increases when the timing error increases, since then the linear one fully benefits from its 'extended' observation set (namely both in-phase and in-quadrature components). We observe the same behaviour for the mean squared error. Note, however, that, this time, the final performance is no longer the same. Indeed, whereas the EMA estimator reaches the Cramer-Rao bound, the linear one remains slightly above.

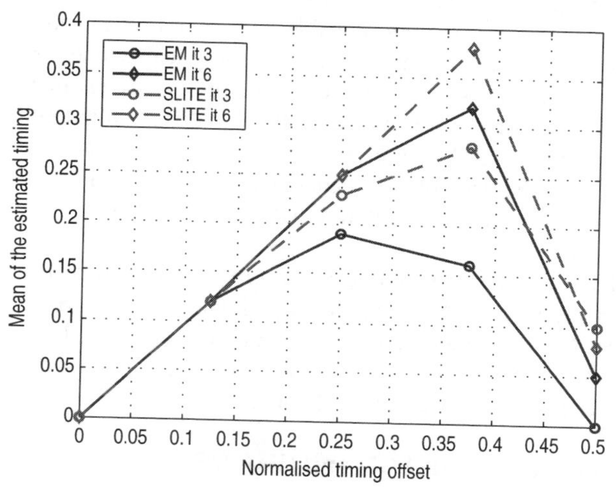

Figure 3.21 *Linear iterative timing estimator mean at $E_b/N_0 = 4$ dB.*

Message-passing synchronisation by combining the SP and EMA algorithms

We consider turbo synchronisation in the factor graph and the Sum-Product (SP) algorithm framework. The SP-algorithm messages related to the synchronisation parameters are approximated by a Dirac function. We show that this approximation requires to solve an ML problem at each iteration, the considered likelihood function being built by considering symbol extrinsic information as an a priori knowledge. We propose to solve these intermediate ML problems by means of the EMA algorithm. We then show in the particular case of a turbo coded transmission that the EMA-based turbo synchronisation framework (proposed earlier in the section) is actually an approximation of the approach proposed here. The proposed method is illustrated by simulations reported in the particular case of carrier phase estimation for turbo coded and bit-interleaved coded modulation transmissions. The resulting synchronisation method is shown to be powerful although very simple to implement [HeVa05].

A subspace method for soft estimation of block fading channels in turbo equalisation

The performance of turbo equalisation strongly relies on the quality of CSI. In block transmission systems, channel estimation is usually performed on a block-by-block basis from the training symbols included in each block. Higher estimate accuracy can be gained from an extended training set obtained by merging Multiple Blocks (MB) and using, within each block, both training symbols and soft-valued data symbols. A soft iterative MB technique based on this approach was developed in [MoNS04] for block fading channels, where the fading is constant within the block and it varies from block to block due to the terminal mobility.

Since turbo processing is usually performed on a set of $L>1$ data blocks (L depending on the interleaver size), the MB approach takes advantage of this latency to improve the estimate accuracy for the slowly varying channel parameters. The MB method relies on the assumption that in multipath channels the delays remain constant over the considered ensemble of blocks, while the fading amplitudes vary from block to block. As a consequence, the subspace spanned by the channel responses over the different paths (temporal subspace) is constant and can be estimated (with high accuracy for large L) by averaging the signals received over the L blocks, while the fast varying parameters have to be updated block by block. When used in turbo receivers the MB estimate can be modified to integrate soft information fed back by the channel decoder. The initial estimate obtained from the pilot symbols is refined in the subsequent iterations by extending the training set with soft-valued information symbols.

The benefits of soft MB estimation with respect to conventional Single-Block (SB) training-based methods are shown in Figure 3.22 for a single-user single-antenna turbo receiver complete with soft channel estimation, MMSE Single Input Single Output (SISO) equalisation and log-MAP SISO decoding. The soft MB method is shown to outperform all training-based and SB methods and to closely approach the performance for a known channel.

Iterative decoding networks with SDD and EMA CSE

Here, we establish a general framework for iterative joint detection and CSE in general iterative decoding networks. Two particular previously described cases (SDD and EMA) of CSE are generalised and put into a common general framework. Both have capabilities for exploiting the iteratively improved backward measure from the decoding network; however, both exhibit different properties and provide different possibilities for iteration scenarios. An example application with simple serially concatenated code with QPSK mapping in AWGN channel with phase rotation is investigated to demonstrate differences between the algorithms in terms of Mean Square Error (MSE), ambiguity resolution, and convergence behaviour.

An SDD approach is an *iterative* implementation of the data elimination principle usually performed directly at the channel symbol's $\mathbf{q}$ level. The factorised form of the expectation using the

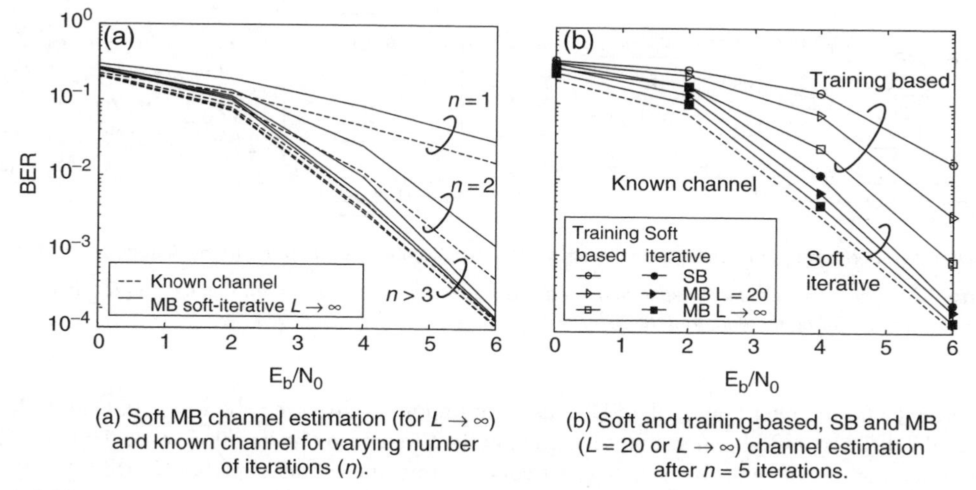

(a) Soft MB channel estimation (for $L \to \infty$) and known channel for varying number of iterations (n).

(b) Soft and training-based, SB and MB ($L = 20$ or $L \to \infty$) channel estimation after $n = 5$ iterations.

Figure 3.22 *BER performance for the turbo equaliser with training-based and soft-iterative SB-MB channel estimation vs. E_b/N_0.*

estimates of backward measure must rely on the *approximation* $\hat{p}^{k,\mathbf{m}}(\mathbf{q}) \approx \prod_n \hat{p}^{k,\mathbf{m}}(q_n)$. The iteration indices k, $\mathbf{m}$ describe iteration in CSE and decoding (Forward-Backward Algorithm) loops respectively. The final iterator for the channel parameter is

$$\hat{\theta}^{k+1,\mathbf{m}} = \arg\max_{\check{\theta}} \sum_n \ln \left(\sum_{q_n} p(r_n|q_n, \check{\theta}) \hat{p}^{k,\mathbf{m}}(q_n) \right) \tag{3.38}$$

Notice that the right-hand side *depends* on the iteration index k, i.e. iterations over the CSE, only through $\hat{p}^{k,\mathbf{m}}(q_n)$, which is fixed for a given fixed $\mathbf{m}$; therefore, it only makes sense to iterate *simultaneously* at one step both k and $\mathbf{m}$ index loops.

An application of the EMA algorithm to the CSE problem in data communication is usually done by setting the unavailable observation equal to the data $\mathbf{y} = \mathbf{d}$. The marginalisation operation $p(q_n|r, \hat{\theta}^k) = \sum_{\mathbf{q}:q_n,\mathbf{d}\mapsto\mathbf{q}} p(\mathbf{q}|r, \hat{\theta}^k)$ is exactly what the SISO module provides, and due to a condition $\mathbf{q} : q_n, \mathbf{d} \mapsto \mathbf{q}$ *correctly respects the code structure* without any approximation. The factorised final CSE iterator is

$$\hat{\theta}^{k+1,\mathbf{m}} = \arg\max_{\check{\theta}} \sum_n \sum_{q_n} \ln p(r_n|q_n, \check{\theta}) \hat{p}^{k,\mathbf{m}}(q_n|r, \hat{\theta}^{k,\mathbf{m}}) \tag{3.39}$$

Compare this result with the SDD case. Unlike the SDD case, the right-hand side depends on k even for fixed $\mathbf{m}$ and the iteration loops can run *independently*. The possibility of iterating over k and improving the estimation can save the number of necessary runs of computationally expensive decoding iteration (Forward-Backward Algorithm). An example network with iterative SDD/EMA CSE is shown on Figure 3.23.

The example simulation results are depicted in Figure 3.24. The most notable is a strong dependence of EMA CSE on the initial estimate (unlike for the SDD case), resulting in high probability of synchronisation failure if the ambiguity is not resolved a priori. On the other hand, the EMA CSE

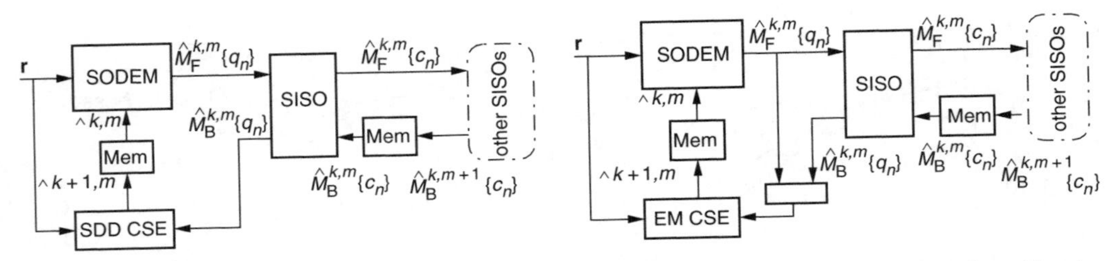

(a) Iterative non-parametric SDD decoding network. (b) Iterative non-parametric EMA decoding network.

Figure 3.23 *Iterative non-parametric SDD and EMA decoding network – multidimensional feedback system. An example of serial concatenated network.*

convergence is smoother than in the SDD case. An important difference between SDD and EMA is also in the different possible iteration scenarios; in the EMA case, the iterations over the CSE with a given fixed iteration step of the decoding network improve the estimate quality [SyBu04].

3.4.2 Channel state estimation and prediction

The fundamental issue concerning adaptive transmission techniques and the demand for a fading forecast has been analysed, the approach being based on the time-variant transfer function. The estimator uses current and past observations to predict channel behaviour. Two-dimensional Unitary-Estimation of Signal Parameters via Rotational Invariance Techniques (ESPRIT)-based prediction of the time-variant transfer function is suggested. In contrast to existing methods, it allows small-scale fading forecast also in the frequency domain.

A coherent detection in mobile communication requires a large number of channel parameters to be estimated. This problem is even magnified in MIMO systems. The problem is attacked here by the subspace-based method that exploits the different varying rates in the structure of the space-time channel form moving terminals. The channel is separated into two subspaces – fast varying (faded amplitudes off the paths) and slowly varying (delays and directions of arrival). A multislot structure of the transmission is utilised. The fast-varying parameters are estimated based on the slot-by-slot approach, whereas the slowly-varying ones are on a multislot basis. The subspace separation is based on the following P path channel impulse response model

$$\mathbf{h}(t,\ell) = \sum_{p=1}^{P} \beta_p(\ell)\mathbf{a}(\alpha_p)g(t-\tau_p) \tag{3.40}$$

where ℓ denotes slot index, β_p is the complex amplitude, $\mathbf{a}(\alpha_p)$ is the angle of arrival dependent antenna array response vector and $g(.)$ is a convolution between the transmitter filter and the receiver matched filter. The ML estimator was applied on these separated subspaces. The performance was verified by extensive simulations for a variety of parameters. These results show that relevant advantage over the single-slot technique is present as far as the quasi-static approximation of the space-time subspace holds true. For additional details, see Section 7.3 and [NiSS03].

A specific channel estimation algorithm for a MIMO transmission based on Alamouti's scheme was developed. We introduce a real-valued matrix transmission model embedding orthogonal space-time block codes as a linear transformation. We analysed the degradation that is due to non-ideal

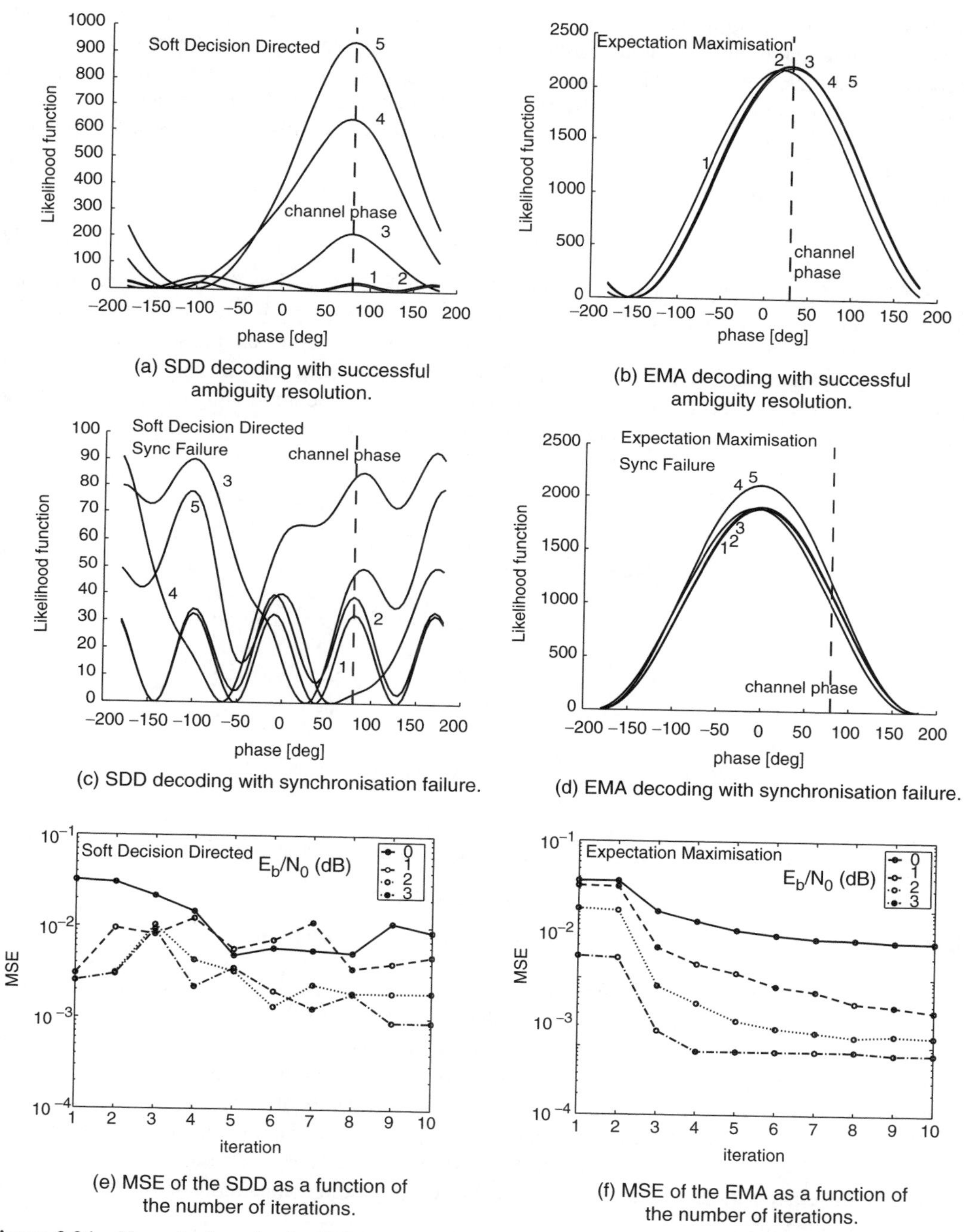

Figure 3.24 *Numerical results for SDD and EMA phase estimator with successful ambiguity resolution and synchronisation failure. True channel phase is shown as a dashed line. The decoding loop iteration number is a parameter,* $\gamma_B = 2$ *[dB].*

pilot-based maximum likelihood channel estimation. We show that Alamouti's scheme does not only provide a simple maximum likelihood estimate of the data, but may as well be deployed to obtain independent estimates of the MIMO channel impulse responses by a simple linear transformation using the detected symbols. This is done by exploiting the mathematical duality between the channel coefficients and the transmitted data. The theory is verified by a simulation for a simple Rayleigh channel model.

The channel estimator is shown to be

$$\hat{\mathbf{H}} = r\mathbf{a}^H \left(\mathbf{a}\mathbf{a}^H\right)^{-1} \tag{3.41}$$

where r is the received signal and $\mathbf{a}$ are the pilot symbols. The estimation errors of individual paths are mutually *independent* and have variance $\sigma_n^2/\|\mathbf{a}\|^2$ where σ_n^2 is the variance of AWGN. Figure 3.25 shows the results for a MIMO system with two transmit antennas and four receive antennas employing a four Phase Shift Keying (PSK) modulation scheme. The channels between all transmit antennas and all receive antennas are assumed to be independent Rayleigh channels resulting in a channel matrix with statistically independent entries [PiLi04].

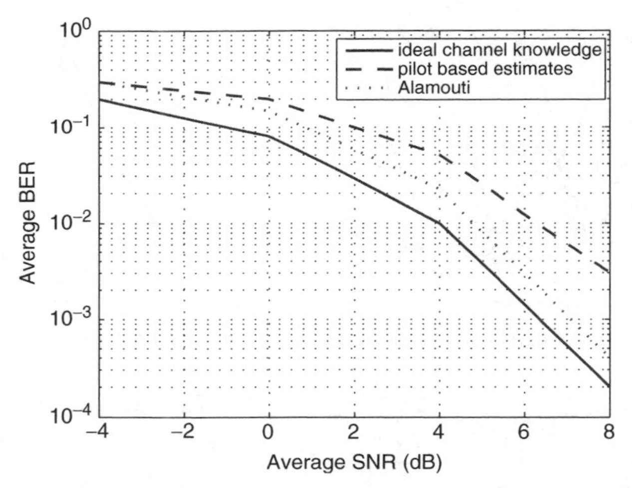

Figure 3.25 *Simulation results for MIMO CSE with Alamouti transmission scheme.*

It was shown that real-valued notation allows to integrate orthogonal space-time block codes in a linear matrix transmission model. This is possible because the complex conjugate operation may be expressed as a matrix multiplication in the real-valued case. Based on the real-valued transmission model, it was explained how to obtain independently disturbed maximum likelihood estimates of the entries of the channel matrix if the transmission is based on Alamouti's scheme. One might think of various ways how these estimates may be used to update the current channel estimates. Due to its low complexity, the scheme is also well suited for iterative schemes possibly jointly carrying out estimation, equalisation, and decoding. Some basic simulation results indicated the potential the introduced scheme may have.

Mobility in a wireless communication system affects how fast the channel fluctuates, and thus, the performance of the Adaptive Coding and Modulation (ACM) scheme is also affected. The ACM scheme must choose the correct codec for a given SNR. Pilot symbol assisted modulation and optimal maximum a posteriori prediction are used to predict the future of the fading envelope and subsequently select the appropriate codec for transmitting information. See Section 3.6 for the application of the predictor. The performance of the predictor is shown on Figure 3.26. The correlation coefficient ρ between predicted SNR and actual SNR is plotted against average expected SNR and velocity [JØHH03].

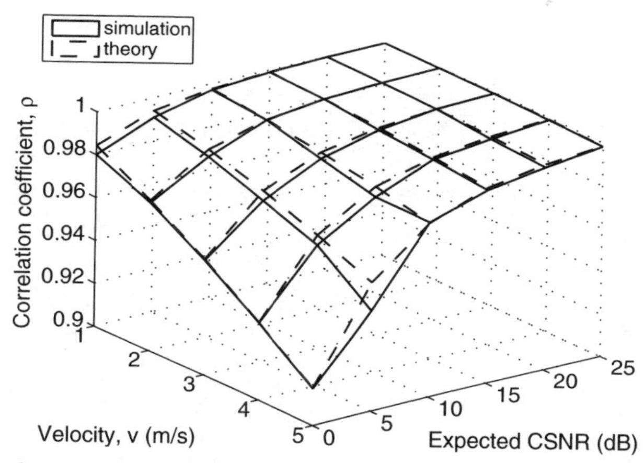

Figure 3.26 *The correlation coefficient of the predictor as a function of velocity and expected SNR.*

3.4.3 Performance evaluation and fundamental limits

We analyse the influence of an imperfect symbol timing estimate on the space-time coded modulation error performance, using the concept of the self-noise to describe this influence. We derive an expression for the white self-noise approximation in a slowly and fast Rayleigh flat fading MIMO channel. The self-noise is shown to cause a substantial performance degradation, which is emphasised by dimensionality of the MIMO channel.

Assume that the detector decision metric ρ is derived with the perfect synchronisation assumption for the θ parameter

$$\rho(r, \check{\mathbf{d}}, \theta) = \|r - \mathbf{u}(\check{\mathbf{d}}, \theta)\|^2 \tag{3.42}$$

At the time of receiver real operation, estimate $\hat{\theta}$ is substituted *instead* of the actual CSI value θ. The detector then operates (instead of the correct one) with the metric

$$\hat{\rho}(r, \check{\mathbf{d}}, \theta) = \|r' - \mathbf{u}(\check{\mathbf{d}}, \theta)\|^2 \tag{3.43}$$

where $r' = r + \mathbf{u}(\check{\mathbf{d}}, \theta) - \mathbf{u}(\check{\mathbf{d}}, \hat{\theta})$. The expression $\zeta(\check{\mathbf{d}}, \theta, \hat{\theta}) = \mathbf{u}(\check{\mathbf{d}}, \theta) - \mathbf{u}(\check{\mathbf{d}}, \hat{\theta})$ is additive to the received signal and it is also random, because of the random nature of channel nuisance parameters and data. Therefore, it is called *self-noise*. It can be conveniently used as a tool for approximate evaluation of the channel nuisance parameter errors on the detector performance. This approximation is based on the idea of replacing the actual self-noise by an equivalent (first and second moments) white Gaussian noise. This replacement is then equivalent to the perfectly synchronised system operating under the new *effective* level of AWGN with spectral density $\mathcal{S}'_w(f) = \mathcal{S}_w(f) + Z_0$, where Z_0 is the white power spectrum density approximation of the self-noise.

This technique was used to evaluate the mean pairwise error probability performance of the Tarokh two-space four-state 4PSK code in the slowly fading channel. Symbol timing errors are assumed to be independent identical distributed (iid) Gaussian zero mean random variables with variance σ_ϵ^2. Numerical results for this code are shown on Figure 3.27. See [Syko02] and Section 7.4 for additional details on the application of this general technique on the particular case of ST coded transmission in the MIMO channel.

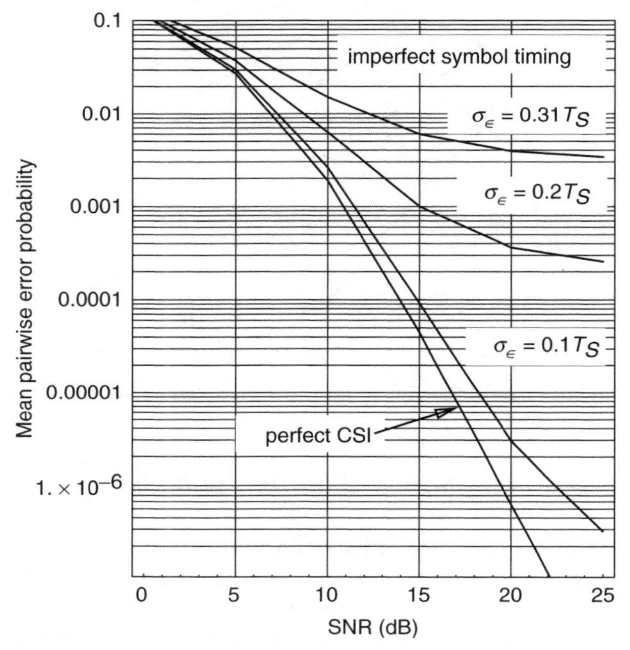

Figure 3.27 *Mean pairwise error probability of the Tarokh two-space four-state 4PSK code in the slowly fading Rayleigh channel.*

The standard iterative approach to the detection of data sequence **d** is to estimate the channel vector **h** using the known pilot symbols $\mathbf{d}_{\mathrm{id}}$ using the least-squares solver. Using this channel estimate, the unknown data symbols are estimated and, in turn, those estimates are then used to improve the channel estimate. Applying this approach to the UMTS FDD-like transmission scheme, it was found that the convergence strongly depends on signal-to-noise ratio. We find that only at very low signal-to-noise ratios three iteration rounds are necessary, e.g. at an SNR value of 0 dB only two iterations are needed. See [Pomm02] and Sections 2.3 and 7.3 for more details.

The channel knowledge on the receiver side seems to be crucial for exploiting the potentials of MIMO channels. The channel knowledge on the receiver side can be obtained, e.g. by channel estimation. We analyse the influence of channel estimation errors on the system performance. For large SNRs at the receiver input, the resulting small channel estimation errors act like additional noise if linear receivers are employed.

We investigate the (N_t, N_r) MIMO system with pilot signals (energy E_p)-based estimator; data symbols energy is E_d. The performance degradation of the zero-forcing receiver expressed in terms of effective noise enhancement is shown to be quadratically proportional to the number of system inputs

$$\delta = 1 + \frac{N_t^2 E_p}{E_d} \tag{3.44}$$

At first sight, the influence of channel estimation errors on data detection seems to be moderate as they act as a noise enhancement factor. Unfortunately, this noise enhancement factor grows quadratically with the number of channel inputs, which limits the number of channel inputs in a practical system. In other words, the potentials of MIMO transmission are limited by the required channel knowledge. See [WeMe04a] and Chapter 7 for additional details.

3.4.4 Sequence synchronisation technique

Analytical approaches to the synchronisation acquisition process and methods for construction of sequences with the best aperiodic autocorrelation properties (alternatively, with minimal total simulation probability) have been the subject of numerous analyses in the early days of digital transmission. Only a few of them investigated the relationship between frame length and synchronisation sequence length and structure. But, since the performance degradation due to imperfect frame synchronisation is small compared to the other types of error, the area was almost neglected, except for some sporadic contributions [BaSD02], [BaSt03]. However, new solutions and high data rates offer more levels of freedom. New design of almost all parameters is required, so more suitable analytical tools had to be developed.

In the background of the sequence analysis lies a process interesting to both mathematicians and engineers – search for a fixed sequence in random equiprobable L-ary data. The historical engineering approach started with an introduction of bifix, a name proposed by Prof. J. Massey. It denotes a subsequence of length $n \leq N$ that is both a prefix and a suffix of an observed sequence of length N. A sequence is completely described with a set of bifix-indicators $h^{(n)}$, $n = 0, \ldots, N$. If a bifix of length n exists, or if $n = 0$, $h^{(n)}$ equals to 1; otherwise, it is equal to 0.

Current techniques perform a search for sequences within the specified distance of the inserted one. At symbol level, this is equivalent to a search for a set of sequences. For the sake of such a search, a new feature is introduced: a cross-bifix. A cross-bifix is a subsequence of length $n \leq N$ that is a suffix of the ith sequence and a prefix of the jth sequence, both chosen from the observed set of M sequences ($i, j = 1, \ldots, M$). A corresponding cross-bifix indicator $h_{ij}^{(n)}$ equals to 1 if a cross-bifix of length n exists, e.g. binary sequences P_i = **0001** and P_j = **001**1 have a 3-bit cross-bifix $h_{ij}^{(3)}$, while obviously $h_{ij}^{(3)} = 0$. If $i = j$, $h_{ii}^{(n)}$ denotes a classical bifix indicator $h^{(n)}$. The default values for the cross-bifix indicators are

$$h_{ij}^{(0)} = 1, \quad h_{ij}^{(N)} = \begin{cases} 0, & i \neq j \\ 1, & i = j \end{cases}, \quad i, j = 1, \ldots, M \tag{3.45}$$

The probability that the search would be accomplished in exactly k steps (a discrete probability density function – PDF – of a search process) equals to

$$\Pr\{k\} = \sum_{i=1}^{M} a_k(i) p^{k+N-1}, \quad p = 1/L \tag{3.46}$$

$$a_k(i) = \sum_{j=1}^{M} \sum_{m=1}^{\min(N,k-1)} \left(Lh_{ji}^{(N+1-m)} - h_{ji}^{(N-m)} \right) a_{k-m}(j), \quad a_1(i) = 1, \quad i = 1, \ldots, M \qquad (3.47)$$

Other statistical parameters (expected duration of a search, variance) can be derived from this [BaSL03], [BaSt04]. A modification of this result [Baji04] yields the PDF of the duration of search in a frame, where a synchronisation sequence is periodically inserted into a stream of usually scrambled and therefore random and equiprobable data, Figure 3.28.

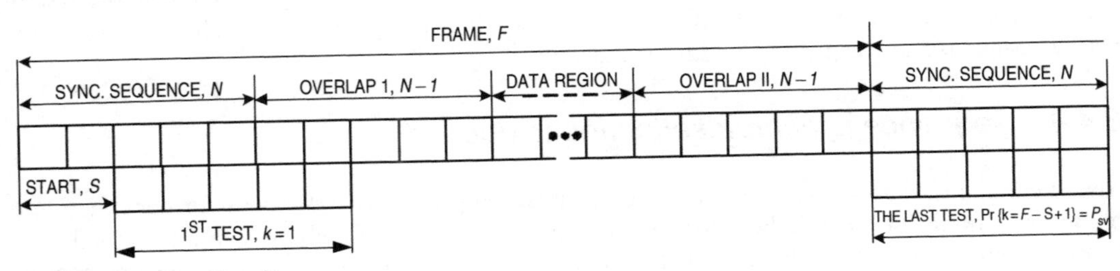

Figure 3.28 *Search in frame.*

One of the results that emerge is survival probability P_{SV}, i.e. the probability that there would be no accidental simulation of synchronising sequence, given that the search has started at the offset S (the worst case is $S = 1$) from the correct synchronisation position. Figure 3.29 shows the dependence of P_{SV} upon the frame length F and sequence length N, if the allowed Hamming distance is $H = 1$ (for this case the number of sequences in set $M = N + 1$). It should be mentioned that the 'all-zeros' sequence, as well as the bifix-free sequence $00\ldots01$, have $P_{SV} = 0$; Figure 3.29 presents the influence of various allowed Hamming distances.

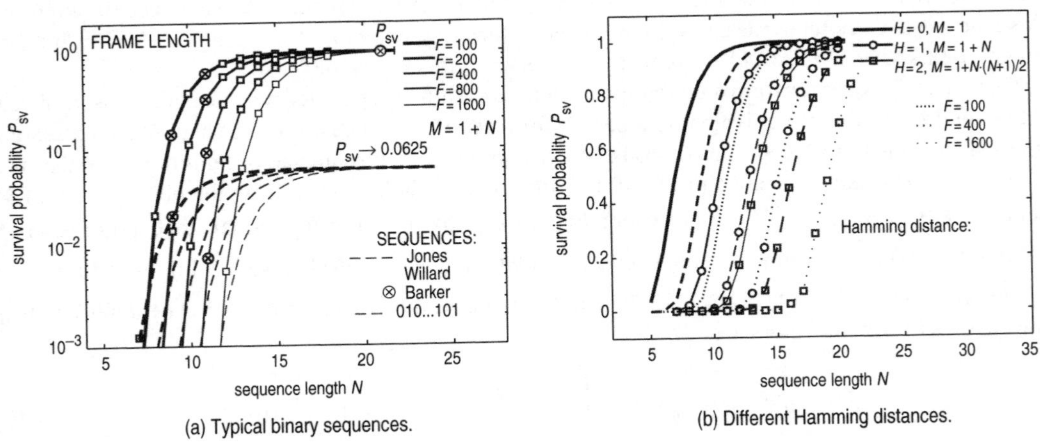

Figure 3.29 *Survival probabilities.*

The survival probability reaches limiting value if sequence length N increases. For good sequences, this value equals to one, i.e. it is not likely that the sequence would be simulated prior to its correct position. Yet, among the sequences with $P_{SV} = 1$ other distinctions can be made, as can be observed in Figure 3.30, which shows an example of $\Pr\{k\}$ (PDF). While all bifix-free sequences have the

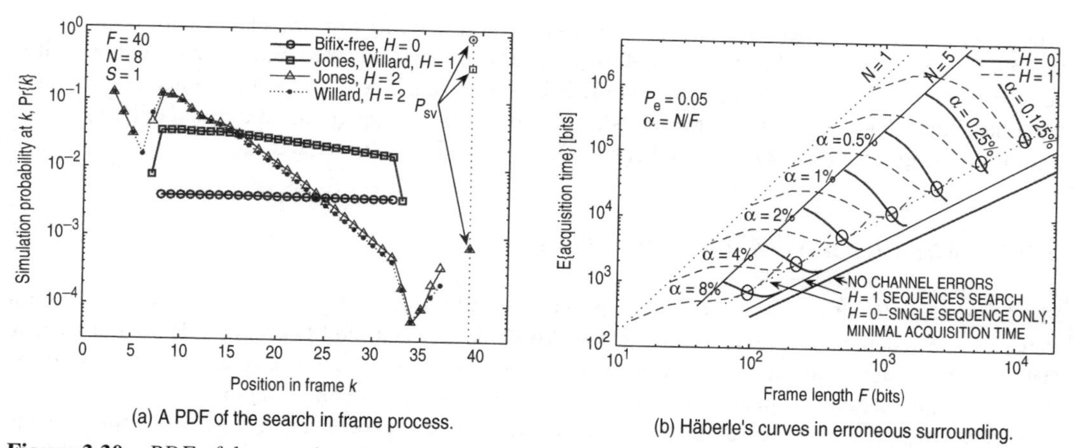

Figure 3.30 *PDF of the search in frame process and Häberle's curves in erroneus environments.*

same PDF if $H = 0$, only the good ones (e.g. Jones' 00011011 and Willard's 00100111) behave alike if $H = 1$. If $H = 2$, Willard's and Jones' sequences have the same survival probability, although their PDFs are slightly different. The derived PDFs, if appropriately averaged [BaNa04], enable reporting of the famous Häberle's curves for the expected value of frame acquisition time, with the redundancy $\alpha = N/F$ as a parameter. The thick curves show the case when sequences at distance $H = 1$ are allowed, while the dashed curves represent acquisition time if only one ($M = 1$) sequence is considered correct. For shorter frames the latter case yields better results (shorter acquisition time), but it is not an optimal solution. When the sequence length increases, reaching approximately the saturation part of P_{SV} curves in Figure 3.29, acquisition time for $M = 1$ saturates, and the multiple search ($M > 1$) improves.

This work might be extended to the multilevel ($L > 2$) case. Further on, the present results show that bifix and cross-bifix indicators are actually the probabilities and the whole analysis could be diverted to the non-equiprobable case! Finally, the derived analytical tool is implemented within the area of biomedical time-series analysis.

3.5 Multi-user systems and multi-user detection

Wireless communication systems need to deal with more and more simultaneous users. Most of the recent approaches tend to allow users to transmit simultaneously on the same frequency band, either using CDMA or SDMA. As a consequence, some MAI will arise in the decoding process and this interference will strongly limit the system capacity. Besides, the need to resist to multipath interference due to the larger frequency bands can be solved in two ways, either by a single carrier transmission scheme (DS-CDMA) or a multicarrier scheme (MC-CDMA) [GCDL02]. In both approaches, but also in SDMA, Multi User Detection (MUD) is an efficient way to deal with the MAI. When complexity is not a problem, single-user performance can be reached by highly loaded multi-user systems. However, complexity is often a limit and some sub-optimal solutions have to be found. Such solutions are considered in this section. In a first part, we present a transmission model for a MIMO multi-user system allowing to evaluate the performance of conventional MUD. We then present several MUD schemes for CDMA based on the turbo principle, neural networks or associated with multiple antenna schemes. The present approaches allow a significant performance enhancement with a limited complexity increase. We finally propose some MUD solutions in MC-CDMA systems with multiple antennas. Joint detection or joint transmission schemes can be implemented to reduce the MAI and a solution using Low Density Parity Check (LDPC) codes is also proposed.

3.5.1 Multi-user MIMO transmission

SDMA is a solution to share the resource among multiple users using some joint detection or joint transmission schemes. However, realising a MIMO link between the MTs and the BS is not an easy task if multiple antennas have to be used at both sides. A solution is to use multiple antennas at the BS and to consider each transmitter as an antenna of a multiple antenna system.

A new concept can be introduced to reduce the correlation between different channels in the uplink. BSs are grouped in service areas and the links between several BSs and several MTs are considered as a MIMO configuration. In [SWCH02] and [WSLW03], this concept is applied to an OFDM system. Joint detection and transmission are performed in order to reduce intra- and interservice area MAI.

If multiple antennas are used to differentiate the users, classical MUD can be implemented at the receiver side. We present a model to evaluate the performance of such a system and of some conventional MUD, eventually excluding some users from the multi-user detection process for complexity reasons.

In [Piet02], [PSTL03b], a discrete-time matrix transmission model is introduced for multi-user MIMO transmission systems. The model includes correlated interference and noise and allows a flexible treatment of users that may or may not be included in a joint detection process. On the basis of the matrix transmission model, conventional MUD is adapted to the multi-user MIMO case.

We consider a system with n_t transmitting and n_r receiving antennas, the former including antennas that are not of interest, i.e. excluded from joint detection. The received signal can be expressed as [PSTL03b]:

$$g(t) = g_{[1,m]}(t) + g_{[m+1,n_t]}(t) + n(t) \tag{3.48}$$

where $g(t)$ is the received signal, which can be divided in the contribution of m desired users $g_{[1,m]}(t)$ and the undesired contributions $g_{[m+1,n_t]}(t)$; $n(t)$ is the thermal noise component.

To exclude undesired users, we basically need a receive filter that maximises the desired signal contributions, while minimising those of the interfering transmitters. Optimum results in terms of signal to interference and noise ratio are obtained with a pre-whitening filter. We introduce the interference and noise correlation matrix (^{H} denotes the complex conjugate transpose):

$$\phi_{[m+1,n_t]}(\tau) = E\left\{\frac{1}{2}\left(g_{[m+1,n_t]}(t+\tau) + n(t+\tau)\right)\left(g_{[m+1,n_t]}(t) + n(t)\right)^H\right\} \tag{3.49}$$

The impulse response of the pre-whitening filter is obtained from the time representation of the inverse correlation filter, the transmit filter and the channel. We can finally derive the discrete-time matrix transmission model at time $t = kT_s$:

$$y_{[1,m]}(k) = r_{[1,m]}(k) * x_{[1,m]}(k) + (y_n(k) + n(k)) \tag{3.50}$$

$y_n(k)$ and $n(k)$ denote the interfering transmitters and thermal noise contributions. $r_{[1,m]}(k)$ determines the correlation of the interference and noise:

$$r_{[1,m]}(k) = \frac{1}{2} v_{[1,m]}^{H}(-t) * \phi_{[m+1,n_t]}(t) * v_{[1,m]}(t)|_{t=kT_s} \tag{3.51}$$

where $v_{[1,m]}(t)$ is the concatenation of the transmit filter matrix and the channel matrix.

However, this complex representation is only valid for proper noises. The statistical properties of the interfering transmitters determine whether properness is satisfied. They depend on the symbol alphabet and, for example, the symbol alphabet of binary phase shift keying does not comply with properness. A more general real-valued notation can be introduced and used for any type of symbol alphabet. This new notation is also valid for space-time block codes for which no complex-valued equivalent exists for $r_{[1,m]}(k)$ due to the complex conjugate operation [PSTL03b].

Finally, the discrete-time transmission matrix is used to estimate the performance of several classical MUD techniques in the case of MIMO systems. The impact of the exclusion of some users from the joint detection by means of pre-whitening is evaluated. For simulation, two users are considered with, each, two omnidirectional antennas; the receiver has four omnidirectional antennas, and a 200 MHz band is considered with a 2.4 GHz carrier frequency. In Figure 3.31, for the matched filter approaches, user 2 is excluded from joint detection for complexity reasons. This leads to unacceptable performance. On the other hand, the use of a pre-whitening filter and maximum likelihood detection of user 1 gets close to the performance of the maximum likelihood detection of user 1 when user 2 is turned off.

The loss of performance due to the exclusion of some users from the joint detection is quite small but leads to an important gain in the complexity.

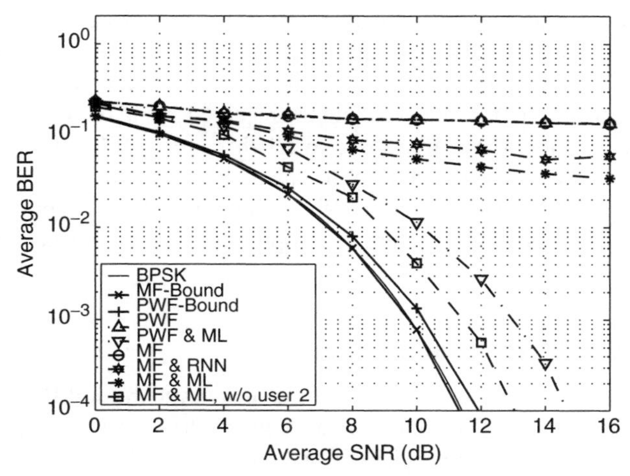

Figure 3.31 *Wideband transmission: user 2 is ignored for all MF curves whereas it is suppressed from joint detection for all PWF curves; ML detection does not include user 2.*

3.5.2 Multi-user detectors for DS-CDMA

To overcome the CDMA's vulnerability to MAI, great attention has been devoted to MUD [KoAg00] that constitute the natural answer to this problem. Such receivers exploit the interference as an additional information source. However, they are generally characterised by a very high complexity.

First, multi-user detection researches have been focused on uncoded systems and on sub-optimal multi-user detection approaches. One of the most promising solutions is the Interference Cancellation (IC) method based on subtracting interfering users from the received signal before making data decision.

However, practical CDMA communications rely on the utilisation of error control coding and interleaving so that, recently, more and more attention has been addressed to coded systems. Optimal joint decoding/detection is an excellent solution to MAI suppression [GiWi96]. However, the prohibitive computational complexity for actual implementation is once again a major difficulty. As a consequence, sub-optimal solutions, which separate the operations of symbol detection and channel decoding, appear more attractive for practical applications.

Particularly, the successful proposal of turbo codes [BeGT93] naturally leads to investigate the feasibility of iterative (turbo) processing techniques in the design of MUD. In the iterative MUD, extrinsic information is determined in each detection and decoding stage and used as a priori information for the next iteration. This detection philosophy is defined as turbo MUD and the advantages due to its introduction are remarkable even for heavily loaded systems.

In this section some promising MUD techniques are described, both for the uncoded and the coded Direct Sequence (DS)-CDMA systems.

Uncoded systems

In literature, IC is generally accomplished by Successive Interference Cancellation (SIC) or Parallel Interference Cancellation (PIC) methods: the former solution aims to successively subtract off the strongest remaining signal while in the latter scheme interference of all the users is simultaneously subtracted off from all the others [KoAg00]. An alternative PIC scheme is the so-called Selective Parallel Interference Cancellation (SPIC) receiver: this technique is based on an SPIC, i.e. on the idea of dividing users' signals into reliable and unreliable according to the outcome of the comparisons of the received signal decision variables with a proper threshold value. Reliable signals are directly detected and (after reconstruction) cancelled from the received signal. Unreliable signals are detected later, after cancellation of the MAI effects due to reliable ones [RFMM01]. In the following, the SPIC strategy is combined with antenna array to enhance the system capacity. Then, a non-linear MUD based on a Recurrent Neural Network (RNN) structure [TeRe01] is presented. This algorithm provides performance which is generally close to the optimum MUD, while keeping the computational complexity low.

Space-time selective PIC receiver

In [RFMM01] the advantages of joint use of antenna array systems and IC schemes for WCDMA communications are highlighted. The focus is on a receiver scheme that is based on the application of the SPIC method in conjunction with an adaptive antenna array for WCDMA systems. The proposed detector, named ST-SPIC, exploits angle of arrival information in order to obtain optimum space-time IC. In the considered propagation environment the number of resolvable paths is assumed to be four. For simulations a QPSK modulation is considered and two coding levels are used (Orthogonal Variable Spreading Factor (OVSF) and scrambling sequences).

Figure 3.32 illustrates the performance of the ST-SPIC receiver, single sensor, in comparison with the rake receiver, in the case of 16 simultaneous users. The curves are referred to the in-phase data stream at 60 kb/s and to the quadrature control sequence at 15 kb/s. From this figure, we can note that the ST-SPIC receiver outperforms the rake while the increase of the implementation complexity is low. The numerical results have been obtained for an optimum threshold value nearly independent of parameter E_b/N_0 [RFMM01].

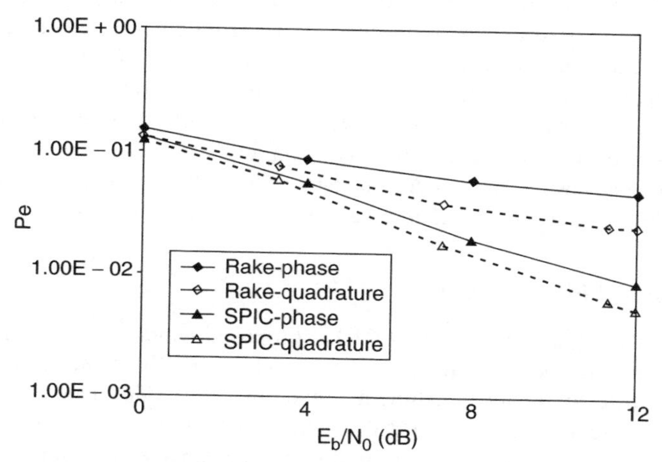

Figure 3.32 *BER comparison between ST-SPIC and rake receivers.*

Recurrent neural network-based MUD receiver

In [TeRe01], a TD-Synchronous Code Division Multiple Access (SCDMA) system is considered. In order to increase the capacity of the system, multiple antennas at the BS and a more accurate time synchronisation in the uplink have been included in the standard. In this system, a linear decorrelating detector [KoAg00] completely eliminates interchannel and ISI but suffers from a substantial noise enhancement which leads to a degradation of the performance.

A non-linear MUD based on an RNN has been proposed [TeRe01]: it has been shown that the energy function which can be defined for these RNN structures has the same structure as the log-likelihood function of the MUD problem. The principle of RNN-based MUD is an iterated non-linear feedback of tentative decisions: a partial interference subtraction on symbol level is made where soft decisions variables are fed back. For moderate channels, linear and non-linear MUD show similar performance while for severe channel conditions, the non-linear MUD based on RNN structures outperforms the linear MUD. In this case a proper choice of the soft feedback is critical to the performance of the non-linear MUD.

Coded systems

Iterative receivers that can achieve near-optimal performance [RSAA98] have been investigated but a major drawback is their complexity, still exponential in the number of users. As a consequence, important attention has been addressed to iterative IC schemes [KoBC01], [WuWa01], [QiTe00],

[HsWa01]. As it is known, as the number of decoding iterations increases, the coding gain offered by a turbo decoder becomes larger. However, the performance improvement obtained by turbo codes is remarkable in the first iterations, and more and more negligible in the successive ones. Hence, it is better to concentrate the significant part of interference cancellation in the first iterations: for the same reason many IC-based iterative receivers with a first linear stage have been proposed [QiTe00], [HsWa01], [TaMB01], [WaPo99].

Besides, in the case of a frequency selective channel, a classical CDMA scheme can be combined with multiple transmit and receive antennas and Space-Time Trellis Codes (STTC). Nevertheless, the conventional MMSE detector, generalised to accommodate multiple antennas and multiple paths [MaVU01], [PaHu01], is not sufficient to cope with MAI, ISI and Inter Antenna Interference (IAI) and cannot take advantage of multipath diversity in highly loaded systems. Conversely, the generalisation of this receiver by means of the turbo principle permits to obtain an iterative multi-user MMSE detector that completely removes MAI, ISI and IAI and achieves the single-user performance, even in a fully loaded system, at the cost of a high computational complexity.

Some iterative schemes using turbo codes, the benefit of multiple antenna schemes and the choice between MMSE receiver and MF are all addressed in the following section.

MAP decoding aided PIC receiver

In [MRFB03], [MBRF03], a new iterative MUD is proposed which is based on the utilisation of a PIC and a bank of turbo decoders. In the proposed structure the PIC detector is broken up so that it is possible to perform IC after each constituent convolutional decoder. Due to the tight relationship between the proposed receiver and the MAP decoders, it is defined as *MAP decoder aided PIC (MPIC)*: this solution aims to profit by IC introduction from the first iterations. Moreover, the variance of the noise-plus-(residual) interference is determined by a new algorithm: particularly, only the most reliable symbols are used in variance determination, neglecting all the others. This solution affords performance improvement for all the considered systems.

In a synchronous AWGN channel, the quantised log-MAP algorithm using the enhanced variance estimator leads to an important performance improvement in comparison with the basic estimator. We illustrate with Figure 3.33 the performance of the proposed receiver in an asynchronous three-path fading channel with 20 users and perfect power control: the replica amplitudes are assumed to be Rayleigh distributed, with relative attenuations equal to 0, -1, -9 dB. The spreading operation is performed by a set of pseudo-noise short codes and processing gain is $G = 15$. The frame length is 500, a maximum ratio combining rake receiver is used as path combiner and the max-log-MAP algorithm is used for the decoding. The channel has a normalised Doppler frequency $f_d T_b = 0.0002$ and just two receiver iterations are performed. We compare the proposed iterative MPIC receiver with the conventional PIC [WuWa01]. With a system load $\beta = N/G = 1.33$ greater than one, the MPIC outperforms the conventional PIC: a gain of 3.5 dB for the MPIC versus the PIC receiver at a BER of 10^{-3} is observed. We can also notice the great benefit obtained for both receivers by using a correct estimate of the noise-plus-(residual) interference variance rather than approximating it by the ambient noise, assumed perfectly known as in [WuWa01].

Iterative MMSE-PIC receiver

An iterative MUD based on the combination of linear MMSE blocks with IC schemes and a bank of turbo decoders is proposed in [MRFC05]. First, the MMSE outputs are used to reconstruct the

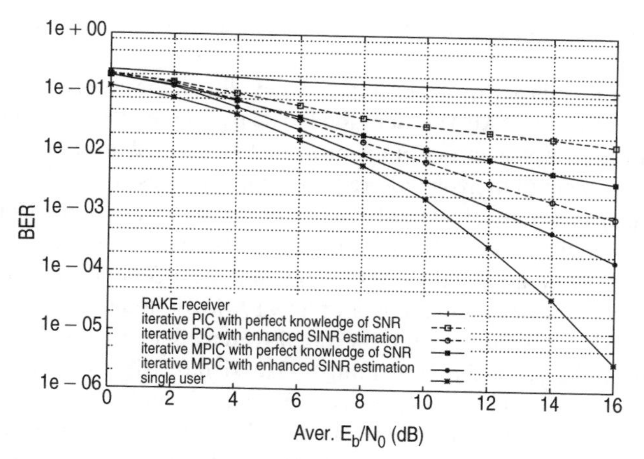

Figure 3.33 *Performance of the iterative IC receivers in a multipath fading channel with 20 equal-power users.*

interferences to be cancelled from the received signals: this solution, anyway, is not optimal due to the negative effects of noise enhancements caused by MMSE filtering; hence, after some cancellation iterations, the detector begins considering the MF outputs, for the cancellation: this solution permits to retain the positive effects of the MMSE introduction in the first iterations.

Figure 3.34 shows the BER performance of the association of the MMSE filter and the PIC

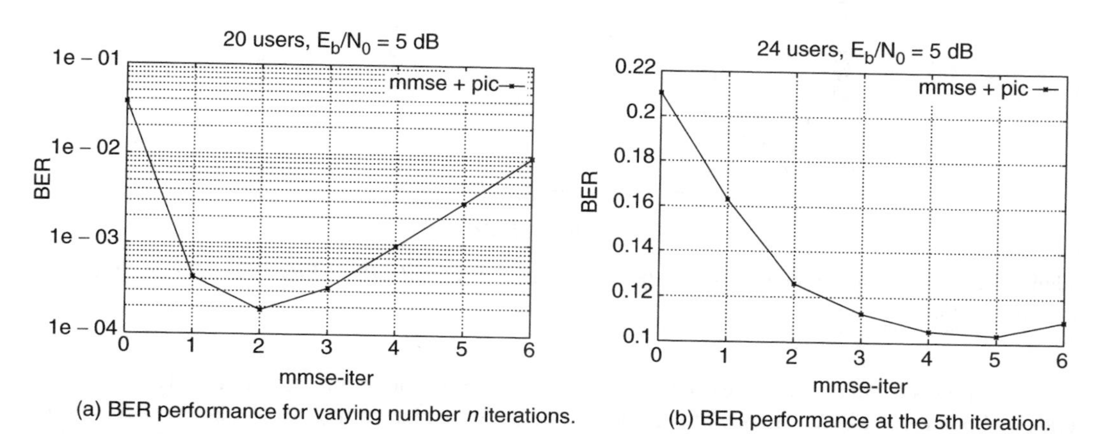

(a) BER performance for varying number *n* iterations. (b) BER performance at the 5th iteration.

Figure 3.34 *BER performance for different number of iterations using the MMSE filter.*

receiver for a different number of iterations made with the MMSE detector: the optimum value of the parameter increases as the number of users grows. This behaviour can be explained because a system with a higher number of active users is characterised by a greater MAI. A higher number of MMSE iterations is then required to perform effective IC. On the contrary, an underloaded system does not need too many MMSE iterations: particularly, performance gain due to one

MMSE iteration introduction is weak while more iterations cause noise enhancement and performance losses.

Though this receiver affords performance improvement for all the considered systems, it is difficult to identify the optimal number of MMSE iterations to be performed before switching to the ordinary IC. Due to the very low error probability of the proposed receiver, definition of the optimal structure by means of Monte Carlo simulations would require a huge processing time.

So, the study of the proposed receiver has been performed by means of the Density Evolution (DE) technique [RiSU01], [DiDP01]: though the result of DE analysis holds for the asymptotic regime, i.e. after some turbo-MUD iterations, it allows to compare the different MUD approaches and draws general conclusions about the optimum number of linear MMSE iterations to be performed. We need to estimate the SNR_{in}-SNR_{out} relations both for the SISO decoder and the SISO MUD. For the SISO decoder this relation can be derived by assuming a Gaussian distribution with mean $\mu = E(\lambda)_{in} = 2.SNR_{in}$ and variance $\sigma^2 = \mathrm{Var}(\lambda)_{in} = 4.SNR_{in}$ and obtaining an SNR_{out} estimate by averaging the soft outputs $SNR1_{out} = \overline{\lambda_{out}}/2$. On the other hand, the SISO MUD relation is dependent on the value of E_b/N_0; hence, we have to generate a proper Gaussian distribution $N\ (c_i * 2SNR2_{in},\ 4SNR2_{in})$, where $c_i \in \{\pm 1\}$ is the coded bit and $SNR2)_{out} = \overline{c_i * \lambda_{out}}/2$.

From Figure 3.35, some interesting characteristics of the proposed receiver can be deduced. When

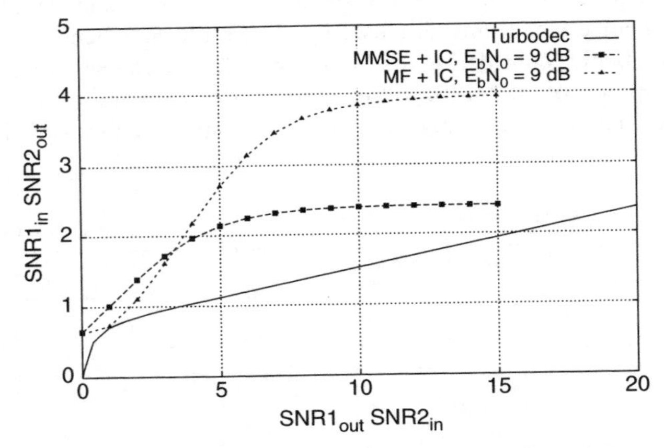

Figure 3.35 *Performance comparison between MMSE + PIC and MF + PIC in a 32-user system with $E_b/N_0 = 9$ dB.*

SNR_{in} is low, DE analysis proves how useful the MMSE introduction is: particularly, the MMSE ends up boosting the convergence of the iterative detection and decoding. Nevertheless, MF solution is characterised by a higher asymptotic value of the SNR_{out}, so confirming that switching to MF IC after some iterations is beneficial.

Turbo PIC receiver for multi-antenna systems

In [ShBu02] an MMSE-based PIC detection approach is also proposed. Unlike the structure in [WaPo99], the linear MMSE filter is used only in the first iteration, which is regarded as the weakest link of the whole detection process because of the unsatisfactory performance of the matched filter.

This structure has been used on a slow, flat Rayleigh fading channel, with relatively poor performance, limited by the single-user bound. Accordingly, an improved turbo MUD that can be used in systems with multiple receive antennas has been proposed. The simulation results demonstrate that this diversity technique allows the turbo PIC receiver to recover its ability to enhance the system performance, even in the presence of severe multipath fading.

The performance of the joint turbo decoding and MMSE-based PIC detection approach has been analysed by some simulation examples for both the single antenna and receive diversity cases. An uplink CDMA system with 10 equal power users has been considered. All users are assumed to employ the 1024-bit, 1/2 rate, 16-state turbo codes (with generator 23, 37). The BS uses four separate antennas to receive the users' signals. The performance of the turbo receiver in a slow Rayleigh fading channel is shown in Figure 3.36 with a significant improvement using spatial diversity rather than a single antenna system. A BER of 10^{-6} is obtained at an acceptable value of SNR. Note, however, that the receiver reaches its performance limit within only five iterations. This suggests that further suppression of the multipath fading may bring out the full potentialities of the turbo PIC detector.

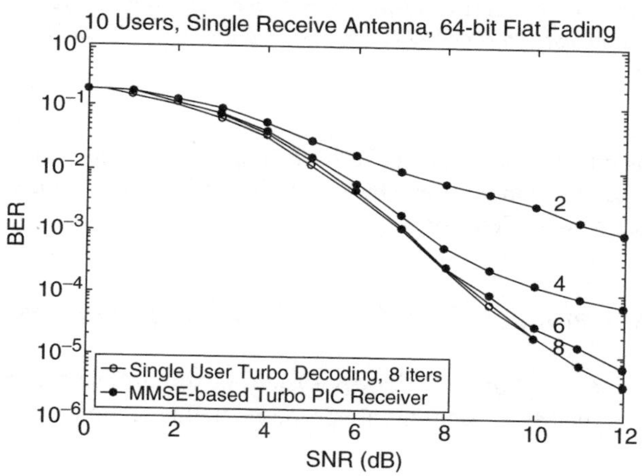

Figure 3.36 *Performance of receive diversity MMSE-based turbo PIC receiver in slow flat fading channel.*

Iterative multi-user MMSE receiver for space-time trellis coded CDMA systems

A sub-optimal multi-user iterative MMSE receiver for STTC CDMA systems over multipath fading channels is derived (SO-MMSE) [OsNA05]: it represents a tradeoff between computational complexity and performance and it is compared with the optimal iterative and non-iterative multi-user solutions.

Simulation results have been achieved by assuming QPSK modulation, OVSF spreading codes, gold scrambling codes (different for each transmit antenna), L-paths Rayleigh slow fading channel, independent resolvable paths for each subchannel, maximum delay spread equal to the spreading factor, two transmit and two receive antennas, n active users, frame length equal to 260 information bits (130 QPSK symbols) and perfect channel state information at the receiver.

The proposed receiver suffers from the ISI, MAI and intra-antenna interference in a highly loaded system, Figure 3.37. The SO-MMSE performance is highly degraded when the number of active users is increased and the optimal receiver is much better. But the complexity is linear with the product

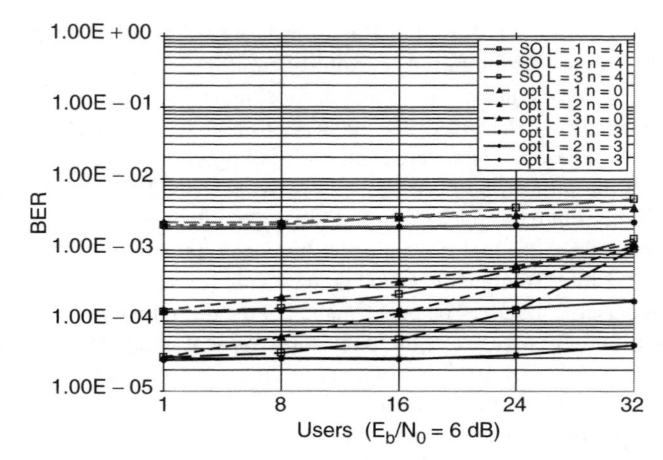

Figure 3.37 *BER performance comparison of optimal iterative multi-user MMSE, non-iterative optimal MUD and SO-MMSE with two antennas 16 states STTC, $E_b/N_0 = 6$ dB, $N_s = 32$ and $n_R = 2$ for different number of users (n) and paths (L).*

of the number of transmit antennas, number of users and iterations and it consistently outperforms the conventional non-iterative single-user MMSE receiver and closely approaches the conventional non-iterative multi-user MMSE detector.

3.5.3 Multi-user detection for MC-CDMA with spatial diversity

To increase the performance of an MC-CDMA system or to enhance the uplink capacity, spatial diversity can be used. Some MUD solutions are proposed in this section for the downlink and uplink situations.

Downlink

As far as the downlink is concerned, we can improve the system possible load by applying the principle of joint transmission with transmit antenna arrays. The user capacity of MC-CDMA systems is mainly limited by MAI arising from the different users sharing the same subcarriers. To avoid costly interference mitigation at MTs, antenna arrays are employed at the BS for interference mitigation through joint Space-Frequency Transmit Filtering (SFTF). In [SaMo03], assuming channel state information at the BS, two SFTF criteria are considered: a single-user Maximum Ratio Transmission (MRT) and a multi-user Maximum Signal to Interference plus Noise Ratio (MSINR). This second approach attempts to minimise MAI with a constraint on the transmit power. If it is too complex to maximise jointly for all MTs the signal to interference ratio, the same Short-Time Fourier Transform (STFT) can be applied to all users. In this way, we allow simple detection techniques at MTs. Moreover, we can increase the user capacity by reallocation of orthogonal spreading codes, thereby efficiently combining CDMA and SDMA. As a simple approach, when the number of active users exceeds the spreading factor, we reuse spreading codes without taking into account the spatial signatures of users' signals. However, to reduce the interference occurring between MTs with the same spreading code and similar spatial signatures, a different random scrambling code is applied prior to OFDM modulation on the top of the spreading code. Figure 3.38 illustrates the possible

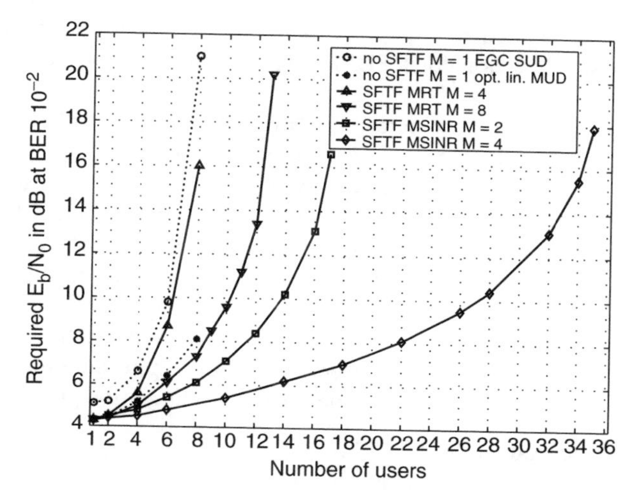

Figure 3.38 *Capacity increase due to SFTF and combined CDMA/SDMA.*

capacity increase through SFTF with M antennas and code reallocation with a spreading factor of 8 for a HIPERLAN/2-like scenario. For comparison, performance of a single antenna MC-CDMA system with single-user (Equal Gain Combining (EGC)) or optimum linear multi-user detection is plotted. For a given transmit power, we show that the multi-user MSINR SFTF approach basically allows multiplying the system load by the number of transmit antennas M.

To further improve a centralised network, several antennas can be used at the BSs. In [WeMe03] and [WeMe04b] several linear techniques are presented for joint transmission and compared with a non-linear scheme. If the transmit match filtering approach maximises the energy efficiency of a single data symbol it results in an important MAI. Transmit zero forcing allows to suppress this MAI but the resulting energy efficiency is poor. Some partial zero forcing allows to reduce MAI between certain selected symbols and is a compromise between the first two approaches. Finally a novel non-linear scheme is proposed and allows a significant performance gain. Those techniques are detailed in Section 3.3.

In [TQMJ02] multiple antennas are also used at the mobile terminal and joint transmission is realised. The choice of the modulation matrix is made in order to reduce the intercell MAI by minimising the transmitted power. The energy transfer is optimised using the space-time eigensignatures of the mobile radio channels.

Uplink

In [ReCF03], a MUD scheme applied to non-spreading modulation is proposed. One encountered problem is the frequency impairment caused by the low-cost local oscillators, which occurs when multiple subscribers simultaneously transmit to the access point. The frequency impairment dramatically reduces the performance in the SISO system.

The proposed algorithm is the MMSE MUD in a version adapted to scenarios where frequency impairments arise. The key to this adaptation is to make at the transmitter the inverse discrete Fourier transform operations at different frequencies according to the considered transmitter. Then, frequency offsets clearly appear on the mathematical expression of the received signal, and the MMSE criterion can be calculated taking into account those frequency shifts. This enables the system to operate multi-user detection for OFDM transmissions even in the presence of frequency impairment.

The performance of the multi-user detector is evaluated under HIPERLAN/2 conditions. In the first scenario, two users transmit to an access point with two antennas. Simulation results, including the channel estimation, show that the system receiver is not sensitive to the frequency offsets and that the performance achieved is similar to the ones obtained by a SISO system in Rayleigh fading conditions. However, when trying to reduce the MMSE receiver complexity using fewer subcarriers, the degradation due to frequency impairment arises needing some more accurate receivers.

In [GoCG05], a new scheme for MUD combining MC-CDMA and LDPC codes is proposed. The users are grouped in clusters. The different users in a cluster share the same spreading code. The clusters are identified by different spreading codes. The decoding principle is as follows:

- the cluster signal is despread;
- the noise and multi-access interference power is estimated;
- in-cluster joint detection is performed thanks to the LDPC codes;
- interference cancellation is realised (PIC receiver principle).

This scheme has several advantages. The number of spreading sequences does not limit the number of users any more and the transmission scheme keeps the same whatever the number of clusters or users in the clusters. The use of LDPC codes over Galois Field (GF)(q) and the association with a PIC receiver structure lead to better performance than a classical receiver because non-binary codes exhibit better performance than the binary ones. However, the best codes are not built in the higher Galois field and no simple link exists between the detection performance and the Galois field order. Figure 3.39 shows the joint detection performance for a three-user cluster. The best performance is obtained with the GF(4) codes. A special case where users have codes in different Galois fields does not improve the system performance.

The proposed approach allows to circumvent the loss of code orthogonality in an MC-CDMA-based uplink transmission. The conjunction of PIC and joint detection can provide substantial improvement without any complexity growth of the transmitter.

3.6 Link adaptation and rate-adaptive systems

3.6.1 Introduction to link adaptation

Link adaptation is a prerequisite for close-to-optimal capacity exploitation in temporally and spatially varying radio channels, both in the single-link and multi-user (network) case. The principle is to adapt throughput (transmit spectral efficiency), power, and/or possible other transmission scheme parameters dynamically to predicted channel conditions, in order to come as close as possible to the ergodic channel capacity while adhering to some given design quality constraints. In this way one

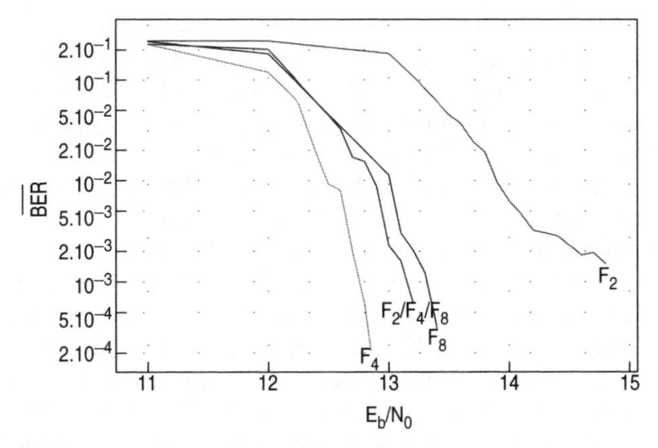

Figure 3.39 *Joint detection for three users with equal power.*

avoids the 'worst-case' choice of transmission schemes that must be made in non-adaptive systems, which therefore typically can only exploit a fraction of the theoretically available capacity. However, some sort of feedback mechanism from receiver to transmitter is needed in order to perform the adaptation at the transmitter side.

In a typical adaptive transmission scheme, the receiver estimates the channel state, and the obtained CSI is fed back to the transmitter via a feedback (also called return) channel. On a flat fading channel, the CSI fed back will typically be an index indicating in which subrange (out of a finite number of possible subranges) the instantaneous Channel Signal-to-Noise Ratio (CSNR) resides.

The transmitter subsequently uses the CSI to update the transmission parameters such as modulation constellation, error-control code, and/or transmit power level, to fit the instantaneous channel quality. A given combination of these parameters is called a transmitter *mode*. This principle is illustrated in Figure 3.40. Here, n is the index of the transmitter mode to be used, typically indicating that the channel is in state n, which corresponds to the nth subrange (out of N possible subranges or intervals) of CSNR levels.

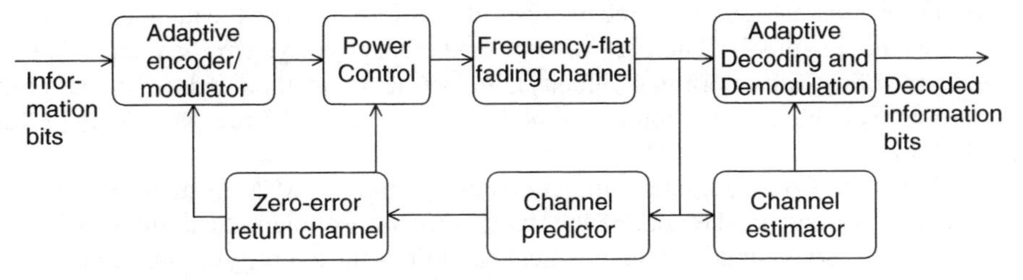

Figure 3.40 *Block diagram of a generic link adaptation scheme.*

The choice of transmitter settings can, for example, be done subject to fulfilment of a chosen target BER and an average transmit power constraint, so that the transmitter attempts to transmit information at the highest instantaneous spectral efficiency possible, while simultaneously fulfilling the BER constraint and the power constraint. Alternatively the PER may be used instead of the BER. Thus, for a high CSNR the transmitter will choose its mode corresponding to a high spectral efficiency, whereas it will decrease the spectral efficiency dynamically as the channel quality deteriorates. Other design criteria are also possible, depending on the application at hand.

The idea of link adaptation dates back at least to the early 1970s. An exhaustive overview of most research contributions up to 2002 can be found in [HaYW02]. In 1997, link adaptation was put on a firm information-theoretic grounding when Goldsmith and Varaiya showed that the ergodic capacity of a flat fading channel can be attained by a certain adaptive scheme in which transmit rate and power are continuously and instantaneously adapted to the channel conditions if these conditions were assumed perfectly known at the transmitter side [GoVa97]. Practical link adaptation schemes constitute discrete-rate, discrete-power approximations to this idealised and practically non-realisable capacity-achieving scheme. It is fair to say that the advantages, design choices, available degrees of freedom, and practical limitations of link adaptation techniques have now become well known, and such techniques are gradually being included in several upcoming wireless communications standards. Link adaptation can also, as will be seen in subsequent sections, be combined with other state-of-the-art wireless technologies such as MIMO and OFDM.

3.6.2 Impact of imperfect channel state information

In COST 273 some emphasis has been devoted to analysing ACM systems with imperfect CSI, in the presence of flat fading. These ACM schemes can, however, also be applied to frequency selective channels where multicarrier modulation (OFDM) is used, as long as each subchannel in the multicarrier scheme can be well modelled as flat fading.

We focus here on ACM schemes where only the *modulation constellations* and *channel codes* (jointly termed *codecs*) are dynamically adapted to the channel conditions. The transmit power is kept constant throughout. In such *rate-adaptive* systems, the transmitter can choose between multiple codecs in order to adapt the Spectral Efficiency (SE) to the variations in the CSNR. Both the transmitter and receiver in a rate-adaptive system will of course need CSI available, assumed to be transmitted from the receiver to the transmitter on a separate return channel.

Theoretical performance analysis has previously been performed for rate-adaptive systems assuming certain idealised conditions such as zero return channel delay, noiseless return channel, and perfect channel estimation. However, unless the forward and return channel are reciprocal, as might be the case in TDD systems, there will be a non-zero return channel delay, and the CSI must be predicted for future transmissions at the receiver based on information on the fading process extracted from received channel symbols. The performance of the adaptive system then depends on several factors: the degree of mobility (terminal velocity), the prediction method and (if a linear predictor is used) filter order, available information about the fading process at the receiver, delay and errors in the return channel.

In [JØHH03] results from computer simulations of a complete ACM system on a Rayleigh flat fading channel are presented. The goal of the work is to confirm theoretical analysis by means of simulations, and to ascertain under which practical conditions the simplifying assumptions normally used during theoretical analysis actually hold. Five channel codes from the very promising class of block codes called Gallager codes (or low-density parity check codes [Gall63]) are used as component codes. The coded information is modulated using Gray mapping onto M-QAM or M-PSK constellations. Parameters further describing the simulated codecs are given in [JØHH03].

The system uses Pilot Symbol Assisted Modulation (PSAM) [Cave91] to provide the receiver and the transmitter with information on the fading envelope. Under the assumption of known fading correlation properties, linear channel prediction and estimation is used. The predictor and estimator filter coefficients are optimised in the MAP (maximum a posteriori) sense.

The relevant CSI for the transmitter is the index of the selected fading region. The receiver predicts the future CSI after each received block of 220 channel symbols, where every 11th symbol is a pilot symbol. The delay between the receiver prediction and the transmitter update is referred to as the prediction lag. Because of the block structure of the Gallager codes employed by the ACM scheme, it is only feasible to update the codec used between two successive transmitted blocks. Thus, an underlying assumption during system analysis is that the fading is slow enough to stay within the same fading region over one transmitted block.

After the prediction the fading region index is transmitted from the receiver to the transmitter. The return channel is assumed free of errors, but with a transmission delay known at the receiver. When the CSI information changes, the transmitter and receiver perform a codec update in order to maximise the instantaneous spectral efficiency subject to a BER constraint.

The above system has been evaluated both through analysis and simulations on Rayleigh fading channels with a Jakes correlation profile, for HIPERLAN/2-type system parameters and moderate mobile speeds (walking speed). The codec switching thresholds used are those designed for perfect CSI, which means that the system is expected not to fulfil the BER requirements at very low average CSNRs and/or long feedback delays. This is confirmed by the simulations, and in general the correspondence between the theoretical predictions and the simulation results is excellent as long as walking speeds are considered, both with regard to BER performance, Average Spectral Efficiency (ASE), and measured correlation between the true and predicted fading. This implies that the simplifying assumption made, of the fading within a symbol block staying within the same fading interval throughout the block, is valid at such speeds. A comparison between simulated and theoretical ASE at a velocity of 1 m/s is provided in Figure 3.41. However, at higher speeds the simplifying assumption is no longer true, and the simulation results start to deviate from the theoretical predictions.

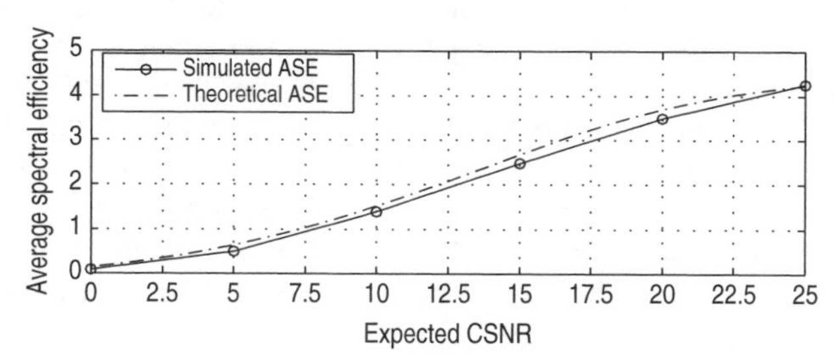

Figure 3.41 *Comparison of simulated and theoretical ASE for the adaptive Gallager coded modulation system, at a mobile velocity of 1 m/s.*

In another contribution [ØiHH04], a set of eight QAM-based *four-dimensional trellis codes*, with individual spectral efficiencies 1, 5, . . . , 8.5 bits/s/Hz, are used instead of Gallager codes, and the effect of *spatial receive diversity* on an ACM scheme is discussed, again under the assumption of flat

Rayleigh fading and pilot-assisted MAP-optimal prediction of each subchannel, with perfect channel knowledge in the receiver (i.e. perfect coherent detection). Receive diversity is obtained through the use of MRC of H receive antennas. Numerical examples are given for a Jakes fading correlation profile under average BER and transmit power constraints.

A main conclusion is that even a moderate amount of receive diversity (combining of only $H =$ 2–4 receive antennas) can significantly improve the robustness of an ACM scheme with respect to channel prediction errors, allowing it to fulfil a given BER constraint over a much wider range of CSNRs and feedback delays than a SISO system can for the same codec switching thresholds, without any loss of ASE (indeed, receive diversity provides an additional ASE improvement due to the array gain). This is because the receive antenna combining effectively stabilises the channel (decreases the probability of deep fades), turning a Rayleigh fading into a Nakagami-H channel. Still, the channel fluctuations are still severe enough for there to be an important ASE gain associated with link adaptation.

There is always an inherent tradeoff between ASE and robustness towards channel prediction errors and return channel delay, since the CSNR switching thresholds used to switch between different transmitter modes must be increased if we want to ensure that each code operates at low enough BER under imperfect CSI. For a given channel model, this will mean that the outage probability will be higher and that lower-rate codes will be used with a higher probability, and thus the achievable ASE will decrease. The contribution [Jetl05] shows that the lowest CSNR thresholds are most affected, which means that the ASE decrease is more noticeable for bad average channel conditions (low expected CSNR) than for good. This can be traced back to the fact that it is harder to estimate and predict the channel when there is a lot of noise in the received signal.

However, another contribution (TD(05)009) demonstrates that with careful system design and optimal parameter choices, the ASE penalty associated with imperfect CSI need not be too severe. In this paper, the same set of $N = 8$ four-dimensional trellis codes as in [ØiHH04] is still used, in conjunction with receive antenna diversity implemented by means of an MRC combiner. Also, pilot-assisted MAP-optimised linear prediction is still used. In addition, the assumption of perfect receiver CSI is relaxed, and is replaced by an assumption of pilot-assisted MAP-optimised linear channel estimation also in the receiver. In this setting, the pilot symbol period and the distribution of transmit power between information symbols and pilot symbols (under an average transmit power constraint) are jointly optimised to maximise the ASE under an instantaneous BER constraint.

The results, which generalise previous conclusions for an uncoded SISO system [CaGi05], show that there is a significant gain to be achieved by performing this parameter optimisation. A numerical evaluation, assuming a carrier frequency of 2 GHz, a channel bandwidth of 200 MHz, a mobile velocity of 30 m/s, and a return channel delay of 1 ms (corresponding to 20% of a Doppler period), shows that the BER constraint (BER $< 10^{-5}$) is fulfilled over the whole CSNR range, while the ASE is roughly 3 bits/s/Hz lower than the ergodic channel capacity, with a loss of about 1 bit/s/Hz compared to the perfect CSI case. The use of multiple receive antennas allows for the use of a longer pilot period and less power allocated to pilot symbols, which contribute to the ASE gain in this case.

3.6.3 Link adaptation in MIMO systems

Recently a lot of research works have been concerned with MIMO systems, evaluating the pros and cons of systems where multiple antennas are used both at the transmitter and receiver side [CaEG02],

[FoGa98], [Fosc96], [FGVW99], [TaSC98], [Alam98]. The quality of the available CSI is crucial for efficient decoding of transmitted information. In the majority of MIMO papers it is assumed that CSI is known only at the receiver. However, calculations on MIMO system capacities show, as previously discussed for SISO and Single Input Multiple Output (SIMO) systems, an increase of system throughput if the channel information is known also at the transmitter. In this case the transmitted signal for each transmit antenna, characterised by its power, modulation, and coding scheme, can be adjusted to the estimated or predicted channel conditions. MIMO systems which select a subset of transmit and receive antennas, adjust the coding and modulation scheme, or allocate transmit power adaptively to the channel conditions, are known as adaptive rate MIMO systems [CaEG02].

The main research activities presented during COST 273 dealing with link adaptation have been focused to a large extent on the physical layer in the point-to-point single-user communications system. Several approaches exist for link adaptation of MIMO systems:

- link adaptation applied to MIMO eigenmode transmission schemes,
- active antenna selection and coding modulation scheme adjusting of active antennas,
- link adaptation using space-time block codes, and
- link adaptation in MIMO OFDM systems.

Since practical mobile communication systems are multi-user systems, some research efforts are also devoted to the link adaptation in multi-user mobile communication systems studying different adaptive resource allocation techniques, optimising the overall system throughput.

Link adaptation in MIMO eigenmode transmission schemes

Theoretically the MIMO Eigenmode Transmission Scheme (EMTS) with optimal power allocation maximises the system capacity [FoGa98], [Tela99]. The MIMO eigenmode transmission scheme uses the left and right eigenvectors of the channel matrix as eigenbeamformers in the receiver and transmitter respectively. The EMTS theoretically outperforms other MIMO systems and furthermore, it can easily be combined with the adaptive coding and modulation principles known from single-input-single-output systems, like adjusting transmission power and coding modulation scheme of individual transmit antennas, as well as new space adaptation techniques like selecting the set of reliable eigenmodes, combining two or more eigenmodes, and allocating the available transmit power among transmit antennas.

A straightforward method of adapting a MIMO system to the channel conditions is by allocation of the available transmitter power in order to achieve maximum system throughput. The capacity of MIMO systems has been calculated, with the optimal power allocation using the waterfilling algorithm in the space domain. However, there exist several constraints which make it impractical to implement power allocation by means of standard waterfilling. The standard waterfilling algorithm, with an infinite number of power levels, requires a large amount of data in the return channel. Therefore, discrete power allocation is introduced, which reduces the amount of data in the return channel on one hand, and decreases the system performance on the other. Other constraints which have to be considered in a real MIMO system are the limited maximum power in each individual eigenmode, and the total amount of transmit power available in the communication system.

The theoretical analysis and simulation results in [Fise04], where the performance of discrete power allocation in a flat block MIMO channel were studied, show that a communication system with $L = 4$ energy levels in each eigenmode is sufficient for not significantly decreasing the performance of a

MIMO system with a low number of transmit and receive antennas, $N_T \leq 4$. An increase in number of transmit and receive antennas means that a higher number of discrete power levels is required.

As noted above, in real transmission systems constraints in the total power available, and the maximum power per individual RF amplifier, exist. The influence of the total instantaneous peak power (defined as a sum of the instantaneous powers over all transmit antennas) and the maximum instantaneous power per transmit antenna on the outage probability is shown in Figure 3.42 [FiSy03]. The parameter K is defined as the ratio between the maximum allowed power per individual eigenmode and the total peak energy available. As expected a decrease of the parameter K causes a degradation of the MIMO system performance.

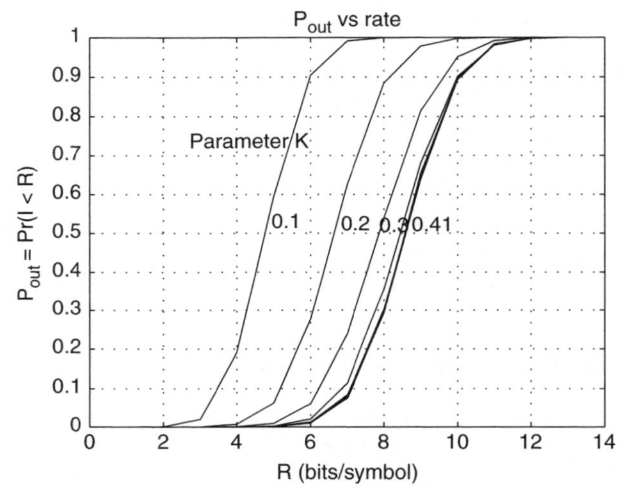

Figure 3.42 *Outage probability of a MIMO system with power constraints for $N_T = N_R = 3$, SNR = 10.*

Assuming a quasi-static flat fading channel and a MIMO eigenmode transmission scheme, the eigenmodes can be classified into *reliable* subchannels, where a certain demand on the Quality of Service (QoS) parameters can be guaranteed, and *unreliable* subchannels, where the QoS demand cannot be met. Two approaches for how to use the reliable and unreliable subchannels to achieve the desired QoS, for example to keep the system BER below a certain target value, are proposed in [JaKP04]. In basic channel mapping strategies no data is transmitted in unreliable subchannels, while in reliable subchannels the coding and modulation scheme is dynamically adapted to the instantaneous channel conditions. The same signal is transmitted in one reliable subchannel and in one or more unreliable subchannels, using advanced channel mapping strategies. At the receiver the reliable and unreliable subchannels are combined, applying equal power or maximum ratio combing. When no reliable channel is available, the same signal is transmitted in two or more unreliable subchannels, to establish an overall reliable communication link with the required QoS. The simulation results, where M-QAM modulation schemes were applied for data transmission, show that the basic channel mapping strategy does not differ significantly from the advanced one for high signal-to-noise ratios, while at low signal-to-noise ratios, only advanced channel mapping strategies will guarantee an overall communication link supporting a BER below the target value.

As mentioned previously it has been proved in the literature that optimal power allocation and coding/modulation scheme selection in MIMO eigenmode adaptive transmission systems can maximise the system capacity. However, in practical implementation many problems occur. Some main problems are

- non-linearity of high power amplifiers,
- channel estimation (prediction),
- timing recovery,
- power and modulation scheme adaptation algorithm, and
- accurate return channel.

A MIMO adaptive eigenmode transmission scheme was implemented to study its performance in typical non-line-of-sight office environments, and quantify the gap between the theory and practice [SaTA04]. When no Power and Modulation Adaptation (PMA) algorithm is implemented in a MIMO system with four transmit and four receive antennas, the experimental results show good agreement with the results obtained by simulation of a corresponding MIMO system with implemented Time and Channel Estimation (TCE), 1–2 dB difference. The main reason for the 1–2 dB differences between measured and simulation results is due to slow channel variations. The comparison between the simulation results with perfect TCE and the results obtained by measurements is the evidence for the excellent performance of the implemented TCE algorithm for a MIMO system without PMA. The modulation scheme shown is QPSK; however, similar results are obtained for 16-QAM signals.

The PMA algorithm allocates the power and selects the modulation scheme for each eigenmode to maximise the system throughput. Now, either 16-QAM, QPSK, or BPSK can be transmitted in each eigenmode. The simulation results, plotted in Figure 3.43, show that the measured throughput of the MIMO system with PMA is in fact worse than the one without PMA. This highlights the problems and constraints encountered when implementing PMA: non-linearity of high power amplifiers, imperfect timing recovery, erroneous channel estimation, power and modulation scheme allocation algorithm,

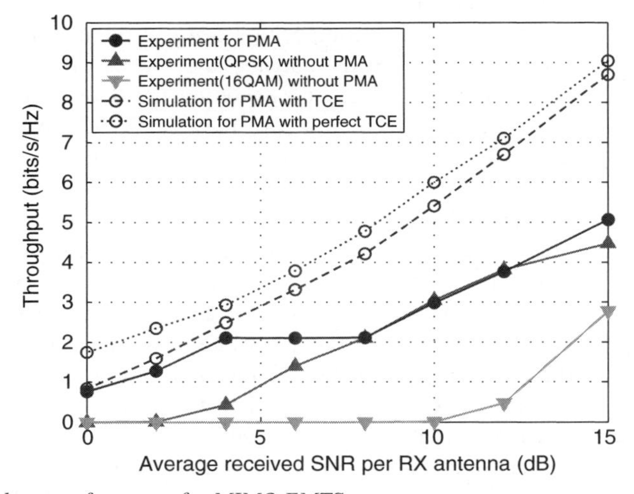

Figure 3.43 *Throughput performance for MIMO EMTS.*

discrete modulation order for the PMA, and feedback errors. The non-linearity of high power amplifiers can be accounted for by backing off the operating point of a high power amplifier. Even though the time and channel estimation algorithm is accurate enough for a MIMO system without PMA, the system with PMA requires a more accurate algorithm. The misestimation of the CSI, in addition to the interference among eigenmodes, causes misselection of the coding/modulation scheme, and the wrong power allocation. However, the algorithm for power allocation is not optimal. In the future, the implementation should take into account the increase of noise due to imperfect eigenbeamforming. The effects of discrete modulation scheme order can be compensated by continuous power allocation. The system throughput is also significantly lower due to the Doppler frequency shift inherent in the measurement environment. All above-mentioned effects must be included in design of future PMA algorithms in order to achieve better properties of adaptive MIMO systems.

Active antenna selection and coding/modulation scheme adjustment approach

Early approaches, which study adaptive MIMO systems mainly from a theoretical viewpoint, suggest that link adaptation on eigenmodes offers the maximum theoretical system throughput. However, practical implementation problems, among them erroneous channel state information estimation, high signal peak-to-average power ratio in high power amplifiers, and the cost of RF chain implementation significantly decrease the measured system throughput. An alternative approach for link adaptation in MIMO systems is to adjust the transmission mode, i.e. the coding and modulation scheme and power on each transmit antenna. In that case the sensitivity of the system on the CSI errors and non-linear distortion is lower. The approach can be combined with the one which considers the implementation cost of MIMO systems, where the RF chains, comprised of digital-to-analogue converters, low noise amplifiers and downconverters, are a significant cost factor. The cost of the system can be reduced applying hybrid-selection schemes, where the L best out of N antennas are selected for the transmission or reception of the MIMO signal.

A simple but efficient algorithm, which determines the set of active transmit antennas observing self-generated noise at the detection, and adjusts the coding modulation scheme of each selected antenna to the instantaneous channel conditions, is proposed in [PlJK05], [JaPK04]. The input to the algorithm is the estimated or predicted channel matrix. The self-generated noise for each transmit antenna is calculated using singular value decomposition of the channel matrix. If the target BER cannot be achieved by the weakest antenna, the antenna is switched off and the available power is allocated to active antennas. The procedure of calculating the SNR of remaining antennas is repeated. By switching off the weakest antenna the diversity gain of the system is increased, and consequently the self-generated noise of active antennas is significantly reduced. The increased SNR may enable the transmission of more efficient coding and modulation schemes from the active transmit antennas. The whole procedure of finding the optimum modulation scheme with pseudo-inverse calculation is recursively repeated with a new deflated channel matrix until all unreliable antennas are switched off. When the target BER can be achieved for all active antennas in the system, the coding/modulation scheme is determined for each antenna separately, from the pre-calculated thresholds. An additional optimisation loop can be added to maximise the throughput of the system. Note that the algorithm proposed is not an optimum algorithm for selecting the transmission scheme. The optimum set of active antennas at the transmitter and their transmission modes can be determined only by an exhaustive search through all available possibilities. This exhaustive search algorithm is time consuming, and thus inappropriate for implementation. The bandwidth efficiency of an adaptive MIMO system with four transmit and four receive antennas and target BER $= 10^{-3}$, for both the optimum search algorithm and the proposed algorithm with and without additional throughput-optimisation loop [PlJK05], is shown in Figure 3.44. Both the throughput of the MIMO eigenmode transmission system and the MIMO Shannon capacity is shown.

The analysed algorithm gives nearly constant BER for zero-forcing MIMO detection over the observed SNR range. The additional bandwidth efficiency optimisation loop in the communication mode selection algorithm improves bandwidth efficiency for high signal-to-noise ratios, and the system bandwidth efficiency over the observed SNR range does not differ noticeably from results for the optimal search algorithm. Due to the use of non-orthogonal channels, 20% of system capacity is lost by comparison with the throughput of an optimal adaptive EMTS system.

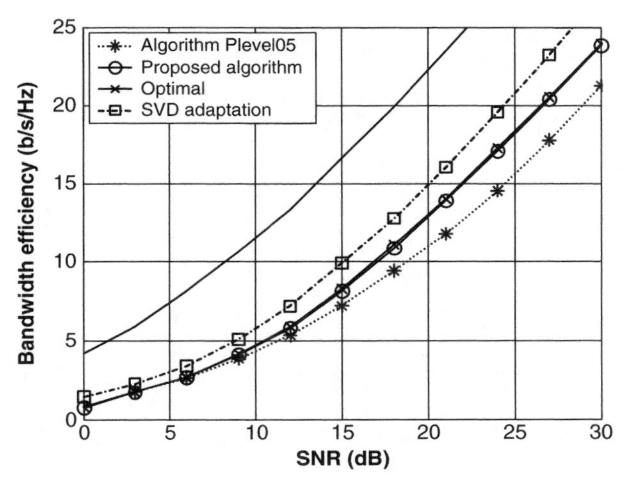

Figure 3.44 *Bandwidth efficiency of an adaptive MIMO system with* $N_T = 4$ *and* $N_R = 4$ *and target BER* 10^{-3}.

The impact of channel estimation errors due to channel variations, simulated as a Gaussian process with zero mean value and standard deviation as a parameter, was also analysed. The simulation results for $\sigma = 0.05$ show that the proposed adaptive MIMO system is insensitive to simulated channel error for zero-forcing MIMO signal detection. A negligible increase of BER is observed regardless of the number of transmit and receive antennas and the choice of target BER. It can also be seen that at a lower target BER, a SISO system is more sensitive to channel variations.

Link adaptation applying space-time block codes

Since the work of Alamouti [Alam98] several Space-Time Block Codes (STBCs) have been designed, most of them assuming no available CSI at the transmitter. As orthogonal full-rate design offers full diversity of an arbitrary complex symbol constellation for only two transmit antennas, a quasi-orthogonal design has been proposed for more than two transmit antennas. The knowledge of CSI at the transmitter can improve the outage performance of the system. A simple method to improve the system performance by combining two or more extended Alamouti space-time codes over four transmit antennas was suggested in [BaRW03], [BaRW04]. Two systems were proposed: the simple one where the information in the return channel is represented with only one bit, and the extended one, where two bits are transmitted in the return channel. Depending on the information received via the return channel the transmitter switches between two predefined STBCs in the first case, and four in the second. The predefined Space-Time Block Code (STBC) is selected to maximise the diversity of the system, to minimise bit error rate, and to improve the orthogonality of the code.

The BER as a function of E_b/N_0 for the zero-forcing receiver is shown in Figure 3.45. The results are obtained for QPSK signals and a flat fading quasi-static MIMO channel. A substantial improvement of the BER is achieved by providing only one bit of information in the feedback channel, enabling the transmitter to switch between two space-time codes and choose the one with the higher diversity. The maximum of fourth order diversity is achieved with four space-time codes and two bits sent to the transmitter via the return channel.

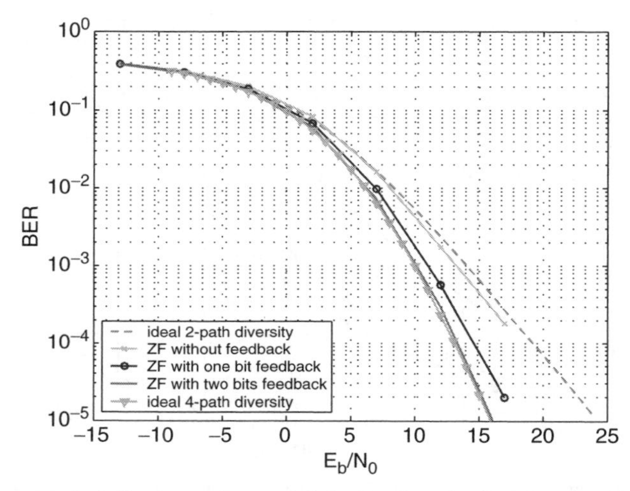

Figure 3.45 *BER for extended Alamouti scheme with feedback and zero-forcing receiver.*

A second promising technique to reduce complexity and cost of MIMO systems is channel adaptive Angular Spread (AS). AS at the transmitter, combined with quadrature space-time codes for four transmit antennas, has been investigated in [BaFW05]. The simulation results for $4 \leq N_t \leq 7$ available transmit antennas and $n_r = 1$, where four transmit antennas are selected according to the best optimisation rule – that is, maximising $h^2(1 - X^2)$ – is presented in Figure 3.46. For a BER $= 10^{-3}$ and $N_t = 5$, the coding gain is about 3 dB compared to $N_t = 4$, the case without antenna selection. Increasing the number N_t of available transmit antennas by one, the coding gain increases by about 1 dB. Most important, Figure 3.46 shows that the system diversity increases substantially with the number N_t of available transmit antennas.

Link adaptation of MIMO systems in a multi-user environment

The majority of works dealing with link adaptation in MIMO systems consider the single-user point-to-point communication. In mobile communication systems multi-user communications are more common. The resource allocation strategies have important influence on the system performance in a multi-user environment. Adapting the transmit power, the number of active antennas at the transmitter, the rate allocation per slot, and the allocation of slots itself (user scheduling) may significantly increase the performance of a MIMO system.

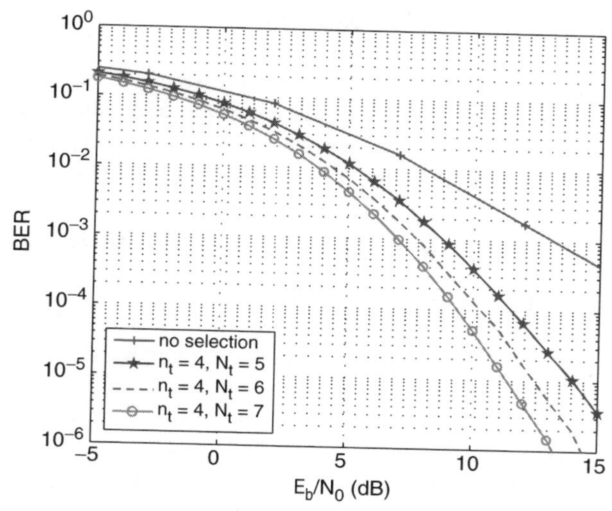

Figure 3.46 *Transmit antenna selection, iid MIMO channels, $n_t = 4$ out of $5 \leq N_t \leq 7$.*

Different resource allocation strategies for multi-user wireless SDMA/TDMA systems with multiple antennas at the transmitter and receiver using V-BLAST architecture were presented [MaMT03]. Joint adaptive allocation of slots, rate, and power in the uplink of a single cell system was considered, with transmission organised in slots and frames. The frame can have fixed length (FF) or variable length (VF).

Three resource allocation algorithms trying to maximise the number of data packets per slot sent through the system are proposed, namely:

- T-based Max Fit: chooses the slot with the maximum per-slot expected throughput, normalised to the number of users in the slot,
- T-based Best Fit: chooses the slot with the maximum per-slot expected throughput increment due to the allocation of the new user,
- S-based Best Fit: chooses the slot with the maximum value for the minimum post-detection SINR over each slot.

Two possible versions of the proposed algorithms are simulated: unconstrained and default. In the unconstrained version the new users are added until all resources are filled, while in the default version new users are added up to predefined thresholds.

The transmit power and rate are dynamically assigned to each antenna. The rate adaptation is obtained by selection of active antennas. The non-active antennas are switched off. The power allocation is obtained by setting the power to each antenna in order to achieve the same post-detection signal-to-noise ratio for each user.

The simulation results show that the greedy algorithms based on throughput estimation outperforms the classical SNR-based algorithms. The combination of rate and power adaptation techniques significantly improves system performance.

3.6.4 Link adaptation in OFDM systems

The flat fading channel models assumed in previous discussions are realistic only for narrowband communications, while high data rate communication requires a broadband radio channel, which is usually frequency selective. A popular technique to convert the frequency-selective broadband radio channel into a parallel collection of frequency-flat subchannels is OFDM.

Low complexity and overhead bit-loading schemes for OFDM

Conventional OFDM modems use fixed constellation size and power level allocation of all subchannels. If the receiver can provide the transmitter with CSI using a robust feedback channel, bit/power-loading techniques can be adopted to increase the throughput or the performance [Bing90], [ChCB95]. In the case that bit-loading techniques are considered for a wireless scenario [Czyl96], particular attention must be paid to the effects of channel estimation and CSI update rate on the performance [YeBC02]. However, *waterfilling*-based techniques require a large overhead for CSI feedback, making them suitable only for static or very slowly time varying channels.

The contributions [Dard04a], [Dard04b] propose and analyse a very simple adaptive bit allocation technique to improve system performance. Only the K most reliable subchannels among N are selected, and higher-order modulations are used to compensate for the reduced subcarrier utilisation by keeping the total bit rate and transmitted power unchanged. In contrast to other bit-loading algorithms, both the power level and the constellation size are kept constant over the selected set of subchannels, thus drastically reducing the modem hardware complexity and signalling overhead. The optimal value for K (optimum loading configuration), which minimises the average bit error probability for the uncorrelated fading channel model, is analytically derived. In addition, it has been verified by simulation that the same result holds also for typical correlated fading scenarios. For performance evaluation, system parameters are taken from the IEEE 802.11a physical layer specifications [IEEE99a]. The channel is assumed constant during a packet transmission and a Reed Solomon (RS) ($n = 112$, $k = 56$, 8 bits/symbol) coding scheme is considered.

Figure 2.15 shows the average packet error probability P_{ep} as a function of E_b/N_0 for the 12 Mbit/s mode (4-QAM). The European Telecommunications Standards Institute (ETSI) channel model 'C' for the 5 GHz band (typical open space environment) is considered [MeAS98]. In this case, the best loading condition results for $K = N/2$ (i.e. half of the more reliable subchannels and quadrupled constellation size). For example, by fixing $P_{ep} = 10^{-2}$, the gain obtained in terms of SNR, compared to the reference scheme ($K = N$, no loading), is around 5–6 dB. The distance from the optimal Campello's algorithm [Camp98] curve is less than 2 dB. It is worthwhile noting that this good result can been obtained with a considerable lower hardware complexity compared to that required by the optimal solution. The robustness of this technique has been addressed in the presence of imperfect CSI updating for time varying channels, concluding that an update time, normalised to the Doppler shift, of less than 0.1 does not degrade the system performance significantly.

Link adaptation in MIMO-OFDM systems

When an OFDM approach is applied to each transmit antenna and each receive antenna in a MIMO system, a frequency selective MIMO channel is transformed into a set of N flat fading MIMO channels, where N is the length of the FFT block used in the OFDM algorithm. A strategy of mapping the information bits to antennas and tones, to achieve both spatial and frequency diversity, is *space-frequency* encoding. The space-frequency code may be concatenated with an outer encoder. These codes are known as Space-Time-Frequency (STF) codes.

The link adaptation in a MIMO-OFDM system may be implemented in three domains: time, frequency, and space. The outer encoder controls time-domain link adaptation. The popular options for outer codes are convolutional codes, trellis codes, and, recently, LDPC codes. Bit and power-loading algorithms on a tone-by-tone basis are applied for frequency-domain link adaptation, while the inner space codes are used for link adaptation in the space domain. Space-time codes can be classified into three groups: space-time block encoding as discussed previously, space-time encoding based on MIMO channel eigenmodes, and transmission selection diversity.

In the contributions [GBVM03], [BGVC03], a concatenated coding scheme for a 2 $\times$ 2 MIMO-OFDM system is tested with different outer/inner code options and in the presence/absence of link adaptation. It is shown that in this scheme adaptivity can be easily implemented adjusting the constellation size/energy on a subcarrier-by-subcarrier basis by means of standard loading algorithms for SISO channels, independently of the selected codes. The resulting procedure does not require the application of the Singular Value Decomposition (SVD) technique for channel diagonalisation and consists of a waterfilling operation over a SISO- OFDM system with the same number of subchannels as the MIMO one. Consequently, well-known bit/power-loading algorithms for SISO multicarrier systems can be adopted. Simulations are performed considering a High Performance Radio Metropolitan Area Network (HIPERMAN) compliant scenario with perfect CSI and different options for the outer code (64 states rate $1/2$ binary convolutional code, LDPC code with rate $1/2$ and random parity check matrix) and inner coding (Alamouti's orthogonal STBC [Alam98], symbol precoding via the SVD and Transmission Selection Diversity (TSD)). The FFT order is 256 and the SISO channels associated with different couples of transmit-receive antennas are statistically equivalent and independent. Numerical results show that, in the absence of outer coding, bit loading in a 2 $\times$ 2 MIMO-OFDM system can bring a small advantage over an unloaded system operating at the same rate. When outer coding is adopted, bit loading offers a negligible energy gain due to finite granularity of the constellation set.

The contribution [MuDa04] considers a Bit-Interleaved Coded Modulation (BICM) OFDM system, where a bit-wise ideal interleaving is assumed performed after the encoding. The CSI at the transmitter can be used for Adaptive Coding (AC) and/or Adaptive Bit Loading (ABL) and/or Adaptive Power Loading (APL). The interleaved bits are mapped to a sequence of vectors, each containing the coded bits conveyed by one OFDM symbol. The bit vectors are split up into N_{sc}-bit tuples, where the size m_k of the kth tuple, $k = 1, \ldots, N_{sc}$, is chosen by the ABL subject to a bit-rate constraint. The Gray encoded tuples are then mapped onto complex signals, where the kth signal is weighted by the square root $\sqrt{e_k}$ of the kth subcarrier energy in the APL scheme, subject to a power constraint. The coded bits in the tuples are assumed uniformly distributed and independent due to the preceding ideal interleaving. The transmitter design relies on given bit-rate and power constraints. Finally, after an IDFT and a guard period insertion, the signal is fed to an antenna array with N_{TX} antennas and transmitted over a frequency selective channel where the adaptive front ends at the transmitter and receiver transmit each subcarrier signal over the strongest corresponding spatial eigenmode. The receiver employs a DFT, a zero-forcing Frequency-domain EQualiser (FEQ), a Hard-Decision Demapping (HDD) and deinterleaving and a Viterbi decoder provides the final bit estimates. HDD turns out to be a good compromise between complexity reduction at the receiver and performance degradation as compared to a design based on a soft-decision demapping. Adopting the equivalent Binary Symmetric Channel (BSC) model introduced for a BICM system in [CaTB98], the exact uncoded Bit Error Probability (BEP) can be computed for each of the parallel independent binary input channels and the average BEP P_b is derived as shown in [HuDa03], [MuDH03]. Since the BEP of a coded transmission over a memoryless BSC decreases with the transition probability of the channel, the subsequent loading procedures are based on a minimisation of P_b w.r.t. m_k and e_k. As a result of the ABL, subchannels with large (resp. small) channel gains employ higher (resp. lower) order modulation [MDHF03]. Once the values e_k have been found, an APL can be employed providing values e_k which can be expressed by Lambert's W-function [MuDH03], [MuDa04].

For the simulation of the BER obtained from the ABL and APL, uncorrelated scattering channel coefficients with a Rayleigh distribution [Hunz02] are assumed. A BICM-OFDM scheme with $N_{sc} \gg 1$ is used, employing a rate-1/2 convolutional code with generators 133_{oct} and 171_{oct}. Using a random number generator, new channel and bit interleaver permutations are generated for each burst comprising 20 OFDM symbols. The achievable performance gains are shown in Figure 3.47. The figure displays the average BERs at the decoder output as a function of the mean SNR $\overline{\mathrm{SNR}}$ for Uniform Power Loading (UPL), APL, ABL and combined ABL + APL, where SISO and $N_{TX} \times N_{RX}$-MIMO systems are considered for different N_{TX} and N_{RX}. For an average BER level of 10^{-6}, the largest relative gain results for the ABL SISO system with a gain of 5.5 dB, which is further improved in the ABL + APL scheme. MIMO systems can be used for a further reduction of the required average SNR where the relative gain decreases for an increasing number of antennas. Further improvements are to be expected from adaptive coding [MuDD04] at the expense of a higher overall system complexity.

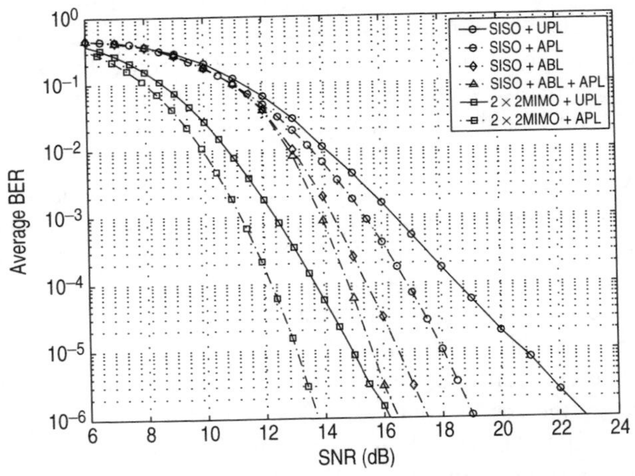

Figure 3.47 *Average BER values for BICM-OFDM using different adaptation schemes for SISO and 2 × 2-MIMO systems.*

3.6.5 Link adaptation in current and upcoming wireless standards

In this subsection, the status and use of link adaptation techniques in some modern wireless communication system standards are considered.

UMTS-HSDPA

A major innovation within Release 5 of the UMTS standard is the High Speed Downlink Packet Access (HSDPA) channel. It provides packet-oriented downlink connections for streaming, interactive and background services at high data rates. HSDPA is designed to reuse as much of the existing UMTS functionality as possible and intended primarily for urban/indoor scenarios and low to medium user speed. The use of multiple codes in conjunction with Adaptive Modulation and Coding (AMC) allows peak data rates of 14 Mbps. Additionally, fast scheduling and fast Hybrid Automatic Repeat reQuest (HARQ) are used [DöMR02].

A simple, backward compatible but versatile implementation of different HARQ techniques, like Chase Combining, Partial and Full IR, is achieved by a two-stage rate matching algorithm [DöMR02]. The gain offered by IR compared to Chase Combining increases with code rate. Full IR outperforms Partial IR for code rates greater than 1/2. For 16-QAM further performance gain is achieved by mapping of systematic bits to high reliable bit positions and by equalising the bit reliabilities after retransmissions [DöGS03].

In HSDPA, the modulation (QPSK, 16-QAM) and code rate ($R = 0.25 \ldots 0.98$) may change every 2 ms. For this link adaptation Channel Quality Feedback (CQF) is required. The user terminal reports the highest AMC level, which it can decode with an average frame error rate of 10% according to the current channel conditions and its specific receiver performance. In an operational system this CQF is subject to errors due to measurement inaccuracy, limited signalling bandwidth, delay, and decoding errors. With increasing link adaptation errors, HARQ becomes more and more important and IR can provide from 5% up to 20% additional throughput compared to Chase Combining [DöMR02]. The allowable errors in channel quality feedback scale with the separation of the AMC levels. CQF signalling with 1 dB granularity in AWGN is sufficient for a system operating with a feedback error of 0.3 dB or more (TD(03)009).

CQF schemes that consider the burstiness of packet-oriented data can provide higher efficiency than a simple periodic sampling of the channel. Investigations are performed in accordance with the 3rd Generation Partnership Project (3GPP) Technical Specification Group (TSG) Radio Access Network (RAN) 1 simulation assumptions [3GPP01] for the Vehicular A channel model at 30 km/h user velocity. The HSDPA traffic model of [3GPP01] includes effects of bursty data traffic, e.g. realistic modelling of the size of packets and packet calls, as well as the packet interarrival time. Furthermore, the delays involved in Acknowledgement (ACK)/Negative Acknowledgement (NACK) transmission and CQF reporting have been considered in detail [DöRM04]. Figure 3.48(a) compares the downlink data and control channel usage for identical throughput and delay performance. The activity-based CQF scheme simply increases the feedback rate during data transmission, whereas the on-demand/NACK-based feedback allows dedicated CQF requested by the transmitter and sends additional CQF when a packet has not been decoded. The autonomous CQF scheme sends a feedback whenever the current channel conditions deviate from the previously transmitted value [DöRM04]. Of the investigated schemes, the on-demand/NACK CQF variant provides highest efficiency in uplink and downlink. It allows reducing the periodic CQF sampling rate during reading times and increases the CQF rate during activity on the data channel in an intelligent way, i.e. it also avoids excessive CQF during data transmission. While maintaining the same data throughput and delay performance, it reduces the downlink channel usage by up to 9% and therefore provides higher system capacity accordingly. Furthermore, by reducing the overall CQF feedback overhead per transmitted bit (cf. Figure 3.48(b)), it saves the scarce power at the MT and reduces the interference due to the uplink control channels [DöRM04].

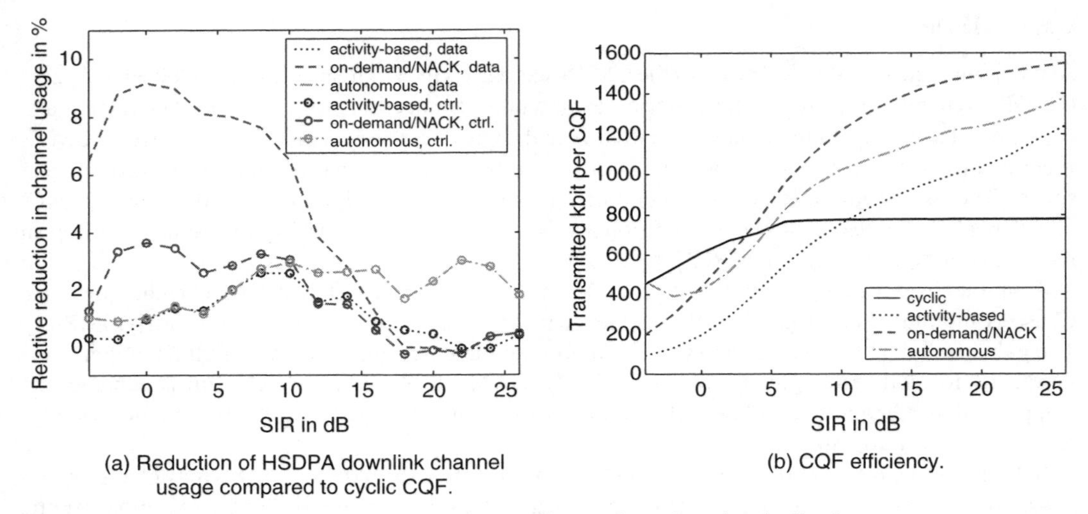

(a) Reduction of HSDPA downlink channel usage compared to cyclic CQF.

(b) CQF efficiency.

Figure 3.48 *HSDPA performance increase due to efficient channel quality feedback.*

For maximum Signal-to-Interference-Ratio (SIR) scheduling and ideal channel quality feedback, the sector throughput of HSDPA increases from 2.3 Mbps to 6.5 Mbps by increasing the number of concurrently active users from 1 to 50 (TD(03)009), highlighting the enormous potential of multi-user diversity in HSDPA.

Wireless LANs

Link adaptation in WLANs has been used from the very beginning of their standardisation. The first IEEE 802.11 WLAN standard defined two physical layer transmission rates with 1 and 2 Mbit/s, respectively, in the 2.4 GHz band [IEEE99d]. Many products at that time offered an additional proprietary 5.5 Mbit/s transmission rate. With 802.11b and 802.11g [IEEE99b, IEEE99c], higher speed transmission rates between 11 and 54 Mbit/s have been standardised as an extension of the existing 2.4 GHz Direct Sequence Spread Spectrum (DSSS) modulation scheme. The IEEE 802.11a Phy was defined from the start to support Location Area (LA) with its eight transmission rates [IEEE99a]. The HIPERLAN Type 2 (H2) standard [ETSI01] defines seven transmission rates and is very similar to the IEEE 802.11a standard.

All standard documents describe different modulation and channel coding schemes for LA, but none of the standards defines criteria or even mechanisms to select the right transmission rate for a given situation. This is intentionally left to the discretion of the implementor. Different from voice networks, the criteria for good and bad state in data networks are not given by fixed packet error rates but rather by throughput on higher layers. Therefore, most performance studies on LA in WLAN systems are not restricted to the Phy but consider Phy jointly with Medium Access Control (MAC) and ARQ protocols.

A number of interesting publications about LA schemes in IEEE 802.11 networks have been published, see, e.g. [QiCS02], [CiSh02], [WuHYJ01], [QCJS03]. Most of these schemes are based on previous knowledge about PER and throughput behaviour for certain SIR, so they rely on selection of the transmission rate for maximum throughput. None of them, however, contains a mathematical justification or performance analysis but all are empirical schemes, investigated with computer simulation methods.

Investigations of LA in HIPERLAN Type 2 are mostly based on the PER versus SIR results from [KSWW99], [KMST00]. A comparison of IEEE 802.11a and H2 Phy layer performance can be found in [DABN03]. Early empirical schemes for H2 considering Phy, MAC and ARQ are described in [LiMT00], [SiBa01]. For detailed investigations of LA in H2 networks, the reader is referred to Section 8.6 of this report.

Note that joint consideration of LA with higher layer protocols is often very simplistic with regard to effects of Rayleigh fading in conjunction with OFDM, as explained in [LaRZ02]. Rayleigh fading in OFDM systems may in good cases result in flat fading or it may, in bad situations, result in frequency-selective fading and consequently make a number of subcarriers entirely unusable. Therefore, bit or packet error rates do not only depend on SIR but also on the current channel transmission function. A possible solution to estimate the PER, given the SIR and the channel transmission function, has been published in [LGRZ03].

Bluetooth

Bluetooth (BT), also known as IEEE 802.15.1 [Blue99], is an emerging technology aiming at providing a wireless connection for MTs, printers, headphones, and other personal devices. The BT paradigm is based on the concept of piconets, where one device plays the role of master and several terminals are slaves polled by the master.

The BT specifications define a collection of protocol stacks based on, at the physical and radio link levels, an RF and a baseband layer. At baseband, six different packet types are defined, having different payload sizes and coding protection and, in principle, a different packet type could be used in each master-slave link; BT specifications, however, give no indications about how to select the proper packet type for each master-slave link.

LA is a well-known technique providing vertical integration within the protocol stack: in this case, it may consist in selecting the proper packet type (i.e. the payload size and the channel coding scheme) as a function of the mean value of the received power level.

The contributions [PaCV02], [PaTV03] show that a proper choice of the packet type can provide advantages in terms of overall throughput. The performance is evaluated through a simple analytical model in a reference scenario, which is also integrated with some results obtained via a ray tracing tool, to characterise the propagation channel in a realistic scenario (an office environment of about 600 square metres). The results obtained allow suitable comparison between the cases of LA and no LA. They show that a significant improvement can be achieved by properly choosing, on the basis of the knowledge of the channel status experienced, the packet type to be adopted in each link.

References

[3GPP01] 3GPP. Physical Layer Aspects of UTRA High Speed Downlink Packet Access. 3GPP, 2001.

[Alam98] S. M. Alamouti. A simple transmit diversity technique for wireless communications. *IEEE J. Select. Areas Commun.*, 16(8):1451–1458, Oct. 1998.

[BaBu04] D. Bajic and A. Burr. A simple suboptimal integer code. In *Proc. ISITA 2004 - IEEE Int. Symp. on Information Theory and its Applications*, Parma, Italy, Oct. 2004. [Also available as TD(03)080].

[BaFW05] B. Badic, P. Fuxjaeger, and H. Weinrichter. Optimization of coded MIMO - transmission with antenna selection. In *Proc. VTC 2005 Spring - IEEE 61st Vehicular Technology Conf.*, Stockholm, Sweden, May to be published 2005. [Also available as TD(05)003].

[Baji04] D. Bajic. On survival probability of alignment sequences. In *Proc. ISITA 2004 - IEEE Int. Symp. on Information Theory and its Applications*, Parma, Italy, Oct. 2004.

[BaNa04] D. Bajic and M. Narandzic. Häberle's acquisition curves revisited: Part II - averages. TD(04)184, COST 273, Duisburg, Germany, Sep. 2004.

[BaRW03] B. Badic, M. Rupp, and H. Weinrichter. Adaptive channel matched extended alamouti space-time code exploiting partial feedback. In *Proc. CIC 2003 - 8th Int. Conf. on Cellular and Intelligent Communications*, Seoul, Korea, Oct. 2003. [Also available as TD(04)102].

[BaRW04] B. Badic, M. Rupp, and H. Weinrichter. Adaptive channel matched extended alamouti space-time code exploiting partial feedback. *ETRI Electronics and Telecommunication Research Institute Journal*, 26(5):443–451, Oct. 2004. [Also available as TD(03)102].

[BaSD02] D. Bajic, V. Senk, and M. Despotovic. Subsets of the STM-1 frame alignment signal: a monitoring analysis. *IEE Proc. Commun.*, 149(5):242–248, Oct. 2002.

[BaSL03] D. Bajic, J. Stojanovic, and J. Lindner. Multiple window-sliding search. In *Proc. ISIT 2003 - IEEE Int. Symp. on Information Theory*, Yokohama, Japan, June 2003.

[BaSt03] D. Bajic and J. Stojanovic. Frame-alignment procedures for STM-1 frame. *IEE Proc. Commun.*, 150(1):37–44, Feb. 2003.

[BaSt04] D. Bajic and J. Stojanovic. Distributed sequences and search process. In *Proc. ICC 2004 - IEEE Int. Conf. Commun.*, Paris, France, June 2004.

[BCJR74] L. Bahl, J. Cocke, F. Jelinek, and J. Raviv. Optimal decoding of linear codes for minimizing symbol error rate. *IEEE Trans. Inform. Theory*, IT-20:284–287, Mar. 1974.

[BeGT93] C. Berrou, A. Glavieux, and P. Thitimajshima. Near shannon limit error-correcting coding and decoding: Turbo codes. In *Proc. ICC 1993 - IEEE Int. Conf. Commun.*, Geneva, Switzerland, May 1993.

[BGVC03] E. Bizzarri, A. S. Gallo, G. M. Vitetta, and E. Chiavaccini. Space-frequency bit-power loading in space-time block coded MIMO OFDM. In *Proc. 12th IST Summit on Mobile and Wireless Commun.*, Aveiro, Portugal, June 2003.

[Bing90] J. A. C. Bingham. Multicarrier modulation for data transmission: An idea whose time has come. *IEEE Commun. Mag.*, 28(5):5–14, May 1990.

[Blue99] Specification of the Bluetooth System. Core, Version 1.0 B, 1999. http://www.bluetooth.com.

[Brän04] F. Brännström. *Convergence Analysis and Design of Multiple Concatenated Codes*. PhD thesis, Chalmers University of Technology, Gothenborg, Sweden, 2004. [Also available as TD(04)098].

[BrRG04] F. Brännström, L. K. Rasmussen, and A. Grant. Optimal puncturing for multiple parallel concatenated codes. In *Proc. ISIT 2004 - IEEE Int. Symp. on Information Theory*, Chicago, IL, USA, June/July 2004. [Also available as TD(04)098].

[BuSh02] A. Burr and J. Shen. A turbo-PIC multiuser receiver with receive diversity for CDMA systems. In *Proc. PIMRC 2002 - IEEE 13th Int. Symp. on Pers., Indoor and Mobile Radio Commun.*, Lisbon, Portugal, Sep. 2002. [Also available as TD(02)073].

[CaEG02] S. Catreux, V. Erceg, and D. Gesbert. Adaptive modulation and MIMO coding for broadband wireless data networks. *IEEE Trans. Commun.*, 40(6):108–115, June 2002.

[CaGi05] X. Cai and G. B. Giannakis. Adaptive PSAM accounting or channel estimation and prediction errors. *IEEE Trans. Wireless Commun.*, 4(1):246–256, Jan. 2005.

[Camp98] J. Campello. Optimal discrete bit loading for multicarrier modulation systems. In *Proc. ISIT 1998 - IEEE Int. Symp. on Information Theory*, Cambridge, MA, USA, Aug. 1998.

[CaTB98] G. Caire, G. Taricco, and E. Biglieri. Bit-interleaved coded modulation. *IEEE Trans. Inform. Theory*, 44(3):927–946, May 1998.

[Cave91] J. K. Cavers. An analysis of pilot symbol assisted modulation for rayleigh fading channels. *IEEE Trans. Veh. Technol.*, 40(6):686–693, Nov. 1991.

[ChCB95] P. S. Chow, J. M. Cioffi, and J. A. C. Bingham. A practical discrete multitone transceiver loading algorithm for data-transmission over spectrally shaped channels. *IEEE Trans. Commun.*, 43(2-4):773–775, Feb./Mar./Apr. 1995.

[CiSh02] K. G. S. Ci and H. Sharif. An link adaptation scheme for improving throughput in the IEEE 802.11 wireless LAN. In *Proc. LCN 2002 - 27th IEEE Conf. on Local Computer Networks*, Tampa, FL, USA, Nov. 2002.

[CVAV04] M. J. Canet, F. Vicedo, V. Almenar, J. Valls, and F. Angarita. An FPGA-based baseband transceiver for HIPERLAN 2. In *Proc. InOWo 2004 - 9th International OFDM-Workshop InOWo*, Dresden, Germany, Sep. 2004. [Also available as TD(04)109].

[CWVM03] D. Cassioli, M. Z. Win, F. Vatalaro, and A. F. Molisch. Effects of spreading BW on the performance of UWB rake receivers. In *Proc. ICC 2003 - IEEE Int. Conf. Commun.*, Seattle, WA, USA, May 2003. [Also available as TD(03)076].

[Czyl96] A. Czylwik. Adaptive OFDM for wideband radio channels. In *Proc. Globecom 1996 - IEEE Global Telecommunications Conf.*, London, UK, Nov. 1996.

[DABN03] A. Doufexi, S. Armour, M. Butler, A. Nix, D. Bull, J. McGeehan, and P Karlsson. A comparison of the HIPERLAN/2 and IEEE 802.11a wireless LAN standards. *IEEE Commun. Mag.*, 40(5):172–180, May 2003.

[Dard04a] D. Dardari. Ordered subcarrier selection algorithm for OFDM based high-speed WLANs. *IEEE Trans. Wireless Commun.*, 3(5):1452–1458, Sep. 2004. [Also available as TD(03)155].

[Dard04b] D. Dardari. A uniform power and constellation size bit-loading scheme for OFDM based WLAN systems. In *Proc. PIMRC 2004 - IEEE 15th Int. Symp. on Pers., Indoor and Mobile Radio Commun.*, Barcelona, Spain, Sep. 2004. [Also available as TD(03)155].

[Dejo02] A. Dejonghe. A unified view of linear MMSE turbo-equalization considering bit-interleaved modulation. TD(02)034, COST 273, Guildford, UK, Jan. 2002.

[DeLR77] A. P. Dempster, N. M. Laird, and D. B. Rubin. Maximum likelihood from incomplete data via the EM algorithm. *Journal of the Royal Statistical Society*, 39(1):1–38, Dec. 1977.

[DiDP01] D. Divsalar, S. Dolinar, and F. Pollara. Iterative turbo decoder analysis based on density evolution. *IEEE J. Select. Areas Commun.*, 19:891–890, May 2001.

[DöGS03] M. Döttling, T. Grundler, and A. Seeger. Incremental Redundancy and Bit-Mapping Techniques for High Speed Downlink Packet Access. In *Proc. Globecom 2003 - IEEE Global Telecommunications Conf.*, San Francisco, CA, USA, Dec. 2003. [Also available as TD(02)041].

[DöMR02] M. Döttling, J. Michel, and B. Raaf. Hybrid ARQ and Adaptive Modulation and coding Schemes for High Speed Downlink Packet Access. In *Proc. PIMRC 2002 - IEEE 13th Int. Symp. on Pers., Indoor and Mobile Radio Commun.*, Lisbon, Portugal, Sep. 2002.

[DöRM04] M. Döttling, B. Raaf, and J. Michel. Efficient channel Quality Feedback Schemes for Adaptive Modulation and Coding of Packet Data. In *Proc. VTC 2004 Fall - IEEE 60th Vehicular Technology Conf.*, Los Angeles, CA, USA, Sep. 2004.

[ETSI01] ETSI. BRAN; HIPERLAN Type 2; Physical (PHY) Layer Specification. Technical Specification 101 475, 2nd edition, 2001.

[FeRa02] S. Feldmann and M. Radimirsch. A novel approximation method for error rate curves in radio communication systems. In *Proc. PIMRC 2002 - IEEE 13th Int. Symp. on Pers., Indoor and Mobile Radio Commun.*, Lisbon, Portugal, Sep. 2002. [Also available as TD(02)083].

[FGVW99] G. J. Foschini, G. D. Golden, A. Valenzuela, and P. W. Wolniansky. Simplified processing for high spectral efficiency wireless communications employing multi-element arrays. *IEEE J. Select. Areas Commun.*, 17(11):1841–1852, Nov. 1999.

[Fisc02] R. F. H. Fischer. *Precoding and Signal Shaping for Digital Transmission.* John Wiley & Sons Ltd., New York, NY, USA, 2002.

[Fise04] R. Fisera. *Adaptive Modulations Ű Adaptation Algorithms under Specific Constraints.* PhD thesis, Czech Technical University in Prague, Prague, Czech Republic, 2004. [Also available as TD(04)131].

[FiSy03] R. Fisera and J. Sykora. Lower-upper energy constrained waterfilling in MIMO channels with delay limited transmission. In *Proc. SCVT 2003 - 10th Symp. on Communications and Veh. Tech. in the Benelux*, Eidhoven, Netherlands, Nov. 2003. [Also available as TD(03)063].

[FoGa98] G. J. Foschini and M. J. Gans. On limits of wireless communications in a fading environment when using multiple antennas. *Wireless Personal Communications*, 6(3):311–335, Mar. 1998.

[Fosc96] G. J. Foschini. Layered space-time architecture for wireless communication in a fading environment when using multiple antennas. *Bell Labs Tech. J.*, pages 41–59, autumn 1996.

[Gall63] R. G. Gallager. *Low-Density Parity-Check Codes.* M.I.T. Press, Cambridge, MA, USA, 1963.

[GBVM03] A. S. Gallo, E. Bizzarri, G. M. Vitetta, and M. Marciniak. A comparison of inner coding options for adaptive MIMO OFDM systems. In *Proc. ICTON '03 - International Conference on Transparent Optical Networks 2003*, Warsaw, Poland, July 2003. [Also available as TD(03)149].

[GCDL02] C. Garnier, L. Clavier, Y. Delignon, M. Loosvelt, and D. Boulinguez. Multiple access for 60 GHz mobile ad hoc network. In *Proc. VTC 2002 Spring - IEEE 55th*

Vehicular Technology Conf., Birmingham, AL, USA, May 2002. [Also available as TD(02)042].

[GiWi96] T. R. Giallorenzi and S. G. Wilson. Multiuser ML sequence estimator for convolutionally coded asynchronous DS-CDMA systems. *IEEE Trans. Commun.*, 44(8):997–1008, Aug. 1996.

[GoCG05] A. Goupil, M. Colas, and G. Gelle. Multiple access receiver based on hierarchical clusters/users detection using non-binary LDPC codes. TD(05)031, COST 273, Bologna, Italy, Jan. 2005.

[GoVa97] A. J. Goldsmith and P. P. Varaiya. Capacity of fading channels with channel side information. *IEEE Trans. Inform. Theory*, 43(6):1986–1992, Nov. 1997.

[Guég03] A. Guéguen. Comparison of suboptimal iterative space-time receivers. In *Proc. VTC 2003 Spring - IEEE 57th Vehicular Technology Conf.*, Jeju, South Corea, Apr. 2003. [Also available as TD(03)086].

[HaMi72] H. Harashima and H. Miyakawa. Matched-transmission technique for channels with intersymbol interference. *IEEE Trans. Commun.*, 20:774–780, 1972.

[HaYW02] L. Hanzo, M. S. Yee, and C.-H. Wong. *Adaptive Wireless Transceivers*. John Wiley & Sons Ltd., New York, NY, USA, 1st edition, 2002.

[HeVa05] C. Herzet and L. Vandendorpe. Message-passing synchronization by combining the SP and the EM algorithms. TD(05)014, COST 273, Bologna, Italy, Jan. 2005.

[HoJo85] R. A. Horn and C. R. Johnson. *Matrix Analysis*. Cambridge University Press, Cambridge, UK, 1985.

[Honk96] M. Honkanen. Modelling of narrowband high power amplifier in radio communication system simulation. In *Proc. URSI/IEEE/IRC - XXI Convention on Radio Science*, Espoo, Finland, Oct. 1996.

[HsWa01] J. M. Hsu and C. L. Wang. A low-complexity iterative multiuser receiver for turbo-coded DS-CDMA systems. *IEEE J. Select. Areas Commun.*, 19(9):1775–1783, Sep. 2001.

[HuDa03] T. Hunziker and D. Dahlhaus. Optimal power adaption for OFDM systems with ideal bit-interleaved and hard-decision decoding. In *Proc. ICC 2003 - IEEE Int. Conf. Commun.*, Anchorage, AK, USA, May 2003.

[Hunz02] T. Hunziker. *Multicarrier Modulation Techniques for Bandwidth Efficient Fixed Wireless Access Systems*. Ph.D. thesis, Swiss Federal Inst. of Technology (ETH), Zurich, Switzerland, 2002.

[IEEE99a] IEEE. Wireless LAN Medium Access Control (MAC) and Physical Layer (PHY) Specifications. 1999.

[IEEE99b] IEEE. Wireless LAN Medium Access Control (MAC) and Physical Layer (PHY) Specifications: Further Higher-Speed Physical Layer Extension in the 2.4 GHz Band. 1999.

[IEEE99c] IEEE. Wireless LAN Medium Access Control (MAC) and Physical Layer (PHY) Specifications: High-speed Physical Layer in the 5 GHz Band. 1999.

[IEEE99d] IEEE. Wireless LAN Medium Access Control (MAC) and Physical Layer (PHY) Specifications: Higher-Speed Physical Layer Extension in the 2.4 GHz Band. 1999.

[IHRF03] R. Irmer, R. Habendorf, W. Rave, and G. Fettweis. Nonlinear multiuser transmission using multiple antennas for TD-CDMA. In *Proc. WPMC 2003 - Wireless Pers. Multimedia Commun.*, Yokosuka, Japan, Oct. 2003.

[Jaff01] E. Jaffrot. Iteractive techniques for channel estimation and equalization. TD(01)005, COST 273, Brussels, Belgium, May 2001.

[Jaff02] E. Jaffrot. Turbo channel estimation for OFDM systems on highly time and frequency selective channels. TD(02)25, COST 273, Espoo, Finland, May 2002.

[JaKa02a] T. Javornik and G. Kandus. A 2MSK receiver based on the regeneration of the larger MSK signal component. *Electrotechnical Review*, 69(1):34–39, Apr. 2002. [Also available as TD(02)023 and TD(02)015].

[JaKa02b] T. Javornik and G. Kandus. Solid state power amplifier impact on the satellite systems performance. In *Proc. EMPS 2002 - 5th European Workshop on Mobile/Personal Satcoms*, Baveno, Italy, Sep. 2002. [Also available as TD(02)134].

[JaKP04] T. Javornik, G. Kandus, and S. Plevel. Dynamic channel mapping strategies in adaptive MIMO systems. In *Proc. SOftCOM 2004 - Intl. Conf. on Software, Telecommunications and Computer Networks*, Dubrovnik, Croatia, Venice, Italy, Oct. 2004. [Also available as TD(03)033].

[JaPK04] T. Javornik, S. Plevel, and G. Kandus. A recursive link adaptation algorithm for MIMO systems. In *Proc. MELECON 2004 - 12th IEEE Mediterranean Electrotechnical Conference*, Dubrovnik, Croatia, May 2004. [Also available as TD(04)112].

[Jetl05] O. Jetlund. *Adaptive Coded Modulation: Design and Simulation with Realistic Channel State Information*. PhD thesis, Norwegain University of Science and Technology, Trondheim, Norway, Apr. 2005. [Also available as TD(04)034, TD(03)040, TD(03)127 and TD(02)108].

[JØHH03] O. Jetlund, G. E. Øien, B. Holter, and K. J. Hole. Adaptive gallager coded modulation scheme on rayleigh fading channels: Comparison of simulated and theoretical performance. In *Proc. NORSIG 2003 - 5th IEEE Nordic Signal Processing Conf.*, Bergen, Norway, Oct. 2003. [Also available as TD(02)108 and TD(02)127].

[KaMa03] K. Kansanen and T. Matsumoto. Turbo equalisation of multilevel coded QAM. In *Proc. SPAWC 2003 - Sig. Proc. Advances in Wireless Commun.*, Rome, Italy, June 2003. [Also available as TD(03)091].

[KaMa04] K. Kansanen and T. Matsumoto. Frequency-domain MIMO turbo equalization. TD(04)143, COST 273, Gothenburg, Sweden, June 2004.

[KaVM05] Y. Kai, N. Veselinovic, and T. Matsumoto. Space-time weighted nonbinary repeat-accumulate codes with turbo equalization for frequency-selective MIMO channels. In *Proc. VTC 2005 Spring - IEEE 61st Vehicular Technology Conf.*, Stockholm, Sweden, May 2005. [Also available as TD(04)155].

[KMST00] J. Khun-Jush, G. Malmgren, P. Schramm, and J. Torsner. HIPERLAN type 2 for Broadband Wireless Communication. *Ericsson Review*, pages 108–119, 2000.

[KoAg00] D. Koulakiotis and A. H. Aghvami. Data detection for DS/CDMA mobile systems: a review. *IEEE Personal Commun. Mag.*, 7(3):24–34, June 2000.

[KoBC01] M. Kobayashi, J. Boutros, and G. Caire. Iterative soft-SIC joint decoding and parameter estimation. In *Proc. 7th Int. Workshop on Digital Signal Proc. Techn. for Space Commun.*, Tirrenia, Italy, Oct. 2001.

[KoMa00] F. Kowalewski and P. Mangold. Joint predistortion and transmit diversity. In *Proc. Globecom 2000 - IEEE Global Telecommunications Conf.*, San Francisco, CA, USA, Nov. 2000.

[KSMTar] K. Kansanen, C. Schneider, T. Matsumoto, and R. Thomä. Multilevel coded QAM with MIMO turbo-equalization in broadband single-carrier signalling. *IEEE Trans. Veh. Technol.*, pages 954–966, to appear. [Also available as TD(04)028].

[KSWW99] J. Khun-Jush, P. Schramm, U. Wachsmann, and F. Wenger. Structure and performance of the HIPERLAN/2 physical layer. In *Proc. VTC 1996 Fall - IEEE 50th Vehicular Technology Conf.*, Amsterdam, The Netherlands, Sep. 1999.

[LaRZ02] M. Lampe, H. Rohling, and W. Zirwas. Misunderstandings about link adaptation for frequency selective fading channels. In *Proc. PIMRC 2002 - IEEE 13th Int. Symp. on Pers., Indoor and Mobile Radio Commun.*, Lisbon, Portugal, Sep. 2002.

[LGRZ03] M. Lampe, T. Giebel, H. Rohling, and W. Zirwas. PER-prediction for PHY mode selection in OFDM communication systems. In *Proc. Globecom 2003 - IEEE Global Telecommunications Conf.*, San Francisco, CA, USA, Dec. 2003.

[LiMT00] Z. Lin, G. Malmgren, and J. Torsner. System performance analysis of link adaptation in HiperLAN type 2. In *Proc. VTC 2000 Fall - IEEE 52th Vehicular Technology Conf.*, Boston, MA , USA, Sep. 2000.

[MaMT03] N. Maretti, S. Mistrello, and V. Tralli. Resource allocation techniques for wireless packet networks based on V-BLAST architecture. In *Proc. WPMC 2003 - Wireless Pers. Multimedia Commun.*, Yokosuka, Japan, Oct. 2003. [Also available as TD(03)159].

[MaVJ04] T. Matsumoto, N. Veselinovic, and M. Juntti. A pdf estimation-based iterative MIMO signal detection with unknown interference. *IEEE Commun. Lett.*, 7(8):422–424, July 2004. [Also available as TD(03)150].

[MaVU01] S. Marinkovic, B. Vucetic, and A. Ushirokawa. Space-time iterative and multistage receiver structure for CDMA mobile communication systems. *IEEE J. Select. Areas Commun.*, 19(8):1594–1604, Aug. 2001.

[MBRF03] S. Morosi, A. Bernacchioni, E. Del Re, and R. Fantacci. Improved iterative parallel interference cancellation receiver for DS-CDMA 3G systems. TD(03)025, COST 273, Barcelona, Spain, Jan. 2003.

[MBWL00] M. Meurer, P. W. Baier, T. Weber, Y. Lu, and A. Papathanassiou. Joint transmission: advantageous downlink concept for CDMA mobile radio systems using time division duplexing. *Elect. Lett.*, 36:900–901, 2000.

[MDHF03] C. Mutti, D. Dahlhaus, T. Hunziker, and M. Foresti. Bit and Power Loading Procedures for OFDM Systems with Bit-Interleaved Coded Modulation. In *Proc. ICT 2003 - 10th Int. Conf. on Telecommunications*, Papeete, French Polynesia, Feb. 2003.

[MeAS98] J. Medbo, H. Andersson, and P. Schramm. Channel models for Hiperlan/2 in different indoor scenarios. TD(98)070, COST 273, COST 259 report, 1998.

[MKSS04] T. Matsumoto, K. Kansanen, C. Schneider, and M. Säreströniemi. Core matrix inversion techniques for conditional MMSE problems. In *Proc. VTC 2004 Spring - IEEE 59th Vehicular Technology Conf.*, Los Angeles, CA, USA, May 2004. [Also available as TD(03)163].

[MoBi04] S. Morosi and T. Bianchi. Comparison between RAKE and frequency domain detectors in ultra-wideband indoor communications. In *Proc. PIMRC 2004 - IEEE 15th Int. Symp. on Pers., Indoor and Mobile Radio Commun.*, Barcelona, Spain, Sep. 2004. [Also available as TD(05)067].

[MoNS04] B. Moschini, M. Nicoli, and U. Spagnolini. A subspace method for soft estimation of block fading channels in turbo equalization. TD(04)090, COST 273, Duisburg, Germany, June 2004.

[MRFB03] S. Morosi, E. Del Re, R. Fantacci, and A. Bernacchioni. Improved iterative parallel interference cancellation for wireless DS-CDMA communication systems. In *Proc. 2nd Workshop COST 273*, Paris, France, May 2003.

[MRFC05] S. Morosi, E. Del Re, R. Fantacci, and A. Chiassai. Design of turbo-MUD receivers with density evolution in overloaded CDMA systems. TD(05)068, COST 273, Bologna, Italy, Jan. 2005.

[MRST03] M. Magarini, L. Reggiani, A. Spalvieri, and G. Tartara. The benefits of the MMSEDFE feedforward filter in reduced-complexity turbo equalization. In *Proc. ICT 2003 - 10th Int. Conf. on Telecommunications*, Papeete, French Polynesia, Feb. 2003. [Also available as TD(02)161].

[MuDa04] C. Mutti and D. Dahlhaus. Adaptive Power Loading for Multiple-Input Multuple-Output OFDM Systems with Perfect Channel State Information. In *Proc. Joint COST 273/284 Workshop on Antennas and Related System Aspects in Wireless Communications*, Gothenburg, Sweden, June 2004.

[MuDD04] C. Mutti, D. Dahlhaus, and D. Destefanis. Adaptive Coding Based on LDPC Codes for OFDM Systems with HD Decoding. In *Proc. 13th IST Summit on Mobile and Wireless Commun.*, Lyon, France, June 2004.

[MuDH03] C. Mutti, D. Dahlhaus, and T. Hunziker. Adaptive Procedures for OFDM Systems with Bit-Interleaved Coded Modulation. In *Proc. 2nd Workshop COST 273*, Paris, France, May 2003.

[NHDL03] N. Noels, C. Herzet, A. Dejonghe, V. Lottici, H. Steendam, M. Moeneclaey, M. Luise, and L. Vandendorpe. Turbo synchronization: an EM algorithm interpretation. In *Proc. ICC 2003 - IEEE Int. Conf. Commun.*, Ottawa, Canada, June 2003.

[NiSS03] M. Nicoli, O. Simeone, and U. Spagnolini. Multi-slot estimation of fast-varying space-time communication channels. *IEEE Trans. Signal Processing,* 51(5):1184–1195, May 2003.

[ØiHH04] G. E. Øien, H. Holm, and K. J. Hole. Impact of imperfect channel prediction on adaptive coded modulation performance. *IEEE Trans. Veh. Technol.*, 53(3):758–769, May 2004. [Also available as TD(02)054].

[OsNA05] F. S. Ostuni, M. R. Nakhai, and H. M. Aghvami. Iterative MMSE receivers for space-time trellis-coded CDMA systems in multipath fading channels. *IEEE Trans. Veh. Technol.*, 54(1):163–176, Jan. 2005. [Also available as TD(04)025].

[PaCV02] G. Pasolini, M. Chiani, and R. Verdone. Performance Evaluation of a Bluetooth Based WLAN Adopting a Polling Protocol Under Realistic Channel Conditions. *Int. J. of Wireless Information Networks on Mobile Ad Hoc Networks (MANETs): Standards, Research, Applications*, pages 141–153, Apr. 2002.

[PaHu01] C. B. Papadias and H. Huang. Linear space-time multiuser detection for multipath CDMA channels. *IEEE J. Select. Areas Commun.*, 19(2):254–265, Feb. 2001.

[PaTV03] G. Pasolini, M. De Troia, and R. Verdone. Throughput Evaluation for a Bluetooth piconet with Link Adaptation. In *Proc. PIMRC 2003 - IEEE 14th Int. Symp. on Pers., Indoor and Mobile Radio Commun.*, Bejing, China, Sep. 2003. [Also available as TD(02)127].

[PeHS03] C. B. Peel, B. M. Hochwald, and A. L. Swindlehurst. A vector-perturbation technique for near-capacity multi-antenna multi-user communication. In *Proc. 41th Annual Allerton Conference on Communications, Control and Computing*, Monticello, IL, USA, Oct. 2003.

[Piet02] C. Pietsch. On capacity and linear processing for multiuser MIMO systems. TD(02)085, COST 273, Espoo, Finland, May 2002.

[PiLi04] C. Pietsch and J. Lindner. Real-valued modeling and channel estimation for transmissions based on orthogonal STBCs. In *Proc. ISSSTA 2004 - IEEE 12th Int. Symp. on Spread Spectrum Techniques and Applications*, Sydney, Australia, Aug. 2004.

[PlJK05] S. Plevel, T. Javornik, and G. Kandus. A recursive link adaptation algorithm for MIMO systems. *AEUE: Archiv für Elektronik und Übertragungstechnik - Int. J. of Electron. and Com.*, 59(1):52–55, Mar. to be published 2005. [Also available as TD(03)113].

[PNTL02] C. Pietsch, M. Nold, W. G. Teich, and J. Lindner. Optimum space-time processing for wide-band transmissions with multiple receiving antennas. In *Proc. ITG - 4th Int. Conf. on Source and Channel Coding*, Berlin, Germany, Jan. 2002. [Also available as TD(01)023].

[Pomm02] C. Pommer. Convergence of a semi-blind Least-Squares-Algorithm for the UMTS FDD uplink with adaptive antennas. TD(02)072, COST 273, Helsinki, Finland, May 2002.

[Proa95] J. G. Proakis. *Digital Communications.* McGraw-Hill, New York, NY, USA, 3rd edition, 1995.

[PSTL03a] C. Pietsch, S. Sand,W. G. Teich, and J. Lindner. Modeling and performance evaluation of multiuser MIMO systems using real-valued matrices. *IEEE J. Select. Areas Commun.*, 21(5):744–753, June 2003. [Also available as TD(02)125].

[PSTL03b] C. Pietsch, S. Sand,W. G. Teich, and J. Lindner. Modeling and performance evaluation of multiuser MIMO systems using real-valued matrices. *IEEE J. Select. Areas Commun.*, 21(5):744–753, June 2003. [Also available as TD(02)125].

[PTVF92] W. H. Press, S. A. Teukolsky, W. T. Vetterling, and B. P. Flannery. *Numerical Recipes in C.* Cambridge University Press, Cambridge, UK, 2nd edition, 1992.

[QCJS03] D. Qiao, S. Choi, A. Jain, and K. G. Shin. MiSer: an optimal low-energy transmission strategy for IEEE 802.11a/h. In *Proc. MobiCom 2003 - Int. Conf. on Mobile Computing and Networking*, San Diego, CA, USA, Sep. 2003.

[QiCS02] D. Qiao, S Choi, and K. G. Shin. Goodput Analysis and Link Adaptation for IEEE 802.11a Wireless LANs. *IEEE J. Mobile Computing*, 1(4):278–292, Oct.-Dec. 2002.

[QiTe00] Z. Qin and K. C. Teh. Iterative multiuser detection with gauss-seidel soft detector as first stage for asynchronous coded CDMA. *Elect. Lett.*, 36(23):1939–1940, Nov. 2000.

[ReCF03] A. Renoult, M. Chenu-Tournier, and I. Fijalkow. Multi-user detection for OFDM transmission in presence of frequency impairments: channel estimation and performance. WP(03)008, COST 273, Prague, Czech Republic, Sep. 2003.

[RFMM01] E. Del Re, R. Fantacci, D. Marabissi, and S. Morosi. Low complexity selective interference cancellator for a WCDMA communication system with antenna array. TD(01)051, COST 273, Bologna, Italy, Oct. 2001.

[RiSU01] T. J. Richardson, M. A. Shokrollahi, and R. L. Urbanke. Design of capacity-approaching irregular low-density parity-check codes. *IEEE Trans. Inform. Theory*, 47:619–637, Feb. 2001.

[RoVH95] P. Robertson, E. Villebrun, and P. Hoeher. A comparison of optimal and suboptimal MAP decoding algorithms operating in the log domain. In *Proc. ICC 1995 - IEEE Int. Conf. Commun.*, Seattle, WA, USA, June 1995.

[RoWi04] J. Romme and K. Witrisal. On transmitted-reference UWB systems using discrete-time weighted autocorrelation. In *Proc. VTC 2005 Spring - IEEE 61st Vehicular Technology Conf.*, Stockholm, Sweden, May 2004. [Also available as TD(04)153].

[RSAA98] M. C. Reed, C. B. Schlegel, P. D. Alexander, and J. A. Asenstorfer. Iterative multiuser detection for CDMA with FEC: Near single-user performance. *IEEE Trans. Commun.*, 46:1693–1699, Dec. 1998.

[SaMo03] T. Salzer and D. Mottier. Downlink strategies using antenna arrays for interference mitigation in multi-carrier CDMA. In *Proc. MC-SS 2003 - 4th Workshop on Multi-Carrier Spread Spectrum and Related Topics*, Oberpfaffenhofen, Germany, Sep. 2003. [Also available as TD(02)137].

[SaTA04] K. Sakaguchi, S. H. Ting, and K. Araki. Initial measurement on MIMO eigenmode communication system. *IEICE Trans. Commun.*, J87-B(9):1454–1466, Sep. 2004. [Also available as TD(04)027].

[ScRe03] D. Schreurs and K. Remley. Use of multisine signals for efficient behavioural modelling of RF circuits with short-memory effects. In *Proc. ARFTG 2003 - 61st Automatic RF Techniques Group Conference*, Philadelphia, PA, USA, June 2003. [Also available as TD(03)105].

[ShBu02] J. Shen and A. J. Burr. A turbo multiuser receiver for receive diversity CDMA systems over flat rayleigh fading channel. In *Proc. PIMRC 2002 - IEEE 13th Int. Symp. on Pers., Indoor and Mobile Radio Commun.*, Lisbon, Portugal, Sep. 2002. [Also available as TD(02)073].

[SiBa01] S. Simoens and D. Bartolome. Optimum Performance of Link Adaptation in HIPERLAN/2 Networks. In *Proc. VTC 2001 Spring - IEEE 53rd Vehicular Technology Conf.*, Rhodes, Greece, May 2001.

[STTM03] C. Schneider, R. Thomä, U. Trautwein, and T. Matsumoto. The dependency of turbo MIMO equalizer performance on the spatial and temporal multipath channel structure - a measurement based evaluation. In *Proc. VTC 2003 Spring - IEEE 57th Vehicular Technology Conf.*, Jeju, Korea, Apr. 2003. [Also available as TD(03)109].

[SWCH02] A. Sklavos, T. Weber, E. Costa, H. Hass, and E. Schulz. *Spread-Spectrum and Related Topics, Fazel, K. and Kaiser, S. (eds.)*. Kluwer Academic, Boston, MA, USA, 2002. [Also available as TD(01)020].

[SyBu04] J. Sykora and A. G. Burr. Iterative decoding networks with iteratively data eliminating SDD and EM based channel state estimator. In *Proc. PIMRC 2004 - IEEE 15th Int. Symp. on Pers., Indoor and Mobile Radio Commun.*, Barcelona, Spain, Sep. 2004. [Also available as TD(04)117].

[SyKn04] J. Sykora and M. Knize. Linear diversity precoding design criterion for blockfading delay limited MIMO channel. In *Proc. Globecom 2004 - IEEE Global Telecommunications Conf.*, Dallas, TX, USA, Dec. 2004. [Also available as TD(04)054].

[Syko01] J. Sykora. Constant envelope space-time modulation trellis code design for Rayleigh flat fading channel. In *Proc. Globecom 2001 - IEEE Global Telecommunications Conf.*, San Antonio, TX, USA, Nov. 2001. [Also available as TD(02)016].

[Syko02] J. Sykora. Self-noise in MIMO space-time coded systems with imperfect symbol timing. In *Proc. PIMRC 2002 - IEEE 13th Int. Symp. on Pers., Indoor and Mobile Radio Commun.*, Lisbon, Portugal, Sep. 2002.

[Syko03] J. Sykora. Spatial inter-branch interference equalization in MIMO frequency flat fading channel with mutually unequal path delays. TD(03)064, COST 273, Barcelona, Spain, Jan. 2003.

[Syko05] J. Sykora. Multicomponent phase discriminator for multichannel CPM modulation in MIMO channel based on nonlinear geometric approach. TD(05)057, COST 273, Bologna, Italy, Jan. 2005.

[TaMB01] A. Tarable, G. Montorsi, and S. Benedetto. A linear front end for iterative soft interference cancellation and decoding in coded CDMA. In *Proc. ICC 2001 - IEEE Int. Conf. Commun.*, Helsinki, Finland, June 2001.

[TaSC98] V. Tarokh, N. Seshari, and R. Calderbank. Space-time codes for high data rate wireless communication: Performance criterion and code construction. *IEEE Trans. Inform. Theory*, 44(2):744–765, Mar. 1998.

[Tela99] E. Telatar. Capacity of multi-antenna gaussian channels. *European Transactions on Telecommunications*, 10(6):585–596, Nov. 1999.

[TeRe01] W.G. Teich and M. Reinhardt. Multiuser/multisubchannel detection based on a recurrent neural network structure for the mobile communication system TDSCDMA. In *Proc. ISCTA 2001 - 6th Int. Symp. on Communication Theory and Applications*, Ambleside, UK, July 2001. [Also available as TD(02)069].

[Toml71] M. Tomlinson. New automatic equalizer employing modulo arithmetic. *Elect. Lett.*, 7:138–139, 1971.

[TQMJ02] H. Troger, W. Qiu, M. Meurer, and C. A. Jotten. A channel oriented joint transmission scheme for MIMO multi-user downlinks. In *Proc. 2nd Workshop COST 273*, Espoo, Finland, May 2002.

[TrMa04] U. Trautwein and T. Matsumoto. Turbo MIMO equalization for real-valued modulation signals. TD(04)191, COST 273, Duisburg, Germany, Sep. 2004.

[Vand02] L. Vandendorpe. Low-complexity fractional turbo receiver for space-time BICM over frequency-selective MIMO fading channels. TD(02)117, COST 273, Lisbon, Portugal, Sep. 2002.

[VcSy04] J. Vcelak and J. Sykora. Analytical error performance analysis for reduced complexity detection of general trellis code with parametric uncertainty. TD(04)132, COST 273, Gothenburg, Sweden, June 2004.

[VeMa03] N. Veselinovic and T. Matsumoto. Iterative signal detection in frequency selective MIMO channels with unknown co-channel interference. WS(03)006, COST 273, Paris, France, May 2003.

[VeSM04] N. Veselinovic, C. Schneider, and T. Matsumoto. Interference suppression and joint detection for reduction of sensitivity to timing offset and spatial correlation in space-time coded MIMO turbo equalization. WS(04)008, COST 273, Bologna, Italy, Jan. 2004.

[WaPo99] X. Wang and H. V. Poor. Iterative (turbo) soft interference cancellation and decoding for coded CDMA. *IEEE Trans. Commun.*, 47(7):1046–1061, July 1999.

[WeMe03] T. Weber and M. Meurer. Optimum joint transmission: Potentials and dualities. In *Proc. WPMC 2003 - Wireless Pers. Multimedia Commun.*, Yokosuka, Japan, Sep. 2003. [Also available as TD(03)008].

[WeMe04a] T. Weber and M. Meurer. Imperfect channel state information in MIMO-transmission. In *Proc. VTC 2004 Spring - IEEE 59th Vehicular Technology Conf.*, Milan, Italy, Sep. 2004.

[WeMe04b] T. Weber and M. Meurer. Low complexity energy efficient joint transmission for OFDM multiuser downlinks. In *Proc. PIMRC 2004 - IEEE 15th Int. Symp. on Pers., Indoor and Mobile Radio Commun.*, Barcelona, Spain, Sep. 2004. [Also available as TD(04)008].

[WeMZ04] T. Weber, M. Meurer, and W. Zirwas. Low complexity energy efficient joint transmission for OFDM multiuser downlinks. In *Proc. PIMRC 2004 - IEEE 15th Int. Symp. on Pers., Indoor and Mobile Radio Commun.*, Barcelona, Spain, Oct. 2004.

[WhBJ03] G. P. White, A. G. Burr, and T. Javornik. Modelling of nonlinear distortion in broadband fixed wireless access systems. *Elect. Lett.*, 39(8):686–687, Apr. 2003. [Also available as TD(02)134].

[WiJo03] T. J.Willink and Y. L. C. de Jong. Iterative trellis search detection for asynchronous MIMO systems. In *Proc. VTC 2003 Fall - IEEE 58th Vehicular Technology Conf.*, Orlando, FL, USA, Oct. 2003. [Also available as TD(04)114].

[WKTM05] R. Wohlgenannt, K. Kansanen, D. Tujkovic, and T. Matsumoto. Outage-based LDPC code design for SC/MMSE turbo equalization. In *Proc. VTC 2005 Spring - IEEE 61st Vehicular Technology Conf.*, Stockholm, Sweden, May 2005. [Also available as TD(04)209].

[WSLW03] T.Weber, A. Sklavos, Y. Liu, and M.Weckerle. The air interface concept JOINT for beyond 3G mobile radio networks. In *Proc. WIRELESS 2003 - Proc. 15th Int. Conf. on Wireless Commun.*, Calgary, Canada, July 2003. [Also available as TD(04)084].

[WuHYJ01] J.-L. C. Wu, L. Hung-Huan, and L. Yi-Jen. An Adaptive Multirate IEEE 802.11 Wireless LAN. In *Proc. ICOIN 2001 - 15th Int. Conf. on Information Networking*, Warsaw, Poland, June 2001.

[WuWa01] K. M. Wu and C. L. Wang. An iterative multiuser receiver using partial parallel interference cancellation for turbo-coded DS-CDMA systems. In *Proc. Globecom 2001 - IEEE Global Telecommunications Conf.*, S.Antonio, TX, USA, Nov. 2001.

[YeBC02] S. Ye, S. Blum, and L. J. Cimini Jr. Adaptive modulation for variable-rate OFDM systems with imperfect channel information. In *Proc. VTC 2002 Spring - IEEE 55th Vehicular Technology Conf.*, Birmingham, AL, USA, May 2002.

[ZhBu01] Li Zhang and Alister Burr. Phase estimation with the aid of soft output from Turbo decoding. In *Proc. VTC 2001 Fall - IEEE 54th Vehicular Technology Conf.*, Atlantic City, NJ, USA, Oct. 2001.

4

Propagation modelling and channel characterisation

Chapter editor: *Pierre Degauque*
Contributors: *Emmanuel Van Lil, Vittorio Degli-Esposti, Pertti Vainikainen, Filipe Cardoso, Wim A. Th. Kotterman, and Luís M. Correia*

4.1 Introduction

The development and application of new signal processing algorithms and, more generally, the optimisation of mobile communication systems are strongly related to the statistical properties of the propagation channel. New coding techniques, such as space-time coding algorithms, have been developed as a result of the fast growth in indoor communications coupled with a never-ending need for higher and higher data transmission rates. At the time that the previous COST 273 Action ended, the existing knowledge of channel characteristics was essentially based on data pertaining to path loss, delay spread and coherence bandwidth. However, because this knowledge proved to be incomplete for new systems, intensive research has been conducted concerning the deterministic propagation modelling and the estimation of channel parameters based on actual measurements.

The approach used for electromagnetic propagation modelling depends on the zone under study. For instance, if only the immediate vicinity of the antenna is to be considered, an integral method can be employed, while the coverage area of a transmitting antenna is usually determined via a ray method. Section 4.2 describes several recently proposed improvements to various existing techniques: integral methods, differential equation methods, parabolic equations, diffraction theory and Gaussian beams. If, as in the example mentioned above, the ray theory is applied to determine the coverage area of a BS, a large database is needed in order to accurately model the environment's electromagnetic and geometric characteristics. Section 4.3 discusses such input database issues, with an emphasis on the complexity that is often cited as the principal obstacle to the widespread adoption of deterministic propagation models. Since decreasing the computation time of ray models is also an important issue, techniques for speeding up ray launching models have also been developed. Simplified or hybrid models, which introduce statistical elements into the deterministic approach, have also been proposed, although they may produce less precise results.

Examples of applications are used throughout the chapter to highlight the advantages and drawbacks of the different methods. Given an assumption of an ideal and complete description of the environment, the hypothesis of smooth surfaces is often implied. However, in many cases, and especially at high frequencies, the roughness or the size of irregularities of a wall surface is on the same order of magnitude as the wavelength, and thus diffuse scattering must be taken into account. This scattering, which plays a fundamental role in time and angle dispersion, is extremely difficult to model. Thus, part of Subsection 4.3.5 deals with recent investigations into this topic, in an effort to incorporate the scattering phenomena into the propagation modelling tools.

Modelling as well as experiments providing complementary information about the channel, channel measurement techniques and parameter estimation are detailed in Section 4.4. Since the development of MIMO systems implies obtaining complete information about the field characteristics in the vicinity of both the receiving and transmitting antennas, double-directional channel sounding is usually needed. New channel sounder architectures, including antennas, have been proposed to provide precise angular information on the rays' Direction of Departure (DoD) and DoA and to ensure a better compromise between measurement accuracy, acquisition time and the number of array elements. To increase the accuracy of the channel transfer matrix's spatial and temporal structure, high-resolution parameter estimation techniques must be used. Improvements in the dedicated algorithms (Multiple Signal Classification (MUSIC), ESPRIT and Space-Alternating Generalised Expectation (SAGE)) are discussed in Section 4.4 and examples of applications are presented. Measurements are then processed to extract the channel's statistical properties.

Section 4.5 introduces improved empirical path-loss models that are valid for rural and suburban environments as well as office buildings. It is usually assumed that long-term fading is due to a shadowing process and different approaches have been studied in an attempt to improve the well-known model based on randomly distributed Gaussian variables. The rationale behind another well-established assumption – that long- and short-term fading should be modelled separately – is called into question, and a new model based on multiple scattering is described. Lastly, extending the models to the wideband case, which is necessary when the relative delays of arriving waves are large compared to the symbol duration, is also discussed. Clearly, fading statistics are quite important for optimising the communication scheme and for predicting the performance of the link. Still, because numerous advantages would result if the channel fading behaviour could be accurately forecast, the potential of prediction techniques, algorithms and applications has been analysed. Extensive measurement campaigns have been conducted over the last few years in order to obtain statistical information on temporal and angular dispersion, including the characteristics of clusters and of diffusely scattered power, for a variety of propagation scenarios or environments. The main results and conclusions are outlined in Section 4.5.

4.2 EM theory and diffuse scattering

This section deals with the progress made in the field of the electromagnetic behaviour of communication systems or parts thereof such as antennas, especially taking into account obstacles: terrain (including forests), (complex) buildings and moving objects. Essentially three kinds of method will be discussed:

1. integral methods (moment methods, physical optics and modal expansion);
2. differential equation methods (Finite Difference Time Domain (FDTD) and parabolic equation);
3. ray methods (tracing/launching and Gaussian beams).

4.2.1 Integral methods

This subsection deals with the most accurate methods available in electromagnetics to describe waves. Unfortunately, some of those methods require very large amounts of computer resources (both memory and processing time), when applied to scenarios that are large compared to the wavelength, so that they are not yet within the reach of the telecommunication engineer. The work within this COST Action has tried to reduce this gap.

Full moment methods

Of course those methods have been applied to the computation of structures, which are limited in size, such as (elements of smart) antennas [VeLC02b], [TrSi02], [TrSi03] and [NdHH04]. Here the emphasis was placed more on the optimal design of specific antennas, with as the main goal the optimisation of the discrimination of the waves impinging on those antennas by numerical methods described mostly in Section 4.4. Paper [VeLC02b] describes in detail the design of 3D array antennas, which allow to pick up signals and discriminates between those signal 'directions' in an unambiguous way, which is not possible with merely planar arrays. This spherical array composed of 12 capacitively compensated monopoles has been designed with a Galerkin piecewise sinusoidal basis function wire-grid program (Numerical Electromagnetics Code (NEC2)) making use of delta-gap excitations. Therefore, the pentagonal parts of the dodecaeder have been approximated by two concentric pentagons and radially extending wires. The emphasis was on the minimisation of the effects of the feedlines on both network parameters and radiation patterns as well as the coupling between the elements. The realised dodecahedral antenna, Figure 4.1, has coupling values below −23 dB, while the simulations predict −22 dB. The matching, however, is not perfectly predicted (−18 dB in the 4900–5100 MHz band, while the program predicts a reflection loss of less than −23 dB).

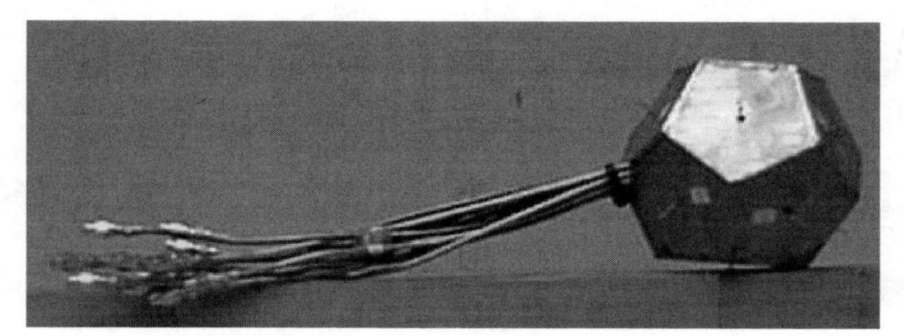

Figure 4.1 *Practical implementation of the dodecahedral antenna.*

Papers [TrSi02] and [TrSi03] use another commercial MoM solver (WIPL) to compute both coupling and impedances of planar circular arrays of monopoles designed around 5 GHz (4.74 GHz), with two, three or four elements. Here the main purpose was to compute the mean capacity of the MIMO channel. The influence of the coupling is described in Section 5.3. Paper [NdHH04] develops an own MoM Galerkin solver and compares the results of a 40 piecewise linear (triangular) basis functions for a half wavelength dipole with the classical sinusoidal distribution. The final goal is the same as in the previous two papers, namely determining the capacity of a (2×2) MIMO system.

Moment methods have not only been applied to antennas, which are of limited spatial extent (a few wavelengths in size), but also to the computation of propagation effects, which usually deal with objects that are of significant spatial extent with respect to the wavelength (in fact a few 1000 wavelengths). A lot of work has been devoted in this COST Action to 'correcting' 2D methods to allow computations of 3D structures with nearly the same amount of computational effort as for the 2D case, or even to reduce the effort for the 2D case. The paper dealing with reduction of computational efforts while using full moment methods [KeCu03] is using a Wiener-Hopf technique to find analytical expressions for obtaining the look-up tables for the Combined Field Integral Equation (CFIE). Those are combinations of relatively easy to compute Fresnel integrals. Only for nearly grazing incidence are there some differences between the full integral equation methods and the analytical formulas of the truncated half planes. For a practical case of a German terrain profile computed at 970 MHz, virtually no difference between both methods is seen, even with plate lengths of about 30λ. Another paper [BrCu01] is introducing an integro-differential equation including the gradients of the surface transverse to the direction of propagation. By approximating the phase on the transverse parts of the surface by a second order polynomial, the 2D Magnetic Field Integral Equation (MFIE) for Perfect Electric Conducting (PEC) bodies can be recast into a one-dimensional integral but involving relatively easy to compute Fresnel integrals. This method has been applied to the computation of both ideal theoretical cases like a smooth sloping wedge at 100 MHz and 1 GHz (for both co- and crosspolar fields), as well as to more complex data of real terrains in both Denmark and Ireland. For those cases measured data was available. The conclusions were that the copolar components did agree quite well, but that significant cross-polar components were present, which are justifying accurate modelling of polarisation diversity effects. This is illustrated in Figure 4.2.

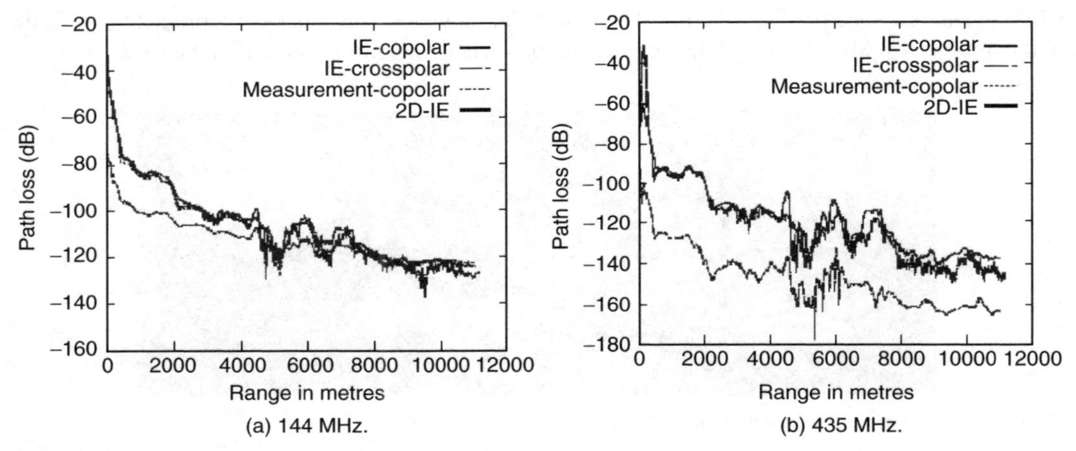

(a) 144 MHz.

(b) 435 MHz.

Figure 4.2 *Co- and crosspolar fields for Danish terrain.*

Physical optics models

Paper [DMSW03] is comparing the standard Kirchhoff formulation with a novel approach called stochastic scattering approach. Here the surface is subdivided into smooth parts with different orientations. Assuming a normally distributed randomness and a Gaussian autocorrelation coefficient, the absolute value of the gradient is Rayleigh distributed with a mean of $\frac{\sqrt{\pi}}{L}\sigma_h$, where L is the correlation distance and σ_h is the root mean square of the roughness. While the Kirchhoff formulation leads to scattering patterns that are more dependent for the coherent scattering part from the size of the plane (in this case an $8 \times 8\lambda^2$ plate is considered), the stochastic scattering approach does show mostly the scattering due to the roughness, which may be much larger than the classical stationary phase assumption in the Kirchhoff case. For this case, with a correlation distance of 4λ and a roughness of 0.35λ, the significant scattering areas are reduced significantly with respect to the Kirchhoff case, which shows a full diffraction pattern. All values are computed at a far-field distance of 2000λ from the plate.

This touches one of the problems of Physical Optics (PO) for large objects or close transmitters or receivers: the plane-wave approximation is not valid any more, leading to fields that continue to grow if the size of the scatterer is increased. Therefore, other solutions have to be found. Paper [LCDC02] is using a parabolic phase approximation for this purpose, which has been developed in a previous COST Action, namely COST Action 259 – Wireless Flexible Personal Communications (COST 259). This paper investigates many further refinements of this Fresnel method such as

1. the use of near-field formulas;
2. using internal reflections in a reflecting plate of finite thickness and material properties (permittivity and permeability);
3. optimal choice of the integration centre, which is critical for the efficient application of this method.

For this last case, seven different methods (starting with the trivial solution) have been investigated. It was shown that the optimally fitted method (optimally means that the parabolic phase behaviour is fitted to osculate the real spherical one as efficiently as possible, i.e. up to and including second degree terms), gave excellent results with only four times the number of FLOating Point operations per Second (FLOPS) of the trivial solution. Both simulations and measurements in an anechoic room on a flat plate also agree very well, both in amplitude, Figure 4.3, and in phase, Figure 4.4.

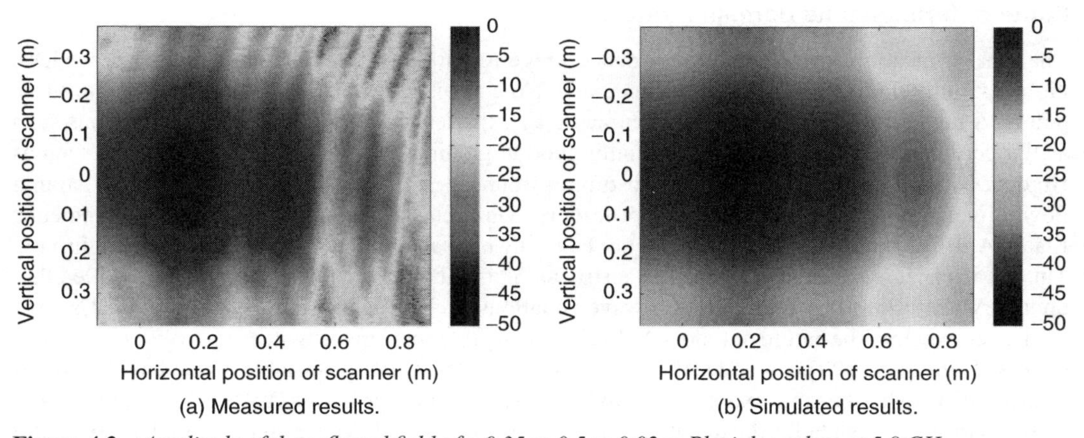

Figure 4.3 *Amplitude of the reflected field of a 0.35 × 0.5 × 0.02 m Plexiglass plate at 5.8 GHz.*

Modal expansion models

Since this method is specially suited for closed areas, one contribution [Kyri03] has applied this method for the evaluation of the K-factor of a hallway. Indeed, the ray-tracing methods are less appropriate to this kind of problem, since the number of reflections on the boundaries of the structure may become very high. The contribution of many different propagating modes is evaluated, taking into account the losses (both dielectric and conduction) with an equivalent reflection coefficient. The constant component estimation proved the most stable, better than the statistical method or the ESPRIT algorithm. The intuitively expected behaviour is confirmed, namely that the K-factor increases with increasing losses, and with decreasing size.

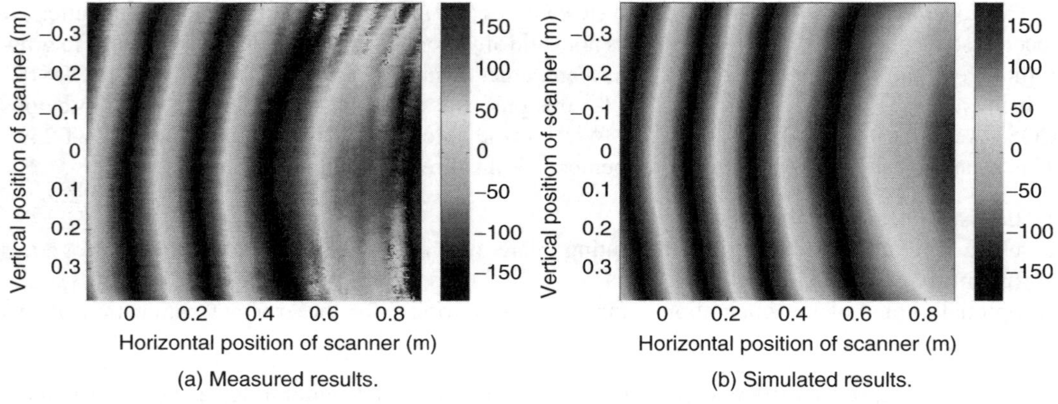

Figure 4.4 *Phase of the reflected field of a 0.35 × 0.5 × 0.02 m Plexiglass plate at 5.8 GHz.*

4.2.2 Differential equation methods

Finite difference time domain models

The ideal truncation method has a low data storage requirement, an easy construction procedure, high flexibility, high stability and high accuracy. No current truncation method satisfies all of these points. A method which shows better accuracy may experience lower stability. Some methods such as the perfectly matched layer are especially good at gaining accuracy and some methods such as Higdon, Mur 1st and 2nd, and Liao's Absorbing Boundary Condition (AbBC) are good at gaining flexibility and have simple construction procedures with relatively small data storage requirements. Liao's AbBC is interesting in that it does not have any media parameter such as relative permittivity. This means Liao's AbBC can be easily constructed in the inhomogeneous environment, while other boundary conditions based on one-way wave equations such as Mur 1st and Mur 2nd AbBCs have media parameters. The nature of Liao's AbBC has been previously investigated [RaAr95], [ChWa92], [PrSh97], and [Rama99] concludes that Liao's higher order AbBC is unstable and even a 2nd order Liao's AbBC exhibits some extent of instability. The higher order Liao's AbBC also requires more memory to store the data for the update of the electromagnetic field values at the boundary. [Cost03] therefore has studied Liao's second order boundary condition with an emphasis on obtaining a reduced memory requirement and improved stability. The results of this work can, however, be applied to higher order Liao's AbBCs. The detail of the numerical formulation is given in [Cost03]. This novel AbBC reduces the data storage requirement for the boundary calculation of the original AbBC by 33%. [WaCh95] pointed out that the high stability is realised when the poles of the artificial reflection coefficient are within the unit circle of the complex plane. Based on this, the possible instability caused by the proposed boundary condition is investigated, compared with the original 2nd order Liao AbBC. The study showed another merit of the proposed AbBC, that is, the proposed AbBC gains its stability with $\alpha \leq 0.5$ while the original AbBC has instability with any α. The reflection coefficient from the proposed AbBC in the time domain was calculated with the marginally stable damping parameter

$\alpha = 0.5$, compared with Mur 1st and 2nd AbBCs which are originated from one-way wave equation like Liao's AbBC. The artificial reflection coefficients from the proposed AbBC and Mur 1st AbBC were in a comparable order, while the Mur 2nd AbBC showed the best absorption among these three boundary conditions. On the other hand, the analysis of the frequency reflection coefficient proved that the proposed method was a better uniform absorber than Mur 2nd AbBC, which is another merit of the proposed AbBC.

In recent years, the Alternating Direction Implicit (ADI) method [Smit65] has been introduced to the FDTD [Nami99], [ZhCZ00]. The standard explicit FDTD [Yee66] has to satisfy the Courant-Friedrichs-Lewy (CFL) bound [Tafl95] to be stable. Cases requiring a small spatial discretisation lead to excessive time sampling and increase the total Central Processing Unit (CPU) time required for the FDTD. The maximum Δt possible in FDTD under the CFL condition is defined as Δt_{CFL}. On the other hand, ADI-FDTD is unconditionally stable when taking sampling times Δt greater than the limit, keeping the total elapsed time within a practical range. $\Delta t / \Delta t_{CFL}$ is called the Courant-Friedrichs-Lewy Number (CFLN) N_{CFL}. Thus ADI-FDTD is attractive when, for example, studying the influence of skin effects in conductive materials or signal reflections in lossy media for microwave imaging with UWB systems. These UWB systems typically have frequency dependent lossy media parameters but ADI-FDTD and FDTD have media parameters which are constant across frequencies. The adaptation of ADI-FDTD to such dispersive environments is required to accommodate UWB systems. Holloway *et al.* [SHBP03], [HMDA02] accounted for material dispersion in the magnetic field and mentioned the possibility of dealing with material dispersion in the electric field. However, [SHBP03] and [HMDA02] do not give an implementation of ADI-FDTD which deals with material with Frequency Dependent (FD) permittivity ϵ and ohmic losses. Chen *et al.*'s [YuCh03] concern focuses on conductive materials. Garcia *et al.* [GRBM03] deal with the Debye model utilising the electrical polarisation vector field. The COST 273 project [Cost04], [ThCo04], [CoTh04a], proposes an alternative technique using the electric flux density $\boldsymbol{D}$, similar to the work by Lazzi *et al.* [ScLa03], but with FD materials for UWB systems. This runs parallel to [GRBM03] and provides convergent evidence for the general approach to FD materials in ADI-FDTD class problems. The novelty of this work is in proposing an FD-ADI-FDTD which can deal with Debye dispersive materials with ohmic losses, along with the solutions to the problems on the $\boldsymbol{D}$ source excitation. [CoTh04a] detailed a method of dealing with the CFL stability condition in a lossy medium for UWB systems, utilising the frequency spectrum of the source excitation. The error in the lossy case was proved to equal the lossless case with an appropriate N_{CFL} adjustment. It is shown that FD-ADI-FDTD is applicable for problems which can accommodate error higher than 0.1 with $N_{CFL} \geq 6$, gaining a simulation time benefit over FD-FDTD. Although the merit of FD-ADI-FDTD is the absence of a limitation on the value of Δt for stability, which allows the reduction of elapsed time of FD-ADI-FDTD, compared with FD-FDTD, the accuracy reduces this merit by a certain amount because accuracy largely depends on the temporal discretisation and spatial discretisation. Therefore, a method to set an appropriate Δt in UWB systems is also inductively proposed [CoTh04b].

Parabolic equation methods

The Parabolic Equation (PE) method is based on a paraxial approximation to the wave equation. Assuming that the fields propagate in a preferred propagation direction, this is a full-wave method accounting simultaneously and accurately for all wave phenomena like diffraction, refraction and scattering. Provided the field is known on an initial plane and adequate boundary conditions on the scattering objects and at the outer boundaries of the integration domain are given, it is easy to proceed (march) in range in the propagation direction. Here, no heavy integrals and matrices are involved. This explains the success of this method for propagation purposes, specially if the equation is reduced to its 2D version. It is very accurate at angles within 15° of this propagation direction. Several contributions apply the 2D version of this method to ducting phenomena on microwave propagation. Indeed, ducting is present in 15% of the time all over the world, especially over humid environments like coastal zones, and hence being able to compute the propagation in those circumstances is important. In [SiMi03] it is shown that surface-based ducts affect the propagation in short distances (up to 3 km) by provoking the shift of the location of the interference minima. At those minima, the differences between the case with the standard troposphere and the ducting case may exceed 20 dB for a ducting layer of 50 m thickness. Paper [SiMi03] discusses the influence of the changes in the evaporation duct thickness on two links at 2.5 and 5.8 GHz. Examples of range dependent ducting (corresponding to refractivity profiles along a mixed land-sea path) are also reported. For this last case between 5 and 10 dB variation in the losses can be accounted for. In both above-mentioned papers a smooth perfectly conducting underlying surface is assumed. The influence of the terrain is introduced in [SiMi04] and [SiMi05]. Assessment of the combined effect of terrain and ducting on UMTS path losses at UMTS frequencies (2.1674 GHz) is reported in [SiMi04]. Ducting can help to increase the signal in valleys. This is seen when comparing the standard case, Figure 4.5, with the ducting case, Figure 4.6. Relying on the precision of the PE and its ability to provide quantitative assessment of the losses, in [SiMi05] the importance of possible inaccuracies in the terrain profile data for the Sofia region is discussed. The influence of the ground type (for vertical polarisation) is also discussed. Another kind of problem that is very well suited for solutions based on PE methods is the case of wave propagation over a forest edge. The results for horizontally polarised links at 1.3555 and 1.5995 GHz are compared in [WaHo03] with Geometric

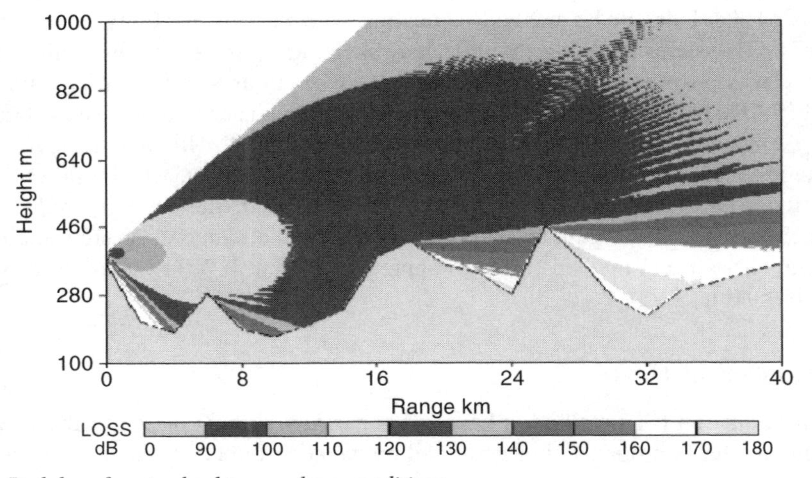

Figure 4.5 *Path loss for standard troposphere conditions.*

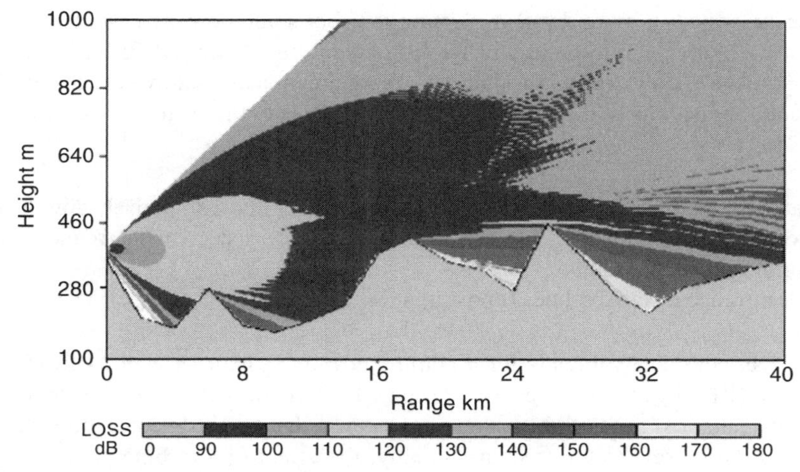

Figure 4.6 *Path loss under a 100 m ducting layer between 450 and 550 m.*

Theory of Diffraction (GTD) approximations, which are in general up to 20 dB worse than the measured cases for low receiver heights. The 2D PE solution with a Leontovitch boundary condition for the ground and an absorbing layer above the region of interest agrees in general within ±5 dB with the measured values.

4.2.3 Ray methods

Here two fundamentally different methods can be distinguished. One is using beams that are not localised in space, but that occupy the whole extent of space (and hence satisfy Maxwell's postulates). The Gaussian beams are one possible implementation. This will be discussed in the first subsection. The second uses the concept of rays, which are cuts of phenomena that do satisfy as a whole Maxwell's equations (plane, cylindrical or spherical waves). Some methods do trace the rays completely and compute the different reflection and diffraction points, which is a cumbersome geometric procedure. Some others do launch rays, and use a receiving sphere to determine the ones relevant to the actual receiver position.

Paper [TaLH02] uses a beam launching method to compute the fields in indoor environments. One first decomposes the source fields into a sum of beam field by the Gabor frame decompositions. This procedure is fully equivalent to the decomposition of antenna fields into plane or cylindrical waves. The expressions for reflection of a Gaussian beam on a plane boundary is then derived, and applied to computations of indoor channels at mm wave frequencies (57–59 GHz). Indeed at those frequencies, the diffractions are negligible and hence no diffraction is taken into account. Measurements in a small laboratory confirm the validity of the procedure, even if only the dominant path is predicted accurately.

Many papers are treating this subject and provide applications in the world of (mobile) communications. Some just use Uniform Theory of Diffraction (UTD)/PO for further investigations, which will be described in more detail in Section 4.3, but some do provide treatments of specific wave phenomena like diffraction and will be discussed here. Paper [GLBM02] uses geometrical optics to compute the effects of wind turbines on communication and safety systems (such as radars). Different effects, depending on the distance from the turbines to the system under investigation, are noticed:

1. a deep shadowing, but of limited geographic extent, turning into a smooth shadowing that reduces the range of the system. This effect decreases considerably with decreasing frequency and is hence only important for radar systems;
2. zones of occurrence of false echoes (ghost images) for radars.

In the same paper, the delay spreads for a communication system as well as Resource Description Framework (RDF) accuracy are also discussed. Both turned out to be negligible for the turbine parks investigated and for the system considered, but might become important if the turbines are placed too closely to the systems. Finally, the effect of a turbine park on a Differential Global Positioning System (DGPS) system (working at 312 kHz) is investigated with a moment method. It turned out that the transmitted wave towards the sea was even improved for this particular park, which was acting as a parasitic antenna at that low frequency. The three-dimensional extent of the areas, where the four different kinds of false echo can occur, were further quantified with spherical-wave approximations in [LiTC05]. Paper [HHBW01] uses a simple ray-optical prediction tool to evaluate the path loss in urban areas. This information is used for UMTS load planning. One of the few papers that deal with extensions of the classical UTD theory is paper [ElVa03]. Here, the theory for a lossy dielectric wedge is derived, by taking into account both the angle of incidence and diffraction into the multiplication factors. This improves significantly the diffracted fields for lossy dielectric wedges over just using the plane wave Fresnel reflection coefficients.

Another [TrLC04] deals with the implementation of curved surfaces in classical 3D ray tracing programs. The most difficult part is not so much the electromagnetic treatment of the diffraction, but to find a fast way to determine the reflection point of a wave on this curved surface, in this case a cylinder. An iteration process converging in all special cases within six steps within an accuracy of 0.1% of the distance transmitter-receiver. Paper [RoMJ04] deals with a solution for multiple edge diffraction with a spherical incident wave for antenna location above, level and below the knife edge plane. The results are in good agreement with the Xia-Bertoni results, while requiring much less computer time, and are accurate within at most 2.27 dB. A general program using a full 3D GTD theory, but tuned to MIMO analysis, is described in [DeHS03]. Here, the efficiency is increased by simplifying the geometry to a discrete number of single-bounce reflectors, called scatterers. Comparison with real measurements at 1.95 GHz and a bandwidth of 120 MHz yield good accuracy for the single-bounce echoes in a half-open environment, just by tuning the size of the scattering clusters, the number of scattering coefficients and the Rician K-factor.

Ray-tracing methods can also be used to determine the permittivity of building materials. Indeed, [JKVW05] is using simple amplitude-only reflection measurements at different incidence angles to fit the permittivity of flat walls. It is even possible to introduce diffuse radiation into the reflection model, giving a more realistic behaviour of the scattered field, by introducing a scattering coefficient, which increases as expected from a small value (0.05) for a nearly flat hangar wall, up to 0.4 for more irregular rural building walls [EFVG05].

4.3 Deterministic propagation modelling

4.3.1 Introduction

Deterministic propagation modelling is aimed at reproducing or studying the actual physical radio propagation process for a given environment. Such approach is particularly suitable for manmade environments, where radio waves interact with relatively simple, geometric obstacles such as buildings and streets. In this case, the geometric and electromagnetic characteristics of the environment and of the radio link can be described and stored in files (databases) and the corresponding propagating field can be computed through analytical formulas and/or computer programs. The degree of accuracy and of 'determinism' varies depending on the chosen approach. While some approaches can be properly considered 'deterministic' others may be defined 'hybrid' or 'semi-statistical', as long as they leave off some deterministic aspects and/or describe them through average, statistical parameters, thus decreasing complexity and computation time. The degree of determinism is also related to the accuracy of environment representation. For example, the adoption of a very accurate, full wave FDTD method in an environment described in terms of soil usage and building density would be a nonsense. Commercially available urban databases are not very accurate and detailed. Therefore, some propagation effects which are critically sensitive to database accuracy (e.g. fast fading, diffuse scattering, phase, polarisation, etc.) must be necessarily neglected or treated in a statistical way. Deterministic propagation models have advantages and drawbacks with respect to statistical models such as Hata-like models or statistical channel models. Deterministic models are physically meaningful, accurate and flexible. Statistical models on the other hand are simpler, faster and more synthetic because their output can describe an entire class of problems instead of a single one.

Generally, statistical channel models are used in the link level design phase while deterministic models are used in the planning phase of mobile radio systems. However, despite the use of deterministic field prediction may in principle lead to great advantages in terms of deployment cost reduction and service quality increase, the widespread adoption of deterministic models is still very limited due to the high cost and the low reliability of input databases and to the high complexity and computation time of the corresponding computer programs. For these reasons, most research within COST 273 addressed topics such as database handling [ScWi03], [CoLA03], [DeFA04], sensitivity to database accuracy [NaCB04] and CPU time reduction [HoWW03], [CoLA03], [DeFA04], [WWWW04], [BCFF02]. In principle, deterministic models such as ray tracing allow a multidimensional characterisation of propagation, i.e. may give information on the distribution in time and space (angle of departure/arrival) of the multipath field. This may be very valuable for the design and planning of present and future systems using array antennas and/or space-time coding techniques such as MIMO systems. However, due to input database inaccuracies and limitations in the modelling of diffuse scattering and of the multipath propagation process in general, multidimensional performances of today's models are still quite poor. Some authors have therefore worked on the improvement of deterministic modelling capabilities [DDGW03], [DESGK02], [FuMa03], [BHHT04a], [JePK05], [GhTI04], [Bert02]. Finally, deterministic models and their realistic propagation simulation capabilities have been applied to system assessments such as network planning issues, system performance assessments, positioning techniques, etc. [CoWS04], [WHZL02], [ZBLL04], [Fusc04]. Deterministic and semi-deterministic field prediction models and techniques, their performance, their pros and cons are described in this section. Applicative aspects and comparative overview of these models include such issues as database accuracy sensitivity, computation time, diffuse scattering modelling, and performance metrics.

Since reliability and cost of input databases represent a major concern, Subsection 4.3.2 is fully devoted to this topic. The various models are then treated starting from the more rigorous, complex (4.3.3) to the more simplified ones (4.3.7).

4.3.2 Input database issues

Deterministic propagation models require as input a proper description of the environment. In the case of urban propagation modelling, which is of interest here, both a geometrical and an electromagnetic description of each object (building, wall, terrain) must be provided and stored in the database. Outdoor databases can be extracted from aerial views using stereo-photogrammetry, from digital cadastre data or by digitising analogue maps. If the space domain is discretised into pixels, and for each pixel a geometrical or electromagnetic parameter is stored, then a *raster* database is obtained. Otherwise, if the environment is decomposed into basic geometrical elements (triangles, polygons, segments, etc.) then a *vectorial* database is obtained. Vectorial databases are usually more compact and accurate than raster databases and are therefore adopted by default in the present section, except where otherwise stated. In vectorial databases, buildings are usually represented as polygonal prisms with flat tops. In all cases, the intrinsic cost of urban databases is high, ranging from some hundred to some thousand euros per square kilometre. Moreover, the attainable accuracy is necessarily limited. The standard deviation of the error in urban databases is of the order of 0.5 m for horizontal coordinates and much higher for vertical coordinates. Vertical dimensions are often unavailable or particularly subject to errors, due to deviations in rooftops from flat polygons, with an evident impact on the reliability of *over-rooftop* propagation models. Moreover, environment cluttering data (vegetation, vehicles, street signs, etc.) are often unavailable or impractical to model. While reliability must be considered a very advisable quality, the 'level of detail' is not necessarily so. A great level of detail (sometimes confused with 'accuracy') means the presence of a great deal of building elements (internal yards, indentations, edges, roof structures, etc.) which, although relatively irrelevant, can still greatly increase the size of the database, with a more than proportional increase in the computation time of the prediction. Therefore, geometric databases must often be properly 'purged' or simplified before they can be input to deterministic propagation models.

In the indoor case, the description of floor and room partitions of the buildings in AutoCAD's native file format for CAD models (DWG) or Data Exchange Format (DXF) can often be extracted from the architectural design files of the building. In this case the accuracy of the database can be good. However, furniture is of course not included: the reliability of the field prediction with respect to measurements on a real, furnished building is therefore questionable no matter the accuracy of the adopted model. The electromagnetic description of both outdoor and indoor environments is also inaccurate because of the lack of related information. Compound materials are often treated adopting *effective* electromagnetic parameters, i.e. parameters representative of the overall behaviour of the compound in terms of reflection, transmission and diffraction of an incoming wave. A usual choice is to use literature or empirical values for the permittivity (ϵ_r) and the conductivity (σ) of each class of materials. For example, $\epsilon_r = 5$ and $\sigma = 10^{-2}$ are usually adopted for typical building walls in European cities.

Each environment type (indoor, dense urban, suburban, etc.) has appropriate database formats. Each database format is associated to appropriate propagation models. Therefore mixed environments pose a number of problems to both database handling and field prediction modelling. The different types of data must be harmonised by means of proper conversion, interpolation, over- or undersampling and/or the different propagation models must be somehow combined. Although few authors within COST 273 worked on database issues, it is evident that problems such as high cost, low reliability and high complexity of input databases (and the corresponding high handling and prediction time) probably represent the main limits to the widespread adoption of deterministic propagation models. Studies related to database extraction, handling and simplification carried out within COST 273 are briefly reported in the remaining part of this subsection.

Electromagnetic parameters determination

In [ScWi03], ray-tracing simulation of wave propagation in hospitals in the frequency range from 600 MHz to 5.2 GHz is considered. What makes wave propagation in hospitals different from that in other buildings is the special construction of their walls. Some of these walls contain metallic layers, e.g. operating rooms with CrNi-steel faced walls, X-ray rooms with lead shielded walls or Magnetic Resonance Tomography (MRTo) rooms with Electro-Magnetic Compatibility (EMC) shielded walls made of a copper foil. If the real conductivity of those metals (copper, lead, CrNi-steel) were used in the simulations, this would lead to an infinite shielding of the walls, because slots and openings in the metallic layers cannot be taken into account. For this reason, the *effective* constitutive parameters of each kind of wall were determined through multiple measurements of single wall attenuation in different hospitals and double checked by comparison between measurement and simulation of attenuation between two adjacent rooms. The attenuation of single walls was measured by positioning horn antennas on both sides of the wall and measuring the attenuation of the transmitted power directly through the wall. For most of the walls it was possible to obtain constitutive parameters that are valid over the whole frequency range from 600 MHz to 5.2 GHz (see Table 4.1). However, this was not possible for the walls of the MRTo room, because the copper EMC shielding of this room shows a sensible frequency dependent behaviour. Therefore, its effective conductivity was determined separately for all five frequencies, Table 4.2.

Similar measurements on electromagnetic parameters were performed in [JePK05] and [JKVW05] by performing both reflected and transmitted power measurements on different kinds of walls with directive antennas. An IEEE 802.11b transmission system between two Power Controls (PCs) was used. In order to separate the desired reflected or transmitted signal from spurious signals coming from surrounding objects a time-domain resolution method using the Pseudo-Noise (PN) sequences of the IEEE 802.11b was adopted. The electromagnetic characteristics were then determined by a 'feedback' iterative technique where the Electro-Magnetic (EM) parameters in input to a 3D ray-tracing tool (see Subsection 4.3.4) were varied until the measured value was reproduced with sufficient accuracy (see Figure 4.7).

Mixed database handling

A variety of radio communication systems such as DAB and DVB, 3G public networks and WLANs must coexist in urban and suburban environments. To assure high-quality in-building coverage for the different systems in different environments (indoor, microcellular, macrocellular, etc.), or to control the limits and interferences between the various cell layers in a complex network, deterministic simulation tools are required in conjunction with appropriate multi-environment database handling techniques. In [LoCo02a], [CoLA03] and [LoCo05] a multi-environment prediction tool based on a 2.5 ray-launching technique for outdoor and for outdoor-indoor propagation and a 3D ray tracing for indoor propagation is presented (see also Subsection 4.3.4). An optimised indoor database format was adopted to speed up indoor prediction without significant loss in accuracy. Only the main structures

Table 4.1 *Composition of walls in hospitals and constitutive parameters.*

Type of room	Material	Thickness	ϵ'	ϵ''	σ (*S/m*)
sick room	plasterboard	2.5 cm	2.4	0.01	0.0
	rock wool	7.5 cm	1.2	0.02	0.03
	plasterboard	2.5 cm	2.4	0.01	0.0
X-ray room	plasterboard	2.5 cm	2.4	0.01	0.0
	lead	1.0 mm	1.0	0.0	70
	rock wool	7.5 cm	1.2	0.02	0.03
	plasterboard	2.5 cm	2.4	0.01	0.0
operating room I	CrNi-steel	0.8 mm	1.0	0.0	1.0
	plasterboard	1.8 cm	2.4	0.01	0.0
	rock wool	7.5 cm	1.2	0.02	0.03
	plasterboard	1.8 cm	2.4	0.01	0.0
	CrNi-steel	0.8 mm	1.0	0.0	1.0
operating room II	*Trespa*™	0.5 cm	3.0	0.1	0.0
	plasterboard	2.0 cm	2.4	0.01	0.0
	lead	0.5 mm	1.0	0.0	20
	rock wool	7.5 cm	1.2	0.02	0.03
	plasterboard	2.0 cm	2.4	0.01	0.0
	Trespa™	0.5 cm	3.0	0.1	0.0
MRTo room	chipboard	1.3 cm	2.5	0.1	0.0
	styrofoam	2.0 cm	1.1	0.0	0.0
	copper shield	0.16 cm	1.0	0.0	$\sigma(f)$ (see Table 4.2)
	rock wool	7.0 cm	1.2	0.02	0.03
	plasterboard	2.5 cm	2.4	0.01	0.0
	rock wool	7.5 cm	1.2	0.02	0.03
	plasterboard	2.5 cm	2.4	0.01	0.0

Table 4.2 *Frequency dependent conductivity $\sigma(f)$ of EMC shielded MRTo rooms for ray-tracing simulations.*

Frequency (MHz)	σ (S/m)
600	750×10^2
900	400×10^2
1800	230×10^2
2450	120×10^2
5200	70×10^2

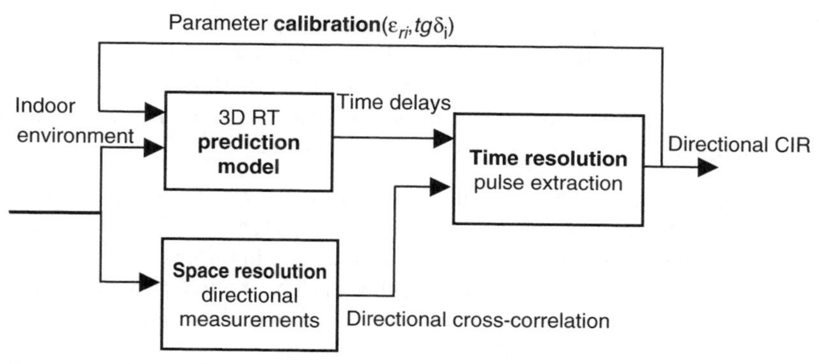

Figure 4.7 *Iterative technique for EM parameter determination.*

that significantly infer on the propagation mechanisms (typically the floor surfaces, the facades and the concrete partitions) are used in the computation of the multipath ray trajectories, the light partitions are only taken into account to determine path attenuation. The distinction between the main structures and the light partitions is incorporated in the building database. The analysis of the hybrid radio networks (coverage and QoS) requires the outdoor and indoor predictions to be done in a unique geographical system. Hence both kinds of terrain modelling, the outdoor map data and the 3D building representation have to be perfectly adjusted. Land usage data is also considered where available. The indoor representation that is derived from architect plans is generally available in a local coordinate system. Successive isometric transformations must be used to adjust the local representation to the building contours in the outdoor map data. Mixed database handling issues are also addressed in [KüMe02] and [KEGJ03]. In the former paper, empirical models for outdoor to indoor penetration are presented and validated with measurements. Of course, since the two different environments must be handled in a uniform way, proper conversion procedures must be adopted (see Section 9.2). In the latter, an automatic propagation model for the automatic planning of UMTS networks is presented which is one of the results of the European Project Information Society Technologies (IST) models and Simulations for Network Planning and Control – IST Project (MOMENTUM). Since the model must work in different deployment scenarios (macro-, microcellular and indoor), algorithms to handle transition between areas with digital terrain and urban databases of different resolution have been developed. A complete description of this work is given in Section 9.2.

Geometric database simplification

Urban field prediction tools are always run over a discrete set of test spots, a route or a limited area such as a cell or a portion of a city. Since computation time strongly depends on the size of the input database, it is necessary to minimise its size by selecting only the *active* area of the topology, i.e. only the buildings or obstacles that participate in the propagation process. As far as field strength prediction is concerned, the active area is limited to the obstacles *around* and *between* the radio terminals, therefore the active area can easily be 'guessed' using rule-of-thumb considerations. On the contrary, when the time and angle distribution of the multipath is to be determined, even far, but prominent, obstacles can have a significant impact on prediction results, and therefore a scientific criterion is needed. A method to attain this aim is presented in [DeFA04]. The method is based on the idea of running a rough, heuristic field prediction model aimed at identifying the 'active' building set.

The basic assumption is that paths experiencing a single (or two) interaction(s) with building walls (often referred to as 'first (second) order rays') are likely to be more relevant than rays experiencing multiple interactions, since more interactions correspond to both a higher number of concentrated losses and a higher distributed path loss. Thus, the simplification method is based on the selection of the buildings belonging to three categories (A, B, C) as follows.

(A) buildings located around and between the radio terminals, buildings located within an ellipse of focuses TX and RX, and of given eccentricity are selected. Usually the eccentricity value is chosen so that the terminals are completely surrounded by buildings.
(B) buildings which can be directly 'seen' from either the TX, the RX or both terminals, therefore generating first- or second-order rays. This step is performed by running a simple geometric visibility tool de-embedded from a Ray Tracing (RT) tool.
(C) buildings not belonging to category A and B, but which are likely to have an impact on propagation because of their height (greater than average) or orientation, i.e. generating low-order rays.

Referring to the ellipse mentioned in (A), its centre is chosen as the origin O of a cylindrical reference system and the map is divided into angular sectors of given amplitude $\Delta\Psi$. For each sector the average building height H_a is computed and a height tolerance Δh is set. Then, all buildings well above average height (with $H > H_a + \Delta h$) are selected. Buildings of average height (i.e. with $H_a - \Delta h < H < H_a + \Delta h$) are selected only if they have at least one wall almost perpendicular (within a given angular tolerance to the radial coordinate). This means that they are selected if they ideally could back-scatter energy toward the ellipse (hence from the TX to the RX). Once classes A to C are identified, the corresponding buildings are collected into the simplified map and duplicates discarded. If more than one RX (TX) is present (ex. receiver path) the whole algorithm is performed once for every RX (TX) and each time the new buildings are added into the simplified map.

In Figure 4.8 a simple case in the city of Helsinki is shown. It is shown in [DeFA04] that in this case the CPU time of an RT prediction can be reduced from 2131 minutes (complete map) to 300 minutes (simplified map) virtually without degradation in Delay Spread (DS) and AS prediction. On the contrary, wideband results obtained with the guessed map are unacceptable.

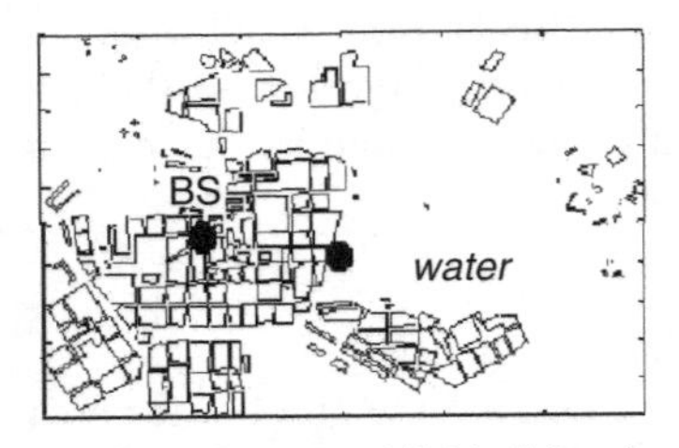

(a) Complete map (183 buildings).

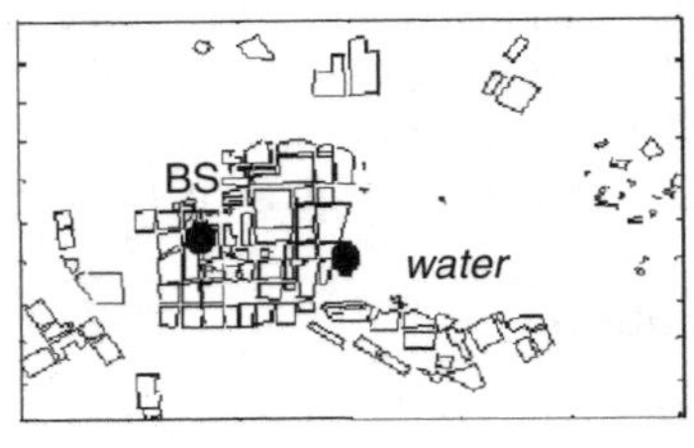

(b) Simplified map (118 buildings).

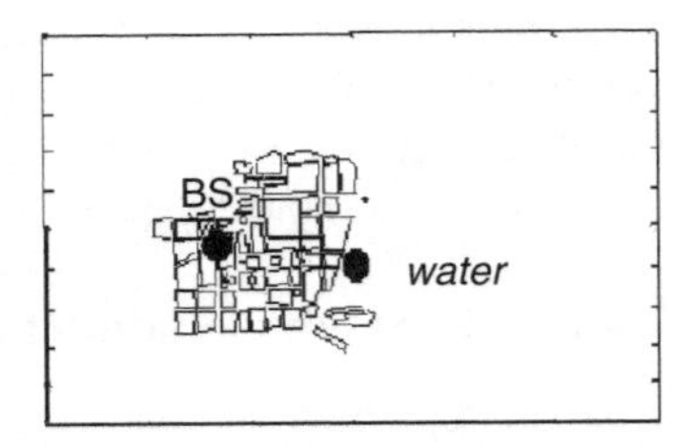

(c) Guessed map (69 buildings).

Figure 4.8 *Central Helsinki maps; black dots = radio terminals.*

4.3.3 Electromagnetic models

Electromagnetic methods are aimed at directly solving Maxwell's equations in some sort of discretised way (see Section 4.2). Depending on the adopted form of Maxwell's equations and on the kind of discretisation of the domain, various methods can be identified such as Finite Element Method (FEM),

FDTD, Method of Moments (MoM) and others. Electromagnetic methods are not commonly used for field prediction at radio frequencies, the main reason being the small wavelength with respect to the dimensional scale of the environment in this kind of problem. In fact, since the space discretisation step must be a fraction of the wavelength, the overall grid dimension would be huge, and consequently memory occupation and computation may become unbearable. Moreover the overall level of accuracy both available in input databases and required in output prediction is usually not so high as to justify the adoption of sophisticated numerical methods. Electromagnetic methods require a detailed description of both the geometry and the electromagnetic parameters of the environment. Real (as opposed to effective) electromagnetic parameters must be used for each single structure composing the environment.

Only FDTD has been used for deterministic field prediction within COST 273. In [Schä03], an FDTD method is used to investigate wave propagation between adjacent rooms in hospitals in the frequency range from 42.6 MHz to 300 MHz. For an accurate simulation of propagation in such complex environments as hospital rooms, with metallic layers in walls and with structures that are small compared to the wavelength such as slots in metal layers, numerical tools based on the solution of Maxwell's equations and a detailed description of each layer composing each room wall are necessary. Wave propagation between two adjacent rooms is investigated for different frequencies and polarisations and results are compared with measurements. The upper simulation frequency is limited due to the limited Random-Access Memory (RAM) resources of the computer, because the calculating area has to be divided in cuboids that are much smaller than the wavelength. With about 4 Giga Byte (GB) of RAM and a calculation area of about 10 m $\times$ 5 m $\times$ 3 m, it is possible to simulate frequencies up to about 500 MHz. In order to determine the so-called 'room attenuation', a transmitter was located inside a room and the pathloss to the adjacent rooms was simulated or measured.

In Figure 4.9 room attenuation with vertical polarisation for different kinds of rooms is reported. It is evident that the simulations are in good agreement with the measurements. Moreover, room

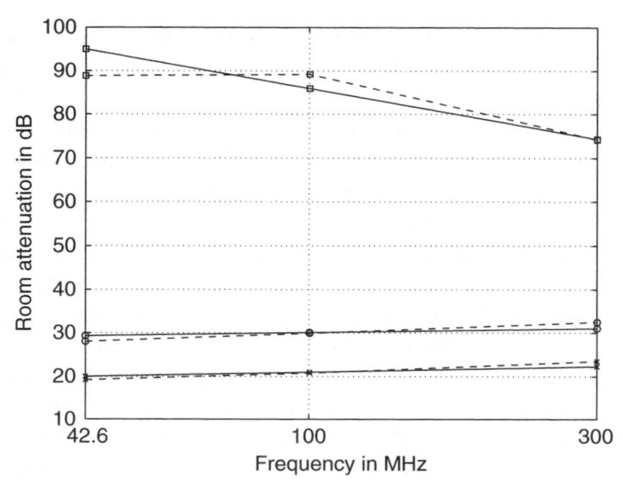

Figure 4.9 *Comparison of the measured and simulated room attenuations. Drawn through: measurement, dashed: simulation. Squares: MRTo room, circles: operating room, x-marks: X-ray room.*

attenuations for different polarisation are obtained: results show that, due to the orientation of slots inside walls, room attenuation and the field pattern strongly depend on polarisation.

4.3.4 Ray models

Ray models are based on Geometrical Optics (GO) theory and its extensions to treat reflection and transmission on plane surfaces and diffraction on rectilinear edges (see Section 4.2). Geometrical optics theory is based on the so-called ray approximation, which is valid when the wavelength is sufficiently small compared to the dimensions of the environment obstacles. This is usually the case in urban radio propagation problems. Under this assumption, the electromagnetic field and its multipath propagation can be expressed in terms of a set of *beams* or a set of *rays*, depending on the transverse extension of them. Rays have a null transverse dimension and therefore can in principle describe the field with infinite resolution. Beams (tubes of flux) have a finite transverse dimension because a space discretisation is usually adopted. Gaussian beams, however, have an infinite extension, but still the field is concentrated around an axis [TaLH02]. If beams are adopted, a limit to space resolution is set. Models adopting rays are usually referred to as RT models. Models adopting beams are usually referred to as Beam Launching (BL) models or sometimes beam tracking models. In RT algorithms the position of both terminals is specified from the beginning at the generic iteration. Then, by means of geometric considerations, all the possible rays that reach the RX from the TX according to GO rules and up to a given maximum number of successive reflections/diffractions (often called *prediction order*) are found. In BL algorithms the position of only one terminal (the TX) is usually specified at the beginning. Then the space is discretised into angular sectors, each one corresponding to a beam which is launched forward in the space. Each time a beam encounters an obstacle it is reflected/diffracted according to GO rules. In BL models unfortunately, space resolution generally decreases with distance from the TX. For what was previously described, RT is suitable for accurate, point-to-point prediction while BL is more suitable for coverage prediction over large areas. Both RT and BL reproduce to a certain extent the multipath propagation process. Multidimensional phenomena related to multipath propagation (time and angle dispersion, space decorrelation, fast fading, etc.) can in principle be reproduced by ray models. However, due to limitations in environment description and/or in the modelling of the propagation process, multidimensional performances of today's models still need to be improved [Bert02].

High computation time, which depends nearly exponentially on both input database size and prediction order, represents the main drawback of ray models. Therefore a great effort has been made in trying to speed up ray models either reducing the dimension of the problem from 3D to 2D or 2.5D, or somehow limiting ray/beam tracing to a subset of dominant rays. In 2D models, propagation is assumed to take place on a horizontal plane almost parallel to the ground and containing the radio terminals. This is strictly true, however, only if the radio terminals are located at the same height and if the Over Roof Top (ORT) paths are negligible. In 2.5D models instead of performing a full 3D prediction, ray tracing is on two or more two-dimensional planes. Usually two planes are considered: the horizontal plane and the vertical plane, the latter being a plane perpendicular to the ground, containing the radio terminals and the main ORT radio paths. An interesting overview of different kinds of ray models is reported in [Bert02]. In particular an original 2.5D method called Vertical Plane Launch (VPL) is described in the paper. Since rays satisfying GO rules belong to folded vertical planes (see Figure 4.10), and the projection of these planes onto the ground plane results in piecewise linear lines also satisfying GO rules as rays in 2D, the basic idea is to perform a 2D ray tracing in

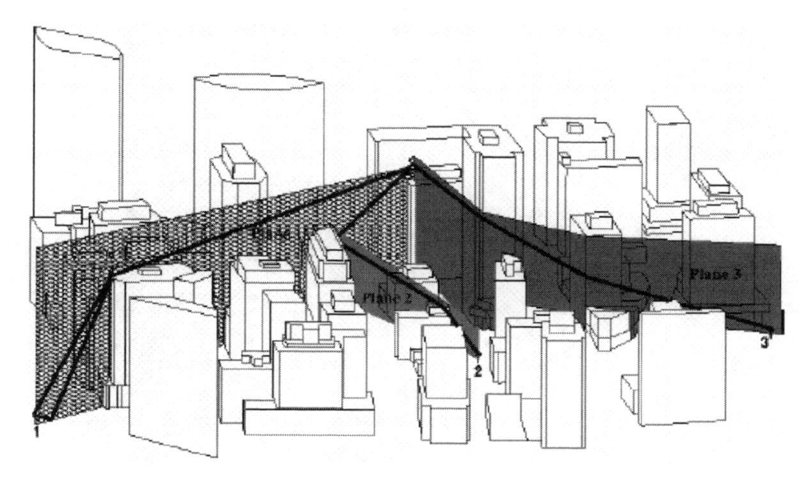

Figure 4.10 *Example of VPL ray tracing.*

the horizontal plane and then to find the actual 3D paths of the rays in the unfolded vertical planes by analytical treatment. VPL is shown to perform well in field strength predictions when compared with measurements carried out in Helsinki. In particular the mean error is of a fraction of dB and the standard deviation of the error is of about 4 dB for all measurement routes. Performance is not as good as regards delay spread prediction.

A sophisticated ray tool for simultaneous prediction of both outdoor and indoor radio coverage is presented by Lostanlen *et al.* in [LoCo02a], [CoLA03] and [LoCo05]. The model is based on 2D BL in the horizontal plane combined with a multiple knife edge diffraction Deygout method to include the radial ORT path in the vertical plane [Deyg66]. The method is therefore of the 2.5D kind. The adoption of ray launching in the horizontal plane saves CPU time w.r.t. RT for predictions over areas. Moreover, the model can treat outdoor to outdoor penetration and predict indoor coverage for outdoor BSs as well. Building penetration and path loss inside buildings can be computed in two different ways. If no accurate indoor database is available, each dominant ray impinging on an outer wall of the building is then prolonged inside the building according to the following empirical-statistical formula:

$$L_{in\text{-}building}[\text{dB}] = L_{outside}[\text{dB}] + L_{interface}[\text{dB}] + d_{indoor}[\text{m}] \times L_{linear}[\text{dB/m}] \qquad (4.1)$$

where $L_{outside}$ is the path loss calculated outside before penetration, $L_{interface}$ is the average loss due to the transmission through the facade, d_{indoor} is the path length inside the building and L_{linear} the average loss due to the in-building propagation taking into account furniture and partition attenuation. An example of outdoor-to-indoor coverage prediction derived using formula 4.1 is reported in Figure 4.11. It is evident that coverage in the fifth floor is better than in the ground floor, as can be easily experimented in the real world.

Outdoor-to-indoor propagation is especially important in public mobile radio systems (GSM, UMTS) and Digital Video Broadcasting systems, where most users are often located indoors. In [LoCo02b], [CoLA03] and [LoCo05] also a quasi-3D RT tool is used for indoor prediction in case of outdoor and/or indoor BSs (WLAN networks). The quasi-3D RT tool, which is perfectly integrated with the 2.5D outdoor one, adopts a 2D approach in the horizontal plane (a plane parallel

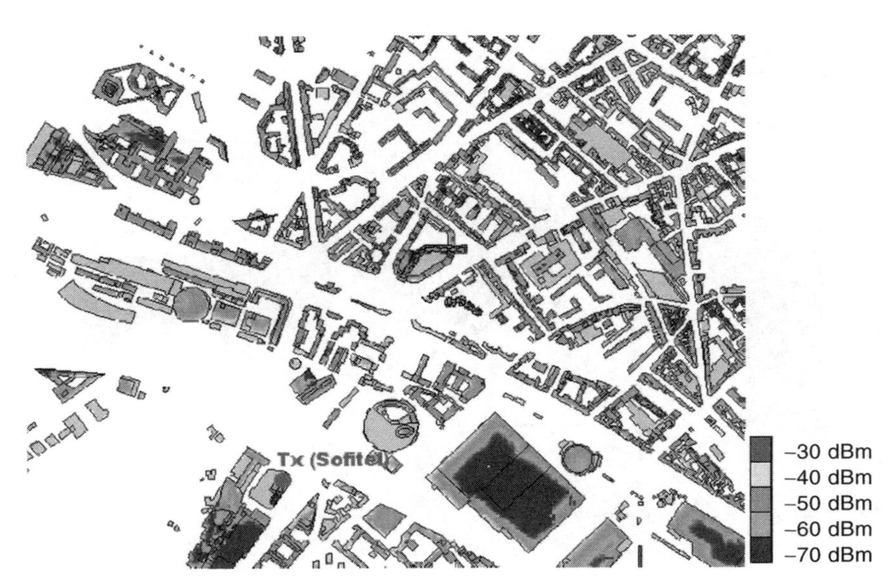

Figure 4.11 *Examples of outdoor-to-indoor coverage.*

to each floor) plus reflections on the floor and the ceiling. Of course the indoor RT model allows the detailed computation of indoor multipath effects. In [LoCo05], however, it is shown that the empirical-statistical method (formula 4.1) outperforms 3D RT as regards outdoor-to-indoor coverage prediction. This is probably due to the great degree of environment cluttering indoors (furniture, etc.) not taken into account by the RT model, which only relies on architect Computer Aided Design (CAD) plans.

Also [JePK05] considers 3D RT for indoor propagation prediction. In this work, a full 3D RT tool, which considers multiple reflections, edge diffractions using UTD coefficients (see Section 4.2) and transmissions through walls, is used in combination with measurement data to determine the electromagnetic parameters of walls (see Subsection 4.3.2). This is a very important application of ray models, which thanks to their adherence to the actual physical process can be used also to model and study multipath propagation.

In [TaLH02] an interesting method based on a Gaussian beam launching technique is presented as an alternative to ray tracing for 3D physical modelling of multipath propagation. The method is applied to the indoor case. By adopting the Gabor frame-based decomposition of source fields it is possible to represent radiated fields as a superposition of elementary Gaussian functions which are translated in the spatial and angular-spectral domains. Gaussian beam tracking through multiple reflections and transmissions is performed in a similar way to conventional BL methods. Fields can then be evaluated by summation of analytical terms representing transformed Gaussian beams. Diffraction is not taken into account, but due to the paraxial concentration of the field in a Gaussian beam, diffraction should have less influence with respect to the case of a plane or spherical wave. It must be noted that the parameters of a given transformed Gaussian beam have to be calculated only once for all observation points: computation time is therefore independent of the number of points where the field has to be evaluated. Simulations of amplitude-delay profiles are performed at 60 GHz, the considered indoor environment being a small laboratory area ($3.5 \times 6.5 \times 2.8$ m^3).

In order to validate the simulated results, measurements were performed in the 59–61 GHz frequency band with a network analyser. Both for simulations and measurements, the transmit and receive antennas were open ended waveguides. The measurement campaign consisted of amplitude/phase measurements at 10 to 30 different sampling points within four spots, chosen to illustrate different types of channels (Line-of-Sight, LoS, Non-Line-of-Sight, NLoS). Root Mean Square (RMS) delay spread values computed from simulated amplitude-delay profiles and from measurement results are shown in Figure 4.12. For clarity, only rectangles containing the values obtained for all points in a given spot are shown in the figure, together with the mean value for each spot. Not surprisingly, simulations seem to underestimate channel dispersion, but the relative difference between measured and simulated results, which is minimum for the LoS case (RX1), is coherent with previously published results in similar environments and frequency range.

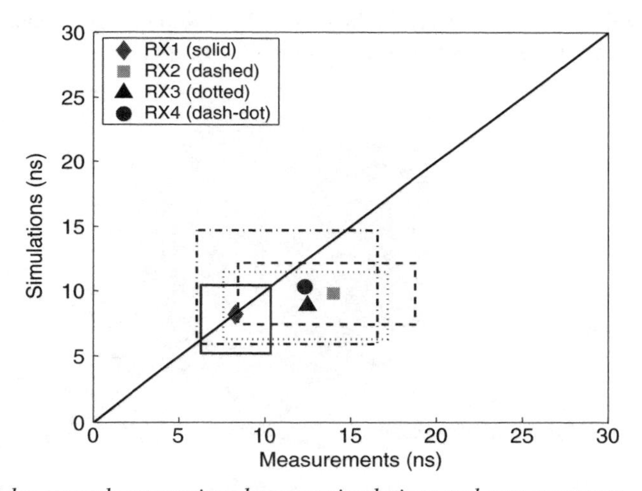

Figure 4.12 *RMS delay spread: comparison between simulations and measurements.*

4.3.5 Advanced techniques for ray models

As highlighted in previous subsections, ray models are generally characterised by a high computation time, especially when large input databases are needed, as for the case of macrocellular urban environments. Moreover, it is known that conventional ray models are not able to fully reproduce a correct space and angle distribution of rays, with a consequent detriment to wideband and multidimensional prediction performance. Beside studies described in previous subsections, such as database simplification studies, fast BL models, etc., several recent studies within COST 273 specifically dealt with the development of techniques aimed at decreasing computation time and improving capabilities or performances of ray models.

Speed-up techniques

In [HoWW03] a novel method for the acceleration of 3D ray models is presented. As the database of the considered urban topology remains the same during one or more ‘runs’ and only the position of the TX or of the RX changes, most part of the rays remains unchanged: only the rays connecting terminals change. Therefore, a sort of 'database preprocessing' can be performed which only depends on the particular topology. In a first step building walls are divided into tiles and the edges into horizontal and vertical segments (see Figure 4.13). After this, the visibility conditions between these different elements (possible rays) are determined and stored in a file. The visibility relations between all tiles and segments in the database are computed in this preprocessing phase, because they are independent of the terminal locations. For this purpose all elements are represented by their centres, which leads to the discretisation of the problem of path finding, Figure 4.13. Based on this preprocessing the path finding can be done similar to the ray launching algorithm by recursively processing all visible elements and checking if the specific conditions for reflection and diffraction are fulfilled. The ray search is stopped if a receiving point or a given maximum number of interactions is reached. If successive runs over the same topology must be carried out, then most part of the computation is already done and saved in a file. The model is shown to perform very well in three different routes in the city of Munich, which were considered in the COST 231 blind test: the mean error is a of fraction of dB and the standard deviation of the error is of about 7 dBs.

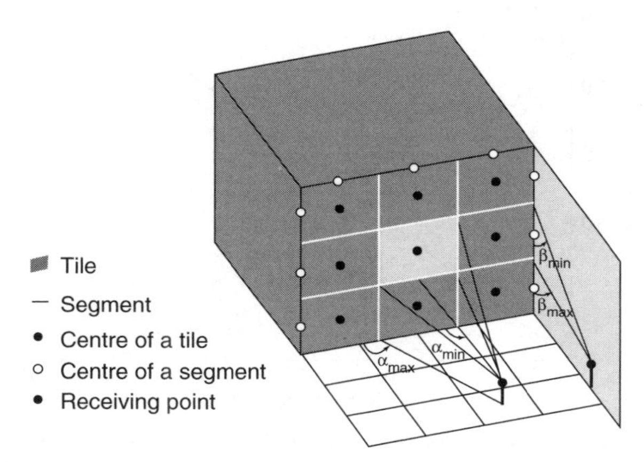

Figure 4.13 *Tiles and segments of a wall.*

A similar performance is achieved in indoor environments. In [WZWH03] the same method as above is used for outdoor prediction, then another BL model is used for enclosed spaces with a proper interface between the two models enabling the computation of the transition from an urban scenario to an indoor scenario and vice versa, thus allowing an accurate computation of the field strength or received power inside vehicles or buildings. As a border for the interface, a polygonal cylinder including the whole indoor database is automatically determined. As an adequate way to handle the interface between the two propagation models the average Power Delay Profile (PDP) is used. It includes field strength values, delays and angles of incidence of the electromagnetic waves impinging on the outer walls of the vehicle or building. The reason for the need for a transition between the two propagation models is that for the indoor or vehicle objects a smaller discretisation has to be used than for the urban scenario. Prediction results in terms of field intensity have been compared with measurements: the agreement is very good, comparable to what is obtained in [HoWW03]. In Figure 4.14 an example of prediction of indoor coverage by an outdoor BS is shown. It is evident that the field penetrates the car mainly through the windows, therefore providing a relatively good coverage in the passenger compartment.

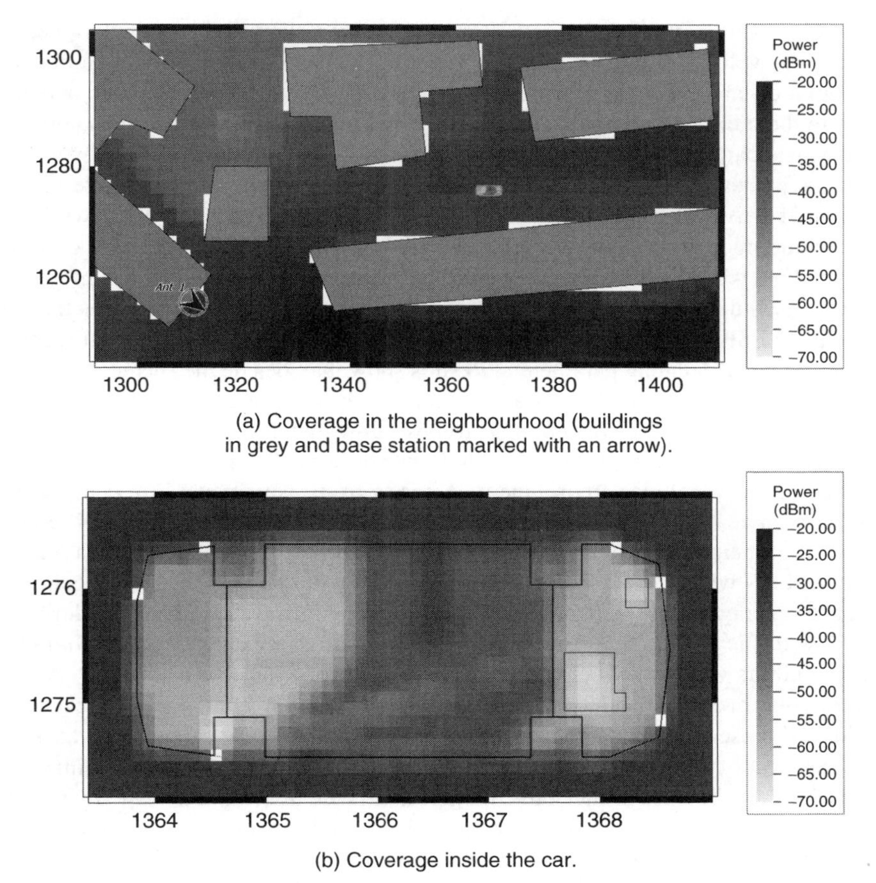

(a) Coverage in the neighbourhood (buildings in grey and base station marked with an arrow).

(b) Coverage inside the car.

Figure 4.14 *Penetration into a car of field from a GSM BS.*

Diffuse scattering studies

Conventional ray models only account for rays that undergo specular reflections or diffractions, but neglect diffuse scattering phenomena which can have a significant impact on propagation. As intended here, diffuse scattering refers to the signals scattered in other than the specular direction as a result of deviations (surface or volume irregularities) in a building wall from a uniform flat layer. The modelling of diffuse scattering from building walls is quite a difficult problem, since building wall irregularities cannot always be modelled as Gaussian surface roughness, as assumed in most theoretical studies. Moreover, objects *next to* (instead of *on to*) building walls such as street signs, traffic lights, advertising boards, trees, etc. can all contribute to non-GO scattering in the urban environment. A growing interest has been focused recently on the possibility to develop some sort of ray theorisation (or characterisation) of diffuse scattering phenomena. In [Bert02] an interesting study using an electromagnetic method to determine the characteristics of diffuse scattering from typical building facades is reported. The study demonstrated that diffuse scattering is concentrated on Keller's cones corresponding to diffractions from horizontal and vertical edges of external decorations and periodical structures such as indentations, windows, etc.

A recent experimental investigation [DFVG05a], however, showed that a single-lobe, continuous scattering pattern well describes the average scattering behaviour of typical building walls when observed from a distance. This is probably due to the fact that a considerable percentage of power penetrates into the building, interacts with furniture and inner structures, and then comes out again in a chaotic way apparently without following diffraction/reflection rules.

In [BHHT04a], the authors used a sophisticated 'synthesised array' method to determine intensity, angle and time of arrival of waves scattered from a brick building wall with windows. The wall was illuminated with a microstrip antenna, then a spatial scanning in front of the wall on a rectangular grid was carried out with the RX antenna (a microstrip antenna as well) so as to configure a 10×10 element synthesised uniform rectangular array. The output of the virtual array was then stored and processed with an ESPRIT superresolution technique in order to identify each single scattered wave and to extract the corresponding parameters. Results show that, beside the specular contribution, the wall generates a multitude of scattered waves: many weak waves are ascribed to 'brick roughness' and a few strong waves are observed coming from window metal frames.

A similar investigation is carried out in [GhTI04] with the aim of identifying the dominant scatterers in a typical urban microcellular environment. A series of measurements in a dense urban area in Yokohama city, Japan, has been carried out. The transmitter (TX) and receiver (RX) antennas were both mounted at a height of 3 metres in an LoS street scenario. The street width was 26 metres and both TX and RX were located 5.5 metres from walls on the same side of the street. Surrounding buildings had an average height of 20 metres. The measurements were accomplished during midnights with a very low traffic in the street. Employing a rotating directive antenna and a sliding-correlator at the RX the authors were able to obtain a Power Delay Profile (PDP) for every DoA. Analysis of PDPs revealed that most contributions arrive in time clusters. By comparing a precise map of the area including all present objects with the spatial distribution of received power it was observed that a significant amount of received clusters are scattered waves with DoAs corresponding to a variety of objects such as signboards, traffic signs, lampposts, traffic lights and almost any metallic object in the environment. Power considerations suggest that in a dense urban area scattering from these objects sums up to around 20% of the NLoS received power.

Recent studies have shown that diffuse scattering plays a fundamental role in determining time and angle dispersion of radio signals in the urban environment [DeBe99], [LKTH02]. Several authors have therefore tried to include diffuse scattering effects into ray models to improve multidimensional and wideband prediction performance. The first such study within COST 273 is reported in [DDGW03], where a model for diffuse scattering from surface roughness to be integrated into deterministic BL models is proposed. It is known that some portion of the energy impinging on building walls is scattered in directions other than the specular one due to the roughness of the surface. The scattered energy is generally split into the so-called *coherent* and *incoherent* components [UlMF96]. The coherent component decreases with increasing surface roughness, whereas the incoherent (or diffuse) component becomes more significant. Since rough surface scattering is generally a stochastic process (with the exception of a deterministic – and therefore known – surface structure), conventionally only its mean components can be included in deterministic ray-based propagation modelling. These mean components can be determined by the Kirchhoff models with scalar and stationary phase approximation or by more complex integral equation methods. The technique proposed in [DDGW03] expands the deterministic modelling by a 'stochastic' component resulting in instantaneous realisations of the scattering process, not only its mean values. Similar to the Kirchhoff formulations, the approach is based on a tangential plane approximation, i.e. it is applicable to surfaces with gentle undulations, whose horizontal dimensions are large compared with the incident wavelength. However, in contrast to the Kirchhoff models, which are only valid for either slightly rough or very rough surfaces, the proposed stochastic scattering approach includes both the coherent and incoherent components at the same time. Using ray launching, each locally plane wavefront is actually represented by multiple discrete rays instead of only one ray. Instead of reflecting all these discrete rays at the same boundary plane, the orientation of the plane (i.e. its normal vector) and its position (i.e. its height) are varied statistically for each discrete ray and for each reflection. In that sense each discrete ray is seen as a representative of an elementary wave for the locally plane incident wavefront. The variations of the local tangential planes are directly related to the statistical behaviour of the roughness of the corresponding surface. These tangential planes are 'local' in the sense that for each discrete ray a different tangential plane is generated.

The new technique is compared to the analytical Kirchhoff model in Figure 4.15. A relatively rough surface with a standard deviation of the surface height $\sigma_h = 0.710$, correlation length $L = 4\lambda_0$, relative permittivity $\epsilon_r = (10 - j3)$ and relative permeability $\mu_r = 1$, of size $Lx = Ly = 8\lambda_0$ is considered in this example. For the ray-optical stochastic scattering approach, the isotropic source is situated at $\theta_i = 60°$ and $\phi_i = 0°$, at a distance of $r_i = 2000\lambda_0$ from the plane. The rays impinge separated by $\lambda_0/10$ in the x- and y-direction on the surface, resulting in $N = 81 \times 81 = 6561$ rays per realisation. The resulting field is determined in the upper hemisphere at a distance $r_s = 2000\lambda_0$ from the centre of the plane, corresponding to approximately eight times the far-field distance, with a quantisation of $\Delta\theta = \Delta\phi = 1$ degree. A total of $M = 10^5$ realisations is taken for the ensemble averaging.

Figure 4.15 depicts the components of the Radar Cross Section (RCS) matrix per unit area of the diffuse (or incoherent) scattering component for the Kirchhoff model with stationary phase approximation, Figure 4.15(a), and for the stochastic scattering approach, Figure 4.15(b). The broken line in Figure 4.15(a) indicates the limit of the validity region of the Kirchhoff model. The absence of colour in the lower corners in Figure 4.15(b) indicates that no ray reaches these regions in the stochastic scattering approach. The absolute value in the specular direction of, e.g. $\sigma^o_{\phi\phi}$ is in both cases 6.3 dB. The coherent scattering component is negligible for this type of surface (more than 50 dB below the incoherent component). Both methods show the same mean scattering behaviour: a more distinct main-lobe for the co-polar *$\phi\phi$ component* compared to the *$\theta\theta$ component*, a maximum of the scattered energy in a region near the direction of the specular direction, and a minimum of the cross-polar components in the specular f-direction. The comparison of coherent scattering from less rough surfaces shows a good agreement between the Kirchhoff model and the stochastic scattering approach as well. By applying this stochastic scattering approach, the inclusion of random (but directive) surface scattering into ray-optical propagation modelling becomes possible. The purely deterministic GO modelling is expanded by a 'stochastic' component, resulting in varying prediction results; thus allowing for the first time to account for non-deterministic scattering in ray-optical wave propagation modelling.

In [DGMA04] and [DSGK02a] an approach for the integration of the diffuse scattering contribution from both surface and volume inhomogeneities into RT models is proposed. The so-called Effective Roughness (ER) approach first proposed in [DeBe99] and [Degl01] is implemented in a 3D RT model and results are compared with measurements. According to the ER approach diffuse scattering, although mainly due to surface and volume irregularities (windows, balconies, indentations, decorative elements, rain-pipes, internal reinforcements, power lines, heating pipes, etc.), is assumed originating from surface roughness scattering. An *effective roughness* is associated to each wall in order to take into account not only real surface roughness but also the mean effects of the above-mentioned elements. A Lambertian scattering pattern is ascribed to each wall (if the wall is far

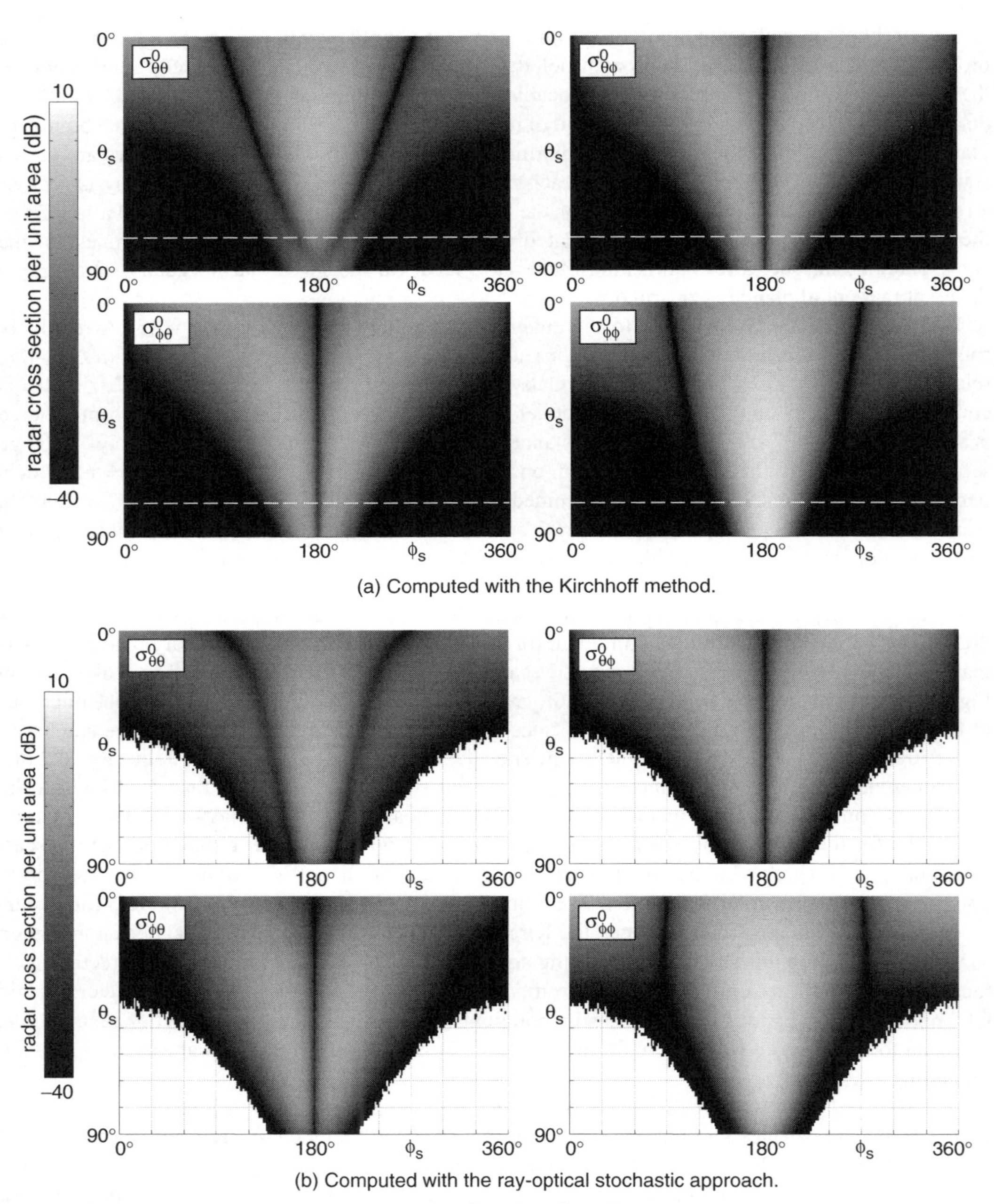

(a) Computed with the Kirchhoff method.

(b) Computed with the ray-optical stochastic approach.

Figure 4.15 *RCS matrix elements per unit area ($\sigma^o_{\theta\theta}$, $\sigma^o_{\theta\phi}$, $\sigma^o_{\phi\theta}$, $\sigma^o_{\phi\phi}$) of a square rough surface. Direction of illumination: ($\theta_i = 60°$, $\phi_i = 0°$).*

enough from TX and RX) or to each surface element (if the wall is close). The scattering contribution of each wall is computed directly from wall distance and orientation with respect to the TX and the RX using simple, analytic formulas derived from power balance considerations, which depend on only one parameter, the scattering parameter S. S is defined as the ratio between the amplitude of the scattered field and the amplitude of the incident field in the vicinity of the surface. In case of a far wall, the square of the total scattered field amplitude is:

$$E_s^2 = K_o^2 S^2 \frac{A\cos\theta_i \cos\theta_s}{\pi} \frac{1}{r_i^2 r_s^2}; \quad K_o = \sqrt{60 G_t P_t}, \tag{4.2}$$

where G_t and P_t are the gain and the input power of the TX antenna, respectively, and A is the area of the wall. Angles and distances in (4.2) correspond to that depicted in Figure 4.16. In order for the overall power balance to be satisfied, specular reflection and diffraction are consequently attenuated according to the Rayleigh factor R. Realistic S and R values for typical buildings have been found to be S = 0.4 and R = 0.6 [DeBe99]. For simplicity, only first-order scattering is considered. Results reported in [DGMA04] show comparisons of measured and simulated power delay profiles relative to a macrocellular scenario in the city of Stockholm. The diffuse component is shown to be of major importance, especially for wideband assessments in macrocells. In Figure 4.17, for example, the agreement is quite good both with S = 0.4 and with S = 0.6. On the contrary, results without scattering obtained with an ordinary 2D RT model are unacceptable. Note that 2D ray-tracing results do not include the first peak, which must of course be computed using an ORT model. Therefore, using a traditional *2D RT + ORT* model the overall received power would indeed be predicted with a good accuracy while wideband parameters such as DS would come out heavily underestimated due to the strong underestimation of the 'tail' of the power-delay profile. Similar results but relative to a more complete characterisation in terms of both delay spread and angle spread of the multipath channel in the city of Helsinki are reported in [DSGK02a] (see Section 4.5).

An interesting study on the role of scattering in ORT propagation is reported in [DeFG03]. In dense urban environment, where the average building height is comparable to or even greater than street width, multiscreen diffraction only partially accounts for the overall ORT propagation mechanism. In particular, the last part of the ORT path, the so-called Roof-to-Street (RtS) part, is a complex phenomenon also involving multiple reflections and scattering on the buildings surrounding the MT. Several authors proposed 2D RT in the vertical plane with multiple reflection/diffractions to model RtS propagation. Reflections and diffractions alone, however, are quite inefficient for RtS propagation in dense urban environment since the former requires a large number of bounces and the latter involves deep-shadow diffraction, which is very weak. According to preliminary studies reported in [DeFG03], where simulations of the RtS mechanism in typical cases are carried out using RT

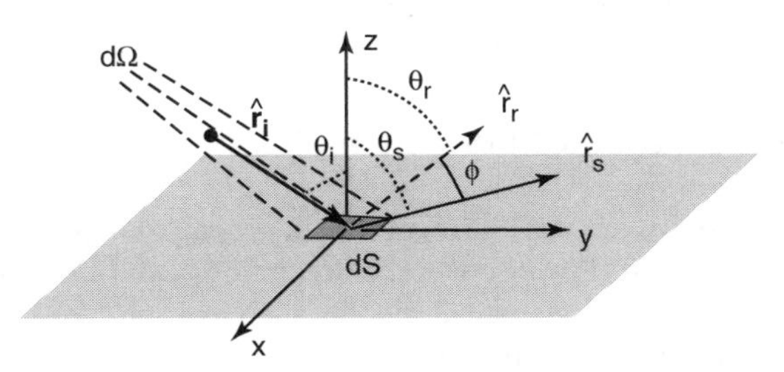

Figure 4.16 *View of the generic surface element with incidence (θ_i) and scattering (θ_s) angles.*

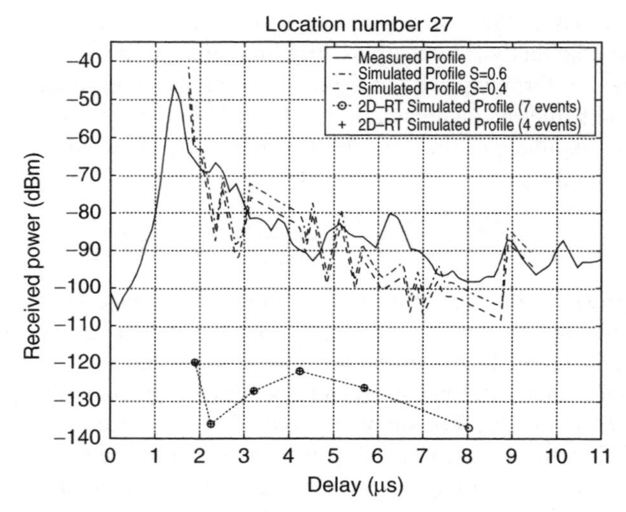

Figure 4.17 *Measured and simulated power-delay profile with and without scattering.*

including ER scattering in the vertical plane, diffuse scattering from building tops and upper parts, which are often directly illuminated by the BS, plays a key role in determining the RtS contribution.

UWB propagation modelling by ray tracing

UWB is a very promising technique for high-data rate, short range transmission and is claimed to permit very low power consumption and therefore very high battery duration in portable terminals. In [LoGC05] a 3D RT model is applied to the modelling of UWB propagation in an indoor environment. The application of ray models to UWB propagation prediction is quite straightforward. Since the geometry of the multipath pattern does not depend on frequency, the rays are generated by the RT algorithm in a similar manner to the narrowband case. On the contrary, the amplitude and the phase/delay of each ray do depend on frequency because of the dependence on frequency of reflection and diffraction coefficients and of the emission/reception characteristics of the antennas. A UWB time-domain-pulse emitted by the TX is therefore not only replicated due to the multipath but also distorted due to the above-mentioned characteristics. The effect of multiple bounces inside the dielectric slab on a UWB pulse transmitted through a wall is also shown in the paper. Since the antenna patterns must be considered frequency dependent, a time-domain antenna diagram must be input to the RT simulator. An example of the effect on a transmitted pulse of the overall indoor propagation process is shown in Figure 4.18.

4.3.6 Applications of ray models

Since ray models can potentially reproduce the multipath propagation process, the range of possible applications is virtually unlimited. Power coverage predictions can of course be achieved with ray models. However, simpler empirical-statistical models such as Hata-like models, multiwall models for indoors, etc. can provide a nearly similar level of accuracy in most cases with a lower computing time. Ray models are therefore worthwhile if a higher insight into the multipath process is needed.

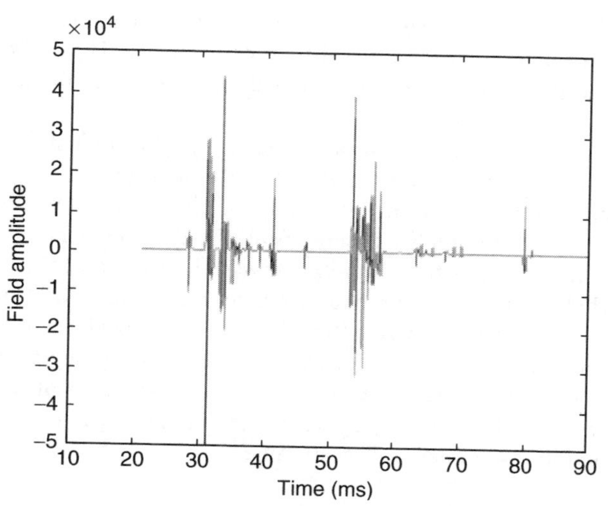

Figure 4.18 *Received UWB impulse with TX and RX in NLoS on the same floor in a simple office environment.*

This is the case of fading statistics assessments, wideband assessments, multidimensional (space, time, frequency, Doppler frequency) channel characterisation, and any other problem involving these aspects. In [Bert02] and [ChBe01] a VPL ray-tracing model is used to perform radio channel characterisation. Channel statistics are needed for advanced system design, but measurements are costly. Moreover, measurements carried out in one environment may not apply to a different environment. The use of ray models is a relatively inexpensive substitute for measurements, with the advantage of allowing the simulation of a large number of cases in a very short time. In [Bert02] it is shown how it is possible to derive channel statistics on the base of the output of a large number of RT runs by means of a Monte Carlo approach. In particular, DS, AS, finger life statistics in rake receivers and other statistics are obtained and reported vs. environment characteristics, building height statistics, etc.

In [TLVD01] fading statistics are obtained by RT simulations. In particular, starting from the multipath information at one single point, it is possible to estimate the signal statistics in a local area of that point, therefore an efficient site-specific channel model might be feasible. Using the ray-tracing technique, the Doppler power spectral density $S(f)$ for a receiver which is assumed to be moving in a certain direction with a given velocity can be computed from direction of arrival and intensity of each path. Moreover, making use of the phase of each wave ϕ_k, the complex function $R(f)$, where $|R(f)|^2 = S(f)$, can be derived and the complex envelope of the received signal in the time domain $r(t)$ can be obtained by simply performing the inverse Fourier transform of $R(f)$. This process is obviously much faster than the simulation of the received signal in a great number of locations (denoted as method 1 in this work). Finally, once the complex envelope of the received signal in the time domain $r(t)$ is known, the spatial distribution of the fades is obtained by considering that time and space are obviously related by the velocity (see formula (4.3)). The fast fading statistics in a local area around the point where the Doppler spectrum has been calculated are obtained by suitably sampling the curve $\tilde{r}(l)$

$$\tilde{r}(l = vt) = |r(t)| = \sqrt{r_I^2(t) + r_Q^2(t)} \tag{4.3}$$

It is important to note that the velocity of the mobile is used here as a parameter whose actual magnitude is not relevant, so that it can be considered as a differential velocity. The spatial distribution of fades depends only on how the power density is distributed along the Doppler frequencies, but not on what the maximum Doppler frequency is.

A wide range of comparisons between measurements and RT simulations has been carried out [TLVD01]. Part of the measurement campaign was carried out on the 5th floor of the Communications Engineering Department of the University of Cantabria, a very complex environment. Three RX trajectories, T1, T2 and T3, are analysed here. Both the data obtained from the measurements and those resulting from the simulations were processed in order to obtain the Rician probability distribution function that best fitted the curve in each case, according to the criterion of the minimum root mean square error. Figure 4.19 shows the Rice K-factor of the local fast fading estimated for these three trajectories. Figure 4.20 presents the normalised level crossing rate for the trajectory T1, this case being representative of all the others. It can be observed that the results obtained, from both the measurements and the simulations, closely approximate the theoretical ones since this propagation environment is close to the case of isotropic scattering.

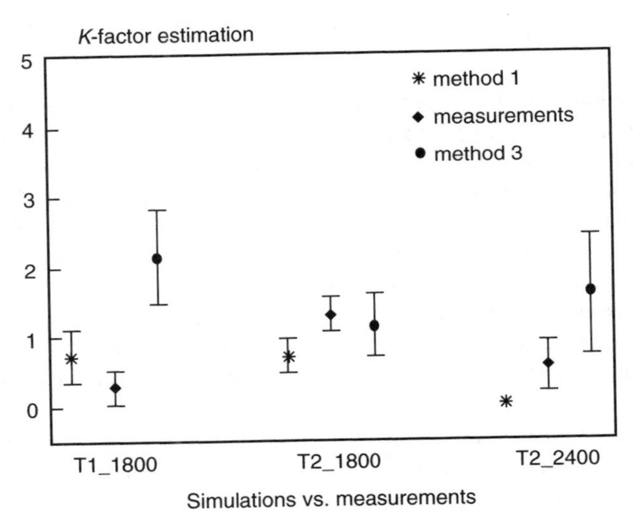

Figure 4.19 *Measured and simulated Rice K-factor for the three routes.*

A different approach, which can be used to derive local fading statistics (including the expected mean level), is presented in [NaCB04]. For network planning purposes, it is of interest to estimate the expected local mean level. Due to multipath fading (often referred to as Rayleigh fading), it is necessary to average large blocks ($10\lambda \times 10\lambda$) of hundreds of points separated by a fraction of wavelength (e.g. $\lambda/4$) to estimate the local mean level, and this can require enormous computational resources. An alternative approach to estimating the expected local mean level, based on artificially modulating the wall surfaces to cover all the possible signal phases, is the proposed wall imperfection model. The proposed model is implemented by randomly varying the position of each wall using a uniformly distributed random variable over a number of 3D RT runs in a simple indoor environment. The standard deviation of the random variable can be varied to study different degrees of wall displacement. The uniform distribution is selected to calculate the local mean field strength by averaging

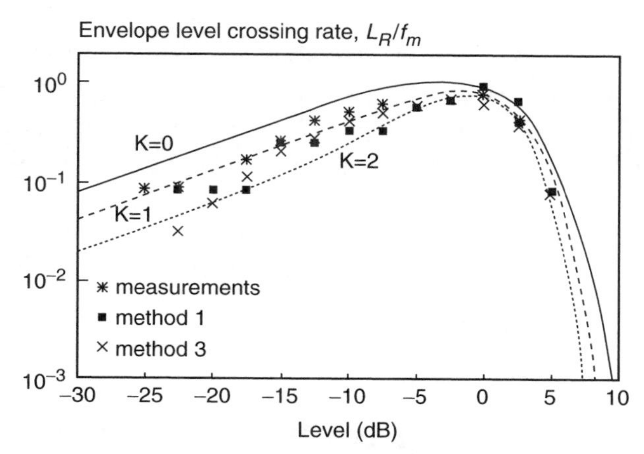

Figure 4.20 *Measured and simulated Rice K-factor for the three routes.*

out all phases with equal weighting. A similar methodology can be used to study imperfections of wall material properties by adding a random variable with an arbitrary distribution (uniform or Gaussian) to the electromagnetic wall property values to account for uncertainties in these values. This proposed approach in estimating the average local value of the signal strength has the advantage of decreasing the number of computer runs compared to the approach of averaging blocks of points with high resolution ($\lambda/4$) grids. The wall imperfection model was tested for several environments. For example, the model is applied to estimate the signal level at a central RX point (15 m, 15 m) in an arbitrary indoor environment of 30 m × 30 m. The procedure is described as follows. During each computer run, all the walls are moved simultaneously and randomly to give uncorrelated wall movements and the phasor sum of all the rays arriving at the receiver is obtained. An optional additional random variable with a 10% standard deviation (selected arbitrarily) is added to the values of the electromagnetic properties of the walls (permittivity and conductivity) to account for uncertainties in these values. The values of n computer runs are then averaged to obtain an estimate of the local mean field strength at the RX point. The estimated level is compared with the value obtained by averaging a block of points around the receiver separated by less than $\lambda/4$. The obtained averaged field value is compared for different computer runs (for the wall imperfection model) and different number of points forming the averaging block (for the spatial sampling approach). The results are summarised in Table 4.3. The results from averaging blocks of points following the spatial sampling approach suggest that averaging 400 points ($5\lambda \times 5\lambda$) seems to be enough for indoor environments to reach a stable average value after minimising the effect of small signal variations. This contrasted with the traditional approach used in outdoor environments of averaging larger blocks of ($10\lambda \times 10\lambda$) or ($20\lambda \times 20\lambda$). The average field level obtained from the wall imperfection model seems to have reached its final value after 50 to 100 runs, therefore with a reduction of 87.5% in the number of computer runs.

In [FuMa03] a full-3D RT tool is used to perform a wideband characterisation at 2 GHz in a typical urban macrocellular scenario, Figure 4.21. The model considers multiple reflections, multiple diffractions and combinations of both. Here the maximum reflection order is set to six and the maximum diffraction order to two. The simulation environment is a 719 × 539 m section of the city of Karlsruhe and 'BS' marks the place of the BS (transmitter), which is located on a 12.5 m high building on a pole 13 m above the roof. The receiver (mobile station – MT) has an antenna height of 1.7 m

Table 4.3 *Comparison of the average local field value around the receiver obtained from the wall imperfection model and the spatial sampling approach for different computer runs.*

(a) Wall imperfection model		(b) Spatial sampling approach	
Number of computer runs	Average field value [dB]	Number of runs (separation: $\lambda/4$)	Average field value [dB]
10	−18.92	100 ($2.5\lambda \times 2.5\lambda$)	−20.52
50	−19.39	400 ($5\lambda \times 5\lambda$)	−19.32
100	−19.34	1600 ($10\lambda \times 10\lambda$)	−19.30
200	−19.22	6400 ($20\lambda \times 20\lambda$)	−19.34
300	−19.45		

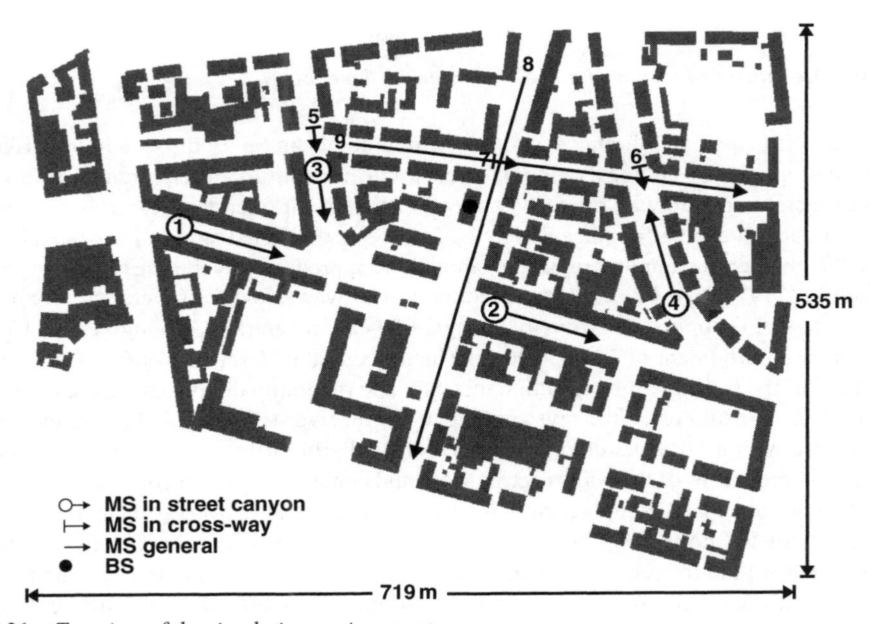

Figure 4.21 *Top view of the simulation environment.*

above ground. A $\lambda/2$ dipole is used as both transmitting and receiving antenna. The channel impulse responses are collected in nine different simulation routes, separated in three different propagation classes: [RX1]–[RX4] capture street canyons, [RX5]–[RX7] intersection and [RX8]–[RX9] capture all propagation classes. Considerations regarding the behaviour of the multipath pattern in different route classes show a good agreement with that reported in [LKTH02]. Table 4.4 summarises the statistical parameters for all simulation routes. Comparing scenario I (street canyon) with scenario II (intersection), one can see that at intersections time and angle spreads are higher than in the latter. Probably, at intersections waves are not influenced by street guiding effect and thus energy is spread over a much larger area in the azimuth plane. Comparing delay spread and angular spread a high correlation can be observed. Typically correlation is higher at the BS than at the MT. The obtained parameters can be used for the parametrisation of spatial channel models.

Table 4.4 *Summary of delay spread, azimuth spread and correlation coefficients between delay and azimuth spread for routes [RX1]–[RX9].*

Route RXN	Scenario	Delay spread (ns)	BS azimuth spread [deg]	MT azimuth spread [deg]	BS corr. coeff.	MT corr. coeff.
1	MT in street canyon	29.3	28.5	3.0	0.23	0.80
2		160.9	18.5	14.7	0.33	0.99
3		75.0	52.2	12.9	0.09	0.65
4		72.6	56.8	13.8	0.57	0.58
Mean value		84.3	39.0	11.1	0.30	0.76
5	MT in intersection	303.5	48.5	31.7	0.78	0.59
6		106.5	55.6	12.7	0.73	0.90
7		32.6	20.4	7.9	0.08	0.61
Mean value		147.5	41.5	17.4	0.53	0.70
8	MT in street canyon and in intersection	72.7	29.4	16.1	0.74	0.84
9		112.3	38.4	14.9	0.40	0.73
Mean value		92.5	33.9	15.5	0.55	0.79
Mean value of all routes (distance = 25–50 m)		108.1	38.1	14.7	0.46	0.75

In [CoWS04] a 2.5D RT model [RiWG97] is used as a propagation engine in a UMTS simulation program. The environment is a dense urban area in central Paris. The same UMTS simulation runs have also been done using traditional Hata-like empirical propagation models such as the well-known and tested COST-231-Hata model. Field prediction and system simulation outputs obtained with the two propagation models are then compared and discussed. In Figure 4.22, for example, field strength coverage with multiple BSs is shown as predicted by the Hata-like model (a) and by the RT model (b). It is evident that the empirical model, although taking into account the pattern of the antennas, is not able to reproduce the interaction of the propagating field with the urban layout, which is on the contrary evident on (b). Of course, such a huge difference reflects on system performance assessments as well. It is shown in the paper that a UMTS system simulation with the empirical propagation engine leads to less than 1% rejection rate R_{jR}, with

$$R_{jR} = (\textit{number of rejected calls/total number of calls})$$

while the same simulation with RT leads to $R_{jR} = 20\%$.

A group of researchers from the University of Stuttgart and from AWE Communications, Germany, have worked on advanced localisation methods using ray models [WHZL02], [ZBLL04]. In particular, the ray-tracing model described in [HoWW03] is used to implement the so-called 'database correlation method' for localisation in an urban environment. Field prediction results obtained with RT in a given service environment are used to define a look-up table for every BS. By evaluating the measured path losses between the MT and the BSs and by correlating these losses with the entries of the look-up table, the determination of the location of mobile terminals is obtained. In order to show the achievable performance in real environments, measurements in a GSM network are used for the localisation. The measurements represent a mobile which receives simultaneously four BSs.

(a) COST 231-Hata model.

(b) 2.5 RT model.

Figure 4.22 *Field prediction over a section of central Paris.*

The received power values lead to a pattern which is compared to the database entries to determine the location of the MT. Figure 4.23 shows the achieved results for the test scenario. The initial effort for the location technique (i.e. the determination of the pixel matrices and the computation of the entries of the look-up table) is higher than the effort for conventional approaches. But the achievable localisation accuracy is significantly better if really accurate prediction models like 3D ray tracing or other accurate models (like the dominant path approach, see Subsection 4.3.7) are used.

Ray models can be of great help in the optimisation phase of an existing mobile radio system. In [LBLC02], for example, the authors tried to understand the reason of a coverage gap in a 400 MHz MPT1327 trunking system by using a 3D ray tracing tool (EPICS). A full 3D representation of the environment, including terrain height, was adopted and the study was carried out by comparing measurements and RT simulations. The study confirmed that bad coverage was due to the effect of the buildings and suggested possible solutions to the problem.

4.3.7 Simplified or hybrid models

Simplified or hybrid models may represent a very attractive compromise between the complexity of ray models and the oversimplification of Hata-like models. The term 'simplified' refers to the process of progressive simplification which usually leads to the development of a simplified model starting from a complete ray model or from a set of measurement data. The required input database is often simplified too, with a consequent acquisition and handling cost reduction. The term hybrid refers to the fact that, while simplifying, statistical elements are necessarily introduced in the deterministic approach. Unfortunately, the simplification of an intrinsically complex propagation process can

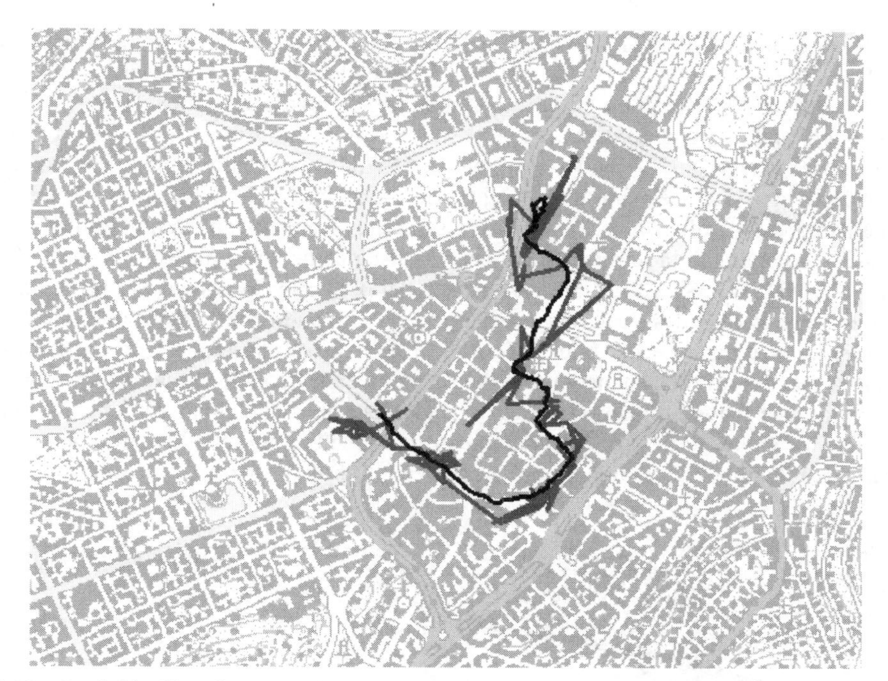

Figure 4.23 *Real (black) and estimated (grey) route of a walk through the city centre of Stuttgart, depicted area: 2.5 km × 2 km.*

indeed yield simple models, but at the expense of a proliferation of arbitrary parameters, which often need to be determined using measurements or RT simulations. It is interesting to notice that some simplified models for indoor environments [LoCo05], [WWWW04] often yield better accuracy in field strength predictions than sophisticated 3D RT tools. This is probably due to the inherent calibration with measurements present in simplified models which permits them to take into account the effect of environment cluttering (furniture, etc.) as opposed to conventional ray models which rely only on architect CAD plans. In principle, a good simplified model could guarantee a good accuracy with a drastic reduction in computation time with respect to conventional ray models, although the output is often limited to field strength and fading statistics. Therefore, these models are suitable for simple field predictions involving a large number of trial runs, such as those required for mobile radio network planning and deployment assessments.

A simplified model for field prediction in the urban environment is described in [BCFF02]. The model is derived from the observation that propagation in the urban environment may occur owing to different mechanisms, namely LoS propagation when there is a direct path between the two terminals, *lateral* NLoS propagation along the streets and around the building corners, and *vertical* NLoS over the building rooftops. For LoS propagation the authors resort to the well-known two-ray propagation model (including the direct ray and the ground reflected ray), while for the two NLoS components they developed new models which benefit from a statistical characterisation of the urban layout. The overall path-loss model is then defined as a weighed sum of these components, and the model yields estimates of the field fluctuations as well. The obstruction probability, which is defined as the probability that a link between a transmitter and a receiver is obstructed, defines the boundary between the LoS and NLoS regions. Then, the contributions to the field strength propagating in the two planes (lateral and vertical) are assumed independent and each of them can be modelled with a Gaussian random variable with known mean and standard deviation. With this hypothesis the field strength and its standard deviation vs. distance result from simple formulas involving some environment parameters (q, μ_{LAT} and μ_{VERT}) which depend on synthetic topological characteristics of the considered area (e.g. mean building height, mean street width, etc.) and, as usual, are expressed in dB. The parameters' values must be extracted from the city map and from measurements or RT simulations. A comparison in terms of path loss vs. distance between measurements performed in Munich at 947 MHz (COST 231 blind test) and model prediction is shown in Figure 4.24(a): the agreement is quite good and we can observe that the vertical and lateral contributions, as expected, equal each other at a distance of about 650 m. Figure 4.24(b) reports the received field standard deviation as a function of the distance. From the figures we may conclude that the developed model is able to reproduce the experimental values of mean path loss and standard deviation with good accuracy once the values of the involved parameters are tuned to the actual environment. It is worth highlighting that the model does not require a detailed building database.

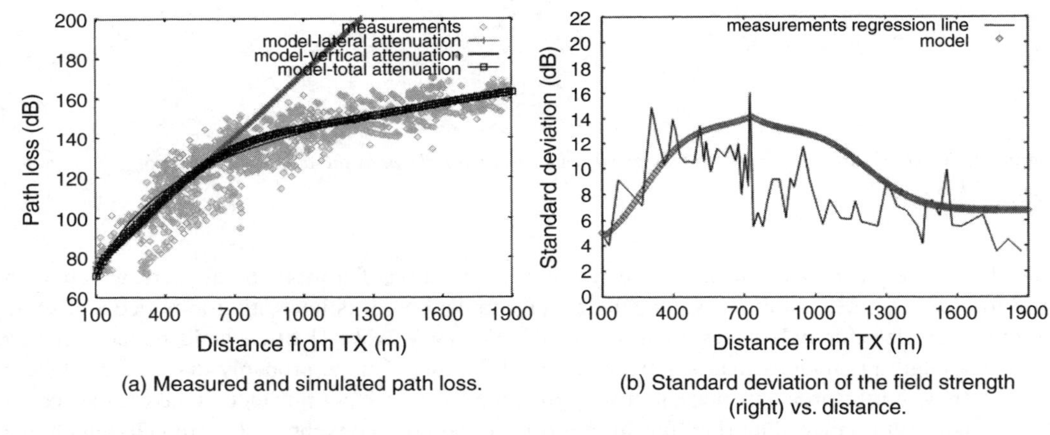

Figure 4.24 *Measured and simulated path loss and standard deviation of the field strength vs. distance.*

In [WWWW04] a new semi-deterministic model for field prediction in both the indoor and outdoor environment is proposed. The model is aimed at taking into account only the dominant paths in each scenario and reducing dependency on the accuracy of the input database with respect to complete ray models. The model requires a simple calibration with measurements or reference data. The model itself can be subdivided into two steps.

(a) Determination of the dominant paths

Determining the dominant paths is the most critical task. For indoor scenarios the algorithm is explained more in detail in [WLGB97] and [WoLa98] and a comparison with ray tracing is shown in Figure 4.25. The model is able to determine the 'less obstructed path' which usually transfers most of the power to the RX. The same principle can also be used for urban scenarios. Figure 4.26 shows an example for urban scenarios (Hong Kong, incl. topography). The dominant paths describe the main direction of propagation (the corridors or rooms passed (indoor)), i.e. the path of the main

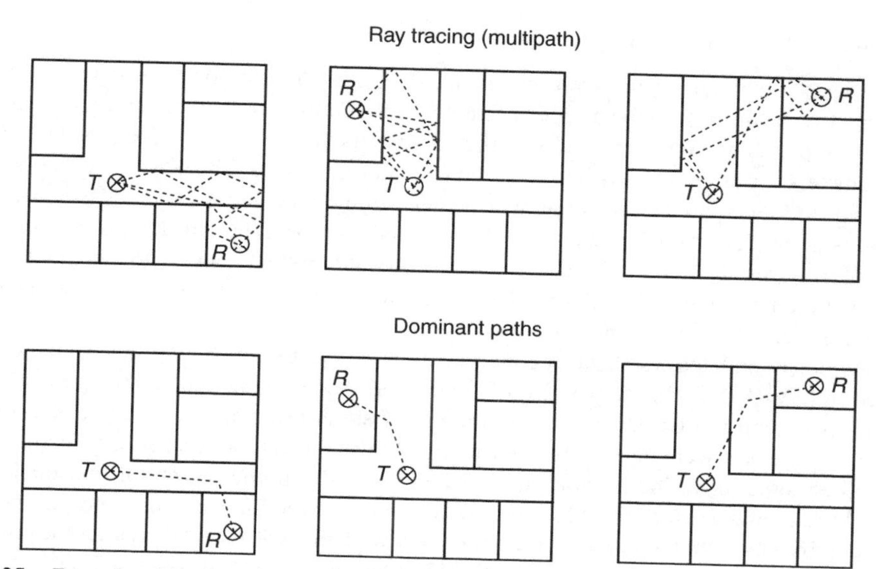

Figure 4.25 *Example of dominant paths tracing as opposed to ray tracing in an indoor scenario.*

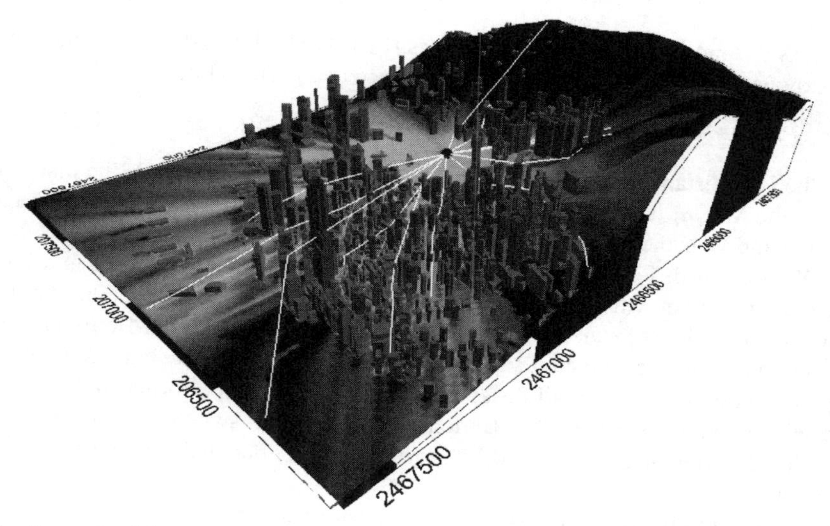

Figure 4.26 *Example of dominant paths tracing in an outdoor scenario (Hong Kong city).*

energy flow is determined. No individual reflection or diffraction coefficients need to be computed. These interactions are included in the 'parameter' of the paths (e.g. waveguiding factor, depending on the reflection loss and distance of walls in the scenario). The paths can either be determined in a horizontal plane (2D mode) in 2.5D (urban case, Figure 4.26) or 3D (indoor). So in urban scenarios the wave can also propagate over the rooftops as well as in the street canyons.

(b) Prediction of the path loss along the paths

The (empirical) prediction of the path loss along the paths depends on path length d (path loss proportional to $n \times log(d)$), on the number of interactions, optionally also on the angles in the path at the interaction points and on the model-specific parameters (waveguiding effect, explained above). These parameters can automatically be calibrated with measurements. The new model has been tested in indoor and outdoor environments by comparing field strength predictions with measurements. Once calibration is done, the attainable accuracy is very good, comparable with RT prediction (standard deviation of the error of about 4 dB indoors and 6–8 dB outdoors) with a very low computation time. The computation time for the Munich COST 231 blind test scenario (2000 buildings, 10 m resolution, 8.7 km^2 area) was below 30 seconds on a 2 GHz PC.

A formulation of path loss vs. distance for urban microcells into simple empirical statistical models is reported in [ZVRK04]. On the base of a 5.3 GHz MIMO measurement campaign carried out in central Helsinki with the BS located in the middle of street crossings at a height of about 10 m, two path-loss formulas were derived by measurement fitting using the least-square method for both LoS and NLoS situations. The BS was equipped with a uniform linear array of four dual polarised elements. Measurements from the MT were performed over rectilinear routes and fast fading was averaged out from measurements using a sliding window of about 80λ, and thereafter the powers over the MIMO subchannels were averaged in each MT location. The LoS formula is a simple Hata-like formula:

$$PL(d_1) = 40.3 + 23.4 \times \log(d_1)[\text{dB}]; \quad \sigma_{LoS} = 2.6[\text{dB}]$$

where d_1 is LoS distance and σ_{LoS} is the Path Loss (PL) standard deviation. It is interesting to note that the PL exponent (=2.3) is greater than in free space (=2). The NLoS formula is:

$$PL(d_1, d_2) = PL(d_1) + 10n_j \times \log(|d_2| - W/2)[\text{dB}]; \quad \sigma_{LoS} = 3.1[\text{dB}]$$

where d_1 is the LoS distance from BS to the closest street intersection, d_2 is the distance from the intersection to the MT, n_j is a randomly drawn path-loss exponent with mean = 2.33 and standard deviation = 0.3 and W is the main street (where the BS is located) width.

In [HeKu05] several different deterministic simplified models are compared in a UMTS urban macrocellular environment with ultra-high sites, i.e. with BSs mounted at a height of at least 100 m. Namely, the models under comparison are COST-WI, flat edge, Maciel-Xia-Bertoni and a reference free-space formula. Since the elevation angle of the LoS path can be quite high, Maciel-Xia-Bertoni shows large prediction errors at small distances, where COST-WI performs well. A hybrid model is proposed which combines Maciel-Xia-Bertoni and COST-WI, switching from the former to the latter where the elevation angle exceeds 0.15 degrees. A more detailed description of [HeKu05] is given in Section 9.2.

The adoption of simplified models is almost mandatory when general results must be derived which are not specific to a particular topology. In [Fusc04] a simplified Hata-like model is used to study innovative coverage solutions for DVB systems. The capability of providing high quality reception by handheld terminals is a very important target in digital video broadcasting. Unfortunately, preliminary assessments and field trials indicate that this target cannot be satisfactorily achieved by conventional network planning, based on the reuse of existing transmission sites (in order to limit the economic effort in the initial deployment of new digital broadcasting networks). Hence new deployment strategies must be investigated. In particular this work shows the possibility of strongly improving coverage probability in urban environment by adding some urban and suburban gap filler to conventional DVB network deployment. This is possible exploiting the flexibility of the single frequency, OFDM transmission technique. Results show that, with a conventional planning adopting extra-urban transmitters only, the minimum power at the TX stations should reach up to 95 kW to achieve a coverage probability $PC \geq 0.95$ all over a circular urban area with a radius of 5 kilometres. By adopting a new, more diffused coverage solution composed of 24 urban TXs (irradiating 8 W each), 10 suburban TXs (30 W) and 3 extra-urban TXs (5.845 kW) the same coverage probability can be obtained saving nearly 90% of the total transmitter (TX) power. The resulting coverage map is shown in Figure 4.27.

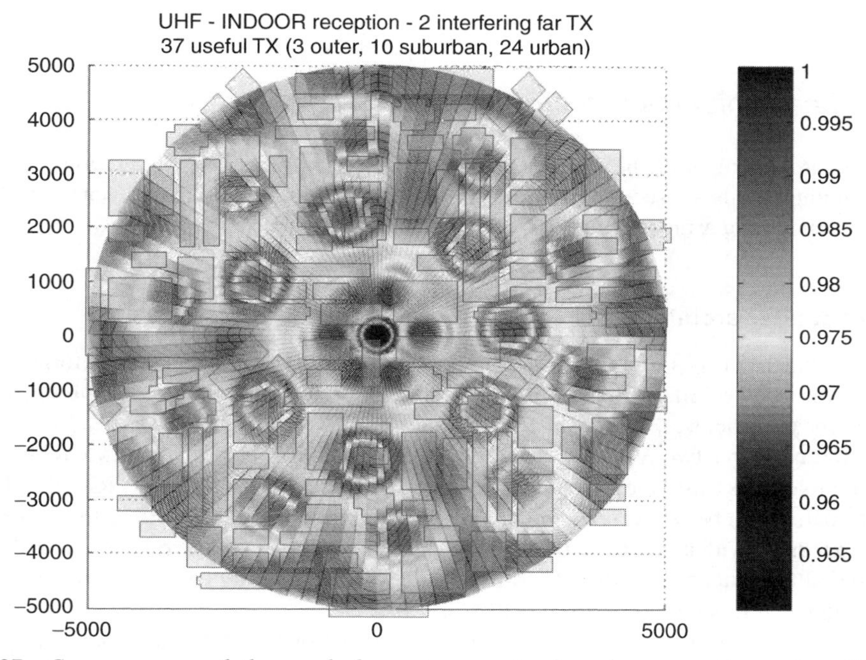

Figure 4.27 *Coverage map with the new deployment over a circular urban area with a radius of 5 km.*

4.4 Channel measurements and parameter estimation

4.4.1 Measurement scenarios

The purpose of measurements and their analysis is typically twofold. In the first place, propagation measurements are still needed to gain deeper understanding on the very complicated combination of propagation phenomena and their mutual significance in mobile radio systems. Due to the progress of transmission technologies, more dimensions of the channels must be taken into account in the measurement. In comparison with COST 259 [Corr01], double-directional channels are extensively focused for the MIMO systems within COST 273. As a very important new aspect, also the polarisation characteristics of the channel have to be known for polarisation diversity or MIMO systems. For all these aspects, one should obtain statistically adequate information in all relevant environments to be able to identify the most significant features of the propagation channel to first understand what is happening in the propagation-physics point of view and further to be able to create realistic channel models for system studies.

On the other hand, there are lots of measurements related to the systems aspects of the current and future mobile radio systems. In a radio system, the radios have to deal with several simultaneous connections between BSs (or access points) and mobiles. Thus one should have information not only on a single link but also on several radio connections and their mutual relationships. Networks are probably the ultimate example of the complexity of the propagation scenarios in the future mobile radio systems.

4.4.2 Channel sounding techniques

Several new architectures of the channel sounders have been proposed for the double-directional channel sounding. This subsection provides an overview of the radio channel sounding hardware architectures including array antennas for directional measurements.

Channel sounder architectures

A channel sounder consists of a TX to repeatedly transmit a fixed wideband waveform and an RX connected to a data recorder to record a waveform distorted due to the propagation channel. For short range measurements, a Vector Network Analyser (VNA) can be used instead of a dedicated sounder. Alternatively, two VNAs can be used as TX and RX respectively [MCRJ04]. A cost-effective approach for SISO measurements is the use of a spectrum analyser as RX [PLVP02]. The phase information can be recovered from the amplitude spectrum by using Hilbert transform.

Various kinds of wideband waveforms are used in the channel sounders, such as a PSK signal modulated by PN sequence [Elek05], [KSPH01], [ZeTS04], a multitone signal with low crest factor [Meda05], and a chirp signal [SFIH02], [LPEO02].

All the signals are designed to have small envelope fluctuation to reduce the backoff of the TX high power amplifier. Vector network analysers use step frequency Continuous Wave (CW) signal. Hybrid stepped frequency and PN sequence is used to realise very wide bandwidth [KaWe00], [VLSK03].

The phase and timing at both sides must be synchronised by some means. When TX and RX are close in distance, the local signal can be shared via coaxial cable or optical fibre. Otherwise, high stability frequency standards such as rubidium oscillators are usually used for the purpose. For high accuracy timing synchronisation, use of Global Positioning System (GPS) and Transmission Control Protocol (TCP)/Internet Protocol (IP) connection have also been considered [LPEO02]. Frequency division multiplexing of the reference signal and the sounding signal is an alternative technique [MATB02].

To obtain the single-directional information, an array antenna is deployed at the receiver side. Time domain multiplexing using a high speed RF switch and a single RF front end is the most popular architecture. Alternatively, parallel receivers can be deployed [SFIH02]. This is advantageous for the measurement of large Doppler shifts due to the shorter measurement time. A multiport network analyser can also be considered as a parallel receiver [MCRJ04]. Yet alternatively, a synthetic array approach can be used [MATB02], [VLSK03], [HaTa03b]. Here the measurement must be performed in a static environment due to the time consuming physical motion of the antenna elements. However, the SISO sounder hardware can be used without modification, and no array antenna calibration is necessary.

For the double-directional measurements, array antennas must be introduced at both TX and RX. Time domain multiplexing can be deployed at both the TX and RX [KSPH01], [Elek05], [Meda05]. The switch timing must be synchronised between the TX and RX multiplexers. As the TX multiplexer handles high power, it operates slower than the RX multiplexer in general. Therefore, the switch timing of the multiplexers is designed so that the RX multiplexer scans the array for each TX antenna element. Some hybrid techniques are also possible solutions, such as time domain multiplexing at TX and parallel RXs [RaSa04], and synthetic array at TX and time domain multiplexing at RX [Herd04]. A full MIMO transceiver is also possible, by using orthogonal signals such as code division multiplexing [ZeTS04], [ChSA04] and frequency division multiplexing [SaTA02].

Antenna architectures

For the non-directional measurements, omnidirectional antennas have been preferred, such as dipole antennas. The antenna element of the synthetic array may also be an omnidirectional antenna due to its original non-directional use. Slot antennas can be used as horizontally polarised omnidirectional antennas [Rich05].

Patch antennas are often used as directional antennas deployed in the base station side, or as elements of conformal arrays to avoid the influence of the support structure of the array. Patch antennas are also advantageous for polarimetric measurements, as the dual polarised element can be easily realised by using two orthogonal feeds in a single patch element [KSPH01], [Rich05], [SLRT03].

Array geometries are designed so that necessary angular information can be obtained. To satisfy the sampling theorem, the element spacing should normally be designed as less than half wavelength. Otherwise, there may appear grating lobes causing estimation errors of directions of arrival/departure.

Azimuth-only measurements can deploy either a uniform linear array [MATB02], [VLSK03], [SLRT03], [RaSa04], [Herd04], or a uniform circular array [SFIH02], [Rich05], [RaSa04]. With the same number of elements, the former has higher resolution in the broadside and is suitable as the BS antenna, while the resolution of the latter is insensitive to the direction and is suitable as the mobile station antenna. The linear array has another disadvantage that it suffers from cone ambiguity and the elevated direction of arrival results in an error of the azimuth angle estimate. To reduce the angular dependency of the resolution, use of a cross array antenna is also considered [VeLC03a].

For joint azimuth and elevation measurements, use of a planar or a solid structure is needed. A uniform circular array can be used also for this purpose [HaTa03b]. Alternatively, a uniform rectangular array may be used [HaTa03b], [Herd04]. To avoid the ambiguity between angles below and above the horizontal plane, a solid array structure is necessary. A spherical array [KSPH01], [KKVV04], a circular cylindrical array [TFBN04], and a rectangular solid array [TsHT04] have been deployed.

Array calibration is necessary for the real array antenna, and the array antenna response must be measured in the anechoic chamber in advance. A continuous reconstruction technique of the array antenna response from discrete data has been proposed in [LaRT03]. On the contrary, the theoretical array response can be used for the synthetic array antenna since there is neither mutual coupling between the antenna elements nor variation of properties for individual elements. When the uniform linear or rectangular array is deployed, ESPRIT algorithm can be applicable.

There has been a trial to optimise the array geometry by using a genetic algorithm [KaKC02].

4.4.3 Parameter estimation

As explained in the previous sections, the design of adaptive arrays or smart antennas brings several technical challenges. It is clear that the knowledge of the DoA parameters is very important. High resolution techniques can resolve these parameters very accurately, which is a prerequisite for the design of antennas. In this subsection, promising methods for determining the full double-directional and polarimetric parameters of the mobile channel are given.

The high resolution parameter estimation techniques are divided into three different types. Conventional methods are based on classical beamforming techniques and require a large number of antenna elements to achieve high resolution. The second class of methods are the subspace methods and are high resolution sub-optimal techniques, which exploit the eigenstructure of the input data matrix. The MUSIC technique for instance uses the eigenstructure of the covariance matrix. In contrast, the ESPRIT exploits the rotational invariance of the underlying signal subspace. The last class of methods is the ML technique, which performs well also under low signal-to-noise ratios. The SAGE algorithm is a way of ordering the maximum likelihood approach in an appropriate way.

This subsection is organised as follows. First, new results of the MUSIC algorithm will be shown, followed by the reporting on the use of the ESPRIT technique. The most common technique with high resolution is of course the SAGE algorithm. Theoretical and experimental results of the implementation of the algorithm in the time domain will be shown. Also the implementation of the SAGE in the frequency domain with important adaptations will be demonstrated with simulations and measurements.

The MUSIC algorithm is a high resolution signal parameter estimation technique. It exploits the eigenstructure of the input covariance matrix, based on a geometric view of the signal parameter estimation problem. Hence, it provides information about the number of incident signals and the DoA of each significant signal.

In order to decrease the computational complexity and to increase the resolution, some improvements to the previously described MUSIC algorithm can be made. The theory of the Smart MUSIC algorithm is described in [Taga97] and some practical simulations in [TaSh96]. Here, the eigenvalue decomposition is interpreted in terms of a Hermetian mapping. Using the Gram-Schmidt orthogonalisation, the speed of the data processing is ameliorated, resulting in a possibility of following the rapid change of the radio environment. This MUSIC algorithm is not only used as Direction-of-Arrival (DoA) estimation in mobile communication, but can also be used in other scenarios; [SICW01] describes the use of the MUSIC algorithm as a signal processing tool for a three-dimensional optical imaging system, based on stepped frequency radar techniques.

Another subspace-based DoA high resolution technique is the ESPRIT. It reduces the computational requirements significantly when compared with MUSIC as an extensive search through all possible steering vectors is not needed. A practical implementation of the ESPRIT is made in [BHHT04b], where the spatial scanning is configured to resemble an antenna array, also called a synthesised uniform rectangular array. With the described measurement set-up, the multipath characteristics of non-specular wave scattering from three-dimensional building surface roughness can be determined, showing that multiple paths can be detected at many scatterers. Here the delay, directly estimated by using ESPRIT, yields close agreement with the delay of the propagation path determined by using DoA estimated by ESPRIT and free space velocity.

In [SeKa01], two-dimensional fading forecast of time-variant channels is explained. Because of the implementation of a direction-of-arrival scheme as the ESPRIT technique, the channel degradation due to small-scaling fading effects must be available. Taking into account the processing delay caused by the information feedback between receiver and transmitter and the adaptation process itself, the channel variations must be predicted reliably into the future as far as possible. In this paper, the feasibility of a two-dimensional fading forecast is studied in the context of the ESPRIT.

Next, the SAGE algorithm is investigated. The SAGE optimisation algorithm is used to replace the high-dimensional optimisation procedure necessary to compute the joint maximum likelihood estimate of the parameters by several separate maximisation processes, which can be performed sequentially.

Here, a technique derived from the ML principle is proposed. It allows for high resolution determination of the complex signals. The EM algorithm updates all parameters simultaneously, which implies a slow convergence and a difficult maximisation. The SAGE method updates the parameters sequentially by replacing the high-dimensional optimisation process necessary to compute the estimates of the parameters, by several separate, low-dimensional maximisation procedures, which are performed sequentially. These separate maximisation procedures are linked up in the final algorithm. The implementation of the algorithm can be made in the time domain and in the frequency domain. The first papers will deal with the time domain implementation.

In [VeLC04a], the simulations are done in three dimensions, resulting in the calculated parameters: the relative delay, the azimuth angle, the elevation angle and the complex amplitude of the waves. Based on these numerical simulations, convergence, resolution and performance analysis are calculated. The ambiguity of detecting different users in a three-dimensional case with a two-dimensional antenna configuration can be resolved by using for instance two parallel planes. This solution suffers from the spatially limited radiation pattern of the array elements. By using spherical arrays, there will always be some elements radiating into the direction of the signals, hence increasing the signal-to-noise ratio of the subsequent digital processing. To minimise the coupling and hence to increase the decorrelation between all the antenna elements, they should be spaced uniformly over the sphere. This configuration will also increase the resolution of the antenna array.

Of course, using different antenna elements causes the radiation from one element to couple to its neighbours. The real current on each array element is the sum of the values due to the excitation plus all contributions from the various coupling sources from each of the neighbours. This mutual coupling is in most cases not desired, but will often be a significant factor in the total radiation characteristics. In [VeLC02a], this mutual coupling is integrated in the SAGE algorithm and the convergence and performance analysis are calculated.

Measurements are often performed in the frequency domain. To use the time domain SAGE algorithm, a transformation of the transfer function in the frequency domain to the time domain received signal, with the training sequence included, is given in [VLSK03]. The DoA parameters of the indoor measurements in a classroom are extracted with the time domain SAGE algorithm and compared with a ray-tracing tool. The same approach – evaluating measurement results in the frequency domain with a time domain algorithm – can be found in [MoRo04], where the results are also compared with a ray-tracing tool.

In fact, most measurements are broadband and the Broadband Time Domain-SAGE (BTD-SAGE) in [VeLC04b] combines the signals in another way, adapting the algorithm to be capable of detecting different paths in a broadband environment and with large antenna arrays, which is the case in the earlier described measurements. The extraction of the measurements is compared with results of a ray-tracing tool, proving that the adaptation of the algorithm works fine.

Another tool is described in [HaTa03a]: the UWB-SAGE algorithm. This algorithm divides the measured data into individual ray paths and estimates the directions of arrival, the propagation time and the variation of the amplitude and phase during the propagation for each signal. Simulations prove the quality in simple indoor propagation environments.

Also the signal model can be extended. In [VeLC03b], for the evaluation of the extracted values, the results can be compensated for the influence of spherical waves. Different simulation parameters are discussed: the extension of the measured bandwidth, the frequency resolution, the geometrical antenna array size and the number of antennas. With an appropriate choice of the measurement geometry and parameters, one can conclude that the SAGE algorithm can compensate for the properties of spherical waves. This knowledge is taken into account in [VeLC03a], where measurements with different antenna array configurations are investigated and compared to each other. Measurements of one linear antenna array that is perpendicular to the LoS and measurements of another one that is parallel to the LoS are combined to virtual measurements with a mill's cross antenna array, resulting in better extraction results of the measurements.

In most measurements, not only the type of antenna, but also the position and the polarisation of the antenna array, are very important. By taking into account both position and polarisation in the radiation pattern of an antenna array, another adaptation to the commonly used time domain SAGE algorithm is made. Hence the direction-of-arrival parameters can be determined very accurately. Simulation results with a cubical or hexahedral antenna array are shown in [LVTC04]. Results with a dodecahedral configuration are given in [VeLC05].

Some ameliorations are suggested for the initialisation and the search procedure in the SAGE algorithm. In [StFJ02] the Initialisation and Search Improved SAGE (ISIS) algorithm is presented, which applies a bin-search-based approach in the initialisation of the parameter estimates. This modified scheme can estimate propagation paths with small amplitudes, and thus the overall profile of the environment can be thoroughly extracted. Paper [FlSJ02] reports the performance of the ISIS algorithm in experimental investigations of dispersions in delay, direction of departure and direction of incidence in a typical non-LoS environment (see Figure 4.28). The results demonstrate the high potential of the technique for getting a detailed insight into the mechanisms by which the electromagnetic energy propagates between the transmitter and receiver sites. In [FYSJ03], the ISIS algorithm is extended to include estimation of the polarisation matrix of individual propagation paths. This new scheme allows for estimation of the directions of departure, directions of arrival, propagation delays, Doppler frequencies, and polarisation matrices of individual paths propagating between the transmitter and the receiver sites. Experimental investigations in a non-LoS pico/microcellular environment show that the polarisation characteristics of individual propagation paths can be related directly to the types of interaction that the waves experience along their paths, such as reflection, diffraction and scattering.

An ambiguity problem in estimating Doppler frequency and directions is analysed in [YiFS03] and [PPYF04]. This problem occurs when high-resolution channel estimation methods are implemented in the channel sounding using a switched timing scheme. Theoretical analysis clarified that the switched timing schemes adopted in such systems can be constructively incorporated into the ISIS algorithm. This attribute allows performing of the Doppler frequency estimation in a large scope

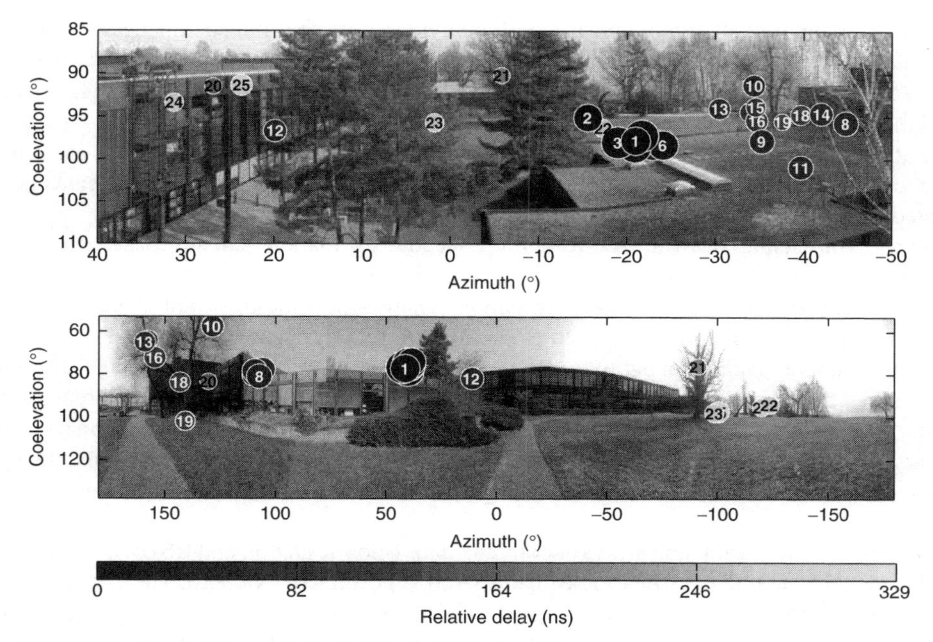

Figure 4.28 *Experimental data obtained with the ISIS algorithm.*

and high resolution without the ambiguity problems. Experimental investigation using PROPSound illustrated this merit in mobile radio environments. Report [HaTa03b] presents a MIMO sounding system dedicated to short range wireless communication. Using the system, a double directional measurement campaign in an indoor NLoS environment was performed, where the channel was estimated in a deterministic way with the above-mentioned ISIS algorithm. Estimation of polarisation characteristics for identical paths was proposed.

An approach leading to lower complexity and faster convergence of the SAGE algorithm is derived in [RiOK04]. The proposed method employs the Von Mises distribution model defined in the angular domain. This models the directional data observed in channel measurement campaigns using only a few parameters. The mixture model allows for representation of scenarios where multiple scatterer clusters are present, and also for representation of each cluster as a finite mixture of Von Mises distributions. The benefits of this method are due to smoother likelihood function and significantly lower dimensionality of the model, leading to lower variance of estimates and reduced computational complexity.

Another approach with the SAGE algorithm is the implementation in the frequency domain, which is more straightforward with frequency domain measurements. An overview of various maximum-likelihood algorithms with different implementation methodologies to estimate multipath parameters can be found in [TaBN04]. Due to the iterative nature of the algorithms, and with the requirement of estimating multidimensional multipath parameters from a vast measurement database, a number of simplified solutions has been proposed in this paper. The performance of the proposed algorithms is compared with that of the classical SAGE algorithm. One of the described algorithms is the Hybrid-Space SAGE (HS-SAGE) algorithm, which is a combination of the element-space and beamspace processing. In [PaTB04], a capacity-based analysis of directional MIMO channel measurements is presented. It is not only suitable for use with a circular array, but it also enhances the processing speed without sacrificing the accuracy and the resolution. A comparison of directly measured MIMO channels and MIMO channels generated from the multipath parameters, extracted using the HS-SAGE algorithm, is described to prove the previously mentioned statements.

In [ThLR04], an ameliorated version of this frequency domain SAGE algorithm is implemented. Instead of the normal serial interference cancellation, the parallel interference cancellation is suggested. The theory is checked with measurement results.

Another implementation of a high-resolution spectral estimation scheme, based on the SAGE algorithm, is suggested in [SeKa02a], and is compared with the subspace-based ESPRIT algorithm. By applying high-resolution parameter estimation schemes on time-invariant channels, the actual and past channel information is analysed to predict the future channel situation. For time-variant situations, parameter tracking or parameter estimation takes the evolutional nature of the radio channel into account. Applied on ray-tracing simulation data, a fading forecast using these different estimation schemes is calculated. A toolbox for spectral analysis and linear prediction of stationary and non-stationary signals is described in [Semm04a]. It includes the generation of test signals, the spectral analysis and parameter estimation of the test signals and measurements and the linear prediction of the signals.

Polarisation-sensitive antenna arrays at both link ends (mobile station and BS) of a MIMO channel are included in the parameter estimation with the SAGE algorithm of [Rich05]. The joint estimation of the four complex polarisation transmission path weights (the directions of arrival, directions of departure, time delays and Doppler frequencies) of measurements showed the polarisation dependence of wave propagation in a mobile radio channel.

An extended data model for high resolution channel parameter estimation and parametric channel modelling is introduced in [TLRT05b]. This is totally different from the well-known ray-optical-based data models, which contain only discrete specular propagation paths. In this paper, distributed diffuse scattering components are also included. For these parameters an estimator is developed and their Cramer-Rao lower bound is derived. Finally, the integration of this extended channel model into the SAGE algorithm is discussed.

A newly introduced multidimensional maximum likelihood parameter estimator is described in [ThLR04] and is called the RIMAX algorithm. It estimates jointly the parameters of the specular components (propagation paths) as well as the parameters of the distributed diffuse scatterers. The algorithm is based on the conjugate gradient optimisation strategy. In comparison with the SAGE algorithm, the complexity is drastically reduced, especially if the number of paths (components) is small compared to the number of observations. Additionally, the algorithm provides an estimate of the variance of the calculated parameters, yielding reliability information of the channel parameters estimated. This tracking algorithm has been tested in a static SIMO scenario in an urban environment and described in [AlRT04].

In a MIMO indoor propagation environment, the knowledge of cluster characteristics is very important. Here also the SAGE algorithm proves its qualities. First, identification of clusters was done visually using the double-directional angular power spectrum. The characteristics of each cluster were calculated with the new algorithm, based on SAGE estimation of eigenmodes. The next step in the development of RIMAX is proposed in [RiEK05]. A state-space-based method, describing the dynamics of the channel parameters in the time, for parameter tracking in order to estimate the propagation paths is suggested here. This results in a lower computational complexity and allows the tracking of the desired time varying parameters as well. The performance of the proposed technique is investigated using channel sounding measurements.

4.4.4 *Device performance and reliability of channel structure estimates*

A thorough investigation of the multidimensional wave propagation mechanisms is a prerequisite for understanding the spatial and temporal structure of the channel transfer matrix. The accuracy and reliability of the estimated channel parameters from a field measurement are dependent on the channel sounding system design. In case of the spatial domain the design of the antenna arrays [TLAP02], [KaKC02] is one parameter which determines the reliability of the DoA estimates. Due to practical aspects, e.g. mechanical, electrical construction and imperfections of the antenna arrays, the performance of real antenna array differs from the designed and simulated one. Therefore methods for calibration and evaluation were developed [LaRT04a], [LaRT04b].

The proposed design and the evaluation methods of antenna arrays are based on techniques using the Cramer-Rao Lower Bound (CRLB) and beamformer optimisation. The CRLB defines variance limits of the DoA estimates independent from any parameter estimator. It results from the calculation of the inverse covariance matrix of the matrix of the first-order derivatives of the observed data with respect to the channel parameters (e.g. DoA). Therefore a derivable data model of the antenna array is required. For this purpose analytic functions of the array manifold are often used in case of design and simulations [KaKC02]. These approaches often neglect the influence of the antenna beam patterns, mutual coupling and other characteristics or imperfections of real antenna arrays for simplicity of the derivatives. Therefore a derivable data model based on calibration measurement data of real antenna arrays is proposed [LaRT04a].

The methods using the CRLB to design an antenna array assume that the estimation algorithm will find the global minimum of the cost function of the optimisation problem. Most of the estimation algorithms are sequential; this means that they estimate the channel parameters path by path and parameter by parameter. This can result in a solution which is just a local minimum of the cost function (e.g. estimation of phantom paths). Therefore also an optimisation of the beamformer is considered [TLAP02]. Both the CRLB minimisation step and the beamformer optimisation step should be combined to achieve satisfying results using the designed antennas in channel measurements and analysis.

CRLB minimisation step

One approach for the optimisation of antenna arrays used in mobile radio channel characterisation devices is to apply a Genetic Algorithm (GA) [KaKC02]. It is proposed to design the array of a channel characterisation system, in order to minimise the CRLB (minimum variance) of the DoA estimates. A critical issue is to choose the optimal array geometry, namely to find the positions where the array elements should be placed, in order to obtain measurements that reduce the DoA estimation error to a minimum, assuming a minimum variance unbiased estimation algorithm, e.g. ML algorithm. Each sensor is supposed to have an isotropic beam pattern. It is shown that the GA determines the position of the sensors to minimise the mean CRLB of the DoAs for the unknown number and position of scatterers and also in case of prior knowledge of the clusters from which the signals originate.

Beamformer optimisation step

To avoid estimation results which are local minima of the cost function in [TLAP02] it is proposed to suppress the side lobes of the beamformer $c_k(x, \Theta) = x_k^H a(\Theta)$, where x_k is the complete data of the kth path and $a\,(\Theta)$ is the normalised steering vector of the antenna array with respect to the DoA Θ. In the initialisation step, from most ML algorithms such as EM and SAGE this function is calculated. For an example of a Uniform Circular Array (UCA) with omnidirectional beam patterns the best side lobe level suppression is achieved by a radius of $r_{opt} \approx N\lambda/16 < \lambda/4 \sin\left(\pi^2/N\right)$. Further suppression is possible through the joint optimisation of the beam pattern of the antenna elements and the array radius. It is found that the radius of the UCA should be as big as possible in terms of a minimum CRLB of the DoA. In terms of side lobe suppression of the beamformer a smaller radius is preferred. Therefore it is necessary to find a compromise between the minimum CRLB and the optimum beamformer.

Evaluation of real antenna arrays

Once the antenna array is realised it can be measured in an anechoic chamber. The antenna response is described by the complex polarimetric beam patterns of all elements which result from vertical and horizontal polarised excitation. Recording of the complete spherical beam pattern requires precise rotation of the array around a suitable defined pivot point located in the phase reference centre of the array and excellent phase stability of the set-up. The measured beam patterns are discrete in azimuth and elevation. Due to the periodicity of the beam patterns in 2π the DFT transforms beam pattern to the Effective Aperture Distribution Function (EADF) domain [LaRT04a], [LaRT04b]. The EADF matrix is calculated by a 2D Fourier transform by applying the Fourier matrices of the azimuth and the elevation to the discrete beam patterns. The EADF concept allows a considerable data compression since it is distinguished by a limited support area as shown in Figure 4.29 (b). On the other hand, it allows a simple analytic calculation of the derivatives with respect to the angular parameters.

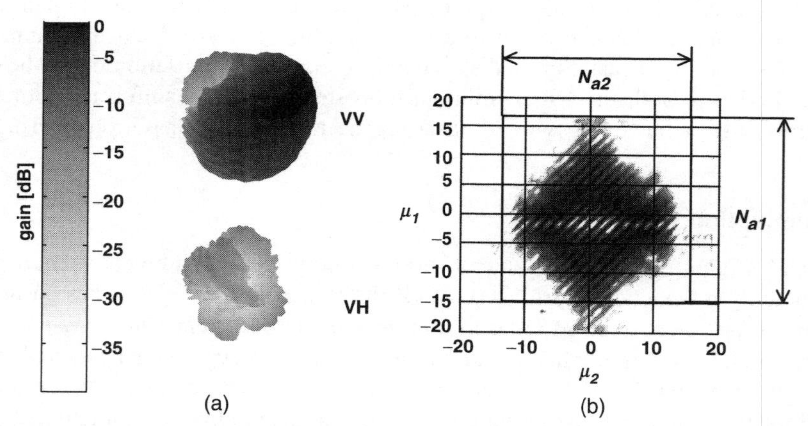

Figure 4.29 *(a) Polarimetric beam pattern for vertical (V) and horizontal (H) excitation; (b) corresponding EADF for the vertical excitation (VV) of an SPUCPA.*

Once one has the first-order derivatives with respect to the DoA the Fisher Information Matrix (FIM) and its inverse the CRLB of an arbitrary test scenario can be calculated. The investigation and exploitation of the FIM structure is essential to design a robust and efficient parameter estimator. The simplest scenario is a single impinging path scenario, which is analysed in the following example. It is related to an SPUCPA with 192 elements. In this scenario the FIM is diagonal. In Figure 4.30 the CRLB of the azimuth angle is compared to the variance which is achieved by an experiment.

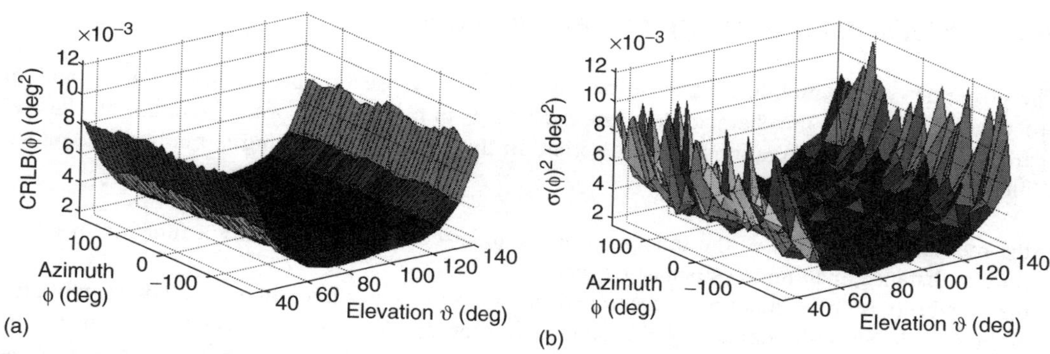

Figure 4.30 *(a) CRLB and estimated variance (b) of estimated azimuth DoA vs. the true azimuth and elevation direction range.*

For any true azimuth/elevation pair φ, ϑ within the coverage sector of the SPUCPA the azimuth was estimated by an ML procedure [Rich05], [ThLR04]. This experiment was repeated 64 times. The noise level for CRLB calculation was adjusted to match the observed device noise level which was held constant according to an SNR in the main beam direction ($\vartheta = 90°$) of 17–18 dB. It is also shown that this evaluation method is valid in the case of more complicated scenarios.

Further, the described method can be used for the ML parameter estimation as clarified in [TLRT05b], [ThLR04], [Rich05]. Here it is used to define the fundamental limit of the achievable variance of the estimated DoA/DoD parameters in terms of the CRLB to check the reliability of the obtained parameters during the estimation process.

Another approach for evaluating the performance of multidimensional channel sounding is presented in [MuSV05]. There the idea is to use existing already parametrised measurement data to create a dataset that resembles closely the original channel information. The approach is based on combining the radiation patterns obtained in calibration measurement of the measurement arrays with the parametrised propagation data. This way a well-known but still realistic test data with multiple signals can be created for studying, e.g. the correctness of estimated angular distributions.

4.5 Channel characterisation

4.5.1 Introduction

This section will mainly deal with aspects of the variability of power received through the mobile radio channel. These aspects comprise the dependence with distance, both on a small and large scale, and its temporal and angular dispersion, all regarded separately. Descriptions of joint behaviour can be found in the chapter on space-time channels and systems, Section 6.3.

The structure will be as follows:

1. a short introduction to the formal requirements on the processing of measured data;
2. a subsection on path loss, its distance dependence and small- and large-scale fluctuations;
3. a subsection on temporal and angular dispersion; and
4. a subsection on fading prediction techniques.

4.5.2 Statistical processing of measurements

The properties of the mobile channel to be treated in this subsection are all modelled as random processes. The bulk of the characteristics are derived, as in the majority of the literature on channel characterisation, from operations on time series. In this way, the following is implicitly assumed [Bult04]:

- the measured variability of the random process under consideration is only a function of either time or distance, making it a stochastic process;
- the variability is only a function of the time difference or distance interval between two samplings, as in a Wide-Sense-Stationary (WSS) stochastic process. As a result, (ensemble) mean values should be independent of time or distance and autocorrelation functions should only depend on the time or distance difference;
- the stochastic process is ergodic, allowing replacing ensembles by isolated sample functions, mostly time series.

Effectively, this means that the mean of the stochastic process should be constant and that the covariance of the process should exist, conditions that could prove difficult to test in practical situations. It is also fair to say that in the literature tests are seldom reported that justify the above assumptions.

One possibility is to assure that the stochastic process is both WSS and Gaussian, as such processes are ergodic. However, Gaussian properties should be inferred through the central limit theorem, as time series analysis inherently assumes ergodicity. Bultitude takes the occurrence of less than 15 multipath components on the radio channel as an indication that Gaussian properties cannot be assumed for the channel transfer [Bult02b]. Depending on the distributions and the independence of the multi-path components, much larger numbers than 15 could be needed to safely assume Gaussian properties.

Bultitude tested on WSS [Bult03] by comparing subsequent Doppler spectra. He used a MUSIC algorithm to extract Doppler parameters from half-overlapping 1 metre segments of measurements and computed the running variance of both the mean power-weighted angle of arrival and the angular spread. Changes of greater than 10% of the maximum value of either variances were taken to indicate non-stationarity. For measurements in a microcellular scenario in downtown Ottawa (Manhattan grid) at 1.8 GHz, approximately negatively exponentially distributed stationarity intervals, 'consistency lengths', were found, but the measurements showed greater probabilities for longer intervals. For Non-LoS (NLoS) streets, the median values of the lengths of these stationarity intervals ranged from 2.6 m to 3.3 m, depending on street orientation [Bult03]. Another technique developed by Bultitude for testing on WSS is inspection of the variation around the mean in spectral lines [Bult02b].

In the process of determining frequency correlation functions, it is common practice to use a Fourier transform on the (average) power delay profile. However, this assumes Uncorrelated Scattering (US). One test on US is to determine frequency correlation functions directly from the spectrum and to test whether the resulting correlation functions are symmetric with respect to arbitrary reference frequencies [Bult02b]. The validity of the assumption of US in indoor environments has been treated by Kattenbach [Katt97]. Also, the presence of dominant coherent components, as in Ricean fading processes, can distort results obtained by Fourier transforms. Salous and Gokalp presented an example where the relatively small differences between urban up- and downlink characteristics at 2 GHz resulted in clear differences between frequency correlation estimates despite similar power delay profiles [SaGo01]. Figure 4.31 shows these differences between up- and downlink for frequency

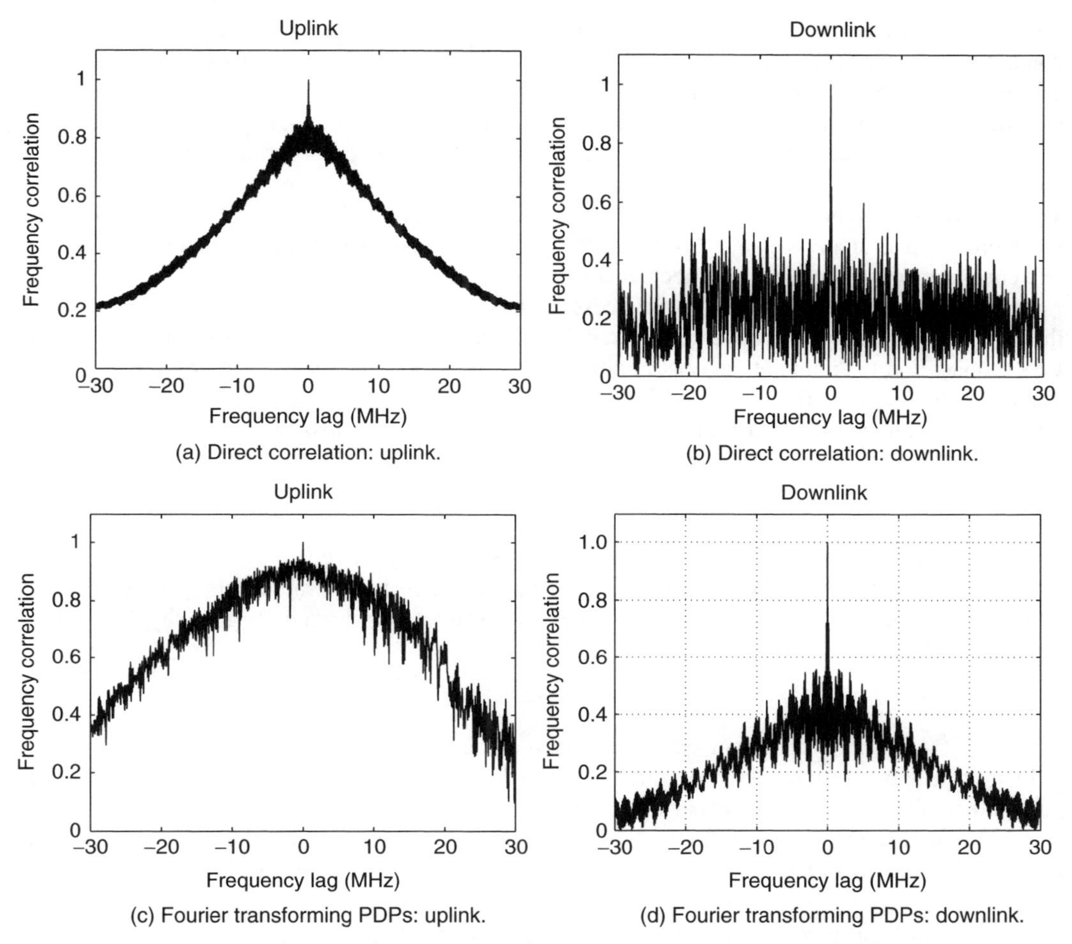

Figure 4.31 *Comparison between frequency correlation estimates at 2 GHz [SaGo01]. (© 2001 IEEE, reproduced with permission)*

correlation estimates by direct correlation as opposed to those derived from Fourier transforming power delay profiles.

4.5.3 Path loss and building penetration

A vast number of different path-loss models have been proposed in the past two decades, ranging from simple empirical to more complex physical ones [Corr01]. Each of these models has it own advantages and drawbacks. Hence, path loss in outdoor macro- and microcells is addressed, including vegetated residential environments. Indoor and building penetration models are also presented.

An approach to extend the validity domain of simple empirical models, e.g. Okumura-Hata [OOKF68], [Hata80], in urban areas is proposed in [GCBV02]. It is assumed that propagation in urban environments occurs by means of reflections and diffractions either along the streets (lateral propagation) or over the buildings (vertical propagation) therefore, two independent models can be combined into an overall statistical one. The two components are treated separately, and each of the submodels can be refined independently of each other. The obtained model leads to the well-known exponential law when BS antenna height and path length are large enough, the Okumura-Hata model is being used in its validity domain, and is being replaced by the proposed extension for distances below 1 km and for BS antennas lower than 30 m. The mix between vertical and lateral propagation, as well as the existence of LoS, is expected to change as a function of the distance. A dominant mechanism exists at any distance from the BS, which allows the identification of three regions, each one characterised by one of these mechanisms.

Once these regions are identified, it is necessary to identify the transition modes between them, and the weighting factors to use for each of the three components. In order to identify the region where LoS propagation is dominant, an obstruction probability factor, p, is used, defining the probability that a link between the BS and the MT is obstructed. Globally, p will be zero in the vicinity of the BS, and it will increase with distance, approaching unity. The path loss in an urban environment is obtained as

$$L_p = (1-p)L_p^{LoS} + pL_p^{NLoS} \tag{4.4}$$

with L_p^{LoS} and L_p^{NLoS} being the path loss in LoS and NLoS, respectively.

The NLoS component results from the combination of two terms, representing the vertical and lateral propagation. For simplicity, it is assumed that the total field strength is given by the strongest component, thus,

$$L_p^{NLoS} = \min(L_p^{LAT}, L_p^{VERT}) \tag{4.5}$$

where L_p^{LAT} and L_p^{VERT} represent the lateral and vertical attenuations, respectively.

In order to apply the model, three components need to be evaluated as a function of the geometrical parameters of the propagation scenario.

Assuming that the received field strengths in the lateral and vertical planes are statistically independent, the total received field strength standard deviation can be evaluated from

$$\sigma(d) = \sqrt{\sigma_{VERT}^2(d)q(d) + \sigma_{LAT}^2(d)\,[1-q(d)] + [\mu_{VERT}(d) - \mu_{LAT}(d)]^2\, q(d)\,[1-q(d)]} \tag{4.6}$$

where q is the dominance probability, i.e. the probability that, at a given distance, d, the dominant component in the received field is the one due to vertical propagation, and the mean value and standard deviation for the two different components are obtained as $\mu_{LAT} = \alpha_1 d + k_1$, $\mu_{VERT} = 10\alpha_2 \log(d) + k_2$, $\sigma_{LAT} = \min(\beta_1 d, 20)$ and $\sigma_{VERT} = \beta_2$.

Parameters α_1, α_2, β_1, β_2, k_1 and k_2 depend on the geometrical and topological characteristics of the considered area, e.g. mean building height, mean street width, BS antenna height, etc. Values of $\alpha_1 = 0.108$, $\alpha_2 = 7.324$, $\beta_1 = 0.027$, $\beta_2 = 6.783$, $k_1 = 64.777$ and $k_2 = -70.000$ were obtained from fitting of measurement data from a small cell at 947 MHz. Additional data for 1800 MHz can be found in [BCFF02].

It is known that path loss depends on the carrier frequency. Wideband measurements at the same environments in two frequency bands, namely 2.45 and 5.25 GHz, and a system bandwidth of 100 MHz, are presented in [LZMJ05]. Rural measurements were performed in a small rural municipality, Tyrnävä, which is located close to Oulu, Finland. The measurement route was about 1 km long and mainly LoS type, thus being representative of flat rural LoS macrocellular environments.

The suburban environment is a typical suburban residential area in the Oulu region with at most four- to six-storey buildings in the vicinity (one- to two-storey buildings in the measurement route), and not so wide streets. The street grid is rather regular, and no open areas can be found in between the houses. An unobstructed LoS condition was observed. Macro- and microcellular cases were considered by positioning the receiver at different heights, 11.7 and 7.6 m being considered, respectively.

Based on measurements, a simple log-distance path-loss model is derived by using linear regression of the scatter plot of the measured data. The results for rural and suburban cases are reported in Table 4.5. Some figures for the rural case are shown as an example, Figure 4.32.

Table 4.5 *Path loss in rural and suburban environments.*

	L_p [dB]	
f [GHz]	Rural	Suburban
2.45	$38.3 + 21.1 \log(d)$, $d \leq 650$ m $-105 + 75.0 \log(d)$, $d > 650$ m	Not available
5.25	$41.8 + 22.0 \log(d)$	'Macrocell' $43.3 + 22.8 \log(d)$ 'Microcell' $41.6 + 23.8 \log(d)$

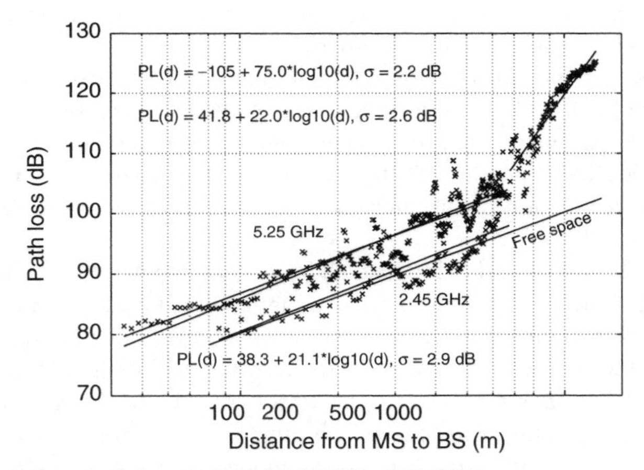

Figure 4.32 *Rural path-loss models at 2.45 and 5.25 GHz [LZMJ05].*

In [ZVRK04], empirical path-loss models for microcells are derived from measurements at 5.3 GHz and a system bandwidth of 100 MHz, performed in an urban microcellular environment in the centre of Helsinki. The environment is characterised by a regular street grid, where the building height varies between four and seven storeys. The BS antenna was placed in the middle of street crossings at a height of approximately 10 m above ground level, i.e. clearly below the rooftop level. At the BS, a horizontally aligned uniform linear array of four dual-polarised microstrip patch elements was used. At the MT, two omnidirectional (dipole) elements and a mockup user terminal with four antennas were used.

Three BS sites were chosen at different street crossings, and in each BS location several BS array orientations were measured. In each BS site, the array was always directed towards a street canyon to measure the received powers along the canyon and the perpendicular streets. Measured streets have different widths within approximately 20 to 30 m.

Empirical path-loss models were derived from measurement data by using the least square error method. For the LoS case, seven streets were measured, a path-loss exponent of 2.34 and a standard deviation of 2.6 dB being obtained. For the NLoS one, nine perpendicular streets were measured. It was observed that the path-loss exponents are within 1.6 to 2.9, with a mean value of 2.34, the standard deviation is within 1.9 to 5.6 dB, with a mean value of 3.1 dB.

Field prediction in vegetated residential environments is not easily accomplished by standard path-loss prediction models [Corr01]. The electromagnetic fields of radio waves propagating through a tree canopy can be decomposed into mean (coherent) and diffuse (incoherent) components. At low frequencies, the diffuse component is relatively small and only mean fields need to be considered [ToBL98]. As the frequency increases, spatial fluctuations of the field must be taken into account. Also, even at low frequencies, the incoherent fields become important as the distance between the transmitter and the receiver increases in a random scattering medium. In [ToLa04], the behaviour of the incoherent component at Very High Frequency (VHF) and Ultra High Frequency (UHF) is examined for a trunk-dominated environment. It is assumed that trunks are symmetric, and that transmitter and receiver are in the same transverse plane. The phase function appearing in the radiative transport theory is obtained from the differential scattering cross-section of a dielectric cylinder, being valid for both low and high frequencies. The exact radiative transport equation is solved numerically by the eigenvalue technique. The solution is used to compute both the coherent and the incoherent attenuation constants in the trunk-dominated environment.

For a complex cross-section permittivity of $10^{-3}j$, a tree trunk radius of 10 cm, and a density of 1000 trees per hectare [ToLa04], it is observed that attenuation expressed in dB decreases linearly with increasing distance, the incoherent component contributing over the coherent one. For example, at 100 m the incoherent intensity adds 10 dB over the coherent attenuation. Globally, the coherent attenuation, in dB, as a function of distance, is given by $0.21d$, whereas the total attenuation is obtained as $0.11d$.

Indoor measurements at 2.45 and 5.25 GHz, and a system bandwidth of 100 MHz, carried out at Elektrobit premises in Oulu during night time in an unpopulated office are presented in [LZMJ05]. By using linear regression of the measured data the path loss was evaluated, $L_p = 48.1 + 17.3\log(d)$ and $L_p = 3.3 + 52.1\log(d)$ being obtained at 5.25 GHz, for the LoS and NLoS cases, respectively. At 2.45 GHz, $L_p = 42.7 + 16.4\log(d)$ and $L_p = 10.8 + 45.4\log(d)$, was evaluated.

A method for modelling path loss inside office buildings is presented in [MeBe02]. The proposed formulation is a combination of a street microcell model with a power law one that is valid in and around straight corridor sections. The model has been validated by comparing it to path-loss measurements in a large office building with a complex topology. Narrowband measurements were performed at 5.2 GHz. The topology of the building was quite complex, consisting of several straight corridor sections, which were connected at crossings and turnings. The building was 70 m wide, 100 m long and five storeys high. The floors were made of reinforced concrete, while the outer walls were made of brick and reinforced concrete, and the inner walls between each room were made of double plasterboard supported by vertical metal studs.

The basic power law model does not account for indoor topologies with corridor segments which are connected at crossings and turnings. The recursive model, however, was designed for such topologies [Berg95]. By combining the power law model, and the recursive model, complex indoor environments can be easily modelled [MeBe02].

By fitting the model parameters to the measurement data, path-loss exponents of 1.55, 2.7 and 3.9 were obtained for the *corridor-corridor*, *room-corridor* and *room-room* classes, respectively.

In Figure 4.33, the received power is shown (solid line) relative to the power measured at 1 m from the BS, and the results of the combined model for the case of *corridor-corridor* class (dashed line). Points B and D are nodes of the recursive model, and at C a loss of 4 dB due to an obstruction is considered.

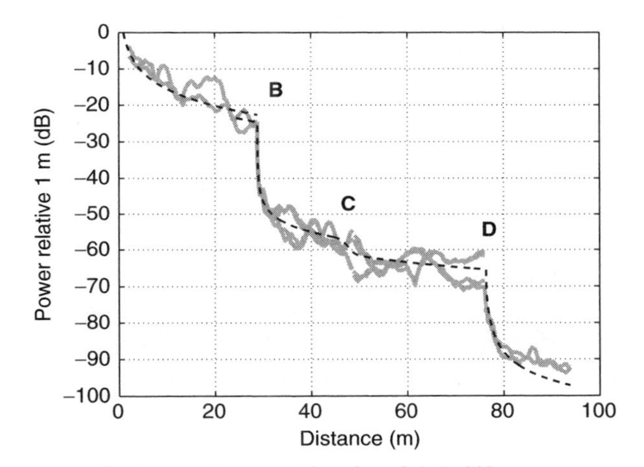

Figure 4.33 *Received power for the corridor-corridor class [MeBe02].*

The millimetre frequency band at 60 GHz is one promising candidate to provide transmission data rates up to several hundred megabits per second, and it has been proposed for future Wireless Local Area Networks (WLANs) [Corr01]. From wideband measurements in rooms and corridors of two different buildings of the Helsinki University of Technology, several results were derived [Geng05], Figure 4.34.

In contrast to usual buildings, hospitals are partly constructed of special walls with metallic layers inside. The wall structure depends on the kind of hospital room. Narrowband measurements in relevant areas of four different hospitals are presented in [ScFW02].

The results of the measurements reveal that it is important to consider the room as a whole, when determining the attenuation to neighbouring rooms. Slots and openings in walls give rise to coupling effects through rooms with metallic layers inside the wall. In order to accurately investigate wave propagation properties, diverse antennas and measurement systems were used. The results show a varying behaviour of the wave propagation in different rooms of the hospital.

The attenuation of single walls was measured by positioning horn antennas on both sides of the wall, and then measuring the attenuation of the transmitted power directly through the wall. Measurements were performed at 2.45 and 5.2 GHz, with the antennas positioned at 50 cm from the wall. In Table 4.6, the results of the measured attenuation of the walls in hospital I at 2.45 GHz are shown.

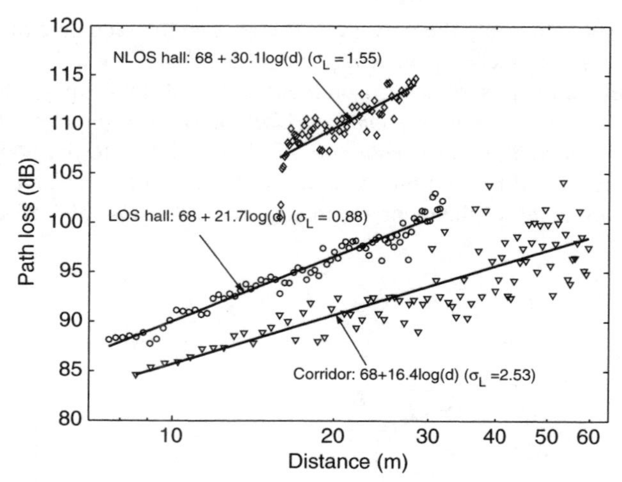

Figure 4.34 *Path loss at 60 GHz [Geng05].*

Table 4.6 *Attenuation in hospital I at 2.45 GHz.*

Room	Attenuation [dB]	
MRTo	window	95.5
	door	73.2
	wall	91.5
X-ray	door	41.5
	wall	51.3
operating	wall	61.6
plasterboard	wall	6.4
concrete	wall	35.5

Table 4.7 shows results of wall attenuations measured in hospital II. Hospital I was newly built in 2000, whereas hospital II was built in 1970. This means that walls are built in different ways, especially those of the rooms for X-ray examinations, which are made of plaster blocks in hospital II. This leads to different results compared to that of Table 4.6.

A common result is that the attenuation increases for higher frequencies, with a difference ranging from about 3 dB for concrete to about 15 dB for Pb-shielded walls (x-ray and operating rooms). Another interesting result is that the wall attenuation is generally lower for horizontal polarisation. A possible reason for this behaviour could be the vertical structure of the walls: all the slots in the metallic shielding and all the steel beams of the basic wall structure are vertically aligned.

There are mainly two different approaches for indoor coverage planning:

1. picocells deployment, where the BS is installed inside the building, e.g. Digital Enhanced Cordless Telecommunications (DECT), and
2. indoor coverage from an externally installed BS (micro- or small macrocells deployment), where, in most cases, propagation takes place over NLoS conditions.

Table 4.7 *Attenuation in hospital II at 2.45 and 5.25 GHz, vertical and horizontal polarisation.*

Room	2.45 GHz Vert. pol.	2.45 GHz Hor. pol.	5.25 GHz Vert. pol.	5.25 GHz Hor. pol.
concrete (10 cm)	7.4	6.8	10.7	10.3
operating	38.2	39.9	52.7	48.3
X-ray wall	42.6	31.1	61.9	46.6
X-ray door	44.2	30.4	61.7	56.9
MRTo wall	91.4	81.2	95.7	95.3
MRTo door	73.7	67.4	93.7	102.9

A great deal of the radio coverage inside buildings is still being provided with the use of BSs located outdoors. The building penetration loss is defined as the difference between the average signal strength in the local area around a building and the average signal strength on a given building floor. For radio planning purposes, the building or room loss factor can be used as an addition to the predicted signal loss for the surrounding local area. At higher floors, the received signal strength will be higher than at the ground floor. This is important for radio planning, because it may cause higher interference levels at higher floors.

Empirical penetration loss models for indoor coverage planning and their parameters for the 1.8 GHz frequency band, experimentally supported on a measurement campaign that has been performed inside 11 different buildings in the city of Lisbon, are presented in [XaVC03]. The models' reference is the propagation loss evaluated at an external wall 1.5 m above the ground level.

The measured buildings were divided into four classes, according to their height and type of integration in the urban structure: High Isolated, Low Isolated, High Integrated and Low Integrated. The difference between integrated and isolated buildings is made by the fact of sharing (or not) walls with the surrounding buildings. An integrated building always shares walls with some others, while an isolated one does not share any. Concerning the height parameters (High/Low), buildings are classified by the number of floors, up to four floors (included) being considered Low, else being classified as High. A description of the measured buildings' characteristics can be found in [XaVC03].

By processing the collected data, it was noticed that all attenuation histograms had a common shape for each building, thus motivating the creation of a new model, the 'Double Sided Gaussian Model'. Histograms had the shape of a kind of 'unbalanced' Gaussian function, with distinct standard deviations for the mean value of 'right' and 'left' sides. The Double Sided Gaussian Model output parameters are the mean value and the standard deviation for both semi-Gaussians, Table 4.8.

Table 4.8 *Double-sided Gaussian parameters per building class.*

Building class	Mean [dB]	Std. dev. right [dB]	Std. dev. left [dB]
High Isolated	2.59	16.59	9.74
Low Isolated	15.79	10.97	12.95
High Integrated	13.27	13.92	12.37
Low Integrated	12.83	9.81	6.28
All classes	10.19	13.85	13.84

For the same buildings, the comparison between 1.8 GHz measurements and 900 MHz measurements was made, a difference of about 6 dB being found, hence verifying the attenuation dependence of 20 log(f). Following the same law, it is expected that the presented results can be extrapolated for the UMTS frequency bands.

In [MaHe03], measurement results taken in office buildings situated in The Hague are presented. Measurements show that long-term fluctuations can be observed between signal levels received in different parts of a building. While on lower floors these fluctuations follow a log-normal distribution, significant differences have been observed at higher floors between LoS and NLoS areas. Therefore, the relationship between the floor height and extra gain with respect to the ground floor level is not linear, depending on factors such as the radiation pattern of the BS antenna and the local urban clutter effect.

Additional results on indoor coverage by outdoor BSs at 1.8 GHz can be found in [KüMe02], empirical factors for building penetration and height gain from extensive measurement campaigns for the 1.8 GHz frequency band being provided.

4.5.4 Long-term fading

In a complex propagation environment, as usually found in mobile communication systems, the received signal results from all possible effects associated with the interactions between the propagating wave and the objects within the propagation space. From the analysis of narrowband measurements in different environments, it is commonly accepted that, in general, the average received signal magnitude is random and log-normal distributed around the mean distance-dependent path-loss value [Pars92].

These variations around the distance-dependent value are usually referred to as slow- or long-term fading, since the magnitude of the received signal remains approximately constant for short periods of time or space; therefore, variations can only be observed on a long-term basis, tenths of wavelengths being usually considered [Pars92]. This type of long-term variation is commonly justified by the changes in the visibility or obstruction of multipath components due to MT or scattering environment changes (movement). This phenomenon is related to the denotation of shadow fading, which is sometimes used when referring to long-term fading.

This distribution is a well accepted model for modelling narrowband fading, being frequently used for system simulation and analytical calculation purposes. Nevertheless, a different long-term fading behaviour can be found in specific environments, e.g. tunnels [Corr01].

Wideband channel modelling is usually based on the channel impulse response of the propagation channel. Contrary to the narrowband case, in which the channel response is described by one single component, in the wideband one, it is described by different components (paths) arriving at different delays, being related to distinct propagation paths. Thus, from the long-term point of view, it is common to model each path as a narrowband channel, i.e. assuming a log-normal distribution for each path [Corr01].

From wideband measurements in urban macrocellular environments [MiVV04], quite large differences in the parameters of the log-normal distribution associated to different clusters were found, depending on the route being considered; therefore, questioning if the assumption of the log-normal distribution is valid when an arbitrary environment, is considered, e.g. all measured routes. In this case, different clusters will contribute to the major part of the total received power, depending on the measured route.

In fact, the assumption that the average received signal magnitude in dB is Gaussian distributed is widespread but not straightforward [Hans02]. From measurements carried out within rooms and corridors of a building with rather solid inner walls made of brick, it is observed that when both are considered, the results will not yield a single Gaussian distribution but rather a superposition of two, i.e. a bimodal Gaussian, which is best described by their two independent means and variances. Therefore, the performed measurements indicate that the assumption of a Gaussian distributed signal magnitude is not valid, when the measured statistics are dominated by a particular scenario, which has for instance much higher amplitudes than all the other parts of the investigated environment. Globally, it is observed that the log-normality of the path loss is not valid for arbitrary scenarios; therefore, it is recommended to separately model each arbitrary scenario by decomposing it into several subscenarios for which the parameters of the potential multimodal distributions should be calculated separately.

In most propagation models, long-term fading is generated independently for different MTs; however, in network simulations it is common to have several MTs operating within a given area. Since long-term fading effects depend on the MT position, which in network level simulations are usually close to each other, a better approximation to reality should be obtained if shadowing effects are not modelled independently for each MT.

Taking into account that long-term fading effects are modelled by the addition of a Gaussian distributed random variable, this does not completely model the shadowing process. An additional aspect should be considered: shadowing is a slowly variant characteristic of the radio channel. This variation rate indicates the existence in the time domain of a non-zero shadowing autocorrelation. As mobility is included, time correlation is intimately related to the space one. In fact physical explanation of shadowing is primarily associated to position [Lee85]. Spatial correlation can be modelled as [ETSI98]

$$R(\Delta r) = 2^{-\frac{\Delta r}{d_{corr}}} \tag{4.7}$$

where Δr is the space shift and d_{corr} the decorrelation distance, for which typical values can be found in [ETSI98].

Computer generation of shadowing data, taking into account (4.7), can be realised in two different ways:

1. The first approach initially generates a set of n independently distributed random Gaussian variables, each corresponding to one of a set of n equally separated positions. Obviously, this results in a sequence that is uncorrelated. Next, the sequence is filtered so that the desired autocorrelation function is obtained.
2. An alternative procedure is described in [ETSI98]. This procedure generates shadowing samples one by one, it is not necessary that they are equally separated. Once a shadowing sample, G_i, is produced, the next one is also a Gaussian random variable whose mean is $R(\Delta r)G_i$, and whose variance is $(1 - R(\Delta r)^2)\sigma^2$, σ being the standard deviation of shadowing values.

One major drawback of the second approach is that the autocorrelation depends only on the distance shift. Considering the case in which a series of propagation maps that account for shadowing needs to be generated, it involves producing a shadowing sample for each location in every map. Since maps are two dimensional, it is not possible to establish an order among its locations; consequently, the second approach cannot be adopted. In this situation, instead, the first one should be used; nevertheless, its application is not straightforward.

A two-dimensional modelling technique, which allows MTs in the vicinity of the simulated environment to undergo related shadowing effects, is presented in [FrLC03], [FGLM04]

$$R(\Delta x, \Delta y) = 2^{-\frac{\sqrt{\Delta x^2+\Delta y^2}}{d_{corr}}} \tag{4.8}$$

where Δx and Δy are the MT shift in horizontal and vertical coordinates, respectively, and d_{corr} is the decorrelation distance.

Assuming that a set of n shadowing maps corresponding to n BSs covering the same area must be generated, the procedure should be as follows:

1. generate $n + 1$ matrices, $g_0, g_1, \ldots, g_n$, every element of which is a Gaussian random variable with zero mean and standard deviation σ;
2. given a correlation coefficient of shadowing from different BSs equal to ρ, produce n shadowing maps, according to

$$G_i = \rho^{1/2} g_0 + (1-\rho)^{1/2} g_i, i = 1, 2, \ldots, n \tag{4.9}$$

3. compute the two-dimensional inverse Fourier transform to obtain $h(x, y)$;
4. use two-dimensional convolution to filter each shadowing map, G_i, to obtain the desired filtered maps.

This scheme for simulating shadowing provides the basis for carrying out more realistic system level simulations, since it provides a link between MT location and shadowing, which is coherent with the environment description.

4.5.5 Narrowband short-term fading

The rapid variations of the received signal over distances of the order of a few wavelengths, or short periods of time, due to the multipath channel behaviour, i.e. the changes in the magnitude and phase of arriving waves, is referred to as small-scale or short-term fading. Globally, the received signal magnitude is described by the superimposition of the distance-dependent path-loss value, long- and short-term fading effects.

Short-term fading results from the path length difference between waves arriving from different paths, i.e. from scatterers located within the propagation environment. These differences lead to significant phase differences.

In the narrowband case, all rays arrive essentially at the same time when compared to the system resolution; hence, the short-term variations of the magnitude of the received signal can be represented by the sum of an LoS component (if it exists) plus several reflected and/or diffracted ones, being usually described by a Rayleigh or Rice distribution, depending on the existence of a LoS component [Pars92].

Under LoS (Rice case), the ratio between the power of the LoS component and the power of the reflected/diffracted ones is usually named the Rice or Ricean factor, K. It should be noted that the value of K is a useful measure of the communications link performance, its proper estimation being of practical importance for an accurate channel characterisation [Corr01].

When the long-term fading effect is modelled by a log-normal distribution, and the Rayleigh one is used to model the short-term fading, the overall signal distribution can be modelled by the Suzuki distribution [Pars92].

It is usually well accepted that long- and short-term effects can be modelled separately. In [Ande02], it is shown that the long-term fading in general may not be due to shadowing, but rather to the slow variation of the coupling between scatterers, when the MT is moving. This means that the long-term fading is just as unpredictable as the short-term one, since it originates from the same scatterers. Shadowing will still exist behind major changes in the environment.

A new model using multiple scattering, replacing the traditional subdivision of the total fading into a slow log-normal and a fast Rayleigh component is proposed in [Ande02]. The physical basis is a model of forward scattering between scatterers, introducing multiple scattered waves defining a new transfer function, H. This function consists of a sum of a small number of terms, where each term is a multiple product of complex Gaussian distributions

$$H = K + H_1 + \alpha H_2 H_3 + \beta H_4 H_5 H_6 + \cdots \tag{4.10}$$

where H_i are complex, independent Gaussian variables (Rayleigh in magnitude), and K is a constant related to the Rice factor. Parameters K, α and β are found by minimising the mean square error against measurements.

The double fading has been described earlier by [EFLR97] as cascaded Rayleigh fading, and it was shown that the probability density for the power, s, is given by

$$p(s) = 2K_0\left(2\sqrt{s}\right) \tag{4.11}$$

where K_0 is the modified Bessel function of the second kind and zero order. The cumulative distribution function is then given by

$$\text{Prob}(s' \leq s) = 1 - 2\sqrt{s}K_1\left(2\sqrt{s}\right) \tag{4.12}$$

with K_1 representing the modified Bessel function of the second kind and first order.

The advantage of this distribution is the insight it gives into the origin of the long-term fading, while its disadvantage is the lack of a simple analytical function for the PDF of the received power except in special cases, e.g. the double product of Rayleigh fading paths.

This new model can be made similar in shape to the Suzuki distribution, but it has a different interpretation. The log-normal (refer to Suzuki) distribution is usually interpreted as a shadowing function, which influences the local mean value, the shadowing being dependent on the local environment. The proposed distribution has a constant mean power for the single scattering for the whole environment, and the variation of the mean of the total power stems from the slowly varying scattering between the scatterers as the MT moves. Thus, there is no need for a shadowing argument to explain the short-term fading. The resulting parameters from the fitting of the distributions may be interpreted as revealing the propagation mechanisms. According to [Ande02], the long-term fading originates from the same random elements as the short-term fading.

Short-term fading can be generated in a simulation by using either ray-tracing or statistical approaches [TLVD01], [Kunn02]. In statistical modelling, the fading processes have given predefined characteristics. Moreover, statistical approaches enable simple modelling of short-term fading in both time and frequency domains. Ray-tracing approaches, being site specific, are numerical intensive techniques.

Since the knowledge of the mean level of the received signal in a multipath environment, and of its variations around the mean, is fundamental when designing a wireless communications system, it should be properly assessed if different techniques are appropriate for evaluating such variations. From the work in [TLVD01] , it is shown that ray-tracing techniques make it possible to estimate with sufficient accuracy not only the mean signal level but also its variations about the mean, therefore allowing to evaluate its statistics, at least in indoor environments.

The capacity to calculate the mean power and the local statistics of the received signal leads to some optimism over the possibility of obtaining site specific channel models, which can account for the channel variations in different locations inside an area of interest. Furthermore, starting from the Doppler spectrum at one single point, obtained using ray-tracing techniques, it is possible to estimate the signal statistics in a local area of that point. This possibility is of great practical importance, since it substantially reduces the local statistics calculation time.

An alternative to such techniques is the statistical modelling approach. In [Kunn02], the statistical modelling and simulation of short-term fading with Rice distributed envelope and the desired temporal, spatial and spectral correlation is considered. The proposed model applies to the simulation of MIMO multicarrier systems. The simulated short-term fading channel gains consist of a deterministic direct part plus a Gaussian one, hence having Rice distributed envelope. The Gaussian part of the channel gains can have desired temporal, spatial and spectral correlation generated by time-correlation shaping filtering and space-frequency correlation transformation. The proposed short-term fading simulator structure enables also the generation of time varying tap weights for the classical tapped delay line model.

The number of Ricean fading channel gains to be generated is given by the product between the number of subcarriers, the number of transmit antennas and the number of receive ones. Concerning the simulator structure, a noise generator outputs uncorrelated white complex Gaussian noise sequences, which are fed to parallel time correlation shaping filters that have identical impulse responses. The uncorrelated output sequences with the desired temporal correlation are fed into a space-frequency correlation transformation that is performed for each set of uncorrelated samples. The complex Gaussian samples with the desired temporal and spatio-spectral correlation are added up with complex samples representing the direct component. Finally, the sequences are interpolated to get to the desired channel sampling rate.

Recently, MIMO systems have gained significant attention due to the potential of achieving high information theoretic capacities. Central to these techniques, multiple antennas are employed at both the BS and the MT in order to exploit the multipath richness of the propagation channel. One way to achieve this is to separate the antenna elements at both the BS and the MT, so that large diversity orders can be achieved. This requires antenna spacings of up to tens of wavelengths at the BS and up to a wavelength at the MT. Employing multiple antennas at the BS does not present a significant problem, but accommodating more antennas at the MT introduces several constraints for practical implementation.

In this regard, the Electromagnetic Vector Sensor (EVS) can be deployed as a compact MIMO receiver antenna capable of measuring six time varying electric fields, and magnetic fields, at a point in space, Figure 4.35. Though the EVS has been widely used for direction finding applications [NePa94], [WoZo00], recent results show that EVS can be applied to mobile communication systems, whereby the use of polarisation diversity can provide capacity improvement over a conventional dual-polarised system [AnMd01].

In [ThYo03], the closed-form expression for the spatial fading correlation function in terms of azimuth and elevation angle, and geometry of the EVS as well as polarisation states of the wave, is derived. Assuming a frequency non-selective directional Rayleigh fading channel model, in a MIMO configuration, the channel impulse response matrix $\mathbf{h}(t)$ is given by

$$\mathbf{h}(t) = \sum_{m=1}^{M} \alpha_m(t) \cdot \mathbf{a}\left(\phi_m, \theta_m, \gamma_m, \eta_m\right) \tag{4.13}$$

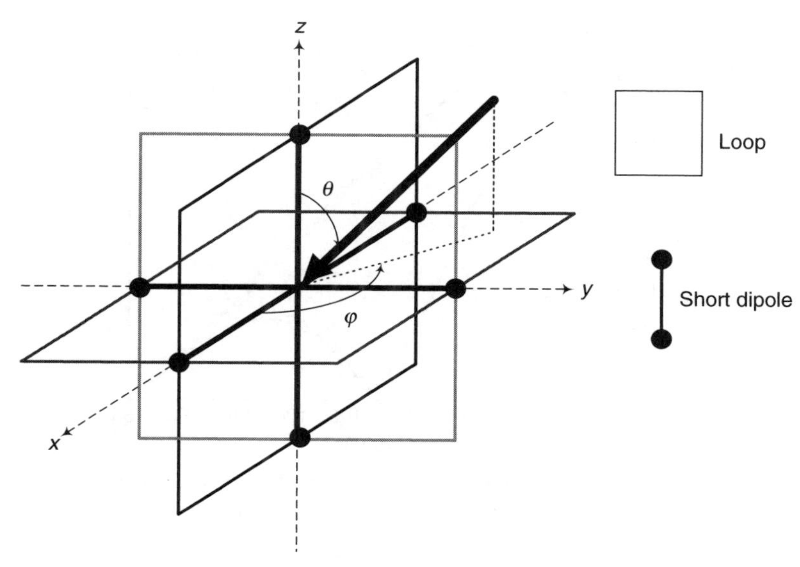

Figure 4.35 *An EVS composed of three identical short dipoles and three magnetically identical small loops [ThYo03].*

where α_m is the complex amplitude of the mth multipath component, M is the total number of components and $\mathbf{a}(\phi_m, \theta_m, \gamma_m, \eta_m)$ is the steering vector. Scalars ϕ and θ are the azimuth and elevation angles, as illustrated in Figure 4.35, while scalars γ and η denote the auxiliary polarisation angle and polarisation phase difference, respectively. In [ThYo03], the angle spread in both azimuth and elevation are modelled by uniform distributions. In addition, the azimuth spread and elevation spread values are the maximum deviation of the angle spread from mean direction of arrival. Let Θ be a spatial vector parameter denoting $\Theta = [\phi, \theta, \gamma, \eta]^T$, where $[\,]^T$ represents transpose; therefore, the spatial fading correlation between any elements (m, n) of the EVS can be expressed as

$$\rho_{(n,m)} = \frac{E\left[a_n(\Theta)a_m(\Theta)^*\right]}{\sqrt{E\left[a_n^2(\Theta)\right]E\left[a_m^2(\Theta)^*\right]}}$$

$$= \frac{\int\limits_{\phi}\int\limits_{\theta}\int\limits_{\gamma}\int\limits_{\eta} a_n(\Theta)a_m(\Theta)^*\sin(\theta)p(\Theta)d\theta d\phi d\gamma d\eta}{\sqrt{\int\limits_{\phi}\int\limits_{\theta}\int\limits_{\gamma}\int\limits_{\eta} |a_n(\Theta)|^2\sin(\theta)d\theta d\phi d\gamma d\eta}} \cdot \frac{1}{\sqrt{\int\limits_{\phi}\int\limits_{\theta}\int\limits_{\gamma}\int\limits_{\eta} |a_m(\Theta)|^2\sin(\theta)d\theta d\phi d\gamma d\eta}} \tag{4.14}$$

where $E[\,]$ denotes expectation, the superscript * signifies the complex conjugate, the scalar $p(\Theta)$ is the joint PDF for all the four parameters, and a_m are the mth entries of $\mathbf{a}(\Theta)$.

Illustrative results for a mean azimuth and elevation angle of arrival of 45° and 90°, respectively, and an elevation spread of 0°, as a function of azimuth spread, are presented in Figure 4.36.

4.5.6 Wideband short-term fading

Contrary to the narrowband case, in the wideband one the relative delays of arriving waves are large compared to the basic unit of information in the channel (symbol or bit); therefore, the wideband channel is usually modelled by a channel impulse response, where the magnitude of the different paths are taken as for the narrowband case.

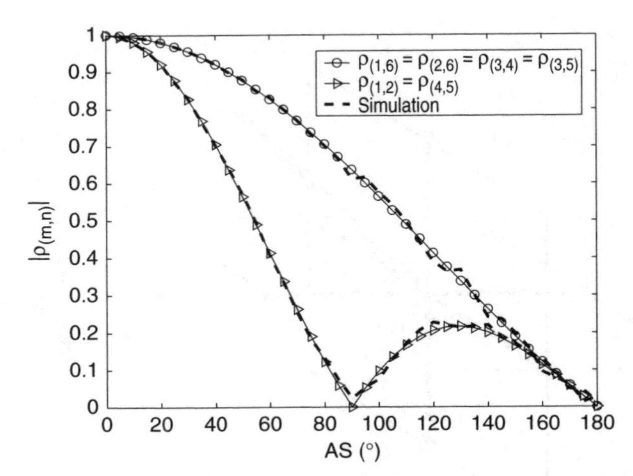

Figure 4.36 *Spatial correlation between elements m and n of the EVS for a mean azimuth of arrival of 45° a mean elevation of arrival of 90° and an elevation spread of 0°, [ThYo03].*

In most cases, under LoS, only the first path is modelled as being Rice distributed, since this path includes the LoS component, all other path magnitudes being modelled by a Rayleigh distribution. This approach was already suggested in [Corr01], where the Rice factor of the first path is derived as a function of the narrowband Rice factor, the system bandwidth, and the RMS delay spread of the propagation channel. Nevertheless, the proposed description of the Rice factor is only valid for a certain structure of the channel impulse response.

A different approach to overcome the bandwidth dependence of modelling parameters is based on the use of the time-variant transfer function instead of the time-variant impulse response, such that all known properties from narrowband models can be applied [Corr01], [Katt02].

A time-domain technique for fading depth characterisation in wideband Rayleigh and Ricean environments, described by its PDPs, is proposed in [CaCo02]. The PDF of the received power, s, is given by

$$p(s) = \int_{-\infty}^{+\infty} \frac{2K}{a_d^2} e^{-\frac{K(2x+1)}{a_d^2}} I_0\left(\frac{2K\sqrt{2x}}{a_d}\right) \sum_{m=2}^{M} \frac{(\lambda_m)^{M-3}\, e^{-\frac{s-x}{\lambda_m}}}{\prod\limits_{\substack{k=2\\k\neq m}}^{M} (\lambda_m - \lambda_k)} dx \tag{4.15}$$

where a_d is the magnitude of the LoS component, K is the Rice factor, I_0 is the Bessel function of first kind and zero order, and λ_m are eigenvalues of a covariance matrix obtained, by generating a matrix whose elements are given by the product between the correlation function between different frequency components, the frequency response of the filter used in the transmitting equipment, and an incremental bandwidth that depends on the system bandwidth.

When K tends to zero the Rice distribution associated to the first path degenerates to the Rayleigh one, therefore corresponding to the NLoS case.

By using this model, the fading depth for different channel models can be easily evaluated for different PDPs. When represented as a function of the Rice factor, and of the product between the system bandwidth and the RMS delay spread of the propagation channel, $B\sigma_\tau$, one gets a graph like the one in Figure 4.37.

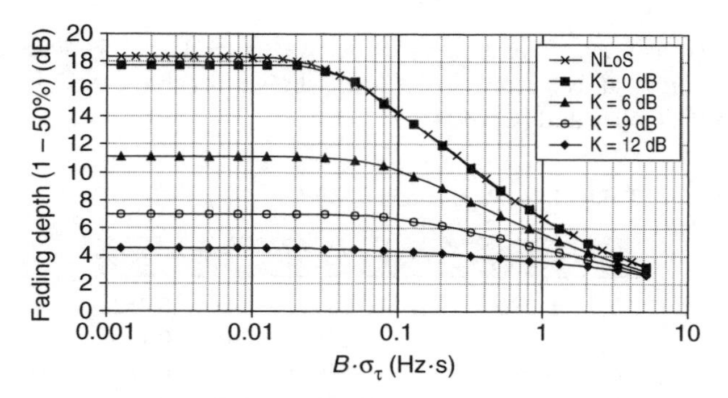

Figure 4.37 *Fading depth, exponential model [CaCo02].*

The results in Figure 4.37 correspond to the case of a continuous exponential decaying function with decay rate equal to the inverse of the RMS delay spread, thus being representative of a large set of channel models, commonly used for simulating the propagation channel. As one can observe, for each value of K, the fading depth remains practically constant for $B\sigma_\tau < 0.02$ Hz·s. This corresponds to a situation where the system bandwidth is below the coherence bandwidth of the propagation channel, defined for a frequency correlation of 90% [Rapp86], thus signals are in a frequency flat environment. For large values of $B\sigma_\tau$, the fading depth decreases for increasing system bandwidths.

A different approach, which is based on environment properties, more precisely on the differences in propagation path length among different arriving components, is presented in [CaCo01]. The proposed approach, accounting for the maximum difference in propagation path length, Δl_{max}, and having also the Rice factor as a parameter, allows evaluating the fading depth, measured between p and 50% of the received power Cumulative Distribution Function (CDF), from a simple mathematical expression

$$FD_p\left(K, B\Delta l_{max}\right)_{[\mathrm{dB}]} = \begin{cases} S_p(K), & B\Delta l_{max} \le w_{b,p} \\ \dfrac{S_p(K) - A_{1,p}(K)}{1 + A_{2,p}(K)\left[\log\left(\frac{B\Delta l_{max}}{w_{b,p}}\right)\right]^{A_{3,p}(K)}}, & B\Delta l_{max} > w_{b,p} \end{cases} \tag{4.16}$$

where $S_p(K)$, $A_{1,p}(K)$, $A_{2,p}(K)$ and $A_{3,p}(K)$ are mathematical functions of K that depend on p, and $w_{b,p}$ is a breakpoint value depending also on p.

For any given value of p, functions $S_p(K)$, $A_{1,p}(K)$, $A_{2,p}(K)$ and $A_{3,p}(K)$ can be evaluated from

$$S_{[\mathrm{dB}]} = \frac{(b_1 - b_2)}{1 + b_3\left(\frac{K_{[\mathrm{dB}]}}{10} - b_4\right)^{b_5}} + b_2 \tag{4.17}$$

$$A_{1[\mathrm{dB}]} = c_{11}\arctan\left(c_{12}K_{[\mathrm{dB}]} - c_{13}\right) - c_{14} \tag{4.18}$$

$$A_2 = c_{21}\left[\frac{\pi}{2} - \arctan\left(c_{22}K_{[\mathrm{dB}]} - c_{23}\right)\right] + c_{24} \tag{4.19}$$

$$A_3 = c_{31}\arctan\left(c_{32}K_{[\mathrm{dB}]} - c_{33}\right) + c_{34} \tag{4.20}$$

where parameters b_i ($i = 1,\ldots, 5$) and c_{jk} ($j = 1,\ldots, 3$; $k = 1, \ldots, 4$) depend on the value of p. Values for the parameters b_i and c_{jk}, for $p = 10\%$, are shown in Table 4.9. Values for different values of p can be found in [CaCo01] and [Card04].

Table 4.9 *Parameters b_i and c_{jk} for $p = 10\%$.*

	S		A_1	A_2	A_3
b_1	8.080	c_{j1}	0.289	0.141	0.452
b_2	0.070	c_{j2}	0.225	0.338	0.180
b_3	0.690	c_{j3}	0.349	2.650	0.689
b_4	−0.410	c_{j4}	0.421	0.610	2.295
b_5	2.943			–	

Since there is a close relation between the physical and geometrical environment characteristics and the PDP of the propagation channel (hence, the RMS delay spread), a simple relationship between the maximum difference in propagation path length among different arriving components and the RMS delay spread of the propagation channel can be established, therefore bridging the gap between the two previous models [CaCo03]

$$\Delta l_{max} = c\sigma_\tau \frac{K+1}{\sqrt{K}} \tag{4.21}$$

where c is the speed of light.

This allows one to use any of the proposed approaches for evaluating the fading depth in a given environment, defined either by physical and geometrical properties or by the PDP. This approach, being simple, is effective for evaluating the fading depth in different environments and for different system bandwidths, while allowing to use any of the described approaches, starting either from the environment characteristics or from the PDP of the propagation channel.

With the emergence of third generation systems, improvements in digital signal processing hardware, and the increasing demand for a larger capacity, the spatial domain appears to be one of the last frontiers for exploiting the possibility of increasing systems capacity. This can be achieved by using different types of antenna, at either the BS, or the MT, or both; therefore, new antenna and signal processing techniques are emerging, including smart antennas, either adaptive or switched beam, spatial diversity combining, and MIMO.

A lot of work has been done concerning the implementation of algorithms for achieving the desired link quality and system capacity using these techniques; however, it is not clear how the values of short-term fading depth observed by the different systems, working in different environments, depend on the antenna arrays being considered, namely on their half-power beamwidth. In [CaCo04], the approach in [CaCo01] is extended, by including the influence of the antenna's radiation pattern. This influence is modelled through the variation of the Rice factor and the maximum possible difference in propagation path length among different arriving components, relative to the case of using omnidirectional antennas.

The variation of the Rice factor, as a function of the antenna half-power beamwidth, $\Delta K(\alpha_{3dB})$, depends on the statistical distribution of AoAs (or AoDs) and on the type of antenna being considered. Assuming an ideal directional antenna (hypothetical antenna with constant gain within the antenna half-power beamwidth and zero outside) and a truncated Gaussian distribution of AoAs (or AoDs) with standard deviation, σ_s, the value of $\Delta K(\alpha_{3dB})$ is evaluated as

$$\Delta K\ (\alpha_{3dB})_{[dB]} = 10\log\left[\frac{1-2Q\left(\frac{\pi}{\sigma_s}\right)}{1-2Q\left(\frac{\alpha_{3dB}}{2\sigma_s}\right)}\right] \tag{4.22}$$

where $Q(\cdot)$ is the well-known Q-function [Carl86].

The variation of the maximum possible difference in propagation path length among different arriving components depends on the scattering model being considered, e.g. Geometry-based Stochastic Channel Models (GSCMs) are usually used, elliptical and circular ones being the most common for simulating micro-, pico- and macrocellular environments. The results in [CaCo04] illustrate such approach.

In the literature on measurements and modelling of time-variant radio channels, there has almost always been a confinement to moving transmitters or receivers in a static environment. For applications of practical interest so far, like mobile cellular phones, this was obviously realistic. Although moving scatterers are certainly always present, their influence is negligible, at least in outdoor environments, due to their scattering cross-section being small compared with fixed scatterers [Cox73].

For indoor environments, the scattering cross-section of moving objects (persons, doors, windows, etc.) still cannot be neglected compared with fixed scattering objects (e.g. walls and furniture). Furthermore, for some applications, e.g. wireless LANs, MTs are usually fixed during operation, whereas quite often some scatterers are moving in the environment [KaFr01].

Some of the few investigations of the influence of moving scatterers that can be found in the literature are related to the temporal power fluctuations of a narrowband channel, e.g. [HMVT94]. For a more comprehensive and more universal treatment, however, wideband measurements are necessary. For this purpose, wideband measurements were performed in a laboratory room with a distance of 6.6 m between transmitting and receiving antenna [KaFr01]. The LoS path was partially obstructed by furniture and laboratory equipment. For most of the measurements, three persons acted as moving scatterers.

Systematic measurements at 1.8, 5.2 and 17.2 GHz have been performed with a bandwidth of 600 MHz. The results from measurements have shown that, for the case of moving antennas, there is a fundamentally different fading behaviour and also a fundamentally different shape of the time-variant impulse response. This fundamental difference leads to the conclusion that it is not allowed (though quite often done) to uncritically adopt results and models found from investigations of channels with linearly moving antennas to applications with moving scatterers in an otherwise static environment, Figures 4.38 and 4.39.

Globally, these results have shown that the influence of moving scatterers cannot be neglected, as it has been done quite often. Concerning the frequency dependence of the influence of moving scatterers and/or moving antennas on the channel behaviour, no fundamental differences have been found from the comparison of the three different frequency bands.

As a major conclusion, it can be stated that, for a statistical modelling, an unmodified application of the well-known models for channels with moving transmitters or receivers to channels with moving scatterers obviously seems to be not very realistic.

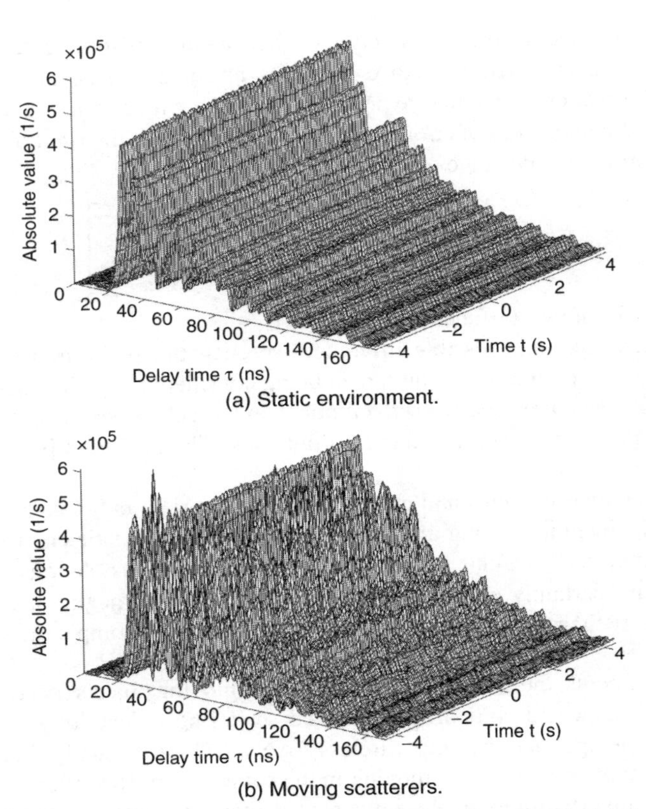

Figure 4.38 *Time-variant impulse response at 1.8 GHz with fixed antennas [KaFr01].*

4.5.7 Temporal and angular dispersion

A number of large measurement campaigns were conducted in the context of comparison of propagation scenarios or environments or of frequency bands. Other contributions intended to cover more rare scenarios or topics. All these contributions contain data on temporal and angular dispersion of mobile radio channels. The choice is made not to remove this data from the context in which it was gathered. That leads to an organisation of this section into subsections on comparisons of environments and frequency bands, respectively, on modelling-based characterisation, on characteristics of clusters, on diffusely scattered power, on presumed static environments and on some special cases.

Comparisons between mobile radio environments

A number of COST 273 participants reported large to very large measurement campaigns that range over diverse propagation environments. This enables the study of the similarities and differences between environments. Salous and Hinostroza conducted extensive measurement campaigns in seven indoor and outdoor-to-indoor environments in Manchester, UK, including a gymnasium, a laboratory, and floor-to-floor and building-to-building links [SaHi03]. The measurements were done at 2.35 GHz in 300 MHz bandwidth with the proprietary UMIST chirp sounder; overall dynamics were stated to

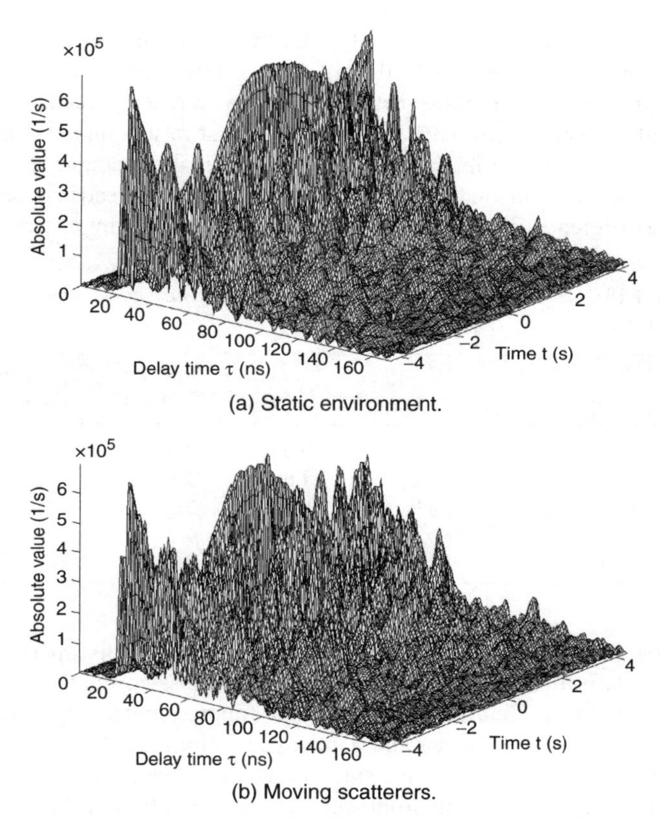

Figure 4.39 *Time-variant impulse response at 1.8 GHz with moving antennas [KaFr01].*

be better than 40 dB. From the measurements, path loss, power delay profiles, delay spread, and coherence bandwidth were determined. The coherence bandwidth ranged from 20 to 250 MHz, the RMS delay spread from 25 to 150 ns for a maximum excess delay of 800 ns. For rooms, halls, and floor-to-floor links, Ricean distributions fitted the delay spread distribution best, for LoS in-building corridors log-normal distributions gave the best fit. The authors noticed a distinct dependence of the delay spread on the clipping level: considering only multipath components stronger than −20 dB under the instantaneous maximum amplitude gave a median in-building delay spread of around 27 ns while for an instantaneous dynamic range of 35 dB this figure became 44 ns [SaHi03].

For an outdoor-to-indoor setting at 5.2 GHz in Lund, Sweden, Wyne *et al.* reported smaller dispersion figures [WAKE04], although this could well be contributed to the geometry of the measurement set-up and not to the different frequency band. Median values for the RMS delay spread were 7–10 ns for rooms facing the TXs, 12–14 ns for rooms at the opposite side of the wing, and 8–12 ns in between, in the corridor. Spreads on Angle of Arrival (AoA) were almost independent of receiver or transmitter location (about 0.24π or 14°) but the spreads of the Angle of Departure (AoD) clearly depended on the TX position, ranging from 0.03 to 0.08π or 1.6 to 4.4° [WAKE04].

A comparison between spreads at 5.3 GHz in different environments was made by Kolmonen *et al.* [KKVV05]. They held an extensive measurement campaign in typical indoor, outdoor-to-indoor, and micro- and macrocellular urban environments (downtown Helsinki). For the microcellular environments, the TX antenna was positioned well above surrounding rooftops. Dual polarised antenna arrays were used and directional information was retrieved separately at the RX and TX side by beamforming, as a fast and robust method. From that, angular spreads, both in elevation and azimuth, were calculated according to 3GGP TR25.966. First results on the angular spreads in the respective environments are given in Table 4.10, showing smaller angular spreads at the TX side than at the RX side, except for the indoor environment where both spreads are comparable in value. The influence of the differences in the antennas' operational angular ranges was not discussed.

Table 4.10 *Angular spreads for various measurement environments [KKVV05].*

Environment	RX azim. spread [°]	TX azim. spread [°]	RX elev. spread [°]	TX elev. spread [°]
Indoor	37.3	40.0	7.0	6.3
Outdoor-indoor	39.2	7.0	5.8	2.2
Microcell (LoS)	28.9	5.1	2.5	1.3
Microcell (NLoS)	40.3	12.6	4.7	2.5
Macrocell	52.3	7.6	7.7	1.7

In grossly the same environments, Kainulainen *et al.* made a large polarimetric survey at 5.3 GHz, determining Cross-polarisation Power Ratios (XPRs) and cross-correlation coefficients between co-polar and cross-polar links. The XPRs are shown to be approximately log-normally distributed in all environments, but with different means and standard deviations, Table 4.11. The cross-correlation coefficients, at both the transmit and receive side, turned out to be low, between 0.26 and 0.45 for NLoS links. In the LoS microcellular environment, the range was 0.51 to 0.62, Table 4.12.

Table 4.11 *Mean cross-polarisation power ratios and standard deviations in diverse environments at 5.3 GHz [KaVV05].*

Environment	$\overline{XPR_V}$ [dB]	σ_{XPR_V} [dB]	$\overline{XPR_H}$ [dB]	σ_{XPR_H} [dB]
Indoor picocell 1	4.5	2.3	5.8	2.9
Indoor picocell 2	4.9	2.3	5.7	2.4
Outdoor-indoor	6.2	3.9	7.3	3.3
Microcell (LoS)	8.6	1.8	9.5	2.3
Microcell (NLoS)	8.0	1.8	6.9	2.8
Macrocell (NLoS)	7.6	3.4	2.3	3.1

The differences between microcellular and macrocellular scenarios in the same area were the subject of an investigation by Hugl *et al.* [HuKL02]. They changed BS antenna heights in order to study the influence on the mean angle of arrival for vehicular channels in a densely built-up urban environment, downtown Helsinki with six- to seven-storey buildings. The BS antenna height was 10 m above rooftop level or around rooftop level, reflecting either macro- or microcellular coverage. The 20 km measurement route covered all accessible streets within a 120° sector from the BS up

Table 4.12 *Mean cross-polarisation cross-correlation coefficients in diverse environments at 5.3 GHz [KaVV05].*

Environment	Transmit correlation		Receive correlation	
	$\overline{\rho(h_{VV}, h_{VH})}$	$\overline{\rho(h_{HH}, h_{HV})}$	$\overline{\rho(h_{VV}, h_{HV})}$	$\overline{\rho(h_{HH}, h_{VH})}$
Indoor picocell 1	0.30	0.31	0.35	0.30
Indoor picocell 2	0.26	0.28	0.29	0.32
Outdoor-indoor	0.35	0.37	0.37	0.45
Microcell (LoS)	0.52	0.56	0.51	0.62
Microcell (NLoS)	0.39	0.40	0.37	0.45
Macrocell (NLoS)	0.30	0.33	0.35	0.31

to about 1.5 km distance. The authors used a Bartlett beamformer for determining the dominant angles of arrival and compared these with the geometrical AoA, the direction of the straight line between TX and RX. For the macrocellular antenna placement, the dominant and geometrical AoA matched quite well, Figure 4.40. However, for the microcellular set-up, large differences were found. The authors assume this to be caused by the difference between over-rooftop propagation and propagation through street canyons. The down-tilting of the receiving antenna could be an additional factor in excluding over-rooftop components at the lower BS position, due to the narrow elevational pattern.

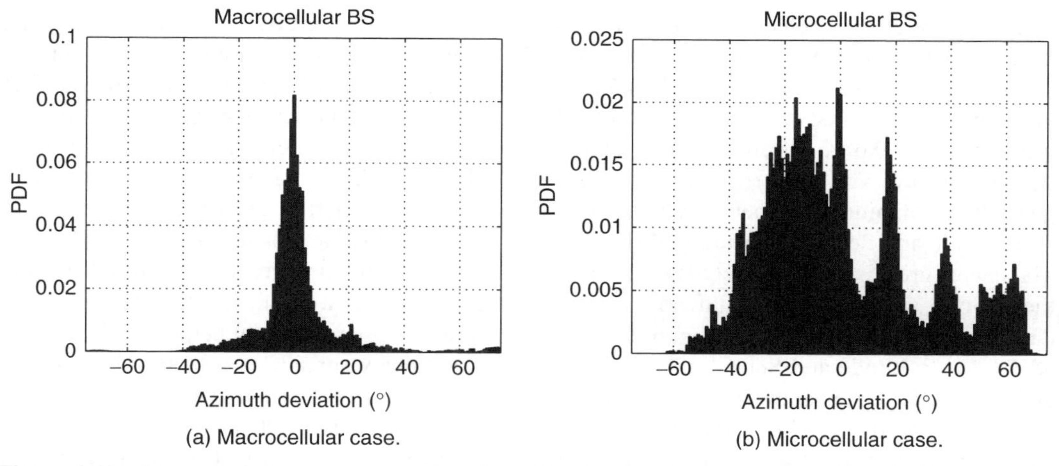

Figure 4.40 *Deviation between geometrical angle and the dominant angle of arrival [HuKL02].*

Frequency-band dependent dispersion

For present system bandwidths, statistical channel properties are generally assumed not to vary over the system bandwidth. For operation over different bands, for instance UMTS FDD operation or multi-band OFDM, and for UWB applications, such invariability is not obvious. Jämsä *et al.* measured delay spreads in a total bandwidth of 10 GHz, centred around 6 GHz, and compared some 500 MHz wide subband results in order to study the frequency dependence of indoor propagation [JäHH04]. Also, a comparison was made between results calculated from 500 MHz-wide cuts from the 10 GHz-wide measurements and from 500 MHz that were individually measured at maximum frequency resolution, rendering 1601 frequency points instead of 81. The full resolution sweeps showed higher dynamic range, but the authors conclude nevertheless that UWB measurements can be used for narrowband modelling with additional benefits from using the same data for different subbands. On the frequency dependence, they report that no real trends with frequency had been seen, apart from the strong dependence of signal strength with the frequency band. Although, in general, delay spread tends to decrease with increasing frequency, Table 4.13.

Table 4.13 *Measured indoor RMS delay spread for different frequency bands, NLoS and LoS in an empty room. UWB results are subband results cut from a full 10 GHz sweep, WB results are swept individually over 500 MHz [JäHH04].*

Centre freq. [GHz]	Band-width [MHz]	RMS delay spread LoS/UWB [ns]	RMS delay spread NLoS/UWB [ns]	RMS delay spread LoS/WB [ns]	RMS delay spread NLoS/WB [ns]
1.25	500	6.3	8.4	10.9	11.3
5.25	500	4.5	7.4	10.4	7.3
10.75	500	3.4	5.9	8.6	–
6	10,000	6.4	6.2	–	–

Spatial correlation behaviour over a similar bandwidth was studied by Liu *et al.*, in this case over 7.5 GHz bandwidth centred around 6.85 GHz. Their aim was to determine practical antenna separations for Multi-Band OFDM (MB-OFDM) applications [LAME05]. The measurements were taken in a single workshop room of $6 \times 6\,m^2$. The distance at which the spatial correlation coefficient was lower than 0.6 sank from 2.2 cm at 4 GHz to 0.4 cm at 10 GHz, roughly inversely proportional with frequency. The differences between LoS and NLoS, the latter achieved by placing a large grounded aluminium sheet between TX and RX, are minor as correlation distance is concerned.

Bultitude and Schenk measured NLoS vehicular metropolitan channels at 1.9 and 5.8 GHz in a Manhattan grid, in downtown Ottawa [Bult02a], [SBAP02]. The measurements were taken from the top of a van in 10 MHz bandwidth and the measurement runs were up to 1 km in length. The data analysis was done per street, in order to be able to separate parallel from perpendicular streets, all NLoS. Bultitude and Schenk found consistently lower delay spreads at 5.8 GHz (16% lower) and larger coherence bandwidths. The median delay spreads were 213 ns and 184 ns at 1.9 GHz and 5.8 GHz, respectively. The corresponding single-sided coherence bandwidths were around 1.8 MHz and around 1.9 MHz. They attribute this partly to higher diffraction losses at higher frequencies. Fresnel zones are smaller too at higher frequencies but it is unclear whether this results in better propagation conditions for multipath components or in fewer available scatterers [Bult02a].

Differences between UMTS FDD duplex bands in urban environments were investigated by Salous and Gokalp, for Manchester city centre [SaGo01], and by Foo *et al.* for Bristol city centre and suburbs [FBKE02]. Salous and Gokalp used the UMIST proprietary channel sounder, to measure pedestrian and vehicular channels at two 60 MHz bands simultaneously, centred at 1950 and 2140 MHz, the UMTS uplink and downlink band respectively [SaGo01]. The measurement results were compared between these two bands and between 5 MHz subbands within the 60 MHz bands. The authors note that frequency dependent differences between subbands or bands are determined by the amount of dispersion on the channel. For the delay spread results, larger variations between subbands were seen in NLoS than in LoS situations and larger differences between uplink and downlink were found in environments with dense scattering than with sparse scattering. It was noted that locally, frequency differences between subbands or bands are larger than apparent from CDFs over pooled data. These differences were largest at largest range or smallest link budget, presumably because dispersion is strongest in these cases [SaGo01].

The latter conclusion was reached by Foo *et al.* too [FBKE02]. In this case, two 20 MHz bands centred at 1920 and 2120 MHz respectively, were measured simultaneously from a moving vehicle. Time of arrival, angle of arrival, and power of multipath components were estimated using a 2D unitary ESPRIT algorithm. Different responses were seen, both temporal and angular, when comparing both bands. The largest difference, between two environments and between both bands, was found for the (azimuthal) angular spread, with strong decorrelation between the two frequency bands. The decorrelation is strongest in urban environment, which has a richer scattering than the suburbs, as measured from the larger angular spreads. The authors conclude from their findings that downlink beamforming based on uplink measurements in scattering-rich environments is sub-optimal [FBKE02].

Modelling-based characterisation

The common approach to characterisation of mobile radio environments is by measurements. Nevertheless, for simple user scenarios and simply structured environments, analytical modelling can still provide insight in fundamental phenomena. Hansen calculated power delay profiles for chambers of a cubic shape for modelling short-range communication as in WLAN or ad-hoc networking applications [Hans03]. Based on ray-optical theory and under the assumptions of well-mixed polarisations, uniformly distributed directions of travel, and random phases of scattered components, he arrives at a power delay profile of the shape of $\exp(-\tau/\sigma_\tau)$ for cubic rooms, in which

$$\sigma_\tau = (2d\xi_\tau)^{-1} \tag{4.23}$$

In turn, $d = -ln(d_E)$, with d_E the mean reflection attenuation, and ξ_τ represents the room dimensions in terms of number of wall interactions per unit of time. As for exponential decay in general, σ_τ equals the delay spread. Hansen claims that delay spread is a robust parameter with respect to modelling inaccuracies and that therefore the derived model will be a good approximation to many propagation scenarios. He compared his model with measurements, at 5.25 GHz with 100 MHz bandwidth, in rooms and a corridor in an office environment. The measured RMS delay spread values were small, ranging from 7 to 13 ns. The analytically derived values could well serve as upper bound as these values generally exceeded the measured ones with one standard deviation or more [Hans03].

In contrast to the approach of Hansen, Fügen *et al.* turned to advanced simulation tools as most of the realistic environments are too complex to use analytical channel models [FMDW03]. They used the proprietary 3D ray-tracing tool of Karlsruhe University in an urban macrocellular environment without high-rise buildings, Karlsruhe, over an area of approx. $500 \times 700\,\text{m}^2$. The simulation incorporates multiple reflections, multiple diffractions, and combinations of both. The BS was virtually placed at 25.5 m height, well above the average building height of 12.5 m and the carrier frequency was 2 GHz. The results show that the angular spreads at the MT are higher on street crossings than in street canyons. Also, the authors model both delay and angular spread, at BS and MT, with an offset negative exponential function of distance between BS and MT. A disadvantage is that the modelled delay spread falls with distance.

Intracluster dispersion

The concept of clusters of scatterers has been brought into radio channel modelling, both to ease and to refine the statistical description of mobile channel behaviour, Section 6.8. The scattering environment is in this concept seen as consisting of a distributed set of concentrations of scatterers, the clusters. Description of the properties of these clusters of scattering objects was undertaken by a number of groups within COST 273. Chong *et al.* took measurements at 5.2 GHz with 120 MHz bandwidth in a large office and a large open space, during office hours [CTLM02]. An FD-SAGE for estimation of multipath components was used and the delay-azimuth estimates were clustered by a 2D Gaussian kernel density estimator; a cluster was defined as a set of multipath components with similar excess delays and AoAs. Chong *et al.* propose a cluster model, to be discussed in the chapter on MIMO modelling, in which the temporal and spatial domain are independent for Obstructed Line-of-Sight (OLoS) and NLoS links, but are coupled for LoS. For LoS, the intracluster marginal power density distributions showed exponential decay with excess delay and were Laplacian for the azimuthal power density, Figure 4.41.

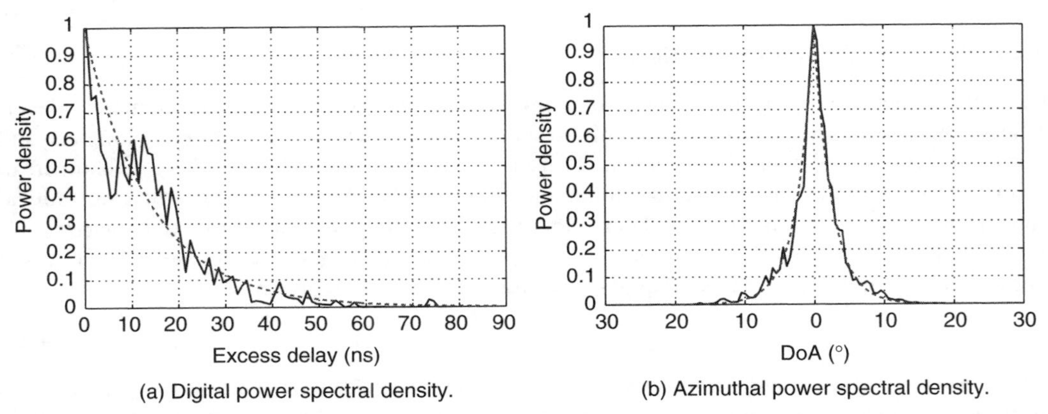

Figure 4.41 *Intracluster marginal power density distributions for LoS, measured (solid line) vs. modelled (dashed) [CTLM02].*

An alternative method by Czink and Yin uses a Bartlett beamformer to coarsely determine clusters in the AoA-AoD domain. SAGE estimates of corresponding AoAs and AoDs derived from instantaneous channel realisations were then selected with this cluster information [CzYi05]. From the clustered estimates, angular spreads were calculated. On synthetic data, spread estimates were shown to be approximately unbiased. The authors observe on measured data different intracluster angular spreads at the TX and RX side [CBYF05], likely due to the position of the TX antenna in the corridor, Figure 4.42. This method is an improvement of an earlier attempt, based on AoA-AoD estimations on eigenmode channels [CHÖB04], that consistently underestimated cluster spreads.

In the UWB band from 3.1 to 10.6 GHz, Haneda *et al.* also determined cluster properties, in a wooden Japanese residential house [HaTK05]. In total 100 paths were extracted, using a SAGE algorithm, that were clustered heuristically in the delay-AoA (azimuth) domain. The results are given in Table 4.14 and are lower than the ones documented above, probably due to the short paths, the light

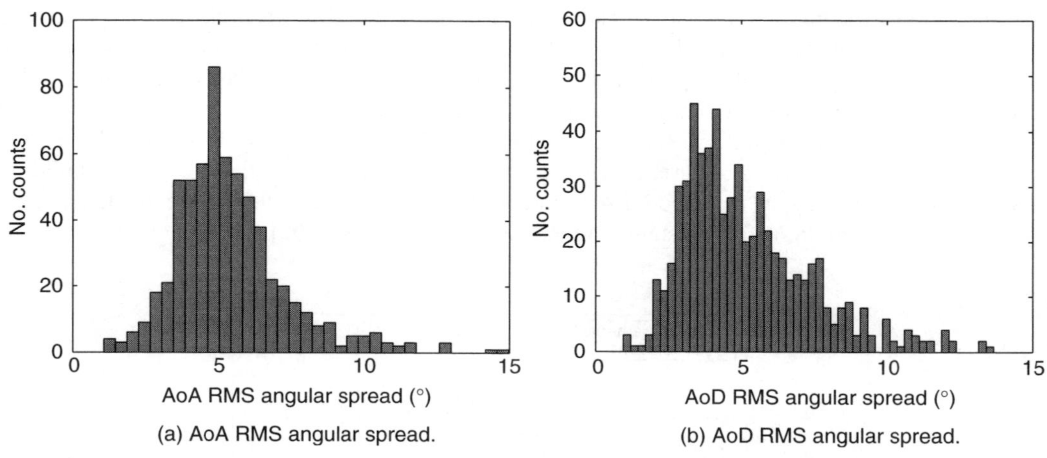

Figure 4.42 *Histograms of cluster angular spread estimates for measured indoor data [CBYF05].*

Table 4.14 *UWB cluster properties in residential environment [HaTK05].*

Scenario	No. clusters	No. paths/ cluster	Delay spread [ns]	Angular spread [°]	Fraction power [%]
Room (LoS)	7	3–24	0.95–1.38	1.9–4.5	73.4
Room-corridor (OLoS)	6	5–23	0.54–1.51	2.4–6.7	50.2
Room-corridor (NLoS)	6	3–27	0.64–1.65	0.9–6.4	50.2
Inter-room (NLoS)	8	6–20	0.21–1.28	1.8–6.3	66.8
Inter-floor (NLoS)	10	3–25	0.07–1.28	3.0–18.2	50.6
Indoor-outdoor (OLoS)	3	15–40	0.25–2.30	3.1–7.6	73.8
Indoor-outdoor (NLoS)	3	21–26	0.10–2.07	2.7–7.1	81.3

construction and the relatively small size of the building. The entries in the last column of Table 4.14 refer to the amount of power that could be recovered by the SAGE algorithm, indicating that roughly a quarter to half the power could not be modelled, see the next section.

Diffusely scattered power

The question of the amount of diffusely scattered power in mobile propagation environments is still largely unanswered. This is of importance for high-resolution parameter estimation applications, where generally a discrete scatterer model is assumed, and from the point of view of MIMO capacity.

Degli-Esposti *et al.* approached the problem from the modelling side by examining to which extent diffusely scattered components determine delay and angular spread in urban environments [DSGK02]. They did this by comparing 3D ray-tracing results with measurements. The measurements were made in the centre of Helsinki, along a waterfront and amid high-rise buildings, at 2154 MHz. The ray-tracing algorithm used is based on a full 3D image ray tracing technique [DDFR94] with diffuse scattering incorporated according to the effective roughness approach [Degl01], [DFVG05a].

The prediction of azimuthal angular spread is reported to be good in general. Delay spread was predicted fairly well too; Figure 4.43 shows the results for a quasi-LoS route. In this case, the authors concluded that modelling of diffuse scattering from far-away buildings across bays helps to reconstruct the delay spread distribution over the measurement route. But, for about 10% of the routes the model does not perform well, especially when the BS is above rooftops and the mobile sunk into a high-rise urban structure. Degli-Esposti *et al.* assume that in these cases the building database could be inaccurate (see Section 4.3) and/or that not enough buildings are considered in the ray tracing so contributions from distant scatterers are missing [DSGK02].

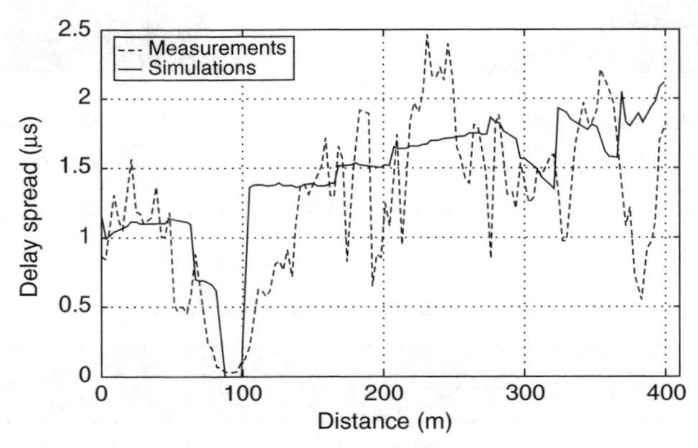

Figure 4.43 *Comparison between measured delay spread and that simulated by ray tracing [DSGK02].*

An approach of analysing large sets of measurements was taken by Trautwein *et al.* [TLST05]. They quantified the amount of diffusely scattered power using the Rimax estimation algorithm pioneered by Richter and Thomä that jointly estimates AoA, AoD, delay, Doppler, and polarisations [TLRT05a], see also Section 4.4. After subtracting the estimated coherent information from the channel transfer, the algorithm parametrises the residual, using the delay domain, into noise and an exponentially decaying diffuse power burst. This power burst is assumed to comprise all Diffuse Multipath Components (DMC). Large differences between environments were found: in an outdoor small macrocell the median value of the ratio between power in specular and in diffuse components was 14 dB while in the large main railway hall with metal support structures in München this ratio was −10.3 dB, Table 4.15. Other environments, such as urban microcell streets and squares, an indoor auditorium lobby, and a highway bridge picocell, showed power ratios in between. In the environments with relatively large amounts of DMC power, the estimator tended to find fewer specular components than in environments with relatively high power in the specular components.

The observation by Trautwein *et al.*, that diffusely scattered power cannot be neglected in a number of environments, was made by others within COST 273 too, i.e. they experienced that there seems to be a ceiling as regards the amount of power that can be captured by high-resolution estimators. Wyne *et al.* retrieved in an outdoor-to-indoor environment a worst case of 60% with 40 multipath components, with a median around 85% and the most probable extraction efficiency of around 90% [WAKE04]. Haneda *et al.* found slightly lower figures indoors, between 50 and 80%, Table 4.14, although they forced 100 specular components to be estimated, despite the high temporal resolution [HaTK05].

Table 4.15 *No. of paths and ratio of power in specular and diffuse components for various environments [TLST05].*

	No. paths		Power ratio specular/DMC				
	Mean	σ	Mean	σ	10%	50%	90%
Environment			[dB]	[dB]	[dB]	[dB]	[dB]
Small urban macrocell	44.4	5.4	13.1	3.4	8.1	14.0	17.0
Urban microcell	20.3	5.5	3.6	4.2	−2.1	3.7	9.0
Urban microcell	17.1	6.0	0.9	4.1	−4.1	0.4	6.8
Urban microcell, square	19.5	4.8	−0.3	5.0	−6.1	−1.2	7.8
Urban microcell, square	27.1	5.0	7.7	3.0	3.6	7.7	11.6
Urban microcell, square	30.2	6.1	2.1	3.8	−2.1	1.7	7.6
Indoor microcell lobby	26.3	7.1	0.8	3.6	−4.6	1.3	5.2
Hotspot railway station	20.4	4.9	−11.3	4.5	−19.0	−10.3	−6.5
Hotspot railway station	21.9	7.4	−9.2	6.9	−19.2	−9.0	0.4
Picocell highway	20.5	6.1	8.5	5.2	1.1	9.2	14.5

Static channels

The influence of moving scatterers on the channel, as compared to moving a receiving antenna, was investigated by Kattenbach and Früchting [KaFr01]. To this end, they made measurements at three different frequencies, at 1.8, 5.2 and 17.2 GHz respectively, in 600 MHz of bandwidth for good temporal resolution, using a proprietary stepped-frequency correlation sounder. People walking around in the room acted as moving scatterers. The effect on static channels was clear from the broadening of the otherwise very narrow Doppler spectrum and the changes in the impulse response. Also, the amplitude variations over time are better noticeable than phase fluctuations, as many multipath components, stemming from for instance the constructional part of the building, do not change at all. The effect of moving scatterers was also distinctly different from time-variant channel behaviour due to receiver movement. Therefore, Kattenbach and Früchting conclude that channels with moving scatterers should be modelled differently from those only showing time-variance due to receiver movement. For modelling temporal correlations, the moving scatterers can be ignored irrespective of their number, but they cannot be ignored for modelling amplitude and phase variations. The authors found no fundamental frequency dependence over the three measurement bands.

Medbo, Berg, and Harrysson specifically measured and modelled temporal variations in a NLoS WLAN scenario without receiver movement [MeBH04]. Overnight, for 15 hours, measurements were taken at 5.25 GHz in 200 MHz bandwidth with a vector network analyser on a 5 s time sweep. The variations seen are believed to be caused by persons moving in the vicinity of the antennas at either end. The terminal was placed in a room next to a corridor, the BS/access point in the corridor. They characterised the measured transfer variations by a deviation parameter p: the total power, as a function of time interval, in the deviation from the mean measured transfer spectrum, the mean determined over intervals of 1000 s, i.e. 100 subsequent measurements. The total received power itself showed to be very constant over the 15-hour period. Without persons around during night time, p was of the order of −50 dB relative to the total received power; during daytime on rare occasions it was higher than −10 dB, Figure 4.44. The authors recommend to use p as an indicator for non-static measurement situations in for instance MIMO measurements [MeBH04]. Two different causes of

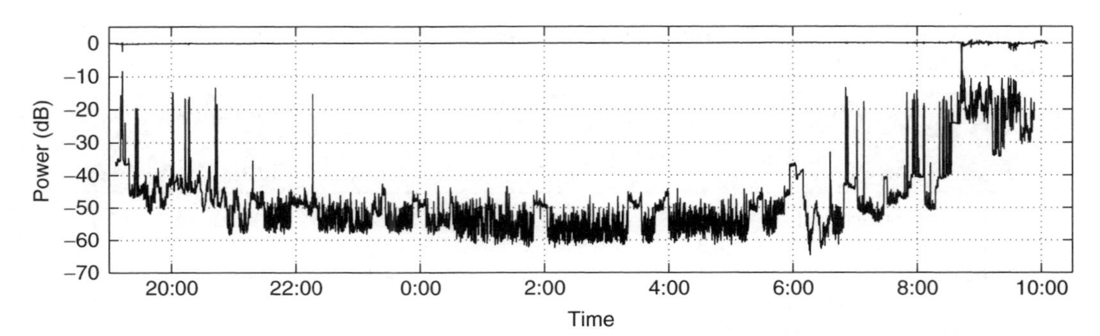

Figure 4.44 *Total wideband power (upper curve) and time-varying power (lower curve) during night time [MeBH04].*

channel dynamics were identified, disturbances with low Doppler shifts and those with high Doppler shifts, with the first most likely related to people standing or sitting close to the antennas, at either side, and the second by people walking by. Medbo, Berg, and Harrysson modelled the variations as the effect of moving persons, consisting of a cylindric scattering and a shadowing component simplified as knife-edge diffraction. Qualitatively good correspondence between measurements and model was reported [MeBH04].

Skentos *et al.* examined static rooftop-to-street scenarios for a wide, busy avenue in Athens under LoS propagation conditions [SKPC04]. The receiver was placed on top of a building at 10 m height, the transmitter at four positions at street level, 1.9 m above ground, between 115 to 236 m away from the transmitter. The carrier frequency was 5.2 GHz, the bandwidth 120 MHz and eight-element uniform linear arrays were used at both ends. The data was tested for wide-sense stationarity by means of a power correlation metric on the power delay profiles, averaged over all 64 SISO links. The threshold for the correlation coefficient with respect to measurement time was arbitrarily set at 0.75; all data passed. Typical values for temporal dispersion were 25 ns mean excess delay and 41 ns RMS delay spread. The measured channel was slowly varying with time, with the maximum observed Doppler shift around 1.5 Hz. The angular dispersion was smaller for the rooftop-positioned RX than for the TX at street level. The clustering of multipath components in the azimuth-delay plane was evident, both for RX and TX, but no cluster parameters were extracted [SKPC04]. Skentos *et al.* also investigated links from rooftop to rooftop as a typical fixed wireless access scenario. In this case, dispersion was found to be very slow, the maximum Doppler shift reported was 0.63 Hz, at 5.2 GHz carrier frequency [SkCK05].

Another experiment under static conditions was conducted by Marques *et al.* [MPKZ01]. They determined the coherence time at 1.7 GHz, bandwidth 30 MHz, in Duisburg, with BS positions above rooftop and the MT stationary in the street, but with path lengths ranging from 20 to 320 m. Coherence was taken to be lost for correlation coefficients under 0.9, between frequency spectra of snapshots averaged over all available measurement-point pairs with the same time offset. The authors found median coherence period lengths between 4.3 and 6 s, depending on the location.

Characterisation of special outdoor environments

Richly scattering environments in an industrial setting were investigated by Kemp and Bryant, with Bluetooth applications in the 2.4 GHz Industrial, Scientific and Medical (ISM) band in mind [KeBr04].

Path lengths were therefore relatively short, about 20 m on a petrochemical plant, 80 m on a car park lined with multistorey buildings, and up to 250 m on an electricity distribution transformer station. Delay spreads and BER were determined from about 120,000 snapshots, the majority of them taken on the petrochemical plant. As the equipment had a higher transmit power than usual Bluetooth appliances, the impulse responses could be clipped at 30 dB instantaneous dynamic range during processing. The average delay spread on the petrochemical plant was 38 ns, that on the transformer station 88 ns, and on the car park 76 ns. The variance of the results on the car park was two orders of magnitude higher than of those on the other environments. The authors attribute this to the presence of two distinctly different reflective processes, those of the cars in the direct vicinity, measured in metres, and those of the buildings at about 100 m distance, measured in hundred metres [KeBr04].

In comparison, indoor environments at 60 GHz would be easier to model, as reported by Geng *et al.* [GKZV05]. From the measured directional patterns at the campus of Helsinki University of Technology, they concluded that in LoS cases, the channel is well described by the direct path and first-order reflections. In NLoS cases, diffraction is the main propagation mechanism.

A special case of non-stationary propagation conditions is encountered in tunnels, just behind the entrance. Molina-García-Pardo *et al.* investigated this transitional regime, called 'excitation zone' [MaLD94], using UTD ray tracing to examine path loss, delay spread, and angular spread over the first hundreds of metres in a rectangular straight tunnel of cross-section $8.5 \times 5.2\,\text{m}^2$ $(w \times h)$ [MoRJL03]. The transmitter, at 2.1 GHz, was placed at the tunnel's centreline at 100 m distance from the entrance at three different heights, corresponding to elevations of 15°, 30°, and 45°. In the excitation zone, delay spread, angular spread and path loss change drastically. The angular spread reaches its maximum transitional value at the beginning of this excitation zone, the delay spread at the end, Figure 4.45. The length of the excitation zone depends on the elevation angle of the transmitter as do the maximum transitional values of the spreads: the smaller the elevation, the longer the excitation zone, and the smaller the maximum transitional spreads.

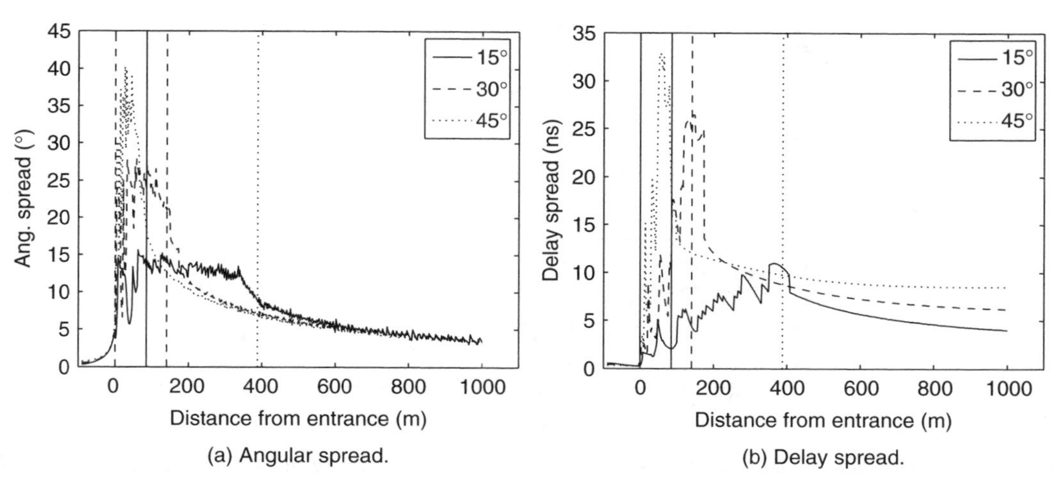

Figure 4.45 *Angular and delay spread in excitation zone behind tunnel entrance for three different TX elevations: 15°, 30°, and 45° (solid, dashed, and dotted). Tunnel entrance at 0 m; vertical lines indicate end of respective excitation zones [MoRJL03].*

4.5.8 Fading prediction

Introduction

In mobile radio system engineering it would be an advantage if channel fading behaviour could be forecast, using measurements of past and present behaviour. Predicting fading behaviour has promising applications in controlling relatively simple combining schemes in cheap receivers where knowing the occurrences of fades beforehand could be used to advantage [AJJF99]. Another envisaged use is that of transmit diversity in situations with insufficient knowledge of the channel or with too low update rates of the channel state information [DuHH00]. Predicting the channel state could then alleviate the need for frequent channel state updates.

This idea has been taken up by several authors. The early publications mainly treat the narrowband case [EyDH98], [HwWi98], [AJJF99] but later ones also address wideband aspects [SeKa02b]. A number of publications showed a limited achievable prediction range, normally fractions of a wavelength to about one wavelength under favourable conditions ([AJJF99], [Teal01], [Ekma02], [SeKa03]).

Principle of fading prediction

As an example of the principle of fading prediction, we will treat here the narrowband case. Extensions to the broadband case, to be discussed later on, were only recently made by Semmelrodt and Kattenbach [SeKa02b]. A common approach of fading prediction is that of frequency analysis on the complex fading signal, for instance, by the use of high-resolution spectral estimation schemes like ESPRIT or root-MUSIC. The constituting sinusoidal signals are easily extrapolated and the predicted signal is formed by the same linear combination of the now extrapolated sinusoids as in the original signal, Figure 4.46. The attractiveness of this approach lies in the fact that numerous frequency analysis tools are available, see for instance [StMo97], and in the physical interpretation

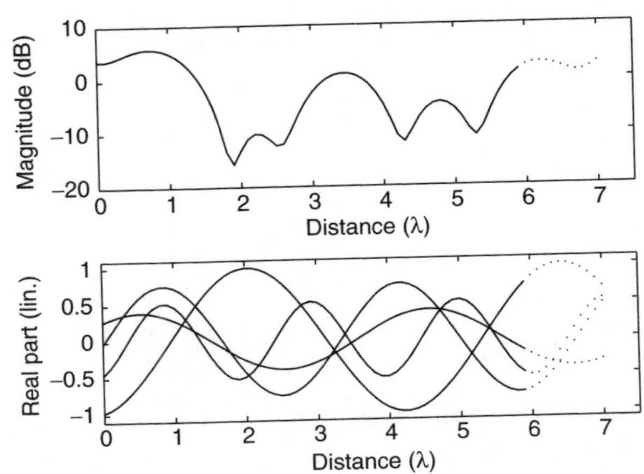

Figure 4.46 *Principle of fading prediction based on spectral estimators. The measured complex fading signal (top, solid, magnitude shown) is decomposed in complex sinusoidal components (bottom, solid, real part shown). These are extrapolated beyond the measurement interval (bottom, dotted) and combined into the fading signal prediction (top, dotted).*

of the results. The sinusoidal components are readily identified as Doppler components [AJJF99], stemming from important reflections in the environment with different angles of arrivals. In the case of narrowband prediction, the individual delays of the components are only visible as additional phase shifts. In practice, this approach does not model the numerous small components attributed to diffuse scattering, the curvature of the wavefronts of the Doppler components, and non-stationary channel states.

Narrowband fading prediction

Semmelrodt and Kattenbach made an extensive survey of spectral estimators for prediction algorithms [SeKa03]. They compared spectral estimators based either on spectral search (MUSIC and Modified Covariance), parameter estimation (EM and RELAXation spectral estimator (RELAX)), and subspace parameter estimation (ESPRIT, Unitary ESPRIT, root-MUSIC) or on adaptive filter algorithms, derived from autoregressive modelling. The adaptive filter algorithms comprised Modified Covariance, Least Mean Squares (LMS) algorithm, Normalised Least Mean Squares (NLMS) algorithm, Affine Projection (AfP), Recursive Least Squares (RLS) algorithm, and Q-R decomposition-based Recursive Least Squares (QR-RLS) algorithm. The comparison was made on synthetic data and measured data. Synthetic data was generated according to either the well-known Jakes model, using nine scatterers giving a small set of deterministic well-separated Doppler shifts, or the stochastic Dersch-model developed for the Research and Technology Development in Advanced Communication Technologies (RACE)-Code Division Testbed – RACE Project (CoDiT) project that only specifies temporal correlation properties, not Doppler profiles [BrDe91]. The measurements were recorded in a hall of $15 \times 15 \times 7\,\mathrm{m}^3$ ($l \times w \times h$) at 5.2 GHz over a distance of 1.84 m. The data was filtered in the Doppler domain in order to suppress out-of-band noise.

Figure 4.47 shows that the best performance is achieved with the Modified Covariance method for the stochastic synthetic channels and the measured channels, not with Jakes channels (not shown). The prediction length shown is defined as the length for which the relative mean-square error between the exact and predicted fading signal reaches −20 dB [SeKa03] and is normalised with respect to the wavelength. The QR-RLS method showed a performance close to that of the Modified Covariance method. For Jakes channels, subspace-based parameter estimation schemes like Unitary ESPRIT and root-MUSIC are preferred. They also perform well with stochastic (outdoor) channels with relatively few and separated scatterers but showed mediocre performance with the indoor measurements. Schemes that work on a fixed delay and Doppler grid, like EM-based methods, do in general not have enough resolution to produce any useful prediction. Therefore, Semmelrodt and Kattenbach suggest that the adaptive filter approach is preferred, giving the best results in realistic scenarios and being of (significantly) less computational complexity than the spectral estimators [SeKa03]. Also, extension to wideband prediction schemes is relatively straightforward.

Error sources in narrowband fading prediction

Kotterman investigated the accuracy of narrowband fading prediction for synthetic channels, consisting of deterministic Doppler components of random shifts, with prediction algorithms built around subspace-based parameter estimator schemes like ESPRIT, root-MUSIC, and some derivatives [Kott04]. His conclusion was that the strongest influence on the achievable extrapolation length for these types of channel came from the relation between the length of the observation interval over which the data is known and the actual number of scatterers. Channels with relatively high densities

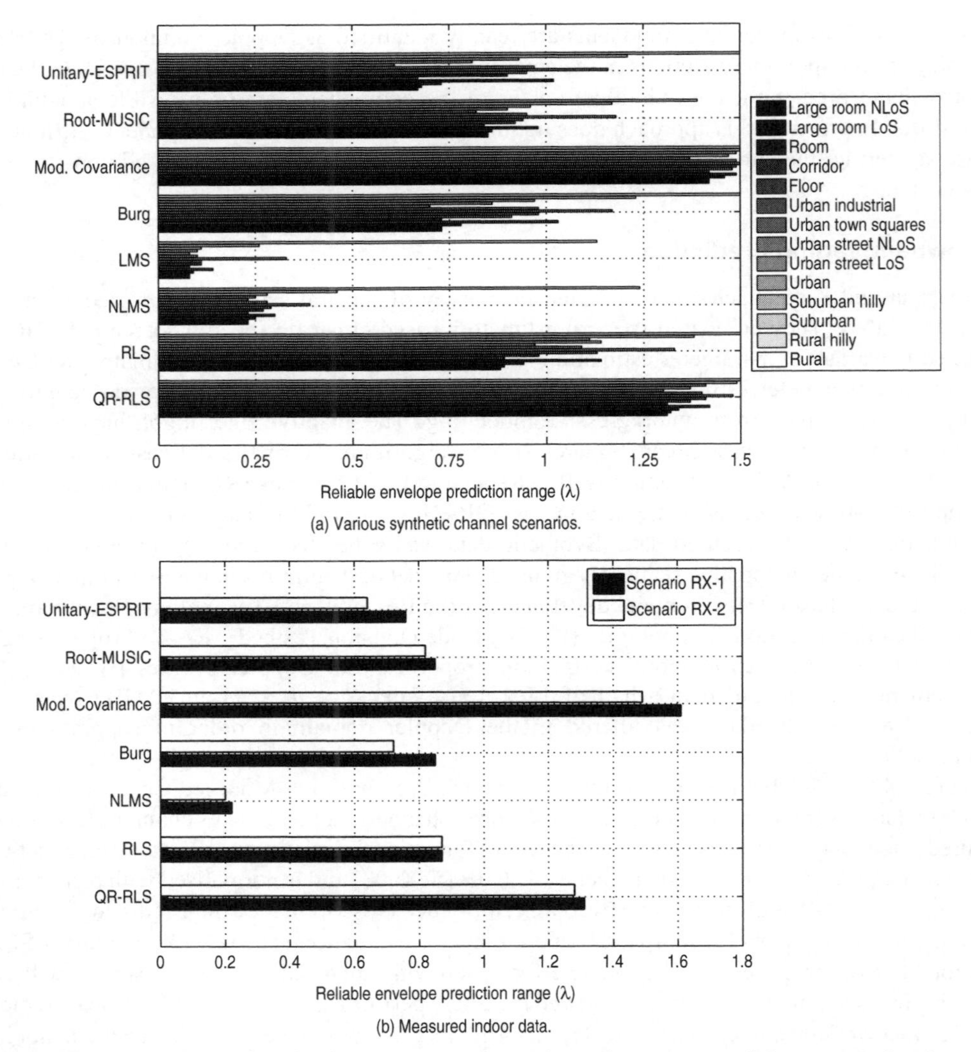

Figure 4.47 *Extrapolation lengths for various synthetic channel scenarios and for measured indoor data, for prediction schemes based on Unitary ESPRIT, root-MUSIC and several AR-based schemes (see text) [SeKa03].*

of scatterers or diffusely scattered components pose severe problems for the subspace-based spectral estimation schemes, see also Besson and Stoica [BeSt00], demanding impractically long data interval lengths over which stationarity of the channel cannot be guaranteed. As a rule of thumb, for a prediction length of one wavelength, the length of the data interval in wavelengths should equal at least half the number of scatterers in the channel, for noise-free cases. The fundamental reason for this is that the well-known reciprocal relation between resolution in time and that in frequency limits the accuracy of the Doppler estimation on short intervals. The maximum achievable prediction length depends mainly on this accuracy. Besides, one of the consequences is that the best prediction schemes for short data intervals render non-physical channel states [Kott04]. As additive noise reduces the achievable prediction length too, practical prediction lengths with prediction schemes based on subspace spectral estimators will be limited to less than a wavelength, a conclusion earlier reached by Teal [Teal01].

In this respect, Jakes models with nine (angularly) well-separated deterministic Doppler components are ideally suited for subspace-based spectral estimators, as already noted by Semmelrodt and Kattenbach [SeKa03], and using these models for evaluation of fading prediction performance [SeKa03], [DuHH00], [Ekma02] is likely to give far too optimistic results as compared to realistic channels with dense scattering [Kott04].

Wideband fading prediction

In case of non-flat fading channels, narrowband extrapolators cannot be used over the entire bandwidth and the wideband character of the channel must be taken into account. Semmelrodt originally pioneered real broadband processing by taking the frequency dependence of the Doppler shifts and the time dependence of the delays into account [Semm04b]. In his two-point analysis, these dependences were estimated by applying a 2D ESPRIT routine twice, at different measurement times. Later on even a 4D SAGE was used on a slightly different formulation, estimating delay, Doppler, and the time gradients of both [Semm04b]. These approaches were abandoned with the success of the much simpler AR-based prediction on narrowband channels described above. Semmelrodt investigated three extensions to the flat fading approach [Semm04b]:

1. using 1D estimation/prediction algorithms in parallel, one per frequency channel;
2. combining information from multiple frequency channels to derive a single extrapolation scheme for the full bandwidth;
3. using 2D Doppler-delay estimation, combining all frequency channels.

Approach 1 is inherently able to treat frequency-dependent Doppler shifts but the frequency resolution per single-frequency prediction is not better than in the narrowband case. The other two approaches sacrifice frequency-dependent information for better resolution. With approach 2, more realisations of the same process are available for the estimation, assuming that the signal bandwidth is much larger than the coherence bandwidth of the channel. Approach 3 makes potentially the best use of the separation between paths in the delay-Doppler domain, but only under stationary channel conditions.

On synthetic channels with five deterministic Doppler components, the Unitary-ESPRIT based method showed better performance than the AR-based Modified Covariance, but in relation to the three wideband approaches the relative results of both methods were comparable: the parallel 1D approach showed the worst performance, similar to the narrowband approximation. The 2D approach showed the best results with approach 2 in between, closer to the 2D than to the 1D results.

Figure 4.48 shows the prediction results for the measured indoor data presented in the section narrowband fading prediction, with a bandwidth of 120 MHz. Again, a mean-square error criterion is used, set at -20 dB. Prediction lengths with root-MUSIC were consistently better than half a wavelength (no 2D algorithm was available for root-MUSIC). Unitary ESPRIT performed worst, only with approach 2 ('Multi-Channel 1D') a prediction length of about one wavelength was achieved whereas the narrowband approximation and the 2D approach failed. In these cases, Semmelrodt expects better results when the model order of the ESPRIT estimation can be made adaptive. Best performance was achieved with the AR-based method; in all three approaches a prediction length

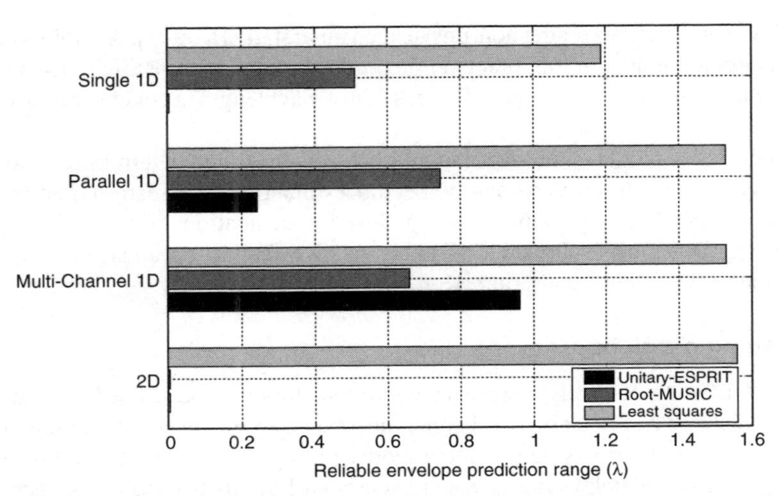

Figure 4.48 *Comparison of prediction performance of three prediction schemes on measured indoor data for three wideband approaches: 'Parallel 1D', 'Multi-channel 1D', and '2D', [Semm04b].*

of about one and a half wavelengths was achieved. The estimation errors were constant over the bandwidth and were comparable for the three approaches [Semm04b].

References

[AJJF99] J. B. Andersen, J. Jensen, S. Jensen, and F. Frederiksen. Prediction of future fading based on past measurements. In *Proc. VTC 1999 Fall - IEEE 50th Vehicular Technology Conf.*, Amsterdam, the Netherlands, Sep. 1999.

[AlRT04] V. Algeier, A. Richter, and R. Thomä. A gradient based algorithm for path parameter tracking in channel sounding. TD(04)124, COST 273, Gothenburg, Sweden, June 2004.

[Ande02] J. A. Andersen. Power distributions revisited. TD(02)004, COST 273, Guildford, UK, Jan. 2002.

[AnMd01] M. R. Andrew, P. P. Mitra, and R. deCarvalho. Tripling the capacity of wireless communications using electromagnetic polarization. *McMillan Magazines*, 409:316–318, Jan. 2001.

[BCFF02] M. Barbiroli, C. Carciofi, G. Falciasecca, M. Frullone, P. Grazioso, and A. Varini. A new statistical approach for urban environment propagation modelling. *IEEE Trans. Veh. Technol.*, 51(5):1234–1241, Sep. 2002. TD-02-044.

[Berg95] J. E. Berg. A recursive method for street microcell path loss calculations. In *Proc. PIMRC 1995 - IEEE 6th Int. Symp. on Pers., Indoor and Mobile Radio Commun.*, Toronto, Ontario, Canada, Sep. 1995.

[Bert02] H. L. Bertoni. Status of ray tracing codes with application to monte carlo simulation of channel parameters. WP(02)016, COST 273, Espoo, Finland, May 2002.

[BeSt00] O. Besson and P. Stoica. Decoupled estimation of DoA and angular spread for a spatially distributed source. *IEEE Trans. Signal Processing*, 48(7):1872–1882, July 2000.

[BHHT04a] H. Budiarto, K. Horihata, K. Haneda, and J. Takada. Experimental study of nonspecular wave scattering from building surface roughness for the mobile propagation modeling. *IEICE Trans. Commun.*, E87-B(7):958–966, Apr. 2004. [Also available as TD(03)200].

[BHHT04b] H. Budiarto, K. Horihata, K. Haneda, and J. Takada. Experimental study of nonspecular wave scattering from building surface roughness for the mobile propagation modeling. *IEICE Trans. Commun.*, E87-B(4):958–966, Apr. 2004. [Also available as TD(03)200].

[BrCu01] R. Bradley and P. J. Cullen. The effect of transverse surface gradients on propagation over undulating terrain, modelled with a PEC surface integral equation. TD(01)006, COST 273, Brussels, Belgium, May 2001.

[BrDe91] W. R. Braun and U. Dersch. A physical mobile radio channel model. *IEEE Trans. Veh. Technol.*, 40(2):472–482, May 1991.

[Bult02a] R. J. C. Bultitude. A comparison of multipath-dispersion-related microcellular mobile radio channel characteristics at 1.9GHz and 5.8GHz. In *Proc. ANTEM 2002 - 8th Int. Symp. on Antenna Techn. Appl. Electromagnetics*, Montreal, Quebec, Canada, July 2002. [Also available as TD-03-015].

[Bult02b] R. J. C. Bultitude. Estimating frequency correlation functions from propagation measurements on fading radio channels : a critical review. *IEEE J. Select. Areas Commun.*, 20(6):1133–1143, Aug. 2002.

[Bult03] R. J. C. Bultitude. Segmentation of measured data and modelling of the non-stationary characteristics of narrowband radio channels in urban microcells. TD(03)135, COST 273, Paris, France, May 2003.

[Bult04] R. J. C. Bultitude. Considerations concerning statistical stationarity in the analysis of measured data and radio channel modelling. COST 273, Duisburg, Germany, Sep. 2004. [Tutorial at COST 273].

[CaCo01] F. Cardoso and L. M. Correia. An analytical approach to fading depth dependence on bandwidth for mobile communication systems. In *Proc. WPMC 2001 - Wireless Pers. Multimedia Commun.*, Aalborg, Denmark, Sep. 2001. [Also available as TD(01)025].

[CaCo02] F. Cardoso and L. M. Correia. A time-domain technique for fading depth characterization in wideband mobile communications systems. In *Proc. URSI -The XXVII URSI General Assembly*, Maastricht, The Netherlands, Aug. 2002. [Also available as TD(02)038].

[CaCo03] F. Cardoso and L. M. Correia. A comparison between different approaches for fading evaluation in wideband mobile communications. In *Proc. VTC 2003 Spring - IEEE 57th Vehicular Technology Conf.*, Jejou, South Korea, Apr. 2003. [Also available as TD(03)046].

[CaCo04] F. Cardoso and L. M. Correia. Short-term fading depth dependence on antenna characteristics in wideband mobile communications. In *Proc. PIMRC 2004 - IEEE 15th Int. Symp. on Pers., Indoor and Mobile Radio Commun.*, Barcelona, Spain, Sep. 2004. [Also available as TD(04)089].

[Card04] F. D. Cardoso. *Short-term Fading Characterisation in Wideband Mobile Communication Systems*. PhD thesis, Instituto Superior Técnico, Technical University of Lisbon, Lisbon, Portugal, 2004.

[Carl86] A. B. Carlson. *Communication Systems - An Introduction to Signals and Noise in Electrical Communication*. McGraw-Hill, Jurong, Singapore, 1986.

[CBYF05] N. Czink, E. Bonek, X.-F. Yin, and B. Fleury. Cluster angular spreads in a MIMO indoor propagation environment. In *Proc. 14th IST Summit on Mobile and Wireless Commun.*, Dresden, Germany, June 2005.

[ChBe01] G. Liang C. Cheon and H. L. Bertoni. Simulating radio channel statistics for different building environments. *IEEE J. Select. Areas Commun.*, 19(11):2191–2200, 2001.

[CHÖB04] N. Czink, M. Herdin, H. Özcelik, and E. Bonek. Cluster characteristics in a MIMO indoor propagation environment. TD(04)167, COST 273, Duisburg, Germany, Sep. 2004.

[ChSA04] H. Y. E. Chua, K. Sakaguchi, and K. Araki. Experimental and analytical investigation of MIMO channel capacity in an indoor LOS environment. TD-04-023, COST 273, Jan. 2004.

[ChWa92] W. C. Chew and R. L. Wagner. A modified form of liao's absorbing boundary condition. In *Proc. IEEE AP-S 1995 - IEEE Int. Symp. On Antennas and Propagation*, Chicago, IL, USA, July 1992.

[CoLA03] Y. Corre, Y. Lostanlen, and S. Aubin. Multi-environment radio predictions involving an in-building WLAN network and outdoor UMTS base stations. TD(03)156, COST 273, Prague, Czech Republic, Sep. 2003.

[Corr01] L. M. Correia. *Wireless Flexible Personalised Communications, COST 259: European Co-operation in Mobile Radio Research*. John Wiley & Sons, Chichester, UK, 2001.

[Cost03] F. Costen. Analysis and improvement of Liao ABC for FDTD. In *Proc. IEEE AP-S 2003 - IEEE Int. Symp. On Antennas and Propagation and USNC/URSI National Radio Science Meeting*, Columbus, OH, USA, June 2003.

[Cost04] F. Costen. A proposal of 3D frequency dependent ADI-FDTD with conductive loss. In *Proc. Int. Conf. Comput. Electromagn.*, London, UK, June 2004. [Also available as TD(04)066].

[CoTh04a] F. Costen and A. Thiry. Alternative formulation of three dimensional frequency dependent ADI-FDTD method. *IEICE Electron. Express*, 1:528-533, 2004.

[CoTh04b] F. Costen and A. Thiry. Temporal discretization for UWB systems in three dimensional alternating-direction implicit finite difference time domain method. *IEICE Electron. Express*, 1:477–483, 2004.

[CoWS04] M. Coinchon, J. F. Wagen, and A. P. Salovaara. Joint channel estimation with array antennas in OFDM based mobile radio systems. TD(01)041, COST 273, Bologna, Italy, Oct. 2004.

[Cox73] D. C. Cox. 910 MHz urban mobile radio propagation: Multipath characteristics in new york city. *IEEE Trans. Commun.*, 21(11):1188–1194, Nov. 1973.

[CTLM02] C.-C. Chong, C.-M. Tan, D. I. Laurenson, S. McLaughlin, M. A. Beach, and A. R. Nix. Indoor wideband directional channel measurement and modelling. In *Proc. 1st Workshop COST 273 - Opportunities of the multidimensional propagation channel*, Helsinki, Finland, May 2002.

[CzYi05] N. Czink and X.-F. Yin. Cluster angular spread estimation for MIMO indoor propagation environments. TD(05)041, COST 273, Bologna, Italy, Jan. 2005.

[DDFR94] P. Daniele, V. Degli-Esposti, G. Falciasecca, and G. Riva. Field prediction tools for wireless communications in outdoor and indoor environments. In *Proc. IEEE MTT-S European Topical Congress -Technologies for Wireless Applications*, Turin, Italy, Sep. 1994.

[DDGW03] D. Didascalou, M. Döttling, N. Geng, and W. Wiesbeck. An approach to include stochastic rough surface scattering into deterministic ray-optical wave propagation modeling. *IEEE Trans. Antennas Propagat.*, 51(7):27–37, July 2003. [Also available as TD(01)007].

[DeBe99] V. Degli-Esposti and H. L. Bertoni. Evaluation of the role of diffuse scattering in urban microcellular propagation. In *Proc. VTC 1999 Fall - IEEE 50th Vehicular Technology Conf.*, Amsterdam, The Netherlands, Sep. 1999.

[DeFA04] V. Degli-Esposti, F. Fuschini, and M. Amorini. Database simplification for field prediction in urban environment. In *Proc. IEEE AP-S 2004 - IEEE Int. Symp. On Antennas and Propagation and USNC/URSI National Radio Science Meeting*, Monterrey, CA, USA, June 2004. [Also available as TD(04)041].

[DeFG03] V. Degli-Esposti, F. Fuschini, and D. Guiducci. A study on roof-to-street propagation. In *Proc. ICEAA 2003 - IEEE Int. Conf. on Electromagnetics in Advanced Applications*, Turin, Italy, Sep. 2003. [Also available as TD(03)143].

[Degl01] V. Degli-Esposti. A diffuse scattering model for urban propagation prediction. *IEEE Trans. Antennas Propagat.*, 49(7):1111–1113, July 2001.

[DeHS03] G. Del Galdo, M. Haardt, and C. Schneider. Ilmprop: a flexible geometry-based simulation environment for multiuser MIMO communications. *Advances in Radio Science - Kleinheubacher Berichte*, page 10, Oct. 2003. [Also available as TD(02)188].

[DESGK02] V. Degli-Esposti, H. El-Sallabi, D. Guiducci, K. Kalliola, P. Azzi, L. Vuokko, J. Kivinen, and P. Vainikainen. Analysis and simulation of the diffuse scattering phenomenon in urban environment. In *Proc. URSI -The XXVII URSI General Assembly*, Maastricht, the Netherlands, Aug. 2002. [Also available as TD-02-036].

[Deyg66] J. Deygout. Multiple knife edge diffraction of microwaves. *IEEE Trans. Antennas Propagat.*, 14(4):480–489, July 1966.

[DFVG05a] V. Degli-Esposti, D. Guiducci, A. De Marsi, P. Azzi, and F. Fuschini. An advanced fieldprediction model including diffuse scattering. *IEEE Trans. Antennas Propagat.*, 14(7):1717–1728, July 2004.

[DFVG05b] V. Degli-Esposti, F. Fuschini, E. Vitucci, and D. Graziani. Measurement and modelling of diffuse scattering from building walls. TD(05)065, COST 273, Bologna, Italy, Sep. 2005.

[DGMA04] V. Degli-Esposti, D. Guiducci, A. De Marsi, P. Azzi, and F. Fuschini. An advanced fieldprediction model including diffuse scattering. *IEEE Trans. Antennas Propagat.*, 14(7):1717–1728, July 2004.

[DMSW03] D. Didascalou, J. Maurer, T. Schäfer, and W. Wiesbeck. An approach to include stochastic rough surface scattering into deterministic ray-optical wave propagation modelling. *IEEE Trans. Antennas Propagat.*, 51(7):27–37, July 2003. [Also available as TD(01)007].

[DSGK02] V. Degli-Esposti, H. El- Sallabi, D. Guiducci, K. Kalliola, P. Azzi, L. Vuokko, J. Kivinen, and P. Vainikainen. Analysis and simulation of the diffuse scattering phenomenon in urban environment. In *Proc. URSI -The XXVII URSI General Assembly*, Maastricht, The Netherlands, Aug. 2002. [Also available as TD(02)036].

[DuHH00] A. Duel-Hallen, S. Hu, and H. Hallen. Long-range prediction of fading signals. *IEEE Signal Processing Mag.*, 17(3):62–75, May 2000.

[EFLR97] V. Erceg, S. J. Fortune, J. Ling, A. J. Rustako, and R. A. Valenzuela. Comparisons of a computer-based propagaqtion prediction tool with experimental data collected in urban microcellular environments. *IEEE J. Select. Areas Commun.*, 15(4):677–684, May 1997.

[EFVG05] V. Degli-Esposti, F. Fuschini, E. Vitucci, and D. Grazian. Measurement and modelling of scattering from building walls. *COST 273*, TD(05)065, Bologna, Italy, Jan. 2005.

[Ekma02] T. Ekman. *Prediction of mobile radio channels, modelling and design.* PhD thesis, Uppsala University, Uppsala, Sweden, 2002. ISBN 91-506-1625-0.

[Elek05] Elektrobit. http://www.elektrobit.ch/products/propsound/index.html. 2005.

[ElVa03] H. El-Sallabi and P. Vainikainen. A new heuristic UTD diffraction coefficient for prediction of radio wave propagation. COST 273, Guildford, UK, Jan. 2003.

[ETSI98] ETSI. Universal mobile telecommunications system (UMTS), selection procedures for the choice of radio transmission technologies of the UMTS. UMTS 30.03, version 3.2.0, Sophia Antipolis, France, Apr. 1998.

[EyDH98] T. Eyceoz, A. Duel-Hallen, and H. Hallen. Deterministic channel modelling and long range prediction of fast fading mobile channels. *IEEE Commun. Lett.*, 2(9):254–256, Sep. 1998.

[FBKE02] S. E. Foo, M. A. Beach, P. Karlsson, P. Eneroth, B. Lindmark, and J. Johansson. Spatio-temporal investigation of UTRA FDD channels. In *Proc. 3rd Int. Conference on 3G Mobile Comm. Technol.*, London, UK, May 2002. [Also available as TD-02-027].

[FGLM04] R. Fraile, J. Gozálvez, O. Lázaro, J. Monserrat, and N. Cardona. Effect of a two dimensional shadowing model on system level performance evaluation. In *Proc. WPMC 2004 – Wireless Pers. Multimedia Commun.*, Abano Terme, Italy, Sep. 2004. [Also available as TD(04)190.]

[FlSJ02] B. Fleury, A. Stucki, and P. Jourdan. High resolution bidirection estimation based on the SAGE algorithm: Experience gathered from field experiments. TD(02)070, COST 273, Espoo, Finland, May 2002.

[FMDW03] T. Fügen, J. Maurer, A. Dallinger, and W. Wiesbeck. Radio channel characterization with ray-tracing for urban environments at 2GHz. TD(03)130, COST 273, Paris, France, May 2003.

[FrLC03] R. Fraile, O. Lazaro, and N. Cardona. Two dimensional shadowing model. TD(03)171, COST 273, Prague, Czech Republic, Sep. 2003.

[FuMa03] T. Fuegen and J. Maurer. Radio channel characterization with ray-tracing for urban evironment at 2 GHz. TD(03)130, COST 273, Paris, France, May 2003.

[Fusc04] F. Fuschini. A study on urban gap fillers for DVB-H system in urban environment. TD(04)201, COST 273, Duisburg, Germany, Sep. 2004.

[FYSJ03] B. Fleury, X. Yin, A. Stucki, and P. Jourdan. High resolution channel parameter estimation for communication systems equipped with antenna arrays. TD(03)132, COST 273, Paris, France, May 2003.

[GCBV02] P. Grazioso, C. Carciofi, M. Barbiroli, and A. Varini. A statistical propagation model for urban environment. TD(02)044, COST 273, Espoo, Finland, May 2002.

[Geng05] S. Geng. Propagation characterization of 60 GHz indoor radio channels. TD(05)015, COST 273, Bologna, Italy, Jan. 2005.

[GhTI04] M. Ghoraishi, J. Takada, and T. Imai. Investigating dominant scatterers in urban mobile propagation channel. In *Proc. ISCIT 2004 - IEEE Int. Symp. on Communications and Information Technologies*, Sapporo, Japan, Oct. 2004. [Also available as TD(04)154].

[GKZV05] S. Geng, J. Kivinen, X. Zhao, and P. Vainikainen. Propagation characterisation of 60 GHz indoor radio channels. TD(05)015, COST 273, Bologna, Italy, Jan. 2005.

[GLBM02] L. Goossens, E. Van Lil, B. Boesmans, and D. Magnus. Wind turbine sites in industrial areas: Specific problems and solutions. In *Proc. Powergen Europe*, Milano, Italy, June 2002. [Also available as TD(02)124].

[GRBM03] S. G. Garcia, R. G. Rubio, A. R. Bretones, and R. G. Martin. Extension of the ADI-FDTD method to Debye media. *IEEE Trans. Antennas Propagat.*, 51-11:3183–3186, 2003.

[Hans02] J. Hansen. Towards a geometrically motivated, analytical indoor channel model. TD(02)063, COST 273, Espoo, Finland, May 2002.

[Hans03] J. Hansen. Analytical calculation of the power delay profile and delay spread with experimental verification. *IEEE Commun. Lett.*, 7(6):257–259, 2003. [Also available as TD-02-010].

[Hata80] M. Hata. Empirical formula for propagation -loss in land mobile radio service. *IEEE Trans. Veh. Technol.*, 29(3):317–325, Aug. 1980.

[HaTa03a] K. Haneda and J. Takada. An application of the SAGE algorithm for UWB propagation channel estimation. In *Proc. UWBST 2003 - IEEE Conf. on Ultra Wideband Systems and Technologies*, Reston, VA, USA, Nov. 2003. [Also available as TD(03)182].

[HaTa03b] K. Haneda and J. Takada. High resolution estimation of the NLOS indoor MIMO channel with a network analyser based system. In *Proc. PIMRC 2003 - IEEE 14th Int. Symp. on Pers., Indoor and Mobile Radio Commun.*, Beijing, China, Sep. 2003. [Also available as TD(03)119].

[HaTK05] K. Haneda, J. Takada, and T. Kobayashi. On the cluster properties in UWB spatiotemporal residential measurement. TD(05)066, COST 273, Bologna, Italy, Jan. 2005.

[HeKu05] A. Hecker and T. Kuerner. Analysis of propagation models for UMTS ultra high sites in urban areas. TD(05)033, COST 273, Bologna, Italy, Sep. 2005.

[Herd04] M. Herdin. *Non-Stationary Indoor MIMO Radio Channels*. Ph.d. thesis, Vienna University of Technology, Vienna, Austria, 2004. [Also available as TD(04)174].

[HHBW01] R. Hoppe, H., Buddendick, G. Wölfle, and F. M. Landstorfer. Dynamic simulator for studying WCDMA radio network performance. In *Proc. VTC 2001 Spring - IEEE 53rd Vehicular Technology Conf.*, Rhodos, Greece, May 2001. [Also available as TD(02)148].

[HMDA02] C. L. Holloway, P. M. McKenna, R. A. Dalke, R. A., and C. L. Perala Devor Jr. Time-domain modeling, characterization, and measurements of anechoic and semi-anechoic electromagnetic test chambers. *IEEE Trans. Electromagn. Compat.*, 44:102–118, 2002.

[HMVT94] H. Hashemi, H. McGuire, M. Vlasschaert, T. Tholl, and D. Tholl. Measurements and modelling of temporal variations of the indoor radio propagation channel. *IEEE Trans. Veh. Technol.*, 43(3):733–737, Aug. 1994.

[HoWW03] R. Hoppe, G. Woelfle, and P. Wertz. Advanced ray-optical wave propagation modelling for urban and indoor scenarios. *European Transactions on Telecommunications*, 14(1):61–69, Jan. 2003. [Also available as TD(02)051].

[HuKL02] K. Hugl, K. Kalliola, and J. Laurila. Spatial channel characteristics for macro and microcellular BS installations. In *Proc. 1st Workshop COST 273 - Opportunities of the multidimensional propagation channel*, Helsinki, Finland, May 2002.

[HwWi98] J. Hwang and J. Winters. Sinusoidal modelling and prediction of fast fading processes. In *Proc. Globecom 1998 - IEEE Global Telecommunications Conf.*, Sydney, Australia, Nov. 1998.

[JäHH04] T. Jämsä, V. Hovinen, and L. Hentilä. Comparison of wideband and ultrawideband channel measurements. TD(04)080, COST 273, Gothenburg, Sweden, June 2004.

[JePK05] J. Jemai, R. Piesiewicz, and T. Kuerner. Calibration of an indoor ray-tracing propagation model at 2.4 GHz by measurements of the IEEE 802.11b pre-amble. In *Proc. VTC 2005 Spring - IEEE 61st Vehicular Technology Conf.*, Stockholm, Sweden, May 2005. [Also available as TD(04)149].

[JKVW05] J. Jemai, M. Kürner, A. Varone, and J.-F. Wagen. Determination of the permittivity of building materials through WLAN meaasurements at 2.4 GHz. TD(05)032, COST273, Bologna, Italy, Jan. 2005.

[KaFr01] R. Kattenbach and H. Früchting. Investigation of the impacts of moving scatterers by wideband measurements of time-variant indoor radio channels. *Frequenz*, 55(7–8):197–203, July 2001. in German. [Also available as TD(01)033].

[KaKC02] P. D. Karamalis, A. G. Kanatas, and P. Constantinou. A genetic algorithm applied for optimization of antenna arrays used in mobile radio channel characterization devices. TD(02)091, COST 273, May 2002.

[Katt97] R. Kattenbach. *Characterisation of time-variant indoor radio channels by means of their system and correlation functions*. PhD thesis, Universität Kassel, Kassel, Germany, 1997. in German, ISBN 3-8265-2872-7.

[Katt02] R. Kattenbach. Transfer function modeling and its application to ultra-wideband channels. TD(02)136, COST 273, Lisbon, Portugal, Sep. 2002.

[KaVV05] A. Kainulainen, L. Vuokko, and P. Vainikainen. Polarization behaviour in different urban radio environments. TD(05)018, COST 273, Bologna, Italy, Jan. 2005.

[KaWe00] R. Kattenbach and D. Weitzel. Wideband channel sounder for time-variant indoor radio channels. In *Proc. AP 2000 - Millennium Conference on Antennas and Propagation*, Davos, Switzerland, Apr. 2000.

[KeBr04] A. H. Kemp and E. B. Bryant. Channel sounding of industrial sites in the 2.4GHz ISM band. *Wireless Personal Communications*, 31(3–4):235–248, Dec. 2004. [Also available as TD-02-111].

[KeCu03] E. Kenny and P. J. Cullen. An analytical formula for obtaining the lookup table in the tabulated interaction method (TIM) for rough terrain. TD(03)139, COST273, Paris, France, May 2003.

[KEGJ03] T. Kuerner, A. Eisenblätter, H.-F. Geerdes, D. Junglas, T. Koch, and A. Martin. Final report on automatic planning and optimisation, technical report IST-2000- 28088-MOMENTUM-D47-PUB. deliverable D4.7 of IST-MOMENTUM, European Commission, Brussels, Belgium, 2003. [Also available as TD(03)172].

[KKVV04] V.-M. Kolmonen, J. Kivinen, L. Vuokko, and P. Vainikainen. 5.3 GHz MIMO radio channel sounder. TD-04-141, COST 273, June 2004.

[KKVV05] V.-M. Kolmonen, J. Kivinen, L. Vuokko, and P. Vainikainen. 5.3GHz MIMO radio channel sounder. In *Proc. IMTC 2005 - IEEE Instrumentation and Measurement Technology Conf.*, Venice, Italy, May 2005. [Also available as TD(04)193].

[Kott04] W. A. Th. Kotterman. *Characterisation of mobile radio channels for small multiantenna terminals.* PhD thesis, Aalborg Universitet, Aalborg, Denmark, 2004. ISBN 87-90834-68-2.

[KSPH01] J. Kivinen, P. Suvikunnas, D. Perez, C. Herrero, K. Kalliola, and P. Vainikainen. Characterization system for MIMO channels. In *Proc. WPMC 2001 - Wireless Pers. Multimedia Commun.*, Aalborg, Denmark, Oct. 2001. [Also available as TD(01)044].

[KüMe02] T. Kürner and A. Meier. Prediction of outdoor and outdoor-to-indoor coverage in urban areas at 1.8 GHz. *IEEE J. Select. Areas Commun.*, 20(3):496–506, Apr. 2002. [Also available as TD(01)013].

[Kunn02] E. Kunnari. Statistical modeling of small-scale fading with temporal, spatial and spectral correlation and rice distributed envelope in the simulation of multi-carrier systems. TD(02)019, COST 273, Guildford, UK, Jan. 2002.

[Kyri03] P. Kyritsi. K factor estimation in a hallway using waveguide mode analysis. TD(03)047, COST273, Barcelona, Spain, Jan. 2003.

[LAME05] J. Liu, B. Allen, W. Q. Malik, and D. J. Edwards. On the spatial correlation of MB-OFDM ultra wideband transmissions. TD(05)015, COST 273, Bologna, Italy, Jan. 2005.

[LaRT03] M. Landmann, R. Richter, and R. Thomä. Performance evaluation of real antenna arrays for high-resolution doA estimation in channel sounding—part 1: Channel parameter resolution limits. TD(03)199, COST 273, Sep. 2003.

[LaRT04a] M. Landmann, A. Richter, and R. S. Thomä. DoA resolution limits in MIMO channel sounding. In *Proc. IEEE AP-S 2004 - IEEE Int. Symp. On Antennas and Propagation and USNC/URSI National Radio Science Meeting*, Monterey, CA, USA, June 2004.

[LaRT04b] M. Landmann, A. Richter, and R. S. Thomä. Performance evaluation of antenna arrays for high-resolution DOA estimation in channel sounding. In *Proc. ISAP 2004 - Intl. Symp. on Antennas and Propagation*, Sendai, Japan, Aug. 2004.

[LBLC02] E. Van Lil, B. Van den Broeck, Y. Van Laer, and A. Van de Capelle. Validation of GO simulations with trunking measurements in leuven. TD(02)014, COST 273, Guildford, UK, Jan. 2002.

[LCDC02] E. Van Lil, I. De Coster, Y. Demarsin, F. Casteels, and A. Van de Capelle. Finetuning the fresnel PO model. In *Proc. URSI -The XXVII URSI General Assembly*, Maastricht, The Netherlands, Aug. 2002. [Also available as TD(02)084].

[Lee85] W. Lee. Estimate of local average power of a mobile radio signal. *IEEE Trans. Veh. Technol.*, 34(1):22–27, Feb. 1985.

[LiTC05] E. Van Lil, D. Trappeniers, and A. Van de Capelle. Simplified formulas for finding the extent of false echo zones in radar systems. TD(05)039, COST273, Bologna, Italy, Jan. 2005.

[LKTH02] J. Laurila, K. Kalliola, M. Toeltsch, K. Hugl, P. Vainikainen, and E. Bonek. 3-D characterization of mobile radio channels in urban environment. *IEEE Trans. Antennas Propagat.*, 50(2):233–243, Feb. 2002.

[LoCo02a] Y. Lostanlen and Y. Corre. A 2.5D model for predicting the in-building propagation penetration in indoor environments. TD(02)052, COST 273, Espoo, Finland, May 2002.

[LoCo02b] Y. Lostanlen and Y. Corre. Studies on indoor propagation at various frequencies for radio local networks. WP(02)013, COST 273, Espoo, Finland, May 2002.

[LoCo05] Y. Lostanlen and Y. Corre. Urban coverage simulations for broadcast (DVB-H, DVB-T) networks. TD(05)048, COST 273, Bologna, Italy, Sep. 2005.

[LoGC05] Y. Lostanlen, G. Gourgeon, and Y. Corre. An approach to model the ultrawideband multipath indoor radio channel by ray-tracing methods. TD(05)047, COST 273, Bologna, Italy, Sep. 2005.

[LPEO02] P. H. Lehne, M. Pettersen, R. Eckhoff, O., and Trandem. A method for synchronising transmitter and receiver antenna switching when performing dual array measurements. URSI General Assembly, Aug. 2002. [Also available as TD(02)048].

[LVTC04] E. Van Lil, J. Verhaevert, D. Trappeniers, and A. Van de Capelle. Theoretical investigations and broadband experimental verification of the time-domain SAGE DOA algorithm. In *Proc. ACES2004 - 20th Annual Review of Progress in Applied Computational Electromagnetics*, Syracuse, NY, USA, Apr. 2004. [Also available as TD(04)142].

[LZMJ05] D. Laselva, X. Zhao, J. Meinilä, T. Jämsä, J. Nuutinen, P. Kyösti, and L. Hentilä. Empirical large-scale characterization of the received power for rural, suburban and indoor environments at 2.45 and 5.25 GHz. TD(05)043, COST 273, Bologna, Italy, Jan. 2005.

[MaHe03] E. Martijn and M. Herben. Radio wave propagation into buildings at 1.8 GHz; empirical characterisation and its importance to UMTS radio planning. TD(03)191, COST 273, Prague, Czech Republic, Jan. 2003.

[MaLD94] Ph. Mariage, M. Lienard, and P. Degauque. Theoretical and experimental approach of the propagation of high frequency waves in road tunnels. *IEEE Trans. Antennas Propagat.*, 42(1):75–81, Jan. 1994.

[MATB02] J. Medbo, H. Asplund, M. Törnqvist, D. Browne, and J.-E. Berg. MIMO channel measurements in an urban street microcell. In *Proc. RVK 2002 - RadioVetenskap och Kommunikation*, Stockholm, Sweden, June 2002. [Also available as TD(03)006].

[MCRJ04] J.-M. Molina-Garcia-Pardo, J. Caldrán-Blaya, J.-V. Rodríguez, and Juan-Llácer. MIMO measurement system based on two network analyzers. TD-04-106, COST 273, June 2004.

[MeBe02] J. Medbo and J. Berg. Simple and accurate path loss modeling at 5 GHz in complex indoor environments with corridors. In *Proc. URSI Commission*

F Open Symposium Radiowave Prop. Remote Sensing, Garmisch-Partenkirchen, Germany, Feb. 2002. [Also available as TD(02)055].

[MeBH04] J. Medbo, J.-E. Berg, and F. Harrysson. Temporal radio channel variations with stationary terminal. In *Proc. VTC 2004 Fall - IEEE 60th Vehicular Technology Conf.*, Los Angeles, CA, USA, Sep. 2004. [Also available as TD-0-4-183].

[Meda05] Medav. http://www.channelsounder.de/. 2005.

[MiVV04] F. Mikas, L. Vuokko, and P. Vainikainen. Large scale behaviour of multipath fading channels in crban macrocellular environments. TD(04)101, COST 273, Gothenburg, Sweden, June 2004.

[MoRJL03] J.-M. Molina-García-Pardo, J.-V. Rodríguez, and L. Juan-Llácer. Angular spread at 2.1 GHz while entering tunnels. *Microwave and Optical Technology Lett.*, 37(3):196–198, May 2003. [Also available as TD-02-043].

[MoRo04] S. Mota and A. Rocha. Experimental results from channel parameter estimation using the SAGE algorithm. TD(04)138, COST 273, Gothenburg, Sweden, June 2004.

[MPKZ01] G. Marques, J. Pamp, J. Kunisch, and E. Zollinger. Wideband directional channel model, array antennas and measurement campaign. IST ASILUM project, Deliverable D4.3bis, Nov. 2001. [Also available as TD-02-095].

[MuSV05] M. Mustonen, P. Suvikunnas, and P. Vainikainen. Reliability analysis of multidimensional propagation channel characterization. TD(05)040, COST 273, Bologna, Italy, Jan. 2005.

[NaCB04] K. Nasr, F. Costen, and S. K. Barton. A study of wall imperfections in indoor channel modelling. TD(04)017, COST 273, Athens, Greece, Jan. 2004.

[Nami99] T. Namiki. A new FDTD algorithm based on alternating-direction implicit method. *IEEE Trans. Microwave Theory Tech.*, 47:2003–2007, 1999.

[NdHH04] H. Ndoumbè Mbonjo Mbonjo, J. Hansen, and V. Hansen. Field theoretical investigations of the influence of mutual coupling effects on MIMO channels. In *Proc. Globecom 2004 - IEEE Global Telecommunications Conf.*, Dallas, TX, USA, Dec. 2004. [Also available as TD(03)145].

[NePa94] A. Nehorai and E. Paldi. Vector-sensor array processing for electromagnetic source localization. *IEEE Trans. Signal Processing*, 42:376–398, Feb. 1994.

[OOKF68] Y. Okumura, E. Ohmori, T. Kawano, and K. Fukuda. Field strength and its variability in VHF and UHF land mobile radio service. *Review of the Electrical Communication Laboratory*, 16(9-10):825–873, Sep. 1968.

[Pars92] D. Parsons. *The Mobile Radio Propagation Channel*. Pentech Press, London, UK, 1992.

[PaTB04] A. Pal, C. Tan, and M. Beach. Comparision of MIMO channels from multipath parameter extraction and direct channel measurements. In *Proc. PIMRC 2004 - IEEE 15th Int. Symp. on Pers., Indoor and Mobile Radio Commun.*, Barcelona, Spain, Sep. 2004. [Also available as TD(04)016].

[PLVP02] I. Pàez, S. Lorendo, L. Valle, R. P., and Torres. Experimental estimation of wideband radio channel parameters with the use of a spectrum analyzer and the hilbert transform. *Microwave and Optical Technology Lett.*, 43(5):393–396, Sep. 2002. [Also available as TD(02)074].

[PPYF04] T. Pedersen, C. Pedersen, X. Yin, B. Fleury, R. Pedersen, B. Bozinovska, A. Hviid, P. Jourdan, and A. Stucki. Investigations of the ambiguity effect in the estimation of Doppler frequency and directions in channel sounding using switched Tx and Rx arrays. TD(04)021, COST 273, Athens, Greece, Jan. 2004.

[PrSh97] D. T. Prescott and N. V. Shuley. Reflection analysis of FDTD boundary conditions-part I: Time-space absorbing boundaries. *IEEE Trans. Microwave Theory Tech.*, 45:1162–1170, 1997.

[RaAr95] O. M. Ramahi and B. Archambeault. Adaptive absorbing boundary conditions in finite-difference time domain applications for EMC simulations. *IEEE Trans. Electromagn. Compat.*, 37:580–583, 1995.

[Rama99] O. M. Ramahi. Stability of absorbing boundary conditions. *IEEE Trans. Antennas Propagat.*, AP-47:593–599, 1999.

[Rapp86] T. S. Rappaport. *Wireless Communication - Principles and Practice*. Prentice Hall, Upper Saddle River, NJ, USA, 1986.

[RaSa04] N. Razavi-Ghods and S. Salous. Semi-sequential MIMO radio channel sounding. In *Proc. CCCT04 - Int. Conf. on Computing, Communications and Control Technologies*, Austin, TX, USA, Aug. 2004. [Also available as TD(04)079].

[Rich05] A. Richter. Estimation of Radio Channel Parameters: Models and Algorithms. Ph.d. thesis, Ilmenau University of Technology, Ilmenau, Germany, 2005. [Also available as TD(02)132].

[RiEK05] A. Richter, M. Enescu, and V. Koivunen. A state space approach to propagation path parameter estimation and tracking. TD(05)053, COST 273, Bologna, Italy, Jan. 2005.

[RiOK04] C. Ribeiro, E. Ollila, and V. Koivunen. Stochastic ML method for propagation parameter estimation using mixture of angular distribution models. In *Proc. PIMRC 2004 - IEEE 15th Int. Symp. on Pers., Indoor and Mobile Radio Commun.*, Barcelona, Spain, Sep. 2004. [Also available as TD(05)024].

[RiWG97] K. Rizk, J. F. Wagen, and F. Gardiol. Two-dimensional ray tracing modeling for propagation prediction in micro-cellular environments. *IEEE Trans. Veh. Technol.*, 46(2):508–517, May 1997.

[RoMJ04] J.-V. Rodríguez, J.-M. Molina-García-Pardo, and L. Juan-Llácer. A multiple-building diffraction attenaution function expressed in terms of UTD coefficients for microcellular communications. *Microwave and Optical Technology Lett.*, 40(4):298–300, Feb. 2004. [Also available as TD(03)153].

[SaGo01] S. Salous and H. Gokalp. Characterisation of W-CDMA channels in FDD UMTS bands. In *Proc. VTC 2001 Fall - IEEE 54th Vehicular Technology Conf.*, Atlantic City, NJ, USA, Oct. 2001. [Also available as TD-02-003].

[SaHi03] S. Salous and V. Hinostroza. Indoor and between building measurements with high resolution channel sounder. In *Proc. ICAP 2003 - 12th Int. Conf. on Antennas and Propagation*, Exeter, UK, Mar. 2003. [Also available as TD-02-064].

[SaTA02] K. Sakaguchi, J. Takada, and K. Araki. A novel architecture for MIMO spatiotemporal channel sounder. *IEICE Trans. Electronics*, E-86C(3):436–441, Mar. 2002.

[SBAP02] T. C. W. Schenk, R. J. C. Bultitude, L. M. Augustin, R. H. van Poppel, and G. Brussaard. Analysis of propagation loss in urban microcells at 1.9GHz and 5.8GHz. In *Proc. URSI Commission F Open Symposium Radiowave Prop. Remote Sensing*, Garmisch-Partenkirchen, Germany, Feb. 2002. [Also available as TD-03-015].

[ScFW02] T. Schäfer, T. Fügen, and W. Wiesbeck. Measurement and analysis of radio wave propagation in hospitals. In *Proc. VTC 2002 Fall - IEEE 56th Vehicular Technology Conf.*, Vancouver, BC, Canada, Sep. 2002. [Also availabe as TD(02)056].

[Schä03] T. M Schäfer. *Experimental and Simulative Analysis of the Radio Wave Propagation in Hospitals.* PhD thesis, University of Karlsruhe (TH), Karlsruhe, Germany, 2003. [Also available as TD(02)193].

[ScLa03] S. Schmidt and G. Lazzi. Extension and validation of a perfectly matched layer formulation for the unconditionally stable D-H FDTD method. *IEEE Microwave Wireless Compon. Lett.*, 13:345–347, 2003.

[ScWi03] T. M. Schäfer and W. Wiesbeck. Effective modeling of composite walls in hospitals for ray-optical wave propagation simulations. In *Proc. VTC 2003 Fall - IEEE 58th Vehicular Technology Conf.*, Orlando, FL, USA, Oct. 2003. [Also available as TD(03)092].

[SeKa01] S. Semmelrodt and R. Kattenbach. Application of spectral estimation techniques to 2-D fading forecast of time-variant channels. TD(01)034, COST 273, Bologna, Italy, Oct. 2001.

[SeKa02a] S. Semmelrodt and R. Kattenbach. A 2-D fading forecast of time-variant channels based on parametric modeling techniques. In *Proc. PIMRC 2002 - IEEE 13th Int. Symp. on Pers., Indoor and Mobile Radio Commun.*, Lisbon, Portugal, Sep. 2002.

[SeKa02b] S. Semmelrodt and R. Kattenbach. Efficient implementation of an extended SAGE algorithm for the extraction of time-varying channel parameters. TD(02)120, COST 273, Lisbon, Portugal, Sep. 2002.

[SeKa03] S. Semmelrodt and R. Kattenbach. Investigation of different fading forecast schemes for flat fading radio channels. In *Proc. VTC 2003 Fall - IEEE 58th Vehicular Technology Conf.*, Orlando, FL, USA, Oct. 2003. [Also available as TD-03-045].

[Semm04a] S. Semmelrodt. *Methoden zur prädiktiven Kanalschätzung für adaptive Übertragungstechniken im Mobilfunk.* PhD thesis, Universität Kassel, Kassel, Germany, 2004. in German, ISBN 3-89958-041-9.

[Semm04b] S. Semmelrodt. Spectral analysis and linear prediction toolbox for stationary and non-stationary signals. *Frequenz*, 58(7-8):185–187, 2004. [Also available as TD(04)019].

[SFIH02] S. Salous, P. Fillipides, I., and Hawkins. Multiple antenna channel sounder using a parallel receiver architecture. In *Proc. SCI 2002 - 6th World Multi-Conf. on Systemics, Cybernetics and Informatics*, Orlando, FL, USA, July 2002. [Also available as TD(02)002].

[SHBP03] S. W. Staker, C. L. Holloway, A. U. Bhobe, and M. Piket-May. Alternatingdirection implicit (ADI) formulation of the finite-difference time-domain (FDTD) method: Algorithm and material dispersion implementation. *IEEE Trans. Electromagn. Compat.*, 45:156–166, 2003.

[SICW01] H. Shimotahira, K. Iizuka, S. Chu, C. Wah, F. Costen, and S. Yoshikuni. Three-dimensional laser microvision. *Applied Optics*, 40(11):1784–1794, 2001. [Also available as TD(02)007].

[SiMi03] I. Sirkova and M. Mikhalev. Influence of tropospheric ducting on microwave propagation in short distances. In *Proc. ICEST 2003, 38th Int. Conf. on Information, Communication and Energy Systems and Technologies*, Sofia, Bulgaria, Oct. 2003. [Also available as TD(02)086 and TD(02)152].

[SiMi04] I. Sirkova and M. Mikhalev. Parabolic equation based study of ducting effects on microwave propagation. *Microwave and Optical Technology Lett.*, 42(5):390–394, Sep. 2004. [Also available as TD(02)152 and TD(05)006].

[SiMi05] I. Sirkova and M. Mikhalev. Digital terrain elevation data combined with the PE method: a sofia region study. TD(05)005, COST273, Bologna, Italy, Jan. 2005.

[SkCK05] N. Skentos, P. Constantinou, and A. G. Kanatas. Results from rooftop to rooftop MIMO channel measurements at 5.2GHz. TD(05)059, COST 273, Bologna, Italy, Jan. 2005.

[SKPC04] N. Skentos, A. G. Kanatas, G. Pantos, and P. Constantinou. Channel characterization results from fixed outdoor MIMO measurements. In *Proc. WPMC 2004 - Wireless Pers. Multimedia Commun.*, Abano Terme, Italy, Sep. 2004. [Also available as TD-04-140].

[SLRT03] G. Sommerkorn, M. Landmann, R. Richter, and R. Thomä. Performance evaluation of real antenna arrays for high-resolution doA estimation in channel sounding - part 2: Experimental ULA measurement results. TD(03)196, COST 273, Sep. 2003.

[Smit65] G. D. Smith. *Numerical solution of partial differential equations*. Oxford University Press, Oxford, UK, 1965.

[StFJ02] A. Stucki, B. Fleury, and P. Jourdan. ISIS, a high performance and efficient implementation of SAGE for radio channel parameter estimation. TD(02)068, COST 273, Espoo, Finland, May 2002.

[StMo97] P. Stoica and R. Moses. *Introduction to spectral analysis*. Prentice-Hall, Upper Saddle River, NJ, USA, 1997.

[TaBN04] C. Tan, M. Beach, and A. R. Nix. Multipath parameters estimation with a reduced complexity unitary-SAGE algorithm. *European Transactions on Telecommunications*, 14:515–528, Jan. 2004. [Also available as TD(03)090].

[Tafl95] A. Taflove. *Computational Electrodynamics*. Artech House, Norwood, MA, USA, 1995.

[Taga97] T. Taga. Smart MUSIC algorithm for DOA estimation. *Elect. Lett.*, 33(3):190, 1997. [Also available as TD(01)001].

[TaLH02] R. Tahri, C. Letrou, and F. V. Hanna. A beam launching method for propagation modeling in multipath contexts. *Microwave and Optical Technology Lett.*, 35(1):6–10, Oct. 2002. [Also available as TD(02)031].

[TaSh96] F. Taga and H. Shimitahira. Proposal of the fast kernel MUSIC algorithm. *IEICE Trans. Fundamentals*, E79-A(8):1232, 1996. [Also available as TD(01)016].

[Teal01] P. D. Teal. *Real time characterisation of the mobile multipath channel*. PhD thesis, University of Sydney, Sydney, Australia, 2001.

[TFBN04] C. M. Tan, S. E. Foo, M. A. Beach, and A. R. Nix. Descriptions of dynamic single-, double-directional measurement campaigns at 5 GHz. TD-04-099, COST 273, June 2004.

[ThCo04] A. Thiry and F. Costen. On the implementation of the frequency-dependent alternating direction-implicit finite difference time domain method. In *Proc.*

IEE Sem. on Ultra Wideband Communications Technologies and System Design - Technical and Operational Development, London, UK, July 2004.

[ThLR04] R. Thomä, M. Landmann, and A. Richter. RIMAX a maximum likelihood framework for parameter estimation in multidimensional channel sounding. In *Proc. ICAP 2004 - 13th Int. Conf. on Antennas and Propagation*, Sendai, Japan, Aug. 2004. [Also available as TD(04)045].

[ThYo03] J. S. Thompson and S. K. Yong. A closed-form spatial fading correlation model for electromagnetic vector sensors. TD(03)106, COST 273, Paris, France, May 2003.

[TLAP02] C. M. Tan, M. Landmann, Richter A., L. Pesik, M. A. Beach, C. Schneider, R. S. Thomä, and Nix A. R. On the application of circular arrays in direction finding part II: Experimental evaluation on SAGE with different circular arrays. In *Proc. 1st Workshop COST 273 - Opportunities of the multidimensional propagation channel*, Espoo, Finland, May 2002.

[TLRT05a] R. Thomä, M. Landmann, A. Richter, and U. Trautwein. *Multidimensional High Resolution Channel Sounding, in Smart Antennas in Europe, State-of-the-Art*. EURASIP Book Series, 2005. [Also available as TD(03)198.]

[TLRT05b] R. Thomä, M. Landmann, A. Richter, and U. Trautwein. *Smart Antennas in Europe – State-of-the-Art*. chapter Multidimensional High-Resolution Channel Sounding. Hindawi, Sylvania, OH, USA, 2005. [Also available as TD-03-198.]

[TLST05] U. Trautwein, M. Landmann, G. Sommerkorn, and R. Thomä. System-oriented measurement and analysis of MIMO channels. T(05)063, COST 273, Bologna, Italy, Jan. 2005.

[TLVD01] R. P. Torres, S. Loredo, L. Valle, and M. Domingo. An accurate and efficient method based on ray-tracing for the prediction of local flat-fading statistics in picocell radio channels. *IEEE J. Select. Areas Commun.*, 19(2):170–178, Feb. 2001. [Also available as TD(01)014].

[ToBL98] S. A. Torrico, H. L. Bertoni, and R. H. Lang. Modelling tree effects on path loss in a residential environment. *IEEE Trans. Antennas Propagat.*, 46:872–880, 1998.

[ToLa04] S. Torrico and R. Lang. Total attenuation through a two-dimensional trunk dominated forest. In *Proc. National URSI Meeting*, Boulder, CO, USA, Jan. 2004. [Also available as TD(04)105].

[TrLC04] D. Trappeniers, E. Van Lil, and A. Van de Capelle. Cylindrical and spherical obstacles in epics-GO. *Revue HF, Belgian J. of Electronics and Communications*, 2004(2):20, Apr. 2004. [Also available in extended version as TD(03)104].

[TrSi02] V. P. Tran and A. Sibille. Inter-sensor coupling and spatial correlation effects on the capacity of compact MIMO antennas. TD(02)128, COST273, Lisbon, Portugal, Sep. 2002.

[TrSi03] V. P. Tran and A. Sibille. MIMO channel capacity and mutual coupling in circular arrays of monopoles. TD(03)099, COST273, Paris, France, Jan. 2003.

[TsHT04] H. Tsuchiya, K. Haneda, and J. Takada. UWB indoor double-directional channel sounding for understanding the microscopic propagation mechanisms. In *Proc. WPMC 2004 - Wireless Pers. Multimedia Commun.*, Padova, Italy, Sep. 2004. [Also available as TD(04)192].

[UlMF96] F. T. Ulaby, R. K. Moore, and A. K. Fung. *Matrix Computations.* Johns Hopkins, Baltimore, MD, USA, 1996.

[VeLC02a] J. Verhaevert, E. Van Lil, and A. Van de Capelle. Influence of coupling in antenna arrays on the SAGE algorithm. In *Proc. URSI -The XXVII URSI General Assembly*, Maastricht, The Netherlands, Aug. 2002. [Also available as TD(02)060].

[VeLC02b] J. Verhaevert, E. Van Lil, and A. Van de Capelle. Study and implementation of a uniform spherical distributed antenna array. In *Proc. URSI Commission F Open Symposium Radiowave Prop. Remote Sensing*, Garmisch-Partenkirchen, Germany, Feb. 2002. [Also available as TD(02)116].

[VeLC03a] J. Verhaevert, E. Van Lil, and A. Van de Capelle. Experimental and theoretical direction of arrival extraction with a mill's cross antenna array. In *Industry-Ready Innovative Research, 1st Flanders Engineering PhD Symp.*, Brussels, Belgium, Dec. 2003. [Also available as TD(03)158].

[VeLC03b] J. Verhaevert, E. Van Lil, and A. Van de Capelle. Extraction of source parameters from broadband measurements with the SAGE algorithm. *Revue HF, Belgian J. of Electronics and Communications*, 3:41–43, 2003. [Also available as TD(03)108].

[VeLC04a] J. Verhaevert, E. Van Lil, and A. Van de Capelle. Direction of arrival (DOA) parameter estimation with the SAGE algorithm. *Elsevier Journal on Signal Processing*, 84(3):619–629, Mar. 2004. [Also available as TD(02)020].

[VeLC04b] J. Verhaevert, E. Van Lil, and A. Van de Capelle. Verification of the BTD-SAGE algorithm with simulated and experimental data. TD(04)022, COST 273, Athens, Greece, Jan. 2004.

[VeLC05] J. Verhaevert, E. Van Lil, and A. Van de Capelle. Applications of the SAGE algorithm using a dodecahedral receiving antenna array. TD(05)013, COST 273, Bologna, Italy, Jan. 2005.

[VLSK03] J. Verhaevert, E. Van Lil, S. Semmelrodt, R. Kattenbach, and A. Van de Capelle. Analysis of the SAGE DOA parameter extraction sensitivity with 1.8 GHz indoor measurements. In *Proc. VTC 2003 Fall - IEEE 58th Vehicular Technology Conf.*, Orlando, FL, USA, Oct. 2003. [Also available as TD(03)035].

[WaCh95] R. L. Wagner and W. C. Chew. An analysis of Liao's absorbing boundary conditions. *J. Electromagn. Waves Applicat.*, 9:993–1009, 1995.

[WaHo03] A. Waern and P. Holm. Wave propagation over a forest edge parabolic equation modelling vs. GTD modelling. TD(03)169, COST273, Prague, Czech Republic, Sep. 2003.

[WAKE04] S. Wyne, P. Almers, J. Karedal, G. Ericsson, F. Tufvesson, and A. F. Molisch. Outdoor to indoor office MIMO measurements at 5.2GHz. In *Proc. VTC 2004 Fall - IEEE 60th Vehicular Technology Conf.*, Los Angeles, CA, USA, Sep. 2004. [Also available as TD(04)152].

[WHZL02] G. Woelfle, R. Hoppe, D. Zimmermann, and F. M. Landstorfer. Enhanced localization technique within urban and indoor environments based on accurate and fast propagation models. In *Proc. WIRELESS 2002 - Proc. 15th Int. Conf. on Wireless Commun.*, Florence, Italy, Feb. 2002. [Also available as TD(02)033].

[WLGB97] G. Woelfle, F. M. Landstorfer, R. Gahleitner, and E. Bonek. Extensions to the field strength prediction technique based on dominant paths between transmitter and receiver in indoor wireless communications. In *Proc. EPMCC 1997 - 2nd European Personal and Mobile Communications Conference*, Bonn, Germany, Sep. 1997.

[WoLa98] G. Woelfle and F. M. Landstorfer. Dominant paths for the field strength prediction. In *Proc. VTC 1998 - IEEE 48th Vehicular Technology Conf.*, Ottawa, Ontario, Canada, May 1998.

[WoZo00] K. T. Wong and M. D. Zoltowski. Closed-form direction finding and polarization estimation with arbitrarly spaced electromagnetic vector-sensors at unknown locations. *IEEE Trans. Antennas Propagat.*, 48(5):671–681, May 2000.

[WWWW04] G. Woelfle, P. Wertz, R. Wahl, P. Wildbolz, and F. M. Landstorfer. Dominant path prediction model for indoor and urban scenarios. TD(04)205, COST 273, Duisburg, Germany, Sep. 2004.

[WZWH03] P. Wertz, D. Zimmermann, G. Woelfle, R. Hoppe, and F. M. Landstorfer. Hybrid ray optical models for the penetration of radio waves into enclosed spaces. In 120 REFERENCES *Proc. VTC 2003 Fall - IEEE 58th Vehicular Technology Conf.*, Orlando, FL, USA, Oct. 2003. [Also available as TD(03)177].

[XaVC03] D. M. Xavier, J. M. Venes, and L. M. Correia. Characterisation of signal penetration into buildings for GSM. In *Proc. ConfTele 2003 - 4th Conference on Telecommunications*, Aveiro, Portugal, June 2003. [Also available as TD(03)069].

[Yee66] K. S. Yee. Numerical solution of initial boundary value problems involving maxwell's equations in isotropic media. *IEEE Trans. Antennas Propagat.*, AP-14:302–307, May 1966.

[YiFS03] X. Yin, B. Fleury, and A. Stucki. Doppler estimation for channel sounding using switched multiple transmit and receive antennas. TD(03)026, COST 273, Paris, France, May 2003.

[YuCh03] C. Yuan and Z. Chen. On the modeling of conducting media with the unconditionally stable ADI-FDTD method. *IEEE Trans. Microwave Theory Tech.*, 51-8:1929–1938, 2003.

[ZBLL04] D. Zimmermann, J. Baumann, M. Layh, F. M. Landstorfer, R. Hoppe, and G. Woelfle. Database correlation for positioning of mobile terminals in cellular networks using wave propagation models. In *Proc. VTC 2004 Fall - IEEE 60th Vehicular Technology Conf.*, Los Angeles, CA, USA, Sep. 2004. [Also available as TD(04)195].

[ZeTS04] R. Zetik, R. Thomä, and J. Sachs. Ultra-wideband real-time channel sounder and directional channel PArameter estimation. In *Proc. URSI 2004 - 18th Triennial Intl. Symp. On Electromagnetic Theory*, Pisa, Italy, May 2004. [Also available as TD(03)201].

[ZhCZ00] F. Zheng, Z. Chen, and J. Zhang. Toward the development of a three-dimensional unconditionally stable finite-difference time-domain method. *IEEE Trans. Microwave Theory Tech.*, 48:1550–1558, 2000.

[ZVRK04] X. Zhao, P. Vainikainen, T. Rautiainen, and K. Kalliola. Path loss models for urban microcells at 5.3 GHz. TD(04)207, COST 273, Duisburg, Germany, Sep. 2004.

5

Antennas and diversity: from narrow band to ultra wide band

Chapter editor: *Pierre Degauque*
Contributors: *Andrés Alayon Glazunov, Alain Sibille, and Jörg Pamp*

5.1 Introduction

This chapter provides new insights on antennas, including diversity schemes and UWB applications. It is important to remember that antenna performances are strongly dependent on the electromagnetic properties of the medium in its immediate vicinity. This is a critical point for handset antennas, for example, since the user's hand and body strongly affect the radiation efficiency. Taken together, the total radiated power in the transmission mode and the total radiated sensitivity in the receiving mode provide a general idea of the handset terminal's behaviour in its environment. However, as outlined in Section 5.2, in order to characterise antenna efficiency precisely, it is first necessary to introduce and define such figures of merit as the mean effective gain and the mean effective radiated power, taking the multipath structure of the communication channel into account. For this, measurement techniques must be reproducible. Two of these techniques have been studied in depth. In the first, which is based on the use of a reverberation chamber, multiple reflections on the walls reproduce ideal Rayleigh channels; in the second, the transmitting antenna is placed in an anechoic chamber, allowing the modification of the antenna radiation pattern, due to a phantom that simulates the user's body, to be clearly demonstrated. One of the main factors influencing the measurement results is, of course, the electrical properties of the phantom material. However, both the position of the phantom hand, or in other words, the way a handset is held, and the position of the antenna handset relative to the head are also quite critical. These issues of measurement repeatability and accuracy were widely discussed during the COST 273 Action, with a focus on updating the standardised measurement techniques.

Antenna diversity was proposed a long time ago as a way to combat fading events and, more importantly, as a way to take advantage of multipath propagation. However, the efficiency of this technique is highly dependent on the correlation between antenna elements, which can be rather high when implementing diversity on small handset terminals. Furthermore, the hand of the user may greatly modify the effective gain of one array element, thus changing the statistics of the mean received power for each element. Section 5.3 addresses this question of antenna diversity as well as polarisation diversity and its possible application to UWB systems. A special case concerns the diversity found in the MIMO techniques, both for the transmitter and for the receiver. Usually, the capacity is calculated by implicitly assuming virtual arrays, since the theoretical approach takes the spatial correlation into account, while ignoring the mutual coupling between elements. However, recent papers have shown that the coupling effect is not at all negligible. The principal results of these papers are presented in the last part of Section 5.3.

The bandwidth constraint in the design of an antenna becomes very important when UWB systems are considered. Indeed their frequency-dependent characteristics distort the shape of the radiated impulse shape, requiring increased effort to detect it. A description of UWB radio channel measurements and models, including the antenna characteristics, is provided in Section 5.4.

5.2 Antenna performance assessment of mobile handsets

5.2.1 Introduction

The handset antennas are required to serve their final purpose, that is making over-the-air wireless communication possible by fulfilling system requirements set upon the User Equipment (UE). In UMTS the performance of the UE will not only affect the QoS to a single user, but the performance of the whole cellular network. Hence, UE performance is even more critical to 3G systems than to 2G systems, since in 2G systems poor UE performance might 'only' result in dropped calls or bad coverage and of course a higher power consumption at the mobile handset.

In [Glaz03] and [Glaz04b] the effect of UE antenna performance on the downlink coverage/capacity trade-off in WCDMA-FDD networks in outdoor urban areas has been investigated. Figure 5.1(a) shows the coverage vs. capacity curves for different UE antenna performance values expressed in terms of Mean Effective Gain (MEG). Is not difficult to see that a 3 dB antenna performance degradation may lead to approximately 40% extra BSs. In turn 6 dB lead to 100% extra BSs to cover the same service area. Moreover, increased capacity and/or extended coverage in downlink WCDMA-FDD systems may be achieved by the joint use of fixed beam smart antennas at BSs and UE with good antenna performance, which is shown in Figure 5.1(b) for the same scenario.

This section deals with uplink and downlink measurements of cellular mobile handset performance including the antenna. Only single antenna devices are addressed; however, some general findings are also applicable to multiple antenna systems on mobile handsets.

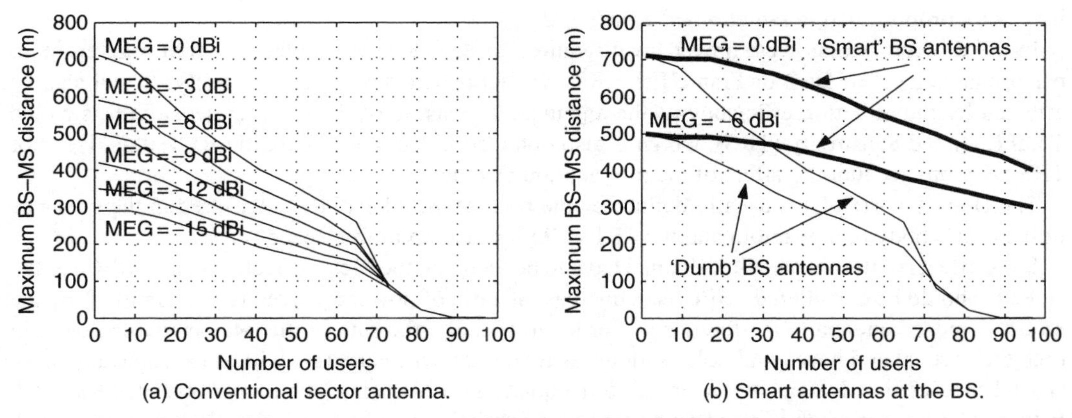

Figure 5.1 *Maximum range in three-sector cell obtained for downlink UMTS-FDD in a urban environment as a function of the number of speech users for different UE MEG values for two different antenna types [Glaz03], [Glaz04b].*

The remainder of this section is organised as follows: in Subsection 5.2.2 the definitions of the figures of merit such as the Total Radiated Power (TRP) and MEG are discussed; the general sketch of the different measurement methods are provided in Subsection 5.2.3; further in Subsection 5.2.4, great attention is given to the many aspects influencing the accuracy of the measurement results, such as the user's hand, head and/or body, tissue-simulating liquids, number of samples, etc.; Subsection 5.2.5 is concerned with the comparison of the different figures of merit and an alternative analysis approach is exposed too; and in Subsection 5.2.6 some antenna design aspects as well as Specific Absorption Rate (SAR) measurement techniques are provided.

Finally, it is worthwhile mentioning that some of the key contributions of the subworking group 2.2 have also been summarised as an input to the 3GPP standardisation body (subworking group RAN 4) [KrGl04] with the objective to incorporate the proposed methodology into the 3GPP standard specifications for mobile handset performance testing.

5.2.2 Figures of merit

The performance of handset antennas in terms of 'body loss' must be included in link budgets of cell plans and system level network simulations for proper capacity and coverage planning. In practice, antenna performance of mobile handsets is far below values usually used in link budgets and can be as low as −15 dB for GSM handsets in talk position. Furthermore, it is more correct to talk about mean effective gain (or mean effective loss) rather than antenna gain minus 'body loss' since in wireless cellular communications multipath rather than single-path propagation prevails. Hence, in this subsection the different terms used for antenna performance characterisation are briefly introduced with focus on the two main modes of operation of the mobile handset, i.e. the transmitting and modes of operation of the mobile handset.

The Total Radiated Power (TRP) is a measure of the antenna performance in transmitting mode and is basically the power with which the MT communicates with the BS. The TRP is assessed by direct measurements in laboratory conditions by means of the Effective Isotropic Radiated Power (EIRP), defined as antenna gain times power accepted by the antenna. The power absorption effects (losses) due to the user's head (phantom) and even hand are also accounted for during measurements. TRP is directly proportional to radiation efficiency.

In the downlink the lowest power level required to decode the received signal at a given radio performance level of BER or Frame Error Rate (FER) is called receiver sensitivity. It will also be affected by the radiation efficiency of the antenna. A measure of the receiver sensitivity is called Total Radiated Sensitivity (TRS), which is also obtained in laboratory conditions. Neither TRP nor TRS take into account the effect of the propagation channel.

The Body Loss (BoL) is usually defined as the ratio of the TRP (TRS) measured in the presence of the user's body or a phantom and the TRP (TRS) measured in free space.

If the effects of the propagation channel have to be included the Mean Effective Gain (MEG) of the antenna should be considered. MEG is defined as the ratio of power received (or transmitted) by the mobile handset antenna to a reference power level. Usually, the sum of the total available power in both the vertical and horizontal polarisations, as it would be measured with ideal isotropic antennas, is used as reference. In [Taga90] a closed form equation for MEG is given for uncorrelated Rayleigh fading. Another approach is measuring performance relative to a physically realisable antenna such as the $\lambda/2$ dipole antenna [OlLa98].

Hence, MEG takes into account the fact that the transmitted RF waves propagate through the radio channel subjected to the spatial dispersion around the receiver, the depolarisation effects and of course the gain (directivity) pattern of the antenna. MEG is therefore not only directly proportional to the radiation efficiency of the antenna, as is TRP, but it also depends upon the angular power distribution of incoming waves and the Cross-Polarisation Ratio (XPR) of the channel. The XPR is a measure of the imbalance of the power in vertical and horizontal polarisations. Hence, MEG is highly dependent on the physical properties of the real propagation environment.

The definition of MEG has been generalised in [Glaz04f] to include Ricean fading (more likely to happen in LOS conditions) with Rayleigh fading as a special case. A closed form formula is provided as well as some numerical examples. The main conclusion is that MEG in Ricean channels may be much lower than in Rayleigh channels depending on the angle of arrival of the strongest wave component. Figure 5.2 shows MEG calculated for the tilted $\lambda/2$ dipole antenna as a function of the inclination angle for (a) Rayleigh channels and (b) Ricean channels.

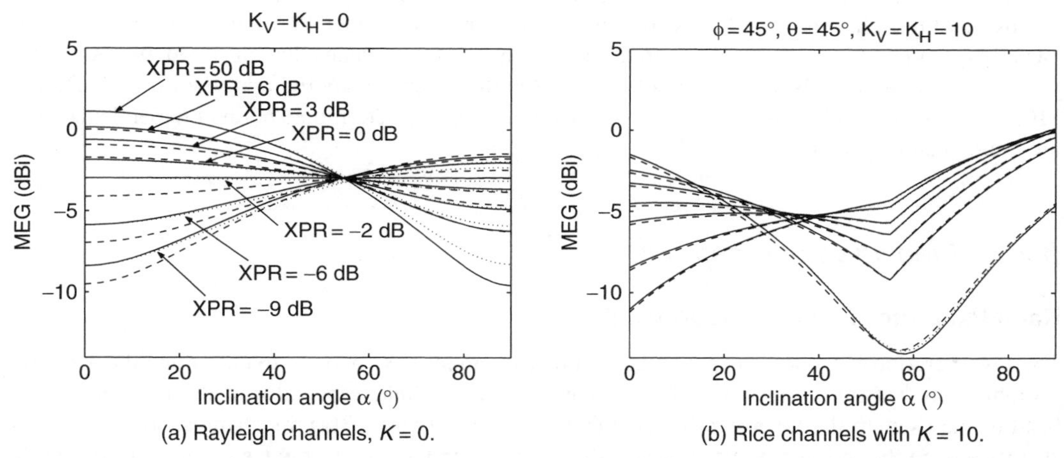

Figure 5.2 *MEG vs. tilt angle relative to the vertical for different XPR in Rayleigh and Rice channels for both the vertically and horizontally polarised components with strongest component coming from azimuth angle and elevation angle both equal 45° [Glaz04f]. (© 2004 IEEE, reproduced with permission)*

In [Glaz04c] a formulation of the MEG in Rayleigh channels is provided that considers the double-directional nature of the propagation channel. It is shown there that the MEG will in general not only depend on the gain pattern of the receiving antenna, but the gain pattern of the transmitting antenna too. The provided closed form equations incorporate the polarisation path gain matrix of the channel, the joint angular power distribution (spatial dispersion) around both the receiver and the transmitter including the double-directional properties of the radio propagation channel. In the same paper, a new parameter denoted average Cross Polarisation Discrimination (XPD) of the antenna was introduced. The XPD is the counterpart to the XPR of the channel but for the antenna (or the handset in general). In other words the XPD is a measure of the power imbalance of the power transmitted (received) with vertical and horizontal polarisation. Based on a numerical example it is further shown that if the horizontal polarisation is 3 dB higher than in vertical polarisation, i.e. $XPD = -3$ dB, the actual MEG difference obtained by the deployment of a BS with vertically polarised antennae and a slant-polarised *antennae* are between 0 and 1.5 dB in favour to the slant polarised BS antennas for $XPR = 6$ dB. In fact many BS are deployed with ±45° cross-polarised antennas, which implies that more power is available in the horizontal polarisation. The most common use of handsets in speech mode is close to the user's head tilted more than 45° from the vertical on average. In this case the antenna polarisation is in favour of the horizontal component [GlPa04]. Hence, more exact assessment of the UE handset antenna performance in real cellular networks is possible.

As stated above MEG is a measure of efficiency. However, in practical applications the effective radiated power is often more valuable. In that case the EIRP may be used instead of antenna gain. This parameter is denoted as Mean Effective Radiated Power (MERP). MERP is the extension of the concept of TRP that takes into account the actual physical propagation channel. Similarly, the concept of receiver sensitivity may be extended to consider the propagation environment, in this case the parameter is denoted as Mean Effective Radiated Sensitivity (MERS) [KrGl04], which is an extension of the TRS concept. It is worthwhile to note that TRP and TRS are deterministic in nature (strictly speaking this is not valid due to measurement errors), it means that once you measure them, you may be sure that this value will not change. On the other hand MEG and therefore MERP and MERS will also reveal a more stochastic behaviour, obviously determined by their dependence upon the radio channel propagation.

5.2.3 Measurement methods

Radiated performance measurement

The reverberation chamber was proposed as a measurement environment to determine the TRP that a mobile handset makes available for the communication link, especially in the presence of the head of the user. In the reverberation chamber shown in Figure 5.3(a), mechanical so-called mode stirrers ensure that a large number of statistically independent cavity modes are present inside the chamber, over all stirrer positions. Hereby, the overall field distribution corresponds well to that in a multipath environment. The reverberation chamber is thus a useful alternative to the traditional anechoic chamber for evaluating mobile handset performance, and the advantages are seen in smaller size and much shorter measurement duration. The chamber presented in [Kild02] is about 2 m^2 in size and contains two metal stirrers. For improved accuracy, platform and polarisation stirring methods have been added. It was shown that the measurement uncertainty for the TRP of a mobile handset determined in this chamber is comparable to that achieved in anechoic chambers. In further work [Mads04] with the reverberation chamber it was shown that it is possible to measure absolute TRP of mobile handsets with a standard deviation of ±0.18 dB at the 900 MHz band and ±0.10 dB at the 1800 MHz band, while the measurement time is only 100 s. It was shown that further decreasing the measurement time results in a larger standard deviation. Still, for a measurement time of 25 s a reasonable standard deviation of ±0.20 dB can be reached at 1800 MHz. The proposed platform stirring enables the simulation of a fast fading effect, which is superimposed onto the homogeneous slow fading environment that the reverberation chamber provides. This represents the real scattered field environment in which mobile handset users are moving.

(a) Reverberation chamber.

(b) Circular multiprobe system.

(c) Spherical multiprobe system.

Figure 5.3 *Reverbation chamber and multiprobe systems.*

The ideal reverberation chamber provides a statistically isotropic environment with uniformly distributed polarisation. Since this seldom is the case for real propagation environments a specialised reverberation chamber for MEG and diversity measurements on mobile handsets have been proposed in [Otte05]. The specialised chamber is a combination of the reverberation chamber and the anechoic chamber where a small anechoic chamber is placed inside a reverberation chamber. The test object is placed inside the anechoic chamber and apertures control the angular distribution of incoming plane waves around the test object. Also the polarisation of the plane waves can be controlled with the apertures. The reverberation chamber is only used as a source of the Rayleigh fading waves that propagate through the apertures of the anechoic chamber.

Another fast method to evaluate mobile handsets uses a compact multiprobe array of measurement antennas to measure radiated near-field fields at 15 elevations simultaneously, Figure 5.3(b) [DuGG04]. The Device Under Test (DUT) needs to be rotated around the vertical axis only. Within a few minutes a full 3D radiation pattern can be measured from 800 MHz to 6 GHz. To evaluate the proposed system, the TRP and radiation pattern of an active mobile handset was measured in free space and in talk position next to a phantom head. The results from the proposed compact system were compared to results obtained in a similar but much larger reference system with 64 measurement probes. The good agreement proves that the measurement radius of 45 cm is sufficient for measurements up to the Personal Communications Service (PCS) band and also in the presence of a phantom head.

Another 3D field measurement system for small antennas was introduced [LOIV03], which is capable of determining the radiation pattern of a mobile handset antenna without rotating the handset. The system operates at 1.8 GHz, and simultaneously measures two orthogonal polarisations of the complex radiated field at 32 locations on a spherical surface with a radius of 1 m, Figure 5.3(c). The spherical wave expansion technique is used to determine the far field from the 64 field samples. The radiated field of an antenna prototype was determined with the presented measurement system. The agreement with the traditionally measured radiation pattern was good. Also the phase of the radiated fields can be determined by using a phase-retrieval network and a spectrum analyser instead of a VNA.

Then can determine the complex radiation properties of, e.g. an active mobile handset without the need to attach a cable to the mobile handset.

In order to be able to perform measurements in an environment similar to an average indoor mobile channel a measurement system is proposed in which the position of several scattering objects in the form of corrugated metal sheets can be arranged [Glaz04e]. Hereby, a Rayleigh fading environment can be created, but also the XPR can be controlled to some extent. The calibration measurements showed that an anechoic chamber was suitable for scattered field measurements at 900 MHz with acceptable repeatability. In the NLoS case the Rayleigh fading was emulated and confirmed with the Kolmogorov-Smirnov goodness of fit test method. The null hypothesis at the 95% significance level that the distributions are Rayleigh was accepted in most cases. No significant variation was noticed in terms of the measured envelope statistics or the measured XPR, which remained within ± 1.5 dB for six test scenarios. The major effect of the sheets was to make fading more severe by controlling the amount of reflections. When the metallic sheets were placed closer to the transmitter to emulate a uniform power distribution around the transmitter, this had no implication in terms of XPR variations. The statistics remained Rayleigh in most cases. The body loss was measured for 13 different mobile handsets in two different inclination angles with respect to the normal of the receiving antenna: 0° and 55°. The DUT were placed on the left side of the phantom. The measured MEG relative to the $\lambda/2$ dipole or SFMG was quite similar for all the handsets and varied within -5.2 to -10.9 dB and -4.4 to -8.1 dB for the 0° and 55° measurements respectively. The average body loss measured at 55° was on average 1 dB lower than at 0°, which was expected.

Radiation testing of mobile handsets in speech mode

The spherical measurement technique has been investigated in [Knud01] using standard dual band GSM handsets. Both uplink and downlink measurements were performed in free space. A GSM tester was used to measure the transmitted power in the uplink. For the downlink the RX level that the handset reports back to the GSM tester was used. There are 64 RX level values specified in GSM, which are mapped to the received signal strength. The measurement set-up has been calibrated in both uplink and downlink by calculating the link loss including cable, switch, connector losses, probe antenna gain and free-space transmission loss. The calibration has been verified by a reference measurement using monopoles connected to a mobile handset. The accuracy in the downlink was improved relative to ETSI specifications by measuring the error in the reported RX level and using that for calibrating the measured downlink data, which was evaluated for 10 commercially available dual band GSM handsets. The RX level as a function of transmit power was measured over the air interface at the centre channel for both the GSM900 and GSM1800 band. A rather large absolute difference was observed, despite the fact the RX levels had been measured at the same spherical position as a function of transmit power for all the handsets.

Hence, the measured value of the transmitted power from the GSM tester relative to the RX level curve is very dependent on the gain of the handset antenna in that particular direction. However, the relative dependency between the reported RX level values for the same handset is independent of the antenna gain. The absolute accuracy of the RX level reporting can be measured via an RF connector, but such a connector is not available for all handsets. Alternatively, the RX level curves can be linked to the receiver sensitivity of the handsets by making a BER reference sensitivity measurement. In GSM it is specified that the reference sensitivity at a class II BER level of 2.0% should be better than -102 dBm for both GSM900 and GSM1800. By making a search for the sensitivity at a class

II BER of 2.0% and finding the RX level that the handset reports back at that level, a direct link between the reference sensitivity and the RX level was obtained.

In [ChRo04] a measurement procedure for 3G UMTSFDD mobile handsets is presented. The procedure involved an anechoic chamber, a Node B simulator (communication tester), a dual polarised antenna, a positioner and a C program that remotely controlled all the components. The measurement capabilities comprised up- and downlink measurements of dual-polarised mobile handset antennas in the far field, which allows the assessment of the TRP, TRS and MEG. In compliance with the UMTS FDD mode, the test procedure is based on a connection between the communication tester and the DUT over the Reference Measurement Channel (RMC) [3GPP03]. Since only the speech application was considered the 12.2 Kbps RMC was used. In the uplink case the maximum power was measured by sending a constant 'up command' to the handset. On the other hand, the downlink measurements were based on the received Common Pilot Channel (CPICH) Received Signal Code Power (RSCP), that is the Common Pilot Channel Received Signal Code Power, and on the BER measurements. In order to evaluate uplink performance the radiation pattern of the mobile handset is measured in both polarisations, which with proper calibration provides the TRP obtained by integrating the EIRP over the sphere of unit radius. For downlink performance measurements the power (target power) needed at the mobile handset to achieve the BER target must be determined. In order to have an efficient procedure, sweeping the power transmitted by the BS cannot be afforded.

A critical moment was the initialisation process that allows performing the needed reference measurements that will be used in the core downlink measurement procedure, that is to say an accurate value of the sent power to obtain a BER equal to the BER target (so-called Pt_{Ref}), which was achieved by the means of dichotomy [ChRo04], which is much less time consuming than a standard sweeping down of the power. Then the reference power received is calculated (so-called $RSCP_{ref}$). This reference measurement allowed measuring the variation between this reference position (the first one, on the first polarisation) and any other position/polarisation. In order to improve the repeatability of measurements, the reference measurement was done at the position giving the maximum CPICH RSCP. Then for every position/polarisation of the handset, the received power variation (at the UE) was calculated by subtracting the value of the CPICH RSCP at a given position (RSCPn) to $RSCP_{ref}$ (the reported CPICH RSCP is not a value but a 1 dB range). This variation (range) is then added to the output sent power of the communication tester (gives a 1 dB range for P_{target}), so that the received power at the UE should remain constant as well as the BER value. Practical experiments have shown that these values are not always inside the range defined by P_{target} limits due to the range of CPICH RSCP and also to the BER instability during measurements. The target power interval is then adjusted so the power associated to a BER of 1% is within its limits. Then the value in the middle of the P_{target} interval is returned – provided that, with a ±0.5 dB accuracy, the power needed by the tested handset to achieve the BER target is known. In the receiver sensitivity procedure, every position on the sphere surrounding the DUT is scanned, and for all these measurement points the RSCP (i.e. the received power at UE) is recorded. Once a full cycle is done, the positioner moves to the position where the strongest RSCP was measured. The transmission power of Node B is then decreased in steps of 0.1 dB until the connection drops. From this, the linearity of the RSCP was investigated, and also the minimal power received at the UE was determined. Finally, in the calibration process, the measurement system's power levels were calibrated using a reference antenna with known gain at European 3G frequency band (1920–2170 MHz).

The effect of an impedance conversion method for the reduction of the measurement equipment drift error in the small antenna measurement is considered in [IdYI01]. In addition to that, sensitivity of the measured input impedance against the fluctuation of the measured reflection coefficients due to the drift of a measurement instrument is investigated in [ITTO03]. The derivatives with respect to the absolute values of the reflection coefficient of the AUT (Antenna Under Test) as well as the phase angle of the reflection coefficient are calculated and used as the index of the sensitivity. As a result, the use of the stub tuner is proved to be very effective to reduce the sensitivity against the drift, especially for the measurement of electrically small antennas. The use of the impedance converter considerably reduces the sensitivity, thus it is very effective for the measurements of the electrically small antennas with large input reactance. This method is especially effective for reducing the effect of the amplitude fluctuation of the reflection coefficient on the input resistance as well as the phase angle fluctuation on the input impedance.

5.2.4 Factors influencing the measurement accuracy

Handset antenna performance measurement is a complex task that involves several moments, one of each being a potential source to measurement uncertainty. In this subsection many of these factors are addressed. The major focus is on the effect of the user's body, including the head and hands, or rather the phantom models used to emulate them. A modal approach for estimating their effect is presented too. Results are also provided for the sampling grid density of the radiation pattern, different channels within the same frequency band, the use of reference antennas and wave propagation (applicable only to MEG). Finally, some general analysis to overall measurement uncertainty and repeatability is presented.

Head, hand and body

One important part of a handset antenna test system is a model of the human body, the so-called phantom. The phantom simulates the effect of the real end user to the radiation characteristics of the handset and antenna. It is basically a structure that is used to simulate electrical properties of human tissue and physical shape of the user's body. Among others, the following three questions arise, when developing the phantom for handset antenna testing:

1. How well do the phantoms model the real human body?
2. Is a standard phantom needed in the antenna test procedures?
3. Are new phantom set-ups or test positions needed for 3G UE and beyond?

Relevance of the phantom for quantification of the influence of human body on η and MEG

In [Boyl02b] and [Boyl03] the efficiency and MEG of several commercial dual-band mobile handsets are evaluated over the DCS1800 and GSM900 frequency ranges respectively. Measurements are performed both with real people, and with phantom heads and hands. The phantom used in this study is the Schmid & Partner Engineering AG (SPEAG) generic head phantom. All measurements are averaged over the frequency ranges concerned (0.88–0.96 GHz for GSM and 1.71–1.88 GHz for Digital Cellular System (DCS)). Particular attention is given to the different loss terms involved, such as mismatch and spatial filtering, and to the relative contributions to the loss due to the different parts of the body.

The method used is that of 3D radiation pattern integration based on fast spherical near-field measurements around the combination of the user and handset. Helical, 'flip' and Planar Inverted F Antenna (PIFA) designs are evaluated. The reported MEG values are the average over seven different propagation environments whose average XPR is 7 dB. Independently of the handset/antenna type it is concluded that efficiency and MEG can be accurately measured using volunteers; further, the mean value of those two parameters converges to within 1 dB and the standard deviation to within 2 dB with 15 users in both GSM and DCS bands. Typical convergence is shown in Figure 5.4. Efficiency and MEG can also be measured using phantoms. However, the loss is very dependent on the hand, so that the position of the phantom hand is critical. In order for the results to be representative, the phantom hand should be placed in a number of typical positions corresponding to real use. However, the way a handset is held is to some extent a function of its mechanical design. Thus, it is difficult to devise a way of randomising the phantom hand position that is suitable for all handsets. At DCS, the frequency-averaged mismatch is not significantly influenced by the presence of the user. However, at GSM, the frequency-averaged mismatch is significantly affected by the presence of the user. Further, at some frequencies the matching can be very bad – with reflection coefficients greater than 0.7. Moreover, the head and the hand are seen to act as a coupled system so that at DCS losses are greater than the additive losses of the head and hand in isolation. At GSM the losses seem to be broadly additive. The losses in the head and the hand are in many cases approximately equal (more or less depending on the handset type). The total efficiency in the presence of the body is low for all of the commercial handsets measured.

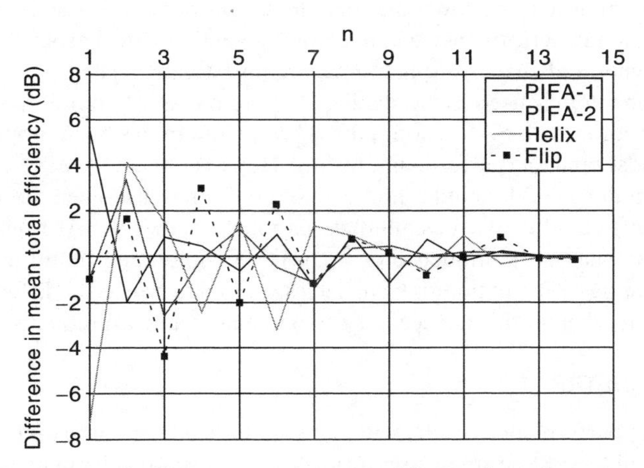

Figure 5.4 *DCS1800 mean total efficiency as a function of the number of users considered for averaging [Boyl03]. (© 2003 IEE, reproduced with permission)*

For GSM the average efficiency is −12.8 dB (−11.3 dB with phantoms). For DCS the average efficiency is −11.1 dB (−10.0 dB with phantoms). The average body losses at GSM and DCS are 11.1 dB (9.7 dB with phantoms) and 8.3 dB (7.1 dB with phantoms) respectively. Clearly, on average, the phantoms are a good representation of the body, but slightly underestimate the loss that occurs with people. The average MEG of the handsets is lower than the average efficiency. At DCS the average MEG is −14.6 dB (a reduction of 3.5 dB compared to the average efficiency). At GSM, the average MEG is also −14.6 dB (a reduction of 1.8 dB).

It is expected that the MEG would be slightly lower at GSM if averaged over all frequencies in the band (as is the case for the efficiency). The main reason for this reduction is that the handsets are predominantly horizontally polarised when in use and radiate in directions that are unlikely to support propagation paths. The different handset/antenna types have different loss characteristics. The handset with the helical antenna has the most interaction with the head, both in terms of loss and mismatch. This is particularly the case at DCS1800 when the local field of the antenna is more significant than the field associated with the phone Printed Circuit Board (PCB). However, relatively low interaction with the hand compensates for this to some degree. The 'flip' handset has the strongest interaction with the hand, since there is a strong probability that the hand covers the antenna. Also, this class of handset does not seem to offer particularly low loss in the head, despite the shielding offered by the PCB. However, it exhibits good MEG performance, particularly at DCS1800. Of the handsets tested, a PIFA-based handset is the best in terms of overall loss in the body. PIFA-based handsets appear to exhibit relatively large losses in the hand but, importantly, low loss in the head. However, they have poor MEG performances at DCS1800 due to high elevation radiation. Using this measure, there is little between all of the commercial handsets tested – all handsets have MEG within 3.3 dB of each other (at both frequencies). Of the handsets tested, a PIFA-based handset appears to have the best overall performance. This handset is the only one that maintains a good match in both bands for all users. This feature will lead to better radio performance.

Phantom head definition and its effect on antenna performance

In the previous subsection it was shown that phantoms can rather accurately be used to represent the human body in antenna performance testing. Now, several practical aspects related to phantoms, including different types and shapes of phantoms, are investigated. The Specific Anthropomorphic Mannequin (SAM) phantom is used as the standard phantom in SAR compliance testing [CENE01]. To increase harmonisation with SAR testing, the SAM phantom has been proposed as the standard phantom also for handset antenna performance testing. However, there have been discussions whether the SAM phantom model should include shoulders or not? It has also been discussed whether some other typical phantoms could be used as alternatives for the standard SAM phantom. Moreover, it is important to know what is the influence of brain tissue-simulating liquid that is used to fill the phantoms. In SAR testing different liquids have been specified to be used at different frequency bands but it has been unclear whether this is necessary also in antenna measurements.

Effect of shoulders on TRP

In [Krog02a] antenna performance results with a phantom head are compared to a similar phantom having head and shoulders. Experiments performed in an anechoic chamber for three active GSM mobile handsets by using a far-field 3D spherical scan system are shown in Figure 5.5(a). The handsets were measured on the middle channels of GSM900 and DCS1800 MT-TX bands. Phantom#1 is a SPEAG generic torso phantom v3.5 (head plus shoulders) and Phantom#2 is a generic phantom head v3.5, see Figure 5.5(b). The used test position was left cheek position, as defined in [CENE01]. The handsets were fixed on the phantom with tape. Both phantoms had a 2 mm thick rubber spacer to simulate the user's ear. The maximum differences in the TRP between phantom head and head plus shoulder phantom were ± 0.2 dB at GSM900 and $+0.3/-0.1$ dB at GSM1800. No clear tendency was observed about which of the two phantoms would give higher TRP values. The peak gain and directivity were approximately 1 dB to 2 dB higher with the Phantom#1 (head plus shoulders) compared to Phantom#2 (head only). In the radiation patterns no significant differences could be seen in the large-scale shapes. In the details of the pattern shapes there are some differences, which in many cases occur at the large elevation angles ($\theta > 140°$).

In [ChKn04] the effect of phantom shoulders was also studied by measuring 3D radiation patterns of two GSM handsets, a candy-bar handset with an internal antenna and a clamshell handset with an

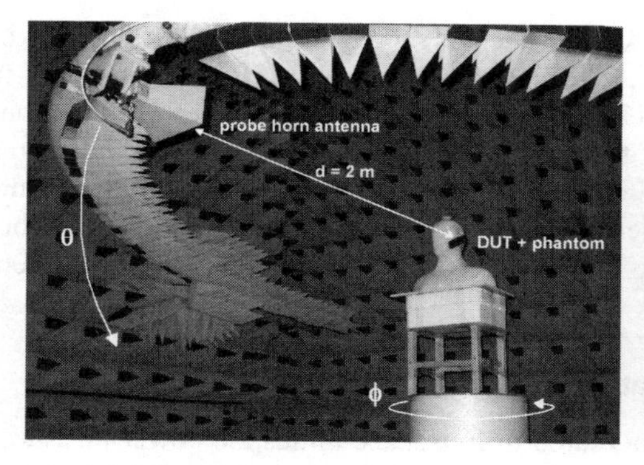

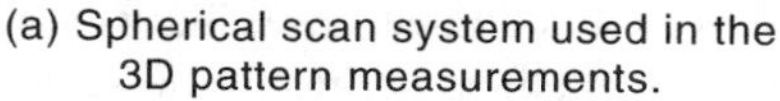
(a) Spherical scan system used in the 3D pattern measurements.

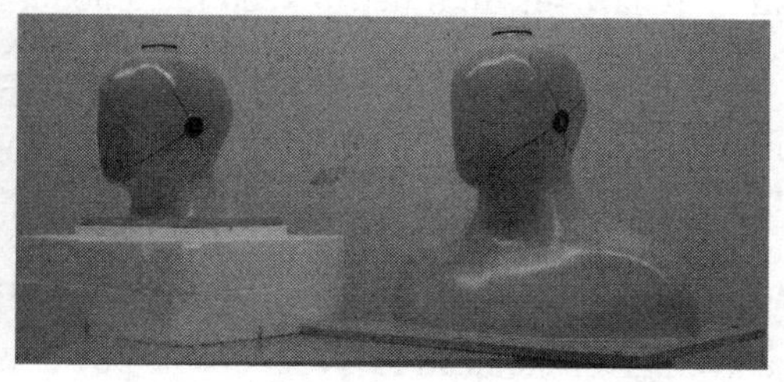

(b) SPEAG generic head and torso phantom used in the tests. Right: Phantom#1, Left: Phantom#2 [Krog02a].

Figure 5.5 *Spherical scan system and SPEAG generic head used in the tests.*

external antenna. The handsets were first measured beside an SAM phantom head and then beside an SAM torso phantom. The handsets were measured on the left cheek position, only for the lowest and highest bands of the handsets: GSM 850/1900 for the candy-bar handset and GSM 900/1900 for the clamshell. Measurements were carried out and three channels were measured per band. The maximum variation of the TRP of the two handsets between the two SAM phantoms was −0.1 dB to +0.4 dB.

Effect of different phantom head types on TRP and TRS

The results for an experimental comparison between SPEAG generic phantom head vs. Microwave Consultants Ltd (MCL) SAM phantom head were reported in [Krog02b]. The comparison of the phantom's shape (SAM head vs. generic head) showed differences up to 1.2 dB in TRS and 1 dB in TRP at 900 MHz. At 1800 MHz the differences were up to 1.4 dB in TRS and 0.4 dB in TRP. In a simulation study, also reported in [Krog02b], two different handset models (internal antenna and external antenna handset model) were simulated with three different head models (only two for the external antenna handset model). The head models differed significantly in shape and size. The average efficiency level with the head models was around 40% and the maximum differences in the efficiency were within 10%. Also the differences in the efficiency results between a phantom filled with tissue-simulating liquid and a solid phantom were experimentally investigated [Krog02b]. The differences in the efficiency between a solid phantom and a liquid-filled phantom were up to 1.5 dB. Not only the material but also the shape of the phantom heads was different.

Effect of different phantom head types on MEG

The effect of the shape of the phantoms on the MEG is expected to be larger than on the TRP since MEG takes into account also the effect of the radiation pattern and polarisation. To study this the MEG relative to a $\lambda/2$ dipole antenna was measured for four different dual-band handsets in passive mode: three GSM and one AMPS handset [Glaz04a]. Two phantoms were considered, the Schmid & Partner V3.5 (phantom head) filled with 5 litres of tissue-simulating SAR liquid for both

900 MHz and 1800 MHz and the Schmid & Partner V2.2 (head plus shoulders) filled with 22.2 litres of water-salt solution with a 1.44 g/l salt concentration. The measurements were performed according to the 'Telia Scattered Field Measurement Method' [OlLa98] in a measurement chamber with fictitious scatters [Glaz04e]. In order to keep the amount of measurements at a reasonable level only three frequencies, low, mid edge and high edge, were considered at each band. Results show that the average difference of body loss measured with the two phantoms is small (~0.3 dB). On the other hand, the difference between the highest and the lowest value was as high as 4 dB for different handsets.

Head loss comparison between different laboratories and phantoms

The average of the head loss (defined as power measured in free space minus the power measured in talk position with a phantom) results obtained at Siemens Mobile Phones in Aalborg (SMP AAL) [ChKn04] are shown in Table 5.1. In the same table, results obtained earlier in the same kind of investigations by Aalborg University (AAU) [PeNi02], [Krog02b] and Nokia RC [Krog02a] are presented. In Nokia RC the phantom was the SPEAG phantom head v3.5. In AAU both the SPEAG phantom head and the MCL SAM phantom head were used, and their results are the average results for the two phantoms. TRP results obtained at SMP AAL include the effects of the holder, which are estimated to be approximately 1 dB higher than the results without the holder. If correcting the SMP AAL results with this expected additional loss, the results are well in line with the results obtained in NRC (the corrected results of SMP AAL are shown in parentheses). The average of the head loss in SMP AAL (corrected results) and Nokia RC at 900 MHz band have nearly 1 dB difference to the results of AAU and at the 1800 MHz band the difference is less than 0.3 dB. Note that different test handsets were used in all the three laboratories. Finally, it seems reasonable to neglect shoulders from the phantom since they do not considerably affect the TRP results of mobile handsets in the 850 MHz to 2000 MHz bands. The differences in TRP were within 0.5 dB. The effect on the MEG is, however, in some cases larger. The comparison between the SPEAG phantom head vs. the MCL SAM phantom head showed that the difference in TRP is not very large either, but is larger than the effect of the shoulders. Hence, the SAM head is recommended to reduce total measurement uncertainty and to advance harmonisation with SAR testing.

Table 5.1 *Average phantom head loss values measured in different laboratories based on TRP measurements. A handset holder was used in the measurements in SMP AAL [ChKn04].*

System	SMP AAL	Nokia RC	UNI AAL
GSM850	5.1 (6.1)	–	–
EGSM900	4.1 (5.1)	5	4.2
GSM1800	1.3 (2.3)	2	2.6
PCS1900	1.3 (2.3)	–	–

TRP measurements with body-mounted phantom set-ups

For mobile handsets that support Internet and video applications, new user phantom(s) and set-ups must be defined in order to evaluate the effect a user has on the antenna performance (e.g. near-field coupling and shadowing). In [Krog02b] an investigation was made on several new body-mounted phantom set-ups. Six active GSM handsets were measured on GSM 900 (channel 62) and GSM 1800 (channel 698) bands in the different body-mounted set-ups. Six commercial GSM handsets were measured on GSM 900 (channel 62) and GSM 1800 (channel 698) bands in the different body-mounted set-ups. Obtained TRP values are presented in Table 5.2. Phantom hand was excluded from the considered set-ups, except in the so-called browsing position, where the extensions of the phantom arms simulated the user's palm. Some of the test set-ups are marked with 'display away from the phantom' set-ups, which describes a set-up where the antenna was towards the body. In fact, these are unintended use situations and here represent the worst-case scenarios. In many of the body-mounted phantom set-ups the body loss is a few decibels larger than in the traditional talk position set-up. However, the talk position set-up did not include here the effect of the user's hand. The MEG is expected to be relatively low in some of the set-ups, in particular in the waist position where the DUT is horizontally oriented. At 900 MHz the spread between the best and worst handsets in body phantom set-ups in terms of TRP was larger than in the traditional talk position and free-space set-up. At 1800 MHz the difference between the best and worst handset did not so largely depend on the phantom set-up. For the six GSM handsets tested, the average body loss due to the phantom head in talk position was about 5 dB at 900 MHz and about 2 dB at 1800 MHz.

Table 5.2 *Average TRP obtained in different body-mounted test set-ups, [Krog02b].*

Measurement description			GSM900	GSM1800
Set-up	DUT position	DUT orientation	TRP [dBm]	TRP [dBm]
phm. head	ear	left cheek pos.	25.6	24.8
phm. head	ear	right cheek pos.	25.8	24.8
phm. body	chest	ver. display in	25.8	25.2
phm. body	chest	ver. display out	24.8	24.0
phm. body	waist	hor. display out	22.8	23.6
box		browse pos.	24.3	23.7
box		ver.	22.0	22.8
free space		ver.	30.6	26.8

Effect of tissue-simulating liquid on TRP

In [Krog02a] the influence of the dielectric parameters of the tissue-simulating liquid used to fill the shell phantoms was studied. Liquid#1 is a sugar-based brain-simulating liquid prepared according to the recipe for SAR measurements for 1800 MHz [Schm99] and Liquid#2 is similar to the first liquid but prepared according to the recipe for 900 MHz. The dielectric parameters for Liquid#1 and Liquid#2 are given in Table 5.3. The SPEAG generic phantom head v3.5 (see Figure 5.5(b)) was used to investigate the influence of the liquid properties. The phantom was first filled with Liquid#1 and then with Liquid#2 and so the measurements were performed both at 'right' and 'wrong' band. The peak to peak differences in TRP between the liquids for the three DUTs (at both bands) was small, +0.2/−0.1 dB. This is close to the repeatability of the measurements. The differences in the peak gain and directivity are also small and the radiation pattern shapes for the two different liquids are nearly similar [Krog02a]. In [Krog05a] the influence of the liquid properties were further studied by FDTD simulations for a few simplified handset antenna models beside SAM head model. The simulated results are in line with the measured results. The simulation results, however, additionally indicate that

Table 5.3 *Dielectric parameters for brain tissue simulating Liquid#1 and Liquid#2. Values between parentheses are target values [Krog02a].*

	900 MHz		1800 MHz	
	ε_r'	σ_{eff} [S/m]	ε_r'	σ_{eff} [S/m]
Liquid#1	48.2	0.7	40.9(41)	1.7(1.65)
Liquid#2	39.7(42.5)	0.8(0.85)	31.0	1.6

the use of non-standard liquids, such as simple salt water, may lead to an error of 1 dB or more in TRP, in particular at 900 MHz. The simple salt water may be a practical liquid, e.g. for handset antenna R&D measurements, but for standardised measurements it is not recommended. It is concluded that the tissue-simulating liquid inside the head phantom does not need to be changed for different frequency bands. For example, one liquid can very well be used at both 900 MHz and 1800 MHz bands. This is opposite for SAR testing where the tissue-simulating liquid parameters (especially the conductivity) have a relatively large effect on the measured SAR and the total measurement uncertainty, which is the reason why the liquid recipes are specified in detail in the testing standards [IEEE03]. It is noted that these conclusions mainly hold for phantom heads. For a phantom hand the electrical properties of the material may need to be more accurately specified than for the phantom head.

The influence of the side of the head on η and MEG and polarisation state

It was found out in [Krog02b] that the differences in TRP between the left and right side of a phantom head in the talk position are typically small. However, this is not generally the case for the MEG. In [KSLK02] the MEG difference calculated from measured 3D radiation patterns was up to 4.5 dB between the left and right side of the phantom head for one of the test antennas at 1800 MHz (an active GSM handset). In order to further study this effect, FDTD simulations were carried out for simplified handset antenna models beside an SAM phantom head model [Krog05b]. The XPR of the model in MEG calculation was 14 dB. For the four test cases, the maximum difference between the left and right side was approximately 2 dB for the MEG and 0.5 dB for the efficiency at 1800 MHz. The maximum MEG difference was found for a monopole mounted on one corner of a metallic box modelling the chassis of a mobile handset. Obviously, a handset model with more complicated polarisation properties, Figure 5.6, would lead to larger MEG differences, as has been the case for the test antenna in [KSLK02]. Therefore if the MEG is used as the performance criterion, the measurement on both sides of the head seems necessary.

Effect of the head and handset orientation on MEG

The user never phones with the same head position and moves during the communication. To investigate the influence of head movements on the MEG evaluation, two numerical models of handsets close to a heterogeneous head have been simulated with an FDTD (Finite Difference in Time Domain) method, they are denoted T_1 and T_2 in Table 5.4 [CoDW03]. The handset-head system was rotated over a large range of angles around the x and y axes, Figure 5.7(a). MEG computation has been performed for all those different configurations. Five environment models have been considered, Figure 5.7(b): the isotropic model; urban area with a vertically polarised (VP) receiving antenna or a cross-polarised receiving antenna and rural area also with the vertically and cross-polarised (XP)

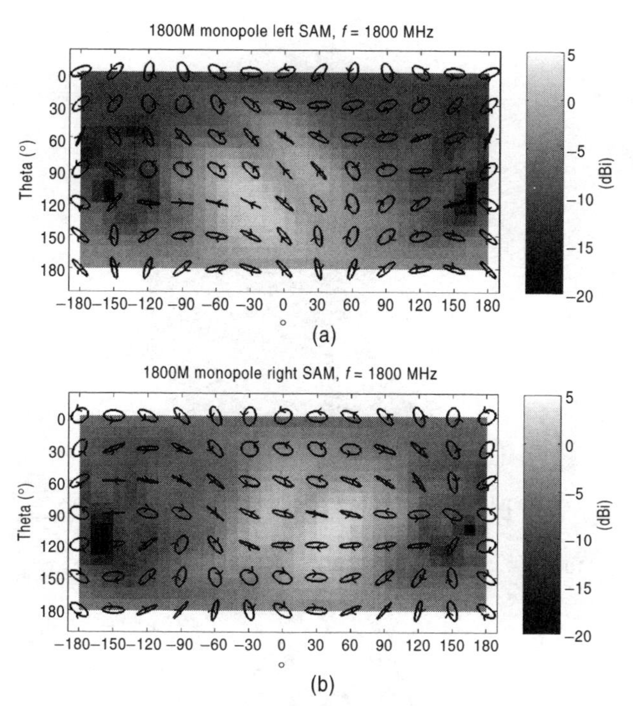

Figure 5.6 *Simulated radiation patterns and polarisation ellipses for a generic handset (monopole antenna on one corner of metallic box) beside the SAM head at 1800 MHz. (a) AUT on left side of SAM, (b) AUT on right side of SAM [Krog05b].*

Table 5.4 *Maximum MEG variation induced by a small head rotation (±20°) around an x or y axes among the different environments for two numerical handset models (T_1, T_2). The frequency is given in MHz and the difference in dB [CoDW03].*

	T_1			T_2		
Rotation axis	900	1750	1950	900	1750	1950
x	−0.1	0.7	−0.3	1.0	1.5	0.3
y	−0.3	0.4	−0.6	0.8	1.5	0.2

receiving antennas. For the urban area the considered angular distribution is defined by a general double-exponential function. The rural area model used is a classical Gaussian distribution. The XPR equals 20 dB for the models with VP antenna, whereas it has been fixed to 0 dB for the models with the XP antenna. The results show a significant variation of the MEG for the different configurations. A large movement (rotation of 90°) can induce an MEG variation higher than 10 dB. Smaller and more realistic head rotations (±20°) still yield considerable MEG variation that can be higher than 4 dB, Table 5.4. Compare this with the environment-induced variation of only 1.4 dB for the conventional measurement set-up (head straight and handset at 55° from the vertical line). Hence, it

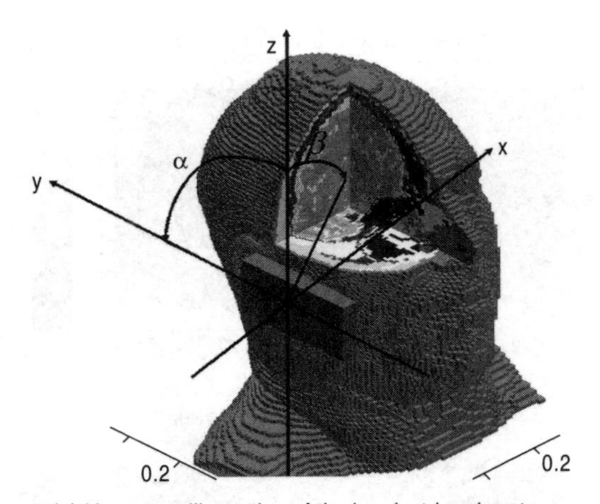

(a) Movement illustration of the handset-head system.

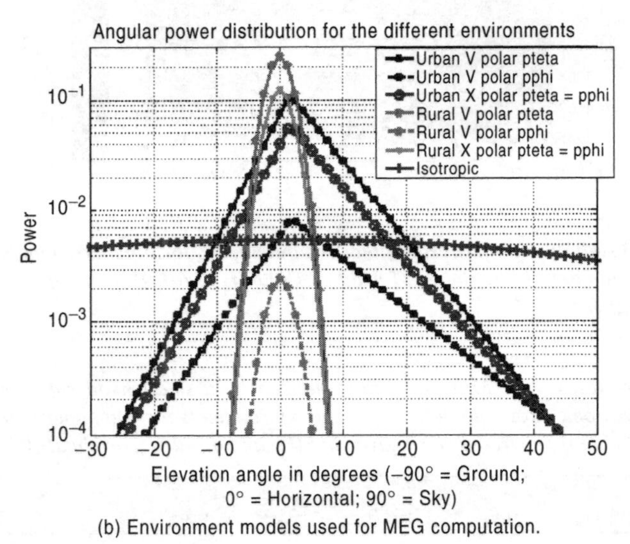

(b) Environment models used for MEG computation.

Figure 5.7 *Movement of the handset-head system and the environment models for the computation. V indicates that a vertically polarised receiving antenna is considered, and X a cross-polarised one [CoDW03].*

seems that several handset positions should be considered in order to provide realistic communication performance in terms of MEG. This point does not automatically imply that more measurements are needed, since the radiation pattern can be moved by post-processing.

Phantom hand definition and effect of hand model

In many cases, the user's hand is responsible for the largest amount of body loss, so that the definition of a representative phantom hand is a key topic. But such a definition is a complicated matter due to the huge number of degrees of freedom of a human hand.

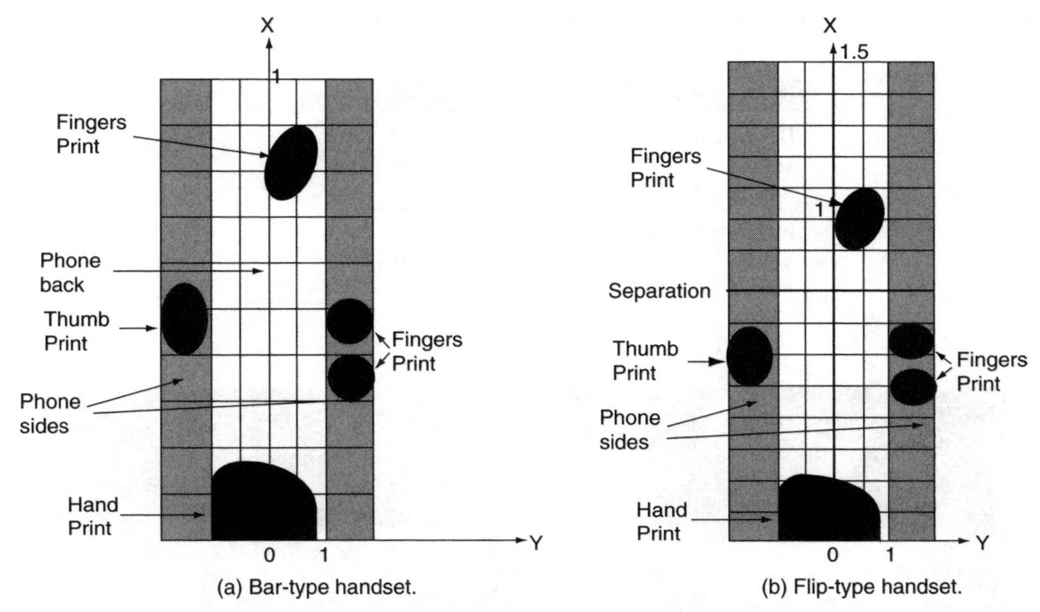

Figure 5.8 *Hand position evaluation grids for different handset types [CoDW04].*

In [CoDW04] and [CDWC04b] a statistical study of the hand position and its consequences in terms of hand modelling has been performed. In order to gain knowledge on the typical hand position in voice applications, an informal survey was carried out with the participation of 27 volunteers. A grid 100 mm long and 40 mm wide was drawn on the back of a commercial 'bar-type' handset, Figure 5.8(a), and a grid 150 mm (80 mm + 70 mm) long and 40 mm wide was drawn on the back of a 'flip-type' handset, Figure 5.8(b). The grids have squares of 10 mm by 10 mm. The grid overall dimensions have been defined in order to be representative of the European commercial handset offers of 2003. The survey results show that most of the people hold their handset in the same way, grabbing the handset between three fingers and using a fourth one to push the top back. The resulting positioning of the fingers and the hand palm has been parameterised as follows:

1. the finger backside x-axis position, which quantifies the height of the fingers touching the back,
2. the finger backside y-axis position, which gives the deviation of the fingers touching the back from the centre,
3. the finger backside print size, which is evaluated from the number of squares covered by the fingers,
4. the thumb position, which quantifies the height of the thumb and
5. the finger side position, which quantifies the height of the fingers touching the side.

Results are summarised in Table 5.5 and should be read together with Figure 5.8.

In 'browse position', the user needs to have access to the keyboard with the thumb. Thus, the hand usually covers the major part of the backside of the handset at the mid section. Following these considerations, TRP and MEG evaluations have been performed in typical talk and browse positions, Figure 5.9. The phantom hand used in those measurements has been made with a flexible material,

Table 5.5 *Statistical characterisation of hand position on bar-type and flip-type handsets [CoDW04].*

Hand position	1	2	3	4	5
Bar-type handset					
Mean	0.8	0.4	2.6	0.6	0.5
Standard deviation	0.1	0.3	1.8	0.2	0.2
Flip-type handset					
Mean	0.9	0.4	2.6	0.7	0.6
Standard deviation	0.2	0.3	1.8	0.2	0.2

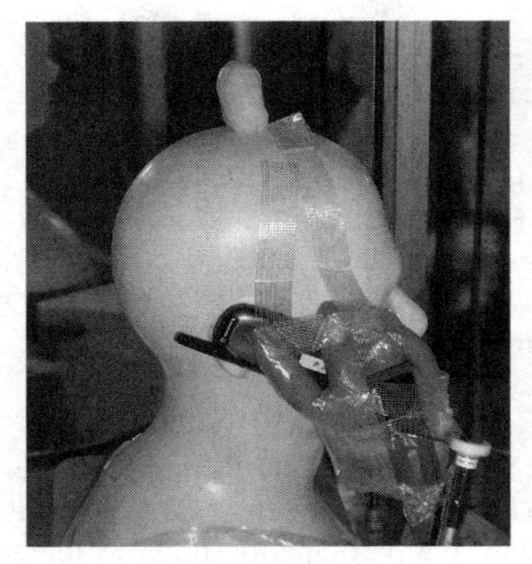

(a) Talk position.

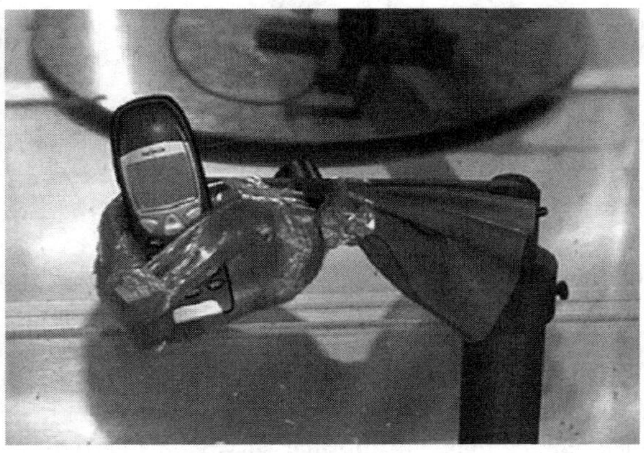

(b) Browse position.

Figure 5.9 *Hand position evaluation [CoDW04].*

in order to keep the possibility to adjust it for each handset. The hand was filled with standardised tissue-simulating liquids used for SAR measurement [CENE01]. In the first study [CoDW04], TRP measurements have been carried out for 12 different handsets in a reverberation chamber. In the second one [CDWC04b], the computation of MEG has been performed on two handsets, whose 3D patterns have been measured in an anechoic room.

The results of the first study are presented in Table 5.6 and show that the hand loss at 1800 MHz for the talk position is 1 to 2 dB higher than the hand loss at 900 MHz, whereas it is the opposite for the head loss. This may be explained by the fact that the radiation pattern is omnidirectional at 900 MHz, whereas it is directional at 1800 MHz, because of the typical size-to-wavelength ratio of current commercial handsets (10 cm long). Therefore, it can be expected that, for 3G handsets, the hand loss will also be dominant and will be an issue of relevance.

The goal of the second study [CDWC04a] was to investigate the difference between the effect of the hand on the TRP and on the MEG for different channel models. As explained before, in order to evaluate the MEG, 3D pattern measurements of two different handsets (T_3 with an helicoidal antenna, and T_4 with a patch antenna) have been performed, with steps of 3° in elevation and 15° in azimuth. The same five environment models as defined earlier have been considered, in order to

Table 5.6 *Hand loss in talk and browse positions, averaged over 12 commercial handsets [CoDW04].*

	GSM900	GSM1800
Talk position		
Mean [dB]	−1.7	−3.7
Standard deviation [dB]	1.4	1.6
Browse position		
Mean [dB]	−1.7	−2.5
Standard deviation [dB]	1.4	2.2

Table 5.7 *Hand losses averaged over realistic head positions (from 40° to 70°) [CoDW04].*

	T3		T4
	GSM900	GSM1800	GSM900
Urban VP			
Mean [dB]	−3.5	−3.0	−3.2
Standard deviation [dB]	0.7	0.5	1.0
Urban XP			
Mean [dB]	−3.5	−3.0	−3.6
Standard deviation [dB]	0.4	0.5	0.6
Rural VP			
Mean [dB]	−3.3	−3.0	−3.0
Standard deviation [dB]	0.7	0.9	1.2
Rural XP			
Mean [dB]	−3.4	−3.1	−3.5
Standard deviation [dB]	0.3	0.9	0.6
TRP			
Mean [dB]	−2.7	−2.1	−3.2
Standard deviation [dB]	–	–	–

compute the MEG. The results are presented in Table 5.7. To analyse the proportion of losses due to the handset polarisation variation and the one due to the directivity distortion, it is possible to compute the MEG in a uniform arrival angle environment. Results are provided in Table 5.8 for the Urban V model. The MEG losses due to the hand are not very different from the TRP losses or body losses, Table 5.8. In fact, the handset polarisation does not significantly change because of the hand presence and even if the pattern directivity is greatly distorted, this modification does not significantly corrupt the MEG, Table 5.8. The hand loss found with the phantom hand in [CoDW04], [CDWC04a] agrees well with the ones found with people in [Boyl02b] and [Boyl03]. With people the hand loss was 2.1 dB at 900 MHz and 3.5 dB at 1800 MHz [Boyl02b], [Boyl03]. With a phantom hand the loss was 1.7 dB at 900 MHz and 3.7 dB at 1800 MHz [CoDW04], [CDWC04b].

Table 5.8 *Hand losses due to the polarisation variation and the one due to the directivity distortion for the Urban V model [CoDW04].*

	T_3	T_4	
	GSM900	GSM900	GSM1800
TRP/MEG difference [dB]	−0.8	0.0	−0.9
Polarisation variation [dB]	0.2	0.4	−0.4
Directivity distortion [dB]	−1.0	−0.4	−0.5

Handset holder possible effect

The SPEAG SAM v4.5 phantom head includes a plastic handset holder as a standard accessory, which can be mounted directly on the phantom and used to hold the handset in the desired position during radiation pattern measurements. In [ChKn04] a total of seven GSM handsets were measured in two bands (GSM 850/1900 or GSM 900/1900) in cheek position (left side) both with and without the holder to investigate its possible influence. Double-sided tape was used to fasten the handset to the phantom, when measuring without the holder. The TRP results obtained in the talk position with the holder deviate with up to 1.2 dB from the results obtained without the holder, when comparing on a channel by channel basis. Unexpectedly, the results with the holder were in most cases higher than the results without the holder. The impedance tuning effect of the plastic holder is thought to be a potential reason for this. Even though the results are based on measurements for only two test handsets, they strongly imply that at least this specific holder affects the results by more than an acceptable degree and should not be used when measuring absolute values like TRP or TRS. The same kind of test is recommended for other handset holders.

Phantom hand for standardised handset measurements

In [KrMo05], an approach and implementation to include the hand effect in the standardised measurement was presented. Conceptually, the approach consists of three steps. The first is to define nominal physical properties (dimensions and dielectric properties) for the phantom hand, the second is to define nominal grip or grips that statistically represent a significant number of end users, and the third is to implement a phantom hand and procedure that utilise these nominal parameters in a repeatable manner for any handset under test. The average values for finger dimensions are given in [KrMo05]. The RF dielectric properties of several human hands were characterised [KrMo05] by performing a comparison between a phantom hand filled with tissue stimulants of varying properties and actual human hands. A human hand and the phantom hand were alternately placed over the open end of the waveguide while measuring its S_{11}. The values for a homogeneous phantom hand thus determined are presented in Table 5.9. The values shown at 2170 MHz are estimated from the measured empirical data at 835 and 1910 MHz.

The specific definition of a hand grip (that is, position of the hand and fingers relative to the handset on the head) during testing is perhaps the most problematic aspect of including the hand in a certification test, since any number of different grips could be said to represent some user. A human factors analysis of hand grip positions (during normal handset usage at the ear) on the two basic types of cellular handset form factors, clam and candy-bar, was conducted and reported in [KrMo05]. The total number of users observed for the analysis was 184, and the number of

Table 5.9 *Dielectric values for a homogeneous phantom hand [KrMo05].*

System	Frequency [MHz]	ε_r'	σ [S/m]
GSM850/900	835	30	0.64
DCS/PCS/UMTS TX	1910	26	1.20
UMTS RX	2170	25 (est.)	1.30 (est.)

handset models was 14. It was found that in both cases, by selecting two grips for each form factor, one encompasses approximately 70 to 80% of the users' grip positions.

Two phantom hand implementations satisfying these requirements were investigated [KrMo05]. In the first, a skeleton of thin-wall fibreglass tubes was constructed to the proper dimensions, with articulated fingers and movable joints between finger segments. A Latex™ glove was installed over the completed hand assembly, and then filled with an appropriate hand tissue simulant solution. This phantom hand can then be posed on the handset and phantom head, in one of the specified grips. With trained test technicians, this scheme was found to have repeatability (in the measurement of TRP of the handset under test) of ± 0.75 dB. The second approach, being investigated at the time of publication, is to mould a phantom hand (of a material that emulates the desired electrical properties) that is fixed into the agreed grip position(s) for each particular handset model to be tested (or perhaps for each range of widths of handset models). It is believed that this would maximise test repeatability. For each new handset model to be certified, a CAD model of a phantom hand gripping the handset in the in-use position at the head, according to the agreed grip(s), is generated. An industry-agreed CAD model would offer the additional benefit of facilitating electromagnetic simulation using the same hand definition that will eventually be used in measurement.

Modal approach for estimating the effect of user's body and phantoms

The method presented in [DeBo05a] provides a more general and analytical way to characterise user-mobile interactions compared to prior work in this field [KuBa92], [VOKK02], [DeBL04]. This technique was first applied to the problem of energy absorption in a phantom, but could also be used to examine the radiated performances of a given source in the presence of a user or user model. Basically, the energy deposition in a phantom is described in terms of power transfer from a finite set of equivalent generators to a microwave junction/multipole load, characterised by its scattering matrix. This equivalent junction/circuit model, derived from an appropriate modal expansion of the electromagnetic field in incoming and outgoing waves of mode order n, allows to introduce the available power and load factor concepts for radiating sources and scatterers. The concepts of available power and load factor in electromagnetic fields provide a new meaningful tool to describe power loss mechanisms in reactive-field regions, by allowing to separate the effect of the source and phantom parameters on the energy absorbed by the phantom.

A calculation example based on the minimal cylinder ($R = 100$ mm) containing the SAM phantom head, exposed to a uniform current line source, is shown in Figure 5.10(a), where the most significant modal contributions to the total dissipated power as function of the distance d between the source and cylinder are represented. Figure 5.10(a) shows that, on the opposite of interactions in radiative-field regions, the lowest order modes are not necessarily the most dissipative ones when the source is held close to the phantom. This can be explained by means of Figure 5.10(b), which represents the load

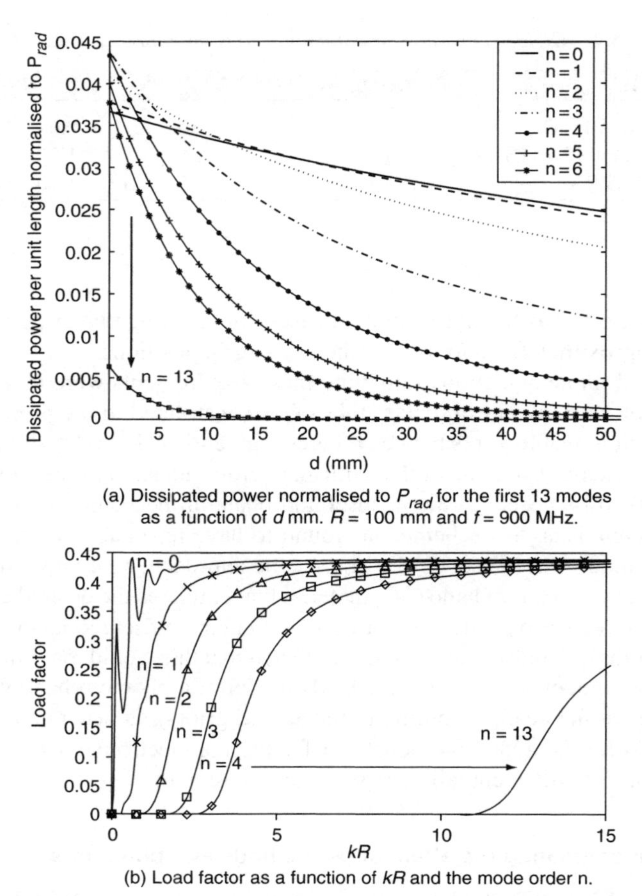

Figure 5.10 *Dissipated powers as a function of the distance and the load factor as a function of the mode [DeBo05a].*

factors of the low-order modes as functions of *kR*. The load factor of a given mode represents the efficiency of power transfer from the corresponding modal equivalent generator to the equivalent junction. It only depends on the phantom geometrical and electrical characteristics and the wave-number k (there is no dependence on any other source parameter). It is observed on Figure 5.10(b) that the higher the mode order, the lower the corresponding load factor. As a consequence, if all the modes were equally excited (with the same available power), modes of lowest orders would be the more dissipative ones. However, it is shown in [DeBo05a] that the available power strongly increases with the order of the mode, and this increase is even bigger when the source gets closer to the phantom. As a result, in the near field of the line source (or the mobile handset), high-order modes can generate a significant absorption. Finally, it is noteworthy that, for a SAM-sized phantom (R = 100 mm, $kR \approx 1.8$ at 900 MHz), the load factors are all smaller than their large radius limit, which corresponds to the power transmission coefficient of a plane wave arriving normally on a half-space filled with tissue-simulating liquid. In the case considered here, the maximal load factor is then approximately equal to 44% independently from the mode order. Current work shows that many ways exist to extend and use this technique to describe more realistic situations, like that of a mobile handset close to a user's head. Moreover, as some work to be published will show [DeBo05a], [DeBo05b], the equivalent junction approach already allows to obtain minimal and maximal bounds of power potentially absorbed by a phantom, for any source-phantom configuration.

Sampling grid

Three methods to characterise as accurately as possible the power of a handset have been investigated, compared and their limitations assessed. The main goal of these methods consists in determining a minimum number of 3D samples that have to be taken to characterise the radiated power or MEG within a required accuracy. In practice there are many issues that need to be taken into account when evaluating the overall error introduced by using a limited number of samples, such as reflections, measurement errors, positioning errors, and calibration errors.

The first contribution [LVKK03] uses an FDTD computation of a handset at 1800 MHz in combination with a realistic model of the human body by combining a phantom head with the upper trunk (shoulder on the side of the headset). This pattern, computed in 984 locations on a spherical surface in the far field was used as a reference model to compare the Fourier transform, the spherical wave expansion techniques as well as a simple numerical integration technique. Some use the knowledge of the full complex radiated field, while others start from the radiated field strength only. The work is based on simulations for four computer models of mobile handsets. The main conclusions from this work are that, in general, the field characterisation techniques utilising both the amplitude and the phase information of the radiated field at each measurement direction, provide, for a certain relatively small number of measurement directions, a more accurate prediction of the field at other directions compared to utilising only the amplitude information. Also, the spherical wave expansion is in general better than the Fourier expansion. The results also give guidelines for choosing a reasonable number of measurement directions for the radiated field of a mobile handset beside a phantom head. When using a spherical wave expansion, 30 measurements give a deviation below 0.05 dB (both amplitude only and full measurements), while Fourier expansion and numerical integration require over 60 measurements.

The second contribution [LiVC04] uses a more coarse model from the headset and the body, by using only a cardioid's pattern generated by an array of two dipole antennas with and without a perfectly electric conducting reflector of size 20 cm $\times$ 50 cm for simulating the body located at a quarter wavelength behind the antenna. This reference pattern was computed with a moment method in 64,800 points at 900 MHz. Since here only a spherical wave expansion is used, but using Fourier orthogonality techniques to determine the expansion coefficients, the conclusions are apparently the same as in the previous case: measurements every 45° in both azimuth and elevation angles (or 32 spherical measurements) will deliver an accuracy of less than 0.05 dB on the power estimation of the full ensemble (headset plus body). In this contribution also a fully exact theory for determining the effects of inaccuracies of the measurement devices on the computation of some parameters like the power, the radiation pattern of arrays, or the received signal on an antenna in the presence of small reflections is presented. This is based on the transformation of inaccuracy shapes in the complex plane.

The last contribution [NiPe05] reports on the computation of the MEG starting from measurements of the spherical radiation pattern measurements for five GSM handsets. The radiation patterns were sampled in a $10° \times 10°$ grid (elevation angle $\times$ azimuth angle). The MEG was computed using different models of the mobile environment, including an isotropic environment resulting in the TRP and TRS for the uplink and downlink, respectively. As expected, the MEG values computed using models, which are essentially isotropic, were found to be influenced less by a reduced sampling density than the MEG values obtained with the two measurement-based, non-isotropic models of the power distribution. Considering all the environment models, a sampling grid of $10° \times 20°$ resulted in a maximum error of 0.4 dB and a standard deviation of 0.1 dB. If the HUT model is not applied, a sampling grid of $20° \times 20°$ may be attractive, in which case the maximum error observed was 0.5 dB and the standard deviation was 0.2 dB. For an isotropic environment the TRP and TRS can be obtained within a maximum error of 0.5 dB if a $30° \times 30°$ sampling grid is used. From measurements it was found that the TRP and TRS were repeatable to within 0.1–0.2 dB. It should, however, be mentioned that the results obtained from the practical measurements are not obtained using spherical wave expansion but rather the simpler numerical integration of the measured values, as specified by CTIA certification document.

Frequency

The computation of the MEG from measurements of spherical radiation patterns of mobile handsets and in particular the variation in the MEG for different measurement frequencies was addressed in [NiPe02a]. The full spherical radiation patterns of five different mobile handsets were measured in an anechoic room. The measurements were performed using a BS simulator for the GSM1800 system at the centre channel as well as the two channels at the band edges. The MEG was then computed for three different environments and for 216 orientations of the handsets. A difference in MEG at the band edges of up to 1.7 dB was found compared to the MEG at the centre channel, and considering all handset orientations. Moreover, in this work, a method is proposed for reducing the total number of measurements. Assuming that the frequency variation is mainly in the total power received or transmitted by the antenna, the radiation pattern is normalised, resulting in an, ideally, frequency independent and normalised MEG. The normalised MEG can then be scaled to any frequency using the total power. Hence, the frequency dependent total power can be estimated within a fraction of a dB by using only a small subset of the full spherical radiation pattern. For the proposed method to be useful the power normalised radiation patterns must result in frequency independent MEG values. From the measurements it was found that the changes in the normalised MEG over frequency for different handset orientations were maximum 0.8 dB. This error introduced in the MEG should be compared to the variation in the MEG for different handset orientations in realistic environments, which may be 3–7 dB [NiPe02b].

Propagation environment

As explained earlier, the MEG of an antenna shall be considered, in order to include the effects of the propagation channel characteristics on the antenna performance. The two main figures introduced by MEG are the power angular distribution of each polarisation and the XPR of the channel. The purpose of this subsection is to give examples of power angular models and their effect on MEG computation.

The angular power distribution of the incoming waves

In [KSLK02], experimental results of angular power distribution in elevation and channel XPR seen from the MT viewpoint obtained in different radio environments at 2.15 GHz are presented. It was found that the double-sided exponential model fitted more accurately the measured distributions. The power distribution is assumed to be uniform in azimuth since the orientation of the mobile antenna in azimuth is considered to be equiprobable. Measurements were performed in five different radio environments: an indoor picocell, an outdoor-indoor, an urban microcell for three BS heights, an urban macrocell and a highway. For each environment, the major part of the measurements was done in a non-line-of-sight situation (77% on average).

The average MEG of three GSM handsets obtained both in free space and in left and right cheek positions with a phantom head with torso was provided for all five environments. It is observed that the effect of the environment on the performance is not as remarkable as the effect of the handset itself, as long as the propagation environment is included in the evaluation. The efficiency (not including the effect of the environment) provides rather different results compared to MEG in a real usage environment. The results were compared with the corresponding MEG computed from the measurement sets. The absolute difference between the model-based results and measurement-based results is in most of the cases less than 0.5 dB. Nevertheless, the difference is smaller for the general double exponential model. Therefore, very precise elevation power distributions models are not needed for accurate MEG estimations.

The cross-polarisation of the channel

In [CoWi05] a model for channel XPR extracted from measurements carried out at 77 GSM/DCS sites in French networks is proposed. The measurement data was collected at 245 locations distributed over the 77 sites (152 locations in urban environments (103 indoor and 49 outdoor), 73 locations in suburban environments (45 indoor and 28 outdoor) and 20 locations in rural environments (12 indoor and 8 outdoor)). Each component of the electric field was measured with a three-axis probe associated with a switch and a spectrum analyser that measured the Broadcast Control Channel (BCCH) signal. Measurements were carried out on a 9-point grid, separated at enough distance to ensure uncorrelated measurements at investigated frequencies (>925 MHz). The resolution bandwidth used was 300 kHz. Each field component was averaged over 50 samples. The results show that the polarisation angles follow a Gaussian law, with an average that depends mainly on the BS antenna tilt. Moreover, the polarisation angle variation is comparable at all locations and was within 15°–20° most of the time.

The average MEG of two numerical handset models placed close to a heterogeneous head model have been calculated, for all the XPR values given in Table 5.10, by weighting each MEG with its corresponding XPR probability density function. The results show that the standard deviation differences yield less than 0.01 dB of variation on the MEG. Moreover, few degrees of variation on

Table 5.10 *Polarisation angle mean and standard deviation for the different configurations [CoWi05].*

Environment		900 MHz Pol. angle	900 MHz XPR [dB]	1800 MHz Pol. angle	1800 MHz XPR [dB]
Rural	VP	67°	7.4	67°	7.4
	XP	38°	−2.1	32°	−4.0
Non-rural	VP	59°	4.4	65°	6.6
	XP	30°	−4.7	26°	−6.2

the mean polarisation angle yield less than 0.2 dB of variation on the MEG. Therefore, an average model is enough.

Reference antennas

Monopole and dipole are often used as reference antennas in mobile handset antenna measurements. In particular, they are commonly used as reference antennas in substitution-type measurements. In these measurements the basic idea is to calibrate the transmission loss of the chamber by substituting the DUT with an antenna with known gain or efficiency characteristics. With the substitution data the raw measurement data obtained with the DUT can be transformed to absolute power (or antenna gain) data. Accurate reference antennas are very important, since the inaccuracy of the gain calibration is typically one of the largest contributions to the total measurement uncertainty in the 3D pattern measurement method. The monopoles and dipoles are often used also as test antennas when characterising the quiet zone of an anechoic chamber or other test rooms used for handset antenna testing.

Monopole reference antennas

In [KrJI03] reference monopole antennas designed for 900 MHz and 1800 MHz bands were presented. These antennas have been designed and constructed for calibration and test system comparison measurements. The monopoles were measured in three different laboratories. Laboratories A and C measured the antennas with the 3D Pattern Integration Method (PIM) and Laboratory B with the Wheeler Cap Method (WCM). Moreover, in one of the laboratories different calibration information sources were tested. The first source was the data sheet gain values from the horn antenna manufacturer and the second source was a calibration by the so-called Three Antenna Method. The differences in the measured total efficiency between the laboratories were approximately 1 dB from 0.83 to 1 GHz for the 900 MHz band monopole and 1 dB from 1.7 GHz to 1.9 GHz for the 1800 MHz band monopole. Below 830 MHz the differences are higher, up to 2.3 dB, and the main reason for this is believed to be the large uncertainty of the specified gain values of the horn antenna, which was used as the gain standard. The advantage of the WCM is that it does not rely on any secondary reference/standard antenna and the measurement with the WCM (minus the mismatch loss) is probably the most accurate way to determine the efficiency of the monopole antennas. It is proposed that the monopoles could be used as 'efficiency standard' antennas. In the calibration procedure, the full 3D pattern of the reference monopole antenna would be measured and the total integrated power would be used as the reference level for the TRP (or efficiency) of the AUT/DUT. This provides higher accuracy than using only the peak gain level of the reference antenna.

Reference sleeve dipoles and magnetic loop antennas

Half-wave dipole antennas are widely used as reference antennas for measurement/calibration of antennas and test ranges for low gain antenna measurements. An example of a reference sleeve dipole is shown in Figure 5.11. The design is based on low loss end-fed sleeve dipole technology minimising cable and feed point interaction. The design includes a choke, which further reduces cable interaction by attenuating the natural return currents from the dipole. Reference antennas for handset antenna measurements were discussed in [GaFo05]. A half-wave dipole is a relatively narrowband antenna with roughly 10% bandwidth and very high efficiency ~95%. A carefully designed and

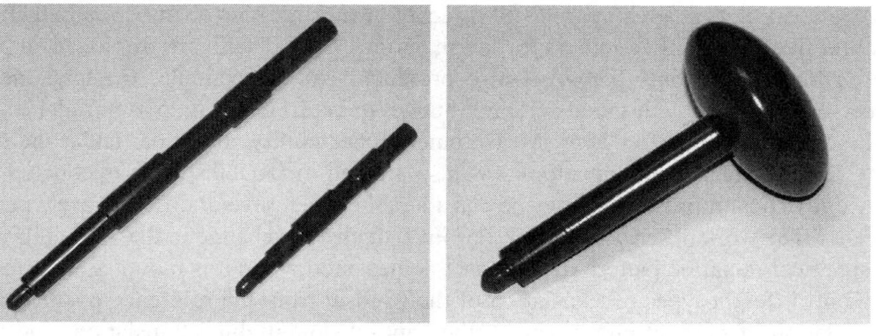

(a) Reference sleeve dipoles. (b) Magnetic loop antennas.

Figure 5.11 *Examples of different types of antenna [GaFo05].*

constructed sleeve dipole is a good choice as the reference antenna for calibration procedures of cellular handset performance measurement standards. For example, the relative bandwidth of UMTS systems is 12% (1920 to 2170 MHz), thus one calibrated sleeve dipole can approximately cover the whole UMTS band. However, different sleeve dipoles are needed for different bands, e.g. GSM900, DCS1800, and UMTS bands. Thus, a set of calibrated sleeve dipoles is usually needed. Usually in the handset antenna test procedures, the measurements are needed with two orthogonal polarisations. Consequently, calibration is also needed for both of these orthogonal polarisations. The magnetic dipole or magnetic loop antenna shown in Figure 5.11 can be used to complement the electrical dipole. The antenna consists of a planar structure generating a loop of current fed by a coaxial cable from below. The cable is orthogonal to the polarisation so any interaction between the two will generate cross-polar radiation. The design includes a choke, to further reduce cable interaction. The radiation pattern of the magnetic dipole is very similar to the electrical dipole but in orthogonal polarisation. The bandwidth and efficiency properties are comparable to those of the electric dipole. For both sleeve dipole and magnetic loop antenna a very high degree of azimuth pattern symmetry can be obtained with a careful design and high precision machining of the antenna components. Theoretical formulas can be used to predict the performance of both of these antennas although these formulas are unable to predict the effect of the choke. Therefore, accurate pre-calibration of the reference antennas is recommended in a laboratory specialised in high accuracy antenna calibrations.

Measurement uncertainty and repeatability

Like in any measurement, in handset antenna performance measurement it is important to identify any possible measurement uncertainty sources and evaluate the overall measurement uncertainty. TRP measurement procedure can be divided into two stages. In Stage 1 the actual measurement of the 3D pattern of the DUT is performed. In Stage 2 the calibration of the absolute level of the DUT measurement results is performed, usually by means of a calibration antenna whose absolute gain is known at the frequencies of interest. The largest uncertainty contributions in TRP (and efficiency) measurement are typically: uncertainty of the absolute level of measurement receiver, absolute antenna gain of calibration antenna, reflectivity level of the anechoic chamber and DUT positioner structures and the uncertainties related to the phantom. More details are discussed in [KDPS04], where an example uncertainty budget is also presented. As an example, the total measurement uncertainty was estimated with the uncertainty budget to be ±1.5 dB at 95% confidence level for a typical measurement system [KrJä05]. Measurement repeatability and in particular the positioning uncertainty of DUT against the phantom was also studied in [KrJä05]. It was concluded that the uncertainty due to positioning errors may be considerable, when several different users perform TRP measurements. The work described in [NiPe05] investigates the change in the TRS, TRP, and MEG when the spherical radiation pattern of a handset is measured while it is mounted incorrectly on the phantom. Four different types of translation of the handset from the reference position were used, namely translation of the bottom/top end of the handset, longitudinal translation, and transversal translation. Six different handsets were measured on both GSM900 and GSM1800 at channel 62 and 698, respectively. The results of this work were obtained with deliberately rather large translations of the handsets on the phantom, usually 15 mm from the reference position. With careful mounting of the handsets smaller deviations from the correct position can be obtained and smaller changes in the results are expected. Generally it was found that TRS and TRP values are correlated so that an increase of the transmitted power due to a translation is usually associated with an increase in the received power. Furthermore, the results show similar influence on the results obtained for the low and high frequency bands. The deviations found for the TRS and TRP values are generally within ±0.5 dB with a maximum deviation of about 1.4 dB. From statistics of the computed MEG values based on data from all handsets, link directions, orientation, and offsets, it is found that the mean MEG deviations due to translations are generally low, about 0–0.2 dB. Furthermore, standard deviations of 0.1–0.5 dB and maximum deviations up to 1.6 dB were found for most handsets, with one exception having a maximum up to 2 dB. The changes due to the incorrect position of the handsets on the phantom should be compared to the uncertainty due to the measurement system and the methods used. Using repeated measurements, the MEG results were found typically to be repeatable within 0.1–0.3 dB. In addition, the changes in the MEG introduced by positioning errors are small compared to the variation in the MEG of 6–8 dB that may be observed for a handset depending on its general orientation in the environment. For the TRP/TRS a difference of 3–4 dB was found between different types of handsets.

5.2.5 Figure of merit comparisons

As exemplified in previous sections a comprehensive assessment of the MEG requires the environment characteristics to be well defined. MEG depends on the propagation channel parameters, for instance XPR and angular power distribution. However, the latter seems to have less effect on the MEG, which has been corroborated by results presented in [KSLK02], [NiPe02b], [Glaz04f], [CDWC04a].

In [KSLK02] the study comprised five different power angular models for 10 different antenna designs. It was concluded that on average MEG will not be highly affected by the used models. However, the calculated MEG was much lower than the radiation efficiency. Another very good example of such variation is provided in [NiPe02b] where TRP and MEG were obtained for five different GSM handsets. Both the up- and downlink were evaluated with and without the phantom (free space). The difference in TRP (TRS) was only 3 dB, which obviously is relevant from a power point of view, but is still low compared to MEG variation in real environments. On the other hand, the difference in average MEG was also 3 dB as in the case of TRP. Moreover, the average MEG for most angular power distribution models was within ± 1 dB around the 3D uniform (isotropic) model, which was a good indication of the fact that 'average' MEG may be estimated through TRP, or more correctly, through the radiation efficiency. Namely, in [Glaz04a], [GlPa04] an approximate assessment of MEG (MERP) was obtained for the radiation efficiency (TRP) of the mobile handset measured assuming a 3D uniform distribution for the angles of arrival distribution was proposed. Still the XPR of the propagation channel must be defined and the XPD of the antenna must be measured. It was further shown that the MEG may be expressed, under the above constraints, as the product of the radiation efficiency and another factor that accounts for the joint polarisation and directivity mismatch denoted as Mean Effective Directivity (MED) [PeAn99], [Plic04], hence,

$$MEG_{[\mathrm{dBi}]} = \eta_{[\mathrm{dB}]} + MED_{[\mathrm{dBi}]}$$

or expressed through TRP

$$MERP_{[\mathrm{dBm}]} = TRP_{[\mathrm{dBm}]} + MED_{[\mathrm{dBi}]}$$

where the mean effective directivity is defined as follows [Glaz04a], [GlPa04]

$$MED_{[\mathrm{dBi}]} = 10 \log \left(\frac{XPR \cdot XPD + 1}{(XPR + 1)(XPD + 1)} \right)$$

where XPR is the cross-polarisation of the channel and XPD is the average cross-polar discrimination of the antenna.

Hence, any improvement in radiation efficiency will result in the same improvement of the MEG provided that the radiation pattern (and MED) is unchanged and vice versa. It is further shown that MEG for the 3D uniform angular distribution is upper bounded by the antenna efficiency independently of the polarisation characteristics of neither the antenna nor the propagation environment. That is $MEG \leq \eta$ since $MED \leq 0$ for all XPR and all XPD. Finally, based on the example of the $\lambda/2$ dipole antenna it is shown in [Glaz04a] that in order to characterise the polarisation effects on MEG the statistical distribution of the XPR of the channel shall be included.

In [GlPa04] passive mode antenna gain measurement results are presented for four dual-band handsets. Three EGSM/GSM handsets were measured, one of which had an external antenna and the other two had embedded antennas. The fourth antenna was operated in the AMPS/PCS bands and had a retractable external antenna. Both left and right cheek positions were measured. The low, mid and high frequency bands were measured in both the uplink and downlink. The measurements were done in the real time spherical near-field antenna test facility (SATIMO) at AMC Centurion. The average XPD of the measured antennas varied between -7.5 dB and 2 dB with an approximate average of -2.7 dB indicating that the received/transmitted power was mainly horizontally polarised. It was shown that in general the radiation efficiency (TRP) overestimates the antenna performance in comparison to the MEG (MERP).

In [CDWC04b] the difference between the MEG computed using five power angular models and its approximation given by equations (above) is investigated. Five environment models have been considered, an urban area with vertical polarised antenna, an urban area with cross-polarised antenna, a rural area with cross-polarised antenna, a rural area with vertical polarised antenna and an isotropic model. For the urban area the considered angular distribution is defined by a general double exponential function. The rural area model used a classical Gaussian distribution. The XPR equals 20 dB for the model considering vertically polarised antennas, whereas it has been fixed to 0 dB for models considering cross-polarised antennas. Numerical models of two mobile handsets have been simulated close to a heterogeneous head with an FDTD code. The first numerical handset, denoted as T_1, is simply a metallic box with a $\lambda/4$ monopole antenna. The second one, T_2, is a numerical model of a commercial handset using a patch antenna. Moreover, measurements of two commercial handsets close to a phantom head have been carried out in an anechoic chamber. One of the handsets is equipped with a helical antenna and is denoted T_3. The second one, T_4, had a patch antenna. The measurement steps were 3° for the elevation and 15° for the azimuth. As a user never uses its handset in the same position, the MEG has been averaged over small head movements. The angle between the handset and the vertical has been moved in 5° steps from 40° to 70°, using post-processing. Table 5.11 shows the difference between the computed and approximated MEG average over realistic head positions. It ranges from 0 dB to 1.7 dB and the mean difference equals 0.7 dB. The approximated average MEG given by equation (above) slightly underestimates the MEG in most cases. This is due to the fact that in uniform arrival angle repartition the antenna lobes are not weighted by the environment model. Nevertheless, this difference is in many cases negligible if compared to usual measurement uncertainty. Therefore, the main parameter in an environment model for MEG computation is the XPR of the channel. The XPR measurements are much easier to perform than the one needed to characterise arrival angle reparations because they can be done with a simple tri-axes E-field probe or a calibrated antenna plug in a spectrum analyser.

Table 5.11 *Difference between computed and approximated MEG for two numerical handset models (T_1, T_2) and two real handsets (T_3, T_4). The frequency is given in MHz and the difference in dB [CDWC04b].*

	T_1		T_2		T_3		T_4	
Model	900	1750	900	1750	900	1750	900	1750
Urban VP	−0.1	0.7	−0.3	1.0	1.5	0.3	1.6	0.3
Urban XP	−0.3	0.4	−0.6	0.8	1.5	0.2	1.6	0.2
Rural VP	0.2	1.2	−0.1	1.6	1.4	0.4	1.6	0.6
Rural XP	−0.3	0.6	−0.6	1.0	1.7	0.3	1.7	0.6

5.2.6 Handset design aspects

In this subsection, several aspects related to handset antenna design are discussed. First, the relation between the handset chassis length and some important antenna parameters are discussed. Second, the influence of materials used in antenna covers is discussed. Next, an antenna concept based on dual-fed PIFA is presented. Then the investigations related to the SAR are discussed. Lastly, adaptive matching systems are introduced and their performance is evaluated.

Antenna design

The electrical characteristics of a mobile handset antenna depend strongly on the size of the ground plane of the device on which the antenna is mounted (chassis) and the position of the antenna on it [TaTs87], which makes the chassis parameters an important part of handset antenna design. Previous results, e.g. in [TaTs87], [VOKK02], indicate that the total radiation bandwidth of the antenna-chassis combination is partly defined by the dipole-type radiation of the chassis currents, whose level further depends on whether the chassis is at resonance or not [VOKK02]. Based on this, it can be expected that the effect of the chassis dimensions is significant also from the handset-user interaction point of view. However, there has been little information available on this issue. In [KOLV04], the effect of the chassis length on the bandwidth, radiation efficiency, and Specific Absorption Rate (SAR) characteristics of the combination of an internal mobile handset antenna and handset chassis is studied with simulations. The SAR and radiation efficiency are determined when the handset models are positioned beside an anatomical heterogeneous head model. In general, when the bandwidth reaches its maximum due to the increased contribution of a resonant chassis, a local increase of SAR and a local decrease of radiation efficiency occur, Figure 5.12. Of particular interest are the dependency of the performance of the studied models on chassis length, the novel information on the connection between the impedance bandwidths, SAR, and radiation efficiencies.

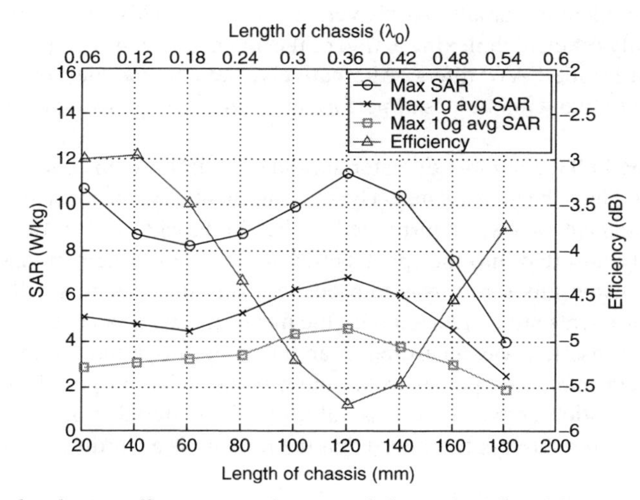

Figure 5.12 *SAR and radiation efficiency as a function of chassis length at 900 MHz. Distance from head to handset is 7 mm.* $P_{in} = 1$ *W [KOLV04]. (© 2004 IEEE, reproduced with permission)*

In a conventional dual-band antenna and RF front-end architecture, a dual-band antenna with a single feed is matched with a circuit that is a compromise between the two bands under consideration. Often this will be achieved with discrete, relatively poor quality components, for example, using simple inductor and capacitor networks, where the inductor Q is unlikely to be greater than 50. The antenna may be self-matching, but the compromise between the two bands will remain. The antenna will then connect to either a diplegia or a band switch. The diplegia, the main function of which is to provide isolation between the low and high frequency sides of the circuit (GSM and DCS, for example), is often fabricated using either discrete components or a multilayer circuit technology such as LTX. Commercial devices typically incur losses of 0.5–0.7 dB. Switches consume power, add a degree of non-linearity and have typical losses of 0.3–0.5 dB. These figures are derived from measurements in a 50 Ω system. Since loss is a function of the load impedance, greater losses may be expected when connected to a typical antenna due to mismatch; mobile handset antennas are typically designed to give a return loss within −6 dB.

The approach taken in [BUGL04] is to optimise the matching and depleting functions by modifying the feed structure of dual-band PIFA. The antenna has two feeds, one for the low frequency band (for example, GSM) and one for the high frequency band (for example, DCS). Having two feeds allows each band to be independently and optimally matched. It also allows each band to be bandwidth broadened. Using double-tuning [Whee75], the bandwidth of each band, measured for a return loss of −6 dB, can be approximately doubled.

The measured average total efficiency (including the effects of mismatch) of the design presented, including the antenna, diplexing and transmit/receive switching, is 81% at GSM and 72% at DCS. The return loss seen at the inputs to the RF module is particularly good in the transmit bands at less than 16 dB in the GSM band and 11 dB in the DCS band. A conventional antenna of the same size achieved a measured return loss of only 6 dB. Since the cumulative losses of subsequent circuit blocks depend on the quality of the match, it is anticipated that the improved match will yield improved efficiency in the subsequent RF chain. Improved performance can be traded for either bandwidth or antenna size. The bandwidth can be approximately doubled using this technique (for a return loss of −6 dB) with only a small loss of average efficiency. This loss of efficiency is more than recovered by the highly efficient diplexing action of the antenna in combination with the out-of-band impedance presented by the SAW filters. Alternatively, the antenna may be reduced in size by a factor of approximately two without loss of performance with respect to a conventional antenna and RF front end.

Finally, in paper [Mikk03] a number of materials and their usability in covers for portable wireless units (DECT handsets) have been evaluated. The measurement method consisting of a patch antenna was outlined. The patch antenna was constructed so that flat blocks (100 × 40 × 3 mm) of different test materials could be inserted into the space between the antenna element and the ground plane. The main reason for using this set-up was to produce a way to measure small losses which is not possible if the measurements are conducted with the material used as a cover. The test antenna with the unknown material inserted was measured in an anechoic room and the loss was calculated by comparing the far-field integrated power to the input power. The proposed measurement method shows a clear differentiation between the loss values of the materials that varied between 0.8 and 4.3 dB. The performed measurements showed that the ranking is according to expectation.

Specific absorption rate

The experimental electromagnetic dosimetry of mobile handsets has been much developed these past years. Most of the existing dosimetric facilities utilise automatic positioning systems to move an E-field detected probe, with the help of robotised arms [ScEK96], or three axes displacement systems in order to achieve SAR measurements. The European Standard prEN50361 [CENE00] details the way to measure the SAR in a head-like phantom. According to the European standard, 12 various configurations (e.g. frequency, handset phantom arrangement) must be considered for a dual-band handset. Consequently, a complete handset test now typically lasts about half a day. Such duration is fully inappropriate for designers of new handset models as well as for pre-compliance testing or systematic picking controls on a production line.

The Parametric Reconstruction of SAR (PARSAR) technique described in [MeBF04] has been developed to overcome the drawback of the standard procedure, while maintaining a good accuracy. It is based on the simple observations that

- the duration of the measurement mainly results from the large number of samples required by the standard approach (about 300),
- this number is largely oversized (factor 10) to accurately account for the field distribution in a homogeneous medium and
- the averaged SAR in 1 or 10 grams does not require an 'exact' description of the field distribution.

This idea that the field distribution in a homogeneous phantom was not so complicated explained the success for correlating the peak SAR value to the averaged SAR in 1 or 10 grams of tissue, in the case of a spherical phantom [MaMa02]. Such a correlation resulted in a fast SAR measurement set-up where the peak SAR can be rapidly found thanks to a mechanical seeker. More fundamentally, the possibility to expand the E-field in a homogeneous medium as a superposition of a finite number of modes allows to claim that the number of measurements could be significantly reduced [Hans88].

The PARSAR technique developed at Supélec since 2001 and now validated on various phantoms and measurement facilities is exploiting those considerations [MeFB01]. The accuracy of the PARSAR algorithm has been assessed with analytical test functions [IEEE03] representing typical SAR distributions and by means of numerical simulations. The global correlation with standard SAR measurements has been evaluated for different phantom shapes and a very wide variety of commercial mobile handsets, first at Supélec and later within the frame of a multicentre evaluation campaign including SAGEM, AMC Centurion and FTR&D. The parametric reconstruction technique has been validated on three different phantoms: truncated sphere, SAM half and full head. In addition, the method accuracy has been evaluated on different sites with different SAR measurement set-ups at Supélec (homemade and SARA2), SAGEM (COMSAR), AMC Centurion (DASY3) and Indexsar (SARA2). The evaluation campaign has allowed to validate the accuracy provided by the PARSAR algorithm, confirming the work done on analytic functions. The algorithm has been implemented on an SARA2 system, and the measurement time has effectively been observed of about 60 s. The number of electric field data is reduced to approximately 30 samples and allows for a reduction in the measurement time (acquisition and post-processing) to about one minute [MeFB02]. Similarly, simple exponential extrapolation techniques [KBDG03] showed too that the number of samples required by the standard procedure is far from strictly necessary.

Adaptive matching systems

Mobile handsets are often used in the vicinity of a human body or nearby other objects. In this condition antenna impedance of the handset is prone to change giving place to considerable loss of the transmitted or received signal's power. Therefore, some quick and simple techniques to recover such impedance mismatch are highly desired. In [ITTO04a], [ITTO04b], [ITTO04c] a detailed comparison between the proposed adaptive impedance matching system and some related works in the past is presented. Some experimental results of the prototype are presented in [ITTO04d]. Figure 5.13(a) shows the configuration of the system. The system is now intended to be used in a transmitting path. It consists of (1) Matching circuit, (2) Mismatch measurement circuit (directional coupler), (3) Switch, (4) Adaptive control circuitry, and (5) Time constant generator (Resistance Capacitor (RC) low-pass filter). The varactors are used for the control elements in the pi-matching circuit. In this configuration, we utilise a test signal, or the perturbational method to determine the

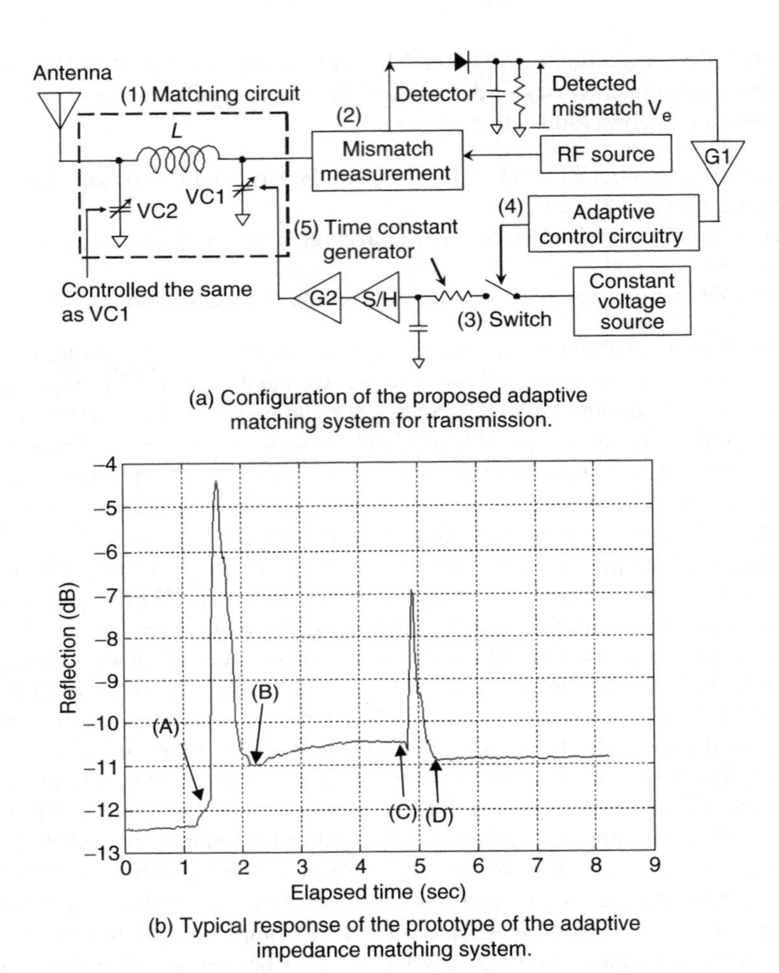

(a) Configuration of the proposed adaptive matching system for transmission.

(b) Typical response of the prototype of the adaptive impedance matching system.

Figure 5.13 *Configuration and typical response of the proposed adaptive matching system [ITTO04d].*

direction of the control, because the phase information is not available from the measured power at (2). The flowchart of the adaptive matching algorithm is given in [ITTO04d]. The basic principle of the algorithm is as follows: the latest value of the mismatch is measured through the detection circuit (2). The mismatch is again measured right after applying the positive voltage (i.e. test signal) to the VC1 by turning on the switch (3). Here, if the mismatch has been increased compared with the previous measurement, then the system begins to give VC1 a negative voltage, and vice versa. After the control frame for VC1 is finished, the control voltage for the VC1 is held with the sample and hold circuit and the VC2 frame is commenced.

The prototype of the proposed adaptive impedance matching system has been produced and measured. In the experiments a small loop antenna was connected to the prototype. The RF source used was a network analyser, which also measures the reflection coefficient between the pi-matching section and the 50 Ω transmission line. The measurement frequency was 2.45 GHz. The varactors are Toshiba 1SV239 (nominal range of the capacitance: about 1.5 pF–5 pF). The inductance in the matching section is 1.2 nH. In this experiment, 3 dB couplers are used as the reflection measurement device shown in Figure 5.13(a).

Typical response of the adaptive control of the prototype is shown in Figure 5.13(b). At $t \approx 0$ s, the reflection is sufficiently small. Then at $t \approx 1.5$ s, the antenna is touched by a finger ((A) in Figure 5.13(b)). After that, the impedance matching is deteriorated. Here, the mismatch loss is increased by about 2 dB, i.e. the power to the antenna from the RF source is decreased by about 2 dB. The adaptive matching system, however, rapidly detects this impedance mismatch, and recovers it at $t \approx 2.2$ s (B). Then, the finger is removed from the antenna at $t \approx 4.8$ s (C). Consequently, the impedance matching is again deteriorated and recovered at $t \approx 5.3$ s (D). Here the operation of the system is held when the reflection is smaller than about -8.0 dB. This holding of the adaptive control is often effective to improve the performance of the system [ITTO04b], [ITTO04c]. Note that the adaptive control does not hold until the end of a frame, even if the reflection becomes smaller than ≈ -8 dB in the middle of the corresponding frame. The speed of the response between (A) and (B) was about 0.6 s on average from 11 trials. Also, the response time between (C) and (D) was about 0.6 s on average from 11 trials. In both cases, the standard deviation values of the samples were about 0.1 s. The response time is sufficiently fast in terms of compensating an impedance mismatch caused by a movement of a human body. The response time, however, can be shortened by using a faster clock timing in the system.

5.3 Diversity techniques

5.3.1 Introduction: diversity, an old concept with renewed interest

The very original idea of diversity probably goes back to the late 1920s [Jake74]. The word diversity itself first suggests that radio systems employing such a technique intend to take advantage of the signal variability from one sensor to another, in order to improve the link robustness. A second simple idea is that by using several antennas at reception, we may try to 'capture' more energy from the ambient medium, thereby improving the Signal to Noise Ratio (SNR) or Signal to Interference Ratio (SIR). Both contribute to performance improvement, be it through the range, the throughput, or more generally speaking through network capacity. However, since the early days of diversity, it has been appreciated how far smart antenna approaches could go, well beyond such simple ideas as expressed above, and could concern all sorts of radio signal processing issues involving multiple antennas.

However, as a result of this sophistication, diversity and smart antenna techniques are currently considered to be limited to the receivers because of cost and also size, which using several antennas implied. In recent years there has been a growing effort to port these techniques to small size/moderate complexity terminals or access points, so that the inherent and attractive advantages of diversity could be widely exploited. Those efforts both involving academia and companies targeted the two main issues: antenna design and antenna integration on the one hand, and signal processing hardware and software development towards practical solutions on the other. COST 273 has regularly contributed to this effort and to the dissemination of results, and the activities are classified in the following subsections.

Subection 5.3.2 recalls a few generalities on diversity, and addresses/summarises the work concerning diversity implemented on receivers. This involves handsets and WLANs but also more specific applications such as digital television mobile reception or ultra wide bands.

Subection 5.3.3 addresses the important and topical subject of Multiple Input-Multiple Output (MIMO) systems. Although Chapter 4 of this book focuses very much on MIMO techniques, only the aspects closest to diversity, such as interelement coupling between antenna elements in relation to signal correlation/decorrelation, are addressed here.

5.3.2 Diversity on receivers

Diversity takes its root in the necessity to combat fading events that may occur when incoming multi-path waves destructively interfere at the antenna. Since this involves the phase of the received signal, simply moving the antenna, or switching to another antenna improves the signal quality. Here lies the seminal idea of diversity. However, it has been known for a long time that optimal diversity performance required decorrelation between the various antenna signals. The famous Clarke's formula states that for an omnidirectional scenario in the horizontal plane, two point source antennas exhibit a complex correlation coefficient given by $J_0(kd)$, where J_0 is the zeroth order Bessel function of the first kind and d the (horizontal) distance between antennas. At small distances, such a formula predicts strongly correlated signals, which is unfavourable. However, a much better decorrelation can be achieved by combining the antenna signals with the proper phasing, in order to construct directional patterns. A simple analysis carried out by Boyle [Boyl02b] showed that a suitable dephasing between two such sources resulted in orthogonal patterns, and minimal correlation for an omnidirectional scenario. In practice, for dipoles, taking into account their mutual interaction and inverting the impedance matrix allows to derive numerically the proper complex voltages which are to be applied in order to achieve the decorrelation.

A comparable approach has been used by Sibille and Fassetta in an attempt to optimise switched beam antenna performance for angular diversity [SiFa03]. They showed that, although for a given scenario the angular spectrum of incoming waves is seldom omnidirectional, it is the case when the averaging is performed over all possible scenarios in which the array might be employed. From the point of view of the antenna designer, achieving the best decorrelation for an omnidirectional scenario is thus the most logical choice. Due to the lack of pattern factorisability for the azimuth and elevation dependencies, the correlation coefficient between antenna beams turns out to depend on the angular spectrum of the scenario in elevation. In addition it was shown that the correlation coefficient in switched beam arrays was very sensitive to an extra spatial diversity, which added to the original angular diversity when the phase centres of the various beams did not coincide.

The result of an indoor measurement campaign at 5.2 GHz for six beam antennas showed that small correlation coefficients could be achieved, even for the angularly closest beams. From the knowledge of the antenna patterns, a fairly good agreement was obtained with computed correlation coefficients, for a Laplacian angular spectrum in elevation.

Diversity on handsets

When diversity antennas are implemented on handsets, the distinct antenna ports exhibit differing gains, due to the antenna technology and placement. Indeed such small terminals are quite small, and the whole handset together with its immediate environment influence the radiators. The concept of mean effective gain, involving not only the antenna pattern but also the received angular power spectrum, allows to take into account both of these effects into a single quantity, expressing the antenna performance in a given context. Takada and Ogawa have extended this concept to handsets implementing diversity with several antenna ports [OgTa00]. Assuming Rayleigh fading they arrived at analytical formulas for the output SNR probability density, $p(\gamma)$, and the average BER for MRC diversity, p_e:

$$p(\gamma) = \frac{1}{\lambda_1 - \lambda_2}(\exp(-\gamma/\lambda_1) - \exp(-\gamma/\lambda_2)) \tag{5.1}$$

$$p_e = \frac{1}{2} - \frac{1}{2(\lambda_1 - \lambda_2)}\left(\frac{\lambda_1}{\sqrt{\frac{2}{\lambda_1}+1}} - \frac{\lambda_2}{\sqrt{\frac{2}{\lambda_2}+1}}\right) \tag{5.2}$$

where λ_1 and λ_2 are the average branch powers and γ the instantaneous output SNR after combining. From these formulas it is possible to give two definitions of the diversity antenna gain, expressed as the required SNR gain compared to an isotropic antenna, either for a given outage probability (slow fading case) or for a given average BER (fast fading). They showed from a simple example that according to the chosen definition, the diversity antenna gain may differ markedly, which requires a clear awareness of the type of improvement expected by the implementation of diversity on a handset.

Experiments have been carried out at 2140 MHz by Kotterman on a handset-like device displaced along an indoor-outdoor route, in order to evaluate the correlation between two close antennas [KPOE01]. From the impulse response measured every 0.3λ with a 10 MHz bandwidth simultaneously on both of them, the short-term received power could be obtained and the fast fading power correlation coefficient extracted from the data. It turns out that this coefficient is generally well below 0.5, Figure 5.14, implying that from the point of view of the diversity performance, the antenna signals can be considered well decorrelated most of the time. As expected the correlation coefficient is still smaller for a pair of slightly more distant antennas; however, it is likely that not only the spatial variance, but also the involvement of the radiation pattern characteristics and their differences for the two antennas explain this reduction. The influence of the user hand is also quite important, as is demonstrated by large branch power differences between the two antennas. Such a difference is obviously detrimental to diversity performance, and a further test should really compare the net BER diversity gain as proposed in [OgTa00].

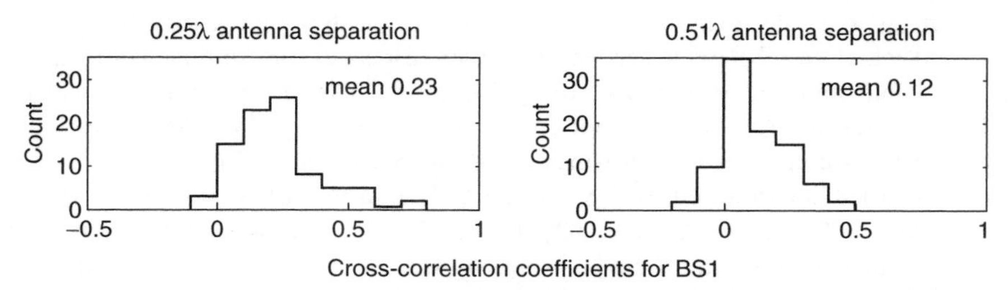

Figure 5.14 *Histogram of the measured correlation coefficient for a two antenna handset, [KPOE01]. BS1 is the BS number.*

More recently, reverberating rooms which have been developed at Gothenburg by the Per Simon Kildal team and Bluetest AB have been used, in order to test the diversity performance of DECT phones in a controlled radio environment [OrBK04]. Due to the highly scattering character of the electromagnetic waves within the chamber and the activation of a mode stirrer, it can be considered that the multipath scenario is basically isotropic and can serve as a reference scenario for diversity performance tests. The measurements were carried out on two different handsets, each equipped with two built-in antennas. It was found that the realised diversity gain at 1% cumulative probability was quite close to the expected 6 dB, Figure 5.15. This value is less than the theoretical gain of 11 dB for the used selection diversity technique under Rayleigh fading and full antenna decorrelation, but it is the maximum gain which can be reached using the measured antenna signals. A moderate loss was observed for the full diversity implementation in the handset, due to the need of thresholding the antenna signal level difference before switching. In addition a very significant reduction of the diversity gain was seen for short antenna separations, due to a worse antenna radiation efficiency caused by loss in the impedance loads.

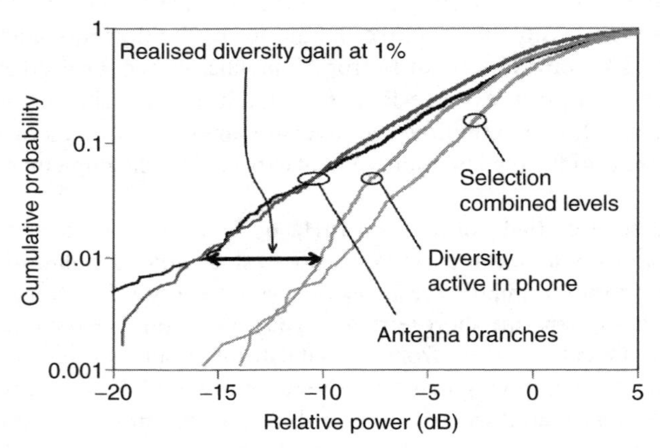

Figure 5.15 *Diversity performance for a DECT phone, [OrBK04].*

Polarisation and mixed diversity

Beyond spatial, angular, or pattern diversity, polarisation diversity is certainly one very interesting and promising way to implement diversity techniques on small receivers. On the one hand, differently polarised antennas generally exhibit low correlation, and on the other it may be possible to co-locate them and achieve a substantial size reduction with respect to conventional co-polarised arrays. However, non-equal branch powers may seriously degrade the diversity performance, which depends not only on the antenna characteristics, but on the scenario as well. In order to investigate these issues, a channel measurement campaign using two orthogonally polarised antennas used successively (vertical/horizontal) was carried out at 1.8 and 2.5 GHz in a few LOS and NLOS scenarios of a university building by Lozano *et al.* [LoMT02]. The results confirmed a good decorrelation between polarisations (smaller than 0.5 in all cases), knowing that in this particular case pattern diversity was superimposed to polarisation diversity due to the nature of the antennas (dipoles). Simulating equal gain combining or MRC with the measured signals showed a slight enhancement of the Rician character after combining (higher K-factor), especially for LOS vertical polarisation. Finally, the simulated diversity gain was indeed quite appreciable, between 7 and 14 dB depending on the scenario.

Another application of mixed spatial-polarisation diversity has been successfully tried for digital TV reception (DVB-T standard). Mobile reception of TV broadcasting signals in particular raises a number of challenges, such as the impossibility of large antenna gains currently achieved with Yagi rooftop antennas, the obstruction of the main LOS paths by buildings in urban areas, and the Doppler effect under fast driving. It has been demonstrated in several past projects that diversity reception brought a significant additional gain, making the reception much more favourable. The paper [GZSP04] describes an experiment carried out in Paris with a full diversity receiver achieving MRC on each subcarrier independently, Figure 5.16, where two antenna arrangements were tried. The first was made of two log-periodic dipoles oriented perpendicularly to each other on the car roof. The second was a pair of orthogonally polarised wideband antennas, placed on opposite windows of the car. Maps of the signal quality (BER and PER) were recorded and compared over several routes in Paris, either with a single or with both antennas of either pair. The results clearly show a strong TV picture quality enhancement under diversity reception, especially with the window antennas. This observation can be interpreted by the worse radiation pattern of the roof antennas, resulting from the uptilted antenna lobes due to the metallic roof reflection combined with horizontal polarisation.

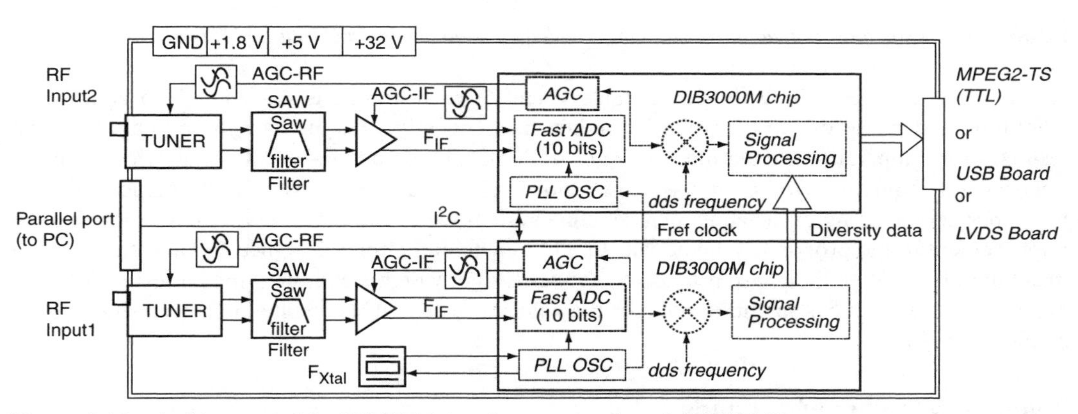

Figure 5.16 *Architecture of the DIBCOM signal processing board, [GZSP04].*

The same authors and others carried out tests of fixed indoor DVB-T reception, in order to evaluate the capability to enhance the coverage in the case of in-home antennas by exploiting diversity [FSCL05]. Systematic measurements were carried out with vertically and horizontally polarised omnidirectional antennas, and vertically and horizontally polarised directional antennas. The measurements thus simultaneously investigated spatial, polarisation and angular diversity in the hope to obtain a significant performance enhancement with two-branch diversity. The results showed that even in the case of deep in-building penetration horizontal polarisation was much preferred, and that in general spatial diversity was somewhat superior to angular diversity. Since the post-processing analysis was carried out only on the total received power rather than on complex received signals on each subcarrier, the conclusions are incomplete and will be completed by the final measurement campaign using the full diversity receiver.

The previously described DVB-T diversity reception employed a diversity receiver implementing MRC on all subcarriers of the DVB-T channel for a 64 QAM-OFDM modulation, based on specially designed chips. Quite another approach has been developed by Tanaka and coworkers in two companion papers ([TITT04], [TITO04]), based on fully analogue electronic circuits in order to provide an adaptation mechanism on the antenna branches. The advantage of this approach is a very simple hardware circuitry at RF level for the diversity operation, avoiding baseband processing with its resulting complexity and consumption. The proposed circuit design makes use of variable capacity diodes as the basic tuning element in each antenna branch, both output signals being subsequently combined in a hardware fixed manner and sent to the receiver input, Figure 5.17.

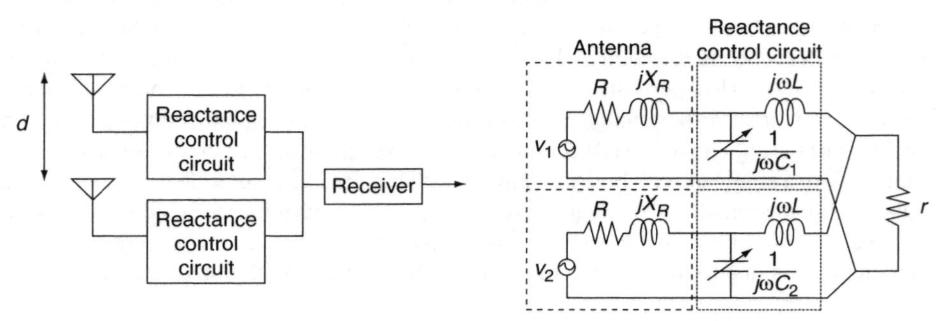

Figure 5.17 *Principle of an adaptive reactance loaded antenna combining circuit, [TITT04].*

It is clear that controlling the capacitance value on each branch is not enough to control the amplitude and phase of the equivalent weights independently, thus being sub-optimal. The best results for a comparative simulation between MRC, SC and the proposed scheme showed the latter to be roughly equivalent to SC. It was shown, however, that a main cause of performance loss was the imperfect matching of the equivalent antenna system seen from the receiver input. Therefore an improvement to the proposed scheme was to add a matching circuit designed to have an optimum matching effect when the phase rotation of the equivalent weights exhibited the largest sensitivity, i.e. close to the practical operating point of the antenna system. This resulted in a diversity antenna gain close to MRC.

Diversity with ultra wide bands

Finally and up to recent times, many of the efforts in the practical implementation of diversity on receivers were in the context of narrowband fading mitigation, since it is where diversity gain is well known and quite beneficial. However, radiocommunications tend to be more and more wideband, in order to deliver high data rates. In this context the advantages of diversity need to be revisited. The case of ultra wide band is especially interesting, on the one hand because the stringent emitted power limitations encourage a strong effort towards any technique able to improve the link margin, and on the other because due to the extreme bandwidth a real comprehensive investigation of the principle of diversity is needed. In particular, as opposed to narrowband diversity there is no unique way to characterise diversity gain since the magnitude of the bandwidth and even of the selected waveform enters the physical layer scheme under consideration. Sibille has defined several correlation functions for spatially variant output signals, taking into account the nature of the receiver principle [Sibi04]. According to the fact, e.g. that the antenna signals are all properly synchronised or not, or are fully equalised or not, the output signals exhibit a very strong spatial correlation or a very fast spatial decorrelation, Figure 5.18. These properties are in direct relation with the very large multipath diversity which can normally be expected in a UWB channel. Both full correlation and fast decorrelation find a direct illustration in the various possibilities and corresponding receiver architectures to exploit UWB diversity. The first one is the achievement of energy gain through the combining of suitably synchronised received signals. The second one is the multiplexing of signals

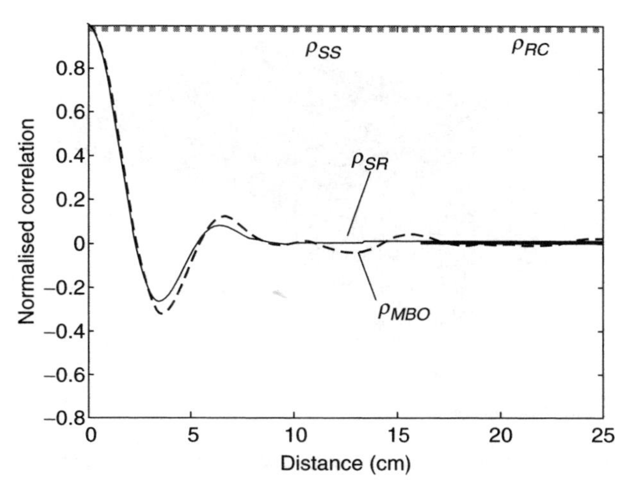

Figure 5.18 *UWB spatial correlation;* ρ_{SS}, ρ_{SR}, ρ_{RC} *and* ρ_{MBO} *respectively denote synchronised sensors, synchronised reference, rake combining and multiband OFDM. Refer to [Sibi04] for further details.*

using several transmitting antennas, and the proper demultiplexing or received data streams made possible by the spatial decorrelation.

Liu *et al.* also considered the specific case of multiband OFDM, and showed from experimental space-variant channel measurements that adequate decorrelation on the various subcarriers could be obtained with an intersensor distance as low as 3 cm in an indoor environment typical of an office [LAME05].

5.3.3 Diversity in the context of MIMO techniques

MIMO techniques are well known to permit the benefit of diversity to the transmitter side in addition to the receiver. However, in most investigations, ideal isotropic antennas are considered, whereas mutual coupling between the antenna elements are reputed to have a significant effect on diversity. For this purpose several works have been carried out in COST 273 in order to assess the influence of mutual coupling on MIMO radio channel capacity.

The spatial correlation and the mutual coupling were first separated in order to analyse their respective effects on the narrowband channel capacity, according to Tran and Sibille in [TrSi02]. Some real and artificial circular array types of two, three and four monopole antennas were defined for this purpose. A real array includes the two phenomena. A virtual array puts emphasis only on the spatial correlation, while an array of distant antennas with reintroduced coupling puts emphasis on the mutual coupling. The angular scenarios being omnidirectional, the spatial correlation can be quantified by the Bessel function of zero order $J_0(2\pi d/\lambda)$, d being the distance between two antennas. The channel capacity employing virtual array appeared to be inversely proportional to this quantity, mutual coupling influencing the capacity only if the mutual impedance Z_{21} is real and resulting in a reduced capacity. The role of the load impedance was also studied in a real array, Figure 5.19. In the case of high loading impedance, the capacity is approximately equal to the capacity with virtual array (no influence of coupling). With small loads, it appears the same capacity reduction as

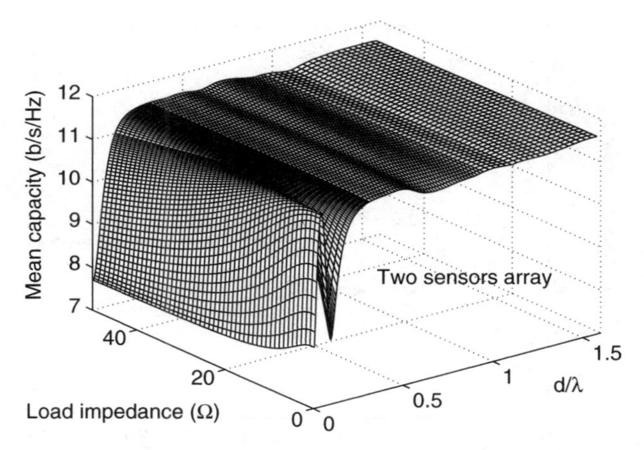

Figure 5.19 *Influence of the load impedance and of the antenna separation, TrSi02.*

for a channel with an array of distant antennas and reintroduced coupling. One main conclusion was that both mutual coupling and spatial correlation are detrimental to the channel capacity, but when both are present this leads to a high capacity. The mean capacity with a real array is always greater or equal to the capacity with a virtual array.

In order to understand these phenomena further, simulated mean capacities have been computed in a later work [TrSi02] with several variants of artificial arrays. Such variants involved artificial radiation patterns (Magnitude Coupled Array: MCA, where only the amplitude of true array radiation patterns is taken into account, and Phase Coupled Array: PCA, where only the phase of the true radiation patterns is taken into account). Some results are plotted in Figure 5.20, and it appears that the phase

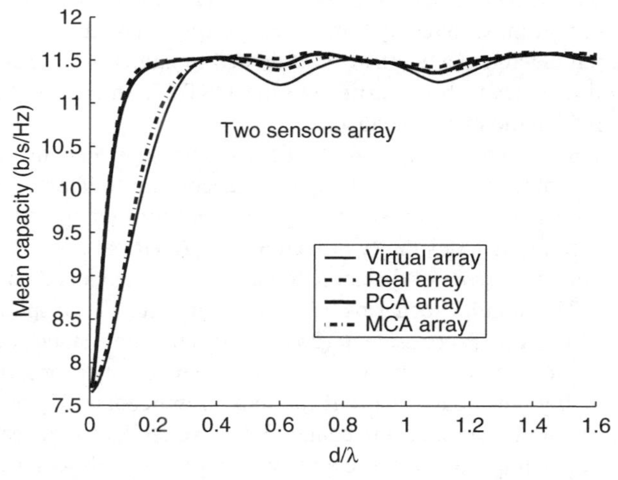

Figure 5.20 *Role of the magnitude and of the phase in the channel capacity, see TrSi03.*

of the radiated fields plays a greater role in MIMO diversity than their magnitude, for small array antennas.

A MIMO channel model based on electromagnetic fields was introduced in [NdHH04a]. The employed antennas again were half-wavelength dipoles operated at their resonance frequency. Coupling was accounted for through an impedance matrix Z of size $(M + N) \times (M + N)$, relating all voltages and currents at both ends of the link, M and N being the number of transmit and receive antennas respectively. Z was decomposed as four submatrices A, B, C and D, where A and D are the mutual impedance matrix of the transmitter and the receiver array respectively, and B and C are the channel transfer impedance matrices verifying $C = B^T$. The problem was formulated in terms of an integral equation and solved by the MoM. A voltage transfer matrix T was obtained for capacity computation. With a two paths channel model (LOS and ground reflection), one no null eigenvalue of the matrix TT^H was found in most cases. Some results were compared to simulated identically independent Gaussian distribution of the transfer impedance matrix B entries. Figure 5.21 shows that the correlation between $T(2, 1)$ and $T(2, 2)$ was increased when mutual coupling was considered for a 2×2 MIMO system, which implies a decrease of the capacity.

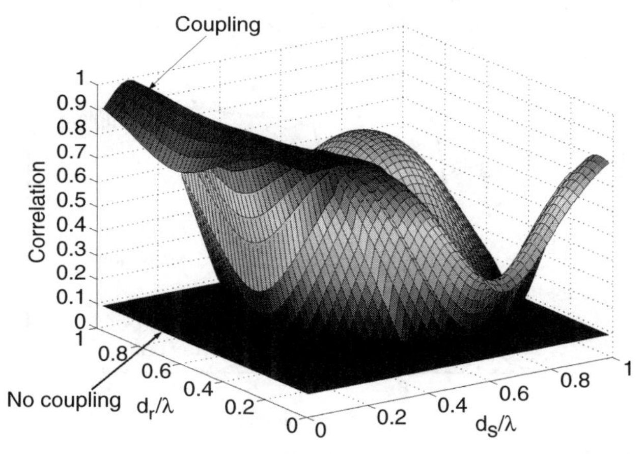

Figure 5.21 *Correlation between T(2, 1) and T(2, 2) for different elements spacing at the transmitter and receiver arrays, NdHH04a. (© 2004 IEEE, reproduced with permission)*

Along the same line, the channel matrix with coupling H_c was expressed by the use of the channel matrix without coupling H_{nc} and of two coupling matrices K_{TX} and K_{RX} at each side, normalised by the self-coupling [NdHH04b]. Expanding this expression results in four terms, one corresponding to the uncoupled channels, two other ones when coupling exists only on one side of the link, and a full coupling term.

The capacity was calculated from the correlation matrix R_H of the channel matrix H in a context of high SNR (20 dB). R_H was also expressed by R_{Hnc} which is the correlation matrix in the absence of coupling. The difference of capacity ΔC between the two cases (coupling or no coupling) was approximately found to be equal to the sum of the logarithms to base 2 of the eigenvalues of the coupling matrices. The channel model used was identically independent distributed (iid) matrix entries, and as usual the antennas were half-wavelength dipoles. Three different normalisations of the channel matrix were used: first, H_{nc} was normalised and H_c was deduced; second, H_c was directly normalised to remove the increasing power due to coupling. A third normalisation was defined to cancel only the power variation at the transmitter side (constant battery life). With antennas spaced by $d_{TX}/\lambda = 0.5$ at the transmitter side, the rank of the channel matrix was reduced due to spatial correlation. Also, reduced antenna spacing increased the transmitted and/or the received power, leading to an improvement of the capacity. As a conclusion, a good trade-off between these two effects has to be found in MIMO antennas designs.

Elaborating on all these previous works, Burr [Burr04] investigated the effect of mutual coupling and antenna element pattern directivity on the steering vector in reception mode Ψ_R and on the correlation matrix R, in the case of a finite scatterers channel model and using a coupling matrix C based on the Steyskal model [StHe90]. In Figure 5.22, the CDF of the channel capacity is plotted for an 8 × 8 antenna system. We see that the in-phase coupling ('c positive') increases the capacity while anti-phase coupling ('c negative') decreases it with array gain (more received power due to coupling leading to an increased SNR). Without array gain, the outcome is reversed, thus the effect of array gain appears to be more significant than that due to enhanced correlation.

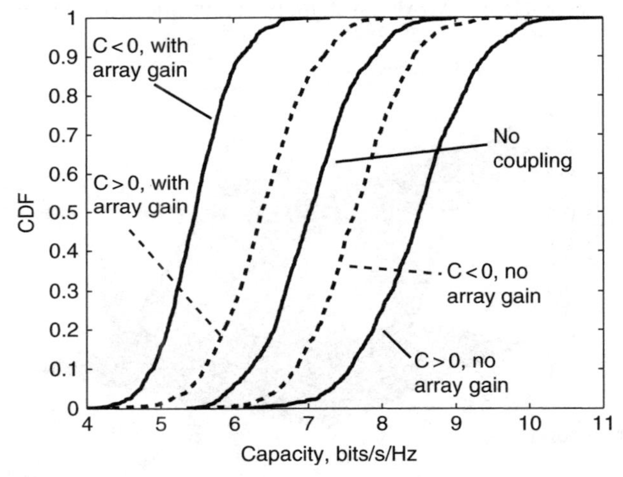

Figure 5.22 *CDF of MIMO capacity for an 8 × 8 MIMO system, Burr04.*

The effects of element directional responses on the entries $\overline{R_{R,ik}}$ of the correlation matrix were also analysed defining a mean Angular Power Density Spectrum (APDS). If the APDS is uniform and all antennas are omnidirectional, the Bessel function of the first kind and zero order is recovered for this average correlation. Coupling can be taken into account in MIMO systems either by considering the modified steering vectors and the isolated elements patterns, or directly by calculating/measuring the directional responses of the elements inside the array.

The correlation coefficient characterising the diversity was also analysed by Derneryd in [DeKr04] for arrays of two dipoles. This coefficient was calculated for near- or far-field parameters, or for both, but it leads to the same results. The mutual coupling, the port termination (50 Ω) and the propagation environment as modelled by a weighting factor S(Ω) which is nothing else than the angular power spectrum, were taken into account. The correlation as computed from the two far-field radiation patterns is depicted in Figure 5.23 for an antenna separation varying along (left) and perpendicular (right) to the incoming waves distributed within a 90° sector. In the near-field method, S(Ω) is equal to the unity and the correlation is obtained from voltages and currents at the antenna terminals.

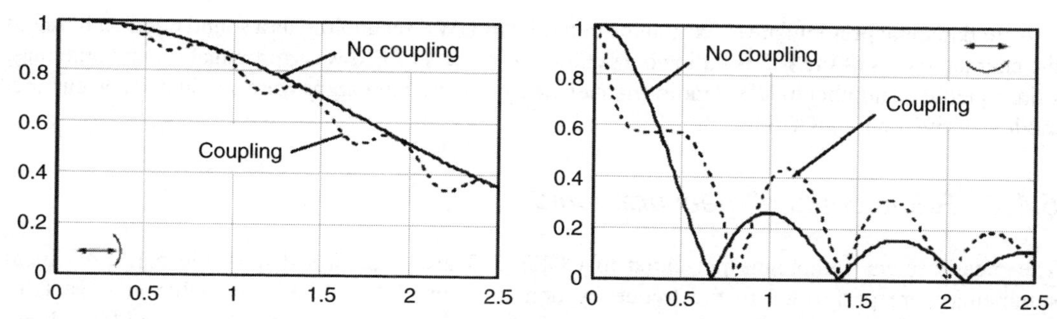

Figure 5.23 *Magnitude of correlation coefficient in the case of two parallel vertical dipoles versus spacing in wavelength, [DeKr04]; the arrows depict the array, and the crescent the angular spread of the scenario. (© 2004 IEE, reproduced with permission)*

With near- and far-field parameters, the correlation matrix of open-circuit voltages $E[V_{oc}V_{oc}^H]$ was first calculated, and the correlation matrix of voltages $E[VV^H]$ with port termination was deduced. A trade-off between low correlation coefficient and high antenna efficiency appeared to be necessary in a four antenna configuration.

Further works were carried out towards the practical implementation of diversity in MIMO array antennas on terminals, e.g. on PDAs [RoHi04], or by exploiting polarisation diversity [WKSW03]. More results on MIMO array antennas are described in Section 6.4 of this book.

5.4 Ultra Wide Band

5.4.1 Introduction

Ultra Wide Band (UWB) radio has recently attracted a lot of attention from both academia and industry as an interesting candidate technology for future consumer communications applications, particularly for low-cost, uncoordinated, and robust operation.

UWB is commonly defined as any radio technology having a spectrum that occupies a bandwidth greater than 20% of the centre frequency, or a bandwidth of at least 500 MHz. Currently, there is a regulatory approval for UWB only in the USA with a frequency allocation between 3.1 and 10.6 GHz for communications and measurement systems.

A number of advantages as well as challenges related to UWB radio technology are frequently discussed. Under ideal conditions, for example, the channel capacity of the band-limited AWGN channel grows faster as a function of bandwidth than as a function of power. Thus UWB radio technology offers potential for short-range high speed data transmission suitable for broadband access to the Internet or connectivity between devices. To take advantage of this, appropriate concepts on the physical and medium access layers are required. Furthermore, UWB offers good accuracy in range determination. The positioning capability enables ad hoc networks of position-aware devices like sensors or RF tags for precise tracking of personnel or assets.

While UWB is expected to be robust against interference by other systems, a way has to be found to apply the technology without causing unacceptable interference to other systems that share the same frequency spectrum.

A fundamental prerequisite to exploit the potential of UWB technology is a sound understanding of the characteristics of the UWB radio channel, i.e. the effects of EM wave propagation and antennas. Consequently, a number of UWB radio channel measurement campaigns were conducted and channel models derived within COST 273.

5.4.2 Radio channel measurements

The measurement campaigns presented to COST 273 are summarised in Table 5.12. All measurements were performed in the frequency domain using VNA readily available at most labs. The major advantage of VNA measurements is the very high measurement bandwidth achievable; the frequencies covered were between 1 and 11 GHz, with bandwidths ranging from 2 to 10 GHz.

The VNA measurement principle requires the environments to be static during a sweep over the measurement band. This is a major disadvantage, therefore the development of a channel sounder capable of assessing time-variability is presented in [ZeST04]. The channel sounder is based on cyclic maximum length binary sequences. Its architecture supports MIMO measurements by synchronous

Table 5.12 *UWB measurement campaigns within COST 273.*

Group	Centre for Wireless Communications, Finland	IMST GmbH, Germany	France Telecom R&D, France
Band (GHz)	2–8	1–11	4–6
Approach	VNA (Agilent 8720ES)	VNA (HP 8719D)	VNA (AB millimetre MVNA 8-350)
Samples	1601	1024	
Environment	Lecture room, small meeting rooms, big auditoria (30 m^2 to 300 m^2)	Office (25 m^2), corridor	Office (16 m^2–75 m^2), corridor, laboratory
Scenario	LoS	'Office-LoS', 'Office-NLoS', 'Office (N)LoS', 'Office-to-Office'	LoS, NLoS
Antenna height	0.6 m, 1.1 m, 2.2 m	1.5 m	TX 2.2 m RX 1.1 m, 1.3 m, 1.4 m, 1.6 m
Type	SISO	SIMO, TX virtual 30 × 150 array (1 cm spacing)	SIMO, virtual 60-UCA (antenna on rotating arm, radius = 27 cm)
TX-RX distance (m)	1.5–13	Office app. 3 corridor <18 m	2.6–16.6
Antennas	Conical 0 dBi	Biconical, app. 1 dBi	Printed dipole
Results	Impulse responses	PDP, S-V parameters, frequency domain power decay, path loss/power law, amplitude statistics	PDP, S-V parameters, RMS delay spread, amplitude statistics
Reference	[HoHä02]	[KuPa02]	[PaVP03]

Group	ENSTA, France	Centre for Wireless Communications, Finland	Tokyo Institute of Technology, Japan
Band (GHz)	2–10	1–11	3.1–10.6
Approach	VNA	VNA (HP 8720ES)	VNA
Samples		1601	801
Environment	Laboratory	Meeting room	Home, out-to-indoor
Scenario	LoS, NLoS	NLoS, LoS	LoS, NLoS
Antenna height			TX 1.3 m, RX 1 m
Type	SIMO	SIMO, virtual ULA	MIMO, TX virtual 10-ULA (4.8 cm spacing), RX virtual 10 × 10 array (4.8 cm spacing)
TX-RX distance (m)	3		5
Antennas	Monocone	Conical, 0 dBi	Monopole
Results	Impulse responses, delay spread CDFs	Impulse responses, PDP, RMS delay spread, subbands	Spatio-temporal analysis, DoA (azimuth/elevation) DoD (azimuth), clusterisation [HaTK04a]; scattering loss, DoA-ToA map, DoA angular spread [HaTK04b]; DoA-ToA map, DoA angular spread, cluster parameters [HaTK05]
Reference	[Sibi04]	[JäHH04]	[HaTK04a], [HaTK04b], [HaTK05]

Group	Lund University, Sweden	Tokyo Institute of Technology, Japan	King's College London, United Kingdom
Band (GHz)	3.1–10.6	3.1–10.6	
Approach	VNA (HP 8720C)	VNA	VNA
Samples	1251	751	1601
Environment	Incinerator hall (13.6 m × 9.1 m × 8.2 m)	Empty meeting room (15.75 m × 6.64 m × 2.46 m)	Workshop (6 m × 6 m)
Scenario	LoS, NLoS	LoS	LoS, NLoS
Antenna height		1.42	TX 1.5 m, RX 1.5 m
Type	MIMO, RX and TX virtual 7-ULA (5 cm spacing)	1. TX fixed, RX virtual 10 × 10 × 7 array (4.8 cm spacing), 2. RX fixed, TX virtual 10 × 10 × 7 array (4.8 cm spacing) V-V, H-H	SIMO, RX virtual 101 × 101 array (1 cm spacing)
TX-RX distance (m)	2, 4, 8	7.86	4.5
Antennas	Conical monopole	Biconical	Discone
Results	PDP, RMS delay spread, DoA vs. delay	Spatio-temporal analysis, DoA, DoD statistics, RMS delay spread	Correlation distance
Reference	[KWAT04]	[TsHT04]	[LAME05]

multichannel operation. In the first phase, it covers the band from near Direct Current (DC) to 5 GHz. Furthermore, its suitability for localisation has been demonstrated.

The presented VNA measurement campaigns address various multiple antenna arrangements covering SISO, SIMO, double-SIMO, MIMO. In all cases virtual arrays have been used, i.e. a single antenna was moved sequentially to the different array element locations.

The measurements presented cover those environments foreseen as the most probable application environments, namely indoor, office, and home environments. One campaign addressed an industrial environment.

A very instructive example for the results achieved [HaTK04a], [HaTK04b], [HaTK05], providing insight into the mechanisms governing indoor UWB propagation, is shown in Figures 5.24 and 5.25 [HaTK04a].

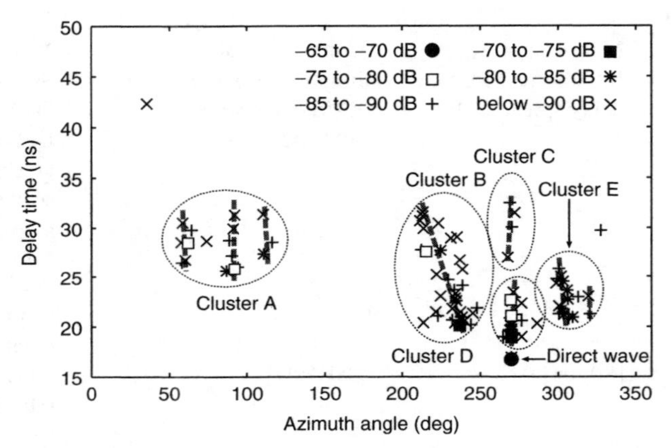

Figure 5.24 *Deterministic components estimated by SAGE and their clusterisation for the LoS measurements [HaTK04a].*

The clustering of propagation paths in the azimuth-delay domain is clearly visible for LoS measurements, Figure 5.24. In this case, specular reflections are identified as the dominant interaction type and can be mapped to physical structures of the room, Figure 5.25.

Based on array measurements [KuPa02], wavefronts may be reconstructed. Figure 5.26 shows such wavefronts passing over a rectangular 150 cm $\times$ 30 cm observation array in an office environment. After approximately 11 ns, the direct wave arrives, Figure 5.26(b). Later, a wave reflected at the right-hand side wall moves over the observation array, Figure 5.26(b) (middle and bottom). Interference between intersecting wavefronts is clearly visible.

The main results are that the UWB radio channel is characterised by strong, individual echoes that exhibit coherent behaviour over large distances compared to the wavelength. These are superimposed on dense multipath clusters with the familiar exponential decay behaviour for large excess delay well known from wideband channels. However, towards low excess delays the dense multipath envelope deviates from exponential decay towards lower power.

Unlike for small fractional bandwidths, frequency trends caused by antennas are not negligible. For example, constant gain antennas at both link ends will impose a $1/f$ trend to the transfer function in addition to possible intrinsic frequency dependencies of the propagation channel.

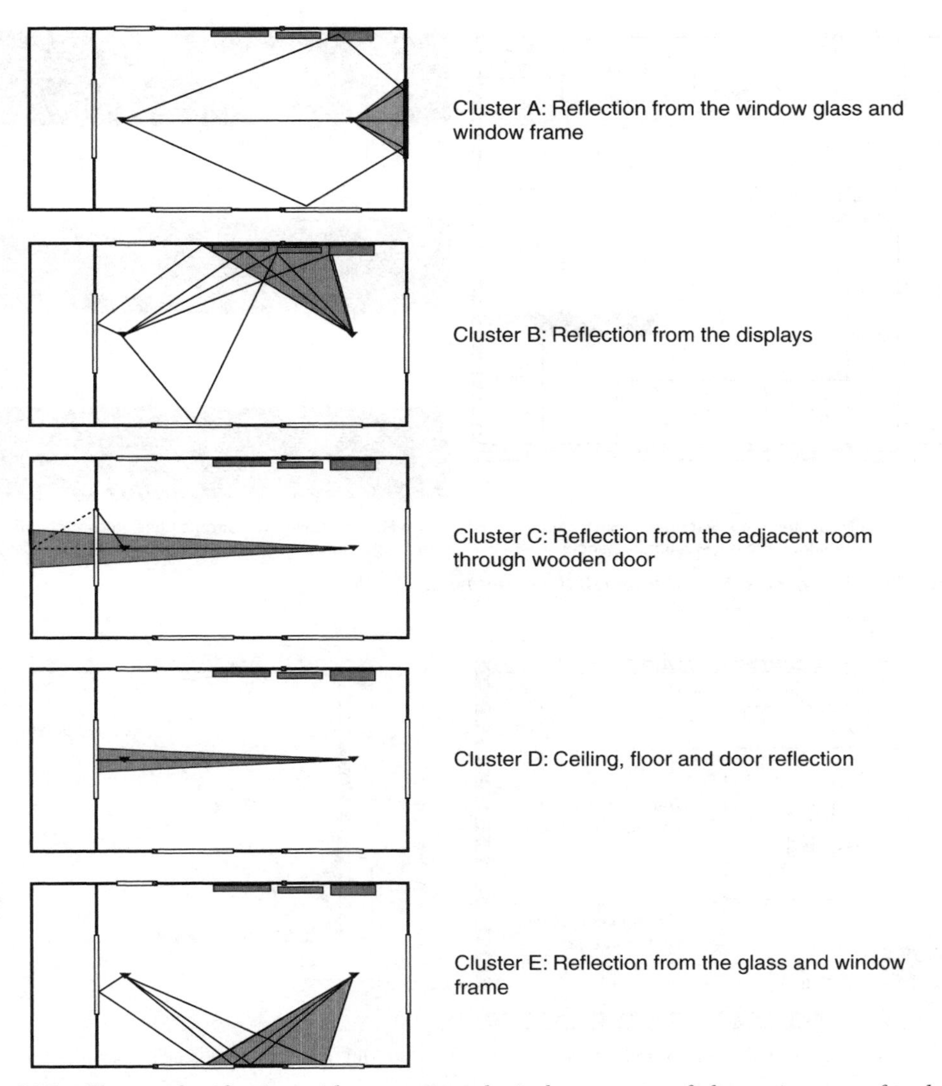

Figure 5.25 *Cluster identification with respect to physical structures of the environment for the LoS measurements [HaTK04a].*

In [ZeST05] results of a combined localisation/imaging experiment using the real-time, multi-antenna channel sounder described in [ZeST04] are presented, demonstrating the capabilities of UWB besides communications. A combination of three receive antennas was used to determine the position of a transmit antenna. Based on this information the multipath components were extracted from the measured channel impulse responses and post-processed to generate an image of the environment, Figure 5.27.

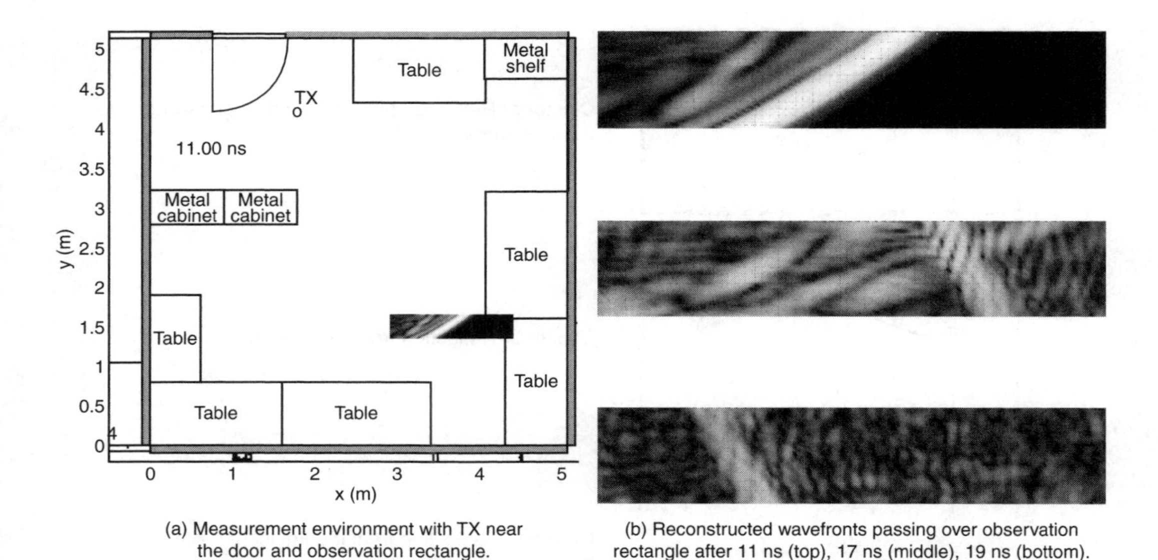

(a) Measurement environment with TX near the door and observation rectangle.

(b) Reconstructed wavefronts passing over observation rectangle after 11 ns (top), 17 ns (middle), 19 ns (bottom).

Figure 5.26 *Wavefronts derived from UWB measurements [KuPa02].*

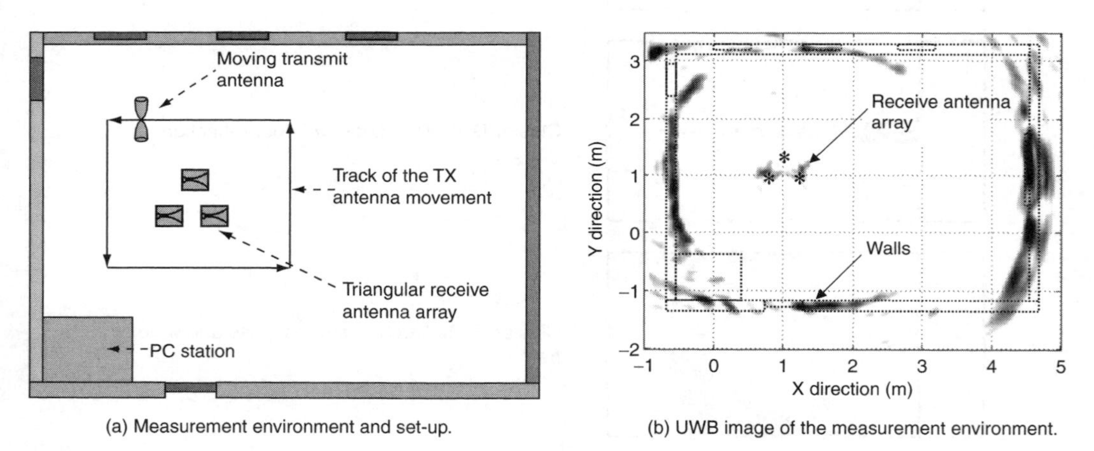

(a) Measurement environment and set-up.

(b) UWB image of the measurement environment.

Figure 5.27 *Results of a UWB imaging experiment [ZeST05].*

5.4.3 Radio channel models

Based on the measurements summarised in the previous section one stochastic and two hybrid channel models are proposed, Table 5.12.

Stochastic models account for the fact that dense multipath cluster(s) have been observed. The corresponding models are variations of the classic Saleh-Valenzuela (S-V) approach, for example with modified path amplitude statistics. The major advantage of this approach is the simple implementation as a tapped delay line.

Table 5.12 *UWB channel models.*

Group	Centre for Wireless Communications, Finland	University of Kassel, Germany	IMST GmbH, Germany
Modelled quantity	ChIR-B	Transfer function	Transfer function, ChIR-B, space-variant
Type	Hybrid	Stochastic	Hybrid
Deterministic	Ray tracing		Simplified ray tracing for surrogate environment (Virtual Sources/Sinks)
Stochastic	Rice fading	Bello's sampling model for delay-spread constraint	Modified S-V
Based on	Measurements [HoHä02]	Theoretical considerations	Measurements [KuPa02]
Parameter sets		'Office-LoS', 'Office-NLoS', 'Office-Office'	
Reference	[HoHä02]	[Katt02]	[KuPa03]

Hybrid: Deterministic + stochastic components
ChIR-B: Complex Channel Impulse Response, Baseband

Hybrid models are composed of (quasi-)deterministic and stochastic components. The deterministic component is included to account for spatial coherence related to the observed dominant echoes, which are considered a distinguishing feature of the UWB radio channel. The employed approaches are highly simplified ray tracing in a generic environment ('virtual sources/sinks') and comprehensive site specific ray tracing. The stochastic component accounts for the dense multipath.

All proposals model the channel impulse response or the transfer function. Some information regarding path-loss characteristics are available in [PaVP03] and [KuPa02].

[LoGC05] presents some thoughts regarding the extension of the ray-tracing technique, well established in radio channel simulations, into the UWB regime by taking the frequency dependence of key parameters and 3D antenna modelling into account.

5.4.4 Antenna aspects

In the ultra wide bandwidth regime the frequency-dependent characteristics of antennas can no longer be neglected. They may alter the spectral content of the transmitted signal, thus distort the impulse shape and increase the efforts for signal detection. Furthermore, antennas required for typical WLAN and WPAN applications need to be small and still efficient which is a main challenge for the design.

[Sibi04] analyses the way antennas affect the performance of an impulse-based UWB system by evaluating the captured energy as measured at the output of a correlator. Several simulations of the radio link performance, both for real and simulated antennas, and for real and simulated channels are performed. It turns out that a low dispersion antenna, i.e. well designed in the UWB sense, also exhibits the smallest degradation of the SNR of the correlator receiver.

[Sibi05] deals with a preliminary assessment of SIMO and MIMO diversity schemes for the improvement of UWB radio link performance. MIMO techniques may bring improvement by synchronising and combining but the characteristics of the radio channel influence the achievable diversity gain dramatically.

References

[3GPP03] 3GPP. Technical specifications group radio access network, UE radio transmission and reception(FDD), release 1999, v3.14.0. TS25.101, June 2003.

[Boyl02a] K. R. Boyle. Mobile phone antenna performance in the presence of people and phantoms. In *Proc. IEE Antenna Measurement and SAR Seminar*, Loughborough, UK, May 2002. [Also available as TD(03)054].

[Boyl02b] K. R. Boyle. Radiation patterns and correlation of closely spaced linear antennas. *IEEE Trans. Antennas Propagat.*, 50(8):1162–1165, Aug. 2002. [Also available as TD(02)018].

[Boyl03] K. R. Boyle. The performance of GSM 900 antennas in the presence of people and phantoms. In *Proc. ICAP 2003 - 12th Int. Conf. on Antennas and Propagation*, Exeter, UK, Mar. 2003. [Also available as TD(04)113].

[BUGL04] K. R. Boyle, M. Udink, A. de Graauw, and L. P. Ligthart. A novel dual-fed, self-diplexing PIFA and RF front-end (PIN-DF2-PIFA). In *Proc. IEEE AP-S 2004 - IEEE Int. Symp. On Antennas and Propagation and USNC/URSI National Radio Science Meeting*, Monterey, CA, USA, June 2004. [Also available as WP(04)022].

[Burr04] A. G. Burr. Multiband MIMO antenna arrays. In *Proc. 13th IST Summit on Mobile and Wireless Commun.*, Lyon, France, June 2004. [Also available as TD(04)107].

[CDWC04a] O. Colas, C. Dale, J. Wiart, G. Christophe, and L. Robert. Comparison between mean effective gain (MEG) and its approximation in a uniform arrival angle environment. TD(04)058, COST 273, Athens, Greece, Jan. 2004.

[CDWC04b] O. Colas, C. Dale, J. Wiart, G. Christophe, and L. Robert. Influence of hand on the mean effective gain (MEG). TD(04)136, COST 273, Gothenburg, Sweden, June 2004.

[CENE00] Basic standard for measurement of specific absorption rate related to human exposure to electromagnetic fields from mobile phones(300 MHz-3GHz). CENELEC TC211 European Standards, June 2000.

[CENE01] Basic standard for measurement of specific absorption rate related to human exposure to electromagnetic fields from mobile phones(300 MHz-3GHz). CENELEC Standard ENS 50361, Brussels, Belgium, July 2001.

[ChKn04] M. Christensen and H. Knöß. Investigation of different phantom head models including holder. TD(04)068, COST 273, Gothenburg, Sweden, June 2004.

[ChRo04] A. Char and M. Roberty. Measurement setup for 3G phones. Master's thesis, Aalborg University, Aalborg, Denmark, 2004. [Also available as TD(04)115].

[CoDW03] O. Colas, C. Dale, and J. Wiart. Influence of user head and the terminal position on the mean effective gain for different environments. TD(03)173, COST 273, Prague, Czech Rep., Sep. 2003.

[CoDW04] O. Colas, C. Dale, and J. Wiart. Influence of hand on the terminal total radiated power (TRP). TD(04)057, COST 273, Athens, Greece, Jan. 2004.

[CoWi05] O. Colas and J. Wiart. Simple environment models for mean effective gain (MEG) estimation. TD(05)054, COST 273, Bologna, Italy, Jan. 2005.

[DeBL04] B. Derat, J.-Ch. Bolomey, and C. Leray. On the existence of a lower bound of SAR value for GSM mobile phone-a resonator based analysis. In *Proc. JINA 2004 - 13th International Symposium on Antennas*, Nice, France, Nov. 2004.

[DeBo05a] B. Derat and J.-Ch. Bolomey. Analytical lower and upper bounds of power absorption in near field regions deduced from equivalent junction model. *J. Electromagn. Waves Applicat.*, pages –, 2005. In press.

[DeBo05b] B. Derat and J.-Ch. Bolomey. A new equivalent junction model for characterizing power absorption by lossy scatterers in reactive field regions. In *Proc. ANTEM 2005 - 11th Int. Symp. on Antenna Techn. and Appl. Electromagnetics*, Saint Malo, France, June 2005. [Also available as TD(04)188].

[DeKr04] A. Derneryd and G. Kristensson. Signal correlation including antenna coupling. *Elect. Lett.*, 40(3):157–157, Feb. 2004. [Also available as TD(04)127].

[DuGG04] Ph. Duchesne, L. Garreau, and A. Gandois. Compact multi-probe antenna test station for rapid testing of active wireless terminals. WP(04)023, COST 273, Gothenburg, Sweden, June 2004.

[FSCL05] A. Fluerasu, A. Sibille, Y. Corre, Y. Lostanlen, L. Houel, and E. Hamman. A measurement campaign of spatial, angular, and polarization diversity reception of DVB-T. TD(05)019, COST 273, Bologna, Italy, Jan. 2005.

[GaFo05] A. Gandois and L. Foged. Reference antennas for calibration and benchmarking of antenna measurement systems. TD(05)073, COST 273, Leuven, Belgium, June 2005.

[Glaz03] A. A. Glazunov. UE antenna efficiency impact on UMTS system coverage/capacity. R4-030546, 3GPPTSG-RAN Working Group 4 (Radio) meeting #27, Paris, France 19th-23rd May, May 2003. [Also available as TD(03)186].

[Glaz04a] A. A. Glazunov. Impact of head phantom models on handset antenna efficiency measurement accuracy in terms of body loss in passive mode. TD(02)144, COST 273, Lisbon, Portugal, Sep. 2004.

[Glaz04b] A. A. Glazunov. Joint impact of the mean effective gain and base station smart antennas on WCDMA FDD system performance. In *Proc. Nordic Radio Symp. 2004*, Oulu, Finland, Aug. 2004. [Also available as TD(04)158].

[Glaz04c] A. A. Glazunov. Mean effective gain of user equipment in double directional channels. In *Proc. PIMRC 2004 - IEEE 15th Int. Symp. on Pers., Indoor and Mobile Radio Commun.*, Barcelona, Spain, Sep. 2004. [Also available as TD(03)187].

[Glaz04d] A. A. Glazunov. On the user equipment antenna performance. WP-04-018, COST 273, Gothenburg, Sweden, June 2004.

[Glaz04e] A. A. Glazunov. Terminal antenna performance measurements in anechoic chamber with fictitious scatters. TD(02)143, COST 273, Lisbon, Portugal, Sep. 2004.

[Glaz04f] A. A. Glazunov. Theoretical analysis of mean effective gain of mobile terminal antennas in ricean channels. In *Proc. VTC 2002 Fall - IEEE 56th Vehicular Technology Conf.*, Vancouver, Canada, Sep. 2004. [Also available as TD(02)090].

[GlPa04] A. A. Glazunov and E. Pasalic. Comparison of MEG and TRPG of practical antennas. In *Proc. PIMRC 2004 - IEEE 15th Int. Symp. on Pers., Indoor and Mobile Radio Commun.*, Barcelona, Spain, Sep. 2004. Also available as TD(03)128 and TD(03)183.

[GZSP04] A. Guena, D. Zapparata, A. Sibille, and G. Pousset. Mobile diversity reception of DVB-T signals using roof or window antennas. TD(04)014, COST 273, Athens, Greece, Jan. 2004.

[Hans88] J. E. Hansen. *Spherical Near-Field Antenna Measurements*. 26. IEE Electromagnetic Wave Series, London, UK, 1988.

[HaTK04a] K. Haneda, J. Takada, and T. Kobayashi. Clusterization analyses of spatio-temporal UWB radio channels in line-of-sight and non-line-of-sight indoor home environments. In *Proc. Joint COST 273/284 Workshop on Antennas and Related System Aspects in Wireless Communications*, Gothenburg, Sweden, June 2004. [Also available as WP(04)012].

[HaTK04b] K. Haneda, J. Takada, and T. Kobayashi. Double directional LOS channel characterization in a home environment with ultrawideband signal. In *Proc. WPMC 2004 - Wireless Pers. Multimedia Commun.*, Padova, Italy, Sep. 2004. [Also available as TD(04)160].

[HaTK05] K. Haneda, J. Takada, and T. Kobayashi. On the cluster properties in UWB spatiotemporal residential measurement. TD(05)066, COST 273, Bologna, Italy, Jan. 2005.

[HoHä02] V. Hovinen and M. Hämäläinen. Ultra wideband radio channel modelling for indoors. In *Proc. COST 273 First Workshop on Opportunities of the Multidimensional Propagation Channel*, Espoo, Finland, May 2002. [Also available as WP(02)002].

[IdYI01] I. Ida, H. Yoshimura, and K. Ito. Reduction of drift error of a network analyser in small antenna measurement,. *IEE Proc. Micro., Ant. and Prop.*, 148(3):188–192, June 2001.

[IEEE03] IEEE. Recommended practice for determining the peak spatial average specific absorption rate (SAR) in the human head from wireless communication devices: experimental techniques. Standard P1528, Brussels, Belgium, Apr. 2003.

[ITTO03] I. Ida, J.-I. Takada, T. Toda, and Y. Oishi. Effective range of drift reduction with an impedance converter for antenna measurements. TD(03)189, COST 273, Prague, Czech Rep., Sep. 2003.

[ITTO04a] I. Ida, J.-I. Takada, T. Toda, and Y. Oishi. An adaptive impedance matching system and appropriate range for control elements. In *Proc. ECTI 2004 - 1st ECTI Annual Conference*, Pattaya, Thailand, May 2004.

[ITTO04b] I. Ida, J.-I. Takada, T. Toda, and Y. Oishi. An adaptive impedance matching system and its application to mobile antennas. In *Proc. IEEE TENCON 2004*, Chiang Mai, Thailand, Nov. 2004. also available as TD(04)121.

[ITTO04c] I. Ida, J.-I. Takada, T. Toda, and Y. Oishi. An adaptive impedance matching system for mobile communication antennas. In *Proc. IEEE AP-S 2004 - IEEE Int. Symp. On Antennas and Propagation and USNC/URSI National Radio Science Meeting*, Monterey, CA, USA, June 2004. [Also available as WP(04)022].

[ITTO04d] I. Ida, J.-I. Takada, T. Toda, and Y. Oishi. Experimental results of the adaptive impedance matching system. TD(04)187, COST 273, Duisburg, Germany, Sep. 2004.

[JäHH04] T. Jämsä, V. Hovinen, and L. Hentilä. Comparison of wideband and ultra-wideband channel measurements. TD(04)080, COST 273, Gothenburg, Sweden, June 2004.

[Jake74] W. C. Jakes. *Microwave mobile communications.* Wiley, 1974. see also [Automatic selection of receiving channels], US patent nr 1747218, Feb. 18, 1930.

[Katt02] R. Kattenbach. Transfer function modeling and its application to ultra-wideband channels. TD(02)136, COST 273, Lisbon, Portugal, Sep. 2002.

[KBDG03] M. Y. Kanda, M. Ballen, M. G. Douglas, A. Gessner, and C. K. Chou. Fast SAR dertermination of gram-averaged SAR from 2-D coarse scans. In *Proc. BEMS 2003 - 25th Ann. Meeting of the BioElectromagnetics Society*, Wailea, HI, USA, June 2003.

[KDPS04] J. Krogerus, B. Derat, S. Pannetrat, H. Shapter, and A. Kruy. Estimation of measurement uncertainty in total radiated power measurements. TD(04)128, COST 273, Gothenburg, Sweden, June 2004.

[Kild02] P.-S. Kildal. Accurate measurements of small antennas and radiation from mobile phones in small reverberation chambers. TD(02)035, COST 273, Espoo, Finland, May 2002.

[Knud01] M. B. Knudsen. *Antenna Systems for Handsets.* PhD thesis, Aalborg University, Aalborg, Denmark, 2001. [Also available as TD(01)143].

[KOLV04] O. Kivekäs, J. Ollikainen, T. Lehtiniemi, and P. Vainikainen. Bandwidth, SAR and efficiency of internal mobile phone antennas. *IEEE Trans. Electromagn. Compat.*, 46(1):71–86, Feb. 2004. [Also available as TD(03)041].

[KPOE01] W. A. Th. Kotterman, D. Prasad, K. Olesen, P. Eggers, I. Z. Kovács, and G. F. Pedersen. Channel measurement set-up for multi antenna handheld terminal and multiple (interfering) base stations. In *Proc. WPMC 2001 - Wireless Pers. Multimedia Commun.*, Aalborg, Denmark, Sep. 2001. [Also available as TD(02)149].

[KrGl04] J. Krogerus and A. A. Glazunov. Measurements of radio performances for UMTS terminals in speech mode. R4-040612, 3GPPTSG-RAN Working Group 4 (Radio) meeting #33, Yokohama, Japan, 15th-19th November, Nov. 2004. [Also available as TD(05)051].

[KrJä05] J. Krogerus and T. Jääskö. Positioning of the DUT against the phantom. TD(05)074, COST 273, Leuven, Belgium, June 2005.

[KrJI03] J. Krogerus, T. Jääskö, and C. Icheln. Comparison measurements of the COST 273 SWG 2.2 reference monopole antennas. TD(03)131, COST 273, Paris, France, May 2003.

[KrMo05] E. Krenz and P. Moller. Systematic approach for hand emulation in radiated measurements. TD(05)062, COST 273, Bologna, Italy, Jan. 2005.

[Krog02a] J. Krogerus. On the phantom and tissue-simulating liquid to be used in handset antenna performance testing. TD(02)024, COST 273, Espoo, Finland, May 2002.

[Krog02b] J. Krogerus. Phantoms for terminal antenna performance testing. TD(02)154, COST 273, Lisbon, Portugal, Sep. 2002.

[Krog05a] J. Krogerus. Influence of the side of head on the mean effective gain of mobile handset antennas in talk position. TD(05)061, COST 273, Bologna, Italy, Jan. 2005.

[Krog05b] J. Krogerus. Influence of tissue-simulating liquid on mobile handset antenna performance. TD(05)060, COST 273, Bologna, Italy, Jan. 2005.

[KSLK02] K. Kalliola, K. Sulonen, H. Laitinen, O. Kivekäs, J. Krogerus, and P. Vainikainen. Angular power distribution and mean effective gain of mobile antenna in different propagation environments. . *IEEE Trans. Veh. Technol.*, 51(5):832–837, Sep. 2002. [Also available as TD(02)028].

[KuBa92] N. Kuster and Q. Balzano. Energy absorption mechanism by biological bodies in the near field of dipole antennas above 300 MHz. *IEEE Trans. Antennas Propagat.*, 41(1):17–23, Feb. 1992.

[KuPa02] J. Kunisch and J. Pamp. Measurement results and modeling aspects for the UWB radio channel. In *Proc. UWBST 2002 - IEEE Conf. on Ultra Wideband Systems and Technologies*, Baltimore, MD, USA, May 2002. [Also available as TD(02)105].

[KuPa03] J. Kunisch and J. Pamp. An ultra-wideband space-variant multipath indoor radio channel model. In *Proc. UWBST 2003 - IEEE Conf. on Ultra Wideband Systems and Technologies*, Reston, VA, USA, Nov. 2003. [Also available as TD(03)154].

[KWAT04] J. Kåredal, S. Wyne, P. Almers, F. Tufvesson, and A. F. Molisch. UWB channel measurements in an industrial environment. In *Proc. Joint COST 273/284 Workshop on Antennas and Related System Aspects in Wireless Communications*, Gothenburg, Sweden, June 2004. [Also available as WP(04)013].

[LAME05] J. Liu, B. Allen, W. Malik, and D. Edwards. On the spatial correlation of MBOFDM ultra wideband transmissions. TD(05)008, COST 273, Bologna, Italy, Jan. 2005.

[LiVC04] E. Van Lil, E. Verhaevert, and E. Van de Capelle. On the influence of the size of objects on the number of power pattern samples and harmonics. TD(04)051, COST 273, Athens, Greece, Jan. 2004.

[LoGC05] Y. Lostanlen, G. Gougeon, and Y. Corre. An approach to model the ultrawide-band multipath indoor radio channel by ray-tracing methods. TD(05)047, COST 273, Bologna, Italy, Jan. 2005.

[LOIV03] T. A. Laitinen, J. Ollikainen, C. Icheln, and P. Vainikainen. Rapid spherical field measurement system for mobile terminal antennas. In *Proc. Instrumentation and Measurement Technology Conf.*, Vail, CO, USA, May 2003. [Also available as TD(03)134].

[LoMT02] S. Loredo, B. Manteca, and R. P. Torres. Polarization diversity in indoor scenarios: An experimental study at 1.8 and 2.5 GHz. In *Proc. PIMRC 2002 - IEEE 13th Int. Symp. on Pers., Indoor and Mobile Radio Commun.*, Lisbon, Portugal, Sep. 2002. [Also available as TD(02)065].

[LVKK03] T. A. Laitinen, P. Vainikainen, T. Koskinen, and O. Kivekäs. Amplitude only vs. complex field measurements for mobile terminal antennas with a small number of measurement locations. In *Proc. Instrumentation and Measurement Technology Conf.*, Vail,CO, USA, May 2003. [Also available as TD(03)051].

[Mads04] K. Madsén. Reverberation chamber for mobile phone radiated tests. TD(04)087, COST 273, Gothenburg, Sweden, June 2004.

[MaMa02] M. Manning and P. Massey. Rapid SAR testing of mobile phone prototype using a spherical test geometry. In *Proc. IEE Antenna Measurement and SAR Seminar*, Loughborough, UK, May 2002.

[MeBF04] O. Merckel, J.-Ch. Bolomey, and G. Fleury. Rapid SAR measurement of mobile phones. *Int. J. of Applied Electromagnetics and Mechanics*, 19(1-4):183–186, 2004. also available as TD(03)098] and TD(04)202.

[MeFB01] O. Merckel, G. Fleury, and J.-Ch. Bolomey. Rapid SAR measurements via parametric modeling. In *Proc. 5th International Congress of the European BioElectromagnetics Association*, Helsinki, Finland, Sep. 2001.

[MeFB02] O. Merckel, G. Fleury, and J.-Ch. Bolomey. Propagation model choice for rapid SAR measurement. In *Proc. EUSIPCO 2002 - XIth Ann. European Signal Processing Conference*, Tolouse, France, Sep. 2002.

[Mikk03] H. V. Mikkelsen. Measurements on materials applied in covers for integrated antennas. TD(03)075, COST 273, Paris, France, May 2003.

[NdHH04a] H. Mbonjo, J. Hansen, and V. Hansen. Field theoretical investigations of the influence of mutual coupling effects on MIMO channels. In *Proc. Globecom 2004 - IEEE Global Telecommunications Conf.*, Dallas, TX, USA, Dec. 2004. [Also available as TD(03)145].

[NdHH04b] H. Mbonjo, J. Hansen, and V. Hansen. Impact of antenna design on the capacity of MIMO systems. WP(04)002, COST 273, Gothenburg, Sweden, June 2004.

[NiPe02a] J. O. Nielsen and G. F. Pedersen. Frequency dependence of the mean effective gain for mobile handsets. TD(02)077, COST 273, Espoo, Finland, May 2002.

[NiPe02b] J. O. Nielsen and G. F. Pedersen. Mobile handset performance evaluation using spherical measurements. In *Proc. VTC 2002 Fall - IEEE 56th Vehicular Technology Conf.*, Vancouver, Canada, Sep. 2002. [Also available as TD(02)021].

[NiPe05] J. O. Nielsen and G. F. Pedersen. Using radiation pattern measurements for mobile handset performance evaluation. TD(05)072, COST 273, Leuven, Belgium, June 2005.

[OgTa00] K. Ogawa and J. I. Takada. An analysis of the effective performance of a handset diversity antenna-proposal for the diversity antenna gain based on a signal bit-error rate. In *Proc. IEEE AP-S 2000 - IEEE Int. Symp. On Antennas and Propagation*, Salt Lake City, UT, USA, Sep. 2000. [Also available as TD(03)142].

[OlLa98] B. G. H. Olsson and S -Å. Larsson. Description of antenna test method performed in scattered field for GSM MS. TD(98-106, COST 273, Duisburg, Germany, Sep. 1998.

[OrBK04] C. Orlenius, R. Bourhis, and P. S. Kildal. Diversity gain of active DECT phones with two built-in antennas measured in reverberation chamber. WP(04)019, COST 273, Gothenburg, Sweden, June 2004.

[Otte05] M. Otterskog. Modelling of propagation environments inside a scattered field chamber. In *Proc. VTC 2005 Spring - IEEE 61st Vehicular Technology Conf.*, Stockholm, Sweden, May 2005. [Also available as TD(05)017].

[PaVP03] P. Pajusco, S. Voinot, and P. Pagani. A study of the ultra-wide band indoor channel: Propagation experiment and measurement results. In *Proc. UWBST 2003 - IEEE Conf. on Ultra Wideband Systems and Technologies*, Oulu, Finland, June 2003. [Also available as TD(03)060].

[PeAn99] G. F. Pedersen and J. B. Andersen. Handset antennas for mobile communications: integration, diversity and performance. *Review of Radio Science 1996-1999*, pages 119–138, Aug. 1999.

[PeNi02] G. F. Pedersen and J. O. Nielsen. Radiation pattern measurements of mobile phones next to different head phantoms. In *Proc. VTC 2002 Fall - IEEE 56th Vehicular Technology Conf.*, Vancouver, Canada, Sep. 2002.

[Plic04] V. Plicanic. Antenna diversity study and applications. Master thesis report, Lund University and Ericsson Mobile Communications, Lund, Sweden, May 2004. [Also available as TD(04)095].

[RoHi04] P. R. Rogers and G. S. Hilton. 3D radiation pattern correlation of PDA-sized MIMO antenna arrays. WP(04)006, COST 273, Gothenburg, Sweden, June 2004.

[ScEK96] T. Schmid, O. Egger, and N. Kuster. Automated E-field scanning system for dosimetric assessments. *IEEE Trans. Microwave Theory Tech.*, 44(1):105–113, Jan. 1996.

[Schm99] Application note:recipes for brain tissue simulating liquids. Schmid & Partner Engineering AG, Switzerland, Mar. 1999.

[Sibi04] A. Sibille. *Spatial aspects of UWB - Section 1: Spatial diversity*, In A. Molisch, I. Oppermann, M. Gabriella di Benedetto, D. Porcino, C. Politano, and T. Kaiser, editors, *UWB communication systems: a comprehensive overview.*

Hindawi Publishing Corporation, 2004. EURASIP Book Series on Signal Processing and Communications. In Press.

[Sibi05] A. Sibille. Time domain diversity in ultra wide band MIMO communications. *EURASIP Journal on Applied Signal Processing*, 3:316-327, 2005. [Also available as TD(03)071].

[SiFa03] A. Sibille and S. Fassetta. Intersector correlations: A quantitative approach to switched beams diversity performance in wireless communications. *IEEE Trans. Antennas Propagat.*, 51(9):2238–2243, Sep. 2003. [Also available as TD(01)002].

[StHe90] H. Steyskal and J. S. Herd. Mutual coupling compensation in small array antennas. *IEEE Trans. Antennas Propagat.*, 38(12):2238–2243, Dec. 1990.

[Taga90] T. Taga. Analysis for mean effective gain of mobile antennas in land mobile radio environments. *IEEE Trans. Veh. Technol.*, 39(2):117–131, May 1990.

[TaTs87] T. Taga and K. Tsunekawa. Performance analysis of built in planar inverted antenna for 800 MHz. *IEEE J. Select. Areas Commun.*, 5:921–929, June 1987.

[TITO04] H. Tanaka, I. Ida, J. I. Takada, and Y. Oishi. Combiner circuit design of two-branch RF diversity antenna controlled with variable capacitors. TD(04)196, COST 273, Duisburg, Germany, Sep. 2004.

[TITT04] H. Tanaka, I. Ida, J. I. Takada, T. Toda, and Y. Oishi. Diversity antenna loaded with variable capacitors for effective combining. In *Proc. International Symposium on Antennas and Propagation (ISAP2004)*, Sendai, Japan, Sep. 2004. [Also available as WP(04)020].

[TrSi02] V. P. Tran and A. Sibille. Inter-sensor coupling and spatial correlation effects on the capacity of compact MIMO antennas. TD(02)128, COST 273, Lisbon, Portugal, Sep. 2002.

[TrSi03] V. P. Tran and A. Sibille. MIMO channel capacity and mutual coupling in circular arrays of monopoles. TD(03)099, COST 273, Paris, France, May 2003.

[TsHT04] H. Tsuchiya, K. Henada, and J. Takada. UWB indoor double-directional channel sounding for understanding the microscopic propagation mechanisms. In *Proc. WPMC 2004 - Wireless Pers. Multimedia Commun.*, Padova, Italy, Sep. 2004. [Also available as TD(04)192].

[VOKK02] P. Vainikainen, J. Ollikainen, O. Kivekäs, and I. Kelander. Resonator based analysis of the combination of mobile handset antenna and chassis. *IEEE Trans. Antennas Propagat.*, 50(10):1433–1444, Okt 2002.

[Whee75] H. Wheeler. Small antennas. *IEEE Trans. Antennas Propagat.*, 23(4):462–469, July 1975.

[WKSW03] C. Waldschmidt, C. Kuhnert, S. Schulteis, and W. Wiesbeck. Compact MIMOarrays based on polarisation diversity. In *Proc. IEEE AP-S 2003 - IEEE Int. Symp. On Antennas and Propagation and USNC/URSI National Radio Science Meeting*, Columbus, OH, USA, June 2003. [Also available as TD(03)107].

[ZeST04] R. Zetik, J. Sachs, and R. Thomä. Ultra-wideband real-time channel sounder and directional channel parameter estimation. In *Proc. URSI 2004 - 18th Triennial Intl. Symp. On Electromagnetic Theory*, Pisa, Italy, May 2004. [Also available as TD(03)201].

[ZeST05] R. Zetik, J. Sachs, and R. Thomä. Imaging of propagation environment by UWB channel sounding. TD(05)058, COST 273, Bologna, Italy, Jan. 2005.

6

MIMO channel modelling

Chapter editor: *Ernst Bonek*
Contributors: *João Gil, Luís M. Correia, Robert J. C. Bultitude, Christiane Kuhnert, Claude Oestges, Mark Beach, Mythri Hunukumbure, Andreas Molisch, and Helmut Hofstetter*

6.1 Introduction

The promise of MIMO (Multiple Input Multiple Output) to overcome the bottleneck in high speed wireless data transmission has fuelled enormous research efforts in space-time channels and systems. The last few years saw conferences that had more than half of their contributions dealing with space-time codes, transceiver algorithms, antenna configurations and channel models for MIMO. Whereas nobody doubts that MIMO will be *the* enabling technology for high speed wireless data, one should not think that it is the only problem to be solved for systems Beyond 3G (B3G). Many questions remain.

The numerous advances gained in COST 273 on MIMO in general can be found in three chapters of this book. Space-time codes and algorithms for MIMO transmission and reception schemes are contained in Chapters 2, 3 and 7 along with corresponding advances in coding and transceiver design in areas other than MIMO. Chapter 4 covers the multidimensional characterisation of MIMO channels together with channels involving fewer variables.

This chapter focuses on measuring and modelling MIMO channels, antenna configurations and the likely deployment scenarios. The paramount characteristic of MIMO, its promise of high capacity, is illuminated from an information theoretic point of view in Chapter 7. Capacity has been the popular yardstick by which models, algorithms, and codes have been measured. As will be seen in the course of this chapter, capacity is not sufficient to capture all aspects of MIMO systems, even considering that there are several different meanings of capacity[1].

The MIMO *channel* prevailing in a given environment determines which of the well-known benefits of MIMO can be exploited, namely

- spatial multiplexing,
- spatial diversity, or
- beamforming gain.

Beamforming, diversity, and multiplexing are rivalling techniques. Full beamforming does not allow diversity or multiplexing; full diversity excludes beamforming and multiplexing; finally, full multiplexing prevents beamforming gain but – note the exception – only *reduces* diversity. Even in the case of full multiplexing, diversity on the signal streams can be achieved [HoTW03]. How to achieve the spatial multiplexing benefit and when it is better to apply beamforming are treated in Section 7.3.

[1] In this introduction, capacity is loosely used also for the more proper terms 'mutual information' and 'spectral efficiency'. For a detailed discussion see Chapter 7.

Models for MIMO are the focus of this chapter. Section 6.3 will provide, along with a general discussion of modelling, details of new physically inspired modelling concepts developed in COST 273. To provide a general basis for the contents, Figure 6.1 shows a MIMO model hierarchy that has been presented in the COST 273 Prague Tutorial [BWMH03].

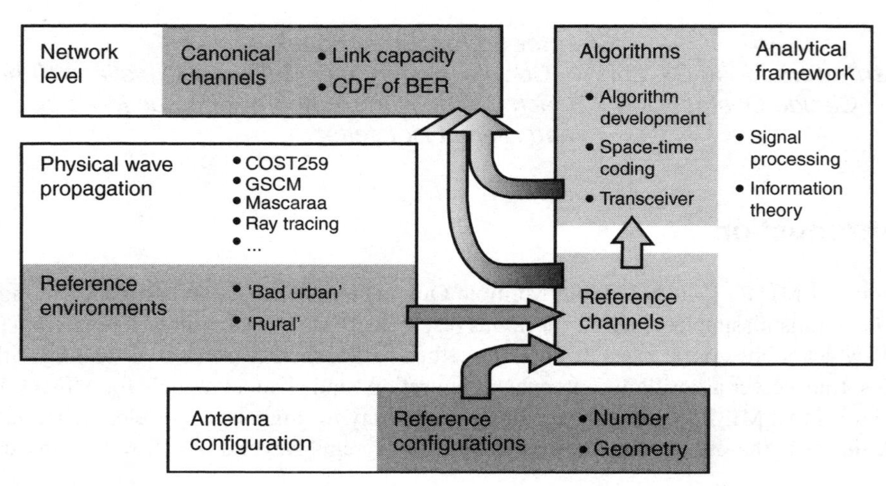

Figure 6.1 *MIMO channel models – an overview.*

Electromagnetic wave propagation provides the basis for *propagation* models. The final result of *physical* modelling is the characterisation of the *environment* on the basis of propagation. Canonical or reference scenarios are agreed-on environments that make comparison of models and their performance much easier[2]. Such scenarios or environments are discussed in Section 6.2. Specifying antenna arrays (Section 6.4) at both link ends by setting the number of antenna elements, their geometrical configuration, and their polarisations turns the propagation model into a MIMO *channel* model. Such a model provides an *analytical* framework for designing transmit and receive techniques for a MIMO link, e.g. space-time codes. Section 6.5 will specifically discuss old and new analytical MIMO models.

How MIMO channel models on the link level may be combined to model a MIMO implementation on the system level is an issue treated in Sections 7.4 and 7.5.

Section 6.8 will present the new COST 273 MIMO Channel Model, mainly based on the numerous and profound measurement campaigns (Section 6.6) reported and discussed in COST 273. Its approach is geometry based on the one hand and stochastic on the other, combining the best of both worlds.

This chapter is also about how to determine whether a MIMO model is 'good'. Creating new models for MIMO has been a popular sport among researchers over the last years. The result is amazing: there seem to be only 'good' models around, if one believes their originators. This observation, of course, immediately raises some questions: Have these models been validated? Are some models

[2]The term 'typical' in conjunction with scenarios is discouraged. Previous use of this expression turned out just to mean the environment of the author's workplace.

better suited to predict a certain aspect of MIMO system performance than others? MIMO models are so important because all the rest of signal processing, coding and deployment hinges on good models. Any model must be a simplification of reality. But how close does one get by using a certain model? Given these facts the discrepancy between efforts to develop new MIMO models and validating them is striking. Still, a lot of the scarce work on model validation has been done within COST 273. So, Section 6.7 of this chapter will deal with what it means to validate a MIMO channel model.

6.2 Scenarios

6.2.1 Introduction

In mobile communications, namely within COST 273, the term *scenario* is applied to many levels, objectives, and fields. Frequently, it concerns a set of parameters adopted within a group of inter-working parties, following the needs and forecast of envisaged situations. Scenarios are often used as foresight exercises, seeking major trends in the overall mobile communications panorama, e.g. [GGFC04], [Poll02]. Several projects in the field define and use scenarios in that sense. Some directly involve propagation issues, while others do not. In another way, scenarios are also used in testing the performance of a system or technology, with near-future objectives within a specific project or group. It is in these cases that scenarios are defined to aid in the study, understanding and measuring of the properties of the propagation channel (several of these are described in Sections 4.4 and 6.6).

Independently of the perspectives, a *propagation scenario* shall be part of an integrated set that determines propagation, interference, correlation and spatial richness. Traffic, services provision, social distributions and physical settings all have an impact on those overall channel conditions. All system layers interrelate, being of the utmost importance to define reference scenarios such as in Section 8.2, Mobile Radio Access Network Reference Scenarios (MORANS). For Section 8.2 this propagation scenario should indirectly reflect some characteristics of such upper layers.

In practical terms, within COST 273, a propagation scenario is defined following two perspectives: considering the location of MTs, subject to certain path-loss conditions, to define bit error rates, allowable services and quality-of-service provision, making use of path-loss propagation models; setting spatial distributions of MTs with their respective spatial propagation channels, in order to cover spatial filtering problems together with directional propagation models. Approaching the physical level means either approximating interference by noise and accounting for path loss, or defining spatial distributions of interference and of desired signals. It is in the latter perspective that activity on MIMO and on spatial filtering has also meant establishing *spatial propagation scenarios*. The inclusion of multi-user settings and resulting non-uniform spatially distributed MAI (seldom referred to as non-spatially white, or spatially coloured interference) have required the generation, evaluation and subsequent analyses as functions of those scenarios. The present section aims at classifying, structuring and combining that activity within COST 273, concerning the physical propagation layer only, dealing with spatial multi-user distributions. Those scenarios should picture situations where the spatial separation of channels may be more or less critical, where the correlation of multi-user MIMO channels is important. There are many practical, real-life situations where these may happen, e.g. in the case that several MTs are grouped together in a bus stop, or near each other in a railway station, spread in a plaza, or more isolated in a car park. Also, several physical environments may be at stake, e.g. indoor or outdoor, micro- or macrocellular, single- or multicellular.

This section is organised as follows: first, it defines the concept at stake, in Subsection 6.2.2. Subsections 6.2.3 and 6.2.4 then describe the propagation scenarios, classifying them into *indoor* and *outdoor spatial scenarios*, relating contributions within COST 273.

6.2.2 Concept

The influence of non-spatially white interference on the capacity of a MIMO system, or array processing and beamforming, is a major concern, as many parts of this chapter will show. The situations that establish such spatially coloured interference are very varied, resulting also in many uses of the term scenario, within the scope of the physical level: a scenario may concern the physical setting where BSs or APs are placed; no MTs or BSs may even be involved; it may characterise a case study, a particularly critical propagation situation; it may also concern a measurement site. Therefore, it is important to define the concept of spatial propagation scenarios.

The term *scenario* has been used as a physical environment where the particular setting of waves incurs important matrix properties and correlation conditions. For example, in [Sibi01], covering the MIMO-keyhole issue, the term refers to a particular situation of reduced transmit/receive diversity. Also related to the MIMO channel correlation issue, [Burr01] and [Burr02] deal with a simple indoor corridor and a rectangular room scenario. Nevertheless, these are not multi-user scenarios, i.e. not defining the spatial distribution of interference.

In many cases the term refers to a measurement *environment*. For example, the work developed in COST 273 Sub-Working Group 2.1 has covered numerous sites, in macro-, micro- and picocellular environments, with or without LoS, indoors or outdoors (see Sections 4.4 and 6.6). Examples are the indoor office scenario in [Özce04], [CHÖB04a], the urban and indoor scenarios in [WWWW04], or the rural, suburban, and urban measurement sites in [CoWi04]. But, again, these are not multi-user scenarios.

The term scenario has also been widely used for system level evaluations, covering several system layers. Though the distributions of MTs, BSs or APs and the type of environment have been established by authors, these studies first aim at evaluating system capacity, service provision, or network deployment, making use of path loss and shadowing, non-spatial models. For that, these are seen as *non-spatial scenarios*. Such is the case of MORANS COST 273 scenarios (see Section 8.2), where the advantage of covering several system layers has the practical drawback of not deeply focusing on the properties of the propagation channel. In fact, the concept of system scenario, initially deviating from the propagation modelling issue, has been first described in [BüNB02], establishing a Reference System Scenario for UMTS Simulations (RSSUS). It was suggested that, in an early stage, either simple modelling (path loss and shadowing) or real world propagation data would be applied. Subsequently more sophisticated propagation modelling, possibly spatial, would then be included in a later stage.

Following this background, the concept of a spatial propagation scenario needs to respect several issues: by being *spatial*, importance is given to the number and distribution of MTs, or of BSs/APs, or even to their physical surroundings; considering spatial properties means accounting for multi-user situations, from which the interference problem naturally arises; the spatial *setting*, the multi-user *situations*, and the *environment* make part of the propagation scenario. Therefore, a spatial propagation scenario sets the corresponding parameters, with direct consequences on the directional propagation and the spatial distribution of interference. Furthermore, it is hereby understood as being independent of any particular propagation model, not being a constituent part of one (as it is the case of the COST 259 Propagation Scenarios [Corr01]). In the referred perspective, for simplicity, these will be hereafter referred to as simply scenarios.

Following this definition, several COST 273 scenario contributions have many points in common, giving rise to the classification into being *indoor*, *outdoor single cell* or *outdoor multicell*. Within some of these, there can be a division according to the cellular environment. As the next section will show, indoor scenarios particularly detail room dimensions and relative positioning. On the other hand, outdoor single cell cases either distribute MTs in space or in angle, possibly randomly distributing MTs within the accepted locations, including their possible grouping. Finally, many of the outdoor multicell scenarios root from the more elementary outdoor single cell situations, dealing with fewer spatial constraints, to encompass larger areas.

As a result, this section contributes to the establishment of a common simulation framework, in the case of dealing with spatially coloured MAI and spatial propagation issues. Several interworking parties may better test their spatial filtering or MIMO technologies in comparable situations, in terms of the way that non-spatially white interference is distributed. Following the classification put forward in [GGFC04], these should then be *organisational-testing* scenarios. These may be a tool to generate harmonising and integrated work, not being too strict (not to limit their use) or too general (establishing enough parameters).

6.2.3 Indoor scenarios

Work towards Wireless Local Area Networks (WLANs) has considered several indoor propagation scenarios for the study of beamforming and related problems [NaCB03b], [NaCB03a], [NaCB04a], [NaCB04b], [CaNB04]. The authors make use of ray tracing in order to generate Directional Channel Impulse Responses (DCIRs) with varied LoS conditions. The scenarios include the size of the covered area, the number and location of MTs or of an AP. These were considered as TXs and RXs, correspondingly. In those studies, the location of MTs and the shape of the environment determine the resulting DCIRs, with several independent interfering sources resulting in different propagation and interference conditions (and in differing spatial filtering performance, as Section 7.3 describes). The so-called *arbitrary* indoor environments cover simple rectangular environments with differing dimensions, e.g. near 40×60 m^2, or 4×9 m^2. Arbitrary partial walls have been placed to generate richer and more varied channel responses. The number of sources is also varied, e.g. 1 AP and up to 5 MTs, or 1 AP and 2 MTs. MTs have been randomly placed, or put in locations to provide specific angular displacement in respect to the AP, e.g. 10° or 140°. Figure 6.2 shows an example of these scenarios.

Focusing more on MIMO, a matrix transmission model is applied together with joint multi-user detection, in [PSTL03]. An indoor scenario, with 1 AP and 2 MTs, is created to analyse performance as a function of the way that interference is accounted for in the system. The scenario consists of a simple room with a given geometry, concrete walls and specified user positions (including angular separation). Ray tracing is used to generate channel responses and matrices. The study in [PNTL02] involves a similar scenario (dimensions, number of MTs and distances to the AP), but a user remains at a fixed location, while the other is placed at multiple different relative angles. In this way, the study is more explicit about the dependence of space-time processing on the angular separation of the two terminals. A partial wall has been placed in the middle of the room, for this scenario, as Figure 6.3 shows.

Concluding, indoor scenarios tend to be detailed regarding the environment, dimensions and angular separation of terminals involved. These have commonly considered simple rectangular rooms,

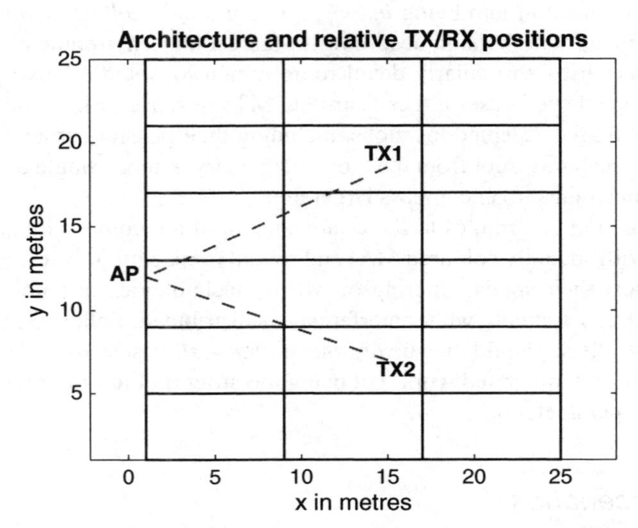

Figure 6.2 *Example of an indoor scenario, with 1 AP (as RX) and 2 MTs (considered as TXs), angularly separated by 45° [NaCB04b].*

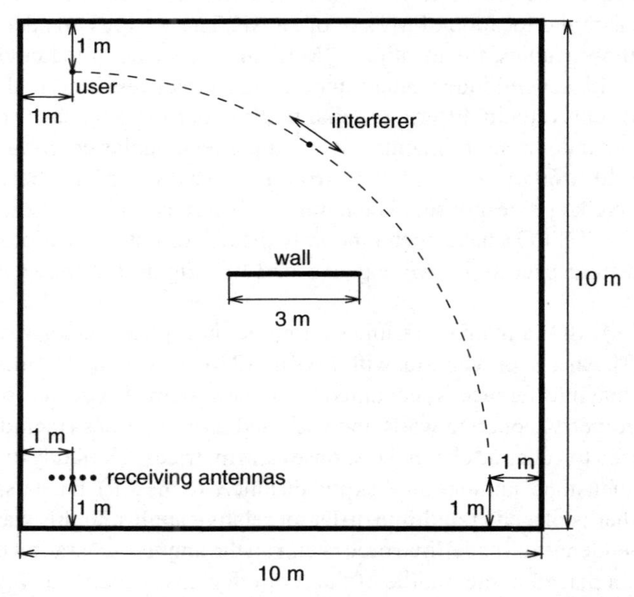

Figure 6.3 *Another example of an indoor scenario, with 1 AP (as RX) and 2 MTs (the TXs), with variable angular separation, and an inner wall [PNTL02].*

possibly with inner walls. The location of MTs (as TXs, in those studies) is defined by their angular placement in respect to the single AP (thereby considered as being the RX). There are indoor scenarios that can further be seen to be a physical repetition of the first, more elementary, ones. Based on the physical layout of all of these, the number of MTs may be varied, forming a large number of possible indoor scenarios. Table 6.1 shows the most important parameters that have been used to define the indoor scenarios, also providing examples of their values and settings.

Table 6.1 *Summary of parameters used to define indoor scenarios.*

Parameter	Examples of values/settings
Room dimensions	Rectangular; 10×10 m^2, 40×60 m^2, or 4×9 m^2
Number of terminals	1 AP, 1 MT, 1 or more interfering MTs
Terminal placement	AP at room corner, or near wall; 1 MT near a corner; MTs in the middle of room
Distance between terminals	8, 15, or 20 m
Angular separation of terminals	MTs with varied angular separations, 10, 45 or 140°, or several angles within [0, 45°]
Size and position of an inner wall	3 m long, at the middle of room

6.2.4 Outdoor scenarios

Going from indoor to outdoor, the definition of scenarios tends to specify the locations and environment with lower detail, e.g. placing MTs randomly. But, besides considering a random spatial distribution of MTs, there can be further restrictions in either angular or spatial domains. Except for [FMKW04], restricting MT location within a regular street grid, most of this sort of scenario concerns situations of a single cell, around which MTs are located. In [Glaz04], for the analysis of smart antennas, desired and interfering MTs are randomly positioned in angle around the BS. This angular distribution is uniform around the BS, with all at the same distance to the BS, within a sector of 120°. The study in [VTZZ04] also applies a simpler scenario, with 3 APs defining a circle within which a certain number of MTs are uniformly spread, to evaluate resource allocation techniques together with SDMA. The related study in [MaVT04] assumes a square service area, defined by the separation of APs. MTs are again randomly distributed, taking numbers up to 16. In [ZhBW04], MTs are angularly separated by 45°, independent of any cellular structure. Following a similar scenario approach, [MaVT04] focuses on a data transmission protocol, in a multicell SDMA/TDMA MIMO system, running a distributed slot allocation algorithm. The considered interference situation involves a number of randomly distributed MTs (up to 16), within the coverage of 2 APs, separated by a certain distance, with differing reuse factors.

Once more, a uniform random spatial distribution of MTs within a single cell setting is applied in [HZWS04], taking up to 32 MTs, and 1 BS in the reported simulations (the work can be extended to a multicell environment). The study focuses on channel estimation in a MIMO system, based on OFDM, and makes use of a geometrically based stochastic channel model to generate channel responses. The scenario is defined by a maximum cell size of 1000 m, and a minimum BS-MT distance of 100 m, it being a macrocellular environment with a single cell.

Also considering 1 BS to focus on intracell interference, several studies on how beamforming performance depends on the wideband and directional properties of the propagation channel have involved a large set of spatial propagation scenarios [GiCo01], [GiCo02], [GiCo03b], [GiCo03a], [GiCo04]. Their concept roots from spatially distributing MTs, within both micro- and macrocell environments. The distributions include grouped MTs, spread MTs, or a single MT separate from the remaining ones, at several distances to the BS. Figure 6.4 depicts the concept where 15 MTs are grouped together at $-\pi/5$ rad, with a single one at $\pi/4$. The number of MTs was varied from 4 to 16. Other parameters have established the scenario, e.g. the dimensions of the respective scattering area imposed by the wideband directional channel model, or the density of clusters. Following the model concept, the microcell environment involves a street with a width of 40 m, with BS-MT distances from 50 to 1000 m. In the macrocell situations, distances range from 1000 to 2000 m, involving scattering circle radii of 50 to 200 m.

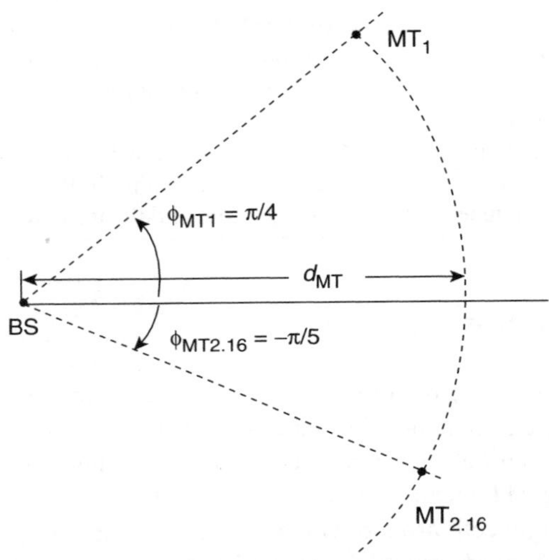

Figure 6.4 *An example of an outdoor scenario with 1 BS, grouped MTs and 1 MT at specific angles [GiCo03a].*

Therefore, the outdoor single-cell scenarios deal with a single cell, within which a certain number of MTs are spread. These may be either macro- or microcellular, depending on the MT-BS distances, and their relative heights. Naturally, some of these single-cell scenarios establish the unitary part of the larger multicellular scenarios. In all of these, varying cell sectorisation, the number of MTs, cell radius, the type of MT distribution, and MT-BS distances have resulted in a large number of scenario contributions within COST 273 (the largest variety among all types of scenarios). Table 6.2 includes the most important parameters used to define these, with examples of their values and settings.

Besides those single-cell outdoor scenarios, others extend from one to several cells. The system level study presented in [ChCz04] accounts for a macrocellular environment, evaluating the impact of a beamforming technique on the outage probability. Its output is the number of MTs for which a certain QoS is fulfilled. The scenario is defined by the number of BSs, cell radii, average and maximum number of paths, as well as fading and DoA statistics. Due to its nature, though this work approaches the propagation channel further (including spatial propagation parameters), the propagation scenario has not been explicitly described. In another case, [HäCC03], system level simulations focus on beamforming, where MTs are uniformly distributed in the cellular area of concern, involving 49 cells. A common characteristic of these system level evaluations is that the

Table 6.2 *Summary of parameters used to define outdoor single-cell scenarios.*

Parameter	Examples of values/settings
Cell radius	Defined, e.g. 500 m; undefined
Cellular environment	Macrocell; microcell
Number of BS antennas	1, 2, or 3; 2 or 3 APs; possibly, include sectorisation
Number of MTs	1, 16, 18, 32 MTs (maximum); a definite number of 4, 8, 16
Spatial distribution of MTs	Random, uniform distribution; random uniform, following a regular Manhattan street grid; random, uniform, within circle defined by APs
Angular distribution of MTs	Grouped MTs; spread MTs; single MT, angularly separated from remaining ones (grouped or spread) by 45°; MTs random, uniformly spread in angle, within angular sectors of 120 or 180°
Distance between BS and MTs	50, 500, 1000, 1500, or 2000 m (used in the case that the angular positioning of MTs is set); 100 m (minimum)
Other spatial parameters	Street width; scattering area (shape and size); number of scatterers; MT mobility; height of MT/BS terminals

number of MTs that achieve certain QoS is an output. Also, in [CzDe01] and [Czyl03], the cellular multi-user propagation scenarios consider that MTs are once more randomly located, following a uniform spatial distribution. The number of MTs is taken up to 10 per cell, in a layout of 55 cells, Figure 6.5, to evaluate downlink beamforming performance with circular arrays or linear arrays with sectorisation. The location of MTs is not a specific one, but still is an input to simulations. Such relative location does determine the performance of the applied beamforming, whose performance analysis is the central issue of [CzDe01]. This is also the scenario approach used in [Glaz04], for network layout simulations, uniformly distributing MTs within a total of 57 hexagonal cells, or in [VTZZ04], with 12 cells. Similarly, in [BrSP03], the scenario is established as a function of the location of MTs, BSs and channel propagation conditions, to study several types of smart antennas and compare the resulting mean network capacity. Up to 600 MTs are again randomly, uniformly placed across the considered area, six sites are hexagonally located around another central one, totallising 21 BSs.

In all of these studies, though the effect of the location of MTs is accounted for in a statistical manner, such uniform spatial distribution is part, in fact, of a spatial propagation scenario. Accordingly, the authors apply simple directional propagation channel models to characterise each MT-BS channel.

Focusing further on MIMO systems, [FMKW04] presents more system level contributions. The macrocellular scenario assumes 1 MT per cell, in order to account for intercell interference. It consists of seven cells, with the respective BSs at their centres, 7 MTs (one per cell), a total area of 2000×2000 m^2, with cell radii of 350 m, Figure 6.6. Instead of beamforming, the issue is evaluating MIMO channel capacity depending on the resulting spatial interference conditions. In this case, the location of MTs is not a specific one, and these are assumed to move within the cell. Their location, at a specific time, is provided within a square grid, following a mobility model. In a related work at the intracellular level, several MTs are assumed to exist within the same macrocell, of radius 500 m [FKWS04]. Up to 18 MTs are considered moving, following a street grid similar to the previously mentioned study. Since the study evaluates how the number of MTs and the number of antennas on each terminal affect the sum capacity within the cell, the situation is close to

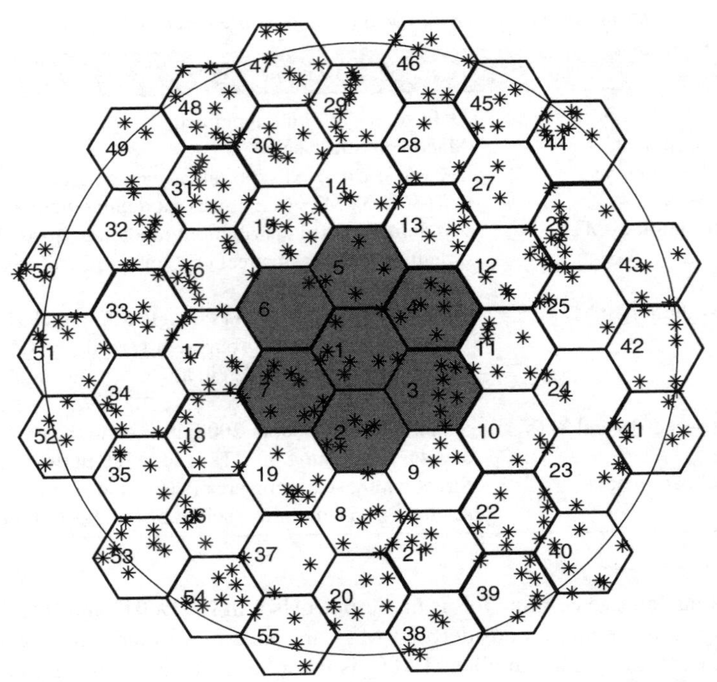

Figure 6.5 *An example of an outdoor scenario, with 55 BSs, 6 MTs per cell, uniform distribution of MTs [CzDe01].*

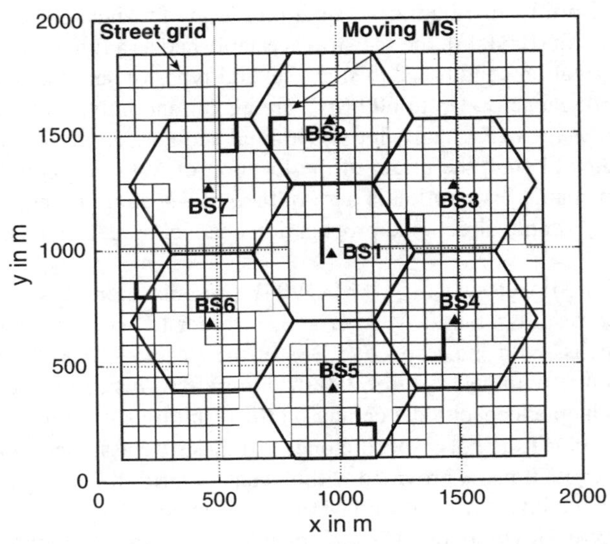

Figure 6.6 *Another example of an outdoor scenario, with 7 BSs, 1 MT per cell, with cell radii of 350 m [FMKW04].*

that where the location of MTs follows a simple, unconstrained random distribution. In this sense, the scenario has similarities to most of the cellular propagation scenarios described so far (though mobility also characterises each MT). Likewise, random distributed MTs are used to evaluate uplink MIMO capacity while using a finite scattering channel model and evaluating the performance of a spatial pre-whitening filter, in [ZhBW04]. MTs are set within 27 cells, with 1 MT per cell, at random spatial positions. The frequency reuse factor is changed, varying the number of MT interferers.

To conclude, there is a considerably large set of scenarios that are used for system level evaluations, with the aim of analysing intercell interference. These involve macrocell, multicellular environments, where the location of MTs is random, and their number or their location are not a specific concern. In COST 273, most of these outdoor multicell scenarios have been set to extract tendencies, extending further than the propagation level. The number of MTs may not be initially set, being a result of the evaluation at stake, e.g. the maximum number of MTs supported for a given service. The restriction in positioning MTs is possible, following a regular Manhattan grid, but still placing MTs randomly over the whole cells. The number of cells, the number and distributions of MTs (total or per cell) and cell radii are some of the most common parameters, also included in Table 6.3.

Table 6.3 *Summary of parameters used to define outdoor multicell scenarios.*

Parameter	Examples of values/settings
Cell radius	350, 500, 1000 m (hexagonal shape)
Cellular environment	Macrocell
Number of cells	E.g. 6, 7, or 57; possibly include sectorisation
Number of BS antennas per cell	1, 3
Number of MTs per cell	A definite number of 1, 6, 10; 10, 600 (maximum)
Spatial distribution of MTs	Random, uniform distribution; uniform, following a regular Manhattan street grid
Total covered area	2000 m × 2000 m; not specified
Other spatial parameters	Number of paths; MT mobility; fading, DoA statistics; height of MT/BS terminals

6.3 Physically motivated MIMO channel modelling and simulation

6.3.1 Introduction

Although physical reasons differ with frequency and environment, the characteristics of signals received over most radio channels vary from instant-to-instant, and location-to-location (i.e. over time and space). If the responsible mechanisms change in an unpredictable manner, received signals resulting from deterministic transmissions, such as channel soundings, exhibit a random nature. Since the randomness is imparted by the channel, radio channels are often modelled as Stochastic Processes (StPs), herein referred to as Channel Processes (CPs). Statistical measures of a CP can be used in a number of ways, including the estimation of achievable performance on the channel of interest, and the generation of realisations of the CP for use in simulations.

On MIMO Links (MLks), where there are M antenna elements at the TX, and N elements at the RX, there are $M \times N$ possible Physical Links (PLks) between different elements of the TX and RX arrays. When there is no motion of the antennas, or in the Critical Region (CrR)[3], randomness among transfer function models for the different PLks results because of the spatial separation of the antennas in the multipath field created by the presence of randomly located Interacting Objects (IOs). Variations in this randomness occur as the RX or TX arrays (or both) move, IOs in the CrR move, or as a result of both occurrences. Six types of randomness can be identified:

- Type I: random changes that result when any of the antennas used in an MLk move continuously in a Physically Stationary (PhS) environment,
- Type II: random changes that result when IOs in the CrR of an MLk move, but all antennas are PhS,
- Type III: random changes that result when Type I and Type II activities occurs simultaneously,
- Type IV: random differences among instantaneous PLk transfer functions between PhS antenna elements,
- Type V: random changes that occur when either the TX array or both are moved in steps to different locations within a local area throughout which shadowing by obstructions remains constant,
- Type VI: random changes that occur when either the TX array or RX array or both are moved in steps beyond the boundaries of a local area[4].

The statistics of CP variations can often be characterised by the same PDFs and correlation functions for any of the above types of randomness. There is frequently, therefore, a tendency to classify the randomness in all such cases as *fading*, and obscure differences in the physical reasons for such behaviour. Thus, for example, characterisation measurements made throughout a coverage area that reflect Type VI randomness might be applied to predict performance on an MLk between PhS RX and TX antenna arrays, which is a case where Type VI randomness could never occur. On such a fixed MLk, the only randomness that could occur are time and space variations of Type II and Type IV respectively, and measurements to characterise such a link must be made under Type II conditions. There are often considerable differences in the characteristics of channel variations, depending on whether IOs temporarily move through a CrR, or whether the CrR changes as a result of changes in antenna locations.

The adjective fading will therefore, in this section of the report, be restricted for use in describing situations in which PLks exhibit variations that influence communications performance as time progresses, regardless of their origin. Thus, Type IV randomness will be referred to as spatial variation, rather than fading. Type V randomness will be referred to as fading if it involves a single communications link (regardless of M and N), or spatial variation if there are multiple users each with antennas in the local area, such that the randomness has no influence on communications over the link to any one user. In MIMO radio engineering, consideration of the source of the required randomness is of paramount importance.

6.3.2 Radio channel modelling

Physical channel models are often based upon parameters estimated from (frequently noisy) radio propagation measurements. In addition to strong Multipath Components (MPCs) that result from

[3]Herein, a CrR is defined as the space surrounding a radio link that contains interacting objects that, when illuminated with energy from the link TX, can re-radiate non-negligible energy towards the link RX.

[4]In such cases there is a probability that the CP exhibits statistically non-stationary characteristics.

specular reflection, radio channel Impulse Response Estimates (IREs) contain lower, diffuse, 'background' energy at a continuum of excess delays which is believed to be the result of electromagnetic scattering. Unless components that contribute to this energy are estimated properly from measurements, not only can they not be modelled correctly for simulation purposes, but their presence also inhibits the accurate estimation of other parameters from measured data. Richter and Thoma [TLRT05] have formulated a solution to this estimation problem.

They conjectured, based on physical arguments, that IRE components (voltages) associated with the background energy within one measurement system delay resolution interval can be modelled as having a complex circular normal distribution (in random selection) with zero mean. Their phases are reasonably modelled as having a uniform distribution on $(-\pi, +\pi)$. The series of components at a continuum of delays is then modelled as an StP, with relative power that decays exponentially as a function of delay, beginning from a starting delay associated with the earliest arrival of energy at the RX. To complete the elements required for modelling an StP, the covariance between components of the diffuse energy at different delays must be represented. Since this is influenced by the bandwidth of the measurement system, a frequency-domain representation for the required covariance is multiplied by the covariance, at the same frequency lags, of the measurement system sounding spectrum, and the result is transformed back to the delay domain. Then, with complete knowledge of the above-described stochastic model for the background components, a statistic is derived for the sampled version of the (frequency-domain) channel transfer function.

Based upon the conjecture that the StP representing the diffuse background energy is Gaussian, the sampled data is modelled as having a multivariate circular Gaussian distribution $d_{bStP} \sim N_C(0, R(\theta))$. The probability density function for the (measured) realisation of the background StP, given the vector, $\boldsymbol{\theta}$, of model parameters, and its log-likelihood function are written. These expressions are then extended to apply to a measured time series of IREs. An equation for the sampled version of the associated (frequency domain) covariance matrix is also written as:

$$\boldsymbol{R}_f(\boldsymbol{\theta}) = toep(\boldsymbol{\kappa}(\boldsymbol{\theta}), \boldsymbol{\kappa}(\boldsymbol{\theta})^H) \tag{6.1}$$

with

$$\boldsymbol{\kappa}(\boldsymbol{\theta}) = \frac{\alpha_1}{M}\left[\frac{1}{\beta_d} \frac{e^{-j2\pi\tau_d}}{\beta_d + j2\pi\frac{1}{M}} \cdots \frac{e^{-j2\pi(M-1)\tau_d}}{\beta_d + j2\pi\frac{M-1}{M}}\right]^T + \alpha_0 e_0 \tag{6.2}$$

where α_1 is the power of the diffuse components at $\tau = \tau_d$, M is the number of frequency points measured within the measurement bandwidth, β_d is the coherence bandwidth of the diffuse components, normalised to the bandwidth of the measurement system, α_0 is the variance of the circular iid normally distributed measurement noise, and $e_0 = [1, 0, \cdots 0]^T$ is a unit vector.

To complete the modelling, the specular components are incorporated as local mean values of the distribution for the background (scattered) component and maximisation of the log-likelihood function for these mean values is treated as a weighted least squares problem. The authors indicate that joint estimation of the diffuse background and the specular components can be incorporated into appropriate high-resolution parameter estimation algorithms like ESPRIT, SAGE and RIMAX to result in considerable improvement to parameter estimation accuracy. They suggest, for example, that in the use of SAGE, or RIMAX, joint optimisation can be effected by simply alternating between the specular and diffuse model optimisation problems in consecutive steps.

Figure 6.7 shows results from the joint estimation of specular and diffuse components from a measurement in a street microcell using the RUSK channel sounder configured with an eight element

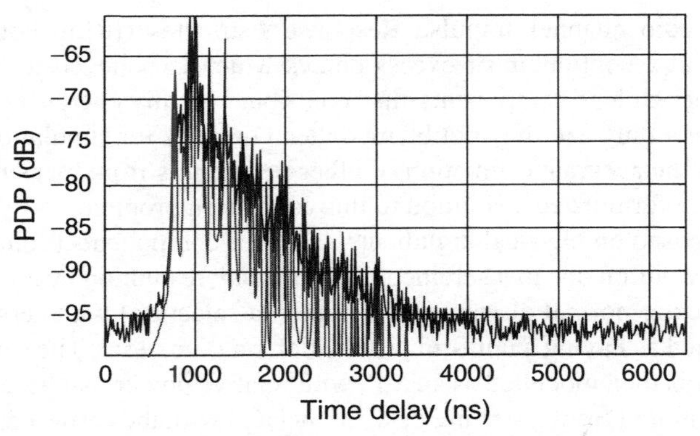

(a) A measured PDP and the related PDP reconstructed from the parameters of the specular propagation paths estimated using SAGE.

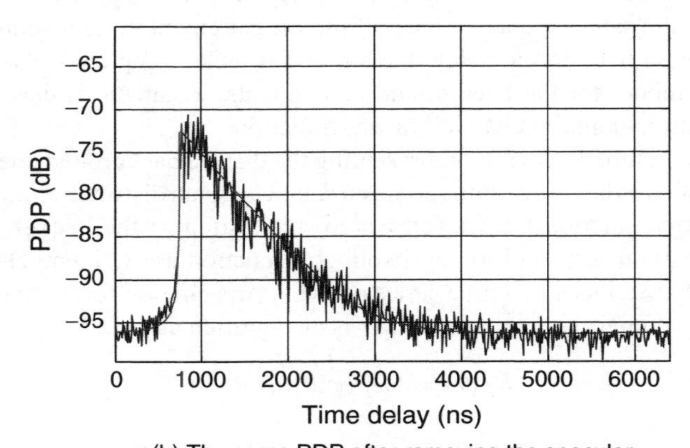

(b) The same PDP after removing the specular components, along with a plot of the main diagonal elements of the estimated covariance matrix (time delay domain) of the dense multipath components.

Figure 6.7 *Results from the estimation of the specular and diffuse content in a measured PDP.*

linear array at the base RX station and an omnidirectional antenna at the TX. The figure clearly shows the advantages of the reported method in the analysis of measured data for modelling of channel parameters. The remainder of this section reports work that was conducted by COST 273 participants in the development and implementation of CP models that are based on pseudo-physical (or approximate) modelling[5] of at least some of the mechanisms involved in the propagation of radiowaves between a TX and an RX, often using channel parameters estimated from measurements. On a real-world mobile radio channel, radiowaves can undergo multiple scattering, diffraction, and reflection from IOs in the CrR. However, a pseudo-physical model might, for example, be developed

[5]This type of modelling can often be verified using measured data and can give good results for a particular application or operating scenario. However, it can also lead to gross inaccuracies when extended analytically to cover new situations. This is particularly true when evolution of channel parameters in time or space is being modelled.

based on the consideration of only specular reflections from IOs having a geometrical arrangement that reflects measured propagation delays between the locations of the MT and BS, whether or not specular reflection is the true mechanism that would cause re-radiation from the IOs involved.

A generic geometrically based MIMO modelling approach was developed by Xu [XCHV02], and tested by comparison with results from the analysis of outdoor macrocellular propagation measurements recorded in New York City. The reported modelling approach is one in which wideband channel realisations are simulated using any given set of channel statistics, including instantaneous PDPs, power azimuth spectra at the TX and RX, and intended antenna array configurations. Based upon the assumption that each element of an MLk matrix, **H**, is the result of the superposition of energy from L waves, elements of **H** corresponding to MT antenna element 'n' and BS antenna element 'm' at delay 'q', and time step 's' are generated in accordance with

$$h_{m,n,q,s} = \sqrt{PDP_q}\sum_{l=1}^{L} A_l\sqrt{G_t(\vec{k}_{t,l})}\sqrt{P_{t,q}(\vec{k}_{t,l})}\exp\{j\vec{k}_{t,l}\cdot\vec{d}_{t,m}\}$$
$$\times\sqrt{G_r(\vec{k}_{r,l})}\sqrt{P_{r,q}(\vec{k}_{r,l})}\exp\{j\vec{k}_{r,l}\cdot\vec{d}_{r,n,s}\} \qquad (6.3)$$

where, $A_l = x+jy$ is a complex Gaussian variable, with $x, y \sim$ N(0,1), G_t, G_r are the gains of the TX and RX antennas, respectively, $k = 2\pi/\lambda$, $P_{t,q}$ is the Power-Azimuth Spectrum (PAS) as a function of DoDs, $P_{r,q}$ denotes the PAS as a function of DoAs, and $\overrightarrow{k_{t,l}}$, $\overrightarrow{k_{r,l}}$ denote the directional vectors to the TX and RX, respectively, at the DoD and DoA of the lth plane wave. The location of the mth transmit antenna element is independent of time, and is denoted $\vec{d}_m$. The location of the nth receive antenna element is a function of time step s, and its velocity, and is given by $\vec{d}_{n,s} = \vec{d}_{n,1} + \vec{v}t$. The phase shift through the propagation medium is modelled by the phase of A_l, which is uniformly distributed in $[0, 2\pi]$. If it is desired to include an LoS component to produce fading with a given Rice factor, K, a strong plane wave is added in accordance with

$$H^{LoS}_{n,m,1,s} = \sqrt{K}\exp\{j\vec{k}_{t,LoS}\cdot\vec{d}_{t,m}\}\exp\{j\vec{k}_{r,LoS}\cdot\vec{d}_{r,n,s}\} \qquad (6.4)$$

A method for extending the foregoing link level model to a system level model that accounts for multiple users and multiple base stations is also reported in [XCHV02].

Model verification results in [XCHV02] include the presentation of Type I correlation data generated from simulations corresponding to an MT PAS width of 35° and a vehicle speed of 10 km/h. Very good agreement with both the intended theoretical correlation function and one derived from Clarke's model is shown for small lags (i.e. small antenna element spacings), such as those applicable to MIMO antenna considerations. A comparison of capacity CDFs is similarly good, and shows degradation, as would be expected, with respect to results for a Rayleigh iid environment. Capacity CDFs from simulations using the model with parameters estimated from measured data were compared with those obtained directly using measured data for 4 × MLks. Agreement between modelling and measurement-based results is shown in the cited work to be excellent.

A dual-polarisation, geometrically based channel model was proposed by Oestges *et al.* [OeEP04] for fixed Local Multipoint Distribution System (LMDS)-type links at 2.5 GHz. Its use begins with the specification of a PDP that is known to be valid for a specific range from a BS, a system bandwidth, and antenna pattern and polarisation characteristics. A physical operating environment is then specified so as to match the reference PDP, by locating IOs on ellipses with dimensions corresponding to the delays of Multipath Groups (MPGs) within the PDP. A tapped-delay-line-type model is then assumed for the channel impulse response, and each tap is allocated a time-averaged power, a delay, and a Rician fading distribution with a specific Rice factor.

The polarisation characteristicsof the energy from each IO are defined by a coefficient matrix defined as in [Oest02], which contains log-normally distributed random elements, with orthogonal and cross-polarisation coupling elements being attenuated and phase-shifted replicas of the co-polarisation coupling elements. Specific attention is paid to the incorporation of gain imbalances among co- and cross-polarised elements. Antenna XPD characteristics are also accounted for in the final apportionment of energy received in an MPG at a particular delay. In addition to energy from incoherent scattering, coherent energy comprised of the sum of an LoS component and coherently scattered components can be modelled. Finally, Doppler characteristics are incorporated using a recombination process in which a fraction of the number of IOs that are considered is removed during each discrete simulation time interval, Δt, and replaced by new IOs, with new properties and new locations. The number of IOs removed from consideration in each recombination is a Poisson random variable ζ, with a mean given by $E\{\zeta\} = 1 - \exp(-n\Delta t)$, where n is related to the desired Doppler spread.

Simulations involving 2×2 MLks with each antenna array having one vertically polarised and one horizontally polarised antenna element were reported in [OeEP04]. PDPs with three MPGs were simulated using a three-tap, tapped delay line model with characteristics in accordance with IEEE 802.16 recommendations for typical Rayleigh and Rician fading channels at a range of 7 km from the BS. PDPs for other ranges were simulated based on scaling from this reference profile, as described in the cited paper. Ergodic capacities were calculated using Type II time averages to replace expectations for different base-subscriber station ranges, antenna polarisation and XPD combinations. Results indicate that channel capacity is only weakly dependent on transmission range, even though instantaneous RMS delay spreads, fading characteristics, and cross-polarisation isolation can be significantly range-dependent.

Molisch [Moli04b] proposed a generic model for MIMO channels in macrocellular mobile radio environments. This model is based substantially on the COST 259 directional channel model [MAHS05b], with an important extension. This is the inclusion of modelling associated with what is frequently referred to as double scattering, even though physical mechanisms do not necessarily have to be limited to electromagnetic scattering. Single scattering models, discussed in [MAHS05b], allow the possibility of:

- direct transmission of energy from a BS to an MT,
- the incidence of energy transmitted from a BS on IOs located close to the MT (usually in a circle about the MT), and subsequent re-radiation to the MT,
- the incidence of energy transmitted from a BS on IOs located remotely from the MT, and subsequent re-radiation to the MT,
- similar interactions on reciprocal propagation paths between the MT and the BS.

The transmission of energy from multiple IOs to the RX terminal is assumed to result in the reception of multiple MPCs, but cases in which there is a predominant, strong specular reflection are also accounted for in the model. The extension to the case of what is termed 'double scattering' involves allowance for the possibility that energy from the BS is first incident on IOs close to the BS, then re-radiated, being received at the MT either directly or via IOs in the vicinity of the MT. Reception of energy via double (i.e. two consecutive) interactions involving remote (referred to as 'far') IOs is not considered as a result of the probability that the received energy would be very weak.

Molisch highlighted a distinction between 'double-directional' and 'vector (matrix)' channel modelling, but emphasised that, although the approaches are different, results must be identical.

Double-directional modelling is based on the consideration of physics, and results in approximations for the impulse response of the propagation path followed by each MPC. The sum of these estimated single-path impulse response functions is then taken as a model for the impulse response of the channel (i.e. a radio link). Since the resulting impulse response contains information on the DoDs, DoAs, delays, and amplitudes of all MPCs, an MLk matrix can readily be calculated. An advantage of using double-directional measurement results is that these parameters can be made almost independent of measurement equipment characteristics. Molisch's generic model was embellished over the duration of the COST 273 Action. Details are discussed in Section 6.8.

Molisch's approach was employed in a model implemented by Hoffstetter and Steinböck, and results were reported in [HoSt04]. One or several BSs as well as remote IOs and IOs in the vicinity of the BS and MT (which move with it) are assumed, as is an MT velocity vector. Specular reflection is assumed at each IO, and a ray-tracing tool calculates the impulse response function associated with the path followed by each MPC. Complete channel impulse response functions as well as MLk matrices are then simulated as described in [Moli04b]. The (software) implementation includes a Propagation Module (PrM), an Antenna Module (AnM), and a Convolution Module (CoM), with appropriate interfaces between the PrM and AnM and the AnM and CoM. The synthesised double-directional channel information (including arbitrary, random information on wave polarisation) is generated in the PrM then applied to the AnM, which accounts for the antenna response to different polarisations, and emerges from the latter in the form of an attenuation and phase shift for each MPC. These results are then summed in the CoM to form the channel impulse response, and MLk matrices. Long-term power variations are accounted for in the normalisation of MLk matrix elements, and true fading correlations as a result of Type I, II and III random variations are estimated in the MIMO analysis. Capacities of simulated MLks are shown in Figure 6.8 to be considerably lower than those that would be achieved under iid Rayleigh fading conditions.

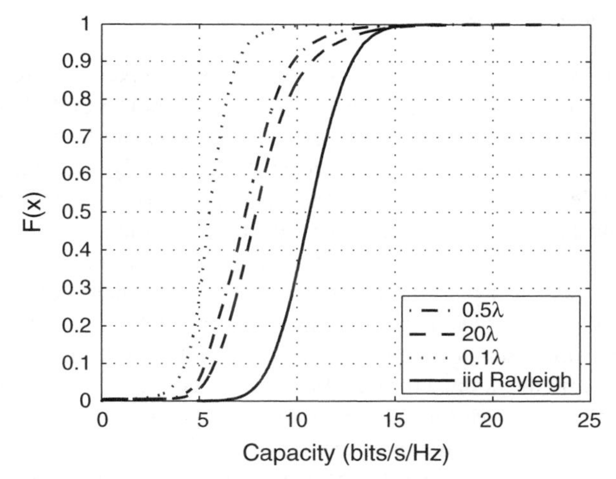

Figure 6.8 *Outage capacity for selected antenna element spacings.*

As well, the figure shows that capacities vary for different antenna element spacings. It was also reported that capacities vary as the number of IOs considered in the simulation is changed.

For the investigation of MIMO systems at the system level, an accurate description of spatially coloured inter- and intracell interference is necessary. This means that for decreasing distance between neighbouring MLks the correlation of channel parameters should increase. This is neglected from consideration in almost all physically based MIMO channel models. The Multiuser Double-Directional Channel Model (MDDCM) presented in [FKMW04], however, enables an accurate characterisation of interference in macrocellular environments [FMKW04]. The conceptual basis for the model is similar to those associated with geometrically based stochastic channel models, but in contrast to them, the position and parameters associated with IOs are not repeatedly calculated stochastically for each channel realisation. The propagation environment is generated in pre-processing and stays fixed during the whole simulation.

The reported MDDCM incorporates new models for interactions with IOs around the MT, and for propagation along street canyons. Known models for far clusters and line-of-sight scenarios are included [AMSM02], [Moli02]. The local cluster is modelled as shown in Figure 6.9. For the sake of

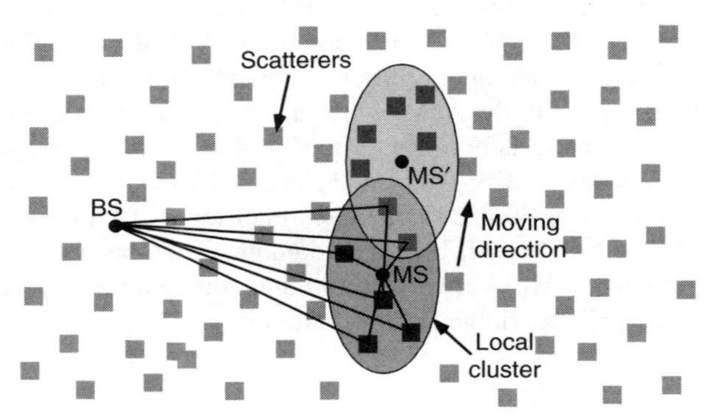

Figure 6.9 *Method for modelling local clusters.*

simplicity the figure shows only the two-dimensional (2D) plane. Fixed IOs are distributed uniformly in the whole 3D simulation environment, which may cover several BSs and MTs. Each MT is located in a circular/elliptic cylinder, which defines the local cluster. This cylinder serves as a search function, since only IOs within it are active. IOs are modelled as ideal, perfectly conducting, rectangular plates. Rays that undergo single interactions between the MT and the BS are calculated according to the model derived in [Svan01]. The advantage of this model is that elementary spatio-temporal properties (e.g. depolarisation of the incident wave field, 'scattering' lobe) are accounted for.

To account for time varying channel behaviour the channel model is combined with a mobility model. When the MT moves to a new location, a number of new IOs contribute energy to the received signal and at the same time, some of the energy from old IOs fades out. As a simulation mode, a Manhattan-like mobility model defined in [ETSI98] is used. The street grid of the urban environment is modelled as a 'chess board'. Mobile stations move linearly along a street and can change their direction at each crossroad. The shape and size of the local cluster is adjusted, depending on the position of the MT. If the MT is located in a street canyon the local cluster is modelled as an elliptical cylinder. If the MT is positioned in a crossroad the local cluster is given the shape of a circular cylinder.

In macrocellular environments, some waves propagate from the BS towards the crossroads. Then, due to street canyon effects, these waves are guided to the MT. Their DoDs (azimuth and elevation) as well as their delays are highly correlated. Street canyons result in the same phenomena as waveguides. To model propagation along a street canyon it is assumed that all buildings are placed along the streets of the Manhattan-like street grid, which therefore represents a waveguide structure. In addition to the waveguide, wave interaction points near the crossroad have to be modelled. For this, additional IOs are placed in a circular cylinder at each neighbouring crossroad. The rays propagate from the BS towards these IOs. If an IO is visible from the MT a reflected ray propagates along the street canyon towards the MT. Associated propagation parameters can be calculated using the 3D image theory discussed in [FMKW04], and exemplified in Figure 6.10.

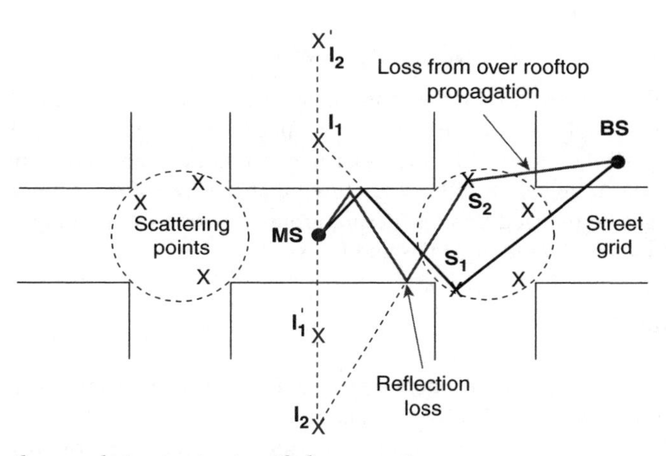

Figure 6.10 *Image theory relating to street-guided propagation.*

A physical model for so-called keyhole channels was conceptualised by Sibille [Sibi01], who formulated a modelling approach based on consideration of coupling between an input wave impinging on a keyhole and a number of output waves produced by diffraction phenomena. The channel transfer function matrix is expressed in terms of steering matrices for the TX and RX arrays and a 'connecting' matrix that describes interrelationships among the complex wave amplitudes at the two arrays. The non-diagonal character of the connecting matrix, which results from splitting incident waves into multiple output waves, leads to a keyhole when the matrix has largely non-zero entries (i.e. when it is full). An example of a 3×3 MLk with TX and RX arrays on either side of a slit in an otherwise perfectly absorbing plane is given. Capacity CDFs are estimated for four different cases involving different array element spacings and two different slit widths. For a very wide (5λ) slit, corresponding to negligible diffraction, and large (1λ) array element spacings, corresponding to little spatial correlation in Type IV random variations at the arrays, capacity was estimated to be 3.54 bits/s/Hz. However, when the slit width was reduced to $1/4\lambda$, the capacity estimate decreased to 2.5 bits/s/Hz. With the narrow slit, it was further shown that reduction of the antenna element spacings in each array to $1/10\lambda$ in order to increase spatial correlation had no effect on capacity. This is indicative that the MLk was already reduced to having a single degree of freedom by the narrow slit and concomitant diffraction effects. It was found that the greatest decrease in capacity occurred when the slit width was reduced from about two to about one wavelength.

Equations are developed in [Sibi01] for both the frequency-domain representation of the MLk matrix, and its delay-domain equivalent, assuming M transmitted waves are coupled into N received waves. The frequency-domain representation is written as

$$\mathbf{H}(\omega) = \mathbf{A}_r(\omega)\mathbf{W}(\omega)\mathbf{A}_t^T(\omega) \tag{6.5}$$

where $\mathbf{A}_r$ and $\mathbf{A}_t$ represent the array steering matrices, and $\mathbf{W}$ is the wave-connecting matrix. In the simple 3×3 example with the diffraction slit, this matrix would have elements

$$w_{ij} = R_{ij}T_{ij}K_{ij}e^{-jkl_i}e^{-jkl_j} \tag{6.6}$$

where R_{ij} and T_{ij} are wave attenuations on either side of the slit, corresponding to propagation lengths l_i, and l_j, and K_{ij} is the Kirchhoff diffraction coefficient associated with path ij. Fourier transformation was applied to give

$$\mathbf{H}(\tau) = \mathbf{A}_r(\tau)\mathbf{W}(\tau)\mathbf{A}_t^T(\tau) \tag{6.7}$$

in which both the steering matrices and the coupling matrix have entries that are Dirac delta functions, dependent on mutual coupling at the arrays, and propagation delays on various PLks between the TX and RX arrays. All parameters are such that they could be estimated from double-directional channel sounding measurements. Sibille conjectured that the size of the connecting matrix in real-world scenarios might be reduced by considering groups of waves impinging at the RX as distinct Rayleigh fading signals with random DoDs and DoAs.

6.3.3 Simulation

In an ideal world, radio propagation phenomena could be measured precisely, modelled analytically, and used to assess the performance of technology and systems under development. However, in reality, the capabilities to do precise measurements and analysis are not often available. Therefore, to allow the repeated testing of prototypes, it is of interest to many in the systems engineering community to develop simulation procedures, based on channel models, some of which can be standardised and used in the comparison and evaluation of different technology on an industry-wide basis. Some participants in the COST 273 Action implemented models in the development of simulators with these objectives, as outlined below.

Kunnari [Kunn02] reported the development of a software package for the simulation of small-scale fading on multiple PLks between TX and RX antenna element pairs in different subbands of an OFDM MIMO system. The number of fading channel gains that can be generated simultaneously is LMN, where L is the number of subbands, M is the number of TX antenna elements, and N is the number of RX antenna elements. Fading with either Rayleigh or Rician temporal envelope fading distributions can be generated, with user-specified time and frequency correlation characteristics for each PLk, and user-specified correlation among fading characteristics on different PLks. The software is also capable of generating time varying tap weights for the fading of MPGs in a tapped delay line model for the impulse response of a frequency-selective fading channel.

It is assumed that plane waves impinge only from the equatorial plane, except in one simulation mode, for which a uniform 3D spectrum is assumed for DoAs. Doppler shifts/spreads are assumed to influence only the carrier frequency, and the time for radio waves to propagate across the TX and RX antenna array dimensions is assumed to be small compared with the inverse of the TX signal bandwidth. The antenna elements in the RX and TX arrays are assumed to be identical, with no electromagnetic coupling. Complex magnitudes of MPCs are assumed to be random realisations from an iid Gaussian distribution and their phases are assumed to be uniformly distributed over $[0,2\pi]$. Gaussian WSS fading characteristics and uncorrelated scattering are also assumed.

The correlation of Type III random variations is imposed by assuming the power spread function $p(\nu, \phi_t, \phi_r, \tau)$ is independent of Doppler characteristics, so that $p(\nu, \phi_t, \phi_r, \tau) = p(\nu)p(\phi_t, \phi_R, \tau)$, where ν represents Doppler shift, ϕ_t, ϕ_r represent DoDs and DoAs, respectively, and τ represents delay. It is also assumed that the corresponding time and spatio-spectral correlations can be represented as a product. Doppler characteristics, and therefore temporal correlations, are generated by applying white noise to filters with either a uniform spectrum, or the well-known fading spectrum proposed by Clarke. Spatio-spectral correlation is imparted to the Gaussian gains for PLks separated from each other in space and frequency by their representation as components of a multivariate normal random vector $\mathbf{z}$ given by $\mathbf{z} = \mathbf{A}\mathbf{x} + \boldsymbol{\mu}$, where $\mathbf{x}$ is multivariate iid normal with zero mean and unit variance, $\boldsymbol{\Sigma} = \mathbf{H}\mathbf{H}^H$ is a covariance matrix, approximated by the Kronecker product of frequency and spatial covariance matrices, $\boldsymbol{\Sigma}_F$ and $\boldsymbol{\Sigma}_S$, respectively, and $\boldsymbol{\mu}$ is the mean. An evaluation of $\boldsymbol{\Sigma}_F$ is obtained from a decaying PDP via Fourier transform. The matrix $\boldsymbol{\Sigma}_S$ is obtained as the Kronecker product of the MLk TX and RX spatial covariance matrices for Type IV random variations, which are specified according to results reported in the literature. Deterministic components with constant amplitude and phases that change in accordance with the Doppler equation can be added to the link gains to simulate Rician fading with specified Rice factors. CDFs resulting from simulation are shown in the cited technical document and show excellent agreement with intended model distributions.

The implementation of a Wideband Directional Channel Model (WDCM) for use in link level simulations to assess the performance of techniques for the mitigation of MAI in UMTS cells was reported by Ferreira *et al.* [FeMC01]. Although the propagation environment that was considered is strongly defined by geometrical considerations, the underlying theoretical propagation model is a statistical one, applicable for simulation of both UMTS micro- and macrocell scenarios. The spatial propagation environment consists of a uniform distribution of randomly oriented groups of IOs, the location of each of which is in turn distributed in the horizontal plane in accordance with a 2D Gaussian PDF. Each IO is assigned a reflection coefficient with an amplitude chosen from a uniform distribution on [0,1] and phase chosen from a uniform distribution with minimum and maximum values given by the maximum Doppler shift associated with a particular MT velocity, multiplied by the simulation step duration of 667 μS. Single interactions and specular reflection are assumed. Only IOs within the CrR surrounding each MT-BS link are activated. Multiple MTs can be activated and signals received and transmitted by them as they move through the propagation environment are simulated sequentially.

A software package, named Mascaraa, for MIMO link level simulations was reported by Conrat and Pajusco in [CoPa03]. The primary modules within this package simulate rays in accordance with a specified channel model, generate a channel impulse response, effect the convolution of an applied signal with the generated impulse response, and simulate mobility.

Mascaraa first simulates rays, each with a specific delay, DoD from a TX array and DoA at an RX array. Polar notation is used, and the field at an RX antenna is represented by the matrix

$$\begin{bmatrix} E_\theta^{RX} \\ E_\phi^{RX} \end{bmatrix} = \begin{bmatrix} G_{\theta\theta} & G_{\phi\theta} \\ G_{\theta\phi} & G_{\phi\phi} \end{bmatrix} \begin{bmatrix} E_\theta^{TX} \\ E_\phi^{TX} \end{bmatrix} \tag{6.8}$$

where $G_{\theta\theta}, G_{\theta\phi}, G_{\phi\theta}, G_{\phi\phi}$ are complex gain values that completely characterise the gain and polarisation properties of each ray. A set of rays with constant amplitude, polarisation, and delay characteristics are used to model the channel in each of a series of WSS intervals within each of which dynamic aspects of the channel are modelled by changing only the phases of impinging rays.

The set of rays can be generated via one of four different approaches. The first of these is to assume a tapped delay line model for the channel impulse response. Approximately 50 rays are then assigned as contributors to each tap in the model. Arrival angles are assigned either to have a uniform distribution in the azimuth plane in accordance with Clarke's model, or to have uniform angles of arrival in 3D to result in a flat fading spectrum. In this mode, only $G_{\theta\theta}$ is non-zero, with relative amplitude defined by the tapped delay line model. Ray directions at the BS are not defined, although it is suggested that elevation and azimuth angles based on information in [Paju98] could be assigned for use in MIMO simulations.

The second approach is by ray tracing based on precise knowledge of the environment and the link terminal locations. A third approach uses a Geometry-based Stochastic Channel Model (GSCM) for defining ray sources, such as clusters of IOs. Finally, rays can be generated in accordance with a directional tapped delay line model, in which the Doppler spectrum associated with the fading of each tap is that associated with power angular distributions that can be specified for the BS or MT. Polarisation modelling is accomplished by assigning randomly selected values to the complex channel gains associated with each ray, which can be either different or the same for all rays associated with a specific tap.

For MLks, an approximation to the impulse response $h_{mn}(k)$ between the mth TX antenna element and the nth RX antenna element at time step k is generated according to

$$h_{mn}(k) = \sum_{i=1}^{NRays} a'_{mn}(i)g(kT_s - \tau(i)) \tag{6.9}$$

where $a'_{mn}(i) = a(i)e^{j\varphi_n(i)}e^{j\varphi_m(i)}e^{jx\cos(\alpha)}e^{jstart(i)}$, a represents the ray's amplitude, φ_n is the spatial phase offset (with respect to the array's phase centre) at the mth antenna element in the TX array, φ_m is the spatial phase offset at the nth element in the RX array, x is the displacement of the MT from its position at the beginning of the interval, α is deduced from the DoA at the MT and the MT's trajectory, $T_s = 1/F_s$ is the simulation clock interval and $start(i)$ is a random phase assumed at the start of simulation for each WSS interval. The function $g(t)$ is the impulse response of a raised cosine filter with total bandwidth equal to $F_s/2$. Detailed consideration of the synthesis of $g(t)$, the accuracy of ray delays, channel impulse response size optimisation, and amplitude and delay normalisation are reported. The processing time needed by the impulse response generation process in Mascaraa is comparable to that of other approaches where tap gains are generated as iid complex variables. That required to simulate a 10 minute-long transmission at 2.2 GHz with a mobile speed of 10 m/s and a signal bandwidth of 5 MHz is reported to be 4900 seconds on a Pentium IV PC, with a 1.4 GHz processor.

Morosi *et al.* reported the software implementation and assessment of a simulator for macro- and microcellular UMTS mobile channels based on the COST 259 WDCM [MAHS05b]. The uplink model used in the simulator is detailed in [MTRF02]. After deterministically specifying the cluster positions in the radio environment of interest, a statistical modelling approach is used. The signal received at the BS is modelled as the superposition of multiple waves, each with its own DoA, (φ_i), propagation distance (d_i), average power (P_i), delay (τ_i), an associated reflection coefficient ($\alpha_i e^{j\phi_i}$) and DoD (γ_i) from the MT. Energy is assumed to arrive at the BS in discrete time delay intervals associated with propagation via clusters, and the signal received from the nth cluster is represented as

$$\vec{\mathbf{y}}_n(t) = \sum_{j=1}^{Q_{cn}} \sqrt{PLW(\tau_j)}\sqrt{PDP(\tau_j)}s(t-\tau_j)\sum_{l\epsilon\tau_j}\alpha_l\sqrt{PAP_l}\sqrt{PEP_l}\,\vec{\mathbf{a}}(\varphi_l) \times e^{-j\phi_l}e^{-j2\pi/\lambda(d_l+vt\cos(\gamma_l))} \tag{6.10}$$

where $\vec{\mathbf{a}}(\varphi_i)$ is the steering vector at the BS antenna, Q_{cn} is the number of taps associated with the nth cluster, PLW represents path loss (including shadowing) associated with the jth tap in the model for the nth cluster, $PDP(\tau_j)$ is the power at delay j in the PDP associated with the nth cluster, PAP_l is a factor associated with the power azimuth profile at the lth DoA in the nth cluster, PEP_l is a factor associated with the power elevation profile at the lth DoA in the nth cluster, v represents the speed of the MT, and L is the number of IOs contributing to energy at delay τ_j. By defining the correlation matrix associated with Type IV (spatial) variations at the BS [3GPP01] as

$$R = \frac{1}{L_{\tau_j}} \sum_{l \epsilon \tau_j} PAP_l(\varphi_l) PEP_l(\varphi_l) \vec{\mathbf{a}}(\varphi_l) \vec{\mathbf{a}}^H(\varphi_l) \tag{6.11}$$

the signal received from the nth cluster can be written more compactly as

$$\vec{\mathbf{y}}_n(t) = \sum_{j=1}^{Q_{C_l}} \sqrt{PLW(\tau_j)} \sqrt{PDP(\tau_j)} s(t - \tau_j) \mathbf{S}_j^H \vec{\mathbf{g}}_j(t) \tag{6.12}$$

where $\mathbf{S}_j$ is chosen such that $\mathbf{R} = \mathbf{S}\mathbf{S}^H$, and fast fading is accounted for by introducing $\vec{\mathbf{g}}_j(t)$, a column vector with independent complex Gaussian entries having unit variance. Thus, each tap within each cluster associated with the signal received at a particular antenna element undergoes a fast fading process, and the correlation matrix $\mathbf{S}$ is derived from the array geometry at the BS, as well as the DoA of energy at delay τ_j, and the corresponding PAP and PEP, in accordance with COST 259 recommendations.

A different geometrical approach for the modelling of indoor radio channels is reported in [HaRe04], [HaLe03]. Unlike many geometrically based channel models, the reported model has no reliance upon a particular distribution of IOs. Instead, it is shown that fundamental statistical parameters associated with spatial variations on the channel are robust, in that they depend only weakly on the geometry of the environment. Thus, in order to estimate these parameters, precise knowledge of the environment, as for instance that required to set up a ray-tracing simulation, is not required. It is also shown that the key parameters that are necessary for geometrical channel characterisation are the volume and the surface area of the domain within which planned radio systems will operate.

Based on the foregoing concepts, a fully analytical model for frequency-selective spatially varying Rayleigh SISO indoor radio channels is developed. Its input parameters are the carrier frequency, antenna characteristics, path-loss exponent, an average reflection coefficient for IOs, the spatial distribution planned for the TX and RX terminals, their minimum distance from each other, and the size of the domain in which the two are to be located. The output of the reported analytical relationships are an upper bound on instantaneous RMS delay spreads and the mean and variance of the log-normal distribution that describes large-scale spatial variations. The distribution of such variations is also reported to provide sufficient information for characterisation of the statistics of small-scale (i.e. local area) spatial variations. The reported model was verified by comparison of analytical results with simulation results and measurements at frequencies in the range of 2 to 60 GHz [HaLe03].

Before moving to consideration of hardware fading channel simulators, attention should be drawn to the work of Lienard and Degauque [LiDe04b], who conducted an interesting investigation into the use of Mode Stirred Reverberation Chambers (MSRCs) for the simulation and testing of MIMO systems. Such chambers have metal walls, and upon their excitation using an antenna, large electromagnetic fields can be generated, associated with numerous propagation modes. A metal paddle can be used for mode stirring, such that for each position of the paddle, different modes persist. If an RX antenna is used to capture energy transmitted through the chamber, each position of the mode stirring paddle results in a different transfer function for the PLk between the RX and TX antennas, including the effects of all propagating modes. The SNR on such PLks can be adjusted through knowledge of the Q of the chamber. Thus, by stirring, different link realisations can easily be simulated under pre-determined average SNR conditions. However, MSRCs, developed initially for EMC testing, result in uniform spatial distributions of electromagnetic fields within the chamber, a desirable property in their original application, but one that prohibits the establishment of realistic MIMO scenarios in which spatial field correlations are, in general, different at the TX and RX antennas. To allow for different field distributions at the TX and RX antennas, two MSRCs can be used, with waveguide coupling between them.

Experiments reported in [LiDe04b] involved the use of two MSRCs of dimension $2.24 \times 2.92 \times 2.0$ m, coupled via either an oversized waveguide, of dimension 20×20 cm or a WR 187 waveguide in which a TE01 mode was excited over an operating bandwidth from 3.95 GHz to 5.85 GHz. Before the experiments began, the chambers were characterised as having Q factors of about 17,000, and when wideband horn antennas were used for transmitting and receiving, a PDP with wide ranging power fluctuations in consecutive delay intervals and a mean power slope of 15 dB/μs resulted. The average of instantaneous RMS delay spreads for 100 different positions of the mode-stirring paddle was determined to be 200 ns. Initial calibration measurements also involved an estimation of spatial correlation of Type IV variations at the TX antenna using a fixed patch antenna in one chamber and a second patch antenna in the other chamber, which was moved in steps of 2 mm. After each step, the channel transfer function matrix was measured for 100 different positions of the mode stirrer using a network analyser configured for transmission measurements. The resulting correlation function decreases nearly linearly from unity to about 0.25 over a distance of 45 mm. MIMO experiments made using MSRC setup, were also reported in the above-cited papers.

6.3.4 *MIMO hardware channel emulators*

Beyond 3G systems use wide bandwidth and multi-antenna technologies. Simulation of these systems is computationally complex and requires significant computation time. HardWare (HW) radio channel emulation provides the necessary processing speed. Other benefits of HW emulation are real-time and repeatable performance evaluation for any device under test. Two COST 273 participants, ARC Seibersdorf Research GmbH and Elektrobit Ltd have developed commercially available hardware channel emulators for MIMO channels. ARC Seibersdorf Research GmbH reported the ARC SmartSim MIMO development platform, a MIMO channel emulator, which includes additional digital signal processing hardware for TX and RX development. Elektrobit Ltd reported the Propsim radio channel emulators [KoJH03], [SJKN05], the Propsound radio channel sounder and the EB4G algorithm development platform.

Elektrobit demonstrated a new time-to-market shortening solution. This system uses the Elektrobit Propsim C8 radio channel emulator, Propsound channel sounder and EB4G algorithm development platform [KoJH03]. It can be used with digital baseband, analogue baseband or RF interface to make it possible to test algorithms without the need to implement RF front ends during testing of early baseband prototypes. The EB4G-algorithm development Software Defined Radio (SDR) platform can be used to run algorithms in real time. Full MIMO configurations can be studied, with freedom to specify the air-interface type. Realistic MIMO channels, with correlation among MLks can be emulated with Propsim C8 using channel data measured with the Propsound channel sounder.

This solution provides unique capabilities to run realistic controllable MIMO testing with real-world channel correlation challenges.

Elektrobit Propsim C8 is a scalable multichannel emulator supporting channel emulation of all known standard and research channel models from 2G to MIMO TGn models with ETSI/Broadband Radio Access Network (ETSI/BRAN) delay profiles and beyond. These models can be defined freely by a user and run within the existing HW configuration. The Propsim C8 capability to run emulations from files provides the possibility for playback of channel measurement data as explained in references [KoNu04], [KKNJ04], and exemplified in Figure 6.11.

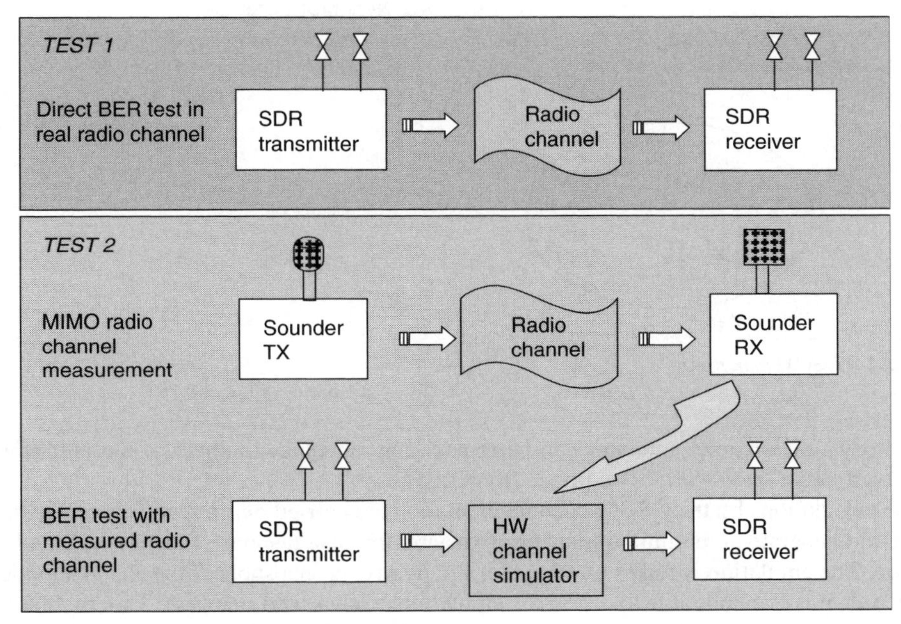

Figure 6.11 *Principle of playback simulation verification.*

The Propsim C8 channel model files can also be examined following testing to debug specific channel states where the system under test encountered problems. Propsim C8 supports 4×4 configurations within one emulator unit and more complex scenarios (e.g. 8×8) can be tested with the synchronisation of multiple units.

The MIMO hardware testbed developed at ARC Seibersdorf Research GmbH consists of a fully scalable channel emulator and additional signal processing units for TX and RX development [KaHP04], [KSKL05]. The channel emulator affords the possibility of using a GSCM based on COST 259 recommendations, which introduces IOs around the BS as well as double scattering (see Figure 6.12) [HoSt04] and is to be enhanced using the latest COST 273 channel model recommendations. Additionally, the user has the possibility of loading impulse response functions from any given channel model or channel sounder measurements. The development platform also features interfaces for digital baseband, analogue baseband, Intermediate Frequency (IF) and Radio Frequency (RF) and supports multiple radio systems as well as multiple frequency bands (e.g. UMTS, WLAN, WIMAX). This allows for seamless rapid prototyping of MIMO systems by starting off with a digital baseband implementation of the system and gradually adding analogue and RF interfaces. For 'real-world'

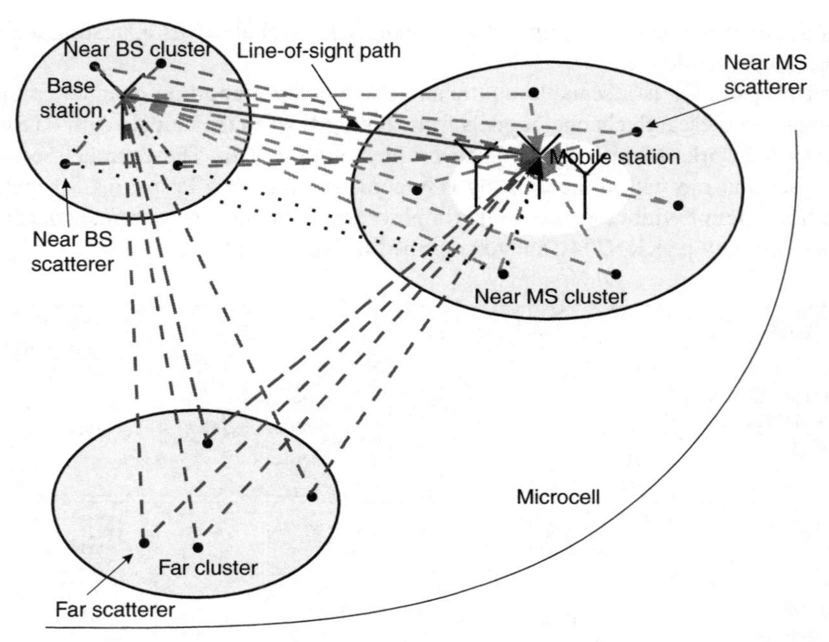

Figure 6.12 *GSCM principle.*

tests, the real-time channel emulator can be replaced by antennas to check if the algorithms also perform well under 'real-world' conditions [KSKL05].

All the calculations by the GSCM channel emulation are carried out in real time on the hardware platform and generate an unlimited number of varying impulse response functions for realistic RX evaluation. The emulation is fully reproducible, and by saving snapshots of the channel model states every second, it is even possible to restart the emulation at any saved snapshot. This enables targeted debugging of RX algorithms and performance evaluation with exactly reproducible channels in the digital emulation domain.

The emulation of a MIMO channel is first partitioned among a set of DSP boards, with each board emulating the radio channel between one pair of TX and RX antennas. Each DSP board consists of a Digital Signal Processor (DSP) and a Field Programmable Gate Array (FPGA), where the DSP calculates the impulse response of the channel and the FPGA convolves the input signal with the impulse response. A novel concept for the convolution in the FPGA with a so-called 'History RAM' was introduced for impulse responses with a long power delay profile in [KSKL05]. The emulation concept is scalable up to 64 channels, to accommodate, for example, an 8×8 MIMO system.

Signal processing hardware is also included for the TX and the RX, which allows for rapid prototyping of MIMO algorithms. For demonstration and evaluation purposes, a smart antenna receiver was implemented and results were reported in [KSKL05].

6.4 Antenna configurations

MIMO is associated with exploiting the spatial domain of the propagation channel and involves sophisticated signal processing. The employment of antenna arrays at both sides of the link is essential for MIMO, hence antennas are an important aspect. Finding feasible antenna configurations is an integral part of enabling the MIMO technology. In this section, different antenna configurations for MIMO are investigated.

MIMO systems exploit the multipath structure of the propagation channel. The antennas are adapted to the propagation channel. Both the antenna arrays and the propagation channel should be treated together and described statistically to take many channel realisations of a propagation environment into account. Correlations among channel coefficients are influenced by the antenna properties. As the antennas are collocated in a MIMO array, mutual coupling effects may occur. All these effects should be considered when designing an antenna array for MIMO systems.

For the evaluation and comparison of different antenna concepts for MIMO systems quality measures are needed. In Subsection 6.4.1, an overview over these specific measures is presented. Different antenna configurations are compared. It becomes clear that power effects should be taken into account, hence power effects are discussed in Subsection 6.4.2. The effects of polarisation diversity and mutual coupling among the antenna elements are addressed. The integration of MIMO antenna arrays into small handsets is a major challenge, because usually a large antenna spacing is needed for the exploitation of the spatial properties of the propagation channel. In Subsection 6.4.3 several examples are given demonstrating how compact MIMO antenna arrays can be integrated into handsets. Finally, a summary is given and future prospects are pointed out.

6.4.1 Quality measures for MIMO antennas

For the development of MIMO antenna arrays some quality measure is needed in order to draw comparisons between different antenna arrays. Obviously, there are several antenna properties which can be assessed by classical antenna measures, e.g. radiation pattern, antenna gain, self and mutual coupling impedances, half-power beamwidth, bandwidth, frequency range, resonant frequency. If an antenna array is employed in a MIMO system, some more specific measures arise. The propagation channel has to be taken into account. The quality measures are statistical measures, which describe the antenna array performance for specific propagation channels. In the following an overview of MIMO-specific quality measures for antenna arrays is given.

As MIMO systems are known for their capacity enhancement for future mobile communications, the instantaneous channel capacity is the most important measure for the evaluation of MIMO systems. It is addressed in the following paragraph. The capacity is influenced by several factors introduced in subsequent paragraphs.

Capacity

The channel capacity depends on the number of transmit and receive antennas n and m, the channel matrix H and the Signal-to-Noise Ratio (SNR). For the sake of simplicity, but without consequences for the quality measures, only MIMO systems without channel state information at the transmitter are considered in the following. With this assumption, the transmit power is equally (and not optimally, see [Ande00]) spread among the transmit antennas. The instantaneous channel capacity of a MIMO system in the presence of spatially uncorrelated Gaussian distributed noise can be calculated by

$$C = \log_2(\det(I + \frac{\text{SNR}}{n} H_F H_F^{\dagger})) \tag{6.13}$$

where I is the identity matrix and $(\cdot)^{\dagger}$ denotes the complex conjugate transpose. To investigate the influence of the correlation properties on the capacity, the channel matrix is often normalised, so that it is independent of the channel attenuation. The capacity is expressed as a function of the SNR at the receiver, as shown by equation (6.13). H_F is the channel matrix, which is normalised with the Frobenius matrix norm, so that the mean attenuation of each channel matrix trace$(HH^\dagger)/nm$ is equal to one. The channel attenuation, which is included in the channel matrix, has to be expressed in the SNR when normalising the channel matrix.

The attenuation is influenced by the antennas and the radio channel. With this normalisation, the influence of the correlation properties on the capacity becomes visible, but any interrelation between the SNR and the correlation properties of H is neglected. As the antennas have an impact on both the SNR and the correlation properties, this normalisation does not predict the behaviour of different antenna arrays properly. In real systems, the effect of antennas is included in the channel matrix. The normalisation of the channel matrix with the Frobenius norm eliminates some of the antenna effects. Therefore another approach is needed to preserve antenna effects allowing for a fair comparison of MIMO antenna arrays.

If H is not normalised, that means the path loss and the gain of the single antenna elements are included in H, (6.13) can be written as

$$C = \log_2(\det(I + \frac{P_T}{\sigma^2 n} HH^\dagger)) \tag{6.14}$$

Equation (6.14) expresses the capacity as a function of the transmit power P_T. The attenuation of the transmission link, influenced by the antennas and the radio channel, is taken into account. σ^2 is the noise power. This formula allows for a comparison of different MIMO systems, including the influence of the transmission gain and therewith of the SNR. Without normalisation the capacity is an appropriate measure for the comparison of MIMO systems employing different antenna arrays.

In Figure 6.13 the capacity calculated using equation (6.14) is shown for several antenna configurations. For the propagation channel, a path-based indoor channel model was used. The capacity is shown with respect to the total size of an array, which is indicated in grey.

It is remarkable that for little array sizes an array with two elements outperforms arrays with three or four elements. Systems employing polarisation diversity are robust against polarisation mismatching, which is demonstrated by a rotation of the transmit array relative to the receive array. With quality measures the influences of an antenna configuration on the capacity are revealed by investigating the effects causing the behaviour. There are two main properties that determine the capacity of a MIMO system. These are the correlation properties and the power level or efficiency in terms of power. Hence measures are needed for the correlation properties and measures for the efficiency in terms of power.

Correlation

The correlation properties of H influence the capacity. The number of correlation coefficients between all elements h_{ij} in H is n^2m^2, thus it is difficult to assess the correlation properties. It is also difficult to show the direct relationship between the capacity distribution and the correlation properties. In [JoBo03] a measure to describe the correlation among all elements h_{ij} is defined, and it is shown that the ergodic capacity of a MIMO system without channel state information increases with decreasing correlation. A simple way to assess whether the correlation is high or low is to consider only the transmit and receive correlation. The complex transmit and receive correlation

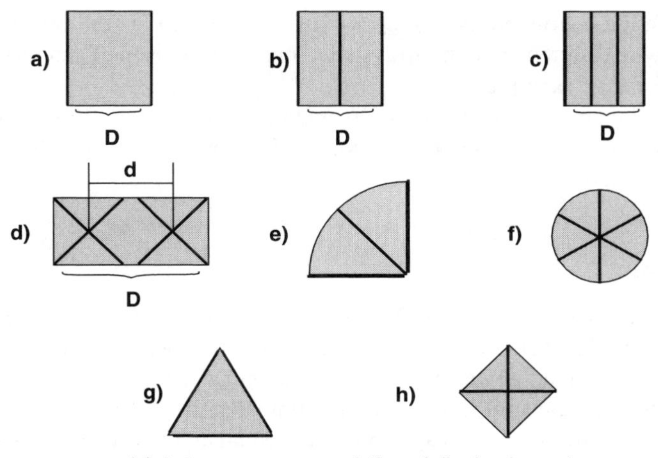

(a) Antenna arrays consisting of dipole elements.

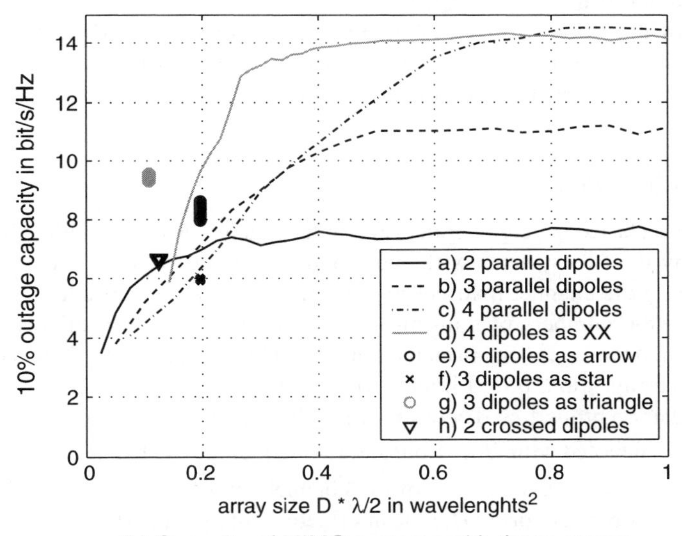

(b) Capacity of MIMO systems with these arrays.

Figure 6.13 *Arrangement of dipoles and their MIMO capacity [WaSW04]. For the systems exploiting polarisation diversity [(e)–(h)] the transmit array is rotated between 0° and 180° relative to the receive array. For the configurations [(a)–(d)], the total array size is changed. (© 2004 IEEE, reproduced with permission)*

coefficients of two zero-mean elements h are defined as

$$\rho_{Tx} = \frac{E\{h_{ki}h^*_{kj}\}}{\sqrt{E\{|h_{ki}|^2\}E\{|h_{kj}|^2\}}} \tag{6.15}$$

$$\rho_{Rx} = \frac{E\{h_{ik}h^*_{jk}\}}{\sqrt{E\{|h_{ik}|^2\}E\{|h_{jk}|^2\}}} \tag{6.16}$$

thus only the correlation among signals transmitted or received from different antennas is considered.

The power correlation coefficient is $\rho_{PTx/Rx} = |\rho_{Tx/Rx}|^2$, given in [PiSt60]. It can be calculated for MIMO systems with different antenna arrays enabling a comparison of antenna arrays for MIMO, see Subsection 6.4.3 for an example.

Correlations also play an important role in diversity systems with multiple antennas at one side of the link only. The investigations and conclusions in Section 5.3 on diversity techniques apply to MIMO antenna arrays, too.

Mean Effective Gain

To assess single antennas in an array the Mean Effective Gain (MEG) can be used, which was introduced in [AnHa77]. The MEG is defined as the ratio of the mean received power of an antenna under test to the mean received power of a reference antenna, when both antennas are used in the same channel with the same transmit antenna. To ascertain that the MIMO system works properly, all antennas need to have an MEG which is approximately equal and as high as possible. If the antenna elements in a MIMO array do not have similar mean effective gains a branch power imbalance may occur deteriorating the system performance.

The definition of the MEG can be extended to assess arrays. The Mean Effective Array Gain (MEAG) is the ratio of the mean received power of an array to the mean received power of a reference antenna in the same channel with the same transmit antenna.

Mean Effective Link Gain

In theoretical analysis based on, e.g. identical and independently Rayleigh fading branches, the mean powers received by the two systems with equal number of antennas are usually normalised to be equal. In such a case the channel matrices are usually normalised according to $E\{||H||_F^2\} = mn$, where m and n are the numbers of transmit and receive antennas, respectively. In the context of antenna comparison, however, such normalisation would ignore the effect of the radiation patterns of antennas as well as array gain if power imbalance occurs between the powers of the antennas caused by the different radiation properties. Therefore we shall normalise the received power with respect to a common reference, denoted with H_{ref}. Consider two sequences of channel matrices, say $\{H_{aut}^{(i)}\}$ and $\{H_{ref}^{(i)}\}$, $i = 1 \ldots N_s$. The received power of the channel can be defined $P = 1/N_s \sum_{i=1}^{N_s} ||H^{(i)}||_F^2$, where $||\cdot||_F$ is the Frobenius norm. The mean effective link gain (MELG) [SSKV04] is a sample mean power over antenna-system-under-test divided by a sample mean power over reference antenna system by

$$G_{e,MIMO} = \frac{P_{aut}}{P_{ref}} = \frac{\frac{1}{N_s}\sum_{i=1}^{N_s} ||H_{aut}^{(i)}||_F^2}{\frac{1}{N_s}\sum_{i=1}^{N_s} ||H_{ref}^{(i)}||_F^2} \tag{6.17}$$

Unlike MEG in SISO systems, MELG defines the average *link* gain of the system. In other words, the MIMO antenna comparison generalises the SISO case to both ends of the link. MELG does not pose any restrictions on array geometry nor does it require equal-power antenna branches, and, in this sense, generalises the notion of MIMO array gain for arbitrary MIMO antenna configurations.

Based on (6.17), the normalised mutual information for the antenna-under-test becomes

$$C_{aut}^{(i)} = \log_2 \left| I + \frac{\rho}{n} G_{e,MIMO} nm \frac{H_{aut}^{(i)} H_{aut}^{(i)H}}{P_{aut}} \right| \tag{6.18}$$

The expression is intuitively appealing, since the SNR of the reference antenna configuration is just ρ, since $G_{e,MIMO} = 0$ dB. The MELG of the test antenna system directly modifies the SNR at which the mutual information is computed. The definition (6.18) is general and makes sense with arbitrary MIMO antenna configurations, e.g. ones where the antennas have different look directions, which is a common situation with MTs.

Power transmission gain

Mutual coupling among closely spaced antennas does not only influence the signal flow and the correlation properties, it can also strongly reduce the efficiency in terms of power of an array which is undesirable since most MTs are battery driven. Due to the fact that the power efficiency of an array depends on its excitation, it is not reasonable to use the power efficiency as a quality measure for arrays in MIMO systems.

The system model, given in [WaSW04], is capable of considering the power transmission gain of the MIMO link. The power transmission gain is a measure for the whole MIMO link including the antenna arrays at the transmitter, the propagation channel, and the antenna arrays at the receiver. The power transmission gain is the ratio of the power received at the signal drain to the power fed into the transmit antennas. The latter is not equal to the power which is radiated from the transmit antennas, if the efficiency of the transmit array is not 100%. Conclusions on the performance of the arrays in terms of power can be drawn by comparing the power transmission gain of MIMO systems with different arrays in the same channel.

In Figure 6.14 the CDF of the power transmission gain is shown for several MIMO systems with two parallel dipole antennas on each side of the link in an indoor channel, simulated using a path-based channel model. When decreasing antenna spacings, the effective gain of the antennas decreases, which reduces the power gain of the transmission channel. For extremely small spacings, the mean power gain of the 2×2 MIMO system can be worse than that of a SISO system due to mutual coupling effects. However, for low outage probabilities, MIMO always outperforms SISO.

Other quality measures

Another important quality measure for an antenna array in a certain channel is the effective diversity order. The effective diversity order combines the measured correlation and effective gain, see [NoTV01]. The diversity antenna gain introduced in [TaOg01] is also a measure that expresses the performance of diversity antennas combining correlation and MEG. These measures focus on diversity at one side of the link only and can thus not be directly applied to complete MIMO systems.

In [RoBK04] the channel capacity measured in a reverberation chamber is introduced as a quality measure for the antenna configurations. The measurements are repeatable allowing for the comparison of different antenna arrays, although the environment does not necessarily correspond to real propagation channels.

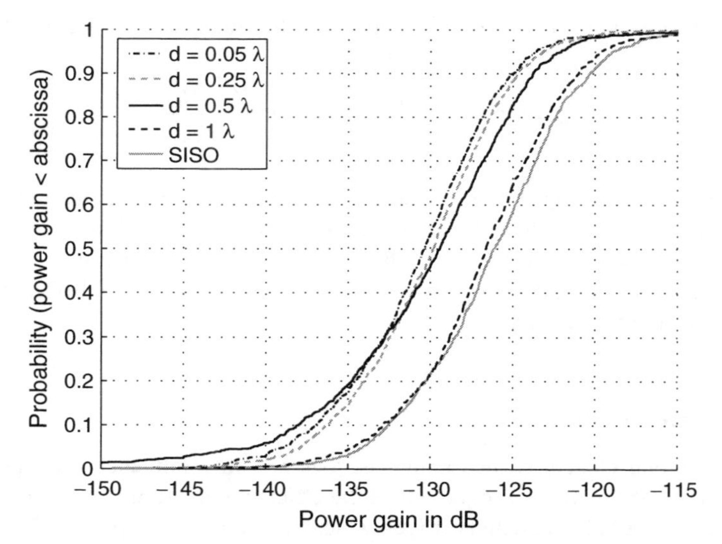

Figure 6.14 *Power gain distributions for different antenna spacings, given in [WaSW04]. The mean attenuation of the physical channel is 132 dB, which is given by the channel model. (© 2004 IEEE, reproduced with permission)*

6.4.2 Power considerations

As mentioned in the previous subsection, the antennas have an impact on the efficiency in terms of power. The mean effective gain, mean effective link gain, and power transmission gain are figures of merit for the efficiency of MIMO antenna arrays. The antenna configuration has a significant influence on the total received power [SSKV04], [WaSW04].

Mutual coupling and radiation efficiency

Mutual coupling effects among the antenna elements in a MIMO array have an influence on power and capacity. For diversity systems with several antennas on one side of the link only, an analysis of mutual coupling effects is given in Section 5.3.

The role of mutual coupling in MIMO is still under investigation. Seemingly conflicting reports on mutual coupling between antenna elements increasing [SvRa01] or reducing [WaSW04] ergodic capacity might not be conflicting at all. Both effects are, in principle, plausible. On one hand, mutual coupling changes the individual antenna patterns. This creates diversity as each antenna 'sees' different portions of the surrounding scatterers (pattern diversity). But the effect seems to be small unless the antennas are located very close to each other. On the other hand, mutual coupling may, by re-radiation of received power, result in higher spatial correlation between antenna signals which is a possible cause for reduced capacity. Let us be clear, however: correlation is not solely caused by mutual coupling and depends on a number of other factors as well. Antennas in close proximity to each other and to lossy material, like human tissue, also suffer from reduced radiation efficiency. This is an effect intricate to measure, but it becomes extremely important if we want to compare the performance of an antenna array with that of a single antenna.

One way to tackle this problem is the method of so-called 'embedded patterns' [KiRo04], which involves full-wave electromagnetic computation of actual antennas and the creation of a well-defined multipath environment. Another possibility is the description of the MIMO communication link including transmit antennas, propagation channel, and receive antennas by scattering parameters, see [WaSW04]. Current findings indicate that spatial correlation has little effect as compared to radiation efficiency.

Impact of polarisation diversity on power and capacity

One important aspect of an antenna configuration is polarisation. Using dual-polarised antennas the correlation can be reduced leading to an improved capacity [WKSW03]. The impact of polarisation diversity on the correlation has been discussed in Section 5.3 on diversity techniques in more detail. The power level is decreased by dual polarisation resulting in a decreased capacity [SSKV04]. As an example, the effect of polarisation on 2×2 MIMO antenna arrays is studied in the following.

The iid capacity [FoGa98] holds for Rayleigh fading channels with zero correlation and for single polarised systems only. With orthogonal polarisations, the channel matrix can be formulated as [ESBP02]

$$H = \sqrt{\frac{K}{K+1}} \begin{pmatrix} e^{j\phi_{11}} & \alpha e^{j\phi_{12}} \\ \alpha e^{j\phi_{21}} & e^{j\phi_{22}} \end{pmatrix} + \sqrt{\frac{1}{K+1}} \begin{pmatrix} X_{11} & \alpha X_{12} \\ \alpha X_{21} & X_{22} \end{pmatrix} \tag{6.19}$$

where K is the Rice factor, α is the cross-polarisation of the channel, X terms are the complex Gaussian random variables, $e^{j\phi}$ terms correspond to the LoS components.

It is shown in [SSKV04] that the normalisation $E\{||H||_F^2\} = mn$ cannot be applied with arbitrary polarisations, because the effect of polarisation on MELG is then ignored. A VH-VH polarised system is used as an example of a dual polarised MIMO system. This can equally be realised, e.g. with opposite handed circular polarisations.

The theoretical analysis gives the following results for 2×2 MIMO systems:

1. In the LoS case ($\alpha = 0$, $K = \infty$), the MELG of the single polarised case is 3 dB higher than the VH-VH polarised case, when the transmitter does not know the channel.
2. If the transmitter knows the channel, the difference is 6 dB, which equals the combined array gain of the system.
3. For high SNR ($\rho > 4$ in case 1 and $\rho > 12$ in case 2) the Shannon capacity of the dual polarised case is higher due to the two eigenvalues.
4. In the Rayleigh case ($\alpha = 0$, $K = 0$) the VH-VH polarised case has smaller eigenvalue spread (4 dB smaller difference in CDF 0.1 point). However, the capacity of a single polarised case is significantly (more than 1 b/s/Hz) higher with low SNR ($\rho = 10$). This is due to the better MELG and better diversity order.

An empirical study was made with channel sounder data [KSPH01], [SSVK03], [SSKV03b] with dual polarised antennas. Using normalisation similar to 6.17, it is observed that the single polarised case has slightly higher capacity with 10 dB SNR (TX does not know the channel).

Here it was assumed that the single polarised case has almost perfect polarisation match. This is not true for handheld devices which can be held in arbitrary position. Employing polarisation diversity becomes important when considering handsets. Due to the random orientation of the handset, a polarisation mismatch can occur. This effect can be overcome by polarisation diversity.

Influence of user's head and hand

The user employing the MIMO terminal has an impact on the MIMO performance as well. As already discussed in Section 5.2 on antenna performance assessment of mobile handsets, every antenna integrated in an MT is strongly influenced by the environment. The power transmission gain is decreased by the influence of the user. Some of the power which should be radiated is absorbed by the human tissue or lost due to mismatching, thus the mean effective gain decreases. The user also has an impact on the balance of the mean effective gains in an array. An imbalance between the mean effective gains occurs causing mean branch power differences. This imbalance deteriorates the MIMO performance of an array. The human tissue blocks parts of the incoming fields, therefore the antenna patterns are deformed. Differences in antenna patterns lead to increased pattern and polarisation diversity. The signals at the antennas become decorrelated due to diversity. Further investigations addressing the influence of the user can be found in [KoPO02], [WKSW04].

6.4.3 *Integration of MIMO antennas into handsets*

For the integration of antennas for MIMO into handsets very compact antenna arrays are needed. The combination of different diversity techniques such as spatial, pattern, and polarisation diversity leads to capable solutions, as shown in [WaSW04]. In the following, the integration of MIMO antenna arrays into handsets is demonstrated by several examples.

Handset with three Inverted-F antennas

The first example handset models a portable device such as a mobile phone. It is equipped with three Inverted-F antennas operating at a frequency of 2 GHz. The aim of the antenna configuration was to combine different diversity technique, i.e. pattern and spatial diversity. To overcome polarisation mismatching effects polarisation diversity is exploited. The simulation model of the small handset consists of a metallic block, representing the battery and the display of the device, and a PVC housing with a wall thickness of 2 mm, see Figure 6.15. The size of the housing is $55 \times 115 \times 27\,\mathrm{mm}^3$. The metallic block is $40 \times 80 \times 10\,\mathrm{mm}^3$ large. Three antennas were mounted onto the metallic block spatially separated and with different orientations to exploit different diversity techniques. Usually Inverted-F antennas require an infinite ground plane, which is not given in the small handset. Thus the metallic block, representing the ground plane, acts as a part of the antennas and influences the shape of the patterns and the mutual coupling impedances. A picture of the handset is shown in Figure 6.16.

The whole device was simulated with a standard EM code based on method of moments. It calculates the pattern of the coupled antenna system as well as the mutual coupling and self-impedances of the antennas, which serve as an input for the model of the MIMO transmission link given in [WaSW04]. On the other side of the link an antenna array set-up, consisting of three half wavelength dipole antennas, which were arranged in a triangle, was used. The channel model is a stochastic, full polarimetric, three-dimensional and path-based indoor channel model.

For a comparison, Table 6.4 presents the quality measures for the MIMO system mentioned above and a much larger reference MIMO system with three vertical $\lambda/2$ dipoles with $\lambda/2$ spacings on each side of the link. Since the correlation is very low for both arrays, the capacity for a constant SNR is equal. But the capacity for a constant transmit power is different, due to the fact that the MEG and the transmission gain are worse for the three small Inverted F-antennas, due to mutual coupling

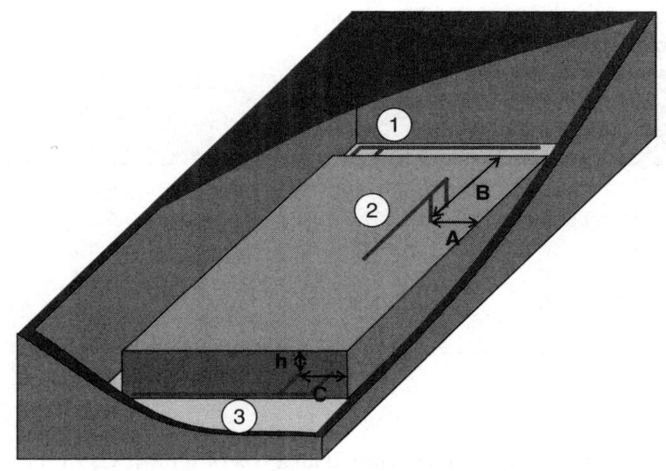

Figure 6.15 *Model of a handset with 3 Inverted-F antennas, [WKSW04]. The positioning parameters are chosen as: A = 13 mm, B = 13 mm, C = 13 mm, h = 5 mm.*

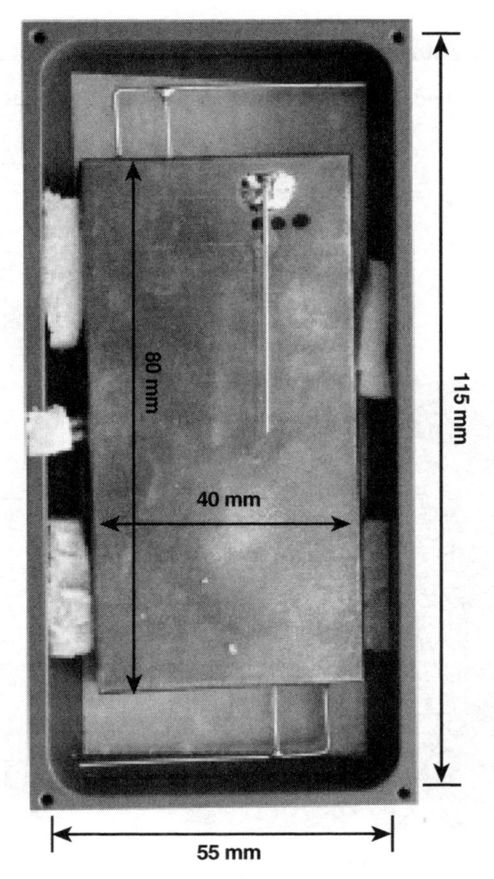

Figure 6.16 *Photograph of the handset with 3 Inverted-F antennas.*

Table 6.4 *Quality measures for three IFAs in the handset compared to three parallel dipoles with half-wavelength spacing, taken from [WaWi04].*

	10% out. Cap. (SNR = 10 dB)	10% out. Cap. (P_T = const.)	Max. power correlation coefficient	MEG for each antenna	50% transmission gain
Reference (3 parallel dipoles)	7.5 bit/s/Hz	11.2 bit/s/Hz	0.1	−0.6 dB −0.6 dB −0.6 dB	−116.4 dB
Handset (3 Inv.-F antennas)	7.5 bit/s/Hz	10.6 bit/s/Hz	0.1	−2.4 dB −2.4 dB −2.5 dB	−121.2 dB

and polarisation mismatching effects. The detailed analysis of the MEGs of the single antennas has allowed for developing and optimising the small array. Further examples and detailed results can be found in [WKSW04].

Handset with four patch antennas

In [KoPO02] a model of a handset is introduced employing four patch antennas at 2140 MHz, Figure 6.17.

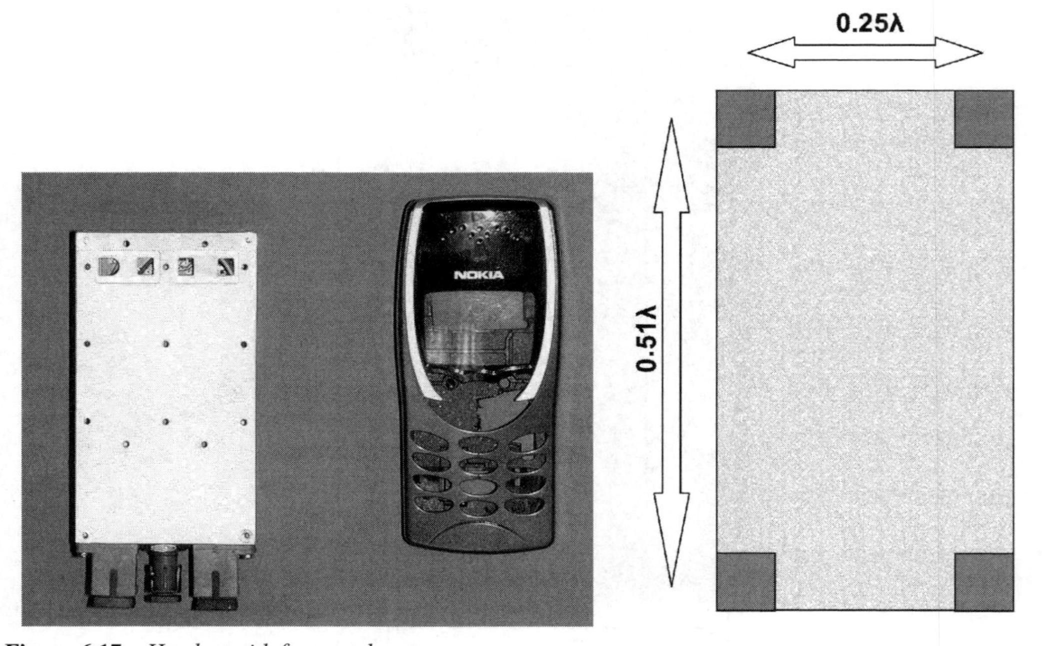

Figure 6.17 *Handset with four patch antennas.*

The handset has the size of a commercial small GSM phone. The four squared patch antennas are located at the corners of the faceplate. The separation between two adjacent patches equals 0.25λ across and 0.51λ lengthwise, respectively.

Outdoor-to-indoor measurements were performed using this handset as the mobile station. Three BSs were distributed surrounding the mobile station at distances between 60 m and 100 m. While moving the mobile station on a straight line 15 m along a corridor, 512 complex channel responses were measured for each of the 12 channels.

The capacity was calculated for every frequency component using formula (6.13) with an assumed SNR of 14 dB. Figure 6.18 shows the bounding curves (the envelope) of all capacity curves for all frequency components. Further results are given in [KoPO02]. It is shown, that integrating four antennas for MIMO in a small handset still makes sense. The channel capacity can be excellent, comparable with fully uncorrelated Rayleigh fading branch signals, while on average the ideal uncorrelated Rayleigh fading case cannot be reached.

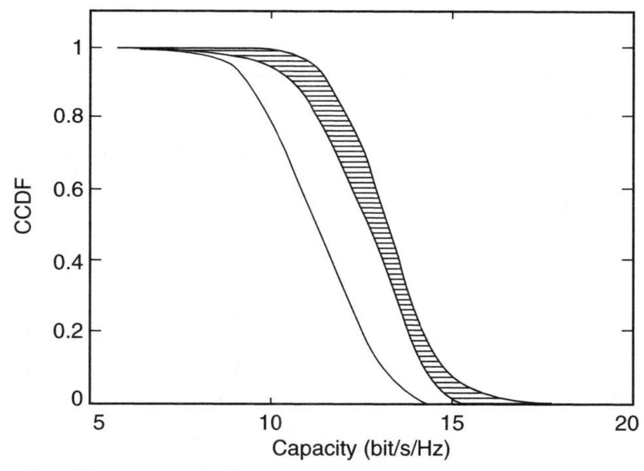

Figure 6.18 *Envelope Complementary Cumulative Distribution Function (CCDF) of the measured capacity with four patch antennas on a handset as the mobile station with test persons as users (open curve) and free-in-air (shaded curve), [KoPO02]. (© 2002 IEEE, reproduced with permission)*

PDAs with four antennas

Three different approaches for the integration of MIMO antennas into handsets have been compared at a frequency of 5.2 GHz in [BHWH04]. The different designs of a four-element antenna array to mount on the surface of a case for a Personal Digital Assistant (PDA) have been considered. Each design uses the same type of element throughout (cavity-backed linear slots, planar Inverted-F or dielectric resonator antenna), although element orientations are different in each case. The size of the case equals $63 \times 113 \times 14\ \text{mm}^3$. The PDA designs are shown in Figure 6.19.

MIMO channel measurements of the three PDAs were conducted in a laboratory environment in a peer-to-peer communications scenario (i.e. PDAs were used at both ends of the link). The transmitter was fixed at one location while the receiver was moved to 21 different sites. The results for the capacity for a fixed SNR of 20 dB are shown in Figure 6.20 as well as for the theoretical Rayleigh case. Further results are given in [BHWH04] and [RoHi04].

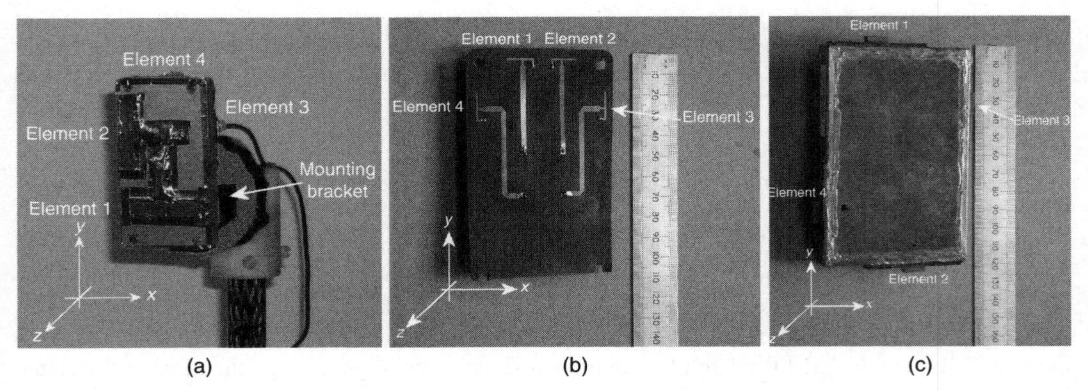

Figure 6.19 *Photograph of the candidate PDA designs, given in [BHWH04].*

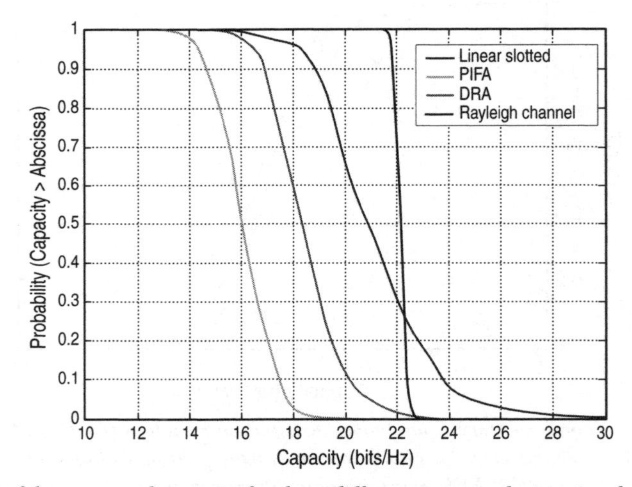

Figure 6.20 *CCDF of the measured capacity for three different antenna designs in a laboratory environment, given in [BHWH04].*

The results indicate that the PDA equipped with linear slot antennas provided the highest overall capacities similar to the ideal Rayleigh case. Again it is shown that MIMO is still feasible even in small handsets.

6.4.4 Summary and future topics

In this subsection, antenna configurations for MIMO have been discussed treating both antennas and the propagation channel at once. Comparisons have been drawn using several quality measures for MIMO antenna arrays. The efficiency in terms of power has to be taken into account to assess arrays for MIMO. Several effects have an impact on the power transmission gain and on the MIMO performance, e.g. the polarisation of the antennas, the user of the MT, and mutual coupling effects among the antenna elements. The challenging task of integrating several antennas for MIMO in handsets has been performed demonstrating that MIMO systems are realisable in compact configurations.

The potential of MIMO systems brings along several topics concerning antenna configuration. The questions arising were only treated to a certain extent by fundamental research. Future research may deal with the realisation of feasible antenna arrays for MIMO in combination with the RF front-end, application-specific antenna arrays for MIMO as well as further MIMO antenna design criteria.

6.5 Analytical MIMO channel models

Physical channel models have been described in Section 6.3. Yet, for system design and simulations, analytical models might be preferred, as they can reproduce the essential characteristics of a variety of channels without being too complicated.

A first possibility is to model the space-time channel without considering the antenna geometries, i.e. describing the multipath channel in both the temporal (or Doppler frequency) domain and the spatial domain. In [LoBF01], [LoFK02], the classical Wide-Sense-Stationary Uncorrelated Scattering (WSSUS) channel formalism is extended to the spatial dimension. Applying a space-time analogy, the signal in the spatial domain is described through replacing time and frequency by spatial location and spatial frequency, respectively. Generalised impulse response and transfer function can then be defined over delay, time and space. Given knowledge of these, discrete sampling filter models are derived [LoFK02] that can be implemented in channel emulators with space-time variable gain multipliers. These multipliers have random properties similar to those of small-scale fading, and Gaussian amplitude resulting from the sum of many multipath components. Integrating this space-time channel model with array geometries finally yields a model of the MIMO channel matrix.

As an alternative, analytical models presented below aim at describing directly the MIMO channel matrix, rather than describing the spatial channel apart from array geometry. In most cases, these models are built upon correlation properties or a decomposition into steering-/eigenvectors, in combination with an iid random matrix with unity-variance, circularly symmetric complex Gaussian entries (in order to obtain different fading realisations).

In this section, analytical MIMO models will be categorised into two classes: correlation-based models, described by means of the channel covariance matrix, and coupling-based models, described by means of a specific coupling matrix and steering-/eigenvectors. Naturally, other classifications are possible, based upon the origin of the model rather than upon its formal description. As a consequence, some coupling-based models can also be considered as correlation based, and vice versa.

6.5.1 Correlation-based models

Correlation-based models describe the (complex) correlation properties between all pairs of receive and transmit antennas. Since the correlations depend on the actual antenna configuration used, these models cannot be generalised easily to other configurations, unless new correlation coefficients are calculated. Yet, these models are particularly useful when analysing the impact of correlations on any performance parameter, since the relationship is explicit in the model.

Full covariance-based model

If zero-mean complex circularly symmetric Gaussian channels are considered, then the covariance matrix (which is equivalent to the correlation matrix in this case) is a sufficient description of the statistical behaviour of the MIMO channel,

$$\mathbf{R}_{full} = E\{\mathrm{vec}(\mathbf{H}) \cdot \mathrm{vec}(\mathbf{H})^H\} \tag{6.20}$$

where $\mathrm{vec}(\cdot)$ stacks the columns of $\mathbf{H}$ into a large vector, $(\cdot)^H$ stands for Hermitian transposition and $E\{\cdot\}$ denotes the expectation operator. From (6.20), any channel realisation is obtained by the full covariance matrix model

$$\mathrm{vec}(\mathbf{H}_{full}) = \mathbf{R}_{full}^{1/2} \cdot \mathrm{vec}(\mathbf{H}_w) \tag{6.21}$$

where $\mathbf{H}_w$ is an $n_R \times n_T$ iid random fading matrix with unity-variance, circularly symmetric complex Gaussian entries and $(\cdot)^{1/2}$ denotes the matrix square root. At this stage, it should be remembered that second-order statistics are only a sufficient and valid representation of the channel statistical behaviour when the entries of the channel matrix are zero-mean Gaussian. Modifications of (6.20) for non-Gaussian channel statistics will be discussed as well in the following.

The representation of (6.20) is thus the most general for circularly symmetric Gaussian channels. Yet, it requires the characterisation of the entire covariance matrix, whose number of elements grows rapidly with the array sizes, while not all entries of $\mathbf{R}_{full}$ have a direct physical interpretation. Finally, the vec formulation of (6.21) is not easily tractable. For these reasons, several models have been developed to simplify (6.21) based on a number of assumptions. These assumptions rely either on a separation between transmit and receive correlation properties, or on a decomposition in the beam-/eigenspace.

Kronecker model

The Kronecker model, introduced by [SFGK00], [ChKT98] and also used in the course of the EU-IST SATURN (Smart Antenna Technology in Universal bRoadband wireless Networks) project simplifies the expression of the full covariance matrix by using a separability assumption,

$$\mathbf{R}_{kron} = \frac{1}{\mathrm{tr}\{\mathbf{R}_{\mathrm{RX}}\}} \mathbf{R}_{\mathrm{TX}} \otimes \mathbf{R}_{\mathrm{RX}} \tag{6.22}$$

where $\otimes$ designates the Kronecker product, $\mathrm{tr}\{\cdot\}$ stands for the trace of a matrix, $\mathbf{R}_{\mathrm{TX}} = E\{(\mathbf{H}^H\mathbf{H})^T\}$ denotes the transmit correlation matrix ($n_T \times n_T$) and $\mathbf{R}_{\mathrm{RX}} = E\{\mathbf{H}\mathbf{H}^H\}$, the receive correlation matrix ($n_R \times n_R$).

Inserting (6.22) into (6.21), the Kronecker channel matrix reads as

$$\mathbf{H}_{kron} = \frac{1}{\sqrt{\mathrm{tr}\{\mathbf{R}_{\mathrm{RX}}\}}} \mathbf{R}_{\mathrm{RX}}^{1/2} \mathbf{H}_w (\mathbf{R}_{\mathrm{TX}}^{1/2})^T \tag{6.23}$$

where $(\cdot)^T$ denotes transposition.

Beside simplified analytical treatment or simulation of MIMO systems, (6.23) allows for independent array optimisation at TX and RX, which is one of the reasons why this model is popular. However, what must be understood is that (6.22) implies the statistical independence between DoD and DoA. Hence, the main drawback of the model lies in its main assumption itself, i.e. it only models separable joint angular power spectra [BöHW03]. Thus, all transmit eigenvectors couple into all receive eigenvectors, and vice versa, with the same profile. This statement also holds true as far as TX and RX directions are concerned.

Diagonally correlated channels

Some particular class of channels cannot be represented by (6.23). As an example, the so-called diagonally correlated channels are obtained when antenna correlations (at both RX and TX) and selected cross- or diagonal correlations (i.e. correlations other than antenna correlations) in $\mathbf{R}_{full}$ are equal to zero, while the remaining correlations have unitary amplitudes. In the case of a 2×2 MIMO Rayleigh fading channel, the absolute value of the full correlation matrix reads thus as

$$|\mathbf{R}_{full,2\times 2}| = \begin{bmatrix} 1 & 0 & 0 & 1 \\ 0 & 1 & 1 & 0 \\ 1 & 0 & 0 & 1 \end{bmatrix} \tag{6.24}$$

where $|\cdot|$ designates the element-wise absolute value of a matrix.

Such channels are shown in [ÖzOe05] to present interesting properties, such as maximising the channel ergodic capacity (above the capacity of iid channels). It is indeed shown in [ÖzOe05] that the ergodic capacity of an $n \times n$ diagonal channel is exactly given by

$$\bar{C} = n \log_2(e) e^{\frac{1}{\rho}} \mathrm{E}_1\left(\frac{1}{\rho}\right) \tag{6.25}$$

where $\mathrm{E}_1(z)$ is the En-Function for $n = 1$ that satisfies $\mathrm{E}_1(z) = \int_1^\infty \frac{e^{-tz}}{t} dt$. In (6.25), the capacity grows exactly linearly with the number of antennas while the capacity of iid channels grows in n only asymptotically.

6.5.2 Coupling-based models

Finite scatterer model

The fundamental assumption of the finite scatterer model [StMB01], [Burr03] is that the transmitter and receiver are coupled via a finite number of paths n_S resulting from discrete scatterers, so that the channel matrix can be written as

$$\mathbf{H}_{finite} = \mathbf{A}_{\mathrm{RX}} \left(\tilde{\Omega}_{finite} \odot \mathbf{H}_w\right) \mathbf{A}_{\mathrm{TX}}^T \tag{6.26}$$

where the operator $\odot$ designates the element-wise Schur-Hadamard multiplication. Matrices $\mathbf{A}_{\mathrm{TX}}$ ($n_T \times n_S$) and $\mathbf{A}_{\mathrm{RX}}$ ($n_R \times n_S$) represent matrices whose columns are steering vectors related to each individual scatterer, and $\tilde{\Omega}_{finite}$ is an ($n_S \times n_S$) coupling matrix whose elements are the independent complex path gains. Note that, except for a scalar scaling, the steering vectors are deterministic in any given stationary scenario. They are fully defined by the DoD/DoA of the radio wave, the array geometry, and the element patterns. However, the DoD/DoA, as well as the number of paths, can be chosen arbitrarily when defining the scenario. It must also be pointed out that this model is linear in the path gains, but not in the DoDs and DoAs. This model is compatible with the single-scattering assumption, and in that case, $\mathbf{A}_{\mathrm{TX}}$ and $\mathbf{A}_{\mathrm{RX}}$ can always be rearranged such that $\tilde{\Omega}_{finite}$ is diagonal. Regarding the application of (6.26) to multiple scattering, two definitions exist. The original version of the approach treats multiple scattering via multiple entries in the respective row/column of the coupling matrix. The approach used in COST 273 [Burr04a], [Burr04b] deals with multiple scattering by considering steering vectors with identical DoDs and/or DoAs. In that case, $\tilde{\Omega}_{finite}$ is again diagonal for adequate DoD/DoA ordering.

The correlation matrix of a MIMO channel as defined by (6.20) is derived for the finite scatterer model in [Burr04a], [Burr04b], together with its relationship to the Kronecker assumption. There, it is shown that when averaging over different fading realisations (i.e. with fixed DoDs and DoAs), the finite scatterer model does not show a Kronecker structure. This is actually the consequence of the one-to-one coupling between DoDs and DoAs, which is precisely the contrary of the Kronecker separability assumption. This holds true even for multiple scattering, which can indeed be accounted for by the finite scatterer model, as detailed in [Burr03]. Yet, unless there is a very large number of scatterers with multiple coupling between DoAs and DoDs, the finite scatterer model will not present a separable structure.

Virtual channel representation

The virtual channel representation models the MIMO channel in the beamspace with predefined steering vectors [Saye02]. It can be expressed as

$$\mathbf{H}_{virt} = \tilde{\mathbf{A}}_{\mathrm{RX}} \left(\tilde{\Omega}_{virt} \odot \mathbf{H}_w \right) \tilde{\mathbf{A}}_{\mathrm{TX}}^T \tag{6.27}$$

with *unitary* response and steering matrices $\tilde{\mathbf{A}}_{\mathrm{TX}}$ $(n_T \times n_T)$ and $\tilde{\mathbf{A}}_{\mathrm{RX}}$ $(n_R \times n_R)$. The $n_R \times n_T$ matrix $\tilde{\Omega}_{virt}$ is defined as the element-wise square root of the power coupling matrix Ω_{virt}, whose positive and real-valued elements $\omega_{virt,ij}$ determine the average power-coupling between the ith virtual transmit angle and the jth virtual receive angle. By contrast to the finite scatterer model of [Burr03], $\tilde{\Omega}_{virt}$ is in general non-diagonal.

Note that the virtual channel representation is linear in the virtual directions, and is characterised by the coupling matrix only. There is a degree of freedom in choosing unitary $\tilde{\mathbf{A}}_{\mathrm{TX}}$ and $\tilde{\mathbf{A}}_{\mathrm{RX}}$. In particular, one steering direction can be selected arbitrarily, the remaining directions result from the orthogonality condition. Note that despite its similarities with the finite scatterer model, the angular resolution of the virtual channel model, and hence its accuracy, depends on the number of virtual angles. The latter cannot be chosen arbitrarily but is related to the actual antenna configuration.

Weichselberger model

In contrast to the prior models, the Weichselberger model treats the MIMO channel in the eigenspace instead of the beamspace. The original idea is to relax the separability restriction of the Kronecker model and to allow for any arbitrary coupling between the transmit and receive eigenbasis.

Introducing the eigenvalue decomposition of the receive and transmit correlation matrices

$$\begin{aligned} \mathbf{R}_{\mathrm{RX}} &= \mathbf{U}_{\mathrm{RX}} \mathbf{\Lambda}_{\mathrm{RX}} \mathbf{U}_{\mathrm{RX}}^H, \\ \mathbf{R}_{\mathrm{TX}} &= \mathbf{U}_{\mathrm{TX}} \mathbf{\Lambda}_{\mathrm{TX}} \mathbf{U}_{\mathrm{TX}}^H. \end{aligned} \tag{6.28}$$

.

Using the assumption that all transmit and receive eigenmodes are mutually uncorrelated, Weichselberger [WHÖB03], [Weic03] proposes the following channel model

$$\mathbf{H}_{weichsel} = \mathbf{U}_{\mathrm{RX}} \left(\tilde{\Omega}_{weichsel} \odot \mathbf{H}_w \right) \mathbf{U}_{\mathrm{TX}}^T \tag{6.29}$$

where $\mathbf{H}_w$ is again an $n_R \times n_T$ iid complex Gaussian random fading matrix, and $\tilde{\Omega}_{weichsel}$ $(n_R \times n_T)$ is defined as the element-wise square root of the power coupling matrix $\Omega_{weichsel}$. This time, the positive and real-valued elements $\omega_{weichsel,ij}$ of the coupling matrix determine the average power coupling between the ith transmit eigenmode and the jth receive eigenmode.

The Weichselberger model mitigates the radical simplification of neglecting the spatial structure of the MIMO channel and describing the MIMO channel by separated link ends by means of the coupling matrix. Hence, the correlation properties at the transmitter and receiver are modelled jointly. Yet, the model includes the Kronecker model as a special case [WHÖB03], [Weic03]. The Weichselberger model parameters are the eigenbasis of receive and transmit correlation matrix and a coupling matrix. It should be mentioned that there is a natural relationship between the eigenvectors and the multipath directions, as outlined in [BuHe03].

Note that this model might be qualified as correlation based as well.

Gesbert model for keyhole channels

All models with complex Gaussian fading statistics, such as those above, are not capable of reproducing the keyhole effect, where each instantaneous channel matrix is rank deficient even though the channel shows low antenna correlation on average. It was predicted theoretically in [Sibi01], [CLWV03], [GeBP02], and demonstrated experimentally in [AlTM03b].

Keyhole channels can be modelled by replacing the iid complex Gaussian fading matrix with a rank-deficient matrix,

$$\mathbf{H}_{keyhole} = \mathbf{H}_{w,\mathrm{RX}}\tilde{\Omega}_{keyhole}^{1/2}\mathbf{H}_{w,\mathrm{TX}} \tag{6.30}$$

where $\tilde{\Omega}_{keyhole}$ denotes the scattering correlation matrix, i.e. for the propagation from the scatterers near TX to the scatterers near RX, and the two other matrices are iid random matrices. This has two important consequences: (i) the amplitude statistics are not complex Gaussian any more, hence the second-order characteristics are no longer sufficient, and (ii) the rank of the instantaneous channel matrix is limited by the rank of the scattering correlation matrix, while the marginal correlation matrices at the link ends are not affected. Verification by inspection of MIMO channel coefficient distribution has also been applied to the first measurement of a true 'keyhole' channel [AlTM03b]. There, it is shown that the 'keyhole' channel shows a double-Rayleigh distribution, as expected from (6.30).

6.5.3 Comparing correlation matrices

Section 6.7 describes how the performance of a model (defined here as the closeness of the modelled channel to the measured data, could be assessed by calculating a metric denoted as Φ, equal to the root mean square difference between the capacity CCDFs of the measured and simulated channels normalised to the mean capacity of the measured channels. Based on results in [McBF02a], Table 6.5 suggests that models employing complex correlation parameters perform much better than those based on power correlation.

Results in [McBF02a] also question the general validity of the Kronecker assumption for MIMO systems larger than 2×2.

6.5.4 Model performance comparison for Rayleigh channels

Since the ultimate test of any model is experimental validation, some of the models described above, viz. the Kronecker model, the Weichselberger model, and the virtual channel representation, are compared with measurements, each for 2×2, 4×4 and 8×8 MIMO systems. These investigations

Table 6.5 *Statistics for the percentage RMS difference between measured and simulated capacity CCDFs, Φ, when (a) complex, and (b) power correlation is employed, in different antenna scenarios.*

(a) Complex correlation

TX/RX array relative orientation	Model	Mean of Φ	Variance of Φ
Parallel	Kronecker	2.76	4.36
Parallel	$\mathbf{R_H}$	3.27	4.51
Perpendicular	Kronecker	5.53	7.91
Perpendicular	$\mathbf{R_H}$	2.91	2.77
Random	Kronecker	3.43	2.27
Random	$\mathbf{R_H}$	3.59	5.78

(b) Power correlation

TX/RX array relative orientation	Model	Mean of Φ	Variance of Φ
Parallel	Kronecker	16.0	31.5
Parallel	$\mathbf{R_H}$	16.5	36.4
Perpendicular	Kronecker	16.6	38.6
Perpendicular	$\mathbf{R_H}$	17.7	40.9
Random	Kronecker	11.1	33.9
Random	$\mathbf{R_H}$	14.0	40.9

are based on an extensive 5.2 GHz measurement campaign in the offices of the Institut für Nachrichtentechnik und Hochfrequenztechnik, Technische Universität Wien, Austria [OzHH04]. For each of the considered arrays, the antenna spacing is always fixed at 0.5 wavelength at the transmit array and 0.4 wavelength at the receive array.

For all scenarios, model parameters are first extracted from the respective measurements. Since the investigated channel models assume that the channel is sufficiently described by its second-order moments, measurements used for the evaluations of these models have to fulfil this requirement, too. In particular, only scenarios where the corresponding MIMO matrices follow a multivariate zero-mean complex Gaussian distribution are considered for validation purposes of these models. The rest, especially scenarios that experience Ricean-fading, are excluded.

From the extracted parameters, channel matrices according to the three models are synthesised by Monte Carlo simulations. Different performance metrics introduced in Section 6.7 (the *mutual information*, the *joint angular power spectrum*, and the *Diversity Measure* [IvNo03]) calculated from the resulting synthesised channels are then compared to the metrics extracted directly from the measurements [ÖzBo04].

The model parameters of the Kronecker model, i.e. the single-sided receive ($\mathbf{R}_{\text{RX}}$) and transmit correlation matrix ($\mathbf{R}_{\text{TX}}$), are estimated by

$$\hat{\mathbf{R}}_{\text{RX}} = \frac{1}{N}\sum_{i=1}^{N} \mathbf{H}(i)\mathbf{H}(i)^H$$
$$\hat{\mathbf{R}}_{\text{TX}} = \frac{1}{N}\sum_{i=1}^{N} \left[\mathbf{H}(i)^H\mathbf{H}(i)\right]^T = \frac{1}{N}\sum_{i=1}^{N} \mathbf{H}(i)^T\mathbf{H}(i)^* \tag{6.31}$$

where N is the number of channel realisations and $\mathbf{H}(i)$ denotes the ith channel realisation.

It is noteworthy that these estimators are, because of the noise, in principle biased [Czin04]. However, since the measurement SNR was in a range of 30 to 40 dB, this bias can be neglected.

From taking unitary steering/response matrices $\tilde{\mathbf{A}}_{\mathrm{TX}}$ and $\tilde{\mathbf{A}}_{\mathrm{RX}}$, the estimated coupling matrix of the virtual channel representation $\hat{\Omega}_{virt}$ can be calculated by

$$\hat{\Omega}_{virt} = \frac{1}{N}\sum_{i=1}^{N}\left(\tilde{\mathbf{A}}_{\mathrm{RX}}^{H}\mathbf{H}(i)\tilde{\mathbf{A}}_{\mathrm{TX}}^{*}\right) \odot \left(\tilde{\mathbf{A}}_{\mathrm{RX}}^{T}\mathbf{H}^{*}(i)\tilde{\mathbf{A}}_{\mathrm{TX}}\right) \tag{6.32}$$

Since the unitary steering/response matrices are not unique, the coupling matrix is not unique either. As already mentioned, one degree of freedom remains in choosing the direction of the first virtual transmit/receive angle. An obvious choice is the broadside direction of the antenna array.

Analogously, applying the eigenvalue decomposition of the estimated correlation matrices, the estimated power coupling matrix $\hat{\Omega}_{weichsel}$ of the Weichselberger model can be obtained by

$$\hat{\Omega}_{weichsel} = \frac{1}{N}\sum_{i=1}^{N}\left(\hat{\mathbf{U}}_{\mathrm{RX}}^{H}\mathbf{H}(i)\hat{\mathbf{U}}_{\mathrm{TX}}^{*}\right) \odot \left(\hat{\mathbf{U}}_{\mathrm{RX}}^{T}\mathbf{H}^{*}(i)\hat{\mathbf{U}}_{\mathrm{TX}}\right) \tag{6.33}$$

Using the extracted model parameters from the measurements, channel matrix realisations are synthesised by introducing different fading realisations of the iid complex Gaussian, unity-variance random fading matrix $\mathbf{H}_{w}$. For the different MIMO systems, the number of realisations is chosen as equal to the respective number of measured realisations.

Joint DoD-DoA spectrum

For the directional evaluations, the joint DoD-DoA Angular-Power-Spectrum (APS) is calculated using the *Capon's beamformer*, described in Subsection 6.7.4. The single-sided marginal spectra (DoD and DoA) are calculated by the one-dimensional Capon's beamformer [ÖzBo04].

Figure 6.21 shows the 8×8 Capon spectra evaluated for an example scenario. Since the receive array is limited to 120° field-of-view, the DoA spectrum is plotted only in the range from $-60°$ to $+60°$.

In the measured channel, specific DoDs are clearly linked to specific DoAs such that the joint APS is *not separable* into a product of the DoD and the DoA APS. The Kronecker factorisation, however, models the joint angular spectrum as separable, thus introducing artifact paths lying at the intersections of DoD and DoA spectral peaks [BöHW03]. The resulting APS is the rank-one product of the two marginal spectra.

In contrast to this, the Weichselberger model exposes this assumption to be too restrictive. Nevertheless, the Weichselberger model does not render the multipath structure completely correct either.

The virtual channel representation should, in principle, be able to cope with any arbitrary DoD-DoA coupling. The joint APS shows that it does not. Because of its fixed and predefined steering directions the virtual channel representation is not able to reproduce any measured multipath components lying *between* two fixed steering directions correctly. Instead, it assumes two independently fading multipath components at the two adjacent directions. In the worst case, a multipath component lying between two fixed DoDs and DoAs is modelled by four equal-powered, independently fading multipath components.

The same spectra as Figure 6.21, but for 4×4 and 2×2, have been analysed as well. All three models suffer from the same types of deficiencies, but, due to the reduced spatial resolution, these are differently pronounced.

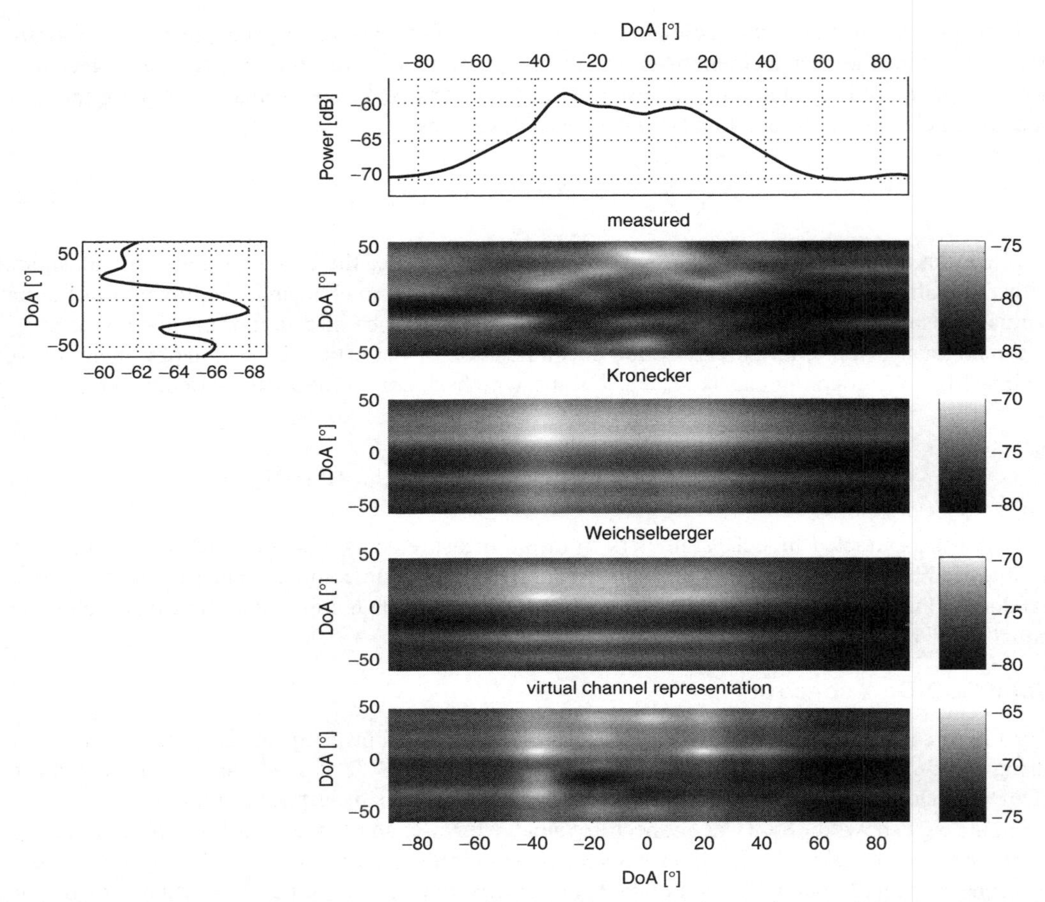

Figure 6.21 *Joint DoD-DoA Capon spectra in dB for 8 × 8 MIMO: the upper right (left) plot depicts the marginal DoD (DoA) spectrum. At the right-hand side, from top to bottom are the joint DoD-DoA spectra of the measured channel, the Kronecker model, the Weichselberger model, and the virtual channel representation.*

In case of the Kronecker model, decreasing antenna numbers comes along with an improved performance. The reduced spatial resolution makes a Kronecker coupling of the measured MIMO indoor channels much more probable.

The same holds true for the Weichselberger model. It also matches reality better with reduced antenna number. For the 4 × 4 system, it captures the spatial characteristics of this channel best of all three models.

On the other hand, the mismatch of the virtual channel model increases since it now provides a decreasing number of steering/response directions to describe the underlying radio channel. In fact, for 2 × 2 it fails completely.

Ergodic mutual information

As far as the ergodic mutual information is concerned, the normalisation is done such that for each scenario the average power of the channel matrix elements h_{ij} is set to unity [HÖHB02]. The average receive SNR for each scenario is then fixed at 20 dB.

In Figure 6.22, a scatter plot of the *average* mutual information when using a synthesised channel vs. the *average* mutual information estimated directly from the measurement data in the case of 8×8 MIMO is shown. For each model, a specific marker corresponds to one of the analysed scenarios. The dashed line corresponds to the identity relationship, whereas the dotted lines indicate different levels of relative errors. As a reference, the average mutual information of an iid Rayleigh fading channel, resulting to 44.0 bits/s/Hz at 20 dB receive SNR, is represented by a black dot. The green diamonds indicate the results of the full covariance matrix model, plotted in order to check whether the channel is sufficiently described by its second-order moments.

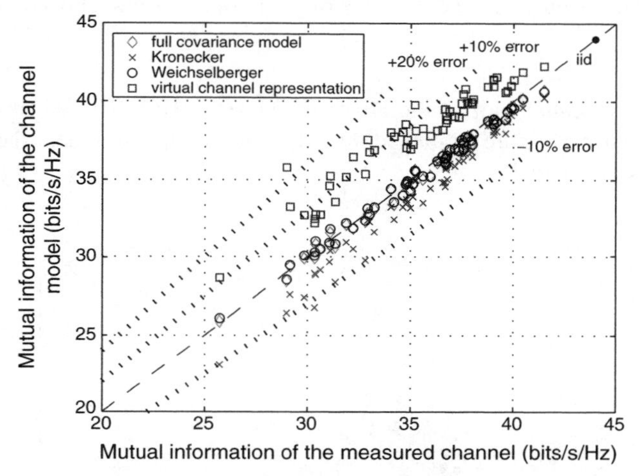

Figure 6.22 *Average mutual information of synthesised vs. measured 8×8 MIMO channels @ 20 dB receive SNR.*

As can be seen, the Kronecker model (crosses) underestimates the ‘true’ mutual information. Moreover, the mismatch increases up to more than 10% with decreasing mutual information. A more detailed analysis in [ÖHHB03] showed that scenarios with low mutual information correspond to high channel correlation. Thus, the Kronecker model, introduced to account for antenna correlation, fails for highly correlated channels.

The virtual channel representation (squares) overestimates the ‘measured’ mutual information significantly. Because of its fixed and predefined steering/response directions, the virtual channel representation is not able to reproduce any measured multipath components lying *between* fixed steering directions correctly. Instead, it assumes independently fading multipath components at the adjacent directions. Thus, it tends to model the MIMO channel with more multipath components than the underlying channel really has, thereby reducing channel correlation and hence increasing the mutual information.

The Weichselberger model (circles) fits the measurements best with relative errors within the range of a few percents.

In analogy to the 8×8 case, the average mutual information of 4×4 and 2×2 MIMO channels has been investigated. The relative model error of the Kronecker model decreases with decreasing antenna number to values of less than 10%. Although for 2×2 channels there exist some exceptional scenarios where the Kronecker model also overestimates the mutual information, a clear trend goes with underestimation of the mutual information.

The relative model errors of the virtual channel representation do not change significantly with the antenna number. The model overestimates mutual information of the measured channel systematically up to 20%.

The performance of the Weichselberger model does not change significantly. It still reflects the multiplexing gain of the measured channel best.

At this point, it is worth mentioning that the mutual information of the Kronecker model has already been investigated in several publications, e.g. in [McBF02a] or [YBOM01], where the performance of the model was found to be satisfactory for 2×2 and 3×3 systems. This is in agreement with the COST 273 results. However, as demonstrated above, when the antenna size increases, thereby improving the angular resolution, the deficiency of the Kronecker model becomes worse.

Diversity order

Figure 6.23 provides a scatter plot of the Diversity Measure [IvNo03] when using a synthesised channel vs. the Diversity Measure estimated directly from the measurement data for each scenario in case of 8×8 MIMO channels.

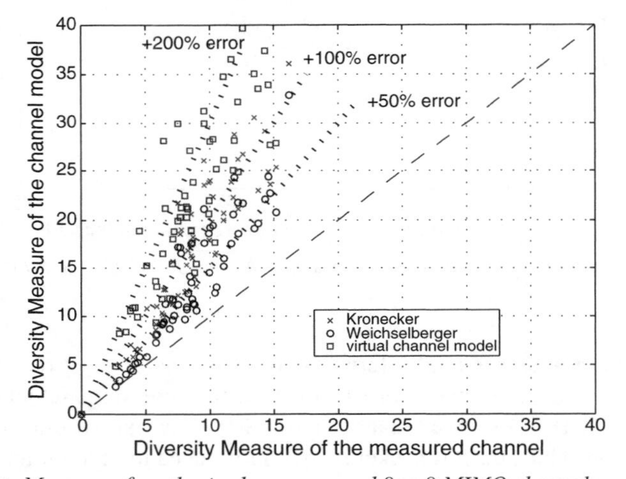

Figure 6.23 *Diversity Measure of synthesised vs. measured 8 × 8 MIMO channels.*

The synthetic channels either match or overestimate the Diversity Measures of the corresponding measured channels. Although the Weichselberger model outperforms both the Kronecker model and the virtual channel representation clearly, it shows relative errors of up to more than 100%. The worst performing virtual channel representation overestimates the diversity metric by a factor of three.

As far as the Diversity Measures for 4×4 and 2×2 MIMO channels are concerned, comparison results show a similar behaviour as 8×8 channels, but decreasing relative errors with decreasing antenna numbers for all three models. Again, the Weichselberger model performs best. For the 2×2 channel, it shows almost perfect match up to some negligible errors for higher diversity values. Also, the match of the Kronecker model is quite tolerable within 10% relative in this case. In contrast, the virtual channel representation systematically significantly overestimates the Diversity Measure, even in the 2×2 case. The reason for the poor performance of the virtual channel representation is, again, its fixed, predefined steering directions.

Although there is no direct relationship between diversity and spatial multiplexing, the overestimation of the diversity order can also be seen in the CDF of the mutual information. The MIMO channel diversity affects the reliability (level of diversity) of the virtual parallel channel paths which is reflected by the slope of the capacity or mutual information CDF curve. Simulations show that the slopes of the Kronecker, Weichselberger and virtual channel representation CDF curves are steeper than the one of the measured channel.

6.5.5 Practical use of analytical MIMO channel models

From the proposed validation procedure we conclude the following. Owing to its simplification in neglecting the spatial structure of MIMO channels, the Kronecker model forces the joint DoD-DoA spectrum to be separable. This yields errors in the considered performance metrics. Therefore, in general indoor environments with interdependent DoDs and DoAs and low antenna spacings, the Kronecker model should not be used for the simulation of capacity (mutual information), beamforming and diversity purposes. Exceptions are 2×2 MIMO systems where the produced error is negligible, because of the reduced spatial resolution of such a system [ÖzBo04]. In the 2×2 case, all three evaluated metrics show a good agreement with the experimental metrics. As a great advantage, the Kronecker model allows for separate optimisation at both link ends, e.g. when designing space-time codes, receive and transmit correlations can be dealt with separately. This is not true for the full covariance matrix model.

The Weichselberger model predicts the average mutual information and the Diversity Measure best of all three models, although it still shows deficiencies in some cases. Regarding average mutual information it shows an almost perfect match with the measurements. Concerning the joint DoD-DoA spectrum it is not so flexible. Although it performs in this sense much better than the Kronecker model, it is not able to render an arbitrary multipath structure without any errors, either.

The virtual channel representation with its fixed and predefined steering directions cannot be used for MIMO beamforming evaluations, except in the limit of very large antenna numbers. If the antenna number increases, the spatial resolution increases enabling simulation of arbitrary DoDs and DoAs. Again suffering from its fixed steering directions it overestimates the mutual information and Diversity Measure significantly.

In general, for capacity evaluations, the Weichselberger model performs thus well. The Kronecker model is an alternative only for MIMO systems with a limited spatial resolution, such as 2×2, maybe 3×3 or arrays with large antenna spacings.

Concerning beamforming evaluations, none of the presented models can reproduce an arbitrary multipath structure accurately, independent of the array size [ÖzBo04]. The Weichselberger model can only cope with systems not larger than 4×4, whereas the use of the Kronecker model should be limited to 2×2. On the other hand, the virtual channel representation improves its performance for very large antenna numbers providing a higher angular resolution, but evidently 8×8 does not seem to be large enough. Therefore, for beamforming purposes with limited antenna numbers in the range from four to at least eight the *finite scatterer model* [Burr03] should be preferred.

For the modelling of the diversity order of a MIMO matrix, again, the Weichselberger model can be used.

Note that the approach discussed in this section is the proper one to validate models that reconstruct realistic MIMO channels, e.g. channels that are measured. This approach, though, is not the only one possible. Should one be interested in a *single* aspect of MIMO only, then models that contain proper parameters (that can be specified more or less freely) might perform better. For instance, the virtual channel representation allows for modelling channels with arbitrary multiplexing orders by choosing appropriate coupling matrices. Similarly, an appropriate choice of the Weichselberger coupling matrix enables the setting of arbitrary multiplexing *and* diversity orders.

6.6 Multi-antenna radio measurements and results

6.6.1 Introduction

Multi-antenna communication techniques (especially MIMO) are a key enabling technology for B3G wireless communications offering high spectral efficiencies and fundamental to the successful deployment of future high data rate communication networks. The current research impetus focuses on applying MIMO concepts to the forthcoming standards in a manner which balances the issues of cost and complexity with performance. Here, two distinct approaches have been taken in order to understand the complex interaction of the channel and the antenna facet. One approach is through a comprehensive characterisation of the radio channel with parameters of DoD, DoA, Time Difference of Arrival (TDoA) and Doppler shift. These are termed as double-directional MIMO (or single-directional in SIMO) measurements and require accurate antenna calibrations and parameter extraction algorithms. The second approach is to record the 'all inclusive' antenna and channel response, without differentiating the spatial parameters. This simpler approach, however, is limited to examining the basics of offered channel capacity and channel correlation effects.

This section begins with a description of single-directional measurement campaigns as a precursor to the double-directional channel measurements by means of both physical and virtual arrays. The second part of this chapter describes measurement campaigns recording the 'all inclusive' antenna and channel response and presents results on issues such as offered capacity in a real physical channel, channel correlation and dynamics of eigenspectrum in wireless deployments.

6.6.2 Directional multi-antenna measurements

Single-directional measurement campaigns and results

A comprehensive single-directional measurement campaign was conducted in Bristol, through the collaboration of the University of Bristol, Telia Research AB and Allgon Systems AB. The main intention was to estimate the correlations between the uplink and downlink bands of the UTRA-FDD in terms of spatial and temporal parameters [FBKE02a], [FBKE02b]. The frequency offset between the uplink and downlink carriers in a Frequency Division Duplex (FDD) system gives rise to a frequency dependency in the channel responses and causes potential problems in downlink beamforming, as the antenna weighting is based upon the spatial parameters extracted from the uplink. This dual-band channel sounding campaign was performed in the 1.92 GHz and 2.10 GHz UTRA-FDD bands, encompassing urban city and suburban residential scenarios. The Medav RUSK BRI vector channel sounder was utilised in these measurements, with hardware modifications to simultaneously transmit a 20 MHz wide probing signal in both 1.9 GHz and 2.1 GHz bands. The mobile transmit unit contained an omnidirectional sleeve dipole antenna, and the BS receiver was equipped with a dual polarised, eight element antenna array. In post-processing, it was observed that the channels spaced apart by 200 MHz (10% bandwidth spacing at 2 GHz) exhibit different responses in both the spatial and temporal domains. The differences were more pronounced in the instantaneous channel responses with very strong decorrelation and different azimuth spreads. The degree of decorrelation also increased for urban environments, compared to suburban settings, due to the increased multipath scattering.

Subsequent analysis with this data has, however, shown that the time averaged DoAs between channels separated by 200 MHz are virtually identical, despite the channels fading independently [FoBe02]. As such, it is desirable to perform uplink-based downlink beamforming in such FDD systems. The performance gain over the conventional Fourier method for DoA estimation with super resolution techniques does not justify their additional computational complexity. In fact, in the presence of grating lobes, the Fourier method offered better performance from the point of view of maximising the array output (lower transmit power requirement for given SNR level). The impact of grating lobes on using the ESPRIT algorithm for estimating the Power Azimuth Spectrum (PAS) is shown in Figure 6.24. The grating lobes at the downlink ESPRIT estimates have caused errors around $-60°$ azimuth spread.

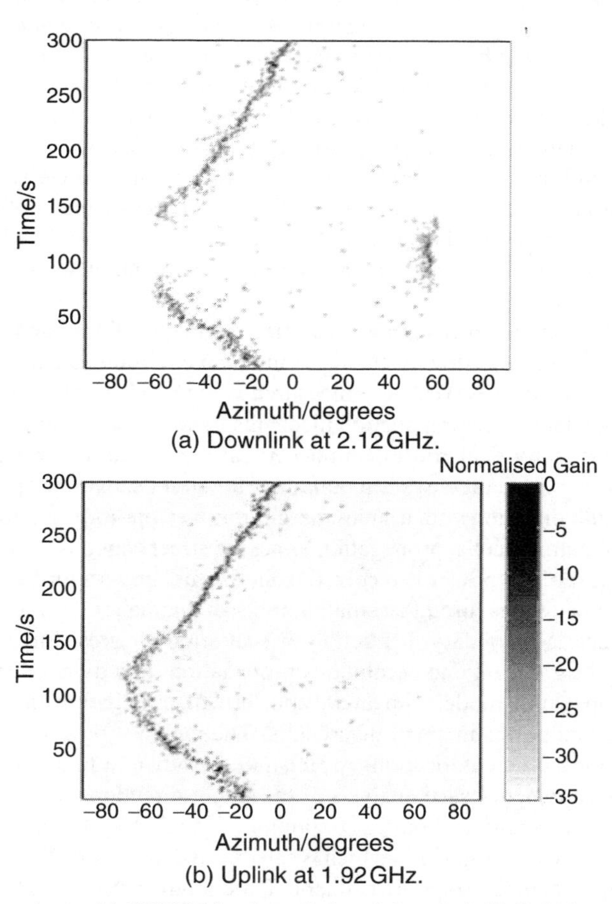

Figure 6.24 *PAS estimations with ESPRIT for the uplink and downlink [FoBe02].*

The design and implementation of a wideband, multichannel sounder by Salous *et al.* for operation in the UTRA-FDD bands is given in [SaFH02]. Using an eight-channel RF receiver in conjunction with an eight signal conditioning unit and eight data acquisition cards, this multichannel sounder has enabled the simultaneous sampling of the radio channel of eight antennas in real time. The potential capabilities of this sounder include double-directional measurements, MIMO measurements (with transmit switching) and angle-of-arrival measurements. Preliminary SIMO results demonstrating the performance of the sounder are included in [SaFH02]. Measurements in outdoor environments with sectored antennas have shown the spatial multipath structure over six antennas each covering 60° azimuth.

A directional channel model for urban microcell environments (specifically for street canyons) and its validation through measurement results is reported in [MaCo01]. The channel model is based on geometrical assumptions, and considers single-specular reflection from scatterers grouped into clusters. The concept of effective street width has been introduced to account for larger spreads in both the time and angular domains. Model simulation results were compared with measurements [MPKZ01], leading to an optimisation of the model's parameter set; an effective street width ratio (effective width considering multiple specular reflections, normalised by actual width) of 12 was achieved. FDD behaviour was also analysed on correlation between up- and downlinks; higher correlation was observed when considering higher resolution. The correlation results have indicated that implementation of adaptive antennas for downlink beamforming is viable. A comprehensive Wideband Directional Channel Model (WDCM) for 3G applications, which utilises the same measurement data [MPKZ01] for validations, is reported in [MTRF02]. This WDCM is based on the results from the COST 259 Action and can be used for UMTS macro-, micro- and Picocells. The model followed a statistical approach based on deterministic considerations and relied on the implementation of clusters and a mobility model for the users. The general street LoS and the general bad urban radio environments have been assessed and excellent agreement between simulation and measurement results has been reported.

Microcell and small macrocell measurements carried out in the 2 GHz band have been analysed to categorise the dominating propagation mechanisms in urban environments [VuSV02]. Both azimuth and elevation angles of incident waves were measured using a spherical antenna array and a wideband radio channel sounder. Three propagation mechanisms were specified: propagation along street canyons, propagation directly over rooftops in a vertical plane, and other mechanisms. The categorisation was based on a constant elevation boundary for street canyon propagation and tracing of the transmitter's azimuth direction with a small margin for over-the-rooftop propagation. The results have shown that while in microcells, propagation along the street canyons is dominating all the time, in small macrocells most of the power is received from the 'unknown' third category. This category can be described as propagations through a small number of anomalies or objects at or above rooftop level. These objects act as secondary diffracting or scattering sources. Street canyon propagation was significant, but no significant and continuous propagation directly over rooftops was detected. Hence a simplified propagation model with lateral and vertical propagation planes would explain only about half of the received power in small macrocells. The received power breakdown for different mechanisms, for two measurement locations in Helsinki, is shown in Figure 6.25.

Theoretical and measurement-based analysis of the single-directional channel has also been conducted at the Catholic University of Leuven, Belgium. In [VeLC04], the authors have proposed an adaptation to the existing SAGE algorithm which is capable of detecting different paths in a broadband environment with large antenna arrays, Broadband Time Domain SAGE (BTD-SAGE) algorithm. By using the indoor channel measurements performed at the University of Kassel with a carrier

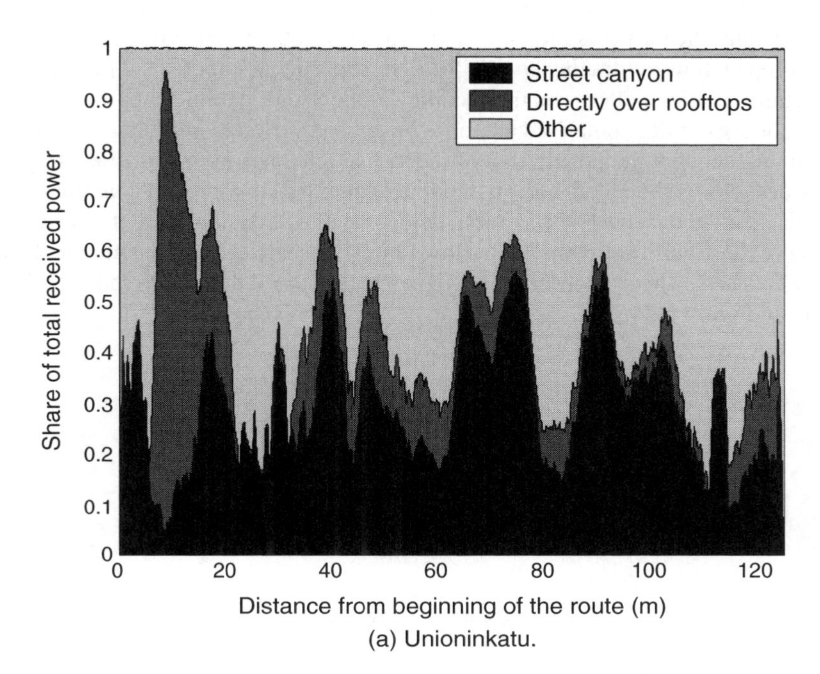

(a) Unioninkatu.

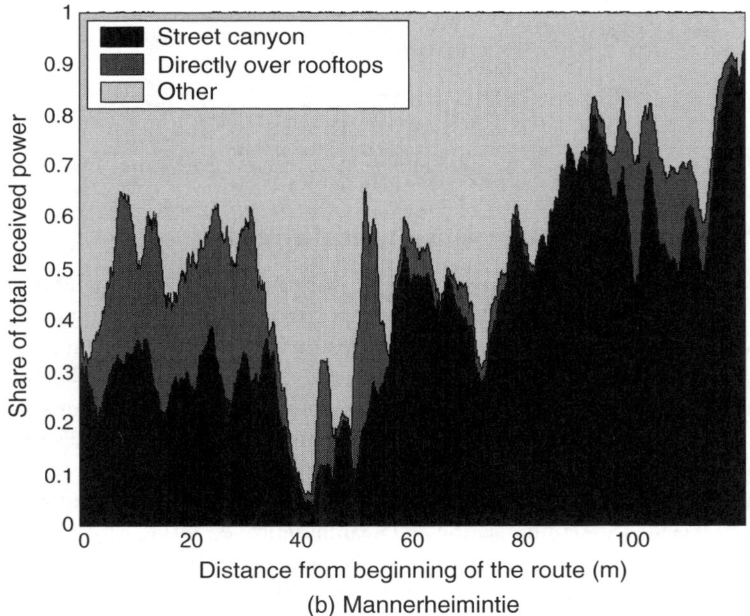

(b) Mannerheimintie

Figure 6.25 *Modes of propagation in two different small macrocell environments [VuSV02]. (© 2002 IEEE, reproduced with permission)*

frequency of 1.8 GHz and 600 MHz bandwidth, the time delay, the azimuth angle and the complex amplitude of the significant paths are extracted. These results are shown to match the parameters given by the propagation ray-tracing tool EPICS. In subsequent papers [LVTC04], [VeLC05] the impact of antenna radiation patterns on the accuracy of the SAGE parameter extraction are discussed. The radiation patterns of dual polarised antennas have been included into the steering vector, which also has taken into account the positioning of individual antennas in the antenna array. A 'smarter' antenna array design was modelled with six monopole antennas in a cubical configuration. Estimates of the time delay (or the distance), the azimuth angle, the elevation angle and the complex amplitude (or the power) of the significant paths have shown high consistency in both simulated (with EPICS) and measured data sets. The stars indicate SAGE extractions and the circles indicate the ray-tracing interpretations in Figure 6.26.

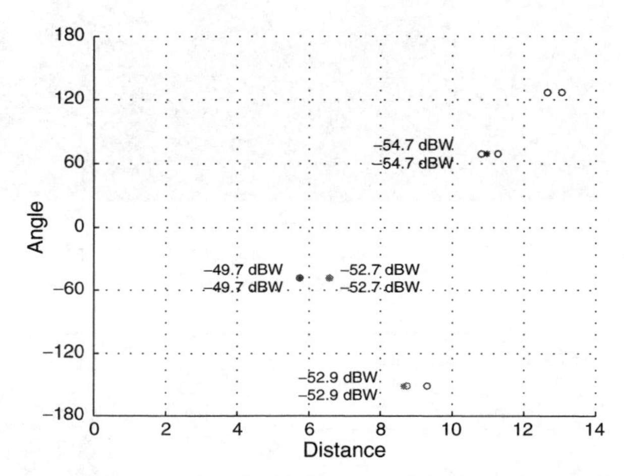

Figure 6.26 *Azimuth angle vs. distance (speed of light × time delay) estimations [VeLC04].*

Double-directional measurement with physical arrays

The Double Directional (DD) representation of the wireless channel is now considered an important aspect in understanding the MIMO channel propagation. Super resolution parameter extraction techniques (like SAGE) are required to accurately estimate the DoD, DoA and TDoA parameters associated with the DD channel description. Hence, the development and validation of these techniques through double-directional MIMO measurements have been a key activity of COST 273. Some of the leading experts in this field have made contributions to COST 273 forums, as detailed in the following paragraphs.

The SAGE algorithm was extensively analysed and improved by Fleury *et al.* at Aalborg University, Denmark. In [StJF02] the ISIS (Initialisation and Search Improved SAGE) algorithm is presented and validated with double-directional measurements in [FYRJ02]. Several test cases and figures of merit have shown in [StJF02] that appropriate, optimised implementation techniques can significantly improve the performance of the SAGE algorithm. More significant paths can be estimated at an improved convergence speed with ISIS than with traditional SAGE. The algorithm provides accurate channel parameter estimates efficiently, using simple and cost-efficient antenna arrays. Relative Squared Estimation Error (RSEE) has been proposed to assess the performance of channel estimation.

This error was found to be already quite small for a low number of paths estimated with ISIS. The signal model can be further extended by taking into account transmit and receive polarisations.

The double-directional measurements to validate the ISIS algorithm were conducted at 2.45 GHz with the Electrobit PropSound channel sounder in an NLoS picocell environment [FYRJ02]. DoA, DoD and TDoA parameters were jointly estimated. A simple ray-tracing method was proposed to reconstruct the propagation paths of the estimated waves under the assumption of one- and two-bounce scattering. The results have shown that most estimated waves can be related to the environment by using the above technique. More specifically, significant objects can be identified in the environment, which coincide with the interaction points (i.e. the edges) of the propagation paths. Estimated waves can also be identified that are likely to interact more than twice with objects/structures along their propagation path. The investigations demonstrate that by extracting these parameters, a detailed insight into the main propagation mechanisms could be achieved.

Recent studies have shown that the ISI-SAGE algorithm combined with a switched multiple-element TX and RX antenna sounding technique makes it possible to estimate the Doppler Frequencies (DF), with an absolute value up to half the rate with which the pairs of transmit and receive elements are switched (switching rate), rather than half the rate with which any fixed pair is switched (cycle rate) as commonly believed. A paper presented by researchers at Aalborg further explores these findings and reports that so-called modulo-type Switching Modes (SMs) used with uniform linear and planar arrays lead to an ambiguity in the estimation of the DF and the directions, when the DF estimation range is extended from minus to plus half the switching rate. The SM of a switched array is the order with which the array elements are switched. Moreover, theoretical and experimental investigations show that the ambiguity problem can be avoided by using some specific SMs, as reported in [PPPB04]. In another of their recent contributions, a deterministic Maximum Likelihood Estimator (MLE) is derived for nominal DoAs of signals originating from Slightly Distributed Scatterers (SDS) [YiFl04]. This estimator is based upon a Generalised-Array-Manifold (GAM) model, which approximates signals originating from SDS. An approximation of the MLE relying on the SAGE algorithm (GAM-SAGE) is also proposed. Results from Monte Carlo simulations of scenarios with a single SDS have shown that when the SNR is high and the azimuth spread of the SDS is less than 8°, the mean square estimation error of the GAM-MLE is 6 dB lower than that of the MLE based on Specular-Scatterers (SS). The GAM-SAGE is shown to estimate nominal DoAs of multiple SDSs more accurately than the SS-SAGE. Figure 6.27 shows the Mean Square Estimation Error (MSEE) of the nominal AoA, plotted against the SNR of the SDS (γ_0) for the different estimators mentioned above.

Extensive double-directional MIMO measurements in indoor environments were carried out by the University of Bristol, using a Medav RUSK channel sounder. The antenna arrays used in the campaigns include 8 element Uniform Linear Arrays (ULA) and 8/16 element Uniform Circular Patch Arrays (UCPA) as depicted in Figure 6.28. A complete technical description of these 5.2 GHz measurements can be found in [TFBN04]. This channel data was utilised in the analysis of maximum likelihood-based super resolution parameter extraction algorithms (primarily SAGE). An overview of different versions of these algorithms, which aim to achieve significant time saving, and reduce both the memory utilisation and processing power of computing resources, is presented in [TaBN04]. The differences of signal processing in the element space and beam space are highlighted in this paper and a hybrid space (HS-SAGE) method is introduced for UCAs. With aid of simulated and measured data, the possible gains in terms of savings in computational effort, while retaining estimation accuracy, are reported.

In a paper presented to the joint COST 273/284 workshop [TPBN04], the 16×16 indoor measurement campaign with identical UCPAs, Figure 6.28(c), was described in detail and parameter

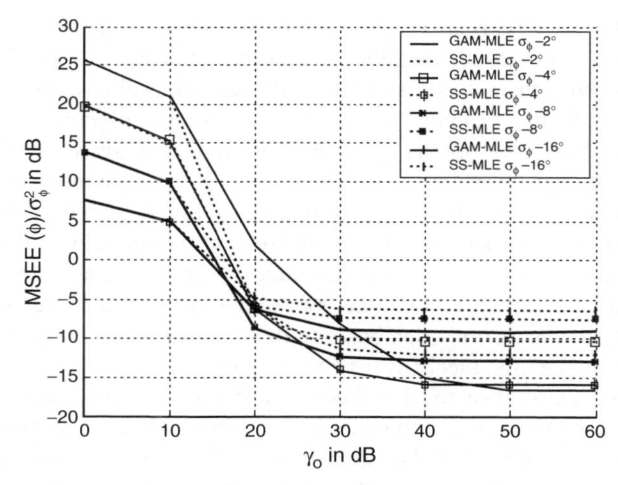

Figure 6.27 *MSEE of nominal AoA vs. SNR of SDS, as a function of azimuth spread [YiFl04].*

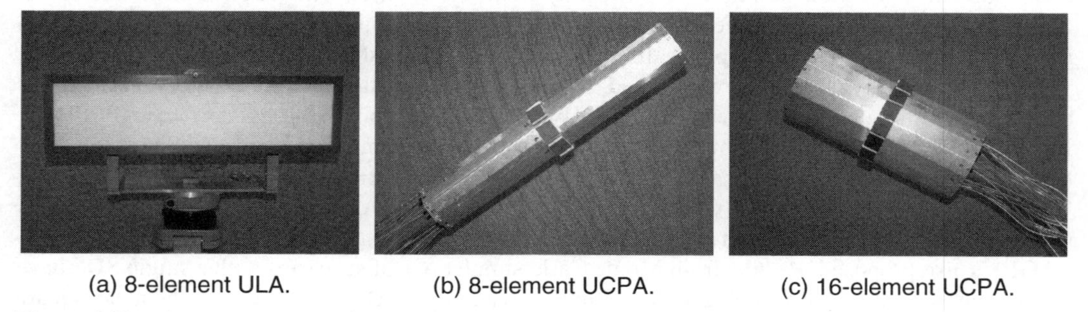

(a) 8-element ULA. (b) 8-element UCPA. (c) 16-element UCPA.

Figure 6.28 *Antenna arrays used in indoor DD measurements [TFBN04].*

extraction results in terms of DoA, DoD and TDoA were presented. Both the Classical SAGE (C-SAGE) and HS-SAGE algorithms were utilised. A comparison of the performance of the two algorithms in joint estimation of TDoA/DoD is presented in Figure 6.29. The results indicate that the distributions of both DoA and DoD are dependent on each other, such that most MPCs concentrate around the direction where both terminals face each other. This effect is observed in both LoS and NLoS propagation conditions. Further, the global distributions of the MPCs are found to exhibit a Laplacian characteristic with its peak located at the region where both terminals face each other.

The above DD channel data was analysed jointly with the University of Edinburgh to investigate a Frequency Domain SAGE (FD-SAGE) approach with Serial Interference Cancellations (SIC) [CLTM02]. The SIC was aimed at replacing the parallel interference cancellation in the standard SAGE, and showed more stable performance in multipath-rich situations. The joint estimation of the number of MPCs, TDoAs, AoAs and complex amplitudes was carried out for the measured indoor data. This performance is evaluated and compared using synthetic data and results using Unitary ESPRIT. A higher number of MPCs were detected by the FD-SAGE and this could be due to the presence of correlated paths that degrade the performance of the 2D Unitary ESPRIT or due to a

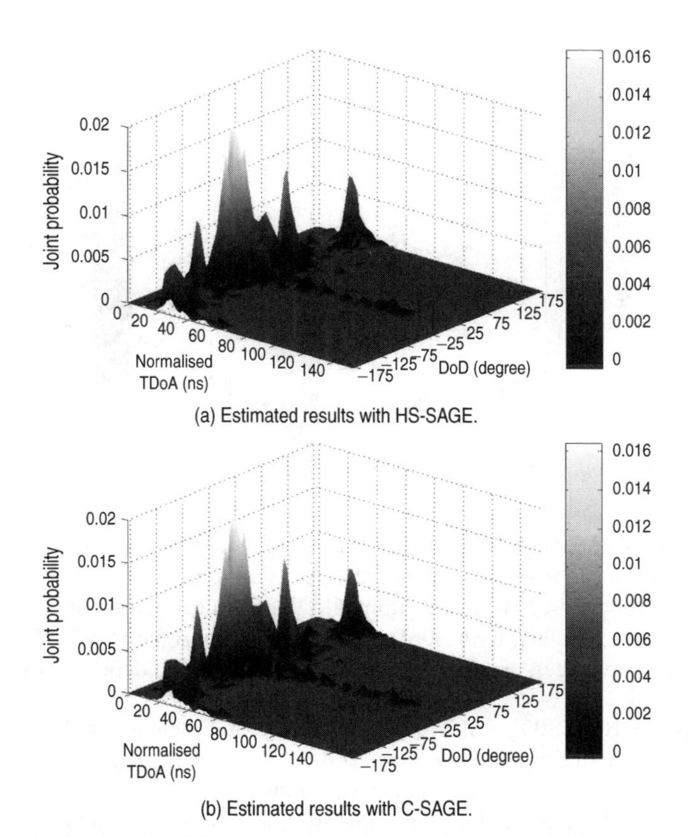

Figure 6.29 *Joint pdfs for TDoA and DoD estimated with classical and hybrid space SAGE [TPBN04].*

leakage in the FD-SAGE. Windowing to account for the limited bandwidth of the channel sounder was also discussed and results have shown that this reduces the leakage in FD-SAGE considerably.

The University of Edinburgh has also developed a wideband dynamic spatio-temporal channel model, which incorporated a Markov chain birth/death process to track the evolution of the multipaths in indoor environments [ChLM03]. This model aims to overcome a major limitation in static channel models, which do not represent the dynamic variations of the multipaths. Analysis of propagation data from [TFBN04] has shown that multiple births and deaths are possible at any instant of time. Furthermore, correlation exists between the number of births and deaths. The joint probability density function (pdf) for the number of births and deaths is depicted in Figure 6.30, where low numbers of births/deaths occur in high probabilities. An M-step, four-state Markov channel model was proposed in this paper to account for these two observed effects. The paths' spatio-temporal variations were also considered by a spatio-temporal vector which was found to be well modelled by a Gaussian pdf. In addition, the methodology used to extract the channel parameters from the measurement data was also discussed.

The Technical University of Ilmenau has also conducted extensive DD MIMO measurement campaigns with a Medav RUSK channel sounder and contributed to the development of super-resolution parameter extraction algorithms. In [Rich05], the authors describe the application of the SAGE

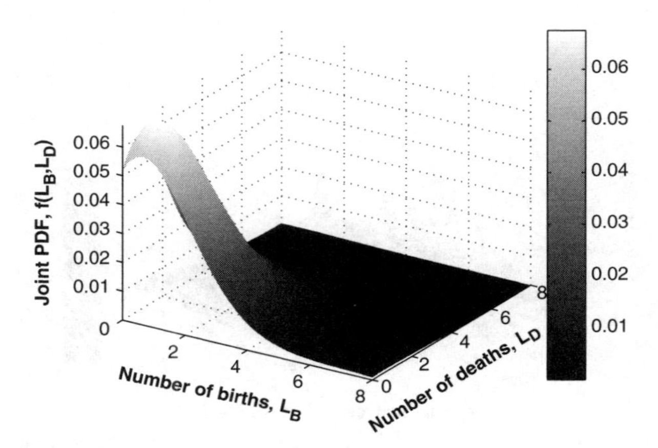

Figure 6.30 *Joint pdf of multipath births and deaths [ChLM03].*

algorithm to measurement data gathered with dual-polarised antenna arrays at both ends of the link. The paper describes the polarisation dependent double-directional time variant complex transfer matrix. Based on this model the SAGE algorithm allows a joint estimation of the four complex polarisation transmission path weights, DoA, DoD, TDoA and DS of radio waves. The estimation results of first measurements have been presented showing the polarisation dependent wave propagation in a mobile radio channel.

In a two-part paper [GaHS03], [SLTR03] University of Ilmenau researches have presented a detailed investigation into the limitations imposed on the accuracy of parameter extraction algorithms. The imperfections of real antenna arrays in terms of mutual coupling and residual calibration errors are noted as causes for these limitations. In [GaHS03], a minimum achievable parameter variance for the unbiased DoA was derived as the Cramer-Rao-Lower-Bound (CRLB). The CRLB can be used to calculate the variance of parameter estimates, especially useful in coherent multipath scenarios. In [SLTR03], experimental results on the performance evaluation of ULAs for high-resolution DoA estimation in channel sounding are reported. In the ULA example, the CRLB determines the minimum achievable estimation variance of the complex path weight and the azimuth angle. The CRLB plots were given for the typical cases 'single-path scenario' and 'coherent two-path scenario' thus demonstrating the performance degradation for closely spaced coherent paths. The standard deviations of the estimated parameter in the two-path scenario are presented in Figure 6.31. As the proposed method relies only on antenna beam pattern measurements, it can easily be applied to any real antenna array and clearly indicates the performance limits imposed by the array.

More recently, a multidimensional maximum likelihood parameter estimator (called RIMAX) to analyse the channel sounding measurements was proposed in [ThLR04]. The algorithm estimates jointly the parameters of the specular components (propagation paths) as well as the parameters of the distributed diffuse scatterers. Depending on the available measurements, the algorithm can estimate the four coefficients of the polarimetric path weight matrix, the azimuth and elevation at the TX site, azimuth and elevation at the RX site, the path delay, and the Doppler shift of the specular components. The algorithm is based on the conjugate gradient optimisation strategy. It has a complexity comparable with SAGE, if the number of paths (components) is small compared to the number of observations. This is typical for multidimensional radio channel sounding measurements. The algorithm additionally provides an estimate of the variance of the calculated parameters, yielding

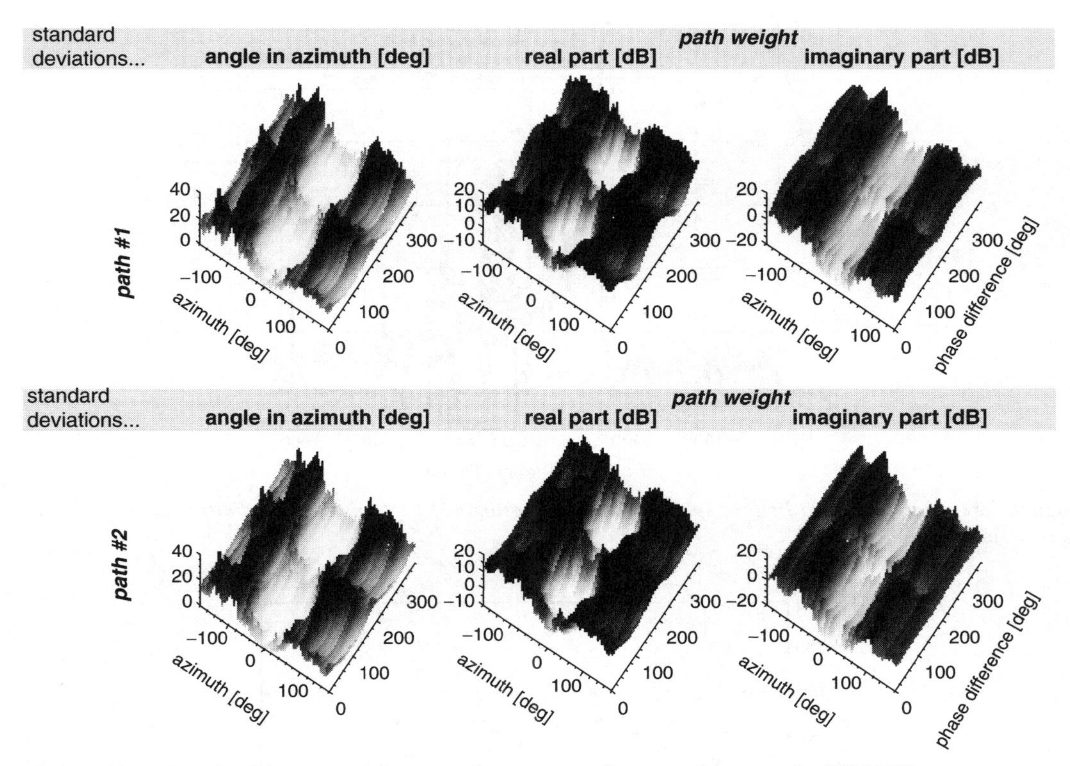

Figure 6.31 *Standard deviations of estimated parameters for two coherent paths [SLTR03].*

reliability information of the channel parameters estimated. The results from an estimation of the PDP are shown in Figure 6.32. Reference [TLST05] gives a comprehensive set of analyses following an extensive propagation measurement campaign under project WIGWAM.

An outdoor DD MIMO campaign was conducted by Ericsson Research and reported in [MATB03]. This was a narrowband channel measurement at 1947 MHz in an urban street microcell environment in both LoS and NLoS conditions. The measured data was used to determine the radio channel characteristics as well as the possible MIMO capacity. The observed capacity was, on average, somewhat less than the ideal. At some LoS locations, a substantial reduction in capacity was observed due to the wave guiding effect of the street canyon. It turned out that the capacity reduction was an effect of both a large Ricean K-factor and a small angular spread. A Maximum Likelihood (ML) method was used to accurately estimate the radio channel in terms of plane waves and their corresponding directions at both ends of the link. The ML estimated channel was successfully used for pathway reconstruction and capacity determination, thereby indicating that the plane wave approach is appropriate for MIMO channel modelling. The channel capacities from measured data and the ML estimated channel has shown very good agreement. Figure 6.33 shows the DoA/DoD estimations for a particular measurement location, with the size of the dots indicating signal power.

An outdoor-to-indoor DD MIMO measurement campaign was recently carried out by researchers at the Lund University [WAKE04]. Measurements were performed at 5.2 GHz between 53 different

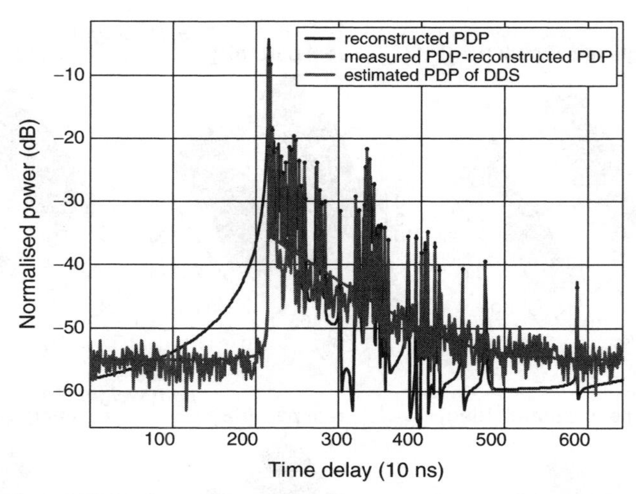

Figure 6.32 *Reconstructed PDP of specular paths and estimated PDP of distributed diffuse scatterers plus noise [ThLR04].*

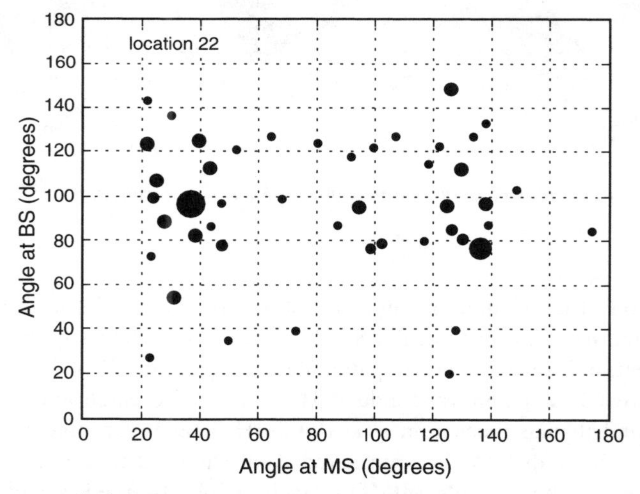

Figure 6.33 *Angular spread of plane wave signal power [MATB03].*

receiver locations in an office building, and three 'BS' positions on a nearby rooftop. The results for angular-delay profiles, RMS angular spread, and other statistical parameters characterising delay and angular dispersion were reported [WAKE04]. The angular spreads at the outdoor link end were found to be rather small. For the indoor link end, they depended on how far the array is from the window facing the outdoor array. The outdoor-to-indoor office wireless propagation channel was characterised with the following parameters: delay spread in the range of 5 to 25 ns, the indoor angular spread 30° to 55°, at the outdoor angular spread of 4° to 20°. By considering 40 multipaths at each measured position, more than 85% of the received power could be estimated in 60% of the 159 measurement locations.

Double-directional measurements with virtual arrays

Virtual arrays are widely used in DD measurements, with a stepper motor mechanism to position the single antenna at different array positions. Although the switching times are higher, this method eliminates the mutual coupling effects and also enables the realisation of large array sizes.

The Vienna University of Technology have conducted extensive MIMO measurement campaigns with virtual arrays, and some of these relate to the analysis of the DD MIMO channel. A detailed analysis of multipath clusters in indoor environments was presented to COST 273 in [CHÖB04b]. The total number of clusters and the dominant clusters are established in a cluttered indoor environment. The analysis was based on DoA/DoD domains and utilised indoor MIMO measurements taken at 5.2 GHz. A single positionable sleeve antenna was used at the transmitter to form a 20×10 virtual rectangular array with 0.5λ spacing. At the receiver, a directional linear array of printed dipoles with eight elements, spaced at 0.4λ, and with a $120° \pm 3$ dB field-of-view was used. The analysis was carried out on a narrowband, 8×8 MIMO data subset. The clusters were identified visually, with DoA/DoD estimates from Bartlett and SAGE algorithms separately and results overlaid. The results have shown the mean number of dominant multipath clusters to be 7.6 with a standard deviation of 2.4, whereas the total number of clusters was found to vary strongly even within a single room. The DoD/DoA estimates for a sample measurement are shown in Figure 6.34, with the white dots pointing to estimates from SAGE and black circles indicating identified clusters.

In [CHÖB04b], cluster characterisation was carried out with a new algorithm based on the SAGE estimation of eigenmodes. Cluster fading statistics were investigated and it was found that strong clusters usually show Rician fading, where clusters with lower power usually exhibit Rayleigh fading. As expected, clusters with Rician fading had few dominant paths, whereas the paths of Rayleigh fading

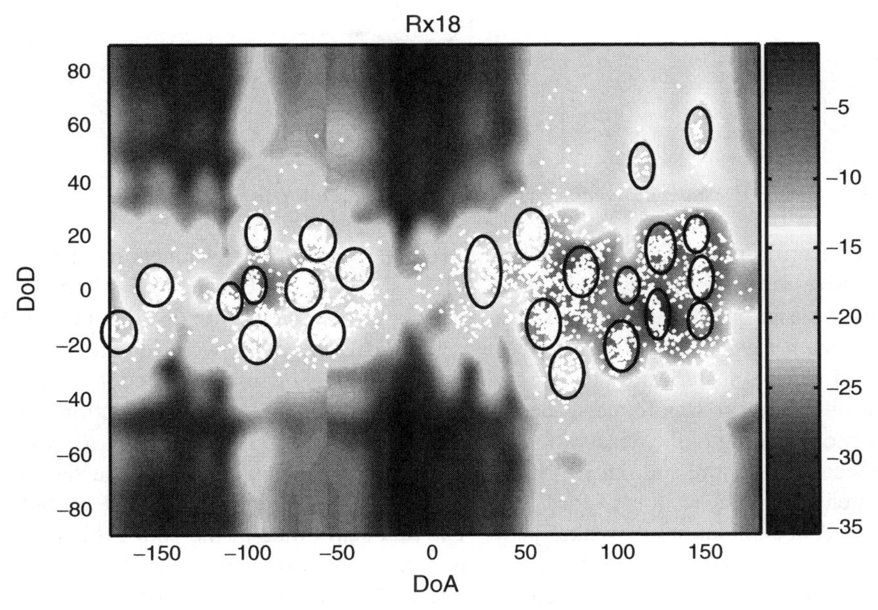

Figure 6.34 *Cluster formations in an indoor environment [CHÖB04b]. (© 2004 IEE, reproduced with permission)*

clusters showed similar powers. The RMS angular spreads for the angles of arrival and departure (AoA, AoD) were also estimated, and values of 2°–5° for AoA and 2°–3° for AoD were reported.

Researchers at Tokyo Institute of Technology have developed a vector network analyser-based system for high-resolution DD MIMO measurements, as detailed in [HaTa03]. This system employs virtual arrays at both the transmitter and receiver ends. Completely automatic measurements are achievable with this set-up, while retaining accuracy. In DD MIMO measurements in an indoor NLoS environment, the channel was estimated in a deterministic way with the ISI-SAGE algorithm [StJF02]. The estimation of polarisation characteristics for identical paths was highlighted. Many waves including dominant paths as well as long delayed paths and many propagation phenomena were effectively detected by ISI-SAGE. The measurement set-up is depicted in Figure 6.35.

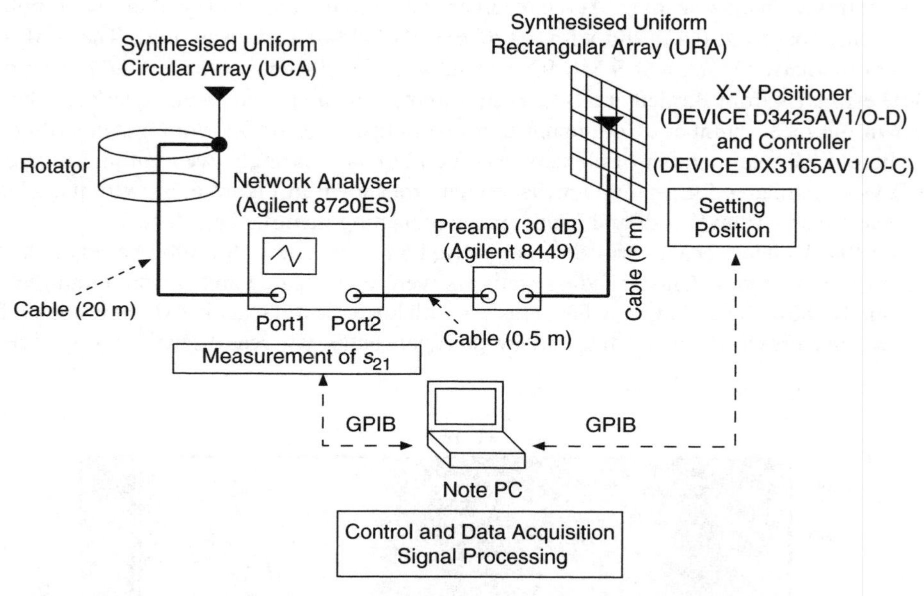

Figure 6.35 *UWB, DD channel sounding system with virtual arrays [HaTa03].*

Some pioneering work on Ultra Wide Band (UWB) DD MIMO channels has been conducted with the above measurement system [HaTK04], [TsHT04]. UWB bands of 3.1–10.6 GHz were considered. Measurements and analysis in a Japanese wooden house are detailed in [HaTK04]. Up to 100 dominant ray paths were extracted using SAGE. The paths were identified with the real environment, where clusterisation behaviour was examined using double directional measurements. The propagating power was found to be concentrated around the specular directions of reflection and diffraction, leading to the observation that the spatio-temporal characteristics of extracted paths highly reflect the structure and size of the building. The power contained in the clusters showed that the estimated 100 paths accounted for 73% of the total received power. The rest existed as diffuse scattering, i.e. accumulation of weaker paths. The practical limits of the path extraction by SAGE were also discussed. The scattering properties and intracluster loss (crucial for deterministic channel reconstruction) were derived for each reflection order. A similar UWB measurement campaign in an indoor office environment and the spatio-temporal analysis are reported in [TsHT04]. After separating the deterministic and diffuse components on both TX and RX positions, the clusters can be coupled based on similar TDoAs by utilising very high temporal resolution. Most of the clusters were found to be determined by physical structures of the experiment environment, i.e. specular reflection, size of reflection objects, and materials. Also, the clusters depended on the wave polarisation and the pattern of the antenna. Intracluster properties were derived based on the moment analysis, and the mean power and spread of the DoDs, DoAs and TDoAs for each cluster were shown. The relationship between DoD and DoA was also investigated.

6.6.3 MIMO measurements of combined channel and antenna responses

At present the use of the Medav RUSK family of sounders is the most common method for MIMO channel characterisation; however, several institutions have developed their own in-house designs based on spread spectrum correlation techniques, chirp methods as well as a dual vector network analyser approach [MoRJ04]. Various types of MIMO channel emulators are now also in use such as the Elektrobit C8 family [KNJY04], [KoNu04], [SJKN05] and MIMO test beds transmitting real MIMO encoded waveforms are also appearing [KaHP04]. Further details on MIMO sounding measurement campaigns and results are now given in the following sections, including special cases such as the 'keyhole' channel and the use of reverberation chambers.

MIMO measurement campaigns at 2 GHz and below

Using a spread spectrum-based sounder at 30 Mchip/s, HUT [KSPH01], [KSVV02] have taken outdoor measurements at a carrier frequency of 2 GHz using a 32 element dual polar spherical antenna array at the mobile station and a dual polar eight element linear array with 0.718λ spacing at 2.154 GHz. An example of the experimental set-up is shown in Figure 6.36.

From the measured data both the directional properties of the channel were derived as described in Section 6.2, as well as MIMO parameter extraction through Singular Value Decomposition (SVD) in order to examine the dynamics of both the eigenvalues and offered channel capacity, Figure 6.37. The results clearly show that the spectrum efficiency gain is highly variable and dependent on the channel characteristics. An investigation of different antenna configurations for a MIMO enabled handset and access point or BS was also conducted based on this experimental set-up [SSKV03a]. It was found that the type of handset antenna had a significant effect on the available MIMO capacity, especially when deployed indoors. Also it was noted that increasing the distance between transmitting antenna elements, or increasing the number of elements, decreases the spread of eigenvalues and, thus, increases MIMO capacity. The smallest eigenvalue spread was found in the indoor picocell.

Given that dual polar data was available from the measurement campaign, a single dual polarised transmitter-receiver configuration (2×2) was also investigated [KSVV02]. It was confirmed that orthogonal polarisations decrease the average signal level, but on the contrary, the eigenvalue spread is smaller, thus yielding higher and more stable MIMO capacities.

Beach and Hunukumbure [HuBe02] have also reported results from an outdoor 2 GHz MIMO measurement campaign with an 8×4 configuration. The transmitting array consisting of two, dual polar UMTS panel antennas spaced 3 m (20λ) apart on a rooftop, overlooking a square surrounded by three- to four-storey buildings, was used to illuminate a vehicular mounted array of four dual polar patches. A Medav RUSK sounder with a customised (in-house) MIMO synchronisation system was utilised. Data was post-processed in order to investigate channel correlation, offered capacity and link level performance of a STBC bearer. The benefits of diversity could be clearly observed

Figure 6.36 *HUT MIMO measurements TX (left) and RX (right) in a microcell [KSPH01].*

for the NLoS deployments. In [HuBe03], this work was extended to include a monopole UCA at RX. The UCA gave better decorrelation in LoS; however, the antenna gain of the directional patches achieved better SNR. In NLoS conditions, the UCA showed higher correlations than the directional patch array, due to the presence of a strong diffracted ray over the rooftops from behind. The spectral efficiencies at fixed SNR for these 8×4 MIMO configurations are shown in Figure 6.38.

FT Wien have also made extensive MIMO propagation measurements at 2 GHz using a Medav sounder with a measurement bandwidth of 120 MHz in urban, suburban and indoor environments. Of particular note is the availability of this data to the scientific community [HMAB02]. An evaluation of the eigenvalue statistics for a subset of this data is given in [ViHU02].

Indoor measurements at 2 GHz with a measurement bandwidth of 60 MHz for a 4×4 system have also been made by Salous [RaSa04], with particular reference time variant nature of the channel. Measurements using devices with equivalent form factors to a mobile phone and a laptop have been conducted by Aalborg University for indoor-indoor and outdoor-indoor deployments at 2 GHz using a spread spectrum-based sounder. Further details can be found in Section 5.2.

MIMO channel characteristics have also been measured by the University of Lille [MDJD03] for railway tunnel deployments using a 4×4 configuration at 900 MHz. Four patch antennas were placed in the train window of the driver's cab as well as deploying an array of horn antennas on the station platform facing the tunnel entrance at various heights. Different antenna and tunnel widths (single and double track) were considered during the measurement campaigns. Again the offered channel capacities calculated for an SNR = 10 dB were calculated and are shown in Figure 6.39. The results indicate that good MIMO capacity gains can be obtained in tunnel deployments, including sections

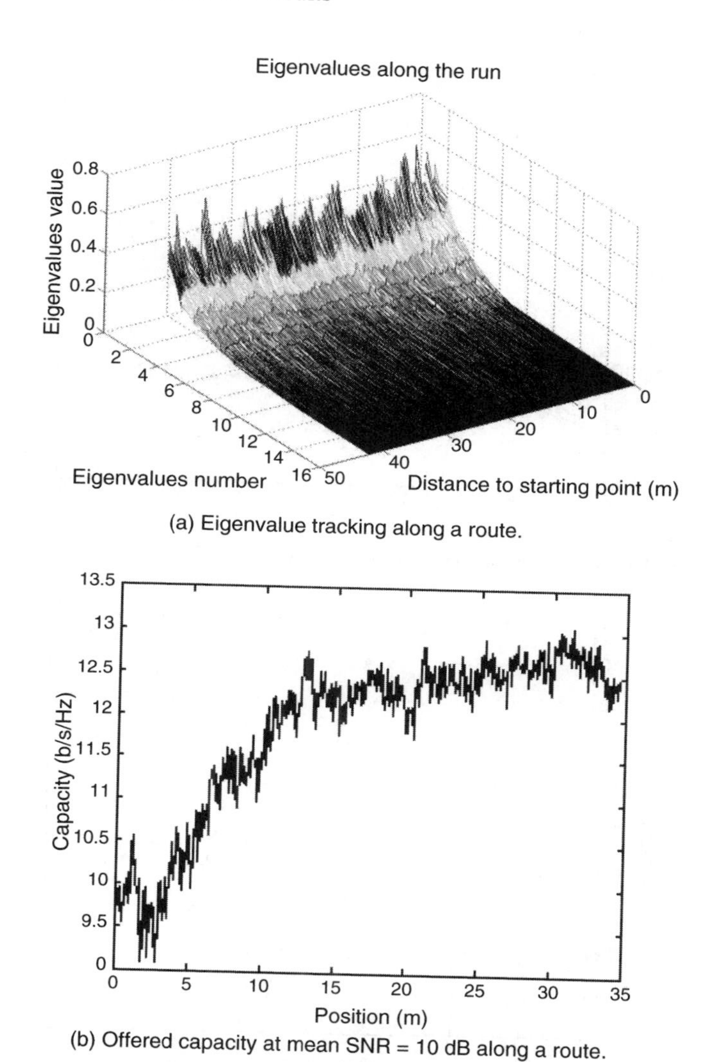

Figure 6.37 *Eigenvalue and offered capacity variations along a route [KSPH01].*

with curved tracks. This work was further extended in [LiDe04a] to consider channel correlations, dynamics of the singular values as well as BER analysis of V-BLAST and Alamouti schemes using the measured channel data.

MIMO measurement campaigns at 5 GHz and above

McNamara *et al.* [McBF02b], [McBF02a] have conducted numerous 8×8 MIMO measurement campaigns in order to extract channel capacities at 5.2 GHz for indoor environments using a Medav RUSK sounder. A simple 0.5λ spaced monopole array on a ground plane was employed at the transmitter, and a cavity backed-dipole array as the receiver. Example results are given in Section 6.3, again illustrating significant capacity gains are attainable as well as sensitivity to variations in channel

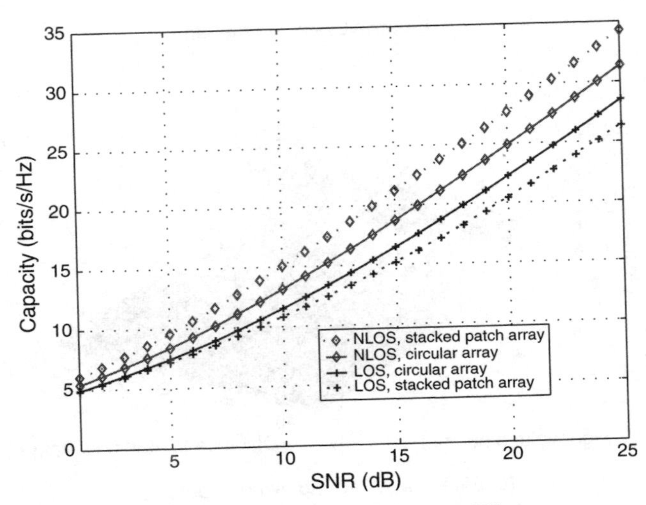

Figure 6.38 *8 × 4 MIMO capacities for two receiver arrays [HuBe03].*

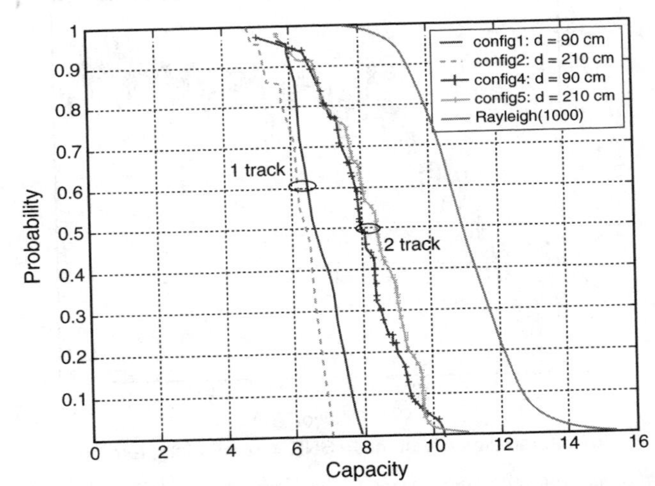

Figure 6.39 *CCDF of channel capacities in 1- and 2-track tunnels [MDJD03]. (© 2003 IEE, reproduced with permission)*

bandwidth and spatial correlation. Kyritsi [KCVW03] has also reported indoor MIMO measurements paying particular attention to channel correlation (as reported in Section 6.7) using a pair of 12 element arrays. Özcelik [Özce04] has used a combination of a linear array (similar to McNamara's) and a virtual array with a 20 × 10 position grid to measure indoor MIMO channel parameters at 5.2 GHz, and also studied the impact of correlation on MIMO performance.

In [ÖHPB03a], Özcelik considers the relationship between the Rician K-factor and SNR to the MIMO capacity, considered in terms of Channel Matrix Coefficient Magnitudes (CMCM), demonstrating that if the CMCMs are strictly Rayleigh distributed (indicating rich multipath) the capacity will be higher than for Ricean distributed CMCMs, at a given SNR. That is, the worse propagation channel is the better MIMO channel. If, however, the comparison is made on the basis of SNR, then there exists a unique relationship between SNR/path loss and MIMO capacity. When the path loss is high (SNR low), as in a bad propagation channel, then MIMO capacity is 'bad', too. Using the same experimental set-up, Herdin [HÖHB02], [HÖHB03] has investigated the capacity as a function of the orientation of the arrays. For his indoor deployment, he found that when path loss is normalised out, MIMO capacity is nearly independent of RX direction and position. A consideration of the impact of human body shadowing on the eigenspectrum is also considered by Herdin in [HeBÖ03], where it was found that the directions of arrival and the beam pattern of the eigenvectors are nearly independent of people movement. Further, these measurements have also been used to benchmark the performance of numerous MIMO transmission techniques by Badic *et al.* [BHGR04].

Investigations to identify the differences between measured and predicted channel capacities for an indoor environment have been conducted at 5.2 GHz by the Tokyo Institute of Technology [ChSA04]. For system dimensions up to 4×4, the capacity has been directly measured as well as the Rician K-factor and channel correlation extracted from the measurements. By comparing measured and predicted capacity, loss associated with both variations in K-factor and channel correlations can be found as shown in Figure 6.40. Measurements and analysis have also been extended to include eigenbeamforming [SaHA04], and further details are given in Section 3.6.

Outdoor 5.2 GHz MIMO measurements [SKPC03] have been conducted by the National Technical University of Athens for 8×8 antenna deployments. The theoretical information capacity has been calculated from the measured data for different MIMO system dimensions and antenna spacings. Investigations of the eigenvalue spread have generated a good alignment to outage capacity values. Further, the measurement databases have been used in conjunction with a genetic algorithm for the selection of the most appropriate antenna configurations as described in [KSKC03], [KaSC04]. Details analysis of Doppler dispersion characteristics, coherence measures, spatial fading correlation and angular power spectra, as well as a discussion on errors induced by the Kronecker product assumption are given in [SkKC04] and [SkKC05].

Recently, HUT [KaVV05] have investigated the benefits of exploiting orthogonal polarisations in a series of campaigns conducted at 5.3 GHz in outdoor-indoor, macro-, micro- and indoor picocell environments. The results contain details on received power levels and Cross-polarisation Power Ratios (XPR) with horizontally and vertically polarised transmission. In addition, correlations between co-polar and cross-polar received components are also given.

In [STMT03], outdoor measured channel data at 5.2 GHz has been used to benchmark the performance of Turbo MIMO Equalisers (TME), illustrating the benefits of schemes based on simple transmit diversity as shown in Figure 6.41. This work has been further extended and a comparison between predicted and measured results given in [ZeST04].

Special MIMO channels – keyholes

It has been shown theoretically (see Chapter 5) that for some environments the capacity of wireless MIMO systems can become very low even for uncorrelated signals. This effect has been termed as 'keyhole' or 'pinhole'.

In practical radio channels, the occurrence of keyholes is very unlikely and hence recordings of keyhole observations in real MIMO channel measurements is extremely rare. One possible example of a MIMO keyhole can be found in the one-way railway tunnel measurements by the University of Lille [LiDe04a]. Evidence of this can be found by studying the instantaneous singular value changes along the tunnels, as depicted in Figure 6.42.

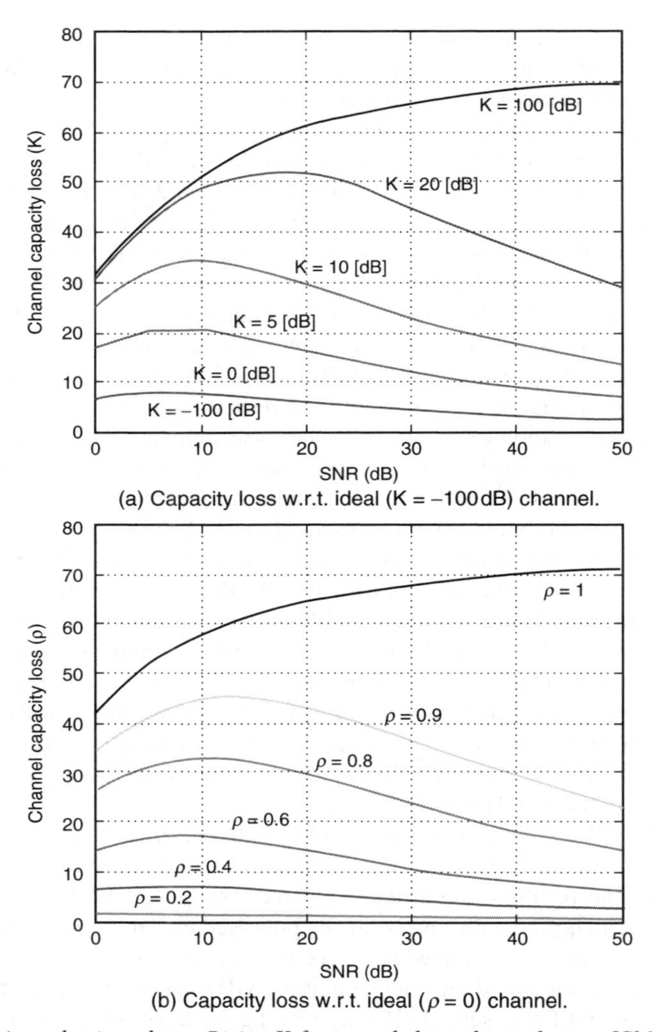

Figure 6.40 *Capacity reductions due to Rician K-factor and channel correlations [ChSA04].*

When the signals were received nearer to the transmit antennas in the curved part of the channel, four significant singular values are present, indicating a full rank channel (H) matrix. But as the reception moved into the one-way tunnel, there is an abrupt drop in three singular values, indicating a rank deficient channel matrix. This segment indicates the behaviour of a keyhole MIMO channel.

In an effort to artificially create keyholes, measurements have been conducted by Lund [AlTM03a] in a controlled indoor environment such that keyholes could be obtained by means of a shielded chamber and an array in an adjacent room. A hole in the chamber wall was the only propagation path between the rooms, as illustrated in Figure 6.43(a). By varying the size of the hole, the 50% outage MIMO capacity was evaluated for both the experimental set-up and several theoretical cases as shown in Figure 6.43(b). It was concluded that the keyhole effect due to real-world waveguides like tunnels or corridors will usually be very difficult to measure.

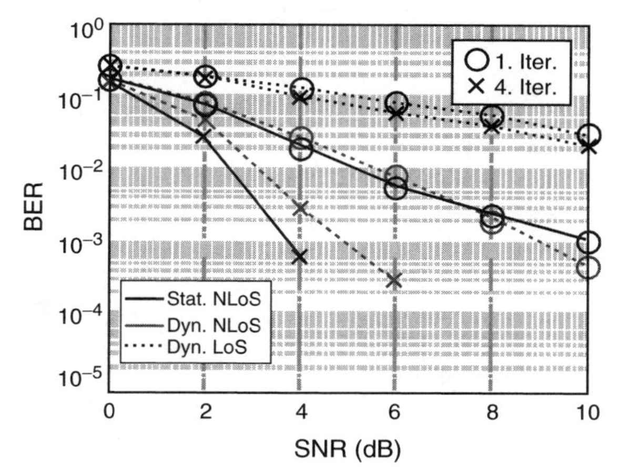

Figure 6.41 *BER performance for Turbo MIMO Equalisers (TME) [STMT03]. (© 2003 IEEE, reproduced with permission)*

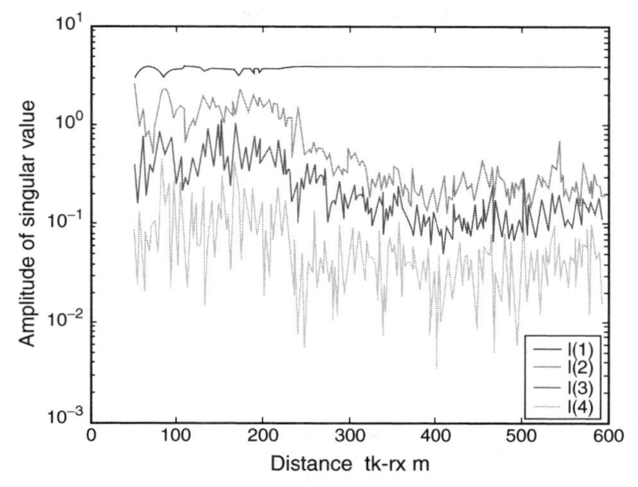

Figure 6.42 *Singular value dynamics indicating a keyhole MIMO channel [LiDe04a]. (© 2004 IEEE, reproduced with permission)*

Mode stirred reverberating chambers in MIMO evaluation

The use of stirred mode reverberating chambers to examine the performance of MIMO enabled devices has been considered in [LiDL04]. The 'emulation' of a MIMO channel is based on a model of room-to-room transmission by coupling two chambers together by means of a waveguide. Figure 6.44 gives an illustration of the results deduced through eigenvalue decomposition of 4×4 MIMO channels. In two experiments, these MIMO channels were generated by connecting the two chambers with a large waveguide and a TE01 mode waveguide, which allows only one mode to transmit. When compared with results calculated from an ideal iid Rayleigh channel, the results with the large

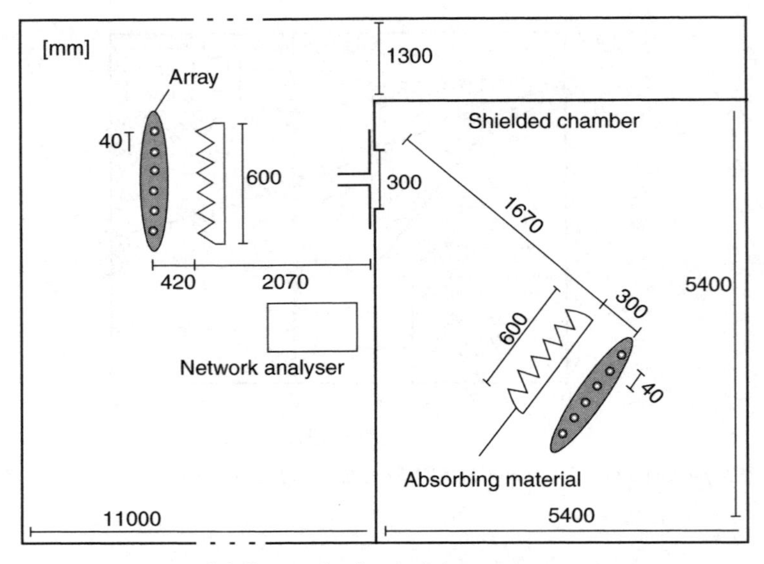

(a) Set-up for 'keyhole' measurements.

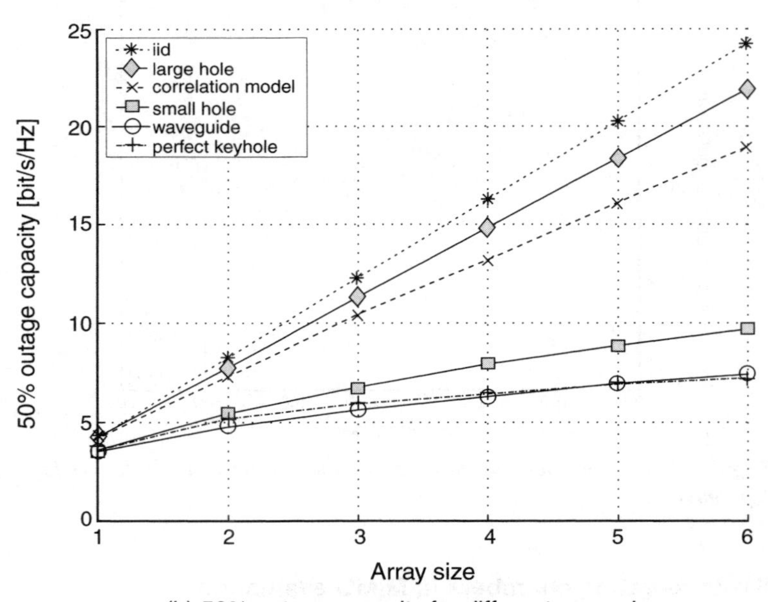

(b) 50% outage capacity for different array sizes.

Figure 6.43 *Measurement set-up and capacity evaluations in a 'keyhole' channel [AlTM03a]. (© 2003 IEEE, reproduced with permission)*

waveguide show very good agreement. When the connection between the chambers only supports one mode (with TE01 waveguide), the keyhole clearly appears since the ratio between the mean first and second eigenvalues exceeds 20. In a further extension of this work, the development and validation of a 3D model of the chamber and associated validation is described in [DDLD04].

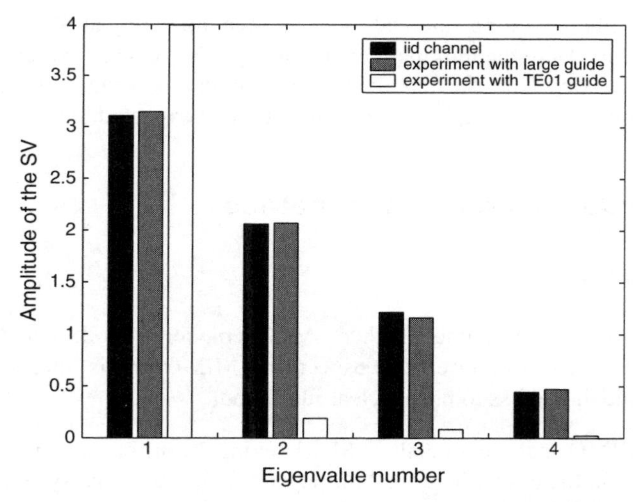

Figure 6.44 *A comparison of eigenvalues of generated MIMO channels with iid Rayleigh channel [LiDL04].*

Comparison between measured MIMO capacities and simulated results

The development of techniques to facilitate the extraction of MIMO capacities for different antenna element types and array geometries from measured channel data will greatly assist in the design and optimisation of antenna components (including element placement) on a MIMO enabled device. One possible approach has been proposed by Bristol [PaTB04] by initially taking high-resolution Double Directional Channel Measurements DDCM (see Section 6.3) of a suitable test environment, and then combining candidate antenna designs through their 3D complex radiation patterns with DDCM data to obtain the complex transmission coefficients of the resultant MIMO system. Figure 6.45 provides

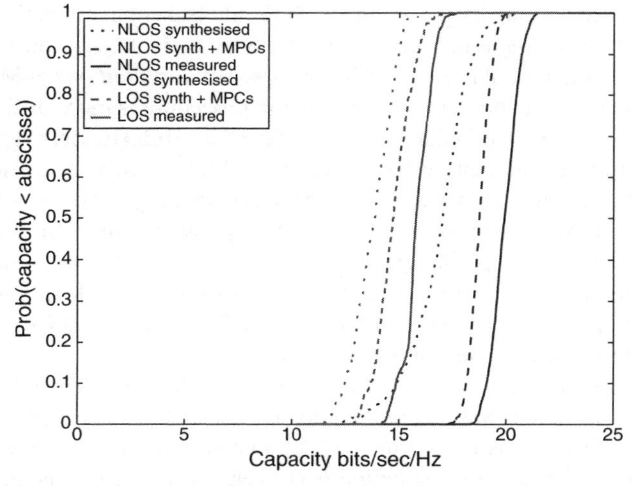

Figure 6.45 *A comparison of MIMO capacities from measured channels and channels modelled with extracted parameters [PaTB04]. (© 2004 IEEE, reproduced with permission)*

an example of results obtained for a candidate 4×4 system operating at SNR $= 20$ dB in both LoS and NLoS environments. Here the multipath component parameters are extracted using SAGE, and in order to reduce processing time, the maximum number of components is limited to 50. Thus, some energy remains undetected, resulting in a lower capacity than expected.

6.7 Model validation and related metrics

6.7.1 *Introduction*

This section is about how to determine whether a MIMO model is good. 'Good' models have been thoroughly validated to reflect precisely the aspect of a MIMO one is interested in. Thus, before adopting a MIMO model, one has to have a clear idea about:

- Which aspect of MIMO shall be modelled? Multiplexing, beamforming or diversity gain?
- At which level should the model function? Propagation, channel, link, system?
- Has the model been validated?
- If yes, by which procedure? Theoretically, or by a single measurement campaign?
- Does the metric chosen for validation correlate with the desired MIMO system performance?

Model validation is so important because how can one otherwise be sure that one is modelling what one wants to model? The result of careful model validation should be twofold:

- Where are the limitations of the model used?
- Are some models better suited to predict a certain aspect of MIMO system performance than others?

Surprisingly, much more effort has gone into the development of new models than into their validation.

One goal of this section is the comparison of measures of quality, in short *metrics*: which is the most suitable to predict a certain specified MIMO property best? Single-number metrics are desirable for model validation because they are easy to compare. It should be kept in mind, though, that any model is a reduction of reality, and single-number metrics are even more so. Obviously, the more complex a metric is, the more information it provides. Metrics based on *distributions* of MIMO characteristics do models more justice, but the outcome of a validation procedure is more difficult to assess.

While capacity has been a very popular metric to verify new MIMO models, it must not be forgotten that capacity is not sufficient to capture all aspects of MIMO systems. It is also not a very 'sharp' measure, i.e. it does not distinguish whether models are accurate or just so-so, as will become evident in this section. As a global metric, neither does capacity tell in which way to improve a model. Further, capacity assumes full exploitation of the channel potential through unlimited flexibility in modulation constellation and adaptation (which is very often not possible in practice), and relies on a reference SNR. And capacity may have a number of different meanings (see Section 7.2). So this section will have a subsection on multiplexing gain instead.

The structure of this section is as follows. The metrics of MIMO models will mirror the measures of *MIMO system performance*. Subsections on spatial correlation and on multipath richness will complement the main three areas of metrics, viz. of spatial multiplexing, of spatial diversity, and of beamforming gain. Although this distinction is convenient, it is not orthogonal, i.e. metrics that are specifically well suited to describe one of the major MIMO benefits may be relevant for others as well. An initial subsection about experimental model validation will weigh the pros and cons of experimental vs. theoretical validation.

6.7.2 About experimental validation

What does it mean to 'validate' a MIMO channel model? The validity of modelling approaches can be verified either theoretically or experimentally.

For a certain modelling approach, theoretical verification means to pitch the results of a certain aspect of the MIMO model against results of the same aspect obtained by an established theory. If the statistical properties of the channel such as shadow fading, delay statistics and angular distribution are known, then a purely stochastic model constructs channel transfer matrices according to these statistics. From the realisations of these channel matrices, a metric such as capacity can be calculated and compared with results from theoretical considerations.

Good agreement between capacity cdfs, if achieved, verifies the usefulness of the new model with respect to an established model, but is not a validation of the new model by itself. Thus, theoretical validation procedures will have to make sure that the reference model is proven solid. Customary has been the comparison with iid uncorrelated Rayleigh fading. Step-by-step validation of intermediate model results will increase the confidence in the model under test. Of course, the validation methodology must be free of circular reasoning, i.e. that the result of the validation has not been a hidden assumption a priori.

The ultimate test of a model is by experiment. This approach is not free of caveats either. For the procedure to be meaningful requires careful planning of the experiment first. Then it should be clear from the beginning which metric will be meaningful to describe which MIMO aspect. As was shown in Section 6.5, even the number of antennas plays a role in validating analytical MIMO models. Some models reflect MIMO performance better if the antenna number is large, others when it is low.

In COST 273, the following experimental validation procedure has been favoured [McBF02a], [ÖHHB03].

- For each environment, appropriately sized subsets of data were selected from the normalised measured channel matrices.
- From this data, the essential model parameters were estimated (as described in more detail in following subsections) and used to
- generate a set of synthesised channels by Monte Carlo simulation.
- The desired model metrics of both sets of channels were then calculated and compared.

The performance of the model was assessed by the closeness of specific metrics of the modelled channels to the measured data. 'Closeness' can be expressed in percentage deviation from measured.

The credibility of the methodology just described has been questioned on the ground of using measured data to derive parameters of models to be validated by just these measurements. Still, it seems to be the best available yet. The fewer the number of parameters extracted or estimated from measured data the better. When these parameters are the crucial ones in the modelling process and no other 'free' parameters can be fitted, the objections against the described approach dwindle.

Another principal objection against experimental validation, done cursorily, concerns the fitting of parameters. Any agreement between measured and modelled testifies that the model parameters have been chosen correctly, not necessarily that the model is correct. And the larger the number of independent parameters necessary to describe a model, the more likely will agreement be obtained.

Gaussian channels

Gaussian channels are a common but tacit assumption in many modelling approaches. Only if zero-mean complex circularly symmetric Gaussian channels are considered, then $\mathbf{H}$ is fully described by its second-order statistics

$$\mathbf{R}_{full} = \mathbf{R}_H = E\{\text{vec}(\mathbf{H}) \cdot \text{vec}(\mathbf{H})^H\}, \tag{6.34}$$

where vec(·) stacks the columns of $\mathbf{H}$ into a tall vector. The validity of this representation is a prerequisite for all analytical models to be valid (Section 6.5). So it is important to see whether it is justified or not. This can be verified by experiment: the entries h_{ij} have to follow strictly a Rayleigh distribution.

6.7.3 Metrics for multiplexing gain

Capacity has been used extensively as a metric for the spatial multiplexing potential of MIMO. Either (i) the ergodic capacity[6] can be estimated – from measurements – by, assuming equal power on all transmit antennas,

$$C_{erg} = E_{\mathbf{H}}\left\{\log_2\left[\det\left(\mathbf{I}_{N_r} + \frac{\text{SNR}}{N_t} \cdot \mathbf{H}\mathbf{H}^H\right)\right]\right\} \tag{6.35}$$

in which $\mathbf{HH}^H$ are measured instantiations, or (ii) the distribution of 'instantaneous' capacity values resulting each from an instantiation of $\mathbf{HH}^H$, or (iii) approximations of the above capacity formula have been used. Further methods, (iv) and (v), obviate capacity and rely on eigenvalue statistics instead.

Practically all new models have been checked on whether they predict correctly ergodic capacity (or mutual information) (method (i)). As an example, Figure 6.46(a) demonstrates excellent agreement between measured capacity and the capacity predicted by the 'Kronecker' model. However, agreement of capacity values synthesised from a model with measured capacity should not be mistaken as a general proof of a model's validity. Figure 6.46(b), taken in another scenario close by, demonstrates an alarming deviation: modelled capacity is *lower* than the measured one [ÖHHB03].

In an investigation of the suitability of a propagation-motivated model to predict range dependence of various MIMO responses in fixed wireless links, [OeEP04] used *median capacity* as a metric, among others (see also Subsection 6.7.8). Median capacity is derived from a capacity distribution (method (ii)).

Besides validating models by the capacity they predict, there are several ways to validate models by *eigenvalue* distributions. It is well known by now that the capacity of MIMO systems is determined by the eigenvalues of $\mathbf{HH}^H$, or equivalently $\mathbf{H}^H\mathbf{H}$, since the two have the same set of non-zero eigenvalues [Burr04b]. Either the (iv) eigenvalue distribution of instantiations of $\mathbf{HH}^H$ or (v) of the average autocorrelation matrix may be used.

If the entries of $\mathbf{HH}^H$ are random, such a form is sometimes called a Wishart matrix. The cdfs (or ccdfs) of capacity calculated for each realisation of this Wishart matrix give an example of validating models by method (ii). Figure 6.47 shows how approximate but analytical expressions for

[6] This section will distinguish five categories of multiplexing gain metrics, numbered (i) to (v).

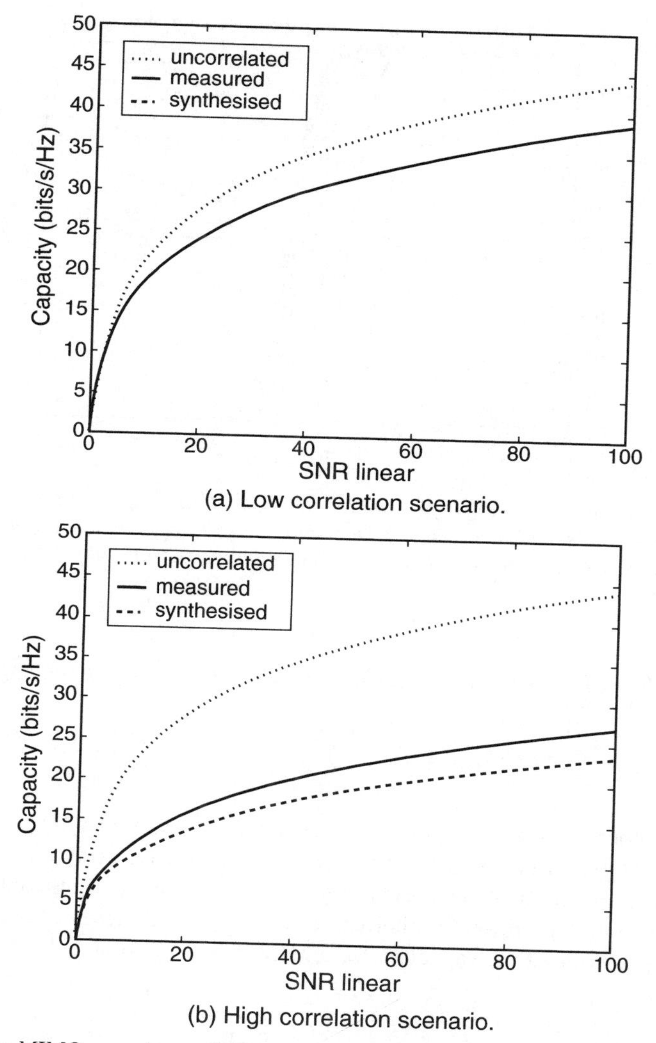

Figure 6.46 *Average MIMO capacity vs. SNR, measured and predicted by the Kronecker model [ÖHHB03].*

the capacity for multiple antenna systems in frequency flat *correlated* Rayleigh fading channels have been validated against Monte Carlo simulations. The proposed analysis allows a fast evaluation of the capacity pdf and of the outage capacity for MIMO systems, so avoiding the necessity of Monte Carlo simulation [ChWZ03].

Eggers [Egge03] has criticised capacity and derived metrics thereof, as they assume full exploitation of the channel potential through unlimited flexibility in modulation constellation and adaptation (which is very often not possible in practice). Also, they rely on a reference SNR, i.e. they are absolute metrics – making it harder to distinguish multiplexing potential from sheer link gain. Instead, he proposes two related metrics. First, the parallelism expressed in normalised cumulative eigenchannel

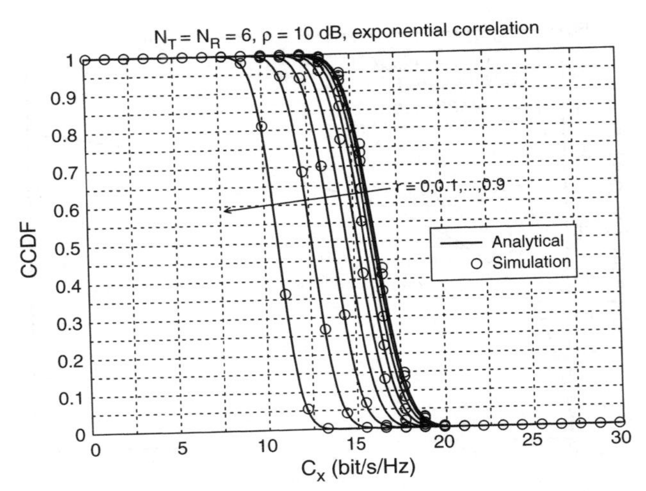

Figure 6.47 *Complementary cumulative probability function of the capacity for a MIMO system with $N_T = 6$, $N_R = 6$, signal-to-noise ratio per receiving antenna $\rho = 10$ dB. Exponential correlation case with r ranging from 0 to 0.9. Comparison between exact analysis and simulations [ChWZ03]. (© 2003 IEEE, reproduced with permission)*

gain is defined as *Normalised Parallel Channel Gain* $(NPCG)_q$, at a given outage level q,

$$NPCG_q = 1/\lambda^2_{\max,q}\sum_{i=1}^{K}{\lambda_i}^2 \qquad \in [1 \ldots K] \tag{6.36}$$

where K is the number of orthogonal spatial subchannels, $K = \min(N_t, N_r)$. This should reflect the multiplexing potential of the environment including the antenna array configuration. However, NPCG will have little sensitivity with respect to eigenpower difference, compared to capacity because of the linear power sum. A proposal for alternative capacity-based metrics, but on relative grounds, can be achieved by looking at the capacity[7] for a MIMO system without feedback, expressed as a sum of eigenvalues:

$$I = \sum_{i=1}^{K}\log_2\left(1 + \frac{\text{SNR}}{N_t}{\lambda_i}^2\right) \tag{6.37}$$

which can be expanded approximately to

$$I \approx \sum_{i=1}^{K}\log_2\left(\text{SNR}/N_t{\lambda_i}^2\right) = K \cdot \log_2(\text{SNR}) - K \cdot \log_2(N_t) + \sum_{i=1}^{K}\log_2\left(\lambda_i^2\right) \tag{6.38}$$

for high SNR and/or eigenvalues. To compare various array constellations on par, the first and second term can be dropped to arrive at a Normalised Parallel Channel Capacity (NPCC). To account for the case of low eigenvalues (<1), which would result in large negative logarithms and throw off the

[7] We better use mutual information, I, here because of omitting the expectation operator.

metric, all eigenpowers are biased with the additive value '1'. Finally, NPCC is approximately

$$NPCC_q \approx 1/\log_2\left(1+\lambda^2_{\max,q}\right)\sum_{i=1}^{K}\left(1+\lambda_i{}^2\right) \in [1\ldots K] \tag{6.39}$$

Both NPCG and NPCC should better reveal the MIMO potential than sheer capacity in cases of constrained modem adaptation or fixed modems.

A verification exercise [LiDe04a] by eigenvalue *statistics* (method (iv)) serves as a good example to carve out the difference between metrics based on distributions and single-number metrics. Comparing distributions is a tricky problem. Figure 6.48 reveals the difficulty in determining what is 'good agreement' when considering distributions. Figure 6.48 compares experimentally obtained eigenvalues with modelled ones. To quantify how good the agreement is one has to resort, again, to single numbers. In this case percentage deviations were calculated to 8, 7, 6 and 10% for eigenvalues 1, 2, 3 and 4, respectively, by

$$\varepsilon_k = \frac{\left\|\lambda_{k,\exp}-\lambda_{k,\mathrm{mod}}\right\|^2}{\left\|\lambda_{k,\exp}\right\|^2} \tag{6.40}$$

Ergodic vs outage capacity

It is important to distinguish between *ergodic* and *outage capacity*. The variance of a capacity distribution (method (ii)) evidences itself as the steepness of the capacity ccdf. The validation of wideband MIMO channel models [McBF02b] highlights this difference particularly well. Figure 6.49

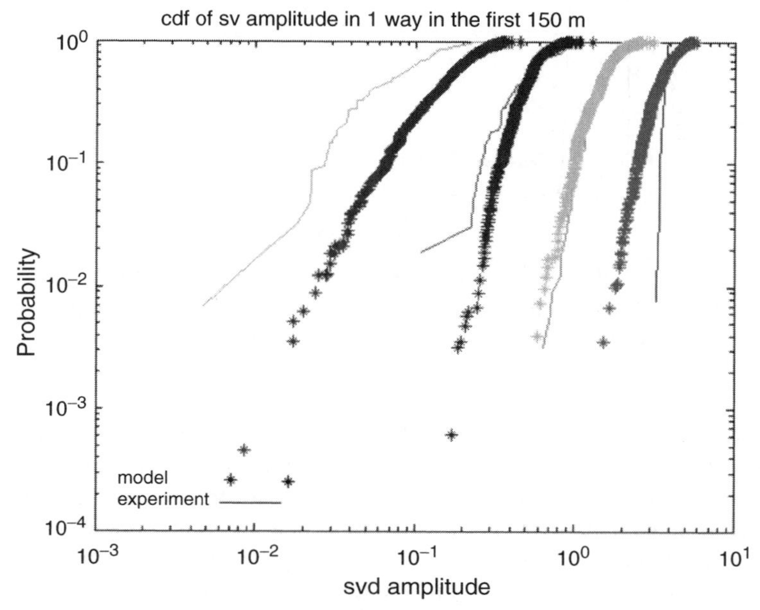

Figure 6.48 *Comparison of measured eigenvalues of a 4 × 4 MIMO system with those synthesised by the Kronecker model [LiDe04b].*

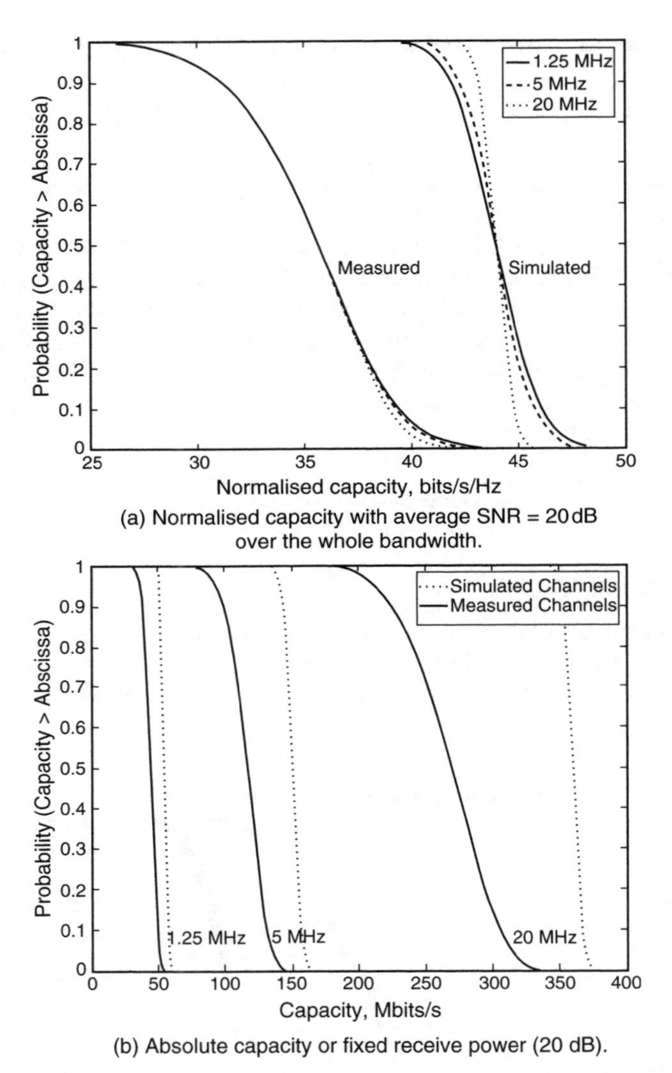

Figure 6.49 *Comparison between capacity ccdfs for 8 × 8 measured and simulated MIMO channels with bandwidths of 1.25, 5, and 20 MHz [McBF02b]. (© 2002 IEEE, reproduced with permission)*

summarises measured and modelled (simulated) capacity results for an 8 × 8 MIMO system with variable bandwidth, i.e. 1.25, 5 and 20 MHz. Note, first, the clear distinction between normalised and absolute capacity: channel response normalisation removes the average path loss in both space and frequency. (It is noted though that normalisation is not always appropriate since the variation in average path loss has a large effect on MIMO performance [MBFK00]. This illustrates that evaluating the performance of channels of varying bandwidth for a constant signal-to-noise ratio and using normalised capacity as a metric can produce misleading results.) Second, in contrast to simulated channels, variation among the plots of normalised capacity for different bandwidths is very small.

As the bandwidth increases almost no change is discernible, only a slight change in the curves' steepness. For lower number of antenna elements, changes in bandwidth effect a larger influence on the normalised capacity. As the bandwidth of these channels increases, the additional frequency diversity is of greater benefit than it would be to a system which exhibits higher spatial diversity. The large difference between measured and simulated is a consequence of the iid Rayleigh fading model. (The channel coefficients were also uncorrelated from one delay to another.) Introducing spatial correlation between array elements would certainly bring the model closer to measurement, but in a wideband channel spatial correlation will be different for different delay taps. Figure 6.50 shows the correlation coefficients for the first six delay taps. Within each plot every 'square' of 8×8 values represents the cross-correlation between one element, h_{ij}, and all others.

The difference in steepness of the modelled and measured capacity curves is a quite common interesting effect and poses the problem of what to match in model validation: the ergodic or the outage capacity?

Several more metrics have been derived from approximations of (6.35). One will be treated in Subsection 6.7.4, another one in Subsection 6.7.5.

6.7.4 Multipath richness and multipath structure

Widely accepted metrics for 'multipath richness' are rare, even though this property of a channel has been identified as important for MIMO early on. In COST 273 several novel concepts have been developed, some of which apply to narrowband models, others to wideband models as well.

An obvious method to test MIMO narrowband models is to study the double-directional multipath structure of a MIMO channel by the joint DoD-DoA Angular Power Spectra (APS) [Özce04]. It may be calculated by *Capon's beamformer*, also known as *Minimum Variance Method (MVM)* [CaGK67], [Capo69],

$$\mathbf{P}_{Capon}(\varphi_{\mathrm{Rx}}, \varphi_{\mathrm{Tx}}) = \frac{1}{(\mathbf{a}_{\mathrm{Tx}}(\varphi_{\mathrm{Tx}}) \otimes \mathbf{a}_{\mathrm{Rx}}(\varphi_{\mathrm{Rx}}))^H \mathbf{R}_{\mathbf{H}}^{-1} (\mathbf{a}_{\mathrm{Tx}}(\varphi_{\mathrm{Tx}}) \otimes \mathbf{a}_{\mathrm{Rx}}(\varphi_{\mathrm{Rx}}))} \tag{6.41}$$

with the normalised steering vector $\mathbf{a}_{\mathrm{Tx}}(\varphi_{\mathrm{Tx}})$ into direction φ_{Tx} and response vector $\mathbf{a}_{\mathrm{Rx}}(\varphi_{\mathrm{Rx}})$ from direction φ_{Rx}. The single-sided marginal spectra (DoD and DoA) are calculated by the one-dimensional Capon beamformer expressed by

$$\mathbf{P}_{DoD,Capon}(\varphi_{\mathrm{Tx}}) = \frac{1}{\mathbf{a}_{\mathrm{Tx}}^H(\varphi_{\mathrm{Tx}})\mathbf{R}_{\mathrm{Tx}}^{-1}\mathbf{a}_{\mathrm{Tx}}(\varphi_{\mathrm{Tx}})} \quad \text{and} \tag{6.42}$$

$$\mathbf{P}_{DoA,Capon}(\varphi_{\mathrm{Rx}}) = \frac{1}{\mathbf{a}_{\mathrm{Rx}}^H(\varphi_{\mathrm{Rx}})\mathbf{R}_{\mathrm{Rx}}^{-1}\mathbf{a}_{\mathrm{Rx}}(\varphi_{\mathrm{Rx}})} \tag{6.43}$$

A sample application is shown in Section 6.5 for validating analytical 8×8 MIMO models.

Driven by the quest for a metric for multipath richness of MIMO channels that contains additional information to the single-number 'capacity', Andersen [Ande04] explored the capacity equation (6.35) in the limit of large SNR or large eigenvalues (vanishing contribution of the '1's on the $\mathbf{I}$)[8].

[8] So this metric is of the kind (iii) as categorised in Subsection 6.7.3.

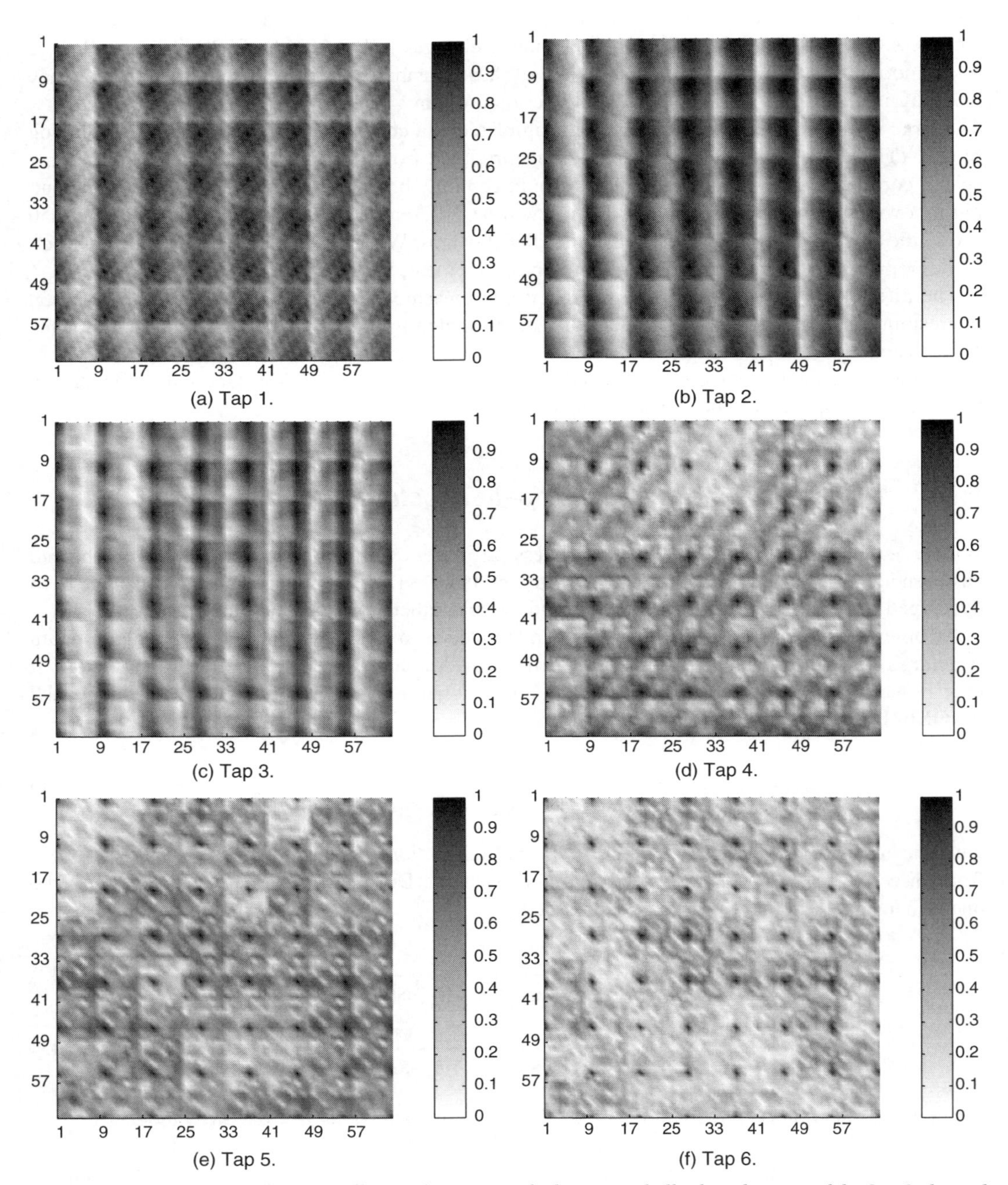

Figure 6.50 *Cross-correlation coefficients between each element and all other elements of the 8×8 channel matrix $\boldsymbol{H}$. (a) to (f) show the matrices of coefficients for the first six taps of the impulse response [McBF02b]. (© 2002 IEEE, reproduced with permission)*

The resulting richness curve (or richness vector) is defined as the sum of the log of the eigenvalues

$$R(k) = \sum_{i=1}^{k} \left(log_2 (\lambda_i) \right). \tag{6.44}$$

Apart from an easily calculated constant term depending on the SNR, the capacity equals the richness. The richness is the same no matter at which end the transmitter is. The eigenvalues are ordered in decreasing order. Figure 6.51 shows a sample application to measurements taken in various indoor environments at Aalborg University with a 16 × 32 array. So the richness definition seems to be a useful tool for comparing environments and MIMO arrays. The richness is plotted as a curve although it is only defined at the integer eigenvalues.

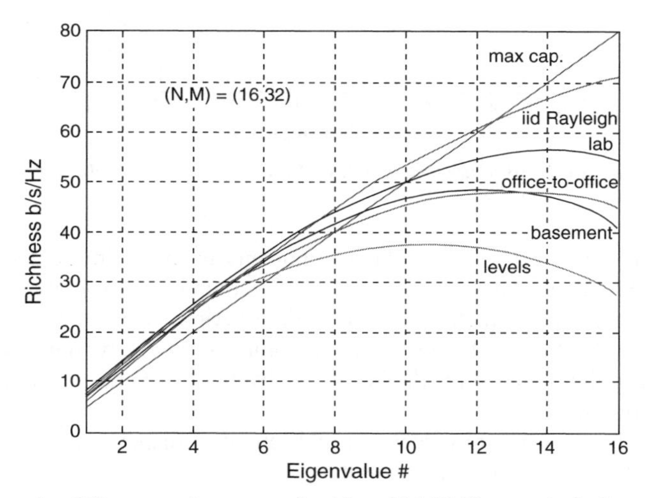

Figure 6.51 *Richness for different environments for 16 × 32 MIMO case, including the iid Rayleigh case, and the maximum capacity case with equal eigenvalues. The experimental environments are: a laboratory full of equipment – lab; a rather empty basement room – basement; the link between two offices at two different building levels – levels; and a link between two offices on the same floor – office-to-office. All signals below −30 dB from the peak are ignored [Ande04]. (© 2000 IEEE, reproduced with permission)*

Since the matrix may be expressed as a function of delay it is possible to extend the definition to the broadband case and evaluate richness also as a function of delay, at a fixed position. Richness *increases* in the measured example of Figure 6.52 for larger delay values.

An entirely different metric for multipath richness was elaborated in [GiCo04]. It is based upon the angular and temporal densities of arriving signals, ρ_ϕ and ρ_τ, respectively, and the observation that these two quantities differ considerably in macro- and microcell environments. Keeping in mind that ρ_ϕ and ρ_τ oppositely reflect richness, a single quantitative measure of richness is introduced by

$$\omega_{DCIR}\,[\text{rad}/\mu\text{s}] = \frac{\rho_\tau}{\rho_\phi} = \frac{\sigma_{\phi,NB}\,[\text{rad}]}{\sigma_\tau\,[\mu\text{s}]} \tag{6.45}$$

in [rad/seconds]. (Given its dimension, one could call ω_{DCIR} also a multipath frequency.) $\sigma_{\phi,NB}$ is the narrowband RMS spread of angles at the receiving end. It might seem paradoxical that definition (6.45) seems to imply larger multipath richness by lower delay spread. One would expect the opposite.

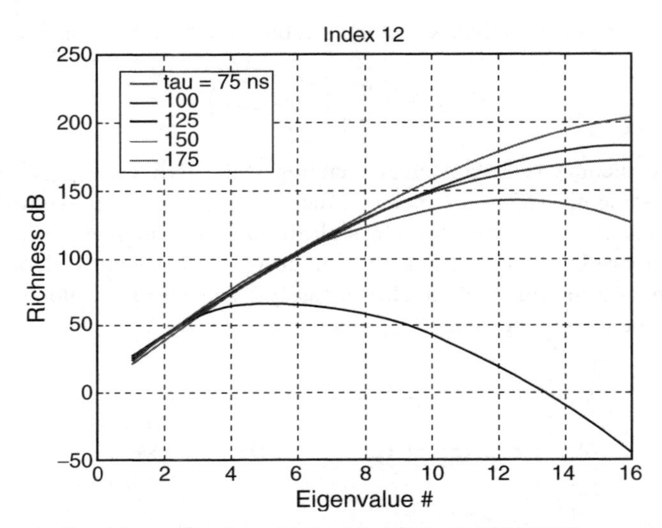

Figure 6.52 *Richness in 'levels' as a function of delay [Ande04]. (© 2000 IEEE, reproduced with permission)*

Higher delay spreads imply relevant angular components in more time units, reflecting a larger multipath richness. But a certain ω_{DCIR} value is an average quantity *per unit* time, If the same significant angular components arrive in a smaller time period, within a lower delay spread, then ω_{DCIR} increases, also reflecting the higher multipath richness, now in both the angular and temporal sense. The multipath richness made available by the channel (6.45) has a bearing on beamforming and interference suppression.

Another effort to link directional dispersion and temporal dispersion into a single metric was undertaken in [Egge03]. The result is the heuristic

$$\mathbf{\Lambda}_{MIMO} = \frac{\mathbf{\Lambda}_{joint} + \mathbf{\Lambda}_{\tau}}{2} \in [0 \ldots 1] \tag{6.46}$$

giving the most compact environment description for expressing parallel channel support of 'randomness' or dual-scattering richness. Here, $\mathbf{\Lambda}_{\tau}$ is a measure of normalised mean delay, τ_{norm}, by

$$\mathbf{\Lambda}_{\tau} = \sqrt{1 - \tau_{inv}{}^2} \in [0 \ldots 1] \tag{6.47}$$

where τ_{inv} is given by

$$\tau_{inv} = 1/\tau_{norm} = \tau_0/\bar{\tau} \tag{6.48}$$

This might seem an odd way of constructing a temporal spread. Note that it does not indicate the degree of time dispersion in the normal sense, i.e. related to frequency-selective fading effects. It only illustrates the scattering-cluster offset from the shortest path (LoS) and thus indicates an average path migration and the possibility for a joint 'centre' of scattering areas between TX and RX. The angular part of the metric, $\mathbf{\Lambda}_{joint}$, is based on the double-directional environment function and relates to a heuristic describing how much 'off-centre' the joint distribution of scatterers is, seen both from TX and RX sides. Its range goes from full dispersive at both ends, $\mathbf{\Lambda}_{joint} = 1$, to fully dispersive at one end but fully directive at the other, $\mathbf{\Lambda}_{joint} = 0$, to fully directive at both ends, $\mathbf{\Lambda}_{joint} = \sqrt{-1}$. The latter situation is one with no parallel channel support, i.e. one realisation of a keyhole channel.

Thus, the imaginary value indicates poor parallel channel support and poor diversity gain, while a 0 value can still have some diversity gain and a 1 of course has the most. For diversity, see also Subsection 6.7.6.

6.7.5 Spatial correlation

In MIMO modelling, spatial correlation is an output of wave propagation on the one hand and an input parameter for capacity calculation on the other. Spatial correlation has been shown early on to impair capacity gain of MIMO systems. So correlation should be an important measure for the purpose of model validation.

Several correlation coefficients

First, it is paramount to distinguish clearly between the different correlation measures, viz. complex, envelope and power correlation, as there exists some confusion in current literature about which of several correlation coefficients are useful for MIMO channel characterisation. Second, the kind of correlation one is talking about needs clarification. Correlation occurs between signals on antenna elements at either the receiver or at the transmitter, or across the link. Special attention must be given to the so-called 'diagonal' correlation, a special kind of across-the-link correlation, as it may provide *higher capacity than iid*, see below. The clarification of correlation coefficients has been done, e.g. in [KCVW03]. Let u and v be two complex variables, e.g. antenna signals in complex baseband. The complex correlation coefficient is defined as

$$\rho_{complex}(u, v) = \frac{E[uv^*] - E[u]\,E[v^*]}{\sqrt{E\left[|u|^2 - (|E[u]|)^2\right] E\left[|v|^2 - (|E[v]|)^2\right]}} \tag{6.49}$$

The envelope correlation coefficient is

$$\rho_{env}(c, d) = \frac{E[cd] - E[c]E[d]}{\sqrt{E\left[c^2 - (E[c])^2\right] E\left[d^2 - (E[d])^2\right]}} \tag{6.50}$$

where $c = |u|$ and $d = |v|$. Finally, the power correlation coefficient is

$$\rho_{power}(f, g) = \frac{E[fg] - E[f]E[g]}{\sqrt{E\left[f^2 - (E[f])^2\right] E\left[g^2 - (E[g])^2\right]}} \tag{6.51}$$

where $f = |u|^2$ and $g = |v|^2$.

The complex correlation is a complex number, whereas the envelope and the power correlations are real numbers. However, they are all limited in absolute value by 1. For Gaussian random variables [LaZo96], the following approximations hold

$$\rho_{env} \approx \rho_{power} \tag{6.52}$$

and

$$\rho_{power} \approx \left|\rho_{complex}\right| \tag{6.53}$$

In the context of MIMO systems, where the quantity of interest is the channel capacity and the *phase* of the variables affects the result, the complex correlation coefficient will be of more interest. (In the MIMO case, the complex random variables are the entries h_{ij} of the channel transfer matrix $\mathbf{H}$.)

Comparing correlation matrices

Spatial correlation is determined by both the propagation environment and the array configuration, through the Power-Azimuth Spectrum (PAS) of multipath signals impinging on each array and the element types and their spacing. Although the array architecture can be controlled, spatial correlation among the elements of the array is a function of the particular environment in which the arrays are located and has been the subject of much research in single array systems.

Various metrics have been introduced for the comparison between modelled and measured correlation data. Employing the method described in [YBOM01], a metric for the difference or the error between an approximated correlation matrix $\hat{\mathbf{R}}$ and the channel correlation matrix $\mathbf{R}_\mathrm{H}$ is

$$\Psi(\mathbf{X}, \mathbf{Y}) = \frac{\|\mathbf{X} - \mathbf{Y}\|_F}{\|\mathbf{X}\|_F} \tag{6.54}$$

$\Psi(\mathbf{X}, \mathbf{Y})$ can be regarded as an indication of the validity of the initial assumption made by using a simplified model such as $\hat{\mathbf{R}}$.

Another interesting distance measure for correlation matrices was recently introduced by Herdin [HCÖB05]. To obtain a metric that is equal to zero when the correlation matrices are identical (apart from at scalar factor) and equal to unity if they differ maximally, the correlation matrix distance between $\mathbf{R}$ and $\hat{\mathbf{R}}$ is defined as

$$d_{corr} = 1 - \frac{\mathrm{tr}\left\{\mathbf{R}\hat{\mathbf{R}}\right\}}{\|\mathbf{R}\|_F \left\|\hat{\mathbf{R}}\right\|_F} \tag{6.55}$$

Here, $\mathrm{tr}\{\cdot\}$ denotes the trace of a matrix, while $\|\cdot\|_F$ stands for the Frobenius norm. The meaning of this metric can easily be understood from a different formulation of it

$$d_{corr} = 1 - \frac{\left\langle \mathrm{vec}\{\mathbf{R}\}, \mathrm{vec}\left\{\hat{\mathbf{R}}\right\}\right\rangle}{\|\mathrm{vec}\{\mathbf{R}\}\|_F \left\|\mathrm{vec}\left\{\hat{\mathbf{R}}\right\}\right\|_F} \tag{6.56}$$

The Correlation Matrix Distance (CMD) is directly related to the inner product between the vectorised correlation matrices, in the present context the measured and the modelled ones. If the correlation matrices are equal, the CMD becomes zero; the more they differ from each other, the larger the CMD becomes. Finally, if they differ to a maximum amount, the CMD becomes 1. Figure 6.53 shows the CMD of a synthetic, time-evolving propagation scenario with two distinct directions. The respective correlation matrices were calculated for each time instant and compared with the very first one. The CMD reflects the difference between the correlation at the beginning and the end precisely (CMD $= 0.5$), as only half of the scenario changes.

The advantage of the CMD is that it measures how the *structure* of the matrices differ, independent of changes in power or system parameters like number of used eigenmodes. A metric utilising the latter is discussed in Subsection 6.7.7.

Capacity metric for channels with cross-link correlations

A further new metric accounting for the cross-link correlation in full evolves from an approximation of ergodic capacity for *correlated* channels [OePa04], and as such falls into the model category (iii) from above. Here, model validation had worked the other way around: simple geometrical scattering

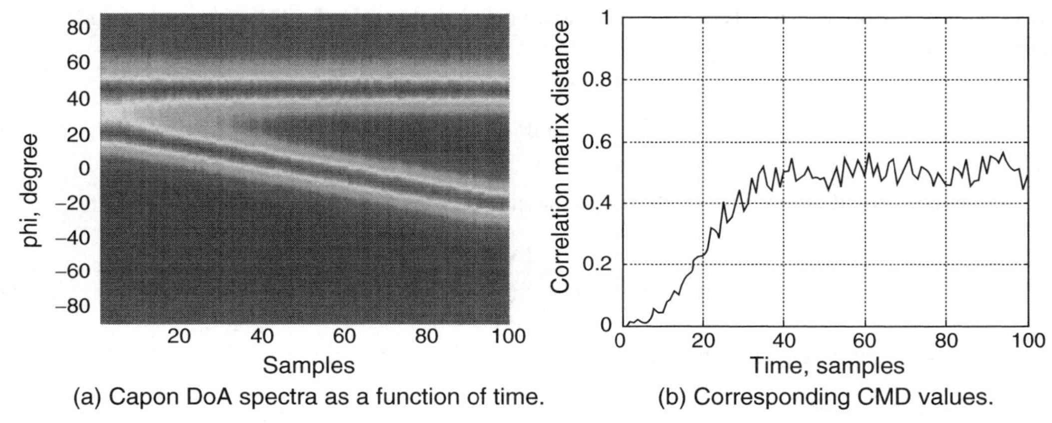

Figure 6.53 *Correlation matrix distance d_{corr} of a time-varying synthetic propagation scenario [Herd04].*

models served to verify the significance of the new metric. Starting from (6.35), the metric reads as ($\overline{C} \equiv C_{erg}$)

$$\overline{C} \approx \overline{C_w} + \log_2\left[\frac{\bar{\kappa}}{\bar{\kappa}_w}\right] \tag{6.57}$$

where the subscript w refers to the uncorrelated case, and the average determinant, $\overline{\kappa}$, is defined as

$$\bar{\kappa} = \mathrm{E}\left\{\det\left[\mathbf{I}_{n_R} + \frac{\rho}{n_T}\mathbf{H}\mathbf{H}^H\right]\right\} \tag{6.58}$$

The metric exploits the fact that $\overline{C}$ and $\log_2\overline{\kappa}$ behave similarly, since $\log_2$ is a monotonic function. The use of $\overline{C}_w$ and $\overline{\kappa}_w$ allows to remove the fixed bias introduced by reversing the expectation and $\log_2$ operations. Closed-form expression of the capacity $\overline{C}_w$ abound in literature and can be easily obtained by a one-time Monte Carlo simulation. On the other hand, $\overline{\kappa}$ can be shown to be uniquely related to the full correlation matrix, $\mathbf{R}_{full}$, (6.34).

Numerical results of the determinant for a so-called diagonally correlated 2×2 MIMO channel [OePa04] lead to the amazing conclusion that the *ergodic capacity is not maximised by independent fading*! So spatial correlation is a dazzling quantity and provides very interesting metrics indeed.

6.7.6 A metric for MIMO channel diversity

The eigenvalues λ_i of the MIMO channel autocorrelation matrix $\mathbf{R_H}$ describe the average powers of the *independently fading matrix-valued eigenmodes* [Weic03]. The degree of diversity offered by this channel is determined by the eigenvalue profile. The *complete* information of such a profile is difficult to capture. A useful single-number metric for Rayleigh fading MIMO systems, the so-called *Diversity Measure* $\Psi(\mathbf{R_H})$,

$$\Psi(\mathbf{R_H}) = \left(\frac{\mathrm{tr}\{\mathbf{R_H}\}}{||\mathbf{R_H}||_\mathrm{F}}\right)^2 \tag{6.59}$$

was recently introduced by [IvNo03]. It is noteworthy that, since the following matrix identities hold [ÖzCB05],

$$\mathrm{tr}\{\mathbf{R_H}\} = \sum_{i=1}^{K} \lambda_i \quad \text{and} \quad ||\mathbf{R_H}||_\mathrm{F} = \sqrt{\sum_{i=1}^{K} \lambda_i^2} \tag{6.60}$$

(6.59) can be equivalently written in terms of the eigenvalues

$$\Psi(\mathbf{R_H}) = \frac{\left(\sum_{i=1}^{K} \lambda_i\right)^2}{\sum_{i=1}^{K} \lambda_i^2} \tag{6.61}$$

Evidently, the more uniform the eigenvalues are distributed the higher the Diversity Measure, as desired.

6.7.7 A metric for beamforming

The F-eigenratio is a measure to describe MIMO channels by a limited number of eigenvalues of correlation matrices [ViHU02]. Specifically, it expresses the degradation of MIMO beamforming schemes due to changes in correlation matrix $\mathbf{R}_H$. So it is another kind of metric based on eigenvalue statistics (method (v)). For MIMO, the eigenvalue decomposition of the correlation matrix $\mathbf{R}_H$ reads as

$$\mathbf{R}_H = \mathbf{W} \cdot \mathbf{\Lambda} \cdot \mathbf{W}^H \tag{6.62}$$

where $\mathbf{W}$ contains all eigenvectors. Define a reduced version $\mathbf{W}_F$, of $\mathbf{W}$ to contain the eigenvectors corresponding to the F largest eigenvectors of $\mathbf{R}_H$. Then, $\mathbf{W}_F$ is used for a low-rank approximation of $\mathbf{R}_H$. Hence,

$$\mathbf{\Lambda}_F = \mathbf{W}_F^H \cdot \mathbf{R}_H \cdot \mathbf{W}_F \tag{6.63}$$

is a diagonal with the F largest eigenvalues as entries. If we use an outdated estimate $\hat{\mathbf{W}}_F$ instead of $\mathbf{W}_F$ for the low-rank approximation of $\mathbf{R}_H$, we get

$$\mathbf{R}_{\hat{\mathbf{W}}_F} = \hat{\mathbf{W}}_F^H \cdot \mathbf{R}_H \cdot \hat{\mathbf{W}}_F \tag{6.64}$$

which in general is not a diagonal. The traces of the matrices $\mathbf{\Lambda}_F$ and $\mathbf{R}_{\hat{\mathbf{W}}_F}$ are a measure for the collected power applying low-rank transforms $\mathbf{W}$ and $\hat{\mathbf{W}}$. Hence, the quotient of the traces provides knowledge about the power of having $\hat{\mathbf{R}}$ instead of $\mathbf{R}$. The F-eigenratio is defined as

$$q_{eigen}^{(F)} = \frac{tr\left\{\mathbf{R}_{\hat{\mathbf{W}}_F}\right\}}{tr\left\{\mathbf{\Lambda}_F\right\}} = \frac{\sum_{k=1}^{K_a} \sum_{f=1}^{F} \lambda_k \cdot |\vec{w}_k^H \cdot \hat{\vec{w}}_f|^2}{\sum_{f=1}^{F} \lambda_k} \tag{6.65}$$

The second form is the element notation. The term $|\vec{w}_k^H \cdot \hat{\vec{w}}_f|^2$ accounts for the mismatch of both eigenbases.

It is noted in passing that (array) gain metrics play an important role in comparing various antenna arrays, MIMO channel models and systems in a fair manner. The meaning of Mean Effective Link Gain (MELG) and Power Transmission Gain (PTG) has been elucidated in [SSKV04] and [WaWi04].

For instance, by comparing the power transmission gain of MIMO systems with different arrays in the same channel, conclusions on the performance of the arrays in terms of power can be drawn (compare Section 6.4).

6.7.8 Miscellaneous

So far the discussion has been focused on channel models. Propagation models have been validated also in COST 273. One question concerns the proper representation of *polarised waves* and cross-polarisation discrimination in MIMO propagation models. Several studies linked dual-polarisation models with capacity (in fixed wireless links [OeEP04], in indoor environments [KCVW03], and in reverberating chambers [DDLD04]) to prove their plausibility. One surprising result was that orthogonal polarisations decreased the average signal level, but also reduced the eigenvalue spread [SSKV04]. This observation was confirmed with measurement data. The influence of cross-polarised antennas in MIMO channel models, however, still seems to be awaiting extensive experimental validation.

An interesting concept to be applied in experimental validation of propagation models is the Multipath Component Cumulative (MCC) power [ZeST04], which represents the number of MCPs that are necessary to achieve 95% (or any other threshold value) of the total power.

Section 6.8 deals with the COST 273 MIMO Channel Model. This model incorporates novel concepts, the very latest measurements and derived model parameters, but papers to validate it as a whole are still lacking. A key concept is the Geometry-Based Stochastic Modelling (GSCM-) philosophy (see, e.g. [Moli04a]). An extension to MIMO, by double scattering and polarisation, of the COST 259 modelling approach, which has been amply validated [Corr01], has been developed [HoSt04]. First validation efforts show encouraging results on the fading statistics and capacity cdf of 4×4 MIMO channels, Figure 6.54.

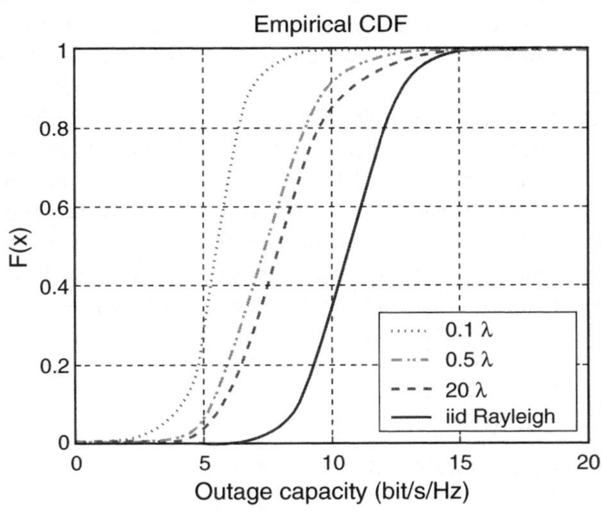

Figure 6.54 *An extension of COST 259 GSCM channel model to MIMO: modelled 4×4 capacity cdfs at SNR = 10 dB [HoSt04]. Antenna distance is the parameter.*

Summarising, one can say that, whereas many MIMO channel models have recently appeared in literature, much less effort has been made to validate these models rigorously. Among these efforts, COST 273 has spearheaded experimental validation and has born many ideas for novel metrics still to be applied. To promote the case for MIMO it would be helpful for MIMO model researchers to focus their ambition on the *cross-validation* of their models with measurement data from other groups.

6.8 The COST 273 MIMO Channel Model

6.8.1 Introduction

For the development, simulation, and testing of any wireless system, it is necessary to know the wireless propagation channel between the transmitter and receiver. The amount of information required for a 'sufficient' description of the channel depends on the system that is analysed. While for narrowband systems a characterisation of the channel attenuation was sufficient, wideband systems require the characterisation of the delay dispersion (impulse response), and smart antennas require directional information of the incident radiation at one link end, usually the BS. For all of these cases, previous COST Actions have made important contributions that are in worldwide use now: COST 231 extended the well-known Okumura-Hata model for the channel attenuation to frequency ranges and environments of interest to modern cellular systems; COST 207 derived models for the delay dispersion in cellular macrocells; these models were instrumental for the development of the GSM system. COST 259 finally derived a model for the directional characteristics in macro-, micro- and picocells, and merged this approach with a description of the attenuation and delay dispersion that is more general than the previous models.

Despite these important advances, there was a lack of accepted channel models for MIMO systems. While the 3GPP standardisation effort derived a model that is now being used for the selection of MIMO approaches in third-generation cellular systems, the description is limited to a small number of environments that is of most interest for cellular communications, but leaves out the important areas for fixed wireless access and wireless LANs; furthermore, this model also makes several simplifications that – while useful in the context of system simulations – restrict the general applicability of the model. COST 273 thus decided to derive a new, general, channel model for MIMO systems. This activity was performed by the subworking 2.1 of COST 273, and the outcome of its deliberations is presented in this section. Note that the model is based both on measurements and modelling approaches developed within COST 273, and on results presented in the open literature. In this section, we only reference the results from either of those sources; more details about specific measurement campaigns within COST 273 can be found in the other sections of this book. Due to space limitations, the description is somewhat terse.

6.8.2 Considered environments

As a first step, the environments for which the parametrisation is to be done had to be determined. Scenarios marked with (*) are mandatory scenarios for system tests, while the other ones are optional.

Macrocells

Macrocellular environments are generally defined as environments where the BS is placed above the rooftop height of the surrounding buildings. It is the most 'conventional' scenario for cellular applications.

1. Small macrocells in city centre (*): this environment described densely built-up areas, with buildings mostly flush next to each other (though parks and empty spaces surrounding major buildings can occur), and the street width typically smaller than the height of the buildings. The building structure is mostly homogeneous in the 'standard' environment; high-rise buildings that are much higher than the majority of buildings give rise to the 'bad urban' environment.
2. Large urban macrocells (*): this is mainly the same environment as above but with a BS far above rooftop level.
3. Suburban: this environment is characterised by a lower building density, as well as lower height of the buildings (typically one–three floors, surrounded by open space).
4. Fixed wireless access: this environment is similar to the small macrocell environment, but differs in the fact that the 'mobile terminal' is outdoors but at a greater height than for the urban scenario. Due to the fact that the 'mobile terminal' is fixed, a different description of the time variance of the channel is required.
5. Outdoor-to-indoor urban (*): this environment is similar to the small macrocells in city centres, but the mobile station is in an indoor, non-line-of-sight location; note that the height of the MT (above street level) can be much higher than in the normal urban environment.
6. Outdoor-to-indoor suburban: same as suburban, but with the MT in an indoor NLoS location.

Note that additional scenarios can occur, especially rural and hilly/mountainous environments. However, due to their relatively low importance for high-data-rate transmission, these scenarios are not included in this model.

Microcells

Microcellular environments are defined as environments where the BS height is at or below the level of the surrounding rooftops, but outdoors. The MT can be located either indoor or outdoor. Note that a cell with the BS at rooftop is sometimes called 'minicell'; it is, however, included in our description of the microcell height.

1. City centre (*): the city centre environment is defined similar to the urban environment. Due to the dense build-up, waveguiding through street canyons is an important phenomenon. Both street canyons and street crossings are in the environment.
2. Bad city centre: this environment is similar to city centre, but is impacted by high-rise buildings that act as far IOs.
3. Open place: this environment is characterised by a large open place (park, square) surrounded by buildings. Also a sports stadium and similar environments fall into this category.
4. BS outdoor – user indoors: this environment is the same as the city-centre environment, but with the MT indoors, and at possibly larger height (in an upper storey).
5. Peer-to-peer: in this environment, there is no BS; rather, two MTs communicate with each other, both of which are at street level.

Picocells

1. Halls (*): this environment contains big enclosed spaces for various usages, e.g. railway stations, airport halls, factory halls. It is assumed that the area covered by such a hall (with no interior walls) is at least 100 m^2, but can be up to several tens of thousands of m^2.
2. Tunnels: this environment contains railway and subway tunnels, car tunnels, and mines.
3. Corridors LoS: this case covers the scenario where both the BS and the MT are in a corridor, and LoS exists between the two.
4. Corridors NLoS: this covers the case where the BS is in a corridor, while the MT can be either in a different part of the corridor (without LoS), or in a room adjacent to the corridor.
5. Office LoS (*): this environment describes an office (covering both the case of cubicles in a larger hall, and a series of self-contained offices along a corridor), where the BS and MT have LoS. A necessary (but not sufficient) condition for LoS is that BS and MT are in the same office; larger distances with LoS can thus occur only in large offices (typically, the cubicle-type environment). Note that an office building is defined to be a concrete/steel/glass building, with a shape and size that is different from a residential environment. It is this building structure that defines the office environment, not the usage (offices put into a residential building are still viewed as 'residential').
6. Office NLoS (*): describes offices where the BS and the MT do not have LoS. They can be either in the same office, with the obstructions like desks and computers, or in different offices.
7. Home environments LoS: this environment describes a residential environment with relatively small rooms (no more than 30 m^2, but typically around 10 m^2 in a brick/plaster/wood structure). There is LoS between the transmitter and receiver.
8. Home environments NLoS: a residential building as defined above but without LoS between BS and MT. The two link ends can be in the same or in different rooms.

Ad hoc networks

Ad hoc network environments are characterised by the following properties: (i) all transceivers are at approximately the same height, (ii) all transceiver stations show nomadic mobility, i.e. remain static for an extended period of time, before being 'dropped' to a new location.

1. Office/residential LoS: ad hoc network in office or residential LoS environment as defined in section 'picocells'.
2. Office/residential NLoS: ad hoc network in office or residential NLoS environment as defined in section 'picocells'.
3. Halls: ad hoc network in hall environment as defined in section 'picocells'.

6.8.3 Generic channel model

In this subsection, we describe the generic channel model that can be used for the generation of Double Directional Impulse Responses (DDIRs) which is the ultimate goal of channel modelling. From these DDIRs, the transfer function matrix can be derived for arbitrary antenna configurations, just by specifying the antenna configurations and patterns of antenna elements at the two link ends. A discussion on how these transfer function matrices can be obtained is given in [Moli04b]; a suggestion for standard antenna configurations is given in [CCGH04].

The COST 273 model uses the same generic channel model for all types of environments. This is an important distinction to the COST 259 model, which used different generic models for macro-, micro- and picocells. While such an identical generic model might not result in the most accurate modelling structure possible, it is a great simplification for the actual implementation. Many of the aspects of the implementation are similar to the macrocellular model of the COST 259 model, which is described in [StMo01], and in greater detail in [MAHS05a] and [AGMP05].

The model distinguishes between external parameters, which are fixed for a simulation run, and describe the simulation environment, and the stochastic parameters, which are chosen according to a certain probability density function that is parameterised for the different environments. The external parameters also enter the parametrisation of the stochastic impulse responses.

External parameters

As mentioned above, external parameters are parameters that remain fixed for a simulation run. They might change according to the system that is simulated, and according to geographical regions (for example, the average rooftop height in city centres can be different in Northern Europe and in Japan).

External parameters for all environments

The following parameters are to be used in all environments:

f_c : *Carrier frequency* [Hz]: for most cellular applications the carrier frequency is in the 2 GHz band; for picocell and ad hoc applications (wireless LAN, etc.) the 5 GHz band is more common.

h_{BS}: *BS height* [m]: for macro- and microcellular environments, height signifies the height above ground. For picocells, it denotes the height of the BS above the floor the MT is in. Negative heights imply that the BS is on a lower floor.

h_{MT}: *Mobile terminal height* [m]: for outdoor environments, h_{MT} signifies the height above street level, and is typically 1.5 m when the MT is at street level; while for outdoor-to-indoor applications, h_{MT} can be much larger. For picocellular applications, h_{MT} signifies the height over the floor the user is on, and is therefore always 1.5 m. For ad hoc applications, the height of the MT is 0.8 m (tabletop height).

$\vec{r}_{BS}$: *BS position* [m]: the distance of the BS from the origin of the coordinate system. Under normal circumstances, the BS is fixed, and at the location $(0, 0, h_{BS})$. However, for ad-hoc scenarios and peer-to-peer communications, a trajectory for the 'BS' can be prescribed.

$\vec{r}_{MT}$: *Mobile terminal position* [m]: this parameter describes the trajectory of the MT through the cell for the duration of the simulation run.

Antenna scenarios (e.g. four-element Uniform Linear Array (ULA)) [no. of antennas, antenna spacing, array shape]: the antenna configuration is prescribed by the system designer, and identifies the number of considered antennas, the spacing between the antennas, the shape of the array, and the pattern and polarisation of the antenna elements.

Antenna orientation [pdf]: this is the pdf of the antenna orientation with respect to the coordinate system in which the movement of the MT is described.

Path-loss model [dB/m]: the path loss model is an external parameter that depends on the distance between the BS and the MT, i.e. $|\vec{r}_{BS} - \vec{r}_{MT}|$. The path-loss law is in general described by a power law (including a possible breakpoint). The exact formulations for the different environments are given (explicitly or by reference) in Subsection 6.8.4. For further use, we also define the excess path loss (EPL) as the difference between the actual path loss and the free-space path loss.

Additional external parameters for macro- and microcells

For macro- and microcells, several parameters describe the building structure. While the model suggests some standard values, they can be adjusted to fit specific cities.
h_B: *Average rooftop height* [m]: suggested values are 15 m for typical urban and 30 m for bad urban; as well as 8 m for suburban.
w_r: *Width of roads* [m]: suggested values are 15 m for typical urban, and 25 m for bad urban.
wb: *Distance between buildings* [m]: suggested values are 25 m for typical urban, and 100 m for bad urban.
ϕ_R: *Road orientation with respect to direct path* [degree]: recommended value is 45°.
These parameters are chosen to obtain maximum compatibility with the COST 259 model.

Additional external parameters for picocells and ad hoc networks

l_l, l_w: *Size of rooms* [m × m]. For residential environments, 3 × 5 m is recommended; for office environments with separate offices, 2 × 4; for offices with cubicle environments 10 × 10; and for halls 50 × 20.
N_{floor}: *Number of floors between BS and MT* [integer]: for standard situations, this number is 0, i.e. the BS and the MT are on the same floor.
Whether there is a building on the opposite side of the building in which BS and MT are located [yes/no]: by default, we assume that there is a building opposite the one in which the BS and MT are located.

Stochastic parameters

The stochastic parameters describe the different locations and radio environments in which the MT might be. Their parametrisation is influenced by the external parameters as reflected in the tables of Subsection 6.8.4.

Following the concepts of [MAHS05b], MPCs arrive in clusters. The total DDIR can thus be written as the sum of the cluster DDIRs, which in turn can be formulated as [AGMP05]

$$P\{cluster\}(\tau, \theta_{BS}, \varphi_{BS}, \theta_{MT}, \varphi_{MT}) = P_\tau(\tau)\, P_\theta^{BS}(\theta_{BS})\, P_\varphi^{BS}(\varphi_{BS})\, P_\theta^{MT}(\theta_{MT})\, P_\varphi^{MT}(\varphi_{MT}) \quad (6.66)$$

Note that this model assumes that *within one cluster*, azimuth spread, elevation spread, and delay spread are independent at the BS and the MT. Note that this is *not* the common Kronecker model that assumes the angular statistics to be independent at BS and MT. In our case here, the independence is per cluster, so that overall, there can still be significant coupling between DOAs and DODs.

Cluster generation – general considerations

The principal idea of the COST 273 Channel Model is to model the mean angles and delays of the clusters by geometric considerations, while the intracluster spreads and the small-scale fading can be represented by either a geometrical approach, or a tapped delay line representation. Also this philosophy is similar to the COST 259 model.

In this context, it is important to distinguish between the situations where the waves propagate from TX to RX via a single interaction (often called single scattering in the literature), as in [Corr01], and those where multiple interactions occur. Single interaction leads to a strong correlation between the delays and the angles at transmitter and receiver. For macrocells, single interaction works quite well whereas for indoor scenarios the correlation between delays and angles does not exist. To cope with the wide range of scenarios, the COST 273 model includes three kinds of clusters: local clusters around BS and/or MT, clusters incorporating single interaction, and finally a twin-cluster concept allowing for multiple IOs. Not all kinds of clusters are mandatory for all scenarios. In macrocells the single interaction cluster is the dominant propagation mechanism whereas in indoor environments multiple interaction processes account for most of the energy of the arriving radiation. The model finally specifies a 'selection parameter' K_{sel} that gives the ratio of single-interaction to multiple-interaction additional clusters. Since no measurements are available at the moment, we put this parameter to unity in macrocells, to 0.5 in microcells, and to zero in picocells. Local clusters show pure single-interaction behaviour.

Cluster generation – local clusters

One cluster always occurs around the MT, whereas a local cluster at the BS occurs only in certain environments. These clusters result in a large angular spread at MT or BS side respectively. For the local MT and BS cluster a pure single-scattering approach is used. The size of the local clusters is given by their delay spreads and the distribution of MPCs inside the cluster, as discussed below.

Visibility region

The concept of visibility regions is explained in [MAHS05b]. Each cluster of IOs is associated with a visibility region. If the MT is in a visibility region, then a cluster is active and contributes to the impulse response; if the MT is outside the visibility region, the cluster does not contribute. The visibility region is characterised by

R_C: size of the visibility region [m]

L_C: size of the transition region [m]

A smooth transition from non-active to active cluster is achieved by scaling the path gain of the cluster by a factor A_m^2. The transition function used is given as [MAHS05b]

$$A_m\left(\bar{r}_{MT}\right) = \frac{1}{2} - \frac{1}{\pi}\arctan\left(\frac{2\sqrt{2}y}{\sqrt{\lambda L_C}}\right) \tag{6.67}$$

with

$$y = L_C + |\bar{r}_{MT} - \bar{r}_m| - R_C \tag{6.68}$$

where $\bar{r}_m$ is the centre of the circular visibility area and λ is the wavelength.

Furthermore, the visibility region is characterised by the probability density function of its location which depends on the distance between the visibility region and the BS. In order to give a constant expectation for the number of clusters that equals N_C, the area density of the visibility regions needs to be [AGMP05]

$$\rho_C = \frac{N_C - 1}{\pi\left(R_C - L_C\right)^2} \quad \left[m^{-2}\right] \tag{6.69}$$

The position of the cluster belonging to one visibility region is discussed below.

Cluster generation – geometric approach for single-interaction clusters

For the single-interaction case, the cluster positions are determined in a geometric way. In a first step, visibility regions for the clusters are distributed throughout the cell. The number of visibility regions in the cell is chosen in such a way that a randomly placed MT sees, on average, the mean number of additional clusters N_C. Each visibility region is associated with one specific cluster. Note that due to this approach, the total number of clusters in the cell can become quite large, but only a small part of them is visible (and thus needs to be taken into account for the computations) at each point in time. The position of the cluster relative to the position of the BS and the cluster centre is determined by the following geometric approach: draw a line from the BS to the centre of the visibility region. The cluster position will be determined relative to that connection line. The radial distance from the BS is determined from an exponential distribution

$$f(r) = \begin{cases} 0 & r < r_{\min} \\ 1/(\sigma_r)\exp(-((r - r_{\min})/(\sigma_r)) & \text{otherwise} \end{cases} \tag{6.70}$$

The angle of the cluster centre is then drawn at random from a Gaussian distribution with a standard deviation $\sigma_{\phi,C}$. This fixes the position of the cluster. The delay (here assumed to be the minimum delay, though this is not exact), azimuth as seen from the BS, and azimuth as seen from the MT, are then computed via simple geometrical relationships. These values are used for further computations.

Cluster generation – angular spectrum approach

For the multiple-interaction clusters, the mean DoA, DoD, and minimum delay are computed as random realisations from the marginal distributions, taken over a large measurement area. Since the variables are drawn from the marginal distributions, i.e. the power angular spectrum at the receiver, power angular spectrum out the transmitter, and power delay profile, this means that delay and angles are independent. However, we stress that this does *not* result in a Kronecker model, i.e. the angular delay power spectrum is not separable.

Let us now consider a more precise formulation. The DoA of a cluster is assigned by computing a uniformly distributed random number x, and then computing $\phi = \text{cdf}_{DoA}^{-1}(x)$, where cdf_{DoA} is the marginal cumulative distribution function of the power-weighted DoA distribution (power angular spectrum), averaged over a large area (not to be confused with the small-scale averaged APS). This is repeated for the DoD and the intercluster delay τ_C. In a next step a delay spread τ and the angular spreads φ_{BS}, φ_{MT}, θ_{BS}, and θ_{MT} are assigned to the cluster. The delay spread to be assigned is a function of the distance between BS and MT. A realisation of clusters in the DoA-DoD plane is shown is Figure 6.55. It can be extended to the delay and/or elevation domain by adding additional orthogonal axes.

Cluster generation – geometric approach for multiple-interaction clusters

In the multiple-interaction case each cluster is divided into a cluster corresponding to the BS side and one at the MT, Figure 6.56. The angular dispersion at BS and MT can therefore be modelled independently. In order to limit complexity, the corresponding clusters look like twins, having the same IO distributions and long-term behaviours. Furthermore each IO at the BS side cluster has exactly one counterpart at the MT side. Therefore the total number of multipath components is equal to the number of IOs. The distribution of IOs within a cluster is discussed in Subsection 6.8.3.

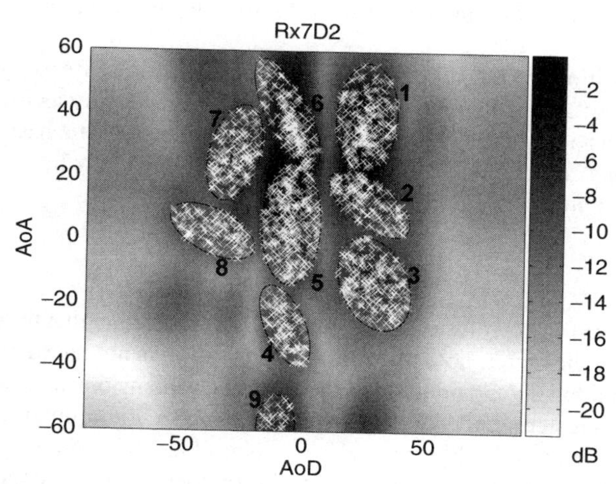

Figure 6.55 *DoA-DoD spectrum of an indoor scenario with nine identified clusters [CHÖB04a].*

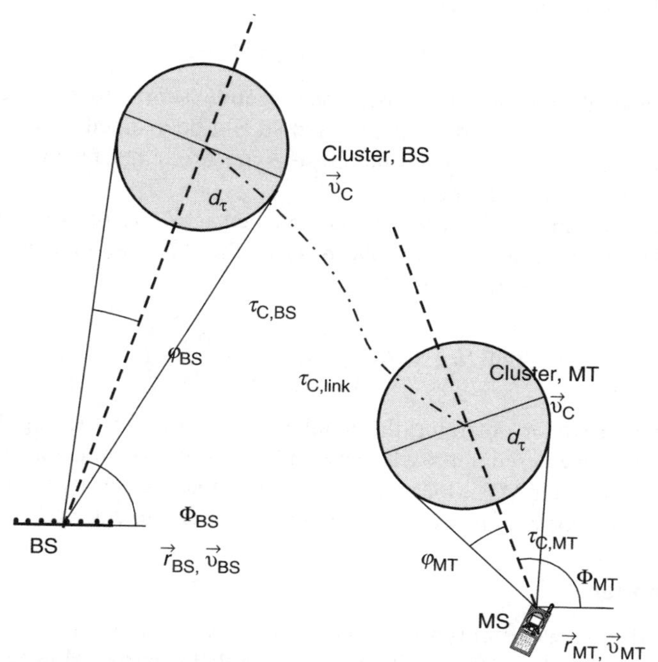

Figure 6.56 *Cluster placement in the xy plane.*

In the following the concept is derived as to how to model a cluster based on the above assumptions. The large-scale behaviour is again an essential part of the model. The model is therefore based on a map in space. This models the movement of MTs and clusters compared to an approach in the delay-azimuth plane.

In a first step the positions of the BS and MT at their start positions are assigned, Figure 6.56. The spreads and directions of a cluster are assigned at random the same way as for the angular spectrum approach above. The distance of the cluster from BS/MT is given as a stochastic parameter $d_{\mathrm{C,BS/MT}}$. If this parameter is not available the following geometric approach may be used. The position of the cluster in space is computed in the following way:

$$d_{\mathrm{BS/MT}} = \frac{d_{\mathrm{C}}}{2\tan(\varphi_{\mathrm{BS/MT}})} \tag{6.71}$$

where d_{C} is derived from the intercluster delay spread. This makes sure that the dimension in the x and y directions is identical; however, note that the parametrisation is not based on a physical argument. Note also that when talking about spreads and the resulting dimensions of clusters, we always mean the σ-RMS spread. A cluster should cover an area of about 3σ for simulations. Figure 6.56 shows the cluster definition for the two-dimensional case. Note that the spreading of the cluster can be represented either in the geometrical plane (as in the figure), or in the delay-azimuth plane. It is obvious that the position of a cluster as seen from the BS is not the same as it is seen from the MT. An additional cluster-link delay $\tau_{\mathrm{C,link}}$ is introduced, which ensures that the total delay of the cluster corresponds to the definitions of the scenario:

$$\tau_{\mathrm{C}} = \tau_{\mathrm{C,BS}} + \tau_{\mathrm{C,link}} + \tau_{\mathrm{C,MT}} \tag{6.72}$$

It may happen that the total delay of a cluster becomes negative (shorter than the delay corresponding to the LoS connection). This is an artificial case and should be avoided by a repositioning of the cluster. The situation occurs mainly when the clusters are originally far away from the MT but subsequently the MT moves close to them.

The distribution of the number of clusters N_{C} is modelled as $N_{\mathrm{C,min}}$ (corresponding to the cluster originating from interactions around the MT, plus possibly the cluster around the BS) plus a Poisson-distributed random variable with parameter N_{P}:

$$\mathrm{pdf}(N_{\mathrm{C}}) = N_{\mathrm{C,min}} + e^{-N_{\mathrm{P}}}\left(\frac{N_{\mathrm{P}}^{N_{\mathrm{C}}}}{N_{\mathrm{C}}!}\right) \tag{6.73}$$

This distribution is used for obtaining the number of clusters for the angular-spectrum-based approach. For the geometry-based approach, the number of clusters immediately follows from the number of visibility regions the MT is located in. Remember that the density of visibility regions was chosen to give the correct mean number of clusters in each environment.

Cluster power model

The power contained in each cluster is a function of the delay (with respect to the LoS or quasi-LoS component). The longer the delay, the smaller is the power that it carries. However, there is a limit to the cluster attenuation – if the attenuation becomes too high, the cluster does not have an impact on the impulse response, and is thus dropped from the considerations. The power of the mth cluster is

$$P_m = P_0 \max\left\{\exp[-k_\tau(\tau_{\mathrm{m}} - \tau_0)], \exp[-k_\tau(\tau_{\mathrm{B}} - \tau_0)]\right\} \tag{6.74}$$

The parameters describing this equation are

k_τ: attenuation coefficient given in units of dB/μs

τ_0: delay of the LoS component given in units of μs,

τ_B: cut-off delay given in units of μs

LoS occurrence

For some environments the occurrence of LoS is modelled stochastically. The modelling approach has a strong similarity to the visibility region for the clusters, and thus need not be described again. The major difference is that the probability for LoS decreases strongly with the distance of the MT from the BS, and is zero after a cut-off distance d_{co}. The model is thus described by the following parameters:

d_{co}[m]: cut-off distance for LoS

R_L[m]: radius of visibility region for LoS

L_L[m]: size of transition region for LoS visibility region

Depending on the existence of an LoS connection, the LoS power factor (power of the first component, compared to the power of all other components) varies. In either case it is modelled as a log-normal random variable, but the mean and variance are different whether there is an LoS or a quasi-LoS. The quasi-LoS case describes an NLoS scenario with Ricean fading of the local cluster and therefore a weak LoS component is modelled. The parameters are thus

μ_K is the mean of the LoS power factor

σ_K is the standard deviation of the LoS power factor

Cluster dispersion

The Double Directional Delay Power Spectra (DDDPS) (i.e. the squared magnitude of the DDIR, averaged over the small-scale fading) can be characterised for each cluster by its dispersion in the following domains: delay, azimuth at the BS, elevation at the BS, azimuth at the MT, elevation at the MT.

In the delay domain an exponentially decaying power-delay profile is used. The exponential profile is characterised by the decay constant σ_τ, which is identical to the well-known RMS delay spread

$$P_\tau(\tau) = \frac{1}{\sigma_\tau} e^{-(\tau-\tau_m)/\sigma_\tau} \tag{6.75}$$

The delay spread is itself a log-normal random variable, with a mean $m_{S\tau}$ (given in ns) and standard deviation $S_{S\tau}$ (given in dB). Note that the mean increases with increasing distance between BS and MT [GEYC97], as

$$m_{S\tau} = \widetilde{m}_{S\tau} d^{-\varepsilon} \tag{6.76}$$

In [Corr01] it was shown that $\varepsilon = 0.5$ fits measurement data for some environments; due to the absence of any other (contradicting) data, it is chosen as the default value for all scenarios. For the angular dispersion at the BS, Laplacian power spectra are used

$$P_{\varphi}(\varphi) = \frac{1}{\sigma_{\varphi}\sqrt{2}} e^{-\sqrt{2}|\varphi - \varphi_m|/\sigma_{\varphi}} \tag{6.77}$$

where the azimuthal spread σ_{φ} is a log-normal random variable with mean $m_{S\varphi}$ (given in degrees) and standard deviation $S_{S\varphi}$ (given in dB). Similarly, the elevation power spectrum is given as

$$P_{\theta}(\theta) = \frac{1}{\sigma_{\theta}\sqrt{2}} e^{-\sqrt{2}|\theta - \theta_m|/\sigma_{\varphi}} \tag{6.78}$$

where the elevation spread σ_{θ} is a log-normal random variable with mean $m_{S\theta}$ and standard deviation $S_{S\theta}$.

Similarly, the angular parameters are also defined for the MT. It is noteworthy that those parameters might depend on the delay of the cluster. For example, the elevation spread is usually different for early-arriving components (over the rooftop in macrocells) and late-arriving components (guided along the street canyons). Also, for some environments, a Laplacian description of the elevation spectrum might not be optimum.

For a geometrical implementation of the model, it is theoretically possible to map the delay-angle distributions onto IO distributions [MoLK98]. However, for a practical implementation of the model the exact distributions are not essential. For the single-IO clusters, approximate IO distributions are given in [LaMB98]. For the twin-cluster approach, a truncated Gaussian distribution of the IOs is preferable (this distribution is related to the Von Mises distribution on the angular domain when the clusters are small ($\theta_{\text{RMS}} < 14°$)) [Fleu00]. We thus propose that the IO distribution in space is drawn from the following probability density function (shown here for one dimension)

$$P_{\text{x}}(x) = \begin{cases} \frac{1}{\sqrt{2\pi\sigma_{\text{x}}^2}} e^{-\frac{1}{2}\left(\frac{x-\mu_{\text{x}}}{\sigma_{\text{x}}}\right)^2}, & \text{for } |x| \le x_{\text{T}} \\ 0, & \text{for } |x| > x_{\text{T}} \end{cases} \tag{6.79}$$

where x_{T} denotes the truncation value. Choosing $x_{\text{T}} = 3\sigma_{\text{x}}$ ensures that nearly the whole Gaussian distribution is kept inside the cluster area. Since the clusters are defined in such a way that the size in the xy plane is circular and identical for the BS and MT side the same relative positions can be used for the BS and MT cluster in the xy plane. To adjust for independent elevation spreads the elevation of the IOs in the MT cluster is refined by $h_{\text{C,MT}}/h_{\text{C,BS}}$ compared to the BS side. $h_{\text{BS/MT}}$ denotes the modified height of the cluster and is computed as

$$h_{\text{C,BS/MT}} = d_{\text{BS/MT}} \tan(\theta_{\text{BS/MT}}) \tag{6.80}$$

For an exponential PDP the cluster has to be rotated at the BS and the MT in such a way that it is orientated towards the BS and away from the MT, for a mathematical formulation see Table 6.6. For the LoS component, the amplitude is characterised through a Ricean distribution

$$\text{pdf}(r) = \frac{r}{\sigma_K^2} \exp\left(-\frac{r^2 + A_K^2}{2\sigma_K^2}\right) I_0\left(\frac{rA_K}{\sigma_K^2}\right) \tag{6.81}$$

where $I_0(x)$ denotes the zeroth-order modified Bessel function of first order, A_K is the amplitude of

Table 6.6 *Implementation of multiple interacting clusters.*

Parameter	Value	Computation
BS/MT position	$\vec{r}_{BS}$, $\vec{r}_{MT}$	
MT/BS velocity	$\vec{v}_{BS}$, $\vec{v}_{MT}$	
Angles of clusters	$\theta_{BS}, \varphi_{BS}, \theta_{MT}, \varphi_{MT}$	$P_\theta(\theta)$, $P_\varphi(\varphi)$, $P_{\theta'}(\theta', \tau)$, $P_{\varphi'}(\varphi', \tau)$
Size of cluster in xy plane	d_C	$P_\tau(\tau_C)c_0$
Distance of clusters from BS/MT	$pdf(d_{C,BS/MT})/d_{BS/MT}$	$d_C/\tan(\varphi_{BS/MT})$
Position of clusters	$\vec{r}_C^{BS/MT}$	$d_{BS/MT}\mathrm{T}(\varphi_{BS/MT}, \theta_{BS/MT})$
Size of cluster in z	$h_C^{BS/MT}$	$d_{BS/MT}\tan(\varphi_{BS/MT})$
Relative IO positions	$\vec{r}$	$\mathcal{N}\left(0, \begin{pmatrix} 1 & 0 & 0 \\ 0 & 1 & 0 \\ 0 & 0 & 1 \end{pmatrix}\right)$
IO positions of add. clusters	$\vec{r}_{C,IO}^{BS}$	$\vec{r}\begin{pmatrix} d_C \\ d_C \\ h_C^{BS} \end{pmatrix}\mathrm{T}(\varphi_{BS}, \theta_{BS})$
	$\vec{r}_{C,IO}^{MT}$	$\vec{r}\begin{pmatrix} d_C \\ d_C \\ h_C^{MT} \end{pmatrix}\mathrm{T}(\varphi_{MT} + \pi, \theta_{MT} + \pi)$
Cluster link delay	$\tau_{C,link}$	$\tau_C - \tau_{C,BS} - \tau_{C,MT}$
IO delay	$\tau_{C,IO}$	$\tau_{C,link} + \left(\mid \vec{r}_C^{MT} + \vec{r}_{C,IO}^{MT} \mid + \mid \vec{r}_C^{BS} + \vec{r}_{C,IO}^{BS} \mid\right)/c_0$
Rotation matrix	$\mathrm{T}(\varphi, \theta)$	$\begin{pmatrix} \cos(\varphi)\cos(\theta) & -\sin(\varphi) & \cos(\varphi)\sin(\theta) \\ \sin(\varphi)\cos(\theta) & \cos(\varphi) & \sin(\varphi)\cos(\theta) \\ -\sin(\theta) & 0 & \cos(\theta) \end{pmatrix}$

T denotes the spherical rotation matrix in space, and $\mathcal{P}(n)$ is a Poisson process.

the LoS component, and σ_K is the variance of the scattered component. The Rice factor K_{MPC} is given by $A_K^2/(2\sigma_K^2)$ and depends on the environment.

Shadow fading

Following a widely used approach, each cluster undergoes shadow fading, which is modelled log-normally distributed with standard deviation σ_{S} [dB]. The mean of the shadowing variance (see below) is correlated with the path loss.

Autocorrelation distances and cross-correlations

The shadow fading, delay spreads and angular spreads are correlated random variables, given as

$$S_{\mathrm{m}} = 10^{s_{\mathrm{S}} X_{\mathrm{m}}/10} \tag{6.82}$$

$$\sigma_{\tau,\mathrm{m}} = m_{\mathrm{s}\tau} \left(\frac{d}{1000}\right)^{\varepsilon} 10^{s_{\mathrm{s}\tau} Z_{\mathrm{m}}/10} \tag{6.83}$$

$$\sigma_{\varphi\mathrm{BS},\mathrm{m}} = m_{\mathrm{s}\varphi\mathrm{BS}} 10^{s_{\mathrm{s}\varphi\mathrm{BS}} Y_{\mathrm{m}}/10} \tag{6.84}$$

where S_{m} denotes the shadow fading of cluster m, $\sigma_{\tau,\mathrm{m}}$ is the delay spread, and $\sigma_{\varphi\mathrm{BS},\mathrm{m}}$ stands for the angular spread in azimuth for the BS side. X_{m}, Y_{m}, and Z_{m} are correlated normal random variables with zero mean and unit variance. Correlated random processes can be computed using the Cholesky factorisation [MoKl99]. Equation (6.84) can also be used for the elevation spread, and for the angular spreads at the MT, by replacing φ_{BS} with θ_{BS}, φ_{MT}, and θ_{MT}, respectively. We thus end up with six log-normal random variables that are characterised by their cross-correlation coefficients ρ_{ij}, with $i = 1..6$, $j = 1..6$.

Furthermore, the shadowing as well as the delay and angular spreads change as the MT moves over large distances and are therefore characterised by a spatial autocorrelation function. This function is assumed to be exponential

$$ACF(x, x') = \exp(-|x - x'|/L_x) \tag{6.85}$$

The autocorrelation distances L_x for the different spreads are thus also important parameters of the model.

Polarisation

For the characterisation of the polarisation it is assumed that orthogonal polarisations suffer from statistically independent small-scale fading, while all other statistical parameters (e.g. shadowing, delay spread, etc.) are identical. The polarisation is thus characterised by the polarisation matrix

$$\begin{pmatrix} P_{VV} & P_{VH} \\ P_{HV} & P_{HH} \end{pmatrix} \tag{6.86}$$

where the entries characterise the powers, averaged over the small-scale fading. We find that the XPD, i.e. the ratio

$$\mathrm{XPD} = \frac{P_{VV} + P_{HH}}{P_{VH} + P_{HV}} \tag{6.87}$$

is a log-normally distributed variable, characterised by its mean m_{XPD} [dB] and the standard deviation S_{XPD}. Following the distribution of the XPD, all other polarisation coefficients are modelled log-normally as well. The XPD does depend on the distance between TX and RX, as well as on the delay and the angles of the MPCs. For some selected environments, such dependencies have been measured [SZMS06], [KLKK02]. However, those results are insufficient to give a general parametrisation relationship, and are thus not considered further here.

Temporal variations in fixed wireless systems and nomadic applications

For fixed wireless systems, we need to define the temporal K-factor, which describes the ratio of the power in the time-invariant MPCs to that of the time-variant MPCs. First of all, the K-factor of the LoS component is given by

$$K_{\mathrm{LoS}[dB]} = 10\log_{10}\left[F_s\left(\frac{h_{\mathrm{MT}}}{3}\right)^{0.46} K_{\mathrm{LoS},0}\left(\frac{d}{1000}\right)^{-0.5}\right] + s_K U_{\mathrm{LoS}} \tag{6.88}$$

where F_s is a seasonal factor, $F_s = 1$ in summer (leaves) and 2.5 in winter (no leaves), $K_{\mathrm{LoS},0}$ is to be estimated (suggested value: 8 to 9), s_K is the standard deviation (6 to 8 dB), and U_{LoS} is normal variable. Note that we use the notation h_{MT} to designate the height of the user terminal, although this one is by definition fixed.

The K-factor K_m of the mth cluster is then expressed as follows:

$$K_{m[dB]} = 10\log_{10}(m_K) + s_K U_m \tag{6.89}$$

where m_K is the average K-factor of the cluster, s_K is the standard deviation (6 to 8 dB) and U_m is a Gaussian variable which must be correlated with X_m, Y_m and Z_m.

The average cluster K-factor can be represented by

$$m_K = F_s\left(\frac{h_{\mathrm{MT}}}{3}\right)^{0.46} K_{c,0}\left(\frac{d}{1000}\right)^{-0.5} \exp\left(-\frac{\tau_m}{\beta_K}\right) \tag{6.90}$$

where $K_{c,0}$ is to be estimated (suggested value: 0.5), and an estimation of β_K is 1.5 dB/100 ns (for bandwidths up to 10 MHz).

Diffuse scattering

Diffuse scattering is the part of the measured signal which cannot be resolved in the temporal domain. First results have been presented in [ThRi05]. The diffuse component can be described by the ratio of power in the discrete components vs the diffuse power with mean level μ_{diff} and standard deviation σ_{diff}. The PDP of the diffuse component is modelled uniformly in azimuth and exponentially in decay.

6.8.4 Parameter values in the different environments

Table 6.7 *External parameters – macrocells (large urban macrocells).*

Parameter	Value	Comments, incl. references
f_c [Hz]	900 MHz–2 GHz	2 GHz is typical because of UMTS
h_{BS} [m]	50 m	From COST 259
h_{MT} [m]	1.5 m	Pedestrian walking (note that MT is outdoors)
$\vec{r}_{BS}$ [m]	(0,0,0)	Origin of coordinate system
$\vec{r}_{MT}$ [m]	Uniform distributed in cell area	
Cell radius [m]	1000	
N_{ant}, d_{ant}, $\vec{r}_{array}$ [.,m,m]	Different arrays possible	Typical case: ULA with 4 elements, $\lambda/2$ spaced at BS; ULA with 2 elements dual polarised at MT
φ_{ant} [pdf]	Uniform	User can be rotated in all directions. Tilt of MT antenna not covered here
P_L [dB/m]	COST 231 Walfish-Ikegami	Identical to COST 259
$h_{rooftop}$ [m]	15 m	From COST 259 (GBU)
w_r [m]	25 m	From COST 259 (GBU)
w_b [m]	50	From COST 259 (GBU)
φ_{road} [°]	45	From COST 259

Table 6.8 *External parameters – microcells (city centre).*

Parameter	Value	Comments, incl. references
f_c [GHz]	1–5	
h_{BS} [m]	3–10	From COST 259
h_{MT} [m]	1.5	From COST 259
$\vec{r}_{BS}$ [m]	$(0,0,h_{BS})$	
$\vec{r}_{MT}$ [m]	Uniform distr. inside cell sector	
Cell radius [m]	Any	
N_{ant}, d_{ant}, $\vec{r}_{array}$ [.,m,m]	Any	
φ_{ant} [pdf]	Uniform	
P_L [dB/m]	$10 \cdot n \cdot \log_{10}(d) + L_0(1m)$ $n = 2.6$, $L_0(1m) = 20 \cdot \log_{10}(4\pi(1m)/\lambda)$	From COST 259 (GSN)
$h_{rooftop}$ [m]	15	From COST 259
w_{road} [m]	20	From COST 259
w_{street} [m]	50	
φ_{road} [rad]	$\frac{\pi}{4}$	

Table 6.9 *External parameters – picocells (halls).*

Parameter	Value	Comments, incl. references
f_c [Hz]	2 GHz–5 GHz	2–5 GHz is typical range of WLAN
h_{MT} [m]	2	
$\vec{r}_{\mathrm{BS}}$ [m]	(0,0,0)	Origin of coordinate system
$\vec{r}_{\mathrm{MT}}$ [m]	Uniform inside each office	
Cell radius [m]	30	
$N_{\mathrm{ant}}, d_{\mathrm{ant}}, \vec{r}_{\mathrm{array}}$ [.,m,m]	Different arrays possible	Typical case: ULA with 4 elements, $\lambda/2$ spaced at BS; ULA with 2 elements dual polarized at MT
φ_{ant} [pdf]	Uniform	User can be rotated in all directions. Tilt of MT antenna not covered here
P_{L} [dB/m]	COST 273	TD02-055
A_{room} [m × m]	2 × 3	
N_{floor}	4–10	
Opposite building [y/n]	y	

Table 6.10 *Stochastic parameters – macrocells (large urban macrocells).*

Parameter	Value	Comments, incl. references
Visibility region		
$\mathrm{pdf}_{\mathrm{cluster}}(n)$	Poisson	From COST 259
R_{C} [m]	100	From COST 259
L_{C} [m]	20	From COST 259
Cluster movement		
$\mu_{C,v}$ [m/s]/$\sigma_{C,v}$ [dB]	0/0	No cluster movement in macro-cells
Cluster power		
k_τ [dB/μs]	1	From COST 259
τ_{B} [μs]	10	From COST 259
Line of sight		
d_{co} [m]	500	
R_{L} [m]	30	
L_{L} [m]	20	
μ_{K} [dB]/σ_{K} [dB]	$((26 - EPL)/6)/6$	
K_{sel}	1	Only single interacting clusters
Average number of local clusters N_{C}	1(MT)	From COST 259 (BU)
Average number of add. clusters $N_{\mathrm{C,add}}$	1.18	From COST 259 (BU)
Marginal DoA pdf	n.a.	
Marginal DoD pdf	n.a.	

Continued

Table 6.10 *Continued*

Parameter	Value	Comments, incl. references
pdf of the delay of clusters	n.a.	
Number of MPCs per cluster N_{MPC}	20	From [Kott05]
Rice factor of additional clusters K_{MPC}	0	gw
Diffuse radiation		
μ_{diff}/σ_{diff}	0.05/3.4	[Kott05]
PDP of the diffuse radiation	Uniform in azimuth, $\exp(-t/\tau)$, $\tau = 0.5\,\mu s$	
Delay spread, goes with $d^{0.5}$		
μ_τ [μs]/σ_τ [dB]	0.4/3	
Angular spread at BS		
$\sigma_{AS} = 10^{\varepsilon x+\mu}$, $x = N(0,1)$		
$\mu_{\varphi BS}$ [°]/$\sigma_{\varphi BS}$ [dB]	0.81/0.34	From 3GPP
$\mu_{\theta BS}$ [°]/$\sigma_{\theta BS}$ [dB]	0.5/3	From COST 259
Angular spread at MT		
$\mu_{\varphi MT}$ [°]/$\sigma_{\varphi MT}$ [dB]	35/0	From 3GPP
pdf(θ_{MT}) [°]	uniform [0, 45]	From COST 259
Shadowing		
σ_S [dB]	6	From COST 259
Autocorrelation distances		
L_S [m]	100	All values set to 100 (COST 259); no separate measurements available
L_τ [m]	100	
$L_{\varphi BS}$ [m]	100	
$L_{\theta BS}$ [m]	100	
$L_{\varphi MT}$ [m]	100	
$L_{\theta MT}$ [m]	100	
Cross-correlations		
ρ	$\rho_{\tau-\phi_{BS}} = 0.5$ $\rho_{\tau-\phi_{MT}} = 0$ $\rho_{\tau-S} = -0.6$ $\rho_{\phi_{BS}-\phi_{MT}} = 0$ $\rho_{\phi_{BS}-S} = -0.6$ $\rho_{\phi_{MT} S} = 0$	Unknown correlation values set to zero; no correlation of elevation angles assumed
$\rho_{\text{BS-BS}}$	0.5	From 3GPP
Polarisation		
μ_{XPD} [dB]/σ_{XPD} [dB]	6/2	From COST 259
μ_{VVHH} [dB]/σ_{VVHH} [dB]	$0/-\infty$	
μ_{VHHV} [dB]/σ_{VHHV} [dB]	$0/-\infty$	

Table 6.11 *Stochastic parameters – microcells (city centre).*

Parameter	Value	Comments, incl. references
Visibility region		
$\text{pdf}_{\text{cluster}}(n)$	Poisson	
R_C [m]	50	
L_C [m]	20	
Cluster movement		
$\mu_{C,v}$ [m/s]/$\sigma_{C,v}$ [dB]	0/0	Clusters assumed static, but the MT moves
Cluster power		
k_τ [dB/μs]	40	From measured mean PDP
τ_B [μs]	0.5	Slope in NLoS-LoS route
Line of sight		
d_{co} [m]	300	
R_L [m]	50	
L_L [m]	50	
μ_K/σ_K	7/2.3	Values from 2 and 5 GHz LoS measurements
K_{sel}	0.5	
Number of local clusters N_C	1	
Average number of add. clusters $N_{C,add}$	3	From median of clusters at 10% significance at 2 and 5 GHz
Marginal DoA pdf	Equal	Random user orientation
Marginal DoD pdf	Equal in $[-\pi/8 - \pi/8]$	Angle 0 towards the sector centre
pdf of the delay of clusters	Probability	Decreasing, zero after 0.5 μs
Number of MPCs per cluster N_{MPC}	7	
Rice factor of additional clusters K_{MPC} [dB]	2	
Percentage of power in the diffuse radiation	10	
PDP of the diffuse radiation [dB/μs]	−40	If $e(-t/\tau)$, $\tau = 0.11$ μs
Delay spread		
μ_τ [μs]/σ_τ [dB]	13/14	
Angular spread at BS		
$\mu_{\varphi BS}$ [°]/$\sigma_{\varphi BS}$ [dB]	2.3/3.4	These are from MT, but same values are used for BS since no other data was available
$\mu_{\theta BS}$ [°]/$\sigma_{\theta BS}$ [dB]	1.3/3.3	
Angular spread at MT		
$\mu_{\varphi MT}$ [°]/$\sigma_{\varphi MT}$ [dB]	2.3/3.4	
$\mu_{\theta MT}$ [°]/$\sigma_{\theta MT}$ [dB]	1.3/3.3	
Shadowing		
σ_{IO} [dB]	2.9	

Continued

Table 6.11 *Continued*

Parameter	Value	Comments, incl. references
Autocorrelation distances		
L_{IO} [m]	5	
L_τ [m]	5	
$L_{\varphi\text{BS}}$ [m]	50	
$L_{\theta\text{BS}}$ [m]	50	
$L_{\varphi\text{MT}}$ [m]	25	
$L_{\theta\text{MT}}$ [m]	25	
Cross-correlations		
ρ	$\rho_\tau - \phi_{BS} = 0.1$ $\rho_\tau - \phi_{MT} = 0.1$ $\rho_\tau - S = 0.04$ $\rho_{\phi_{BS}} - \phi_{MT} = 0$ $\rho_{\phi_{BS}} - S = -0.2$ $\rho_{\phi_{MT}} - S = -0.2$	BS values from MT, no clusterised DoD measurements available
$\rho_{\text{BS-BS}}$		
Polarisation		
μ_{XPD} [dB]/σ_{XPD} [dB]	8.5/1.8	Polarisation parameters are extracted from microcell NLoS and LoS 5 GHz measurements
μ_{VVHH} [dB]/σ_{VVHH} [dB]	0.3/3.2	
μ_{VHHV} [dB]/σ_{VHHV} [dB]	−0.5/1.8	

Table 6.12 *Stochastic parameters – picocells (halls).*

Parameter	Value	Comments, incl. references
Visibility region		
$\text{pdf}_{\text{cluster}}(n)$	Poisson	
R_{C} [m]	n.a.	
L_{C} [m]	n.a.	
Cluster movement		
$\mu_{C,v}$ [m/s]/$\sigma_{C,v}$ [dB]	0/0	No cluster movement in macro-cells
Cluster power		
k_τ [dB/μs]	50–100	ChLa03, Medb05
τ_{B} [μs]	n.a.	
Line of sight		
d_{co} [m]	n.a.	
R_{L} [m]	n.a.	
L_{L} [m]	n.a.	
$\mu_{\text{K}}/\sigma_{\text{K}}$	n.a.	
K_{sel}	0	Only multiple interacting clusters

Table 6.13 *Stochastic parameters – picocells (halls).*

Parameter	Value	Comments, incl. references
Average number of local clusters N_C	1	
Average number of add. clusters $N_{C,add}$	3	Medh05
Marginal DoA pdf (BS)	$\delta(\varphi_{BS}) + \delta(\varphi_{BS} - 180°)$	Medh05
Marginal DoD pdf (MT)	$\delta(\varphi_{MT} - 90°)$	Medh05
pdf of the delay of clusters	Uniform between $\tau_{min} < \tau < \tau_{max} = 30/k_\tau$	
Number of MPCs per cluster N_{MPC}	>20	
Rice factor of additional clusters K_{MPC}	0.5	Medh05
Percentage of power in the diffuse radiation	<10	
PDP of the diffuse radiation	Same as corresponding clusters	
Delay spread		
Cluster decay [dB/μs]	150–800, uniform pdf	
Angular spread at BS		
$\mu_{\varphi BS}$ [°]	$5° + k\exp(-(0.15d)^2)(k = 30° - 70°$, uniform pdf)	
$\mu_{\theta BS}$ [°]	5–10, uniform pdf	
Angular spread at MT		
$\mu_{\varphi MT}$ [°]	30–70, uniform pdf	
$\mu_{\theta MT}$ [°]	5–10, uniform pdf	
Shadowing		
σ_{IO} [dB]	3	MeBe02
Autocorrelation distances	Unknown	
L_{IO} [m]	n.a.	
L_τ [m]	n.a.	
$L_{\varphi BS}$ [m]	n.a.	
$L_{\theta BS}$ [m]	n.a.	
$L_{\varphi MT}$ [m]	n.a.	
$L_{\theta MT}$ [m]	n.a.	
Cross-correlations		
ρ	n.a.	
$\rho_{BS\text{-}BS}$	Unknown	
Polarisation		
μ_{XPD} [dB]/σ_{XPD} [dB]	−9/2	
μ_{VVHH} [dB]/σ_{VVHH} [dB]	$0/-\infty$	
μ_{VHHV} [dB]/σ_{VHHV} [dB]	$0/-\infty$	

References

[3GPP01] 3GPP. Deployment aspects. TS25.943, TSG RAN, June 2001.

[AGMP05] H. Asplund, A. A. Glazunov, A. F. Molisch, K. I. Pedersen, and M. Steinbauer. The COST 259 directional channel model - II. macrocells. *IEEE J. Select. Areas Commun.*, 2005. Submitted.

[AlTM03a] P. Almers, F. Tufvesson, and A. F. Molisch. Keyhole effects in MIMO wireless channels - measurements and theory. In *Proc. Globecom 2003 - IEEE Global Telecommunications Conf.*, San Francisco, CA, USA, Dec. 2003. [Also available as TD(03) 179].

[AlTM03b] P. Almers, F. Tufvesson, and A. F. Molisch. Measurement of keyhole effect in a wireless multiple-input multiple-output (MIMO) channel. *IEEE Comms. Letters*, 7(8):373–375, Aug. 2003. [Also available as TD(03)179].

[AMSM02] H. Asplund, A. F. Molisch, M. Steinbauer, and N. B. Mehta. Clustering of scatterers in mobile radio channels-evaluation and modelling in the COST259 directional channel model. In *Proc. ICC 2002 - IEEE Int. Conf. Commun.*, New York, NY, USA, Apr. 2002.

[Ande00] J. B. Andersen. Array gain and capacity for known random channels with multiple element arrays at both ends. *IEEE J. Select. Areas Commun.*, 18(11):2172–2178, Nov. 2000.

[Ande04] J. B. Andersen. Multipath richness - a measure of MIMO capacity. TD(04)157, COST 273, Duisburg, Germany, Sep. 2004.

[AnHa77] J. B. Andersen. and F. Hansen. Antennas for VHF/UHF personal radio: A theoretical and experimental study of characteristics and performance. *IEEE Trans. Veh. Technol.*, 26(4):349–357, Nov. 1977.

[BHGR04] B. Badic, M. Herdin, G. Gritsch, M. Rupp, and H. Weinrichter. Performance of various data transmission methods on measured MIMO channels. In *Proc. VTC 2004 Spring - IEEE 59th Vehicular Technology Conf.*, Milan, Italy, May 2004. [Also available as TD(03)170].

[BHWH04] M. A. Beach, M. Hunukumbure, C.Williams, G. S. Hilton, P. R. Rogers, M. Capstick, and B. Kemp. An experimental evaluation of three candidate MIMO array designs for PDA devices. WP(04)005, COST 273, Gothenburg, Sweden, June 2004.

[BöHW03] E. Bonek, H. Özcelik, M. Herdin, W. Weichselberger, and J. Wallace. Deficiencies of the kronecker MIMO radio channel model. In *Proc. WPMC 2003 - Wireless Pers. Multimedia Commun.*, Yokosuka, Japan, Oct. 2003. [Also available as TD(03)123].

[BrSP03] L. E. Braten, A. Spilling, and M. Pettersen. A UMTS FDD simulator for smart antennas; general description and preliminary results. TD(03)052, COST 273, Barcelona, Spain, Jan. 2003.

[BuHe03] A. G. Burr and M. Herdin. The relationship between eigenvectors and multipath direction of arrival. TD(03)136, COST 273, Paris, France, May 2003.

[BüNB02] H. Bühler, T. Neubauer, and E. Bonek. RSSUS - reference system scenario for UMTS simulations. TD(02)162, Lisbon, Portugal, Sep. 2002.

[Burr01] A. Burr. Evaluation of the capacity of the MIMO channel in a corridor using ray tracing. TD(01)037, COST 273, Bologna, Italy, Oct. 2001.

[Burr02] A. Burr. Evaluation of the capacity of the MIMO channel in a room using ray tracing. TD(02)022, COST 273, Guildford, UK, Jan. 2002.

[Burr03] A. G. Burr. Capacity bounds and estimates for the finite scatterers MIMO wireless channel. *IEEE J. Select. Areas Commun.*, 21(5):812–818, June 2003.

[Burr04a] A. Burr. On the channel autocorrelation matrix of a MIMO channel. TD(04)108, COST 273, Gothenburg, Sweden, June 2004.

[Burr04b] A. G. Burr. On the full correlation matrix of a MIMO channel. TD(04)197, COST 273, Duisburg, Germany, Sep. 2004.

[BWMH03] E. Bonek, W. Weichselberger, A. F. Molisch, and H. Hofstetter. MIMO channel modeling - revisited. COST 273 Prague Tutorial, Prague, Czech Republic, Sep. 2003.

[CaGK67] J. Capon, R. J. Greenfield, and R. J. Kolker. Multidimensional maximumlikelihood processing of a large aperture seismic array. *IEEE Proc. of the IEEE*, 55:192–211, Feb. 1967.

[CaNB04] H. Cao, K. M. Nasr, and S.K. Barton. Echo domain multiple access (EDMA): A new multiple access technique for impulse radio in a multipath environment. In *Proc. URSI EMTS 2004 - International Symposium on Electromagnetic Theory*, Pisa, Italy, May 2004. [Also available as TD(04)171].

[Capo69] J. Capon. High-resolution frequency-wavenumber spectrum analysis. *IEEE Proc. of the IEEE*, 57(8):1408–1418, Aug. 1969.

[CCGH04] G. Calcev, D. Chizhik, B. Goeransson, S. Howard, H. Huang, A. Kogiantis, A. F. Molisch, A. L. Moustakas, D. Reed, and H. Xu. A wideband spatial channel model for system-wide simulations. *IEEE Trans. Veh. Technol.*, 2004. Submitted.

[ChCz04] B. K. Chalise and A. Czylwik. Uplink user capacity of UMTS-FDD with robust beamforming based upon minimum outage probability. TD(04)179, Duisburg, Germany, Sep. 2004.

[ChKT98] C. N. Chuah, J. M. Kahn, and D. Tse. Capacity of multi-antenna array systems in indoor wireless environment. In *Proc. Globecom 1998 - IEEE Global Telecommunications Conf.*, Sydney, Australia, Nov. 1998.

[ChLa03] C.-C. Chong and D.-I. Laurenson. Spatio-temporal correlation properties for the 5.2-ghz indoor propagation environments. *IEEE Antennas Wireless Propagat. Lett.*, 2:114–17, 2003.

[ChLM03] C. C. Chong, D. I. Laurenson, and S. McLaughlin. A wideband dynamic spatio-temporal markov channel model for typical indoor propagation environments. TD(03)079, COST 273, Paris, France, May 2003.

[CHÖB04a] N. Czink, M. Herdin, H. Özcelik, and E. Bonek. Cluster characteristics in a MIMO indoor propagation environment. TD(04)167, COST 273, Duisburg, Germany, Sep. 2004.

[CHÖB04b] N. Czink, M. Herdin, H. Özcelik, and E. Bonek. Number of multipath clusters in MIMO propagation environments. *Elect. Lett.*, page 1498, 2004. [Also available as TD(04)166].

[ChSA04] H. Chua, K. Sakaguchi, and K. Araki. Experimental and analytical investigation of MIMO channel capacity in indoor LOS environment. TD(04)023, COST 273, Athens, Greece, Jan. 2004.

[ChWZ03] M. Chiani, M. Z. Win, and A. Zanella. On the capacity of spatially correlated MIMO channels. *IEEE Trans. Inform. Theory*, 49(10):2363–2371, Oct. 2003.

[CLTM02] C. C. Chong, D. I. Laurenson, C. M. Tan, S. McLaughlin, M. A. Beach, and A. R. Nix. Joint detection-estimation of directional channel parameters using the 2-D frequency domain SAGE algorithm with serial interference cancellation. In *Proc. ICC 2002 - IEEE Int. Conf. Commun.*, New York, NY, USA, May 2002. [Also available as TD (02)045].

[CLWV03] D. Chizhik, J. Ling, P. W. Wolniansky, R. A. Valenzuela, N. Costa, and K. Huber. Multiple-input - multiple-output measurements and modeling in Manhattan. *IEEE J. Select. Areas Commun.*, 21(3):321–331, Apr. 2003.

[CoPa03] J-M. Conrat and P. Pajusco. A versatile propagation channel simulator for MIMO link level simulation. TD(03)120, COST 273, Paris, France, May 2003.

[Corr01] L. M. Correia, editor. *Wireless Flexible Personalised Communications - COST 259 Final Report*. John Wiley & Sons Ltd., New York, NY, USA, 2001.

[CoWi04] G. W. K. Colman and T. J. Willink. Adaptive array algorithm performance: case studies in different environments. In *Proc. ANTEM 2004 - 10th Int. Symp. on Antenna Techn. and Appl. Electromagnetics*, Ottawa, ON, Canada, July 2004. [Also available as TD(04)044].

[CzDe01] A. Czylwik and A. Dekorsy. System level simulations for downlink beamforming with different array topologies. TD(01)027, COST 273, Bologna, Italy, Oct. 2001.

[Czin04] N. Czink. Optimum Training for MIMO wireless channels. Master's Thesis, Vienna University of Technology, June 2004.

[Czyl03] A. Czylwik. Performance of realistic circular antenna arrays in cellular mobile radio systems. TD(03)010, COST 273, Barcelona, Spain, Jan. 2003.

[DDLD04] O. Delangre, Ph. De Doncker, M. Liénard, and P. Degauque. Effect of 3D antenna parameters on MIMO systems with experimental validation in a reverberating chamber. In *Proc. SCVT 2004 - 11th Symp. on Communications and Veh. Tech. in the Benelux*, Gent, Belgium, Nov. 2004. [Also available as TD(04)100].

[Egge03] P. Eggers. Dual directional channel formalisms and descriptions relevant for tx-rx diversity and MIMO. TD(03)044, COST 273, Barcelona, Spain, Jan. 2003.

[ESBP02] V. Erceg, P. Soma, D. S. Baum, and A. J. Paulraj. Capacity obtained from multiple input multiple output channel measurements in fixed wireless environments at 2.5 GHz. In *Proc. ICC 2002 - IEEE Int. Conf. Commun.*, New York, NY, USA, Apr. 2002.

[ETSI98] ETSI. Selection procedures for the choice of radio transmission technologies of the UMTS - UMTS 30.03 (version 3.2.0). 1998.

[FBKE02a] S. E. Foo, M. A. Beach, P. Karlsson, P. Eneroth, B. Lindmark, and J. Johansson. Frequency dependency of the spatial-temporal characteristics of UMTS FDD links. TD(02)027, COST 273, Guildford, U.K., Jan. 2002.

[FBKE02b] S. E. Foo, M. A. Beach, P. Karlsson, P. Eneroth, B. Lindmark, and J. Johansson. Spatio-temporal investigation of UTRA FDD channels. In *Proc. IEE 3G Mobile Communication Technologies 2002*, London, U.K., May 2002. [Also available as TD(01)028].

[FeMC01] L. S. Ferreira, M. G. Marques, and L. M. Correia. Implementation of a wideband directional channel model for link level simulations. In *Proc. of The IEE Seminar on MIMO Communication Systems*, London, U.K., Dec. 2001. [Also available as TD(02)029].

[FKMW04] T. Fügen, C. Kuhnert, J. Maurer, and W. Wiesbeck. A double-directional channel model for multiuser MIMO systems. In *Proc. URSI EMTS 2004 - International Symposium on Electromagnetic Theory*, Pisa, Italy, May 2004. [Also available as TD(04)003].

[FKWS04] T. Fügen, C. Kuhnert, C. Waldschmidt, M. Schnerr, and W. Wiesbeck. Capacity of the MIMO broadcast channel under realistic propagation conditions. TD(04)103, COST 273, Gothenburg, Sweden, Jan. 2004.

[Fleu00] B. H. Fleury. First- and second-order characterization of direction dispersion and space selectivity in the radio channel. *IEEE Trans. Inform. Theory*, IT-46(6):2027–2044, Sep. 2000.

[FMKW04] T. Fügen, J. Maurer, C. Kuhnert, W., and Wiesbeck. A modelling approach for multiuser MIMO systems including spatially-colored interference. In *Proc. Globecom 2004 - IEEE Global Telecommunications Conf.*, Dallas, TX, USA, Jan. 2004. [Also available as TD(04)004].

[FoBe02] S. E. Foo and M. Beach. Uplink based downlink beamforming in UTRA FDD. TD(02)104, COST 273, Lisbon, Portugal, Sep. 2002.

[FoGa98] G. J. Foschini and M. J. Gans. On limits of wireless communications in a fading environment when using multiple antennas. *Wireless Personal Communications*, 6(3):311–335, Mar. 1998.

[FYRJ02] B. H. Fleury, X. Yin, K. G. Rohbrandt, P. Jourdan, and A. Stucki. High-resolution bidirection estimation based on the SAGE algorithm: Experience gathered from field experiments. TD(02)070, COST 273, Espoo, Finland, May 2002.

[GaHS03] G. Del Galdo, M. Haardt, and C. Schneider. Geometry-based channel modelling of MIMO channels in comparison with channel sounder measurements. *Advances in Radio Science - Kleinheubacher Berichte*, pages 117–126, Oct. 2003. [Also available as TD(03)188].

[GeBP02] D. Gesbert, H. Boelcskei, and A. J. Paulraj. Outdoor MIMO wireless channels: Models and performance prediction. *IEEE Trans. Commun.*, 50(12):1926–1934, Dec. 2002.

[GEYC97] L. J. Greenstein, V. Erceg, Y. S. Yeh, and M. V. Clark. A new path-gain/delayspread propagation model for digital cellular channels. *IEEE Trans. Veh. Technol.*, 46:477–485, 1997.

[GGFC04] J. Gil, G. Galvano, L. S. Ferreira, and L. M. Correia. Propagation scenarios among many others. TD(04)083, COST 273, Gothenburg, Sweden, June 2004.

[GiCo01] J. Gil and L. M. Correia. Combining adaptive beamforming with directional channel modelling for UMTS. In *Proc. PIMRC 2001 - IEEE 12th Int. Symp. on Pers., Indoor and Mobile Radio Commun.*, San Diego, CA, USA, Sep. 2001. [Also available as TD(01)036].

[GiCo02] J. Gil and L. M. Correia. Dependence of adaptive beamforming performance on directional channel macro-cell scenarios for UMTS. TD(02)050, COST 273, Espoo, Finland, May 2002.

[GiCo03a] J. Gil and L. M. Correia. Adaptive beamforming dependencies on wideband and directional propagation characteristics in micro- and macro-cell UMTS scenarios. In *Proc. PIMRC 2003 - IEEE 14th Int. Symp. on Pers., Indoor and Mobile Radio Commun.*, Beijing, China, Sep. 2003. [Also available as TD(04)007].

[GiCo03b] J. Gil and L. M. Correia. The MMSE vs beamforming gain optima discrepancy in adaptive beamforming applied to directional channel scenarios. In *Proc. Conf Tele 2003 - 4th Conference on Telecommunications*, Aveiro, Portugal, June 2003. [Also available as TD(03)053].

[GiCo04] J. M. Gil and L. M. Correia. Fundamental wideband and directional channel parameters ruling adaptive beamforming performance in micro- and macro-cells.

In *Proc. VTC 2004 Spring - IEEE 59th Vehicular Technology Conf.*, Milan, Italy, May 2004. [Also available as TD(04)077].

[Glaz04] A. A. Glazunov. Joint impact of the mean effective gain and base station smart antennas on WCDMA-FDD systems performance. In *Proc. Nordic Radio Symp. 2004*, Oulu, Finland, Aug. 2004. [Also available as TD(04)158].

[HäCC03] L. Häring, B. K. Chalise, and A. Czylwik. Dynamic system level simulations of downlink beamforming for UMTS FDD. In *Proc. Globecom 2003 - IEEE Global Telecommunications Conf.*, San Franciso, CA, USA, Dec. 2003. [Also available as TD(03)018].

[HaLe03] J. Hansen and P. Leuthold. The mean received power in ad-hoc networks and its dependence on geometrical quantities. *IEEE Trans. Antennas Propagat.*, 51(9):2413–2419, Sep. 2003. [Also available as TD(02)010].

[HaRe04] J. Hansen and M. Reitzner. Efficient indoor radio channel modelling based on integral geometry. *IEEE Trans. Antennas Propagat.*, 52(9):2456–2463, Sep. 2004. [Also available as TD(02)063].

[HaTa03] K. Haneda and J. Takada. High-resolution estimation of NLOS indoor MIMO channel with network analyzer based system. TD(03)119, COST 273, Paris, France, May 2003.

[HaTK04] K. Haneda, J. Takada, and T. Kobayashi. Double directional LOS channel characterization in a home environment with ultrawideband signal. In *Proc. WPMC 2004 - Wireless Pers. Multimedia Commun.*, Padova, Italy, Sep. 2004. [Also available as TD(04)160].

[HCÖB05] M. Herdin, N. Czink, H. Özcelik, and E. Bonek. Correlation matrix distance, a meaningful measure for evaluation of non-stationary MIMO channels. In *Proc. VTC 2005 Spring - IEEE 61st Vehicular Technology Conf.*, Stockholm, Sweden, May 2005.

[HeBÖ03] M. Herdin, A. Burr, and H. Özcelik. How human shadowing affects directions of- arrival and eigenvalues at 5.2 GHz. In *Proc. WPMC 2003 - Wireless Pers. Multimedia Commun.*, Yokosuka, Japan, Oct. 2003. [Also available as TD(03)062].

[Herd04] M. Herdin. *Non-Stationary Indoor MIMO Radio Channels.* PhD thesis, Institut für Nachrichtentechnik und Hochfrequenztechnik, Vienna University of Technology, Vienna, Austria, Aug. 2004. Downloadable from http://www.nt.tuwien.ac.at/mobile. [Also available as TD(04)174].

[HMAB02] H. Hofstetter, C. Mecklenbräuker, H. Anegg, E. Bonek, R. Müller, and H. Kunczier. The FTW wireless MIMO measurement campaign at 2 GHz: documentation of the downloadable data sets. TD(02)135, COST 273, Lisbon, Portugal, Sep. 2002.

[HÖHB02] M. Herdin, H. Özcelik, H. Hofstetter, and E. Bonek. Variation of measured indoor MIMO capacity with receive direction and position at 5.2 GHz. *Elect. Lett.*, 38:1283–1285, Sep. 2002. [Also available as TD(02)155].

[HÖHB03] M. Herdin, H. Özcelik, H. Hofstetter, and E. Bonek. Linking reduction in measured MIMO capacity with dominant-wave propagation. In *Proc. ICT 2003 - 10th Int. Conf. on Telecommunications*, Papeete, Tahiti, Feb. 2003. [Also available as TD(02)157].

[HoSt04] H. Hofstetter and G. Steinböck. A geometry based stochastic channel model for MIMO systems. In *Proc. ITG - 6th Int. Conf. on Source and Channel Coding*, Munich, Germany, Mar. 2004. [Also available as TD(04)060].

[HoTW03] A. Hottinen, O. Tirkkonen, and R. Wichman. *Multi-Antenna Transceiver Techniques for 3G and Beyond.* John Wiley & Sons Ltd., New York, NY, USA, 2003.

[HuBe02] M. Hunukumbure and M. Beach. Outdoor MIMO measurements for UTRA applications. In *Proc. 11th IST Summit on Mobile and Wireless Commun.*, Lyon, France, June 2002. [Also available as TD (02)076].

[HuBe03] M. Hunukumbure and M. Beach. Outdoor MIMO measurements and analysis with different antenna arrays. TD(03)007, COST 273, Barcelona, Spain, Jan. 2003.

[HZWS04] H. Hofstetter, T. Zemen, J. Wehinger, and G. SteinbÖck. Iterative MIMO multiuser detection: Performance evaluation with COST 259 channel model. In *Proc. WPMC 2004 - Wireless Pers. Multimedia Commun.*, Abano Terme, Italy, Sep. 2004. [Also available as TD(04)203].

[IvNo03] M. T. Ivrlac and J. A. Nossek. Quantifying diversity and correlation of rayleigh fading MIMO channels. In *Proc. ISSPIT 2003 - 3rd IEEE International Symposium on Signal Processing and Information Technology*, Darmstadt, Germany, Dec. 2003.

[JoBo03] E. A. Jorswieck and H. Boche. On the impact of correlation on the capacity in MIMO systems without CSI at the transmitter. In *Proc. CISS - 37th Conf. on Information Sciences and Systems*, Baltimore, MD, USA, Mar. 2003.

[KaHP04] F. Kaltenberger, G. Humer, and G. Pfeiffer. MIMO/smart antenna development platform. In *Proc. WSR 2004 - 3rd Workshop on Software Radios*, Karlsruhe, Germany, Mar. 2004. [Also available as TD(04)173].

[KaSC04] P. Karamalis, N. Skentos, and P. Constantinou. Comparison of existing MIMO antenna selection algorithms with an evolutionary approach. TD(04)055, COST 273, Athens, Greece, Jan. 2004.

[KaVV05] A. Kainulainen, L. Vuokko, and P. Vainikainen. Polarization behavior in different urban radio environments at 5.3 GHz. TD(05)018, COST 273, Bologna, Italy, Jan. 2005.

[KCVW03] P. Kyritsi, D. C. Cox, R. A. Valenzuela, and P. W. Wolniansky. Correlation analysis based on MIMO channel measurements in an indoor environment. *IEEE J. Select. Areas Commun.*, 21(5):713–720, June 2003. [Also available as TD(02)037].

[KiRo04] P.-S. Kildal and K. Rosengren. Electromagnetic characterization of MIMO antennas including coupling using classical embedded element pattern and radiation efficiency. In *Proc. IEEE AP-S 2004 - IEEE Int. Symp. On Antennas and Propagation and USNC/URSI National Radio Science Meeting*, Monterey, CA, USA, June 2004.

[KKNJ04] J. Kolu, P. Kyösti, J-P. Nuutinen, and T. Jämsä. Playback simulation of measured MIMO simulation. TD(04)110, Elektrobit Ltd, Gothenburg, Sweden, June 2004.

[KLKK02] K. Kalliola, K. Sulonen, H. Laitinen, O. Kivekas, J. Krogerus, and P. Vainikainen. Angular power distribution and mean effective gain of mobile antenna in different propagation environments. *IEEE Trans. Veh. Technol.*, pages 823–838, 2002.

[KNJY04] J. Kolu, J.-P. Nuutinen, T. Jämsä, J., Ylitalo, and P. Kyösti. Playback simulation of measured MIMO radio channels. TD(04)110, COST 273, Gothenburg, Sweden, June 2004.

[KoJH03] J. Kolu, T. Jämsä, and A. Hulkkonen. Real time simulation of measured radio channels. In *Proc. VTC 2003 Fall - IEEE 58th Vehicular Technology Conf.*, Orlando, FL. USA, Oct. 2003.

[KoNu04] J. Kolu and J-P. Nuutinen. Verification of playback simulation in measured MIMO channels. TD(04)169, Elektrobit Ltd, Duisburg, Germany, Sep. 2004.

[KoPO02] W. Kotterman, G. F. Pedersen, and K. Olesen. Capacity of the mobile MIMO channel for a small wireless handset and user influence. In *Proc. PIMRC*

2002 - IEEE 13th Int. Symp. on Pers., Indoor and Mobile Radio Commun., Lisbon, Portugal, Sep. 2002. [Also available as TD(02)149].

[Kott05] W. Kotterman. *Environmental Influences on Variability of MIMO eigen modes at both link ends*. TD(05)119, Cost 273, Lisbon, Nov. 2005.

[KSKC03] P. Karamalis, N. Skentos, A. Kanatas, and P. Constantinou. A measurement-based method for selecting MIMO system array configurations. TD(03)110, COST 273, Paris, France, May 2003.

[KSKL05] F. Kaltenberger, G. Steinböck, R. Kloibhofer, R. Lieger, and G. Humer. A multiband development platform for rapid prototyping of MIMO systems. In *Proc. ITG - 7th Int. Conf. on Source and Channel Coding*, Duisburg, Germany, Apr. 2005. [Also available as TD(04)173].

[KSPH01] J. Kivinen, P. Suvikunnas, D. Perez, C. Herrero, K. Kalliola, and P. Vainikainen. Characterization system for MIMO channels. In *Proc. WPMC 2001 - Wireless Pers. Multimedia Commun.*, Aalborg, Denmark, Sep. 2001. [Also available as TD(01)044].

[KSVV02] J. Kivinen, P. Suvikunnas, L. Vuokko, and P. Vainikainen. Experimental investigations of MIMO propagation channels. In *Proc. IEEE AP-S 2002 - IEEE Int. Symp. On Antennas and Propagation and USNC/URSI National Radio Science Meeting*, San Antonio, TX, USA, June 2002. [Also available as TD(02)078].

[Kunn02] E. Kunnari. Modelling and simulation of small scale fading with temporal, spatial, and spectral correlation. TD(02)019, COST 273, London, U.K., Jan. 2002.

[LaMB98] J. Laurila, A. F. Molisch, and E. Bonek. Influence of the scatterer distribution on power delay profiles and azimuthal power spectra of mobile radio. In *Proc. ISSSTA 1998 - IEEE 4th Int. Symp. on Spread Spectrum Techniques and Applications*, Sun City, South Africa, Sep. 1998.

[LaZo96] R. O. LaMaire and M. Zorzi. Effect of correlation in diversity systems with rayleigh fading, shadowing, and power capture. *IEEE J. Select. Areas Commun.*, 14(3):449–460, Apr. 1996.

[LiDe04a] M. Liénard and P. Degauque. Dual antenna array systems in tunnels: Propagation channel properties June 2004. In *Proc. IEEE AP-S 2004 - IEEE Int. Symp. On Antennas and Propagation and USNC/URSI National Radio Science Meeting*, Monterey, CA, U.S.A., June 2004. [Also available as TD(03)030].

[LiDe04b] M. Liénard and P. Degauque. Simulation of dual array multipath channels using mode stirred reverberation chamber. *Elect. Lett.*, 40(10):578–580, May 2004. [Also available as TD(04)24].

[LiDL04] M. Liénard, P. Degauque, and P. Laly. Mode stirred chambers for simulating MIMO channels. TD(04)024, COST 273, Athens, Greece, Jan. 2004.

[LoBF01] N. Lohse, M. Bronzel, and G. Fettweis. Radio channel characterization using space-time system functions. TD(01)050, COST 273, Bologna, Italy, Oct. 2001.

[LoFK02] N. Lohse, G. Fettweis, and R. Kattenbach. Radio channel modelling using spacetime sampling models. TD(02)141, COST 273, Lisbon, Portugal, Sep. 2002.

[LVTC04] E. Van Lil, J. Verhaevert, D. Trappeniers, and A. Van de Capelle. Theoretical investigations and broadband experimental verification of the time-domain SAGE DOA algorithm. In *Proc. ACES 2004 - 20th Annual Review of Progress in Applied Computational Electromagnetics*, Syracuse, NY, USA, Apr. 2004. [Also available as TD(04)142].

[MaCo01] M. G. Marques and L. M. Correia. A wideband directional channel model for UMTS micro-cells. In *Proc. PIMRC 2001 - IEEE 12th Int. Symp. on Pers., Indoor and Mobile Radio Commun.*, San Diego, CA, USA, Sep. 2001. [Also available as TD(02)012].

[MAHS05a] A. F. Molisch, H. Asplund, R. Heddergott, M. Steinbauer, and T. Zwick. The COST 259 directional channel model - I. philosophy and general aspects. *IEEE Trans. Wireless Commun.*, 2005. in press.

[MAHS05b] A. F. Molisch, H. Asplund, R. Heddergott, M. Steinbauer, and T. Zwick. The COST 259 directional channel model - I. overview and methodology. *IEEE J. Select. Areas Commun.*, 2005. Submitted

[MATB03] J. Medbo, H. Asplund, M. Törnqvist, D. Browne, and J.-E.Berg. MIMO channel measurements in an urban street microcell. TD(03)006, COST 273, Barcelona, Spain, Jan. 2003.

[MaVT04] N. Marchetti, R. Veronesi, and V. Tralli. On the impact of interference on data protocol performance in multicellular wireless packet networks with MIMO links. In *Proc. VTC 2004 Fall - IEEE 60th Vehicular Technology Conf.*, Los Angeles, CA, USA, Sep. 2004. [Also available as TD(04)200].

[MBFK00] D. P. McNamara, M. A. Beach, P. N. Fletcher, and P. Karlsson. Capacity variation of indoor multiple-input multiple-output channels. *Elect. Lett.*, 36(24):2037–2038, Nov. 2000.

[McBF02a] D. McNamara, M. Beach, and P. Fletcher. Spatial correlation in indoor MIMO channels. In *Proc. PIMRC 2002 - IEEE 13th Int. Symp. on Pers., Indoor and Mobile Radio Commun.*, Lisbon, Portugal, Sep. 2002. [Also available as TD(02)097].

[McBF02b] D. McNamara, M. A. Beach, and P. Fletcher. Wideband analysis of indoor MIMO channels. TD(02)026, COST 273, Guildford, UK, Jan. 2002.

[MDJD03] Liénard M., P. Degauque, Baudet J., and Degardin D. Investigation on MIMO channels in subway tunnels. *IEEE J. Select. Areas Commun.*, 21(3):332–339, Apr. 2003. [Also available in TD(02)011].

[MeBe02] J. Medbo and J.-E. Berg. Simple and accurate path loss modeling at 5 GHz in complex indoor environments with corridors. In *Proc. URSI-F-2002 Open Symp. on Propagation and Remote Sensing*, Garmisch-Partenkirchen, Germany, Feb. 2002. [Also available as TD(02)055].

[Medb05] Jonas Medbo. Cost 273 channel model parameters for the office environment. TD(05)050, COST 273, Bologna, Italy, Jan. 2005.

[MoKl99] P. Mogensen and T. Klingenbrunn. Modelling cross-correlated shadowing in network simulations. In *Proc. VTC 1999 Fall - IEEE 50th Vehicular Technology Conf.*, Amsterdam, The Netherlands, Sep. 1999.

[Moli02] A. F. Molisch. A channel model for MIMO systems in macro- and microcellular environments. In *Proc. VTC 2002 Spring - IEEE 55th Vehicular Technology Conf.*, Birmingham, AL, USA, May 2002.

[Moli04a] A. F. Molisch. A generic model for the MIMO wireless propagation channel. *IEE Proc. Signal Proc.*, 52(1):61–71, Jan. 2004. [Also available as TD(02)100].

[Moli04b] A. F. Molisch. A generic model for the MIMO wireless propagation channel. *IEE Proc. Signal Proc.*, pages 61–71, 2004.

[MoLK98] A. F. Molisch, J. Laurila, and A. Kuchar. Geometry-base stochastic model for mobile radio channels with directional component. In *Proc. 2nd Intelligent Antenna Symp.*, Surrey, UK, July 1998.

[MoRJ04] J.-M. Molina-García-Pardo, J.-V. Rodríguez, and L. Juan-Llacer. MIMO measurement system based on two network analyzers. *IEEE Transactions on Instrumentation and Measurement*, 2004. [Also availabe as TD(04)106].

[MPKZ01] M. G. Marques, J. Pamp, J. Kunisch, and E. Zollinger. Wideband directional channel model, array antennas and measurement campaign. Deliverable D2.3bis, Nov. 2001.

[MTRF02] S. Morosi, M. Tosi, E. Del Re, and R. Fantacci. Implementation and assessment of a wideband directional channel model for macro and micro cells. TD(02)147, COST 273, Lisbon, Portugal, Sep. 2002.

[NaCB03a] K. M. Nasr, F. Costen, and S. K. Barton. An optimum combiner for a smart antenna in an indoor infrastructure WLAN. In *Proc. VTC 2003 Fall - IEEE 58th Vehicular Technology Conf.*, Orlando, FL, USA, Oct. 2003. [Also available as TD(03)088].

[NaCB03b] K. M. Nasr, F. Costen, and S. K. Barton. A spatial channel model and a beamformer for smart antennas in broadcasting studios. In *Proc. ICAP 2003 - 12th Int. Conf. on Antennas and Propagation*, London, UK, Mar. 2003. [Also available as TD(02)109].

[NaCB04a] K. M. Nasr, F. Costen, and S. K. Barton. A downlink pattern optimisation algorithm for a smart antenna in an indoor infrastructure WLAN. In *Proc. PIMRC 2003 - IEEE 14th Int. Symp. on Pers., Indoor and Mobile Radio Commun.*, Barcelona, Spain, Sep. 2004. [Also available as WP(04)021].

[NaCB04b] K. M. Nasr, F. Costen, and S. K. Barton. On the angular separation and channel estimation errors on a smart antenna system in an indoor WLAN. TD(04)156, COST 273, Duisburg, Germany, Sep. 2004.

[NoTV01] O. Norklit, P. D. Teal, and R. G. Vaughan. Measurements and evaluation of multiantenna handsets in indoor mobile communication. *IEEE Trans. Antennas Propagat.*, 49(3):429–437, Mar. 2001.

[OeEP04] C. Oestges, V. Erceg, and A. J. Paulraj. Propagation modeling of multi-polarized MIMO fixed wireless channels. *IEEE Trans. Veh. Technol.*, 53(3):644–654, May 2004. [Also available as TD(03)005].

[OePa04] C. Oestges and A. J. Paulraj. Beneficial impact of channel correlations on MIMO capacity. *Elect. Lett.*, 40(10):606–607, May 2004. [Also available as TD(04)002].

[Oest02] C. Oestges. A physical-statistical channel model of macro- and megacellular terrestrial networks and its application to polarisation multiplexing. In *Proc. WPMC 2002 - Wireless Pers. Multimedia Commun.*, Honolulu, HI, USA, Oct. 2002.

[ÖHHB03] H. Özcelik, M. Herdin, H. Hofstetter, and E. Bonek. A comparison of measured 8×8 MIMO systems with a popular stochastic channel model at 5.2 GHz. In *Proc. ICT 2003 - 10th Int. Conf. on Telecommunications*, Papeete, Tahiti, Feb. 2003. [Also available as TD(02)153].

[ÖHPB03a] H. Özcelik, M. Herdin, R. Prestos, and E. Bonek. How MIMO capacity is linked with single element fading statistics. In *Proc. ICAA 2003 - International Conference on Electromagnetics in Advanced Applications*, Torino, Italy, Sep. 2003. [Also available as TD(03)067].

[ÖHPB03b] H. Özcelik, M. Herdin, R. Prestros, and E. Bonek. Is a bad channel a good channel? In *Proc. ICAA 2003 - International Conference on Electromagnetics in Advanced Applications*, Barcelona, Spain, Sep. 2003. [Also available as TD(03)067].

[ÖzBo04] H. Özcelik and E. Bonek. Experimental validation of analytical channel models. COST 273, Duisburg, Germany, Sep. 2004. [Also availlable as TD(04)208].

[ÖzCB05] H. Özcelik, N. Czink, and E. Bonek. What makes a good MIMO channel model? In *Proc. VTC 2005 Spring - IEEE 61st Vehicular Technology Conf.*, Stockholm, Sweden, May-June 2005. [Also available as TD(04)208].

[Özce04] H. Özcelik. *Indoor MIMO Channel Models*. PhD thesis, Institut für Nachrichtentechnik und Hochfrequenztechnik, Vienna University of Technology, Vienna, Austria, Dec. 2004. downloadable from http://www.nt.tuwien.ac.at/mobile.

[OzHH04] H. Özcelik, M. Herdin, and H. Hofstetter. Indoor 5.2 GHz MIMO measurement campaign. TD(04)174, COST 273, Duisburg, Germany, Sep. 2004.

[ÖzOe05] H. Özcelik and C. Oestges. Capacity in diagonally correlated channels. In *Proc. VTC 2005 Spring - IEEE 61st Vehicular Technology Conf.*, Stockholm, Sweden, May 2005. [Also available as TD(04)133].

[Paju98] P. Pajusco. Experimental characterization of D.O.A. at the base station in rural and urban area. In *Proc. VTC 1998 - IEEE 48th Vehicular Technology Conf.*, Ottawa, Canada, May 1998.

[PaTB04] A. Pal, C. M. Tan, and M. A. Beach. Comparison of MIMO channels from multipath parameter extraction and direct channel measurements. In *Proc. PIMRC 2004 - IEEE 15th Int. Symp. on Pers., Indoor and Mobile Radio Commun.*, Barcelona, Spain, Sep. 2004. [Also available as TD(04)016].

[PiSt60] J. R. Pierce and S. Stein. Multiple diversity with nonindependent fading. In *Proc. Institute of Radio Engineers (IRE)*, New York, NY, USA, Jan. 1960.

[PNTL02] C. Pietsch, M. Nold, W. G. Teich, and J. Lindner. Optimum space-time processing for wide-band transmissions with multiple receiving antennas. In *Proc. ITG - 4th Int. Conf. on Source and Channel Coding*, Berlin, Germany, Jan. 2002. [Also available as TD(01)023].

[Poll02] A. Pollard. An operator perspective on smart antennas & MIMO in systems beyond 3G. TD(02)123, Lisbon, Portugal, Sep. 2002.

[PPPB04] T. Pedersen, C. Pedersen, R. R. Pedersen, Bozinovska B, A Hviid, X. Yin, and B. H. Fleury. Investigations of the ambiguity effect in the estimation of doppler frequency and directions in channel sounding using switched tx and rx arrays. TD(04)021, COST 273, Athens, Greece, Jan. 2004.

[PSTL03] C. Pietsch, S. Sand, W. G. Teich, and J. Lindner. Modeling and performance evaluation of multiuser MIMO systems using real-valued matrices. *IEEE J. Select. Areas Commun.*, 21(4):744–753, June 2003. [Also available as TD(02)125].

[RaSa04] N. Razavi-Ghods and S. Salous. Semi-sequential MIMO radio channel sounding. In *Proc. CCCT04 - Int. Conf. on Computing, Communications and Control Technologies*, Austin, TX, USA, Aug. 2004. [Also available as TD(04)079].

[Rich05] A. Richter. *Estimation of Radio Channel Parameters: Models and Algorithms*. PhD thesis, Ilmenau University of Technology, Ilmenau, Ilmenau, Germany, 2005. [Also available as TD(02)132].

[RoBK04] K. Rosengren, P. Bohlin, and P.-S. Kildal. Characterization of antennas for MIMO systems in reverberation chamber and by simulation. WP(04)003, COST 273, Gothenburg, Sweden, June 2004.

[RoHi04] P. R. Rogers and G. S. Hilton. 3D radiation pattern correlation of PDA-sized MIMO antenna arrays. WP(04)006, COST 273, Gothenburg, Sweden, June 2004.

[SaFH02] S. Salous, P. Fillipidis, and I. Hawkins. Multiple antenna channel sounder using a parallel receiver architecture. In *Proc. SCI 2002 - 6th World Multi-Conf. on*

Systemics, Cybernetics and Informatics, Orlando, FL, USA, July 2002. [Also available as TD(02)002].

[SaHA04] K. Sakaguchi, T. Ho, and K. Araki. Initial measurement on MIMO eigenmode communication system. *IEICE Trans. Commun.*, J87-B(9):1454–1466, Sep. 2004. [Also available as TD(04)027].

[Saye02] A. M. Sayeed. Deconstructing multiantenna fading channels. *IEEE Trans. Signal Processing*, 50(10):2563–2579, Oct. 2002.

[SFGK00] D. S. Shiu, G. J. Foschini, M. J. Gans, and J. M. Kahn. Fading correlation and its effect on the capacity of multielement antenna systems. *IEEE Trans. Commun.*, 48(3):502–513, Mar. 2000.

[Sibi01] A. Sibile. Keyholes and MIMO channel modelling. TD(01)017, COST 273, Bologna, Italy, Oct. 2001.

[SJKN05] T. Sarkkinen, T. Jämsä, P. Kyösti, J-P. Nuutinen, and J. Kolu. Future trends in real-time MIMO radio channel simulation. TD(05)046, COST 273, Bologna, Italy, Jan. 2005.

[SkKC04] N. Skentos, P. Karamalis, and P. Constantinou. Results from fixed MIMO channel measurements at 5.2 GHz in urban environment. TD(04)140, COST 273, Gothenburg, Sweden, June 2004.

[SkKC05] N. Skentos, A. Kanatas, and P. Constantinou. Results from rooftop to rooftop MIMO channel measurements at 5.2 GHz. TD(05)059, COST 273, Bologna, Italy, Jan. 2005.

[SKPC03] N. Skentos, A. Kanatas, G. Pantos, and P. Constantinou. Capacity results of MIMO measurements at 5.2 GHz in urban environments. TD(03)056, COST 273, Barcelona, Spain, Jan. 2003.

[SLTR03] G. Sommerkorn, M. Landmann, R. S. Thomä, and A. Richter. Performance evaluation of real antenna arrays for high-resolution doA estimation in channel sounding - part 2: Experimental ULA measurement results. TD(03)196, COST 273, Prague, Czech Republic, Sep. 2003.

[SSKV03a] K. Sulonen, P. Suvikunnas, J. Kivinen, and P. Vainikainen. Study of different mechanisms providing gain in MIMO systems. In *Proc. VTC 2003 Fall - IEEE 58th Vehicular Technology Conf.*, Orlando, FL, USA, Oct. 2003.

[SSKV03b] P. Suvikunnas, K. Solunen, J. Kivinen, and P. Vainikainen. Effect of antenna properties on MIMO-capacity in real propagation channels. WP(03)001, COST 273 Workshop, Barcelona, Spain, Jan. 2003.

[SSKV04] P. Suvikunnas, J. Salo, J. Kivinen, and P. Vainikainen. Comparison of MIMO antennas: performance measures and evaluation results of two 2×2 antenna configurations. In *Proc. Nordic Radio Symp. 2004*, Oulu, Finland, Aug. 2004. [Also available as TD(04)032].

[SSVK03] K. Sulonen, P. Suvikunnas, L. Vuokko, J. Kivinen, and P. Vainikainen. Comparison of MIMO antenna configurations in picocell and microcell environments. *IEEE J. Select. Areas Commun.*, 21(5):703–712, June 2003.

[StJF02] A. Stucki, P. Jourdan, and B. H. Fleury. ISIS, a high performance and effiecient implementation of SAGE for radio channel parameter estimation. TD(02)068, COST 273, Espoo, Finland, May 2002.

[StMB01] M. Steinbauer, A. F. Molisch, and E. Bonek. The double-directional radio channel. *IEEE Antennas Propagat. Mag.*, 43(4):51–63, Aug. 2001.

[StMo01] M. Steinbauer and A. F. Molisch. *Directional channel models*. John Wiley & Sons Ltd., New York, NY, USA, 2001.

[STMT03] C. Schneider, U. Trautwein, T. Matsumoto, and R. Thomä. *The Dependency of Turbo MIMO Equalizer Performance on the Spatial and Temporal Multipath Channel Structure - A Measurement Based Evaluation*, In *Wireless Flexible Personalised Communications - COST 259 Final Report.* John Wiley & Sons Ltd., New York, NY, USA, 2003. [Also available as TD(03)109].

[Svan01] T. Svantesson. A physical MIMO radio channel model for multi-element multipolarized antenna systems. In *Proc. VTC 2001 Fall - IEEE 54th Vehicular Technology Conf.*, Boston, MD, USA, Sep. 2001.

[SvRa01] T. Svantesson and A. Ranheim. Mutual coupling effects on the capacity of multielement antenna systems. In *Proc. ICASSP 2001 - IEEE Int. Conf. Acoust. Speech and Signal Processing*, Salt Lake City, UT, USA, May 2001.

[SZMS06] M. Shafi, M. Zhang, A. L. Moustakas, P. J. Smith, Andreas F. Molisch, F. Tufvesson, and S. H. Simon. Polarized MIMO channels in 3D: Models, measurements and mutual information. *IEEE J. Select. Areas Commun.*, Mar. 2006. Special Issues on 4G.

[TaBN04] C. M. Tan, M. A. Beach, and A. R. Nix. Multipath parameters estimation with a reduced complexity unitary-SAGE algorithm. *European Transactions on Communications*, 515–528:515–528, Jan. 2004. [Also available as TD (03)090].

[TaOg01] J. Takada and K. Ogawa. An analysis of the effective performance of a handset diversity antenna influenced by head, hand and shoulder effects — A proposal for a diversity antenna gain based on a signal bit-error rate and analytical results for the PDC system. *IEEE Trans. Veh. Technol.*, 50(3):845–853, 2001. [Also available as TD(03)142].

[TFBN04] C. M. Tan, S. E. Foo, M. A. Beach, and A. R. Nix. Descriptions of dynamic single-, double-directional measurement campaigns at 5 GHz. TD(04)099, COST 273, Gothenburg, Sweden, June 2004.

[ThLR04] R. S. Thomä, M. Landmann, and A. Richter. RIMAX - a maximum likelihood framework for parameter estimation in multidimensional channel sounding. In *Proc. ISAP 2004 - Intl. Symp. on Antennas and Propagation*, Sendai, Japan, Aug. 2004. [Also available as TD(04)045].

[ThRi05] R. S. Thomä and A. Richter. Joint maximum likelihood estimation of specular path and distributed diffuse scattering. In *Proc. VTC 2005 Spring - IEEE 61st Vehicular Technology Conf.*, Stockholm, Sweden, May 2005.

[TLRT05] R. S. Thomä, M. Landmann, A. Richter, and U. Trautwein. *Multidimensional High-Resolution Channel Sounding.* EURASIP Book Series, 2005. [Also available as TD(03)198].

[TLST05] U. Trautwein, M Landmann, G. Sommerkorn, and R. S. Thomä. System-oriented measurement and analysis of MIMO channels. TD(05)063, COST 273, Bologna, Italy, Jan. 2005.

[TPBN04] C. M. Tan, D. L. Paul, M. A. Beach, A. R. Nix, and C. J. Railton. Dynamic double directional propagation channel analysis with dual circular arrays. WP(04)001, Gothenburg, Sweden, June 2004.

[TsHT04] H. Tsuchiya, K. Haneda, and J. Takada. UWB indoor double-directional channel sounding for understanding the microscopic propagation mechanisms. In *Proc.*

WPMC 2004 - Wireless Pers. Multimedia Commun., Padova, Italy, Sep. 2004. [Also available as TD(04)192].

[VeLC04] J. Verhaevert, E. Van Lil, and A. Van de Capelle. Verification of the BTD-SAGE algorithm with simulated and experimental data. TD(04)022, COST 273, Athens, Greece, Jan. 2004.

[VeLC05] J. Verhaevert, E. Van Lil, and A. Van de Capelle. Applications of the SAGE algorithm using a dodecahedral receiving antenna array. TD(05)013, Bologna, Italy, Jan. 2005.

[ViHU02] I. Viering, H. Hofstetter, and W. Utschick. Validity of spatial covariance matrices over time and frequency. In *Proc. Globecom 2002 - IEEE Global Telecommunications Conf.*, Taipeh, Taiwan, Nov. 2002. [Also available as WP(02)011].

[VTZZ04] R. Veronesi, V. Tralli, J. Zander, and M. Zorzi. Distributed dynamic resource allocation techniques for multicell SDMA packet access networks. In *Proc. WCNC 2004 - IEEE Wireless Commun. and Networking Conf.*, Atlanta, GA, USA, Mar. 2004. [Also available as TD(04)052].

[VuSV02] L. Vuokko, K. Sulonen, and P. Vainikainen. Analysis of propagation mechanisms based on direction-of-arrival measurements in urban environments at 2 GHz frequency range. In *Proc. IEEE AP-S 2002 - IEEE Int. Symp. On Antennas and Propagation and USNC/URSI National Radio Science Meeting*, San Antonio, CA, USA, June 2002. [Also available as TD(02)030].

[WAKE04] S. Wyne, P. Almers, J. Karedal, G. Ericsson, F. Tufvesson, and A. F. Molisch. Outdoor to indoor office MIMO measurements at 5.2 GHz. In *Proc. VTC 2004 Fall - IEEE 60th Vehicular Technology Conf.*, Los Angeles, CA, USA, Sep. 2004. [Also available as TD(04)152].

[WaSW04] C. Waldschmidt, S. Schulteis, and W. Wiesbeck. Complete RF system model for the analysis of compact MIMO arrays. *IEEE Trans. Veh. Technol.*, 53(3):579–586, May 2004.

[WaWi04] C. Waldschmidt and W. Wiesbeck. Quality measures and examples of arrays for hand-held devices. In *Proc. IEEE AP-S 2004 - IEEE Int. Symp. On Antennas and Propagation and USNC/URSI National Radio Science Meeting*, Monterey, CA, USA, June 2004. [Also available as TD(04)178].

[Weic03] W. Weichselberger. *Spatial Structure of Multiple Antenna Radio Channels.* PhD thesis, Institut für Nachrichtentechnik und Hochfrequenztechnik, Vienna University of Technology, Dec. 2003. downloadable from http://www.nt.tuwien.ac.at/mobile. [Also available as TD(03)144].

[WHÖB03] W. Weichselberger, M. Herdin, H. Özcelik, and E. Bonek. A stochastic MIMO channel model with joint correlation of both link ends. *IEEE Trans. Wireless Commun.*, 2003. [Also available as TD(03)144]. To be published.

[WKSW03] C. Waldschmidt, C. Kuhnert, S. Schulteis, and W. Wiesbeck. Compact MIMO-arrays based on polarisation diversity. In *Proc. IEEE AP-S 2003 - IEEE Int. Symp. On Antennas and Propagation and USNC/URSI National Radio Science Meeting*, Salt Lake City, UT, USA, June 2003. [Also available as TD(03)107].

[WKSW04] C.Waldschmidt, C. Kuhnert, S. Schulteis, and W. Wiesbeck. On the integration of MIMO systems into handheld devices. In *Proc. ITG - 6th Int. Conf. on Source and Channel Coding*, Erlangen, Germany, Jan. 2004. [Also available as TD(04)161].

[WWWW04] G. WÖlfle, R. Wahl, P. Wildbolz, P. Wertz, and F. Landstorfer. Dominant path prediction model for indoor and urban scenarios. TD(04)205, Duisburg, Germany, Sep. 2004.

[XCHV02] H. Xu, D. Chizik, H. Huang, and R. Valenzuela. An efficient channel modeling approach for MIMO systems. TD(02)098, COST 273, Espoo, Finland, May 2002.

[YBOM01] K. Yu, M. Bengtsson, B. Ottersten, D. McNamara, P. Karlsson, and M. Beach. Second order statistics of NLOS indoor MIMO channels based on 5.2 GHz measurements. In *Proc. Globecom 2001 - IEEE Global Telecommunications Conf.*, San Antonio, TX, USA, Nov. 2001.

[YiFl04] X. Yin and B. H. Fleury. Nominal direction-of-arrival estimation for slightly distributed scatterers in channel sounding. TD(04)206, COST 273, Duisburg, Germany, Sep. 2004.

[ZeST04] R. Zetik, C. Schneider, and R. Thomä. Correlation between MIMO link performance evaluation results and characteristic channel parameters. In *Proc. 13th IST Summit on Mobile and Wireless Commun.*, Lyon, France, June 2004. [available as TD(04)049].

[ZhBW04] L. Zhang, A. G. Burr, and M. W. Webb. Capacity of MIMO cellular system with finite scattering. TD(04)038, COST 273, Athens, Greece, Jan. 2004.

7

MIMO systems

Chapter editor: *Alister Burr*
Contributors: *Andreas Czylwik, Tricia J. Willink, Misha Dohler, and Hamid Aghvani*

7.1 Introduction

In the previous chapter we considered the underlying MIMO channel, including both the fundamentals of the double-directional electromagnetic propagation which distinguishes the MIMO from the SISO case, and also in particular various forms of channel model. It has also become clear how the MIMO channel is able to offer significantly increased capacity compared to SISO. We now consider the techniques which can be used to achieve this promised capacity, by exploiting these channel characteristics.

The basis of the advantage of MIMO systems arises from a feature which is simultaneously the greatest strength and the greatest weakness of wireless systems. The very absence of conductors to guide the radiation is what makes wireless so attractive as a communication medium, but also means that the signals cannot be restricted in space. This fact is the origin of the large path loss we encounter in wireless systems and also gives rise to both multipath fading and to interference between systems. Multi-Element Antenna (MEAs) arrays allow us to greatly reduce these disadvantages by exploiting the spatial dimension, and targeting the signals we transmit and receive to specific directions, while losing none of the flexibility and convenience which typify wireless communications.

This ability to direct signals in space also means that the phenomenon of multipath propagation, which has been a perennial problem, now becomes an advantage because it provides diverse routes for the signal to reach the receiver. MIMO allows us to exploit this diversity by transmitting redundant information by these multiple routes, or to increase throughput by transmitting independent information by the different paths (or indeed a combination of both). With the continually increasing demand for information bandwidth over what remains a strictly limiting spectrum resource, this is undoubtedly the most promising technology available for fulfilling such demands.

There have traditionally been two ways to exploit multi-element array antennas, although they are of course closely related. The first (the older of the two) is based on beamforming and array signal processing, and the results are commonly known as ‘smart antennas’. The second, and more general, is MIMO (Multiple-Input, Multiple-Output). This, broadly, provides the distinction between Sections 7.3 and 7.4 of this chapter: Section 7.3 describes techniques based on beamforming, while Section 7.4 considers MIMO techniques. To achieve MIMO, of course, we require MEAs at both transmitter and receiver, while ‘smart antenna’ systems need have an array at only one end, usually the BS. The latter may thus be described as Multiple Input, Single Output (MISO) on the downlink, or Single Input, Multiple Output (SIMO) on the uplink.

To confuse the issue, however, transmit diversity schemes, which may also be SIMO, are usually considered along with MIMO techniques. The distinction between these and downlink beamforming is that in the latter case the channel is known to the transmitter, and hence the energy can be directed, while transmit diversity achieves its diversity advantage even in the absence of such channel knowledge. Here transmit diversity techniques, including most space-time codes, have been included in Section 7.4. Note, however, that this distinction vanishes in the case of adaptive MIMO (or MISO) systems, where the channel is known to the transmitter, and hence can also direct the signal.

As mentioned, MIMO systems must have MEAs at both ends of the link. This is also required to achieve the capacity increase, since for this purpose it must be possible to separate the different spatial paths at both transmitter and receiver. MISO and SIMO systems, in contrast, can achieve only a diversity advantage. The focus on 'smart antennas' is usually on increasing the gain of the antenna and/or on reducing interference, either to the receiver from unwanted user signals, or from the transmitter to other users. However, much work is based on the eigenmode theory of MIMO systems: although usually only one mode of the channel can be accessed.

As mentioned in the introduction to Chapter 3, discussion of MIMO systems is also to be found elsewhere in the book, since many of the techniques explored in Chapter 3 are applied to MIMO systems. However, in this chapter the focus is on the techniques that are used specifically to exploit the MIMO channel: the transmission techniques and receiver architectures.

Finally, in Section 7.5 we consider the implications of MIMO techniques for complete wireless systems. Traditionally MIMO techniques have been devised and analysed for single links only, whereas in practice they will be used within networks, where interference can occur between links. Clearly a particularly important case is their use in cellular systems. In this section we begin to consider the question of whether MIMO techniques can in fact increase the capacity of complete networks, rather than just of individual links. We also consider several MIMO or MIMO-like systems which invoke issues on higher levels, especially the network layer.

7.2 Information-theoretic aspects

The following section focuses on information-theoretic aspects of MIMO systems. First, the results on capacity of single links are summarised. The next subsection discusses results on capacity of multi-user MIMO links. Finally, the capacity of relay-assisted MIMO links is considered.

7.2.1 Capacity of single link MIMO systems

The authors of [SSSV04a] and [SSSV04c] decompose a lower bound of the ergodic capacity of a MIMO link (for the high SNR case) as a sum of three terms: (i) a supremum capacity, (ii) the effect of fading, and (iii) the eigenvalue dispersion of the channel. It is shown that the terms in the decomposition are statistically independent under ergodic iid Rayleigh fading. Furthermore, the exact probability distributions of random terms in the decomposition are derived. For a Rayleigh fading MIMO channel, the authors calculate also the probability distribution of an upper bound of the channel capacity thereby obtaining the outage probability of channel capacity [SSSV04b].

The capacity of a multipath channel with correlated fading is investigated in [Fise04]. First, a SISO channel is considered. It is modelled by an impulse response with correlated circularly symmetric zero-mean Gaussian tap coefficients. The tap coefficients exhibit the same variance, thus corresponding to a rectangular power delay profile. The standard method for calculation of the capacity of a frequency-flat MIMO channel is used to calculate the capacity of the frequency-selective SISO channel. The result is extended to a frequency-selective MIMO channel. It is numerically shown that the correlation between channel taps decreases the channel capacity.

The capacity of a so-called diagonal correlation channel is investigated in [ÖzBo04]. Such an $n \times n$ MIMO channel exhibits n orthogonal DoDs and n orthogonal DoAs where each DoD couples into a single DoA and vice versa. Diagonal correlation Rayleigh fading channels show a larger ergodic capacity than iid channels. The authors provide an intuitively appealing explanation for the ergodic capacity enhancement in the case of a 2×2 MIMO channel. For a diagonal correlation channel, fading of the channel matrix elements occurs, in contrast to iid channels, not independently but in pairs of h_{11}, h_{22} or h_{12}, h_{21}. As a consequence, diversity of such diagonal correlation channels is smaller than that of iid channels.

The capacity of a MIMO channel with non-linearly modulated finite alphabet signals is analysed in [Syko04]. Upper and lower bounds for the capacity are determined. It is shown that the capacity scales linearly with the rank of the channel matrix as it is known for an unconstrained (linear) channel.

In order to obtain the maximum capacity, the energy distribution on eigenmodes has to be optimised according to the waterfilling principle. It is shown in [Fise04] that when using discrete energy levels instead of continuously optimised energy levels, the capacity degradation is only marginal. For four-antenna systems, three energy levels are sufficient whereas for systems with more antennas more energy levels are required.

In [KnSy04b] an approach is discussed which aims for a constant capacity per eigenmode. Of course, this is a sub-optimal approach, but it gives some implementation-oriented advantage. There is a non-zero probability that the weakest eigenmode is not present so that the required transmit power tends to infinity. In order to avoid this situation, it is proposed to skip the weakest mode. Numerical evaluations show that both the constant capacity per eigenmode approach and the approach not using the weakest eigenmode do not yield a dramatic reduction of capacity.

In [LeMJ04] the outage probability of the capacity of specific channel models in the 5 GHz range is investigated by means of simulations. The single-user downlink transmission in macrocell, microcell, and picocell environments is considered. It has been found that data rates around 1 Gbit/s for LoS conditions and 500 Mbit/s for NLoS conditions can be theoretically achieved in a microcellular environment with a bandwidth of 100 MHz at a link distance of 250 m. In order to provide high data rates in NLoS indoor environments, only short link distances, under 20 m, are allowed. Data rates around 400 Mbit/s can be theoretically achieved in indoor LoS conditions with a link distance of 40 m.

Because of the time-variant behaviour of the radio channel, its capacity also changes with time. The relative fluctuations decrease with increasing bandwidth since the frequency correlation of the channel vanishes for large frequency differences. In [Czyl04] the temporal fluctuations of the capacity of ultra-wideband SISO radio channels with frequency-selective fading are analysed on the basis of a simplified channel model. Assuming perfect channel knowledge, the capacity fluctuations are calculated by an analytical approach. An equation is presented which relates the variance of the fluctuations of channel capacity with the frequency-correlation function of the radio channel.

In order to investigate which part of the huge channel capacity of MIMO systems can actually be achieved in a practical system, the capacity loss associated with Space-Time Bit-Interleaved Coded Modulation (STBICM) approaches has been calculated in [HiBu04] for quasi-static fading. Example capacity CDFs have been used to illustrate the losses of STBICM over the constrained modulation capacity which itself shows a loss over the unconstrained modulation capacity. Rather than just considering losses in capacity for given SNR and probability of outage, the CDFs were translated to plots of outage probability as functions of SNR for a given rate. For capacity-achieving schemes, the SNR loss due to the modulation constraint was found to be mostly independent of the number of transmit and receive antennas but decreased as the modulation order increased. Whereas the loss due to bit interleaving was found to be relatively independent of the rate, it increased with the number of transmit antennas and/or modulation order and decreased as the number of receive antennas increased. Example losses of the turbo code-based STBICM scheme over the modulation constrained capacity were around 0.6–0.9 dB for eight receive antennas, these being comparable to the loss of the turbo code on the single-input/single-output BPSK/AWGN channel. Example losses of the STBICM scheme over the unconstrained modulation capacity were around 1.0–1.7 dB.

In [DAZL05] a transmit antenna selection scheme operating over a flat MIMO Nakagami fading channel is investigated, where at the transmitting side a single antenna is selected according to an SNR maximising criterion and at the receiving side maximum-ratio combining is performed. Such a scheme minimises the signal processing at the transmitting side at the expense of an additional feedback channel. The ergodic capacity and the outage probability for a given communication rate are rigorously derived for such a scenario.

7.2.2 *Capacity of multi-user MIMO systems*

The capacity of multi-user MIMO systems is investigated in [PiTL02]. In a multi-user MIMO system a simple but practical way for the detection process is to suppress interference by linear filtering at the receiver. The paper analyses the capacity loss with respect to the optimum joint detection process where all transmit signals – including interference – are received simultaneously.

The capacity of a general MIMO system with n_t transmit and n_r receive antennas is given by [Fosc96]:

$$C^{0}_{1\ldots n_t}(\sigma_x^2, \sigma_n^2) = \log_2\left\{\det\left(I_{n_r} + \frac{\sigma_x^2}{\sigma_n^2} H_{n_t} H_{n_t}^H\right)\right\} \tag{7.1}$$

where σ_x^2 denotes the signal power per transmit antenna and σ_n^2 the noise power per receive antenna, respectively. In (7.1) it is assumed that the signals of all transmit antennas are desired signals. Now, assuming that there are desired signals from only m transmit antennas and interfering signals from $n_t - m$ transmit antennas, the capacity can be shown to be:

$$C^{n_t - m}_{1\ldots m}(\sigma_x^2, \sigma_n^2) = C^{0}_{1\ldots n_t}(\sigma_x^2, \sigma_n^2) - C^{0}_{m+1\ldots n_t}(\sigma_x^2, \sigma_n^2) \tag{7.2}$$

The capacity of multi-user MIMO systems is also considered in [Burr02]. By a stochastic simulation example it is shown that the capacity is increased by using a pre-whitening filter to suppress interference. The simulations furthermore show that the capacity advantage of a MIMO system remains also in an interference-limited region. The preliminary conclusion of the paper is that the capacity of MIMO systems is as robust to co-channel interference as that of a SISO system.

The information-theoretic capacity of MIMO systems in co-channel interference is investigated in [WeBN04]. It could be shown that adding an interfering antenna has the effect of eliminating one of the MIMO subchannels (in the limit of high Interference-to-Noise Ratio (INR)) but leaving the others unchanged. This observation has been used to derive bounds on the performance in high INR. It has also been found that MIMO systems perform more efficiently the more spatially coloured interference they experience, and, with many interferers, their overall effect is a reduction of the prevailing SNR. By simulation, the authors examined the effect of having more or fewer receive antennas than transmit antennas.

The authors of [FKWS04] focus on the multi-user MIMO broadcast channel. By means of simulations the cumulative distribution of the sum capacity of a single cell is calculated. The BS as well as the mobile stations are equipped with multiple antennas. The simulations are carried out for a double-directional geometry-based flat fading channel model which is well suited for a macrocellular environment. The simulation results show that the sum capacity is increasing with the number of both transmit and receive antennas as well as the number of mobile stations in the cell. The relative increase of the sum capacity decreases with an increasing number of mobile stations.

7.2.3 Capacity of MIMO systems with relays

For a MIMO channel with iid Rayleigh fading coefficients, the capacity of the MIMO channel increases with the minimum of transmit and receive antennas. In a real propagation scenario the capacity gain with respect to a SISO channel diminishes with increasing correlation between antenna signals. Node cooperation is the natural extension of space-time processing for multiple distributed nodes. In [WiRa03], [RaWi04] the capacity of a corresponding relay-assisted MIMO link is calculated. A concept with two time slots is considered: during the first time slot, the source is transmitting and the relays are receiving, during the second relays as well as the source are allowed to transmit signals. The relays may or may not have multiple antennas and are described by noisy linear amplifiers. For LoS channels (without relays) the capacity of the link increases only logarithmically with the number of transmit/receive antennas due to the high correlation of antenna signals. Using relays, the link capacity can be improved dramatically to an almost linear increase which is usually obtained in a rich scattering environment.

7.3 Array processing and beamforming

High capacity demand and limited spectral resources for wireless communications require solutions that take advantage of opportunities for adaptivity and diversity. The time-varying spatial and temporal channel characteristics of the mobile multipath environment provide many such opportunities.

If the relative time delays experienced by the impinging signals are small compared to the symbol period then the system model is determined to be narrowband. In this case, the sampled baseband system model for the space-time communication system with N_t transmitting antenna elements and N_r elements in the receiver array is

$$\mathbf{r}(k) = \mathbf{H}(k)\mathbf{s}(k) + \mathbf{i}(k) + \mathbf{n}(k) \tag{7.3}$$

where the elements of the length-N_r received signal vector $\mathbf{r}(k)$ are the signals at the output of each antenna element, $\mathbf{s}(k) = [s_1(k) \ldots s_{N_t}(k)]^T$ is the length-N_t vector of transmitted signals of interest, $\mathbf{i}(k)$ is the interference vector and $\mathbf{n}(k)$ is the additive noise. $\mathbf{H}(k)$ is the $N_r \times N_t$ channel response matrix. In general, the channel response is assumed to remain constant over multiple symbol periods, in which case its dependence on k will be omitted.

Antenna arrays, or Multiple Element Antennas (MEAs), serve two main purposes – to exploit diversity gain, thereby mitigating the effects of fading and increasing the SNR, and to reject interference by steering the array gains in favour of the desired signals. This section is concerned with techniques for processing the transmitted and received signal vectors to achieve these goals. As discussed in Section 6.3, the spatial channel can be described in terms of the AoD or AoA of the emitted or impinging signals. Herein, techniques which use directional beams, possibly aligned to the angles of the multipath components, will be termed 'beamforming'. When the spatial characteristics are used without specifically extracting AoD or AoA information, it will be referred to as 'array processing'.

7.3.1 Antenna selection

Antenna selection at the transmitter and receiver, in which a subset of the available antenna elements are used, was considered in conjunction with space-time block codes in [BaFW05]. For a 4×4 quasi-orthogonal code, only four of the N_t transmit elements and one out of N_r receive elements were used. The selection of antenna elements was based on a parameter which takes into account the self-interference caused by the partial decoupling of the data streams. It was seen that the performance improved with increasing $N_t > 4$ or $N_r > 1$, and that even in highly spatially correlated channels, an improvement was possible relative to the case with $N_t = 4$, $N_r = 1$ and iid gains.

7.3.2 Beamforming

The application of different beamforming techniques at the seven BSs of a UMTS FDD network was considered in [BrSP03]. The simple Switched Lobe (SL) technique, in which the predefined beam providing the largest signal power is selected, achieved a doubling of network capacity relative to a fixed sector antenna with a 65° lobe as a result of enhanced interference reduction resulting from the narrower beams. The Paging Area (PA), using a DoA algorithm to continuously track the strongest angular component of the desired signal and direct a beam accordingly, achieved performance marginally better than the SL. Adaptive array processors, using eight antenna elements linearly spaced by one-half wavelength and designed to maximise the SINR (see Subsection 7.3.3), achieved an increase of 235% in network capacity relative to the fixed sector antenna due to their ability to combine signal energy from different directions.

When simulating antenna array performance, it is generally assumed that the antenna elements are omnidirectional. In reality, when an antenna array is mounted on a mast the beam pattern of each element will deviate significantly from the omnidirectional characteristic due to shadowing effects of the mast itself. The impact of these effects on downlink beamforming using a circular antenna array was modelled using sector antennas in [Czyl03]. The average SIR in an FDD CDMA-based cellular radio system was simulated for different sector antenna beamwidths, and it was found that even with beamwidths as narrow as 40°, the performance was equivalent to that achieved using omnidirectional antennas.

7.3.3 Uplink array processing

The array processing receiver applies a complex weighting to the signals arriving at each of the elements in the antenna array. This weight vector, $\mathbf{w}$, can be viewed as a spatial filter, leading to the decision variable

$$\hat{s}(k) = \mathbf{w}^H \mathbf{r}(k) \tag{7.4}$$

The optimum spatial receiver filter with respect to maximising the SINR was derived for wideband signals and anisotropic interference, such as that caused by users in neighbouring cells, in [PNTL02]. It was shown to consist of a pre-whitening filter, which whitens the coloured interference signal, and a filter matched to the concatenation of the transmit filter, the physical channel and the pre-whitening filter. Thus, the impact of the coloured interference is that the overall receiver filter response is increased, resulting in additional ISI. It was noted that, unless the desired signal and interference have the same AoA, it might not be possible to eliminate all the ISI, even if an ML equaliser is employed; however, if no restrictions are placed on the bandwidth, the gain always exceeds the total ISI in the presence of thermal noise; therefore, increasing the SINR should enable complete elimination of the ISI.

MRC was used in [DuOH05] for adaptive coded modulation using a Multi-Element Antenna (MEA). In this technique, the signals at different antenna elements are combined in proportion to their SNRs. It was seen that not only was the overall system performance improved, as expected, but less power was required for the pilot symbols used for channel estimation.

Weight vector calculation

The weight vector $\mathbf{w}$ in (7.4) depends on the optimisation parameter used. For beamforming to minimise the MSE the weight vector is given by

$$\mathbf{w}_{MMSE} = \arg\min_{\mathbf{w}} \mathcal{E}\{|\hat{s}(k) - \mathbf{w}^H\mathbf{r}(k)|^2\} \tag{7.5}$$

which leads to the Wiener-Hopf weight vector for user of interest q

$$\mathbf{w}_{MMSE} = \mathbf{R}^{-1}\mathbf{p}_q \tag{7.6}$$

where the covariance matrix is $\mathbf{R} = \mathcal{E}\{\mathbf{r}(k)\mathbf{r}^H(k)\}$ and the propagation vector for user q is $\mathbf{p}_q(k) = \mathcal{E}\{\mathbf{r}(k)s_q^*(k)\}$.

The SINR performance of the MMSE array processor (7.6) was investigated in [NaCB03a]. As shown in Figure 7.1, it was found that for high SNRs, the interference dominates and the array processor acts as a null steerer, suppressing the interfering signal components. At low SNRs, the

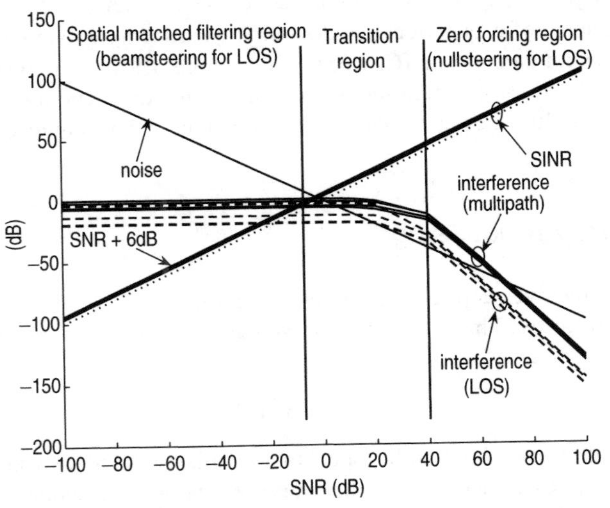

Figure 7.1 *Noise and interference components vs. SNR for $N_t = N_r = 4$ using MMSE array weights.*

array processor acts as a beam steerer, maximising the signal energy from the desired user. A smooth transition region exists in which the MMSE array processor finds the best compromise between null- and beamsteering.

When the channel matrix $\mathbf{H}$ is column rank deficient, for example, when the array is under-loaded ($N_t < N_r$), the covariance matrix of the signal component of $\mathbf{r}(k)$, i.e. $\mathbf{R}_{ss} = \mathbf{H}\mathcal{E}\left\{\mathbf{s}(k)\mathbf{s}^H(k)\right\}\mathbf{H}^H$, is singular and the impact of the noise can be reduced using a subspace-based array processor [Will02]. For $N_t \leq N_r$, there are at most N_t non-zero eigenvalues of $\mathbf{HH}^H$, denoted λ_i, $i = 1, \ldots, N_t$. Since

$$\mathbf{R} = \mathbf{HH}^H\sigma_s^2 + \sigma_n^2\mathbf{I} \tag{7.7}$$

it can be decomposed as

$$\mathbf{R} = \mathbf{U}_s\mathbf{\Lambda}_s\mathbf{U}_s^H + \mathbf{U}_n\mathbf{\Lambda}_n\mathbf{U}_n^H \tag{7.8}$$

where $\mathbf{\Lambda}_s = \mathrm{diag}(\lambda_1, \ldots, \lambda_{N_t}) + \sigma_n^2\mathbf{I}$ and $\mathbf{\Lambda}_n = \sigma_n^2\mathbf{I}$. The columns of $\mathbf{U}_s$ are the eigenvectors corresponding to the eigenvalues in $\mathbf{\Lambda}_s$, and span the signal subspace. The columns of $\mathbf{U}_n$, where $\mathbf{U}_s \perp \mathbf{U}_n$, span the noise subspace. As the desired user's signal must lie in the signal subspace, the MMSE weight vector is

$$\mathbf{w}_{SS} = \mathbf{U}_s\mathbf{\Lambda}_s^{-1}\mathbf{U}_s^H\mathbf{p}_q \tag{7.9}$$

Joint detection

For JD, $\hat{\mathbf{s}}(k)$ can be found using the ZF weight array, given by the pseudo-inverse of the channel matrix, i.e. [WeMe04]

$$\mathbf{W}_{ZF} = (\mathbf{HH}^H)^{-1}\mathbf{H} \tag{7.10}$$

Note that this requires knowledge of the channel matrix, rather than just the covariance matrix of the received signal vector as in (7.6) and (7.9).

Overloaded arrays

When the array is overloaded, i.e. $N_t > N_r$, the ZF array processor fails to perform due to its inability to null out all the interferers. In this case, the ML-JD is the optimum receiver, but, at $O(2^{N_t m})$ for m bits/symbol, the complexity becomes prohibitive for a reasonable number of users. An alternative approach to joint detection for the overloaded array case was proposed in [CoWi04b], using a genetic algorithm with a priori information used to select the initial population provided by a ZF detector. At each generation, the fitness of each member of the population is evaluated, and the next generation is created by a process of reproduction, mutation and elitism. Using the number of fitness evaluations as a measure of the computational complexity, the resulting BER-to-complexity trade-off is illustrated in Figure 7.2, for $N_t = 8$ users and $N_r = 5$ elements in the receiver array. The importance of using a priori information is noted in the improved initial performance as well as the faster convergence. The population size, P, also affects the convergence rate and the asymptotic performance.

7.3.4 Downlink array processing

The BS antenna array can also be used to improve the performance of downlink communications; if the channel response at the BS array for a given MT with a single element antenna is $\mathbf{h}$, then, by

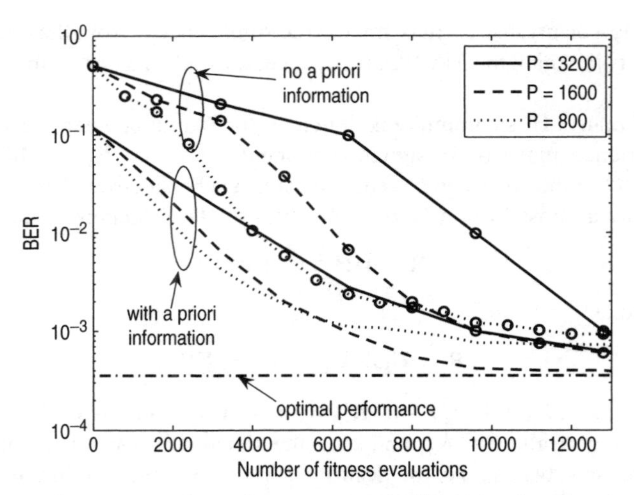

Figure 7.2 *BER vs. complexity curves for eight users using QPSK with a five element array for SNR 17.8 dB.*

reciprocity, the received signal at the MT from the BS is

$$r(k) = \mathbf{h}^T \mathbf{x} s(k) + n(k) \tag{7.11}$$

where $n(k)$ is additive noise and $\mathbf{x}$ is the transmitter weight vector at the BS.

When channel knowledge is available at the receiver, for example in a TDD system, the array processor weights used in the uplink can be applied in the downlink. This situation was investigated in [ZeZe03] for data measured in an indoor environment. It was found that by acquiring phase information at each antenna element on the uplink and applying the phase inverse on the downlink, the fades observed by the MT were significantly reduced.

In [HäCC03], a downlink beamforming algorithm was proposed for UMTS systems, in which the angular and attenuation characteristics of all the multipath components in the downlink were assumed to be known. The objective was to maximise the average signal power at the MT of interest, while keeping the total interference power constant. The weight vector $\mathbf{x}$ which provides the solution to this problem for user q was shown to be the eigenvector corresponding to the largest eigenvalue, λ, solving

$$\mathbf{R}_q \mathbf{x} = \lambda \mathbf{R}_{in} \mathbf{x} \tag{7.12}$$

where $\mathbf{R}_q$ and $\mathbf{R}_{in}$ are the spatial covariance matrices of the desired user and of the interference-plus-noise, respectively. When combined with a simple power control algorithm, this downlink beamformer was demonstrated to provide support for 60% more users with $N_r = 4$ receiver elements relative to a single antenna element. Power control was seen to play a considerable role, especially in the case of MEAs with large numbers of elements. For $N_r = 4$, an increase of 50% in the number of supported users was observed when power control was incorporated.

In [NaCB04a], the objective was to determine downlink weights to achieve a target SINR for each user while minimising the total transmitter power. The problem was formulated as a constrained optimisation, which was solved using a gradient-based algorithm. It was observed that the resulting downlink weights are not always the same as those for the uplink.

For the JT of multiple data streams, the parallel of the ZF joint detector in Subsection 7.3.3, the transmitter weight array is the ZF-JT [WeMe04]

$$\mathbf{X} = \mathbf{H}^*(\mathbf{H}^T\mathbf{H}^*)^{-1} \tag{7.13}$$

7.3.5 Channel estimation

In [MaWe04], a technique was proposed for estimating the channel responses in an OFDM system for multiple MTs, each with a single element antenna, whose signals were assumed to impinge on the BS array from a single AoA, which is known perfectly at the receiver. ML and MMSE channel response estimates were derived and it was seen that the MMSE approach provided better performance due to the inclusion of additional channel state information from the power-delay profile. It was noted that this information is difficult to obtain in real-world systems.

It was observed in [NiSS03] that in time-varying multipath channels the AoAs and arrival delays change slowly compared to the fading amplitudes. A subspace method exploiting this property was proposed for training-based channel estimation in block-by-block transmission systems. The quasi-static nature of the AoAs and delays is converted into the invariance of the spatial-temporal subspaces that contain the channel response. These subspaces are estimated by means of long-term averaging (over several blocks), while the faster-varying parameters are updated in the short term (block-by-block). As the accuracy of the estimate of the subspaces increases with the number of blocks, the parameters that affect the variance of the overall channel estimate reduce asymptotically to the fast-varying features only, as long as the space-time subspace is quasi-static. This reduction of the number of parameters is particularly relevant in radio environments where the angle-delay spread is small compared to the system resolution.

An example is given in Figure 7.3; the multipath structure shown in Figure 7.3(a) is composed of five paths whose AoAs $\{\theta_i\}_{i=1}^5$ and delays $\{\tau_i\}_{i=1}^5$ stay constant over the blocks. The Power Delay Angle (PoDA) diagram for the channel within the ℓth block is given in Figure 7.3(b); Figure 7.3(c) shows a noisy estimate obtained by the unconstrained least squares approach. Since $\theta_1 = \theta_2, \theta_3 = \theta_4$ and $\theta_4 = \theta_5$, the spatial and temporal subspaces have dimension, i.e. diversity order, $r_\mathrm{S} = 3$ and $r_\mathrm{T} = 4$, respectively. The diagrams of the subspace-based estimates are shown in Figure 7.3(d)–(f). The estimates were obtained by exploiting the stationarity of the temporal (T) subspace only, Figure 7.3(d), the spatial (S) subspace only, Figure 7.3(e), or both (ST) the subspaces, Figure 7.3(f), using as diversity orders $\hat{r}_\mathrm{T} = 1$ to $\hat{r}_\mathrm{T} = 4$ and $\hat{r}_\mathrm{S} = 1$ to $\hat{r}_\mathrm{S} = 3$. The comparison shows how the projection onto the spatial-temporal subspaces reduces the estimate error with respect to the unconstrained estimate.

An alternative subspace-based method was proposed in [WeMZ05], in which the channel matrix decomposition was based on the observation that, when the mobile moves only in a small area, the impinging wavefronts do not change but the subspace-based channel vector changes quickly due to fast fading. Therefore, only the subspace-based channel vector needs to be updated at each snapshot, reducing the number of unknowns that need to be estimated compared to the conventional estimator. It was shown in [WeMZ05] that the degradation in SNR decreases with the number of impinging wavefronts because the dimension of the subspace which must be estimated is correspondingly smaller.

Impact of channel estimation errors

In practical systems, errors in the estimation of the channel matrix **H** or its covariance matrix **R** result from noise, non-ideal training sequences, quantisation, time-varying channel conditions, etc.

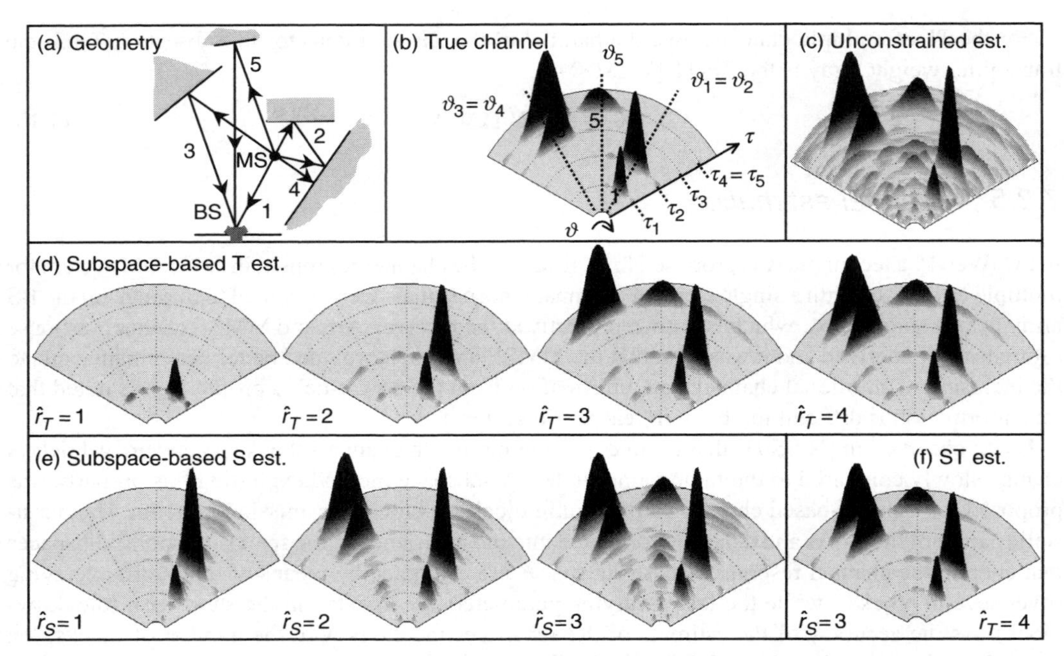

Figure 7.3 *Example of space and/or time projections for a channel with five multipath components with* $r_S = 3$ *and* $r_T = 4$.

Even with good estimation techniques, channel estimation errors are unavoidable. The impact of estimation errors on the performance of a zero-forcing array processor was considered in [NaCB04b]. It was found that even small errors in the channel matrix estimate result in significant degradations in SINR and this effect was more pronounced in the LoS scenario.

An analytical investigation of the impact of estimation errors in array processors was undertaken in [Will02]. The estimation errors in the covariance matrix and propagation vectors were treated as perturbations

$$\hat{\mathbf{R}} = \mathbf{R} + \mathbf{E}_R \qquad \text{and} \qquad \hat{\mathbf{p}} = \mathbf{p} + \boldsymbol{\epsilon} \tag{7.14}$$

The mean increase in the MSE resulting from the estimation errors was derived using a first-order approximation of $\mathbf{R}^{-1}$ and the statistics of $\mathbf{E}$ and $\boldsymbol{\epsilon}$, for both MMSE and MMSE subspace-based array processors, (7.6) and (7.9), respectively.

A perturbation approach to analysing the impact of channel estimation errors on the BER performance of the ZF-JD and ZF-JT array processors was presented in [WeMe04]. The perturbation was defined as

$$\hat{\mathbf{H}} = \mathbf{H} + \mathbf{E}_H \tag{7.15}$$

where the variance of the elements of $\mathbf{E}_H$ was σ_e^2. For small σ_e^2 the data estimation error was expressed in terms of $\mathbf{E}_H$ using truncated Taylor series, and the resulting error floors were derived.

Compensation for estimation errors

As noted above, channel estimation errors are inevitable. The impact of the channel estimation errors was taken into account in determining robust array weights for CDMA in [BoBP04] and [ChCz04]. The 'robust adaptive beamformer' was presented in [BoBP04]. This technique is used to train the array processor with no knowledge of the angular make-up of the desired signal, without the use of a training sequence. This formulation is a low complexity approach based on the worst case optimisation problem, as follows.

For received signal $\mathbf{r}$ consisting of signal, interference and noise components, the corresponding despread output vector of the N_r correlators is $\mathbf{y}$. The weight vector which maximises the SINR is computed using the covariance matrices before and after despreading, i.e. $\mathbf{R}$ and $\mathbf{R}_{yy}$, respectively; this approach is called 'modified code filtering'. Thus

$$\mathbf{w}_{mcf} = \arg\max_{\mathbf{w}} \frac{\mathbf{w}^H \mathbf{R}_{yy} \mathbf{w}}{\mathbf{w}^H \mathbf{R} \mathbf{w}} \tag{7.16}$$

which is equivalent to the Minimum Variance Distortion Response (MVDR) problem, i.e.

$$\min_{\mathbf{w}} \mathbf{w}^H \mathbf{R} \mathbf{w} \qquad \text{s.t.} \qquad \mathbf{w}^H \mathbf{R}_{yy} \mathbf{w} = \sigma_y^2 \tag{7.17}$$

for a fixed constraint, σ_y^2. This optimisation does not have a closed form solution, but $\mathbf{w}$ can be obtained using an iterative algorithm, elaborated in [BoBP04]. It was seen that the proposed algorithm is a more robust solution to the modified code filtering problem than direct matrix inversion.

In a similar way, a robust array processor based on a minimum outage probability criterion was proposed in [ChCz04]. The covariance matrices, $\mathbf{R}_k$, $k = 1, \ldots, N_{MT}$, of all users were assumed to be subject to unknown perturbations, $\mathbf{E}_k$, giving the SINR for user q

$$\gamma_q = \frac{\mathbf{w}^H (\mathbf{R}_q + \mathbf{E}_q)\mathbf{w}}{\sum_{k \neq q, k=1}^{N_{MT}} \mathbf{w}^H (\mathbf{R}_k + \mathbf{E}_k)\mathbf{w} + \sigma_n^2 \mathbf{w}^H \mathbf{w}}. \tag{7.18}$$

For an SINR threshold γ_{th}, the outage probability for user q is then

$$O_q = Pr\left\{ \operatorname{tr}\left((\mathbf{R}_q + \mathbf{E}_q)\mathbf{w}\mathbf{w}^H \right) \leq \gamma_{th} \sum_{k \neq q, k=1}^{N_{MT}} \operatorname{tr}\left((\mathbf{R}_k + \mathbf{E}_k)\mathbf{w}\mathbf{w}^H \right) + \sigma_n^2 \gamma_{th} \operatorname{tr}\left(\mathbf{w}\mathbf{w}^H \right) \right\}. \tag{7.19}$$

For a seven-site UMTS network using $N_r = 4$ elements in the BS array, modelling $\mathbf{E}_k$ as $\mathcal{CN}\left(0, \sigma^2\right)$ with $\sigma = 1^{-10}$, it was seen in [ChCz04] that the robust array processor was able to support approximately 25% more users.

7.3.6 Time-varying channels

In [CoWi04a], two approaches to weight adaptation for tracking changing channel responses were evaluated using measured data in different operating environments. The simulations used a training sequence of length N_T to generate the estimated covariance matrix, $\hat{\mathbf{R}}(n)$, and propagation vector, $\hat{\mathbf{p}}_q(n)$, where n is the frame index.

The two weight adaptation techniques considered were Direct Matrix Inversion (DMI), a block-adaptive algorithm, and LMS, a gradient-adaptive algorithm. For the DMI algorithm, the weight vector for frame n is

$$\mathbf{w}_{DMI}(n) = \hat{\mathbf{R}}^{-1}(n)\hat{\mathbf{p}}_q(n) \tag{7.20}$$

In the LMS algorithm, the weight vector is adapted over each of the N_T training symbols,

$$\mathbf{w}_{LMS}(k) = \mathbf{w}_{LMS}(k-1) + \mu\mathbf{r}(k-1)e^*(k-1) \tag{7.21}$$

$$e(k) = s(k) - \mathbf{w}_{LMS}^H(k)\mathbf{r}(k) \qquad k = 1, \ldots, N_T \tag{7.22}$$

The average SNR gains for different array sizes relative to a single receiver antenna element were determined in [CoWi04a] using measured data from rural, suburban and urban environments, and are shown in Figure 7.4 for $N_T = 12$ training symbols in a frame of length 60. The DMI algorithm maintains similar performance in all propagation environments – as long as the channel response change over a single frame is small, all significant multipath groups are included in the diversity combining process. However, as the number of elements was increased beyond $N_T/2$, the SNR gain decreases, which is attributed to the inadequacy of the channel estimates, as discussed in Subsection 7.3.5.

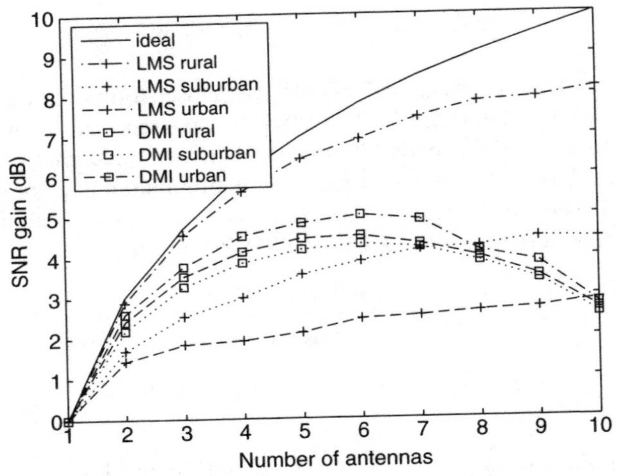

Figure 7.4 *Performance of LMS and DMI algorithms in different time-varying environments for different size arrays.*

The performance of the LMS algorithm was much more dependent on the operating environment. In the rural scenario, where there was an LoS component and the angle of arrival changed slowly with the vehicle's movement, the gradient algorithm was able to track the changes in the channel response well. In the suburban and urban environments, the performance was worse as the adaptive algorithm was unable to track the rapidly changing channel response.

7.3.7 Applications of array processing

UMTS

Many of the techniques presented above can be applied to the UMTS scenario, for example [Pomm02], [ChCz04], [HäCC03]. A detailed investigation of the use of adaptive antennas in different UMTS operational scenarios (see Section 6.2) was presented in [GiCo01], [GiCo02], [GiCo03a], [GiCo03b], [GiCo04]. The weight vector that minimises the MSE was determined using the Conjugate Gradient (CG) method [GoLo96, Section 10.2] to minimise the mean-square error. The gain achieved with the array processor relative to a single omnidirectional antenna element was given by G_{N_r}.

The performance of the array processor was evaluated in numerous macrocellular scenarios, and it was found in [GiCo02] that the distance between BS and MT, the radius of the circle containing the scatterers and the density of clusters and scatterers all lead to similar performance of the array processor. Specifically, when the angular spread of the impinging signals is reduced, the limited resolution of the array processor means that lobes and nulls cannot isolate the desired signals from the interferers. The cluster/scatterer density has the most significant effect. As long as the angular displacement and grouping of the MTs enable sufficient spatial separation at the BS, substantial gains are achievable through the use of an adaptive array processor. In microcellular environments, the signals have greater angular and temporal richness relative to the macrocellular environment, therefore the gains achievable using an adaptive antenna are greater [GiCo03a], [GiCo04].

Observations of the evolution of G_{N_r} through the iterations of the CG algorithm showed that the gain peaked after a small number of iterations, before decreasing to its asymptotic performance [GiCo01]. The evolution of the SINR gain and the MSE with the number of iterations was seen to depend on the number of users and their grouping [GiCo03b]; the degradation after the peak is greater when the impinging signals have similar angular and temporal characteristics. The impact of noise, block size and code length on this effect were investigated and it was observed that longer reference codes provided better performance. An alternate approach would be to use uplink beamforming at the MTs to increase their differentiation at the BS.

A semi-blind least squares algorithm was proposed in [Pomm02] which iteratively improves the array weight vector based on estimated data symbols. For a spreading sequence length N_{SF}, the transmitted signal can be described as the temporal vector

$$\mathbf{s} = \alpha \cdot \mathbf{s}_{DC} + j\beta \cdot \mathbf{s}_{CC} \tag{7.23}$$

where the subscripts DC and CC refer to the data and control channels, Dedicated Physical Data Channel (DPDCH) and Dedicated Physical Control Channel (DPCCH), respectively. The received signal matrix is denoted by $\mathbf{U}_r$, and a flat fading channel is assumed. The initial estimate is derived from the known identification sequence of the control channel using the MMSE criterion as above, and the DPDCH and the unknown part of the DPCCH can then be estimated iteratively by solving, for iteration m,

$$\mathbf{w}_m = \arg\min_{\mathbf{w}_m} \|\mathcal{I}\{\mathbf{U}_r\mathbf{w}_m\} - \beta \cdot \hat{\mathbf{s}}_m\|^2 \tag{7.24}$$

The iterative process is repeated until the estimated data sequence does not change. It was found in [Pomm02] that the algorithm converges in only a few iterations, i.e. one to three, depending on the SNR.

HIPERLAN/2 and IEEE 802.11a/e

The use of antenna arrays at the access points in a WLAN using HIPERLAN/2 was proposed in [NaCB02], [NaCB03a], [NaCB03b], [NaCB04b], [NaCB04c]. The environment of particular interest was television broadcasting studios, in which multiple users, possibly operating at different data rates, must be supported. Adaptive antennas provide a feasible method for reducing interference from other users. The investigation used ray tracing to provide a spatio-temporal model of the studio environment at 5.18 GHz using a ULA with $N_r = 4$. Using a ZF-JD array processor, it was found that in an environment with a coherence bandwidth in excess of 20 MHz the beam patterns varied smoothly across the signal bandwidth indicating that only two or three updates of the array weights would be required across the 52 subcarriers. In an environment in which the coherence bandwidth was considerably narrower, approximately 2 MHz, the weights would need to be updated every two or three subcarriers.

JOINT

The Joint Transmission and Detection Integrated Network (JOINT) was proposed in [WSLW03] as a TDD-OFDM-based system with MEAs at MTs and access points (APs). Each group of APs covering a given service area is linked to a central unit. JD and JT are applied at the central unit in the uplink and downlink, respectively. The system model can be described using (7.3), in which now the block-wise elements of **H** are the channel matrices between each of the N_{MT} MTs and the N_{AP} APs, and the transmitted and received vectors similarly represent the vectors at the corresponding arrays. It was shown in [WSLW03] that, for an iid complex Gaussian channel matrix **H**, the transmit energy required to provide a specified SNR for a given MT is inversely proportional to the number of APs, N_{AP}, for a given N_{MT}/N_{AP}. This means that the total energy required in a service area, and the interference caused to other service areas, is a function of N_{MT}/N_{AP} and therefore the number of users that can be supported for a given QoS is limited by the number of access points, N_{AP}.

7.3.8 Eigenbeamforming

When a wireless communication system has arrays at the transmitter and receiver, beamforming approaches related to those considered above can be used to increase the system throughput. Specifically, it is assumed that the transmitter and receiver terminals both have knowledge of the channel response matrix, **H**, and use this to generate a set of orthogonal spatial filters. The signalling bandwidth is assumed to be much less than the channel's coherence bandwidth – for frequency-selective channels, eigenbeamforming is used in conjunction with OFDM to create multiple parallel narrowband subsystems [Will03], [Will05b].

The singular value decomposition of **H** is given by

$$\mathbf{H} = \mathbf{U}\boldsymbol{\Sigma}\mathbf{V}^H \tag{7.25}$$

where the columns of the unitary matrices $\mathbf{U} \in \mathbb{C}^{N_r \times N_r}$ and $\mathbf{V} \in \mathbb{C}^{N_t \times N_t}$ are the left and right singular vectors of **H**, respectively, and the diagonal elements of $\boldsymbol{\Sigma}$ are the real singular values, σ_i, $i = 1, \ldots, \sigma_N$. Thus, there are at most $N = \min(N_t, N_r)$ orthogonal modes of **H** which have non-zero gains.

The equivalent system model is now

$$\hat{\mathbf{s}} = \mathbf{W}^H\mathbf{H}\mathbf{X}\mathbf{P}^{\frac{1}{2}}\mathbf{s} + \mathbf{n} \tag{7.26}$$

where **X** and **W** are the $N_t \times N$ and $N_r \times N$ spatial filter array matrices at the transmitter and receiver, respectively, and $\mathbf{P} = \mathrm{diag}(p_1, \ldots, p_N)$ is the power weighting matrix. The weight matrices are selected such that

$$\mathbf{W} = \mathbf{U} \quad \text{and} \quad \mathbf{X} = \mathbf{V} \tag{7.27}$$

then (7.26) becomes

$$\hat{\mathbf{s}} = \mathbf{\Sigma}\mathbf{P}^{\frac{1}{2}}\mathbf{s} + \mathbf{n} \tag{7.28}$$

Thus, the spatial filters at the transmitter, $\mathbf{X}$, direct the N data substreams along the orthogonal eigenmodes of the channel so that they can be extracted without interference at the receiver using the left singular vectors of $\mathbf{H}$ as array weights. The channel gain on eigenmode k is $\lambda_k = \sigma_k^2$, where λ_k is the kth eigenvalue of $\mathbf{H}^H\mathbf{H}$.

An important problem in applying eigenbeamforming efficiently is the determination of suitable modulation and coding schemes to use on each eigenmode as well as the corresponding power allocation, $\mathbf{P}$. Approaches to this problem depend on whether it is assumed that the transmitter and receiver have perfect channel knowledge or only an estimate of $\mathbf{H}$.

Perfect channel knowledge

Several approaches to the problem of adapting power allocations and modulation/coding rates on the eigenmodes have been proposed, based on bit error rates and outage capacity.

In schemes whose objective is to achieve specified bit error performance, the power level and modulation constellation on each eigenmode is selected to achieve the target performance with the most efficient allocation of power.

In [SaTA04], the BERs for BPSK, QPSK and 16-QAM were determined for each of ten power levels. The proposed power and modulation adaptation algorithm searches through the set of possible candidate values of power and constellation size to find those which maximise the total throughput.

For the case of MIMO OFDM, it was proposed in [MuDa04] that only the strongest eigenmode on each frequency subcarrier be used. A power loading algorithm was proposed for predetermined constellation sizes and was simulated for IEEE 802.11a physical layer environments. It was seen that at a BER of 10^{-6}, the proposed adaptive power loading scheme provides a gain of over 4 dB for 2×2 MIMO relative to SISO, and approximately 1 dB relative to waterfilling.

The first adaptation strategy based on outage capacity was presented in [Fise04]; in that case, the power was allocated in discrete quantities and the objective was to minimise the outage capacity, which is equivalent to maximising the mutual information. The optimisation problem was defined as a Lagrangian minimisation for which the solution can be found using an iterative local discrete first-order search method. Results using this iterative solution for $L = 2, \ldots, 6$ power levels for $N_t = N_r = 3$, presented in [Fise04], show that low outage probabilities can be achieved even with coarse power resolutions – as the SNR decreases, the degradation for small L also decreases.

A disadvantage of this approach is that an iterative solution is required for each coherence interval. A non-iterative approximation was also indicated in [Fise04] which exploits observations of the solutions obtained using the iterative approach. Figure 7.5 shows the mapping of the solutions onto the two-dimensional eigenvalue space for $N_t = N_r = 2$ with $L = 6$. The grey shades correspond to the different available solutions. The regions are quite well partitioned, and are approximately

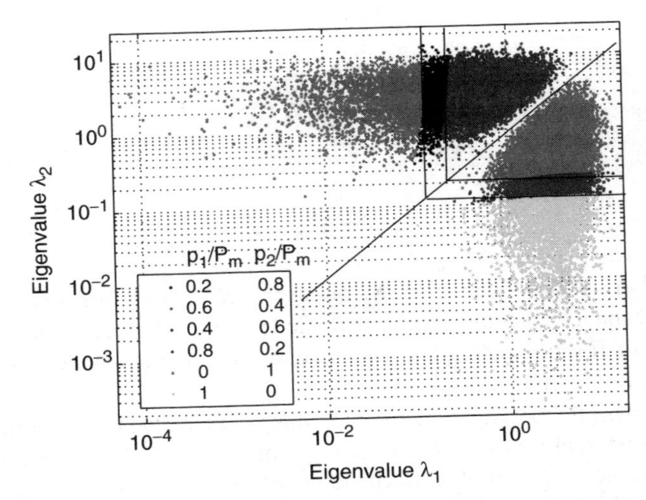

Figure 7.5 *Solutions obtained by iteration mapped to the eigenvalue space for $N_t = N_r = 2$, $L = 6$, SNR 10 dB.*

separated by lines parallel to the axes. The use of such predetermined thresholds would enable a rapid selection of **P** using the eigenvalue ratio.

Another approach to adaptive power allocation based on the outage probability was presented in [KnSy04b]. The objective in this case was to achieve a constant target capacity on each eigenmode, and thereby a constant overall capacity with minimal transmission outage probability. This criterion enabled the use of a single codebook for a given rate on all eigenmodes. One solution to this problem is 'truncated channel inversion', in which eigenmodes with gains less than some threshold are 'switched off'; however, the outage probability then becomes non-zero. The alternative approach proposed in [KnSy04b] is 'subspace total channel inversion', in which the weakest eigenmode remains switched off at all times, and the remaining subspace of dimension $N-1$ is fully inverted with no truncation.

The ergodic capacities for unordered eigenmodes using full space truncated inversion and subspace total inversion are shown in Figure 7.6, along with that achieved using optimal waterfilling over space and time. Note that the gap between the performance of full space truncated and subspace total inversion decreases as the channel dimension, $N = \min(N_t, N_r)$, increases.

Imperfect channel information

It was observed in [SaTA04] that there was considerable deviation between the simulated and experimental performance of the proposed power and modulation adaptation algorithm. It was noted that a high degree of accuracy is required of the channel estimation, feedback and synchronisation techniques in order to support the adaptive power loading. The impact of channel estimation errors is that the transmit and receive weight arrays, (7.27), are mismatched to the channel, resulting in SI [Will05a]. The impact of the SI caused by channel estimation errors can be estimated using perturbation analysis [Will05a], whereby the expected SINR can be determined for each eigenmode taking into account the effect of the estimation errors. This was used in a robust loading algorithm to select the number of eigenmodes that can support transmission at the desired performance, and to

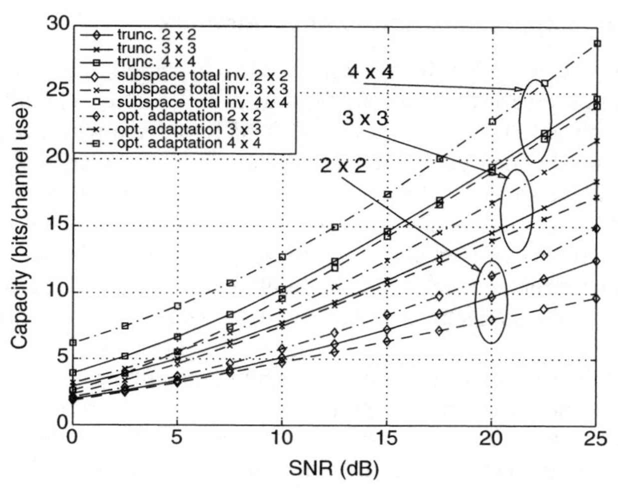

Figure 7.6 *Ergodic capacities for unordered eigenmodes using full space truncated inversion, subspace total inversion and waterfilling.*

allocate the transmitter power accordingly. Figure 7.7(a) demonstrates the resulting simulated SINRs obtained for many realisations of the 8×8 channel matrix $\mathbf{H} \sim \mathcal{CN}\left(0, \sigma_h^2\right)$ with estimation error variance σ_e^2. The specified SINRs are 20 dB for modes 1–3, 10 dB for modes 4–6 and 5 dB on modes 7 and 8, with $\sigma_h^2/\sigma_e^2 = 25.2$ dB. The SINRs resulting from a waterfilling allocation for the same model are shown in Figure 7.7(b). The robust power loading algorithm selects seven or more modes in less than 13% of cases, whereas the waterfilling method uses all $N_t = 8$ modes in over 72% of realisations.

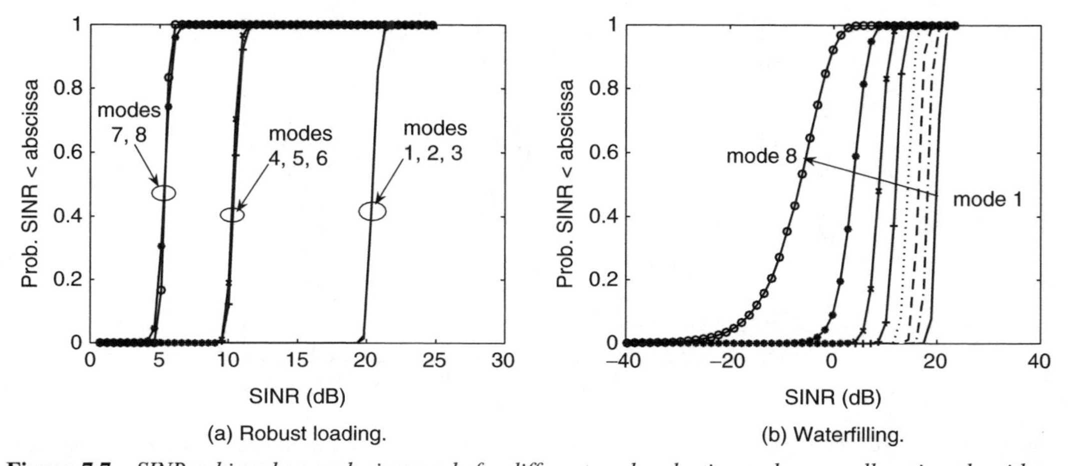

Figure 7.7 *SINR achieved on each eigenmode for different mode selection and power allocation algorithms.*

Eigenbeamforming in sparse scattering environments

Although most of the focus for MIMO systems is on rich scattering environments, useful gains can also be achieved in sparse scattering conditions as demonstrated in [Will03], [Will05b] where eigenbeamforming was used in conjunction with OFDM for a fixed wireless access system. Channel measurements were obtained using ULAs with $N_t = N_r = 8$ at 5.725 GHz at a height of approximately 60 m above street level in Toronto, Canada, with the LoS between the transmitter and receiver blocked by a small cluster of buildings. The relatively static nature of the channel characteristics supported the use of an eigenbeamforming technique. The self-interference resulting from channel estimation errors was controlled by limiting the number of eigenmodes used. Overall, spectral efficiency increases of a factor of three were obtained relative to a system employing single element antennas.

An investigation of the relationship between the impinging multipath groups and the structure of the singular vectors was undertaken in [BuHe03]. For the case of two correlated impinging waveforms with propagation vectors $\mathbf{h}_1$ and $\mathbf{h}_2$, there are two extreme cases: when $\mathbf{h}_1$ and $\mathbf{h}_2$ are orthogonal, the singular vectors in $\mathbf{U}$ are aligned with the two propagation vectors and the singular values are their two norms. If $\mathbf{h}_1^H\mathbf{h}_2 \neq 0$ and $\|\mathbf{h}_1\| = \|\mathbf{h}_2\|$, then the singular vectors contain both propagation vectors, but with opposite phases. Thus, the singular vectors take a 'sum and difference' form. When $\mathbf{h}_1^H\mathbf{h}_2 \neq 0$ and $\|\mathbf{h}_1\| \neq \|\mathbf{h}_2\|$, the singular vectors are part way between these two extremes: both propagation vectors are present in both singular vectors, but at different amplitudes. This effect was observed in measured data, as shown in Figure 7.8 [BuHe03]. It can be seen at AoAs $-55°$ and $20°$ that the two strongest modes generate beams in the same direction, indicating a strong 'sum' relationship. These effects were also observed in [Will05b], where it was seen that the eigenstructure in sparse scattering environments is highly dependent on frequency because the ratio of element separation to wavelength impacts the effective angle of arrival.

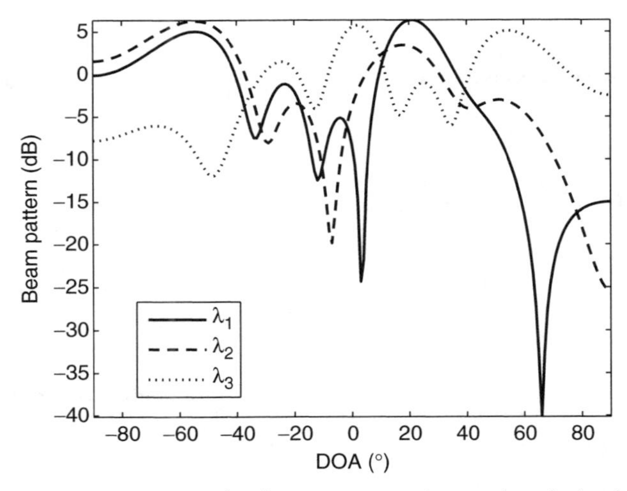

Figure 7.8 *Eigenbeam response patterns for three strongest eigenmodes obtained from multiple antenna measurements.*

7.4 MIMO transmission techniques

7.4.1 *Introduction*

The work conducted in COST 273 with respect to MIMO transceiver techniques has been very diverse, ranging from space-time coding techniques to enhanced channel estimation algorithms. As such, numerous contributions have been made to the areas of STBC, STTC, multiplexing, channel estimation, reduction of transceiver complexity, channel estimation techniques, as well as distributed and cooperative MIMO techniques.

The choice of structure for this section has been as follows: we first dwell on the design of space-time transmitter; second, on the design of space-time receivers; third, on the performance evaluation of various space-time coding schemes; and fourth, on the interesting generalisation of the space-time coding analysis by means of the real-valued notation. Finally, we explore the contributions made in the areas of distributed and cooperative MIMO transceivers.

7.4.2 *Space-time transmitter design*

Space-time block coding schemes

The STBC Alamouti scheme with feedback is a special form of link adaptation and is hence dealt with in greater depth in Section 3.6.

Performance results on the concatenation of high-rate pragmatic Trellis Coded Modulation (TCM) codes with a simple high-rate space-time block code operating on a MIMO channel with four transmit and four receive antennas are presented in [HaGe02]. Four TCM encoders feed four data streams of 32-QAM symbols into a simple rate-two Alamouti – like space-time code spreading the data over four transmit antennas. Perfect CSI at the receiver is assumed for all investigations. In this way an overall data rate of eight information bits per channel use is obtained. Using four receive antennas with a low complexity ZF receiver we get diversity order of approximately six. Comparing this system with coded Vertical-Bell Labs Space Time (V-BLAST) with the same bit rate and complexity, our system performs much better for all types of correlated and uncorrelated MIMO channels under investigation.

Performance results [HaGe02] show that the proposed concatenated coding system is much better than the coded V-BLAST system in the whole range of SNR. Assuming an uncorrelated channel and a BER of 10^{-3}, the concatenated STBC needs a mean SNR of 15 dB and coded BLAST achieves this BER at a mean SNR value of 26 dB.

The proposed low complexity concatenated STBC achieves a high data rate with reasonable BER at low SNR values. The coded V-BLAST system shows a rather gentle slope of the BER curve, due to a lower degree of diversity. Our system illustrates the importance of a certain amount of space diversity in MIMO systems which can be easily obtained by generalising the Alamouti principle. It is shown that simple high-rate STBCs providing only limited spatial diversity improve the system performance dramatically.

Space-time trellis coding schemes

To date, STTC for fast fading channels are known for up to two transmit antennas. This restriction results from the prohibitive search time required to find suitable codes. [BRHR04], [ABRH03] presents novel four-PSK STTCs for three and four transmit antennas operating over fast Rayleigh fading channels. The codes are designed according to the symbol-wise Hamming distance and product distance criteria; Criteria I, using a fast code-search method. This design criterion is optimal for constructing codes when the diversity order of the system is three or less. Otherwise, the symbol-wise Hamming distance and accumulated distance (trace) criteria – Criteria II – are chosen.

The performance of these new codes is evaluated by means of Monte Carlo simulation giving the performance as a function of SNR vs. FER. In each of the simulations, the frame length is 130 four-PSK symbols, which are transmitted simultaneously from every antenna. A maximum likelihood ST-Viterbi decoder is employed at the receiver, and it is assumed that the receiver performs perfect channel estimation, where a fast fading channel has been assumed. The results [BRHR04], [ABRH03] indicate that the novel codes outperform known codes by several decibels; further, a slight performance loss is observed when deploying codes obtained for diversity order of three or less in systems with diversity order greater than three.

Based on the distance evaluation procedure for non-linear modulation with memory, described in more detail in Subsection 2.4.2, we designed a simple trellis code for Rivest's Cipher 2 (RC2) CPM modulation in a 2×2 MIMO Rayleigh channel; for details see [Syko01].

Space-time coding over correlated channels

Space-time codes maximise the diversity advantage whereas Spatial Multiplexing (SM) focuses on the data rate advantage. High rate codes such as linear dispersion codes try to exploit both the diversity and the spectral efficiency advantages of the MIMO channel. A common assumption in the design of such techniques is to consider the fading coefficients between the pairs of transmit-receive antennas as independent and identically Rayleigh distributed (iid).

This is, however, an idealistic situation. In practice, the fading coefficients are correlated and this correlation depends on the antenna spacings and orientation, on the mutual coupling, the richness of scattering and the presence of dominant components. These effects highly influence the capacity as well as the performance of the space-time processing. High-rate codes, such as SM, are significantly more affected by the propagation conditions than low-rate codes, such as space-time orthogonal codes.

Space-time codes designs commonly rely on the assumption of a high SNR. In the first part of [CVVO04], we investigate the impact of this assumption when the channel is correlated. Therefore, we discuss the impact of transmit and receive correlations on the performance of space-time codes as a function of the SNR and the diversity achieved by the codes on independent and identically distributed channels. Full diversity codes are shown not to interact with the channel at high SNRs while at realistic SNRs, interactions occur and affect the coding gain. For non-full diversity codes, interactions with the channel occur whatever the SNR. At realistic SNRs, every space-time code interacts with the channel; we show that on correlated channels, the 'high SNR' assumption is totally unrealistic and may lead to bad code designs.

In the second part of [CVVO04], we formalise the interactions between the channel and the code and derive a new design criterion to guarantee the robustness of space-time codes in correlated channels when the transmitter has no information about the channel. The design criterion is totally general, i.e. it does not depend on the channel gain distribution and can be applied to any kind of space-time code. Based on that criterion, new spatial multiplexing schemes, linear dispersion codes and space-time trellis codes are derived. By simulations they are shown to perform much better on real-world channels than codes only designed for iid channels.

In the third part of [CVVO04], we show how to exploit some statistical knowledge of the channel (i.e. channel correlations) at the transmitter in the design of space-time codes. First, we focus on an SM scheme and design new non-linear signal constellations that exploit the spectral efficiency advantage of SM and the robustness of eigenbeamforming. Then, we derive STTC based on those non-linear signal constellations. Finally, assuming a large number of receive antennas, we derive linear dispersion codes based on the transmit correlation matrix for the classical PSK and QAM constellations.

Space-Time pre-coding schemes

There are important requirements for coding schemes in future wireless networks. These networks will be heterogeneous with regard to node complexity, link level requirements, propagation environments, etc. The choice, if it is better to use spatial multiplexing or TX diversity methods, depends on a number of parameters, such as demanded link reliability, demanded data rate, channel conditions (correlated or uncorrelated fading, Rayleigh or Ricean fading), etc. So it is desirable to use a coding scheme that allows a flexible trade-off between spatial multiplexing and transmit diversity.

Using a class of recently proposed linear scalable space-time codes we demonstrate in [WiKH03] the trade-off between spatial multiplexing and transmit diversity for Rayleigh and Ricean flat fading. The codes are able to use jointly transmit diversity in combination with spatial multiplexing, and they achieve spatial and temporal diversity in Rayleigh and Rice fading environments. Frequency diversity of frequency-selective channels can be utilised by combining the considered codes and OFDM. Simulation results for different scenarios show that a system consisting of the considered codes and decoder meets the requirements of future communication systems, e.g. it is possible to exploit high diversity factors and handle high rates with a reasonable and scalable complexity.

An interesting approach related to the design of linear diversity precoders in block fading delay-limited MIMO channels has been developed in [KnSy04a].

The objective there has been to develop a design criterion for the precoding (inner code) across the finite set of block fading channel observations such that the virtual channel seen by outputs of the outer coders has uniform achievable rates over all blocks in the frame. The design criterion regards the specific frame probability of outage. Outage event occurs whenever at least one block in the frame fails to support the required rate. To put it differently, the goal is to equalise or stabilise the instantaneous capacity development within the whole frame consisting of a finite number of blocks obeying iid block fading channel realisation.

The solution relies on a novel *virtual multiple access capacity region* approach. The specific design criterion demands an equal outage capacity for each codeword at the output of outer codes and the performance of the two-stage approach (outer code with independent codebooks and inner precoder) has to be equal to the joint one-stage coding (joint codebook across the whole frame).

The selected precoders, however sub-optimal regarding the design criterion, are evaluated. It is proved that the precoding offers significantly lower probability of outage for a given desired rate with very simple processing complexity compared with the direct joint codebook design. The temporal-only precoding (Kronecker Product (KP) precoders) is shown to be essential since the additional improvement in the transmission reliability using the precoding in both space and time is negligible in the MIMO channel. Details can be found in [KnSy04a].

7.4.3 Space-time receiver design

Frequency-selective MIMO channel estimation

The time-varying multipath channel has some characteristics that are stationary (or varying over a long term) and other features that are fast varying (e.g. fading amplitudes). This property has recently been used in designing signal processing algorithms for wireless communications. A simple example can illustrate the different varying rates: while the fading amplitudes can vary completely when either end of the communication link moves as little as $\lambda/4$ (λ is the carrier wavelength), the angles (DoAs and DoDs) remain constant with changes in position of several wavelengths (i.e. 10 to 1000λ).

In [SiSp04], we show that entries of the frequency-selective MIMO channel matrix within the kth block can be arranged into a vector $\mathbf{h_k} = \mathbf{T}\beta_\mathbf{k}$ so that the slowly varying term, represented by the matrix T, and the fast-varying fading vector $\beta_\mathbf{k}$, can be decoupled. We consider an estimator that (a) is able to consistently estimate the long-term features of the channel $\mathbf{T}$ so that for a large number of bursts K (ideally $K \to \infty$) these can be assumed to be acquired with any accuracy and (b) performs optimum (MMSE) tracking of the variations of the fast-varying features $\beta_\mathbf{k}$. Notice that many known estimators proposed in the literature under simplified settings have (at least one of) the aforementioned properties.

By deriving the asymptotic ($K \to \infty$) MSE matrix of the estimate for this method, we set a lower bound on the achievable performance of any channel estimation algorithm over the considered channel models.

MIMO turbo-type receivers

When using a frequency-selective MIMO communications system, a number of issues have to be addressed. First of all, to benefit from transmit diversity, appropriate ST coding needs to be used. Second, a receiver structure has to be proposed which is able to handle the ISI, the Co-Antenna Interference (CAI) and the receive diversity capability.

Regarding transmit diversity, we consider an approach based on bit interleaving and named STBICM. The information bits are encoded by means of a convolutional coder. The coded bits are interleaved and distributed to the different antennas where they are mapped on complex symbols and transmitted. The receiver is turbo based. It jointly performs turbo space-time equalisation to counteract ISI and CAI, and turbo demodulation.

Most of the turbo detectors are based on optimal a posteriori probability evaluation and are thus trellis based. This leads to excellent error rate performance but the detector complexity is huge and basically prohibitive if multiple antennas and/or multilevel/phase modulations are used. Less complex solutions have been proposed, like the max-log-MAP detector or reduced-state approaches. Further complexity reduction is possible using Filter-Based (FB) solutions. Ariyavisitakul has used a low-complexity FB detector but this detector does not fully exploit the available a priori information.

In the CDMA context, Wang and Poor have proposed a multi-user FB detector that can be shown to be the exact MMSE solution.

High spectral efficiency is attained by a simultaneous transmission of multiple data streams that are not orthogonal by any of the conventional communication signal dimensions, i.e. by time, frequency, or code. Instead, they are transmitted from different antenna elements at different spatial locations.

Sophisticated and yet efficient signal processing approaches are required in order to separate the transmitted data streams from the received signal mixture, especially in wideband channels. Detection algorithms for MIMO systems in frequency-selective fading channels that have only cubic complexity have already been proposed. The core of these algorithms is a Soft Canceller (SfC), whose aim is to suppress MAI and ISI signal components, followed by an instantaneous MMSE filter (Selection Combining (SC)/MMSE).

This combination enables the approximation of extrinsic information for each receive data stream, which is used as a priori LLRs for the SISO channel decoders. Each of the references above employs BPSK modulation for the transmit signals, because in this case there exists a simple relation between the LLRs of the coded bits and the corresponding value of the (soft) data symbols.

Nevertheless, the impact of using a real-valued modulation is neglected. In [MaTr04], we show that for the choice of BPSK modulation or any other modulation, which can be approximated as a pulse amplitude modulation with a real-valued symbol alphabet (multilevel PAM, MSK, GMSK), the approximation of the LLRs should explicitly consider this fact. Furthermore, the MMSE filter computation is optimised, and the numerical complexity is slightly reduced. Simulations for a multipath Rayleigh fading channel model show an improved performance especially for the first iteration. Simulations using channel sounder measurements from a microcellular environment reveal a slight improvement in performance critical situations.

Iterative-type space-time receivers

At high spectral efficiencies, optimal decoding of STBC MIMO systems has a critical complexity unless the space-time code structure offers simple optimal decoding, as is the case for Space-Time Orthogonal Block Codes (STOBC). Indeed, the complexity of true ML decoding or APP decoding of STBC increases exponentially with the spectral efficiency. As an SfISfO decoding scheme, APP space-time decoding is of special interest since it allows the space-time receiver to efficiently work with a channel decoder. Besides, it can advantageously be used in an iterative manner with an SfISfO channel decoder to approach the optimal joint receiver performance. List Sphere Decoding (LSD) provides a good approximation of APP decoding with reduced complexity and can therefore be considered as a performance reference.

In [Guég03], we introduce a new sub-optimal SfISfO space-time receiver based on MMSE using priors, which has significantly lower complexity than LSD but turns out to perform similarly when used in an iterative receiver. The system under consideration is first presented in [Guég03], followed by a detailed description of the new proposed MMSE-based receiver and a brief description of the reference LSD receiver. These two schemes are then evaluated in iterative receivers in terms of performance and complexity.

The Iterative Tree/Trellis Search (ITTS) detection scheme discussed in [JoWi02], [JoWi03] is a low-complexity MIMO detection technique for STBICM MIMO systems employing turbo processing at the receiver. The ITTS scheme is based on the observation that sequences of spatially multiplexed data symbols can be represented by a tree or a trellis structure, depending on whether the transmissions from different antennas are received synchronously or not. The M-algorithm is employed to perform a breadth-first search for the best paths through the tree or trellis, or equivalently, for the symbol sequences most likely to have been transmitted. QAM signal constellations with block partitionable labelling, referred to as multilevel bit mapping, are used in order to enable the tree/trellis search to be performed in steps of two bits at a time, regardless of the modulation order. As a result, the complexity per bit of the ITTS scheme is only linear in the number of transmit antennas, and nearly independent of the constellation size. This is a major improvement over existing soft-input soft-output MIMO detection schemes. Another advantageous feature of ITTS detection is that its performance can be traded off for lower complexity, by changing the list size parameter of the M-algorithm.

Theoretically, the performance of ITTS detection is identical to that of the optimal MIMO detector only for the maximum list size, which is prohibitively high even for a small number of transmit antennas and low modulation order. However, as has been verified by simulations, good performance can be achieved at very small fractions of the maximum list size. Furthermore, the efficiency of the ITTS scheme increases with decreasing correlation between the spatially multiplexed symbols, in both space and time. In case of synchronous reception, where temporal correlation between symbols is maximum, trellis-based ITTS yields identical performance as tree-based ITTS. For asynchronous operation, however, temporal correlation between symbols is reduced, which improves the performance of trellis-based ITTS detection. For example, [JoWi03] reports a 4 dB performance gain for an 8×8 MIMO configuration employing 64-QAM if the system is operated asynchronously.

For MIMO spatial multiplexing data transmission systems the complexity of the Maximum Likelihood Detector (MLD) can be prohibitively expensive when the number of transmitting antennas and constellation points is high. To simplify the MLD many linear and non-linear detection techniques have been proposed. The V-BLAST architecture, proposed in [GVWF98], is a practical low-complexity receiver where symbols are detected sequentially according to the well-known nulling and successive interference cancellation process. The main drawback of V-BLAST is that the diversity order in the early stages is lower than in the next ones. Moreover, the overall performance may be limited by error propagation that takes place in the first stages of the detection process.

In our proposal [SpMa04] the principle of reduced state sequence detection, based on mapping by set partitioning, is applied to perform detection in spatial multiplexing MIMO systems using QAM constellations. In QAM modulation, the real and imaginary part of each symbol belongs to the integer set $\mathbb{Z}$. The binary partition $\mathbb{Z}/2\mathbb{Z}$ is considered in each dimension of the QAM constellation for each transmitted substream. Let 2^{2k} be the size of the QAM constellation in use, with $k = 1, 2, \ldots$ A list of 2^{2n_T} candidate subsets is generated by considering the 2^{2n_T} combinations of least significant bits for the n_T entries of the transmitted vector. A sub-optimal receiver is applied to perform the detection in each of these 2^{2n_T} subsets containing $n_T 2^{2(k-1)}$ constellation points. At each stage the detector examines the decision statistic for the symbol sent from antenna n and compares it with the candidate symbols that are drawn from the current subset associated to substream n. At the end of this procedure a list of 2^{2n_T} candidate vectors is generated. A final decision is taken by applying the MLD to this reduced set

$$\hat{\mathbf{a}} = \arg \min_{\hat{\mathbf{a}}_r \in \mathcal{A}_r} \|\mathbf{r} - \mathbf{H}\hat{\mathbf{a}}_r\|^2,$$

where $\mathbf{r}$ is the $n_R \times 1$ received signal vector ($n_R \geq n_T$), $\hat{\mathbf{a}}_r$ is an $n_T \times 1$ vector taken from the reduced set $\mathcal{A}_r$ containing the 2^{2n_T} candidate vectors and $\mathbf{H}$ is the $n_R \times n_T$ channel matrix whose elements are iid random variables having uniform distributed phase and Rayleigh-distributed magnitude with average power equal to 1.

We consider the scheme where the received vector elements are linearly weighted to minimise the effects of the total disturbance, that is, interference plus noise. This approach leads to MMSE V-BLAST [Hass00]. The weighting operated by the MMSE linear matrix introduces bias in the decision process [SpMa04]. By removing the bias a better performance is obtained and the resulting approach is denoted as Unbiased MMSE (UMMSE). The performance of the proposed detection algorithm, termed SP, is compared with that of the linear UMMSE and of the UMMSE V-BLAST with ordering. The simulated MLD performance is also reported as a benchmark. We consider the application of the SP detection algorithm both with the linear UMMSE (UMMSE SP) and with the ordered UMMSE V-BLAST (UMMSE V-BLAST SP) as subset candidate vector detectors. Figure 7.9 reports the Symbol Error Rate (SER) vs. SNR (SNR $= 1/\sigma_w^2$) for the different detection algorithms in a 2×2 system using 64-QAM modulation. We observe that the performance of the UMMSE V-BLAST SP algorithm is close to that of the MLD for an SER greater than 10^{-2}. Due to a different slope of the curves, for a lower SER the performance of the UMMSE V-BLAST SP detection algorithm deviates from that of the MLD.

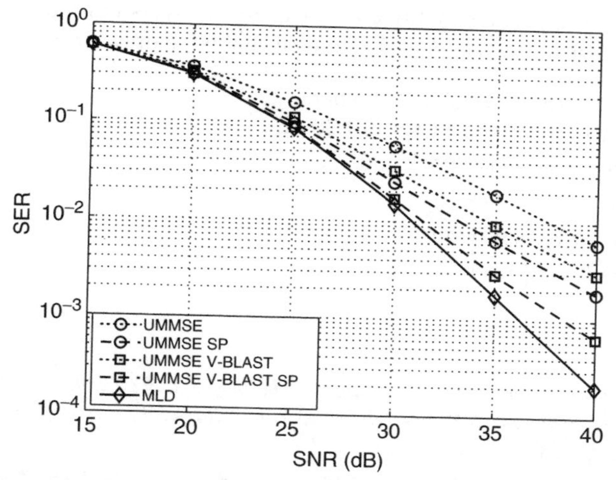

Figure 7.9 *SER vs. SNR for the different detection algorithms in a 2×2 system using 64-QAM.*

7.4.4 Space-time transceiver performance

Influence of imperfect symbol timing

A general procedure for the evaluation of the impact of the self-noise in slowly and fast Rayleigh flat fading is applied to the particular case of imperfect symbol timing in the MIMO channel; see Section 6.3 for detailed treatment of the general procedure.

Performance of STBCs

The BLER performance of a selection of well-known linear STBCs was investigated in [PaSL03]: spatial multiplexing, an orthogonal STBC by Alamouti and linear codes by Sandhu and Damen. The codes were evaluated using a diverse range of radio channel measurements, recorded in two contrasting office environments. The sensitivity of code performance to the mean Ricean K-factor of the channels was examined; and the relative merits of using a dual polar patch array, opposed to a linear array of monopoles, were investigated. The most important conclusions drawn from the study were:

- Spatial multiplexing and Alamouti codes have similar performance, with spatial multiplexing performing slightly better in path-rich, high-capacity channels and Alamouti performing better in LoS channels.
- The design criterion for Damen-style codes is more effective than the Sandhu criterion for 2×2 channels, over the SNR range considered, leading to an improvement of typically 1.2 dB. However, receiver complexity is dramatically increased for both of these code types relative to the Alamouti STBC.
- Patch antennas perform at least as well as the linear array of monopoles, although minor performance gains are probably due to the 22% increase in element spacing and variations between the measured channels.
- Polarisation diversity appears more effective than spatial diversity for a 0.61 wavelength spacing of the patches, although more channels need to be studied to draw firm conclusions.

The results are summarised in Table 7.1, where

Cap	Capacity in bps/Hz at an SNR of 20 dB
Al	Alamouti orthogonal block code
VB	V-BLAST
S1,S2	Sandhu-style codes with $T = 1$ and $T = 2$
Dm	Damen ($T = 2$) code

Table 7.1 *Results from the open-plan office. The performance of the STBCs is given by the SNR required to achieve 1% block-error rate. Ray = Rayleigh, LoS = line of sight, NLoS = non-line of sight, MIX = mixture of LoS/NLoS.*

Channel	Type	Cap [bps/Hz]	Al [dB]	VB [dB]	S1 [dB]	S2 [dB]	Dm [dB]
Ch1	Ray	11.3	15.5	16.7	18.1	16.4	15.3
Ch2	LoS	9.2	14.8	16.0	18.9	16.0	15.8
Ch3	LoS	8.8	16.0	16.3	19.1	16.9	15.7
Ch4	MIX	9.6	16.0	16.0	17.7	16.8	15.6
Ch5	MIX	11.0	13.2	13.1	15.1	13.5	12.5
Ch6	MIX	10.3	14.2	14.0	16.0	14.5	13.8
Ch7	MIX	10.5	11.2	13.1	15.3	12.6	12.2
Ch8	NL	9.7	14.8	14.6	16.9	15.6	14.2
Ch9	NL	9.4	15.9	15.9	18.2	16.7	15.7
Ch10	NL	9.6	14.8	14.7	17.0	15.6	14.3

In [GeHR03], we show in the case of general STBC that it is not only the Pairwise Error Probabilities (PEPs) between nearest neighbouring code words that have to be considered to evaluate the total BER performance of the STBC. For STBCs in general we have to consider all Pairwise Error Probabilitys (PEPs) and not only the nearest neighbour error events in order to adequately describe the BER performance of these codes. The upper bound of the BER curve shows that for low SNR the BER curve is dominated by the nearest neighbour errors (PEP1) whereas for high SNR the PEPs corresponding to multiple symbol errors dominate the error performance. This fact is explained by means of the Euclidean distance profile of pairwise error events. Additionally, we showed that for reasonably low SNR values it is not that important to use full rank STBCs and to consider all possible PEPs. If rank deficiencies occur quite rarely, then they do not influence the total BER in the range of low to medium SNR that is of major interest in practical applications. In contrast to general STBC, the BER performance of Generalised Complex Orthogonal Design (GCOD) codes is dominated by single symbol errors in the whole range of SNR. For such codes it is not necessary to consider all distinct PEPs to characterise the BER performance.

The basic technique to overcome the code rate limitation of orthogonal STBCs for more than two transmit antennas is to introduce a small amount of non-orthogonality in the STBC matrices. As a result, several full rate quasi-orthogonal STBC Quasi Orthogonal STBC (Q-OSTBC) schemes have been introduced. However, some of these codes, e.g. the ABBA code, are very sensitive to spatially correlated channels.

In [RuWB04], Q-OSTBCs are investigated over spatially correlated channels. In our simulations we utilise the so-called Kronecker channel model to analyse the performance of Alamouti-like space-time block codes, designed for four transmit antennas. We present a transmission system where only one channel information bit b per code block is returned to the transmitter. Depending on the value of b the transmitter switches between two predefined Q-OSTBCs and chooses that code matrix which achieves better performance.

In [RuWB04] two sets of STBCs, namely Extended Alamouti Codes (EACs) and ABBA codes for four transmit antennas are combined with very low-rate channel feedback information. The BER performances of all systems are compared in case of spatially correlated and uncorrelated channels. It turns out that this simple transmit scheme, selecting the best transmission code for a given transmit channel, improves diversity and BER performance over the whole SNR range in a spatially correlated and uncorrelated transmission environment. Even a small amount of channel state information helps to decorrelate the transmission system using the ABBA code and the EAC. However, in strongly correlated channels, the transmission performance deteriorates equally in both cases compared to the performance in uncorrelated MIMO channels.

Adverse and FER performance of STTC

[RBAR03] has addressed the joint FER and BER performance of a two antenna space-time trellis code as a function of trellis complexity. The results have shown that increasing code complexity reduces FER, but yields an increased BER for the given frame size. However, when two receive antennas are employed in a high SNR scenario, the effect does not occur.

The analysis has allowed further understanding of the characteristics of space-time trellis codes, and has highlighted the need for improving both the FER and BER simultaneously. Reducing the frame size has shown to improve the FER, but with no improvement in the BER. Furthermore, it has been shown that for an increasing number of states in the trellis, the time to converge from an erroneous instant increases therefore causing an error burst, hence impairing the BER performance. For more details, consult [RBAR03].

Performance of various space-time architectures

Link level or system level simulation methods have the aim of evaluating and comparing the performance of different air interface concepts as well as system concepts under realistic considerations. In contrast to the development of prototypes, multidimensional channel sounding techniques provide efficient possibilities to perform such simulations with a manifold of variations. Furthermore, these techniques open the way to high-resolution path parameter estimation results, and hence face simulation performances with the physical nature of propagation. Realistic conclusions and understandings of the system or algorithm under test can be drawn and consequently used to enhance and optimise the considered concept. Important aspects for performing MIMO measurements and using the measurement data in transmission system simulations are discussed in [MTST02] and [TSSH03]. A BPSK version of the Turbo MIMO Equaliser presented in [KSMT05] was analysed considering broadband channel measurements in a microcell WLAN scenario [STMT03]. Here a transition from NLoS to LoS propagation characteristics could be found. The performance of the Turbo MIMO Equaliser with three transmit as well as receive antennas degrades rapidly if the channel condition changes from NLoS to LoS. A strong correlation between the link performance and the spatial channel characteristics in terms of RMS azimuth spreads on transmit and receive sides was highlighted. Furthermore the realistic MIMO capacity based on the same measurement data corresponds directly to the BER of the Turbo MIMO Equaliser [ZeST04].

Two information-theoretical approaches to linear diversity precoding in a block fading delay-limited MIMO channel are developed and compared. Performance of precoders of different orders and kinds is then evaluated in terms of outage capacity. Beneficial influence of selected precoders is shown. To justify our theoretical approach, the real scheme of precoding with both iid space-time multiplexing and Alamouti code is examined. The cumulative density function of the probability of error and the average BER for optimal and simple sub-optimal detection techniques are compared.

In [KnSy04a], there was developed a very useful mathematical framework to handle outage capacity in a space-time system with precoding across a finite number of iid channel realisations in the frame. The ultimate design criterion, which is optimal from the outage capacity point of view, is also given and serves as as benchmark for linear precoders. The frame length specifies the allowed transmission delay and so the maximal order of the proposed precoder. There was shown that some multiple access ideas can be formally utilised for the case of linear diversity precoders and that the final design criterion has simple definition based on the shape of achievable capacity region.

This chapter provides an extension of the mentioned paper. Two other design criteria are developed and their interpretation is discussed. The so-called *delay-limited approach with independent observations* and the *true multiple-access approach* with precoding are surveyed to clarify the difference from the approach in [KnSy04a]. For both approaches the numerical results are introduced.

In [KaSK04], the efficiency of the evolutionary antenna selection method has been investigated in terms of capacity performance. For problems with moderate search space, the arrays selected by the proposed and existing sub-optimal algorithms are compared to the optimum computed by exhaustive search.

Additionally, for problems with large search space, solutions of both algorithms are compared to reference ULA antenna configurations. Results show that for joint selection the GA outperforms other algorithms. Under flat fading conditions and for single link-end antenna selection, the incremental (2003) Gorokhov *et al.* algorithm performs slightly better in iid channels, whereas in realistic measured channels both offer approximately the same capacity amendment.

Results also indicate that the capacity performance of arrays with selected elements is superior to that of ULAs, sometimes even in cases where the latter employ more antenna elements. Therefore, a reduction of the required RF chains is feasible.

Finally, the authors expect that a hybrid algorithm, i.e. a GA with hints from a sub-optimum algorithm, may offer capacity enhancement in lesser computational times.

An efficient scheme for packet-based data transmission using broadband single-carrier modulation has been proposed in [KSMT05].

The proposed scheme allows for efficient MIMO turbo equalisation at the receiver to be performed and provides reliable data throughput in varying spatio-temporal channel characteristics. Performance evaluation results were obtained through simulations in measured channel conditions and evaluated against spatial channel parameters estimated with super-resolution methods. The proposed scheme was shown to closely match or exceed the throughput efficiency of BICM in all tested channel conditions. It was shown that when the unequal error protection of the modulation is taken into account in the ARQ algorithm design by separating the ARQ processes of the differently protected levels, further performance improvement can be achieved. For further information, consult [KSMT05]. Moreover, the proposed concept was extended by an antenna variable modulation scheme [SGKM04]. The simulation results showed that with significantly changing MIMO channel characteristics the transceiver signalling has to be adopted. Along a measurement track with a transition from NLoS to LoS robust performances considering a predefined throughput threshold could be achieved using the proposed variable modulation scheme at the different transmit antennas.

In [TMWK05], we have introduced a new design concept for LDPC codes with EXIT charts based on outage probability. The EXIT charts for the turbo equaliser are obtained by a computationally efficient semi-analytical procedure that enables the study of random channel realisations. Simulation results show that improvements are possible and an adaption of code rates to channel E_b/N_0 is feasible. However, performance lies below expectations because of the following reasons:

- The Gaussian assumption is not totally fulfilled. Particularly irregular codes with high variable node degrees cause high magnitudes of the LLRs which is in contrast to the Gaussian assumption.
- Due to the short block length, the designed performance is not achieved at the threshold value E_b^*/N_0 but at higher values of E_b/N_0. Simulation results show that code design with EXIT charts lead to performance improvements. However, the optimised codes show the improvements several decades above E_b^*/N_0 and can even perform poorer than unoptimised codes at E_b^*/N_0.
- The assumption of parallel equaliser transfer curves $f_{EQ}()$ does not strictly hold in varying channels. Each equaliser EXIT curve crossing $f_{EQ}()$ fails to decode, and the true outage ratio becomes larger than the design target.

7.4.5 Real-valued notation and its applications

Typically, linear baseband transmission models use complex numbers to represent signals and impulse responses. Unfortunately, certain space-time coding techniques such as, e.g. orthogonal STBCs, [Alam98], [TaJC99], [LaSt03], may not be incorporated in such models, because the complex conjugate operation may not be expressed as a linear transformation. Most commonly, the problem is tackled by introducing two linear transformation matrices, one for the symbols and another one for the complex conjugate of the symbols, see, e.g. [HaHo02]. This way, however, the resulting model is not linear with respect to its input symbols and some of the convenient properties of linear transmission models get lost. A different approach employs real-valued transmission matrices [PSTL03] that are based on a general real-valued description of matrices [NeMa93]. As a matter of fact, we point out that many known STBCs may then be interpreted as a certain type of spreading, i.e. encoding and decoding of the STBC is carried out by a simple matrix multiplication. Although a real-valued notation has been used before in connection with orthogonal STBCs, e.g. [StGa03], [WaTS01], none of these models provides a transmission matrix that maps the input symbols to the output symbols by vector matrix multiplications.

A series of TDs [PSTL03], [LiPi04], [PiLi05b], [PiLi05a] addresses various aspects of the transmission matrices which are based on the real-valued notation. The general rule for transforming a complex matrix into its real-valued equivalent (indicated by the bar) is

$$\overline{\mathbf{H}} = \begin{pmatrix} 1 & 0 \\ 0 & 1 \end{pmatrix} \otimes \Re\{\mathbf{H}\} + \begin{pmatrix} 0 & -1 \\ 1 & 0 \end{pmatrix} \otimes \Im\{\mathbf{H}\} = \begin{pmatrix} \Re\{\mathbf{H}\} & -\Im\{\mathbf{H}\} \\ \Im\{\mathbf{H}\} & \Re\{\mathbf{H}\} \end{pmatrix}$$

where $\otimes$ means the Kronecker product. A transmission model is now defined by

$$\overline{\tilde{\mathbf{x}}} = \overline{\mathbf{U}}^T \overline{\mathbf{H}}_\ell^T \overline{\mathbf{H}}_\ell \overline{\mathbf{U}} \overline{\mathbf{x}} + \overline{\tilde{\mathbf{n}}} = \overline{\mathbf{R}} \overline{\mathbf{x}} + \overline{\tilde{\mathbf{n}}}$$

where $\overline{\mathbf{x}}$ is the transmit symbol vector, $\overline{\tilde{\mathbf{x}}}$ is the received symbol vector after matched filtering, and $\overline{\tilde{\mathbf{n}}}$ is the coloured (due to the matched filter $\overline{\mathbf{U}}^T \overline{\mathbf{H}}_\ell^T$) noise vector. $\overline{\mathbf{x}}$, $\overline{\tilde{\mathbf{x}}}$, and $\overline{\tilde{\mathbf{n}}}$ are all real valued with separate entries for the real and imaginary part of the receive vector. $\overline{\mathbf{H}}_\ell = \mathbf{I}_\ell \otimes \overline{\mathbf{H}}$, where $\mathbf{I}_\ell$ is an identity matrix of size ℓ. This way $\overline{\mathbf{H}}_\ell$ denotes ℓ time slots of a single tap MIMO channel. $\overline{\mathbf{U}}$ denotes a spreading matrix that may carry out any linear transformation of the signal constellation at the transmitter. Due to the real-valued notation the transformations are more general than what could be obtained with complex valued matrices. The matrix product $\overline{\mathbf{U}}^T \overline{\mathbf{H}}_\ell^T \overline{\mathbf{H}}_\ell \overline{\mathbf{U}}$ is combined into a single transmission matrix $\overline{\mathbf{R}}$. As a matter of fact, $\overline{\mathbf{R}}$ fully describes the transmission.

A capacity (mutual information) analysis [PSTL03], [LiPi02] for the real-valued transmission model is as straightforward as it is for the complex case except for a factor $\frac{1}{2}$

$$C = \frac{1}{2} \log_2 \left\{ \det \left(\mathrm{I}_{2n_t} + \frac{\sigma_x^2}{\sigma_n^2} \overline{\mathbf{H}}^T \overline{\mathbf{H}} \right) \right\} = \frac{1}{2} \log_2 \left\{ \det \left(\mathrm{I}_{2n_t} + \frac{\sigma_x^2}{\sigma_n^2} \overline{\mathbf{R}} \right) \right\}$$

Here, we assumed that the transmitter does not have any channel state information. Further note that we use $\mathbf{H}^T \mathbf{H}$ instead of the commonly used expression $\mathbf{H}\mathbf{H}^T$. This way the capacity is a function of the transmission matrix $\overline{\mathbf{R}}$. From this it should be immediately clear that any unitary spreading matrix has the potential to preserve capacity (mutual information).

TD [LiPi04][1] shows how to describe orthogonal STBCs as linear spreading matrices using Alamouti's scheme [Alam98] as an example, i.e. we may find a spreading matrix $\overline{\mathbf{U}}$ such that it carries out the appropriate mapping between the symbols, transmit antennas, and time slots. The model also easily provides access to the properties of orthogonal STBCs. E.g. from the structure of

[1] This paper also explains why orthogonal STBCs are well suited for data directed channel estimation.

the spreading matrices it is clear that they cannot achieve capacity for arbitrary channels because the spreading matrices have orthogonal columns but are not square. Therefore, capacity may only be achieved in certain cases when the channel is rank deficient [PiLi05a]. This is the case for Alamouti's scheme when a single antenna is used at the receiver.

Based on these considerations new STBC (spreading matrices) have been proposed in TD [PiLi05a]. These spreading matrices are based on a layered transmission of orthogonal STBC. A spreading matrix based on Alamouti's scheme, e.g. is

$$\overline{\mathbf{U}}_{2\mathrm{A}} = \begin{pmatrix} \overline{\mathbf{P}}^1\overline{\mathbf{U}}_{\mathrm{A}}^1 & \overline{\mathbf{P}}^2\overline{\mathbf{U}}_{\mathrm{A}}^1 \\ \overline{\mathbf{P}}^1\overline{\mathbf{U}}_{\mathrm{A}}^2 & \overline{\mathbf{P}}^2\overline{\mathbf{U}}_{\mathrm{A}}^2 \end{pmatrix} \quad \text{where} \quad \overline{\mathbf{U}}_{\mathrm{A}} = \begin{pmatrix} \overline{\mathbf{U}}_{\mathrm{A}}^1 \\ \overline{\mathbf{U}}_{\mathrm{A}}^2 \end{pmatrix}$$

denotes the spreading matrix for Alamouti's scheme. $\overline{\mathbf{U}}_{\mathrm{A}}^i$ are the subspreading matrices which carry out the mapping for the ith time slot. The matrices $\overline{\mathbf{P}}^j$ are different for each layer and are chosen such that $\overline{\mathbf{U}}_{2\mathrm{A}}$ is unitary. Among the set of matrices which cause $\overline{\mathbf{U}}_{2\mathrm{A}}$ to be unitary, we may choose those which optimise the distance profile of the received constellation (assuming a MIMO Rayleigh channel). Figure 7.10 compares the distance profile of the proposed codes with existing codes from the literature.

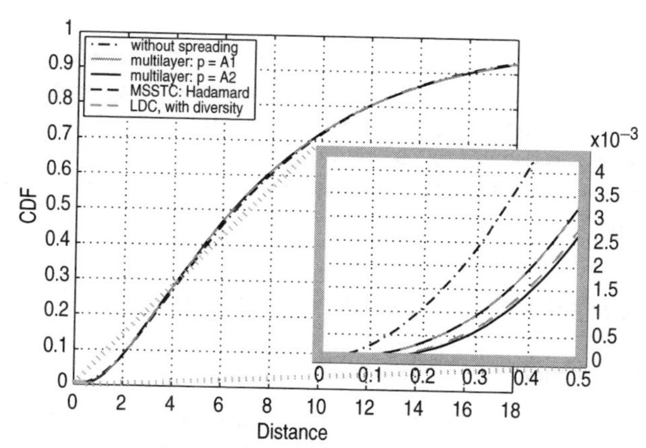

Figure 7.10 *Distance profiles for different full rate spreading matrices ($\ell = 2$); the modulation scheme is 4 PSK; MIMO Rayleigh channel with two transmit antennas and two receive antennas; the LDC is from [HaHo02] and the MSSTC is from [WaTS01]; further, two codes of the proposed scheme are presented with two different choices for the matrices $\overline{\mathbf{P}}^j$.*

TD [PiLi05b] carries out an analytical analysis of how imperfect channel knowledge influences the performance of orthogonal STBCs. Although the real-valued notation may not be necessary, it enables a straightforward analysis. An exact solution is derived by considering that the received symbols are given by the scalar product of two independent non-central chi-square distributed vectors. On the other hand, an approximation is given based on the assumption that the overall interference and noise are Gaussian distributed. For a large number of antennas, this assumption is justified due to the central limit theorem. Figure 7.11 shows the analytical results for one, two, and four transmit antennas. As we can see, the approximation is quite good for four transmit antennas already.

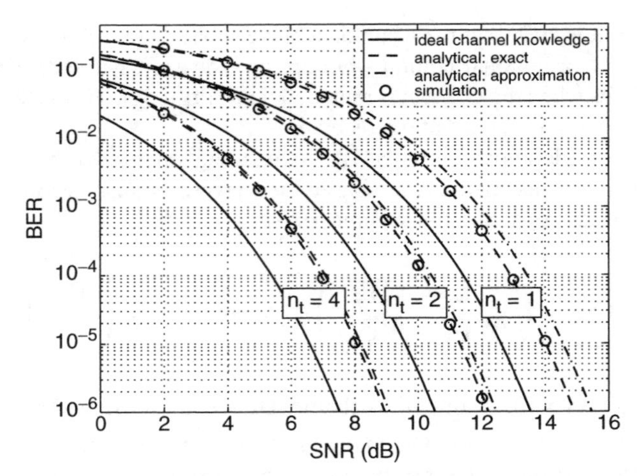

Figure 7.11 *BER performance for a receiver with $n_r = 4$ antennas; the transmitters with $n_t = 2$ and $n_t = 4$ antennas employ the Alamouti code [Alam98] and a rate 3/4 STBC [HaHo02], respectively.*

Another aspect of the real-valued notation is that non-proper random variables [NeMa93] may be taken into account in the transmission model. Certain interferers, e.g. may cause the receiver noise and interference to be non-proper. TD [PSTL03] takes this into consideration and also compares different detection techniques. As many transmission techniques, e.g. CDMA, MC-CDMA, and MIMO, may all be represented by a transmission matrix $\overline{\mathbf{R}}$, it is easy to adapt most detection techniques to any of these schemes.

7.4.6 Distributed and cooperative space-time systems

Virtual antenna arrays

Naturally, the deployment of MIMO techniques seems to be impossible where the number of antennas in the MT is the limiting factor. However, one could view a cell not as a system of single point communication links, but rather as a network with a certain number of antenna elements available in it, which are allowed to communicate among each other. With appropriate precautions, such deployment could emulate a MIMO system but the difference between this system and a traditional MIMO antenna array is that the antenna elements are connected through a wireless link, justifying the term Virtual Antenna Array (VAA). The application of MIMO system techniques, like space-time codes, to such a system is suggested and drastic performance gains have been reported [DoLA02a].

Distributed relaying networks

VAAs constitute a promising candidate for ad hoc-type networks with the following advantages. First, a VAA allows an automatic scaling of the ad hoc network as a higher density of MTs requires more capacity; however, this capacity is easier to provide if more antenna elements are available to form VAAs. Second, they can be deployed in transmitting and receiving mode, i.e. adjacent MTs may form a VAA to enhance the reception of data from another ad hoc MT. Furthermore, the same VAA may be utilised in receiving as well as transmitting mode, i.e. it relays data.

The latter mode of operation leads to the fairly novel concept of distributed-MIMO multistage networks [GADS01], [DoAg04], [GkAg04], [GkAD03], which is depicted in Figure 7.12. Here, a source MT (s-MT) communicates with a target mobile terminal (t-MT) via a number of relaying MTs (r-MTs). Spatially adjacent r-MTs form a VAA, each of which receives data from the previous VAA and relays data to the consecutive VAA until the t-MT is reached. Note that each of the involved terminals may have more than one antenna element; furthermore, the MTs of the same VAA may cooperate among each other. With a proper set-up, this clearly allows the deployment of MIMO capacity enhancement techniques.

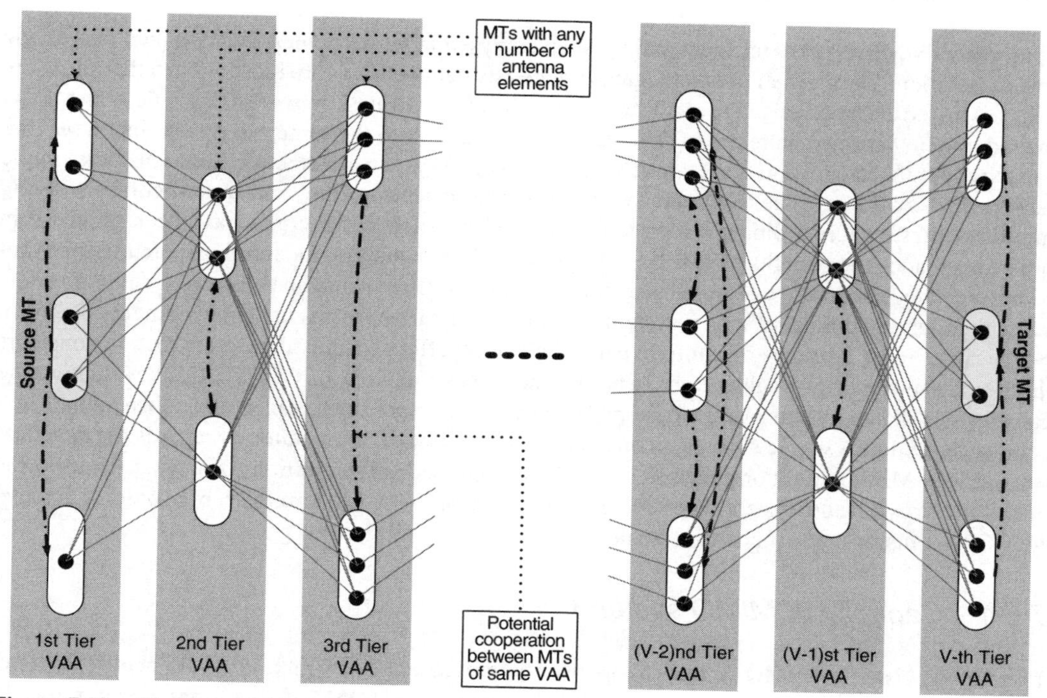

Figure 7.12 *Distributed-MIMO multistage communication network.*

Analysis in the above-mentioned work concentrated on Frequency Division Multiple Access (FDMA)-based relaying, where fractional bandwidths α and fractional powers β were derived. An equivalent TDMA-based system would require fractional frame durations α and fractional powers β/α.

Cooperative relays in rank-deficient MIMO channels

Future wireless MIMO systems that operate at higher frequencies (for example, in the 24 GHz ISM) will offer the possibility to equip transmitter and receiver with many antennas. However, capacity will not grow linearly with the number of antennas due to the high spatial fading correlation which leads to low-rank matrix channels. In [WiRa03], [RaWi04], we study the impact of multiple linear amplify-and-forward relays on the mutual information of rank-deficient MIMO channels. We derive a general system model for wireless networks with one source/destination pair and several linear amplify-and-forward relay nodes which assist the communication between source and destination. All nodes may be equipped with multiple antennas. For a given allocation of gain factors at the relay nodes we give an analytical expression of the mutual information. We compare the performance of a relay assisted MIMO link in a line-of-sight environment with a MIMO link without relay nodes. Our results show that the proposed signalling scheme helps to increase spectral efficiency of MIMO systems in poor scattering environments.

7.5 Network aspects

Until recently MIMO systems have mainly been analysed in terms of their performance on a single link in isolation. The dramatic capacity gains previously demonstrated are most often in that situation. This has raised the question of the applicability of MIMO to a wireless network as a whole. Can MIMO extend system capacity, in terms of the total user throughput, to the same extent as it increases link capacity? A related question is the application of MIMO in system architectures other than the simple hub-based network such as a cellular system or wireless access point, and in particular involving multihop links, or forwarding between terminals. Several such architectures have been proposed in order to enhance the benefits of MIMO in scenarios where fading on the terminal antennas may not be uncorrelated because the terminal size does not allow sufficient antenna element spacing. In some cases the individual links in such a system may not be MIMO (perhaps using only single antennas), but because of the presence of multiple paths the overall effect is equivalent. This section considers these two aspects in two subsections. The first describes the work of COST 273 in evaluating the capacity of MIMO systems first in the presence of co-channel interference and subsequently in a cellular system as a whole. We also consider the viewpoint of an operator as regards the practical reasons why MIMO might or might not be deployed. The second subsection considers various forms of multihop system and link architecture, using either MIMO-enabled terminals, or in order to achieve MIMO benefits in single antenna systems.

7.5.1 Capacity of MIMO systems

The principal issue in MIMO systems as opposed to individual MIMO links is the effect of co-channel or intercellular interference, which in a cellular system using TDMA is due to frequency reuse. In the work of COST 273 this has been assumed to have an effect similar to Gaussian noise: techniques like multi-user detection of intercellular interference have not been considered, nor have those which exploit the distribution of the interference in order to suppress it. Interference suppression techniques are considered in Section 3.3, and multi-user detection in Section 3.5, albeit in the context of CDMA systems, where the interference is mainly intracellular, rather than for co-channel interference in TDMA. However, in a MIMO system this interference does have a characteristic that can easily be exploited, namely its directional distribution. A beamformer at the receiver can thus minimise the effect of interference. The optimum would be a 'Max SINR' beamformer (which maximises the signal to interference-plus-noise ratio at its output). This is also referred to in [Burr02] as a 'spatially whitening pre-filter', by analogy with the optimum matched filter in (frequency-domain) coloured noise. Here the interference is spatially coloured. [WeBN04] derives it as a matrix given by the inverse matrix square root of the covariance matrix of the interference-plus-noise.

[Burr02] and [WeBN04] both consider the effect of interference on the capacity of a MIMO link on an independent Rayleigh fading channel, although the latter includes the case when the wanted channel and the interference channel are known at the transmitter of the wanted signal, while the former covers only the case of no channel knowledge at the transmitter. The most important result is that the capacity of a MIMO link is not disproportionately reduced by interference, compared to a SISO link. The 'pre-whitening filter' gives a significant advantage, and also means that a MIMO system can perform better in the presence of interference than of white noise of the same power. However, this advantage diminishes as the number of interferers increases, since the interference becomes more spatially white. [WeBN04] shows that the effect of an interfering antenna, if very strong, is to eliminate one of the degrees of freedom of the MIMO link; this of course reduces the performance of the MIMO system, but also means that the effect of one interferer is limited, whatever its strength, since once it has eliminated this MIMO mode it can have no further effect. It also shows that in a MIMO system where the number of interfering antennas is less than the number of receive antennas, the capacity will not be interference limited, but may continue to increase indefinitely with increasing signal-to-noise ratio. Comparing the case of known channel and interference at the transmitter (which allows optimum waterfilling power distribution to be employed) for an independent Rayleigh channel gives rise to only a modest increase in capacity. It is, however, equivalent to providing one more receive antenna.

In [ZhBP04] and [FMKW04] these results are extended to a cellular system. In both cases a cellular system model is used like that illustrated in Figure 7.13. Note that shadow fading is included in this model, which means that in some cases the best link from a mobile is not to the nearest BS, but to one in a neighbouring cell. The interference caused by each mobile to each BS other than its own can then be calculated. Both these documents use a double-directional model of the MIMO channel, so that fading correlation can be taken into account: [ZhBP04] uses a simple single bounce model with a scattering cluster around the mobile, while [FMKW04] uses a more sophisticated Multi-user Double-Directional Channel Model (MDDCM), described in [FKMW04] (see Section 6.3),

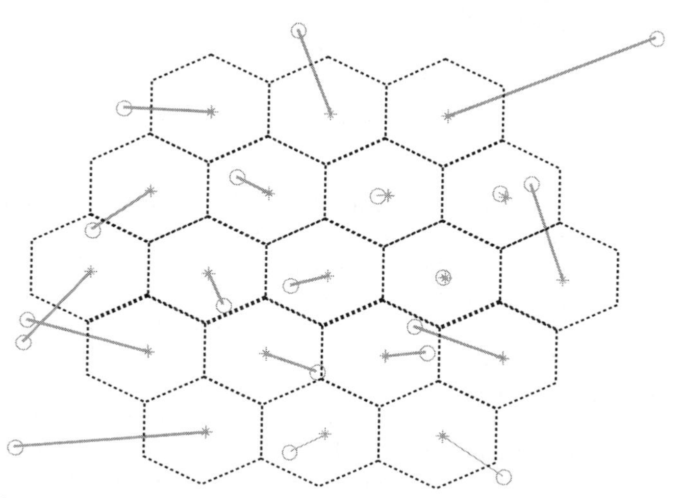

Figure 7.13 *Cellular system model showing position of BSs (stars) and mobiles (circles), the nominal cell boundaries (dotted lines) and the mobile-base links (solid lines).*

which also includes far clusters and is able to take into account the correlation of the interfering signals.

The results again show that there is a significant advantage for MIMO systems in this realistic cellular environment. Figure 7.14 shows results for a cellular system with reuse factor $\frac{1}{3}$ (cellular frequency plan with three cells per cluster). The MIMO system with pre-whitening has mean capacity more than three times that of the SISO system, for a wide range of signal-to-noise ratio. However, compared to a SIMO system with pre-whitening, which in fact is equivalent to a 'smart' antenna system using a Max SINR beamformer, the capacity advantage at high SNR, in the interference-limited region, is less than a factor of two. Note that in this case the interferers are also assumed to be SIMO, which means that the number of interfering antennas is smaller by a factor of four, and hence more degrees of freedom are available at the receiver to suppress the interferers. Nevertheless the MIMO system still has a significant advantage, which diminishes only when the SNR becomes unrealistically high.

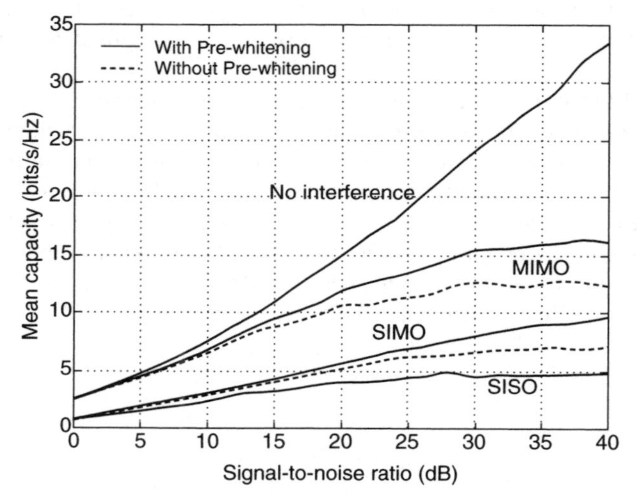

Figure 7.14 *Mean link capacity for 4 × 4 MIMO system compared with SISO and 1 × 4 SIMO, reuse factor $\frac{1}{3}$.*

To obtain system spectral efficiencies in terms of bits/s/Hz/cell one should multiply the abscissa values in Figure 7.14 by the reuse factor, $\frac{1}{3}$. Note, however, that results given in [ZhBP04] show that the system spectral efficiency would in fact be greater for a system with 100% reuse.

From this we may conclude that MIMO maintains its capacity advantage in cellular systems as much as in isolated links. However, [Poll02] gives a view from an operator's perspective about the implementation of MIMO cellular systems. It is seen as unlikely that MIMO will be implemented in a 3G system because the network now being rolled out does not include it, and an 'add-on' approach would require costly upgrades, and would not yield a benefit for the operator until a large number of terminals were MIMO enabled. However, its implementation in a new air interface realising the ITU vision of 'gigabit wireless' [ITU00] seems much more probable, since otherwise it will be difficult to meet these requirements in the spectrum likely to be available. Nevertheless the size of multiple antenna systems is still seen as a barrier to implementation.

7.5.2 Multihop MIMO architectures

Two multihop relaying architectures have been proposed within COST 273 to overcome two different common deficiencies of the conventional MIMO channel. The first, in [DoLA02b] covers the case where multiple antennas cannot be implemented on a terminal because of its size. The proposal, called VAA, uses a group of closely spaced MTs with single antennas to act as a MIMO array. This allows increased capacities to be achieved on the main channel between this virtual array and the BS array. It also requires additional wireless links to be deployed between these mobiles to relay the information to the target node. These links cannot share the same spectrum, so they occupy a bandwidth greater than the main channel by a factor equal to the number of relay nodes. However, because they are required only for very local communication, they can be reused between different groups of relays, and hence for a sufficiently large number of groups will not greatly affect the spectrum occupancy.

[WiRa03] uses cooperative relaying to overcome another potential problem of MIMO channels: that of rank deficiency of the channel due to insufficient scattering. The relays, like the VAA proposal, forward analogue information but this time all relays use the next time slot to signal over the same channel, so that the signals interfere. The effect, essentially, is to provide 'artificial scatterers' to increase the rank of the channel. Results are given for a scenario with a large number of relays where the channel without relays has no multipath (LoS only), showing that the capacity can approach more closely that of the independent Rayleigh fading channel.

[MiBG03] describes a different approach which, while it does not directly involve MIMO, employs some of the principles of MIMO in the context of a relaying network. Here data is transmitted over multiple routes through a multihop network, similar to both the VAA and the cooperative relaying networks. It is encoded at the network level at the transmitter; the coded data (containing additional redundancy) is distributed over the multiple routes, and it is then reassembled and decoded at the receiver. For unreliable links the additional redundancy is more than compensated by the reduced retransmissions over the whole network. When the reliability of the links is unknown to the transmitter, the distribution of the data over multiple routes increases the overall reliability significantly. Unlike the other schemes, this does not assume that relay nodes transmit analogue data, which in many systems will be more realistic.

References

[ABRH03] H. Aghvami, Bilal, A. Rassool, F. Heliot, L. Revelly, and R. Nakhai. 4-PSK spacetime trellis codes with five and six transmit antennas for slow rayleigh fading channels. *Elect. Lett.*, 39(3):296–298, Feb. 2003.

[Alam98] S. M. Alamouti. A simple transmit diversity technique for wireless communications. *IEEE J. Select. Areas Commun.*, 16:1451–1458, 1998.

[BaFW05] B. Badic, P. Fuxjaeger, and H. Weinrichter. Combining quasi-orthogonal spacetime coding with antenna selection in wireless communication. TD(05)003, COST 273, Bologna, Italy, Jan. 2005.

[BoBP04] M. Borgo, M. Butussi, and S. Pupolin. A novel robust beamforming using convex optimization for CDMA space-time receiver. TD(04)122, COST 273, Gothenburg, Sweden, June 2004.

[BRHR04] A. Bilal, A. Rassool, F. Heliot, L. Revelly, R. Nakhai, and H. Aghvami. On the search for space-time trellis codes with novel codes for rayleigh fading channels. *IEE Proc. Commun.*, 151(1), Feb. 2004. [Also available as TD(03)043].

[BrSP03] L. E. Bråten, A. Spilling, and M. Petterson. A UMTS FDD simulator for smart antennas: general description and preliminary results. TD(03)052, COST 273, Barcelona, Spain, Jan. 2003.

[BuHe03] A. G. Burr and M. Herdin. The relationship between eigenvectors and multipath direction of arrival. TD(03)136, COST 273, Paris, France, May 2003.

[Burr02] A. G. Burr. Capacity of MIMO systems in realistic cellular wireless systems. In *IEE Conf. on Getting the most out of the radio spectrum*, London, UK, Oct. 2002. [Also available as TD(03)031].

[ChCz04] B. K. Chalise and A. Czylwik. Uplink user capacity of UMTS-FDD with robust beamforming based upon minimum outage probability. TD(04)179, COST 273, Duisburg, Germany, Sep. 2004.

[CoWi04a] G. W. K. Colman and T. J. Willink. Adaptive array algorithm performance: case studies in different environments. In *Proc. ANTEM 2004 – 10th Int. Symp. on Antenna Techn. and Appl. Electromagnetics*, Ottawa, ON, Canada, July 2004. [Also available as TD(04)044].

[CoWi04b] G. W. K. Colman and T. J. Willink. Genetic algorithm assisted array processing in overloaded systems. In *Proc. VTC 2004 Fall – IEEE 60th Vehicular Technology Conf.*, Los Angeles, CA, USA, Sep. 2004. [Also available as TD(05)026].

[CVVO04] B. Clerckx, L. Vandendorpe, D. Vanhoenacker-Janvier, and C. Oestges. Optimization of non-linear signal constellations for real-world MIMO channels. *IEEE Trans. Signal Processing*, 52(4):894–902, Apr. 2004. [Also available as TD(04)086].

[Czyl03] A. Czylwik. Performance of realistic circular antenna arrays in cellular mobile radio systems. TD(03)010, COST 273, Barcelona, Spain, Jan. 2003.

[Czyl04] A. Czylwik. Fluctuations of the channel capacity of ultra-wideband radio channels. TD(04)139, COST 273, Duisburg, Germany, Sep. 2004.

[DAZL05] M. Dohler, H. Aghvami, Z. Zhou, Y. Li, and B. Vucetic. Capacity and outage of transmit antenna selection schemes over nakagami fading channels. TD(05)023, COST 273, Bolgna, Italy, Jan. 2005.

[DoAg04] M. Dohler and A. H. Aghvami. Outage capacity of distributed STBCs over nakagami fading channels. *IEEE Comms. Letters*, 8(7):437–439, July 2004.

[DoLA02a] M. Dohler, E. Lefranc, and A. H. Aghvami. Virtual antenna arrays for future wireless mobile communication systems. In *Proc. ICT 2002 – 9th Int. Conf. on Telecommunications*, Beijing, China, June 2002.

[DoLA02b] M. Dohler, E. Lefranc, and A. H. Aghvami. Virtual antenna arrays for future wireless mobile communication systems. In *Proc. ICT 2002 – 9th Int. Conf. on Telecommunications*, Beijing, China, June 2002. [Also available as TD(03)039].

[DuOH05] D. V. Duong, G. E. Oien, and K. J. Hole. Adaptive coded modulation with receive antenna diversity and imperfect channel knowledge at receiver and transmitter. TD(05)009, COST 273, Bologna, Italy, Jan. 2005.

[Fise04] R. Fisera. *Adaptive Modulations - Adaptation Algorithms under Specific Constraints*. PhD thesis, Czech Technical University in Prague, Prague, Czech Republic, 2004. [Also available as TD(02)071].

[FKMW04] T. Fügen, C. Kuhnert, J. Maurer, and W. Wiesbeck. A double-directional channel model for multiuser MIMO systems. In *Proc. URSI EMTS 2004 – International Symposium on Electromagnetic Theory*, Pisa, Italy, May 2004. [Also available as TD(04)003].

[FKWS04] T. Fuegen, C. Kuhnert, C. Waldschmidt, M. Schnerr, and W. Wiesbeck. Capacity of the mimo broadcast channel under realistic propagation conditions. TD(04)103, COST 273, Gothenburg, Sweden, June 2004.

[FMKW04] T. Fügen, J. Maurer, C. Kuhnert, and W. Wiesbeck. A modelling approach for multiuser MIMO systems including spatially-colored interference. In *Proc. Globecom 2004 – IEEE Global Telecommunications Conf.*, Dallas, TX, USA, Dec. 2004. [Also available as TD(04)004].

[Fosc96] G. J. Foschini. Layered space-time architecture for wireless communication in a fading environment when using multi-element antennas. *Bell Labs Tech. J.*, pages 41–59, 1996.

[GADS01] A. Ghorashi, H. Aghvami, M. Dohler, and F. Said. Improvements in or relating to electronic data communication systems. Publication No. WO 03/003672, June 2001.

[GeHR03] G. Gerhard, W. Hand, and M. Rupp. Understanding the BER-performance of spacetime block codes. TD(03)085, COST 273, Paris, France, May 2003.

[GiCo01] J. M. Gil and L. M. Correia. Combining directional channel modelling with beamforming adaptive antennas for UMTS. In *Proc. PIMRC 2001 – IEEE 12th Int. Symp. on Pers., Indoor and Mobile Radio Commun.*, San Diego, CA, USA, Sep. 2001. [Also available as TD(01)036].

[GiCo02] J. M. Gil and L. M. Correia. Dependence of adaptive beamforming performance on directional channel macro-cell scenarios for UMTS. TD(02)050, COST 273, Espoo, Finland, May 2002.

[GiCo03a] J. M. Gil and L. M. Correia. Adaptive beamforming dependencies on wideband and directional propagation characteristics in micro- and macro-cell UMTS scenarios. In *Proc. PIMRC 2003 – IEEE 14th Int. Symp. on Pers., Indoor and Mobile Radio Commun.*, Beijing, China, Sep. 2003. [Also available as TD(04)007].

[GiCo03b] J. M. Gil and L. M. Correia. The MMSE vs. beamforming gain optima discrepancy in adaptive beamforming applied to directional channel scenarios. In *Proc. ConfTele 2003 – 4th Conference on Telecommunications*, Aveiro, Portugal, June 2003.

[GiCo04] J. M. Gil and L. M. Correia. Fundamental wideband and directional channel parameters ruling adaptive beamforming performance in micro- and macro-cells. In *Proc. VTC 2004 Spring – IEEE 59th Vehicular Technology Conf.*, Milan, Italy, May 2004. [Also available as TD(04)077].

[GkAD03] A. Gkelias, A. H. Aghvami, and M. Dohler. 2-hop distributed MIMO communication system. *Elect. Lett.*, 39(18):1350–1351, Sep. 2003.

[GkAg04] A. Gkelias and M. Dohler A. H. Aghvami. A resource allocation strategy for distributed MIMO multi-hop communication systems. *IEEE Commun. Lett.*, 8(2):99–101, Feb. 2004.

[GoLo96] G. H. Golub and C. F. van Loan. *Matrix Computations*. Johns Hopkins, Baltimore, MD, USA, 1996.

[Guég03] A. Guéguen. Comparison of suboptimal iterative space-time receivers. In *Proc. VTC 2003 Spring – IEEE 57th Vehicular Technology Conf.*, Jeju, South Corea, Apr. 2003. [Also available as TD(03)086].

[GVWF98] G. D. Golden, R. A. Valenzuela, P. W. Wolniansky, and G. J. Foschini. V-BLAST: an architecture for realizing very high data rates over the rich-scattering wireless channel. In *Proc. ISSSE 1998 – Proc. Int. Symp. on Signals, Systems and Electronics*, Pisa, Italy, Sep. 1998.

[HäCC03] L. Häring, B. K. Chalise, and A. Czylwik. Dynamic system level simulations of downlink beamforming for UMTS FDD. In *Proc. Globecom 2003 – IEEE Global Telecommunications Conf.*, San Fransisco, CA, USA, Dec. 2003. [Also available as TD(03)018].

[HaGe02] W. Hans and G. Gerhard. Concatenation of trellis coded modulation and space-time block code for high data rate on wireless MIMO channel. TD(02)126, COST 273, Lisboa, Portugal, Sep. 2002.

[HaHo02] B. Hassibi and B.M. Hochwald. High-rate codes that are linear in space and time. *IEEE Trans. Inform. Theory*, 48(7):1804–1824, July 2002.

[Hass00] B. Hassibi. An efficient square-root algorithm for BLAST. In *Proc. ICASSP 2000 – IEEE Int. Conf. Acoust. Speech and Signal Processing*, Istanbul, Turkey, June 2000.

[HiBu04] S. Hirst and A. Burr. Capacity of st-bicm qpsk codes in quasi-static fading. In *Proc. WPMC 2004 –Wireless Pers. Multimedia Commun.*, Abano Terme, Italy, Sep. 2004. [Also available as TD(04)198].

[ITU00] ITU. Preliminary draft new recommendation (PDNR): Vision framework and overall objectives of the future development of IMT-2000 and of systems beyond IMT- 2000. 2000.

[JoWi02] Y. L. C. de Jong and T. J. Willink. Iterative tree search detection for MIMO wireless systems. In *Proc. VTC 2002 Fall – IEEE 56th Vehicular Technology Conf.*, Vancouver, BC, Canada, Sep. 2002. [Also available as TD(04)114].

[JoWi03] Y. L. C. de Jong and T. J. Willink. Iterative trellis search detection for asynchronous MIMO systems. In *Proc. VTC 2003 Fall – IEEE 58th Vehicular Technology Conf.*, Orlando, FL, USA, Oct. 2003. [Also available as TD(04)114].

[KaSK04] P. Karamalis, N. Skentos, and P. Constantinou A. Kanatas. Comparison of existing MIMO antenna selection algorithms with an evolutionary approach. TD(04)055, COST 273, Athens, Greece, Jan. 2004.

[KnSy04a] M. Knize and J. Sykora. Linear diversity precoding design criterion for blockfading delay limited MIMO channel. In *Proc. Globecom 2004 – IEEE Global Telecommunications Conf.*, Dallas, TX, USA, Dec. 2004. [Also available as TD(04)054 and TD(04)056].

[KnSy04b] M. Knize and J. Sykora. Subspace inversion symbol energy adaptation in MIMO Rayleigh channel with zero outage probability. In *Proc. VTC 2004 Fall – IEEE 60th Vehicular Technology Conf.*, Los Angeles, CA, USA, Sep. 2004. [Also available as TD(04)134].

[KSMT05] K. Kansanen, C. Schneider, T. Matsumoto, and R. Thoma. Multilevel coded QAM with MIMO turbo-equalization in broadband single-carrier signalling. *IEEE Trans. Veh. Technol.*, 2005. Submitted for publication. [available as TD(04)028].

[LaSt03] E. G. Larsson and P. Stoica. *Space-Time Block Coding for Wireless Communications*. Cambridge University Press, Cambridge, UK, 2003.

[LeMJ04] J. Leinonen, T. Matsumoto, and M. Juntti. Single cell capacity evaluation towards 4G MIMO communication systems. TD(04)046, COST 273, Athens, Greece, Jan. 2004.

[LiPi02] J. Lindner and C. Pietsch. On capacity and linear processing for multiuser MIMO systems. TD(02)085, COST273, 2002.

[LiPi04] J. Lindner and C. Pietsch. Real-valued modeling and channel estimation for transmissions based on orthogonal STBCs. In *Proc. ISSSTA 2004 – IEEE 12th Int. Symp. on Spread Spectrum Techniques and Applications*, Sydney, Australia, Aug. 2004. [Also available as TD(03)100].

[MaTr04] T. Matsumoto and U. Trautwein. Turbo MIMO equalization for real-valued modulation signals. TD(04)191, COST 273, Duisburg, Germany, Sep. 2004.

[MaWe04] I. Maniatis and T. Weber. Joint channel estimation with array antennas in OFDM based mobile radio systems. TD(04)009, COST 273, Athens, Greece, Jan. 2004.

[MiBG03] P. D. Mitchell, A. G. Burr, and D. Grace. Performance of two-branch route diversity over a highly correlated rayleigh fading channel. In *Proc. WPMC 2003 – Wireless Pers. Multimedia Commun.*, Yokosuka, Japan, Oct. 2003. [Also available as TD(03)032].

[MTST02] T. Matsumoto, U. Trautwein, C. Schneider, and R. Thoma. On the use of multidimensional channel sounding field measurement data for system-level performance evaluation. TD(02)164, COST 273, Lisbon, Portugal, Sep. 2002.

[MuDa04] C. Mutti and D. Dahlhaus. Adaptive power loading for multiple-input multipleoutput OFDM systems with perfect channel state information. In *Proc. 4th Workshop COST 273*, Gothenburg, Sweden, June 2004.

[NaCB02] K. M. Nasr, F. Costen, and S. K. Barton. An application of smart antenna systems for archiving networks in TV studios. In *Proc. PIMRC 2002 – IEEE 13th Int. Symp. on Pers., Indoor and Mobile Radio Commun.*, Lisbon, Portugal, Sep. 2002. [Also available as TD(02)006].

[NaCB03a] K. M. Nasr, F. Costen, and S. K. Barton. An optimum combiner for a smart antenna in an indoor infrastructure WLAN. In *Proc. VTC 2003 Fall – IEEE 58th Vehicular Technology Conf.*, Orlando, FL, USA, Oct. 2003. [Also available as TD(03)088].

[NaCB03b] K. M. Nasr, F. Costen, and S. K. Barton. A spatial channel model and a beamformer for smart antennas in broadcasting studios. In *Proc. ICAP 2003 – 12th Int. Conf. on Antennas and Propagation*, Exeter, UK, Mar. 2003. [Also available as TD(02)109].

[NaCB04a] K. M. Nasr, F. Costen, and S. K. Barton. A downlink pattern optimisation algorithm for a smart antenna in an indoor infrastructure WLAN. In *Proc. PIMRC 2004 – IEEE 15th Int. Symp. on Pers., Indoor and Mobile Radio Commun.*, Barcelona, Spain, Sep. 2004. [Also available as WP(04)021].

[NaCB04b] K. M. Nasr, F. Costen, and S. K. Barton. Impact of user angular separation and channel estimation errors on a smart antenna system in an indoor WLAN. TD(04)156, COST 273, Duisburg, Germany, Sep. 2004.

[NaCB04c] K. M. Nasr, F. Costen, and S. K. Barton. An OFDM-MMSE smart antenna for an infrastructure WLAN in an indoor environment. In *Proc. VTC 2004 Spring – IEEE 59th Vehicular Technology Conf.*, Milan, Italy, May 2004. [Also available as TD(03)014].

[NeMa93] F. D. Neeser and J. L. Massey. Proper complex random processes with applications to information theory. *IEEE Trans. Inform. Theory*, 39(4):1293–1302, July 1993.

[NiSS03] M. Nicoli, O. Simeone, and U. Spagnolini. Multi-slot estimation of fast-varying space-time communication channels. *IEEE Trans. Signal Processing*, 51(5):1184–1195, May 2003. [Also available as TD(03)042].

[ÖzBo04] H. Özcelik and E. Bonek. Diagonal-correlation channels: Better than i.i.d? TD(049133, COST 273, Gothenburg, Sweden, June 2004.

[PaSL03] S. Parker, M. Sandell, and M. Lee. The performance of space-time codes in office environments. In *Proc. VTC 2003 Spring – IEEE 57th Vehicular Technology Conf.*, Jeju, South Korea, Apr. 2003. [Also available as TD(03)087].

[PiLi05a] C. Pietsch and J. Lindner. On the construction of capacity achieving full diversity space-time block codes. In *Proc. VTC 2005 Spring – IEEE 61st Vehicular Technology Conf.*, Stockholm, Sweden, May 2005. [Also availale as TD(04)129].

[PiLi05b] C. Pietsch and J. Lindner. On the impact of imperfect channel knowledge on the performance of orthogonal STBCs. In *Proc. ICC 2005 – IEEE Int. Conf. Commun.*, Seoul, South Korea, May 2005. [Also availale as TD(04)036].

[PiTL02] C. Pietsch, W. Teich, and J. Lindner. On capacity and linear processing for multiuser mimo systems. TD(02)085, COST 273, Espoo, Finland, May 2002.

[PNTL02] C. Pietsch, M. Nold, W. G. Teich, and J. Lindner. Optimum space-time processing for wide-band transmissions with multiple receiving antennas. In *Proc. ITG – 4th Int. Conf. on Source and Channel Coding*, Berlin, Germany, Jan. 2002. [Also available as TD(01)023].

[Poll02] A Pollard. An operator perspective on smart antennas and MIMO in systems beyond 3G. TD(02)123, COST 273, Lisbon, Portugal, Sep. 2002.

[Pomm02] C. Pommer. Convergence of a semi-blind least-squares-algorithm for the UMTS FDD uplink with adaptive antennas. TD(02)072, COST 273, Espoo, Finland, May 2002.

[PSTL03] C. Pietsch, S. Sand, W. G. Teich, and J. Lindner. Modeling and performance evaluation of multiuser MIMO systems using real-valued matrices. *IEEE J. Select. Areas Commun.*, 21(5):744–753, June 2003. [Also available as TD(02)125].

[RaWi04] B. Rankov and A. Wittneben. On the capacity of relay-assisted wireless MIMO channels. In *Proc. SPAWC 2004 – Sig. Proc. Advances in Wireless Commun.*, Lisbon, Portugal, July 2004. [Also available as TD(03)124].

[RBAR03] T. Rassool, A. Bilal, B. Allen, R. Roberts, R. Nakhai, and P. Sweeney. Error analysis of optimal and sub-optimal space-time trellis codes: Concatenation requirements. In *Proc. PIMRC 2003 – IEEE 14th Int. Symp. on Pers., Indoor and Mobile Radio Commun.*, Beijing, China, Sep. 2003. [Also available as TD(02)150].

[RuWB04] M. Rupp, H. Weinrichter, and B. Badic. Comparison of non-orthogonal space-time block codes using partial feedback in correlated channels. In *Proc. SPAWC 2004 – Sig. Proc. Advances in Wireless Commun.*, Lisboa, Portugal, July 2004. [Also available as TD(04)006].

[SaTA04] K. Sakaguchi, S. H. Ting, and K. Araki. Initial measurement on MIMO eigenmode communication system. *IEICE Trans. Commun.*, J87-B(9):1454–1466, Sep. 2004. [Also available as TD(04)027].

[SGKM04] C. Schneider, M. Großmann, K. Kansanen, T. Matsumoto, and R. Thoma. Measurement based throughput performance evaluation of antenna variable modulation for broadband turbo MIMO transmission. In *Proc. WPMC 2004 – Wireless Pers. Multimedia Commun.*, Abano Terme, Italy, Sep. 2004. invited COST273 session.

[SiSp04] O. Simeone and U. Spagnolini. Lower bound on training-based channel estimation error for frequency-selective block-fading rayleigh MIMO channels. *IEEE Trans. Signal Processing*, 52(11):3265–3277, Nov. 2004. [Also available as TD(03)084].

[SpMa04] A. Spalvieri and M. Magarini. A suboptimal detection scheme for MIMO systems with non-binary constellations. In *Proc. PIMRC 2004 – IEEE 15th Int. Symp. on Pers., Indoor and Mobile Radio Commun.*, Barcelona, Spain, Sep. 2004. [Also available as TD(04)126].

[SSSV04a] J. Salo, P. Suvikunnas, H. El-Sallabi, and P. Vainikaninen. On the characteristics of MIMO mutual information at high SNR. In *Proc. of 6th IEEE Nordic Signal Processing Symp.*, Espoo, Finland, Jun. 2004. [Also available as COST 273 TD(03)185].

[SSSV04b] J. Salo, P. Suvikunnas, H. El-Sallabi, and P. Vainikainen. On the distribution of mutual information in rayleigh fading mimo channels. In *Proc. NORSIG 2004 – 6th IEEE Nordic Signal Processing Conf.*, Oulu, Finland, Aug. 2004. [Also available as TD(04)043].

[SSSV04c] J. Salo, P. Suvikunnas, H. El-Sallabi, and P. Vainikainen. Some results on mimo capacity: the high snr case. In *Proc. Globecom 2004 – IEEE Global Telecommunications Conf.*, Dallas, TX, USA, Dec. 2004. [Also available as TD(04)042].

[StGa03] P. Stoica and G. Ganesan. Space-time block codes: Trained, blind, and semi-blind detection. *Elsevier Journal on Signal Processing*, 13(1):93–105, Jan. 2003.

[STMT03] C. Schneider, U. Trautwein, T. Matsumoto, and R. Thoma. The dependency of turbo MIMO equalizer performance on the spatial and temporal multipath channel structure – A measurement based evaluation. In *Proc. VTC 2003 Spring – IEEE 57th Vehicular Technology Conf.*, Cheju, South Korea, Apr. 2003. [available as TD(03)109].

[Syko01] J. Sykora. Constant envelope space-time modulation trellis code design for rayleigh flat fading channel. In *Proc. Globecom 2001 – IEEE Global Telecommunications Conf.*, San Antonio, TX, USA, Nov. 2001. [Also available as TD(02)016].

[Syko04] J. Sykora. Symmetric capacity of nonlinearly modulated finite alphabet signals in MIMO random channel with waveform and memory constraints. In *Proc. Globe-com 2004 – IEEE Global Telecommunications Conf.*, Dallas, TX, USA, Dec. 2004. [Also available as TD(03)094].

[TaJC99] V. Tarokh, H. Jafarkhani, and A. R. Calderbank. Space-time block codes from orthogonal designs. *IEEE Trans. Inform.* Theory, 45(5):1456–1467, July 1999.

[TMWK05] D. Tujkovic, T. Matsumoto, R. Wohlgenannt, and K. Kansanen. Outage-based LDPC code design for SC/MMSE turbo-equalization. In *Proc. VTC 2005 Spring – IEEE 61st Vehicular Technology Conf.*, Stockholm, Sweden, May 2005. [Also available as TD(04)209].

[TSSH03] U. Trautwein, C. Schneider, G. Sommerkorn, D. Hampicke, R. Thoma, and H. Wirnitzer. Measurement data for propagation modeling and wireless system evaluation. TD(03)021, COST 273, Barcelona, Spain, Jan. 2003.

[WaTS01] U. Wachsmann, J. Thielecke, and H. Schotten. Exploiting the data-rate potential of MIMO channels: Multi-stratum space-time coding. In *Proc. VTC 2001 Spring – IEEE 53rd Vehicular Technology Conf.*, Rhodes, Greece, May 2001.

[WeBN04] M. W. Webb, M. A. Beach, and A. R. Nix. Capacity limits of MIMO channels with co-channel interference. In *Proc. VTC 2004 Spring – IEEE 59th Vehicular Technology Conf.*, Milan, Italy, May 2004. [Also available as TD(03)166].

[WeMe04] T. Weber and M. Meurer. Imperfect channel state information in MIMOtransmission. In *Proc. VTC 2004 Spring – IEEE 59th Vehicular Technology Conf.*, Milan, Italy, May 2004. [Also available as TD(03)152].

[WeMZ05] T. Weber, M. Meurer, and W. Zirwas. Subspace-based channel estimation. TD(05)004, COST 273, Bologna, Italy, Jan. 2005.

[WiKH03] A. Wittneben, M. Kuhn, and I. Hammerstroem. Linear scalable space-time codes: Tradeoff between spatial multiplexing and transmit diversity. TD(03)038, COST 273, Barcelona, Spain, Jan. 2003.

[Will02] T. J. Willink. On the impact of estimation errors on subspace-based array processor performance. In *Proc. PIMRC 2002 – IEEE 13th Int. Symp. on Pers., In- door and Mobile Radio Commun.*, Lisbon,Portugal, Sep. 2002. [Also available as TD(02)017].

[Will03] T. J. Willink. MIMO OFDM for fixed wireless access. In *Proc. WIRELESS 2003 – Proc. 15th Int. Conf. on Wireless Commun.*, Calgary, AB, Canada, July 2003. [Also available as WP(03)005].

[Will05a] T. J. Willink. Improving power allocation to MIMO eigenbeams under imperfect channel estimation. *IEEE Commun. Lett.*, 2005. Accepted for publication. [Also available as TD(05)025].

[Will05b] T. J. Willink. MIMO OFDM for broadband fixed wireless access. *IEE Proc. Commun.*, 2005. Accepted for publication. [Also available as WP(03)005].

[WiRa03] A. Wittneben and B. Rankov. Impact of cooperative relays on the capacity of rankdeficient MIMO channels. In *Proc. 12th IST Summit on Mobile and Wireless Commun.*, Aveiro, Portugal, June 2003. [Also available as TD(03)124].

[WSLW03] T. Weber, A. Sklavos, Y. Liu, and M. Weckerle. The air interface concept JOINT for beyond 3G mobile radio networks. In *Proc. WIRELESS 2003 – Proc. 15th Int. Conf. on Wireless Commun.*, Calgary, AB, Canada, July 2003. [Also available as TD(04)084].

[ZeST04] R. Zetik, C. Schneider, and R. Thoma. Correlation between MIMO link performance evaluation results and characteristic channel parameters. In *Proc. 13th IST Summit on Mobile and Wireless Commun.*, Lyon, France, June 2004. [available as TD(04)049].

[ZeZe03] E. Zentner and R. Zentner. Down link transmitting array for indoor channel equalization. In *Proc. ICECOM 2003 –17th Int. Conf. Appl. Electromagn. and Commun.*, Dubrovnik, Croatia, Oct. 2003. [Also available as TD(03)157].

[ZhBP04] L. Zhang, A. G. Burr, and D. A. J. Pearce. Capacity of cellular system for the finite scatterers MIMO wireless channel. In *Proc. WPMC 2004 – Wireless Pers. Multimedia Commun.*, Abano Terme, Italy, Sep. 2004. [Also available as TD(04)038].

8 Radio network aspects

Chapter editors: *Roberto Verdone, Narcis Cardona*
Contributors: *Silvia Ruíz Boqué, Velio Tralli, John Orriss, Lúcio Studer Ferreira, Luís M. Correia, Markus Radimirsch, Gianni Pasolini, and Alberto Zanella*

8.1 Introduction

The radio network aspects that have been considered in COST 273 are mainly related to radio network planning, radio resource management, and the design and evaluation of techniques for radio network optimisation. The main focus was on infrastructure-based networks, and in particular cellular systems like UMTS. However, the application of such techniques was also considered for Wireless LANs (WLANs), Wireless Personal Area Networks (WPANs), and infrastructure-less environments. The aspects related to terminal location identification were also investigated. In this field, several good and well-known books are published over the literature. Therefore, the two chapters dedicated to such issues here are focused specifically on the most advanced results obtained in COST 273, with no reference to the basic and more introductory aspects related to the air interface techniques investigated. The next chapter deals with the research results related to UMTS.

This chapter treats topics that are not strictly UMTS oriented and is organised in four parts. The first part (Section 8.2) is dedicated to MORANS, an initiative aimed at harmonising the tools used by several research groups participating in COST 273, to study the performance of radio cellular networks with particular regards to UMTS. This harmonisation was achieved through the definition of common Reference Scenarios, which allowed the comparability of results among groups.

The second part includes sections dedicated to general topics, not specifically related to particular air interface technologies: Section 8.3 describes techniques for radio network optimisation, such as, e.g. packet scheduling for cellular systems, or system capacity maximisation through the use of multiple antennas; Section 8.4 provides methodologies for radio network performance evaluation, including, in particular, the description of theoretical connectivity models that can be very useful to describe physical layer characteristics of wireless networks composed of many radio access ports and mobile nodes, randomly distributed over the plane: several application examples of this modelling are reported in the section; next, Section 8.5 gives a description of service and traffic models suitable for 3G and 4G networks.

The third part includes three sections (8.6, 8.7, 8.8) dedicated to specific network domains, i.e. Wireless Local Area Networks (WLANs), wireless PANs, and wireless ad hoc networks, respectively. In this case, specific air interfaces are considered, such as, e.g. HIPERLAN/2, IEEE802.11 and Bluetooth. Finally, the fourth part (Section 8.9) is dedicated to terminal location identification, a topic closely related both to network and propagation aspects.

8.2 MORANS

Systems B3G will not consist simply of one radio access technology but will contain several wireless technologies such as WCDMA, GSM, General Packet Radio Service (GPRS), WLAN and PANs. It is well known that the integration of these technologies and networks in a global heterogeneous network will benefit both users and operators, allowing a global optimisation of coverage, capacity and quality. As an example, 3G networks are working in environments characterised by the presence of 2G, 2.5G systems and hot spots served by uncoordinated Wireless Local Area Networks (WLANs) in public, corporate or residential environments.

The study of these networks requires the introduction of many parameters, and the definition of much more complex evaluation scenarios than in the past. Some of the areas covered by COST 273 are network planning and optimisation, RRM algorithms design, smart antennas and MIMO techniques, coding, scheduling algorithms, traffic models and QoS, interworking and roaming between standards and so on. All of them will require system simulations. Several researchers from different institutions and countries participating in the COST 273 Action started the MORANS initiative in May 2002, which is oriented to provide a useful tool with a set of scenarios, parameters and models to be used as a common simulation platform.

8.2.1 Goals of the MORANS initiative

In many cases the results of different researchers appeared not to be comparable due to the different assumptions and approaches when describing the scenarios. This problem became evident within the COST 273 Action when dealing with comparison of different RRM techniques and Radio Network Planning (RNP) strategies. To solve this the main goal of MORANS is to facilitate the exchange of information among researchers, through the definition of some common Reference Scenarios [BüNB02], allowing researchers to easily exchange methodologies and compare results and strategies. In its first phase it has been addressed to infrastructure-based networks focusing on UMTS.

The Reference Scenarios have to satisfy a number of requirements in order to be of practical relevance. Some of them are:

- ease of use;
- limited number of cases;
- defined interfaces between different parts of the scenario;
- extendability.

8.2.2 Structure of the reference model

As a single scenario would be too complex, and moreover, depending on the subject under study, too inefficient, it was agreed that two types of scenario elements had to be provided [MuRV04]:

- synthetic scenarios, based on simple and regular geometrical layouts, simplified models, which make it easy to interpret the results;

- real-world-based scenarios, where some data is taken from the real world, thus making its definition (and use) more complex, but providing the possibility to test radio network algorithms under more realistic conditions.

As some system simulations require the use of dynamic scenarios, both static and dynamic approaches should be considered. As will be described in the following subsections each scenario consists of several elements chosen from a set of previously described layers (Geo Data, Traffic, Site Locations, Propagation Models, Terminal and Service Requirements, Node B Parameters and Link Level Performance Indicators).

Layers and elements for the real-world scenarios

Scenarios in our context are a set of assumptions for simulation input, where the assumptions for these types of scenario are derived from nominally planned networks in real environments. It is important to understand that they are still only assumptions and that in principle they do not (necessarily) map real radio networks. This is for three reasons: first, scenarios should contain a collection of challenging network situations which may not appear in real networks in limited areas, second it is not the goal of COST 273 to analyse particular operators, and third network layouts are typically kept confidential by operators.

The main advantages of real-world scenario approaches are:

- scenarios are very close to network reality with lower risk of invalid simplifications;
- the linking of theoretical calculations to the real world leads to practical understanding;
- the study of cross-influence of various technologies is enabled, e.g. influence of antenna design, propagation simulation, geo-data accuracy, etc. on network RRM simulation.

With the richness of the scenario comes the complexity, which makes it difficult to zoom in on certain simulation model details during simulation studies. This applies in particular in the early steps of implementation verification as well as in the later steps of implementation quality assurance. Thus from the beginning it is clear that real-world scenarios are complemented by synthetic scenarios.

Real-world scenarios were first suggested in [BüNB02] and further elaborated in [MuRV04].

For MORANS a layer structure has been chosen, where

a 'LAYER' is a set of coherent types of data that are strictly related, either conceptually or for practical reasons;

a 'SUBLAYER' is a subdivision of a layer;

an 'ELEMENT' is a realisation of data in a layer (or sublayer);

a 'REFERENCE SCENARIO' is a choice of matching elements, taken from each layer, one per layer. Thus, a reference scenario can be defined by choosing a set of elements, for specific numerical evaluation purposes;

an 'ELEMENT ID' is a unique identifier for the elements, and thus for data, provided as general input for the users. The element ID is in the form of 's.s-xx' where s.s is the sublayer number, and xx a sequence number.

Each reference scenario has been organised by dividing data elements into different layers, and then collecting elements belonging to the same layer, in the same folder of the MORANS Compact Disc (CD). So that the implementation of each scenario can be seen as a composition of layers, where each layer can be defined as a number of elements.

The layered structure for MORANS has been chosen to be

- extendable;
- flexible for generation of improved scenarios by exchanging elements in layers;
- well structured and easily understandable.

Table 8.1 reports the layers that have been identified and foreseen.

Table 8.1 *MORANS layers.*

#	LAYER	#.#	sublayer
1	GEO DATA		
		1.1	Basic: DEM, Clutter, Backdrop [raster], Backdrop [vector]
		1.2	Extended: building data [vector]
2	TRAFFIC INFORMATION		
		2.1	Model and parameters
		2.2	Realisations (dynamic)
		2.3	Realisations (static)
		2.4	Erlang traffic maps (per pixel) eventually per service [raster]
3	USER REF PLANE SERVICE REQ		
		3.1	User reference plane per service: conversational, streaming, interactive, background, voice
4	LINK LEVEL LOOKUP TABLES		
		4.1	Terminal performance models
		4.2	Realisations per service, terminal, UL/DL, speed
5	SITE LOCATIONS		
		5.1	Site locations (+ antenna pole locations)
6	PROPAGATION		
		6.1	Path-loss models + parameters
		6.2	Calculated path loss by 2D model(s)
		6.3	Calculated path loss by 3D model(s)
		6.4	Log-norm shadowing models
		6.5	Realisation of log-norm shadowing per user
7	NODE B PARAMETER		
		7.1	Static antenna systems and power budget
		7.2	(Shared) node B resourced
		7.3	Smart antenna systems and power budget
8	RRM ALGORITHMS AND PARAMETER DEFINITIONS		
		8.1	CAC
		8.2	PC
		8.3	LC
		8.4	Handover
		8.5	Scheduling

The real-world scenarios are designed to grow by provision of new elements which are generated by partners. These may be variants, new scenarios and simulation results based on existing elements. Thus partners feed results back to the MORANS scenario data set. Only a basic set of elements is provided by sponsoring.

Elements are provided in the form of raster data files (mostly by sponsors), vector data files (mostly by sponsors), tables as text files or in eXtensible Markup Language (XML) (by partners), documentation as pdf files (by partners), references to standards, publications, reports, etc. as pdf files.

For practical implementation purposes it has been found useful to define a strict and closed form representation of the single scenarios. This structuring has been formulated in XML and is described in the corresponding Subsection 8.2.3.

On the MORANS CD R2.0 two real-world-based reference scenarios are provided, which refer to Turin and Vienna real-world environments. The provided data in R2.0 is shown in Table 8.2.

Table 8.2 *Elements available for the real-world scenarios.*

#	Layer	Real-world Turin	Real-world Vienna
1	Geo data	Dimension of the simulation area – Weighting coefficient of each pixel with respect to the others – Digital Elevation Model – Railways and motorways (vector files).	Coordinates of the simulation and analysis area – Digital Elevation Model – Roads arranged in different layers according to their importance (vector) – 3D data of building contours, including height quota of building (vector) – 3D building information (raster) – Topographical map (raster), provided by ARGE Digitalplan ZT GmbH, Graz, Austria.
2	Traffic information	Service characterisation: Service information and usage on four different services in both uplink and downlink. Usage probability (user density – Typical usage of each one of the four services – Usage probability per service type). Traffic information with 50 different realisations (identifier and coordinates of each user; type of environment the user is placed; type of service the user is requiring; service status).	Service information and usage on three different services, in both uplink and downlink. Usage probability (relative user density – Absolute user density for each service and in four different environments). Traffic raster map with 50 different realisations (identifier and coordinates of each user; type of clutter; requested service; speed of user; current status of the link; class and actual value of penetration loss assigned to the mobile).

Continued

Table 8.2 *Continued*

#	Layer	Real-world Turin	Real-world Vienna
3	User reference plane service requirement	Description, by means of a .pdf file, of the UTRA-FDD link level simulator that has been used for the link level evaluations [OIRG03].	Description, by means of a .pdf file, of the UTRA-FDD link level simulator that has been used for the link level evaluations [OIRG03].
4	Link level look-up tables	Description, by means of a .pdf file, of the most relevant assumptions made in the UTRA-FDD link level simulator. This .pdf file also provides link level look-up tables for different type of services, in uplink and downlink [OIRG03].	Description, by means of a .pdf file, of the most relevant assumptions made in the UTRA-FDD link level simulator. This .pdf file also provides link level look-up tables for different type of services, in uplink and downlink [OIRG03].
5	Site locations	Site locations for the Turin scenario, provided by Vodafone Italy (identifier of the site; abscissa and ordinate of the site; altitude of the site).	Site locations for the Vienna scenario, provided by DI Dr. Hermann Bühler GmbH in cooperation with mobilkom Austria (identifier of the site; abscissa and ordinate of the site; altitude of the site).
6	Propagation data	Path-loss predictions provided by Vodafone Italy to MORANS initiative. Referring to each site (path-loss values; elevations angles).	Path-loss matrices containing the propagation loss from the transmitter antenna to any receiver location within 15 km.
7	Node B parameters	Node B configuration parameters: • Node B identifier • Antenna type • Mechanical tilt • Electrical tilt • Azimuth height • Site identifier. Antenna system configuration contains files storing the horizontal and vertical radiation patterns and some additional information.	Node B information for the network case of Vienna: • Site identifier • Transmitter • Identifier • Name of the antenna • Antenna azimuth • Antenna tilt. Antenna radiation pattern: angle in degrees, starting with 0 at the main beam; attenuation in dB against the main beam.
8	Radio resource management	Reference parameters described in a PDF providing parameters, value range, extensive links to publications, reports, standards, etc. for each parameter. [FeCo04]	Reference parameters described in a PDF providing parameters, value range, extensive links to publications, reports, standards, etc. for each parameter [FeCo04].

Layers and elements for the synthetic scenario

MORANS synthetic scenarios have been designed with the aim of providing simplified simulation models which make it easy to analyse simulation results. Scenario models and parameters have mainly been taken from reference models provided by ETSI and 3GPP, and results from IST funded research project Advanced Radio Resource Management for Wireless Services – IST Project (ARROWS). Synthetic scenarios have been structured, according to the general MORANS structure, using the same layer structure as described in Table 8.1.

Layer 1: Geographic data

Geographic characteristics for three different environments have been extracted from [3GPP02a] and [ETSI98b], hence the definition of three elements for layer 1 [FMCR04a]. The defined environments are summarised in Table 8.3.

Table 8.3 *Parameters of layer 1.*

Environment	Mean building height	Mean building width	Mean street width	Building layout
Urban macrocellular	12 m (4 stories)	50 m	30 m	Undefined
Urban microcellular	Assumed infinite	75 m	15 m	Manhattan grid
Mixed macro/micro	12 m	60 m	30 m	Manhattan grid

Layer 2: Traffic information

Layer 2 elements are defined in [MGFC04]. This layer comprises the following elements:

1. *Traffic model*: Traffic is modelled in terms of user characterisation, mapping between services and bearers (both aspects summarised in Table 8.4) and data source behaviour. Models for data sources are as follows:

   - *Speech*: Poisson call arrival process and exponentially distributed call duration and activity periods duration.

Table 8.4 *Service and user demand description.*

	Speech	Videoconference	Web browsing	Videostreaming
Num. of call attempts during busy hour	1	0.06	0.12	0.12
Call duration/load	120 s	120 s	2 Mb	425 Kb
Traffic per user during busy hour	33.3 mE	2 mE	240 Kb	51 Kb
Downlink (DL) bearer (type/kbps)	Speech 12.2	Conversational 64	Interactive 128	Streaming 128
Uplink (UL) bearer (type/kbps)	Speech 12.2	Conversational 64	Interactive 64	not relevant

- *Web browsing*: Self-similar traffic with Pareto-distributed number of files, file size and OFF states during activity periods and Weibull-distributed inactivity periods.
- *Videoconference*: Markov chain modelling H263 video.
- *Videostreaming*: Moving Picture Experts Group (MPEG)-4 modelled through wavelet coefficients.

2. *Mobility*: Two mobility models are proposed; each one is appropriate for one of the different building layouts included in layer 1 (see above). Both models are derived from the one described in [ETSI98b].
3. *Traffic maps*: Uniform traffic densities for suburban and dense urban environments are defined. Also, a non-uniform traffic distribution is included to model hot-spot situations.

Layer 3: User reference plane service requirement

The simulation of the cellular network at the system level needs results from link level simulations that relate to interference conditions provided by the network to the objective signal quality in terms of BER, BLER, etc. This layer describes the simulation methodology that has been followed so as to obtain such link level results. It is based on the assumptions, simplifications and methodology developed within the ARROWS IST research project [ARRO04].

Layer 4: Link level look-up tables

Link level performance in terms of BER and BLER are provided within this layer. Such performance measures depend on several parameters:

- Radio Access Bearer (RAB) and transport format within the RAB.
- Received signal-to-noise plus interference ratio E_b/N_0.
- Transmitter and receiver structure.
- Propagation channel characteristics (delay profile, Doppler shift).

Look-up tables are provided for several parameter combinations.

Layer 5: Site locations

Different cellular network layouts are included as elements belonging to layer 5 [MGFL04]. These elements provide the possibility of performing simulations in both the geographic environments described in layer 1, macrocellular (see Figure 8.1) and microcellular (see [3GPP02a]) with or without using the wrapping-around technique, hence producing four different layouts; an additional fifth layout, inspired by [3GPP02a], has been added to model the coexistence of micro- and macrocells. Layouts have been designed to make it feasible the simulation of both CDMA and TDMA networks with different reuse factors, namely 1/3 and 1/4.

Layer 6: Propagation data

Propagation modelling for system level simulations includes path loss and shadowing, while the short-term channel characteristics are modelled at the link level. Namely, UMTS 30.03 [ETSI98b] and 3GPP [3GPP02a] path-loss models have been assumed in layer 6 for both macrocellular and microcellular environments. Two versions of the macrocellular model are described in order to adapt it to the geographic environment descriptions.

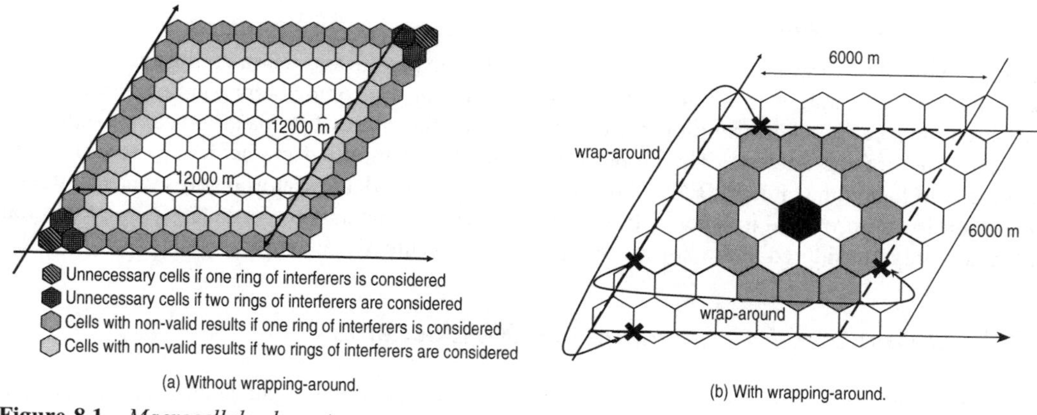

Figure 8.1 *Macrocellular layouts.*

As for shadowing, its statistical properties described in [ETSI98b] have been considered. However, in order to provide coherence with MORANS realistic scenarios, in which specific propagation maps for each cell are known, the concept of shadowing map has been introduced in the synthetic scenario. Thus, layer 6 [FMCR04b] includes bidimensional shadowing maps (see Figure 8.2 for an example) that have been generated using bidimensional filtering and that keep the commonly assumed statistical properties of shadowing, including standard deviation, distance autocorrelation and correlation between cells.

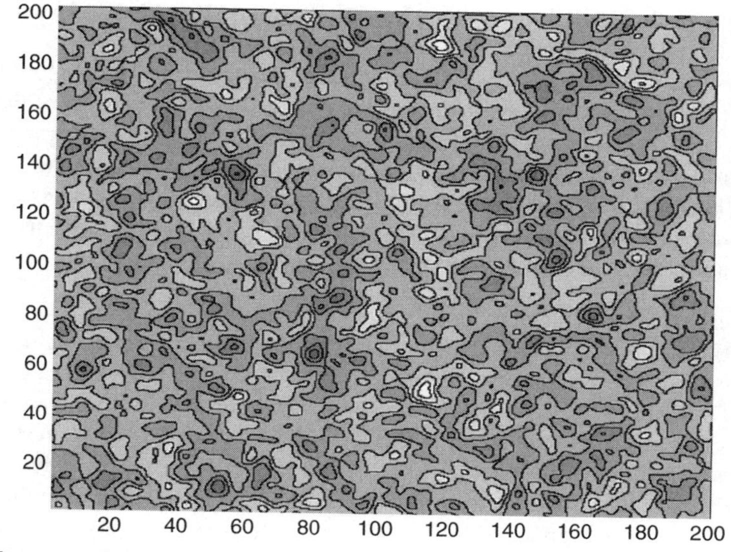

Figure 8.2 *Bidimensional shadowing map.*

Layer 7: Node B parameters

Layer 7 [FMGL04] comprises two aspects: antenna radiation pattern and sector configuration. Specifically, two different antennas are described. The former is an isotropic antenna aimed at modelling omnidirectional cells, hence providing a scenario in which the effect of propagation can be separated from the effect of the antenna. The latter corresponds to a simplified mathematical model of the 90° beamwidth antenna cited in [ETSI98b]. Its purpose is to model trisectorial cells, whose antenna orientation is also provided in layer 7. It must be pointed out that antenna orientation and radiation pattern are only considered in the horizontal plane for the synthetic scenario.

8.2.3 Data model for a scenario and XML schema

This section intends to provide a description of the data model used to organise the scenarios, which lead to the choice of a common platform, XML (eXtensible Mark-up Language), to represent and interchange the MORANS structured data [MuRV04].

The starting point and the main objective of switching towards XML technology was to use a common data model, both to structure the existing data and to be able to host future extensions. Moreover, since large and complex radio network simulation and planning data scenarios had already been published within the European project MOMENTUM (Models and Simulation for Networkplanning and Control of UMTS) [EGTK04] using an XML format, it was decided to harmonise the MORANS data with the existing MOMENTUM data format, believing that this choice would increase the incentive for potential users, since in this way all data provided by MORANS and MOMENTUM could be used at once. To this end, a new XML data structure has been defined: the MOMENTUM format has been enhanced in some places to accommodate the MORANS information, and the MOMENTUM data has been updated accordingly to make all scenarios available in a common format.

In the following an overview on the structure designed to host the Reference Scenarios data in XML format is presented: a more complete description can be found at [MuRV04].

Figure 8.3 provides the general structure designed to host the MORANS and MOMENTUM data, which is distributed over several files.

Scenario.xsd is the anchor schema, specifying the structure of all data. The anchor file *Scenario.xsd* is a tree structure, where the root element creates the link between all elements of the scenario. The name of the root element, *Scenario*, has been chosen to stress the fact that this structure is to be used for several scenarios. This element points to all elements constituting the scenario.

Each of the elements described below in turn imports several other elements, each specifying attributes and attribute types, in order to create the structure of the XML documents collecting data. A short description of the main elements is presented below.

- The name of each particular scenario is collected in the element *ScenarioName*.
- The element *GeoData* refers to all elements defining the scenario from the geographical point of view. The structure of this element has been designed to host data related to Layer 1.
- The element *TrafficModelData* relates to data and parameters used for traffic modelling.
- Specification of propagation parameters is defined in the element *PropagationModelData*.
- The element *TrafficData* concerns the definition of traffic load for the considered scenario. This element gathers all the information and the traffic assumption that can be found in Layer 2.
- The element *ServiceData* presents traffic source models for the services and defines their specifications, with reference to a certain user equipment and in a particular mobility mode.

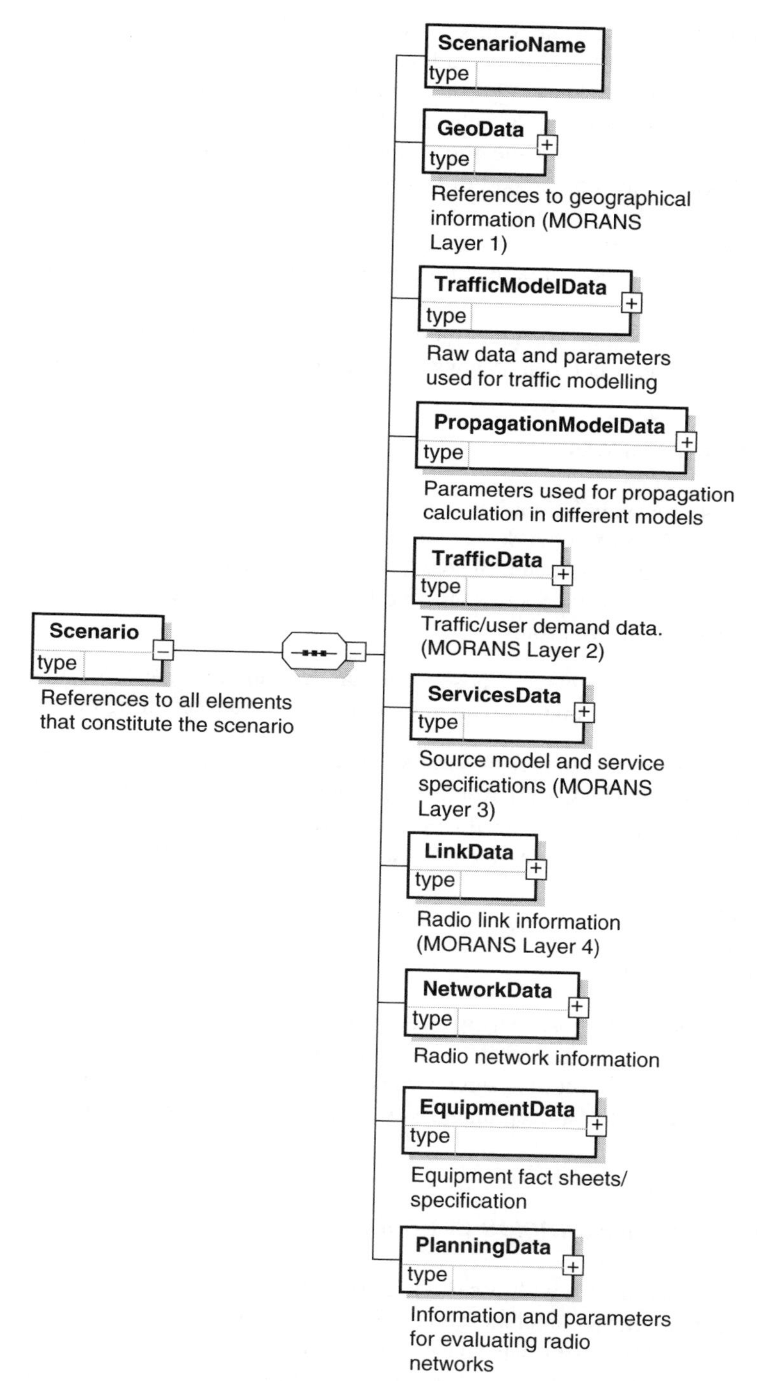

Figure 8.3 *The anchor schema for MORANS and MOMENTUM scenarios.*

- The element *LinkData* contains radio link information such as BER and BLER versus mean E_b/N_0 for different bearers. All this data refers to Layer 4.
- The element *NetworkData* defines fully configured radio network with all transmission parameters. The scenario is defined in terms of cells constituting each site. A complete set of configuration parameters is given for each site and cell of the considered network. This element collects information related to Layer 5 and part of Layer 7.
- *EquipmentData* gives the characterisation of the user and network equipment. In particular, this element contains the specification of the antenna hardware used in each scenario and technical characteristics of specific user equipment, i.e. mobile devices. The antenna information refers to Layer 7.
- Finally, the element *PlanningData* gives guidelines for automatic network planning: at the moment this type of information is not available with respect to the MORANS scenarios.

8.2.4 RRM parameters

In order to perform system simulations it is necessary to provide reference values for the main parameters characterising a WCDMA network. Therefore, one of the goals of MORANS is the identification of that subset of necessary parameters whose knowledge is required for a proper characterisation of the network, but not to provide an exhaustive list of parameters with the optimum values. The parameters are classified in two groups:

- the first group contains parameters whose value can be found in the specification documents,
- the second one contains parameters with unspecified value, and requiring optimisation.

Furthermore, since a large number of parameters deals with system level aspects, a further classification in Node B, Terminal, Call Admission Control, Congestion Control and Soft Handover related parameters is given. Finally for the most important RRM procedures, simplified algorithms that should be taken as reference are given. Some of them are [RCFG04]:

- Call Admission Control
- Congestion Control
- Soft Handover procedure

In [FeCo04] the authors propose to use Call Admission Control and Congestion Control Algorithm parameters as well as Soft Handover decision parameters similar to the one given in [WCFM01]. Also in this work a complete set of system parameters, propagation channel models and transceiver parameters are collected after a deep analysis of what is usually considered in open literature and 3GPP documents.

8.2.5 Results and examples of usability

In [MuVe04] an evaluation of downlink capacity, in terms of users served by the system, in a WCDMA scenario is performed using parameters from the Turin Real World Scenario [MuRV04], over a semi-analytical tool [BaMV03]. Results have been compared to those obtained using a Synthetic Scenario. Mobile Stations (MSs) are randomly generated and uniformly distributed over the simulation area, and at each snapshot a new distribution of MTs has been used. For what concerns propagation aspects, path-loss databases provided by [MORA03] have been used for the Turin Real World Scenario and a simplified model for the Synthetic Scenario. Service parameters have also been extracted from [MORA03]. The purpose of the paper was to investigate the role of the Soft Handover Process and Site Selection Diversity Transmit Power Control (SSDT) Mode in network dimensioning.

In [GMFL04] dynamic downlink simulations to analyse the performance of UTRA for different resource allocation strategies have been done including aspects of handover, load control, packet management, etc. The dynamic scenario includes terminal mobility (with wrap-around), call set-up and release, service demand variations and quality changes in received signal. Cellular layout, as well as propagation models, mobility models and service parameters have been chosen from [MORA03]. Two different handover strategies have been tested: hard handover and soft handover, and each of them for different margins. Results in terms of maximum number of simultaneous users are depicted in Figure 8.4 where it is shown that for similar margins soft handover clearly outperforms hard handover.

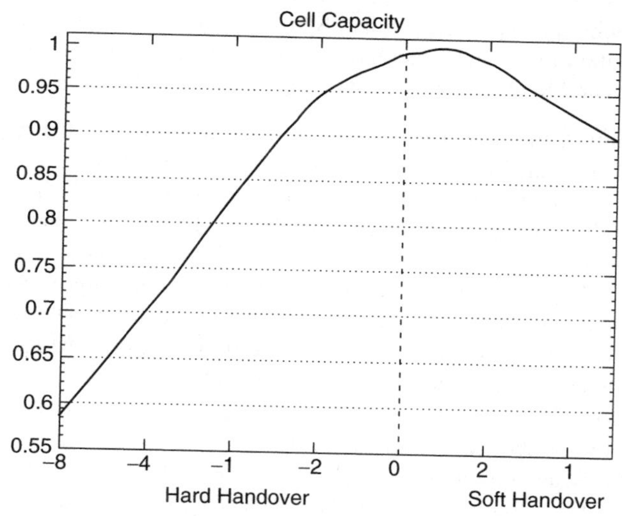

Figure 8.4 *Capacity for different handover configurations.*

In [RGGH04] the real Vienna scenario has been used [MORA03] and adapted to a UTRAN C++ simulation tool and some simple results have been analysed in order to verify that the obtained information is coherent. The results presented should be understood as preliminary simple results (mainly for UL), but could help to understand how to use MORANS scenarios, with the provided layers and files, in order to be able to compare results with other research teams, as well as taking benefit of the use of a detailed scenario description. Some RRM parameters as MTs and Base Transceiver Station (BTS) transmitted powers, cell load and interference factors have been obtained.

In [FGLM04] a new 2D shadowing model has been described, which incorporates spatial information. The performance of the proposed model has been evaluated and a significant impact on the system performance estimation has been identified considering the use of a synthetic scenario as described in [MORA03]. Results on throughput and BLER in a GPRS system have been analysed and compared with other shadowing models (log-normal assumption).

One of the main goals of the MORANS initiative is to perform simulations over the Common Reference Scenarios in XML format facilitating the comparison of results obtained by different tools. The following pages summarise the results obtained by different research teams (Polytechnical University of Valencia – UPV, Konrad Zuse Zentrum für Informationstechnik Berlin – ZIB, DI Dr. Hermann Bühler GmbH – GmBh and Wireless Future – WF) which are described in more detail in [PBCE05].

Macrocellular synthetic scenario

The estimated performance in the downlink of a WCDMA network has been evaluated by means of two dynamic system simulators and results have been compared. The chosen scenario is described by an urban macrocellular environment with regular cell layout (intersite distance of 1 km) with omnidirectional antennas, a macrocell propagation model with 8 dB standard deviation log-normal shadowing and a uniform traffic distribution. Pedestrian (3 km/h) and vehicular (50 km/h) users have been separately simulated for a service mix composed by Speech and the Video Call (VC). A wrapping-around technique has been applied. Some differences between the two dynamic simulators have been identified in the implementation of some RRM strategies and system behaviours. In order to obtain comparable results the following approach has been considered:

- Call Admission Control. In the dynamic simulator developed by WF the required power for the incoming session is estimated in order to take into account the resources needed by this session and the additional resources required by the ongoing sessions in order to face the interference increase caused by the admission of the session. In the dynamic simulator developed by UPV the required power for the incoming session and the additional power required by the ongoing sessions are calculated according to the actual system state and the future state of the system when the incoming session is assumed to be admitted. This means that a perfect estimation is considered: no dropping event happens if the radio channel condition of admitted sessions does not change (i.e. users are motionless and fast fading is not considered).
- Voice Activity Factor. The voice service is characterised by the alternation of silence and activity. For the purpose of this study the activity factor of this service has been set to 0.5, assuming that a user speaks for the same amount of time that he or she listens. In addition, the power needed for supporting the dedicated control channel has to be considered. In the dynamic tool developed by WF this channel is fully implemented while in the dynamic simulator developed by UPV this overhead is taken into account assuming a higher activity factor. For the purpose of this study, it has been set to 0.67.

A summary of some RRM parameters is given in Table 8.5 and results in Table 8.6.

Results obtained by means of the two dynamic simulators are very close to each other in terms of average speech sessions per cell, average video call sessions per cell, mean BS transmitted power and average soft handover sessions per cell. Comparing the blocking rate and dropping rate provided by the tools a difference is recognisable. This is due to the different RRM strategies applied by the tools. In fact, an ideal call admission control is applied by the dynamic simulator developed by UPV. This leads to a higher blocking rate than in the result obtained by WF. Conversely, the dropping rate experienced in the WF simulator is higher than the one obtained by UPV. It means that the sessions that are admitted with an underestimation of the required radio resources in the case of the dynamic tool developed by WF are subsequently dropped in most of the cases. It is worth noting that the higher blocking rate and dropping rate obtained by UPV in the pedestrian case study are mainly due to a

Table 8.5 *RRM parameters for the synthetic scenario.*

Parameter	Value [measurement unit]
Call admission control threshold	75%
Congestion control algorithm time to trigger	5 [s]
Maximum active set size	4
SHO add threshold	3 [dB]
SHO remove threshold	4 [dB]
SHO replace threshold	3 [dB]
SHO time to trigger	0 [s]

Table 8.6 *Synthetic scenario results.*

Performance metric [measurement unit]	UPV		WF	
Speed [km/h]	3	50	3	50
Blocking rate [%]	4.5	13.1	1.4	8.9
Dropping rate [%]	4.7	2.5	2.4	12.2
Average speech sessions per cell	27.3	23.3	27.5	25.2
Average VC sessions per cell	5.7	4.9	6.2	4.4
Mean BS transmitted power [W]	9.6	12.6	10.2	12.5
Average SHO sessions per cell [%]	21	23.1	20.8	21.2

higher offered load. In fact, the same Poissonian process has been considered in both tools but each of them has offered the load according to its own realisation.

Turin Real World Scenario

System evaluations have been carried out in order to compare the estimated performance in the downlink of a WCDMA network by means of a dynamic system simulator and evaluation software that exploit a static approach. The main simulation parameters are reported in Table 8.7.

Table 8.7 *Turin scenario parameters.*

Parameter	Value
Orthogonality factor [0:1]	0.633
Shadowing standard deviation [dB]	0
In car loss [dB]	8
In building loss [dB]	20
CPICH power [dBm]	30
Other common channels power [dBm]	30.5
Maximum BS power [dBm]	43
CPICH E_c threshold [dBm]	−110
CPICH E_c/I_0 threshold [dB]	−18

Two network configurations for azimuth and tilt values have been evaluated in terms of coverage and capacity, named reference and improved configuration.

Coverage analysis

The level (E_c) and the quality (E_c/I_0) of the CPICH of each cell have been separately evaluated. In the former case, results are not dependent on the system load and have been obtained by separately evaluating indoor (additional loss of 20 dB), outdoor (without additional loss) and vehicular (additional loss of 8 dB) users. In the latter case, results depend on the system load so each BS is supposed to transmit the maximum power level (worst case analysis) and are not dependent on the additional loss that a user could experience according to its mobility behaviour. In Table 8.8 the percentage of covered area is reported for the three cases considered for the evaluation of E_c level of the CPICH, and for the case that investigates the quality E_c/I_0 of the CPICH and also for the two network configurations.

Table 8.8 *Coverage results for the Turin scenario.*

		ZIB		WF	
Performance metric	Additional losses [dB]	Reference	Improved	Reference	Improved
Outdoor coverage [%]	0	98.6	99.4	98.5	98.9
Vehicular coverage [%]	8	95.7	96.6	95.2	95.4
Indoor coverage [%]	20	49.0	50.1	47.7	48.0
E_c/I_0 coverage [%]	–	97.8	99.3	97.1	97.8

Results show that in the case of coverage analysis a good agreement between different tools and evaluation methods can be reached. Even if the numerical results differ, the same trend is obtained by means of the two approaches: the improved configuration leads to better system performance in terms of coverage.

Capacity evaluation

A traffic mix composed by 30% vehicular and 70% indoor users is considered. The requirement for the speech service has been set to FER = 1% and for both the video call and the streaming data to BLER = 1%.

Since the evaluation software does not explicitly simulate users the performance comparison has been carried out in terms of average downlink power per cell.

The absolute results differ (Table 8.9). This was to be expected, as the evaluation methods apply a static approach that tends to underestimate the required resources with respect to the dynamic

Table 8.9 *Capacity results related to the WF and ZIB implementations.*

	Average power per cell [mW]	
Network configuration	ZIB	WF
Reference	3660	5415.8
Improved	3534	5300.9

simulator, as explained in Section 9.4. However, it can be seen that the observed trends are similar in both cases. When comparing the two configurations, the average transmit power is reduced in the improved configuration.

Real World Scenario (Vienna)

System evaluations have been carried out in order to compare the estimated performance of a WCDMA network by means of a commercial static system simulator and evaluation software that exploits a static approach. Coverage and capacity evaluations have been carried out and compared.

The service mix is composed of speech and two data services: a symmetric service (64 kbps both in uplink and downlink) and an asymmetric service (64 kbps in uplink and 144 kbps in downlink). As the commercial radio planning tool cannot import the XML files of the MORANS data sets, the propagation files have been recalculated from the propagation model parameters. Figure 8.5 shows the results of the coverage analysis.

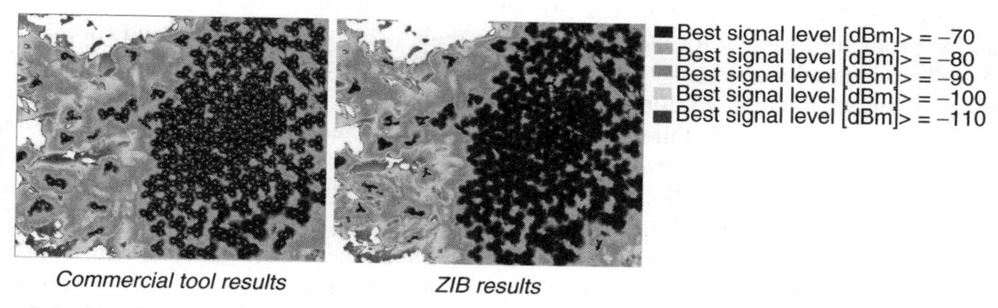

Figure 8.5 *Results of the coverage analysis, based on the pilot power measurement.*

The two plots show very high correlation in absolute and relative signal strength. Since the evaluation software does not explicitly simulate users the performance comparison has been carried out in terms of average downlink power per cell. The difference of the power levels simulated by means of the commercial static tool and the evaluation method for uplink and downlink are reported in Figure 8.6. In the downlink the graph compares the total transmitted power while in the uplink the

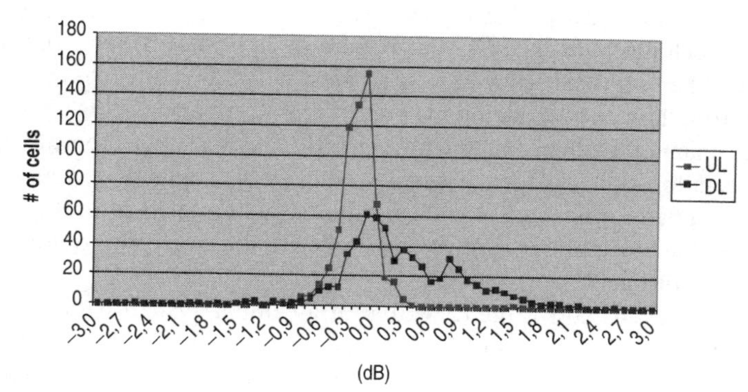

Figure 8.6 *Histogram of power level differences in UL and DL.*

total received power is compared. The graph shows good agreement between the two simulation tools even though the simulation methodology differs.

8.3 Techniques for radio network optimisation

8.3.1 *Resource allocation and packet scheduling techniques in wireless cellular systems*

Packet-switching communication is gaining an ever increasing relevance for wireless networks. On the one hand, releasing radio resources to single communication units, i.e. the packets, is much more efficient than releasing resources to entire connections; on the other, IP is becoming the predominant technology in the backbone network for integration and transport of heterogeneous multimedia and data traffic.

Two important radio resource optimisation problems have to be considered, within this framework, for the design of MAC and RRM functionalities in a wireless packet network.

The first problem is to look for the best way (as a trade-off between capacity and complexity) to dynamically assign radio resources to a set of users having packets to transmit [ZaKi01].

The second problem is to look for the best way (as a trade-off among capacity, user satisfaction, fairness and complexity) to schedule packet transmission in time over the available radio resources for a set of users with one or more packet traffic sources with possible QoS constraint [CaLi01].

Therefore, dynamic resource allocation and packet scheduling techniques in wireless packet networks are relevant aspects for the optimisation of spectrum efficiency and system capacity of next generation radio access networks. In order to achieve efficient use of the radio resources in cellular systems, it is needed that these techniques are channel and interference aware [ZaKi01], [CaLi01]. Moreover, different coordination strategies among the cells have to be considered: a totally centralised algorithm (running on a single common controller or on a network of distributed controllers) would be the optimal choice since it has complete information and therefore can perform optimal decisions. However, it may be complicated and requires a large signalling burden. A distributed algorithm (running independently on each cell) may be simpler, but it requires a smart management of radio power and bandwidth with incomplete information to achieve a large system capacity. An intermediate choice would be a partially centralised algorithm controlling small subsets of cells having the highest reciprocal interference.

The first problem is addressed in [TrVZ02], [VeTr03] which propose and investigate a new dynamic resource allocation technique for TDMA-based (pure or hybrid) cellular radio systems with full frequency reuse. Unlike traditional allocation strategies, which first allocate the time slots and then perform power control, the proposed technique first assigns in advance (static allocation) a maximum power level to each channel and then dynamically performs channel allocation and fine power control. The static power pre-allocation is named 'power shaping' and is implemented by using a set of power profiles, suitably designed for hexagonal cells with three sectors, which limit (or shape) the power transmitted in each slot of the frame with the aim of partially organising intercell and intersector interference in an uncoordinated environment.

This technique is investigated for the downlink of a cellular system with hexagonal cells, each with three sectors. Each sector is assigned a label (A or B or C), and the reuse factor is one. The guidelines for the design of three label profiles were provided in [TrVZ02]. By expressing the power profiles as vectors of power levels, $P_l = [P(0, l), P(1, l), \ldots, P(N-1, l)]^T$ (l is the label), suitable three-label symmetric power profiles can be simply designed through a geometric approach, leading to, for example, the design of Table 8.10. One of the simplest choices for the power levels p_j is the straight line (linear shaping), i.e. $p_j = P_{min} + j(P_{max} - P_{min})/(N-1)$ where P_{max} and P_{min} are parameters that should be suitably set. A more detailed design is proposed in [VeTr03].

Table 8.10 *Three-label power profiles with centre slot.*

Slot	0	1	2	3	4	5	6	7	8	9
P_A	p_0	p_1	p_2	p_6	p_5	p_4	p_6	p_5	p_4	p_3
P_B	p_6	p_5	p_4	p_0	p_1	p_2	p_6	p_5	p_4	p_3
P_C	p_6	p_5	p_4	p_6	p_5	p_4	p_0	p_1	p_2	p_3

The channel allocation algorithm is a greedy algorithm which exploits the knowledge of the conservative power needed by each channel to serve the packet of a given user under an SIR quality constraint. This conservative power is estimated by using the information of power profile used by the interfering cells and the channel state information. The algorithm allocates one (not yet allocated) user at a time, until no more users or resources can be allocated. Each time a user is chosen, its preferred resource is assigned to it. After the termination of the allocation procedure a power control algorithm is applied inside each sector (or group of sector/cells in case of a centralised control).

The results show that this technique is able to increase the capacity of systems with and without centralised resource management control. Moreover, it is also able to fill the efficiency gap between centralised and not centralised strategies.

The resource allocation problem has also been investigated for wireless packet networks with smart antennas in [VTZZ04]. Resource allocation techniques with smart antennas, through SDMA, have been pointed out to be able to increase the capacity of the radio system by allowing a simultaneous transmission to multiple users: on the downlink, each beam is adjusted to serve the desired user and nulls are placed in the direction of interfered users. The still open issue for packet multicell SDMA is how to manage the intercell interference, which is very difficult to be predicted in an uncoordinated environment, due to packet access and downlink beamforming.

The optimal joint power control and beamforming on the downlink was addressed in [RaLT98], where an iterative algorithm that first adjusts weighting vectors based on an equivalent uplink problem and then adjusts transmit powers is proposed. The work in [VTZZ04] proposes an allocation algorithm which works in a distributed SDMA environment and is able to support power shaping. The intercell interference is managed by means of an estimation based on a fraction of the worst case interference, which takes into account multiple antennas and is obtained through pilot tone signalling (the actual intercell interference value is not a priori known because it depends on the allocation performed by the other APs). Power shaping imposes a constraint on each slot of the frame about the maximum transmit power. The allocation algorithm tries to fill each slot with a set of packets, depending on users' spatial separability, channel quality, interference estimation and available power. It is shown (see Figure 8.7) that even with SDMA this technique provides a significant capacity gain over the baseline case of random allocation and can significantly reduce the gap due to the lack of control, with respect to a greedy centralised algorithm.

The second optimisation problem, i.e. packet scheduling, is addressed in [CuMB01] for the uplink of a single cell and in [SaGV05] for the uplink of a multicell MC-CDMA system. Both works consider a channel and traffic adaptive scheduling. The work in [CuMB01] proposes a general framework useful to build a wide class of scheduling algorithms. It is based on the definition of a matrix,

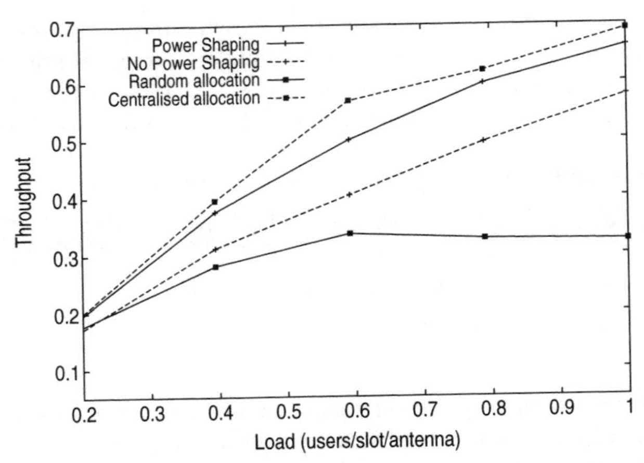

Figure 8.7 *Throughput vs load: comparison between the proposed algorithm, with and without power shaping, and the centralised and random allocation.*

which is updated frame by frame, that includes for each packet the information about the queue state (length, age) and the channel state (channel gain). Based on this framework, the work proposes a new scheduling algorithm for TDMA and CDMA access schemes which is channel dependent and load adaptive. In the numerical results, an extensive comparison with more classic load adaptive algorithms shows that channel dependent algorithms allow a more efficient use of resources in terms of capacity, delay and packet loss.

In [SaGV05], the problem of designing efficient scheduling strategies for the MC-CDMA interface is addressed by taking proper consideration of the key aspects typical of communication on a wireless medium, that is, fading (with its time correlation properties) and the frequency-selective characteristics of the environment. The work proposes and evaluates the performance of a channel-adaptive scheduling algorithm which reconsiders all decisions on a frame-by-frame basis, depending on the channel conditions that are estimated at the BS, where the scheduler is assumed to be implemented. The performance is measured in terms of QoS (throughput, outage, delivery delay, etc.) and of average transmitted power. The channel state is described with two variables: the maximum average channel gain over all the Groups Of Frequencies (GOF) and its increment with respect to the value of the previous frame. The performance figures are compared to those of a static scheduling strategy which assigns the radio resource only by taking traffic aspects into account. Figure 8.8 outlines the different scheduling behaviour in an interval of 100 frames for a traffic class with Linear Decorrelating Detector (LDD): the allocation process in the adaptive scheduler tries to exploit the channel gain, by allocating when at least one GOF presents a high gain.

In both schemes, delay due to signalling plays a key role. Fairness aspects and support of QoS call for future investigations.

8.3.2 *Joint link and network optimisation with MIMO systems*

With the aim to increase system capacity through efficient spectrum exploitation, the use of MEA technology at both transmitter and receiver has been extensively investigated in the last few years.

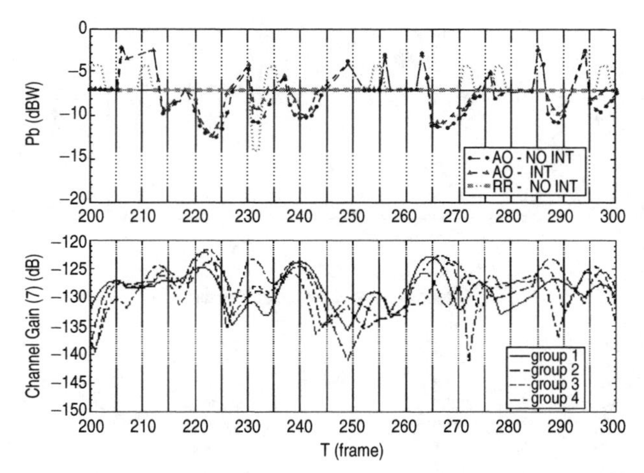

Figure 8.8 *Transmitted power of an LDD bearer during 100 frames and the corresponding fluctuations of GOF gain.*

By transmitting different signals in parallel over different antennas, in an environment characterised by rich scattering and multipath, and by receiving them through a set of receiving antennas (MIMO scheme), it is possible to achieve very large spectrum efficiency. Works by Telatar and Foschini [Tela99], [FoGa98] revealed that MIMO channels have very large theoretical capacity, which can be approached by using appropriate transmission and processing techniques.

The application of MIMO concept goes beyond the simple single-link scenario including only one transmit and one receive terminal. It can also be applied to scenarios where the transmitter and the receiver are composed of multiple distributed cooperating terminals or nodes or relays in a wireless network: a group of nodes can cooperate to transmit (or relay) from multiple distributed antennas the information coming from a single user in an ad hoc or a sensor network [DoAg04]; a set of terminals can cooperate to transmit information from multiple users to a single access point as in the uplink of an SDMA cellular system [MaMT03]. In single-link scenarios, as well as in multilink network scenarios different radio resource optimisation problems can be investigated. The set of parameters or radio resources involved in the problem depends on the scenario, the availability of channel information at the receiver, the availability of a feedback to the transmitters. Hence, we may have power allocation, rate allocation, slot allocation, antenna selection. Two possible approaches have been considered up to now for resource optimisation:

- the first considers the maximisation of mutual information of the channel, which gives an indication of the system capacity achievable with ideal signalling, coding, etc.;
- the second considers the maximisation of the throughput of a protocol at link layer or above in a practical system configuration.

The first approach is the one considered in [WiRa03], [RaWi04] and [DoAg04]. The work in [WiRa03], [RaWi04] investigates the benefit gained by the use of cooperating relay nodes in a single-user communication on a rank-deficient MIMO channel (those channels whose capacity does not grow linearly with the number of antennas due to the high spatial fading correlation). The related scenario is shown in Figure 8.9 and each relay is a simple amplify-and-forward device with

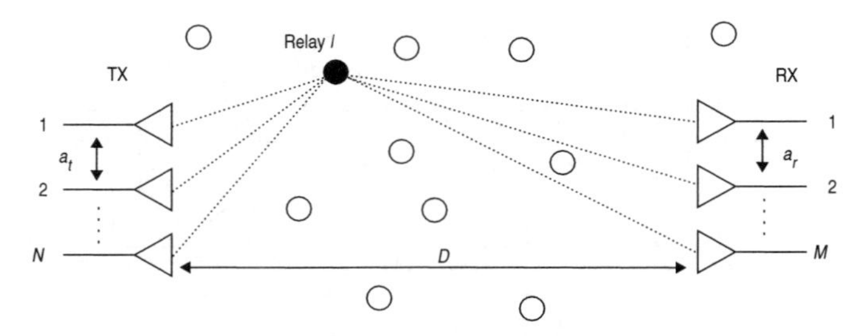

(a) Antenna arrays at source and destination.

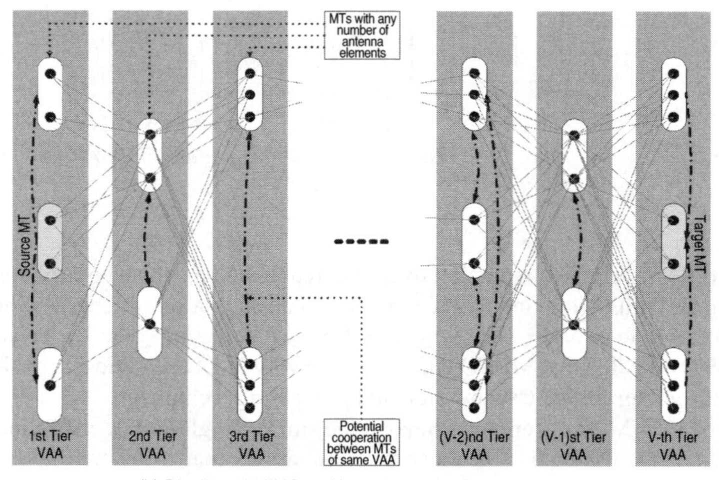

(b) Distributed-MIMO multi-stage communication network.

Figure 8.9 *A two-hop relay network.*

multiple antennas. Here, the optimisation problem is done with a space of variables that includes the slots allocated to source and relay transmission and the power gain assigned to each antenna of the relay nodes. The paper proposes three protocols for slot allocation to the available links. The three choices for a frame of two slots are: source-relays, relays-sink (P1); source-all, relays-sink (P2); source-all, all-sink (P3). The paper also proposes static power allocation at the relay antennas: two power values are selected for each node as a function of source-relay distance. The results illustrated in Figure 8.10 for the basic case of no multipath (rank 1 channel) show the significant increase of capacity and its sensitivity to the slot and power allocation strategy.

The work in [DoAg04] investigates the capacity achievable in multihop distributed MIMO systems where two or more nodes at each hop or stage can collaborate to form a single spatially distributed MIMO node. The scenario is depicted in Figure 8.9. The aim of the paper is to provide a near-optimal analytical solution for an optimisation problem with a space of variables including fractional bandwidth and transmission power to be assigned to each node; therefore, it is also a resource allocation problem. Here, the nodes are modelled as regenerative (decode-and-forward) relays. The improvements obtained through this optimisation are illustrated in Figure 8.11 for a two-stage network

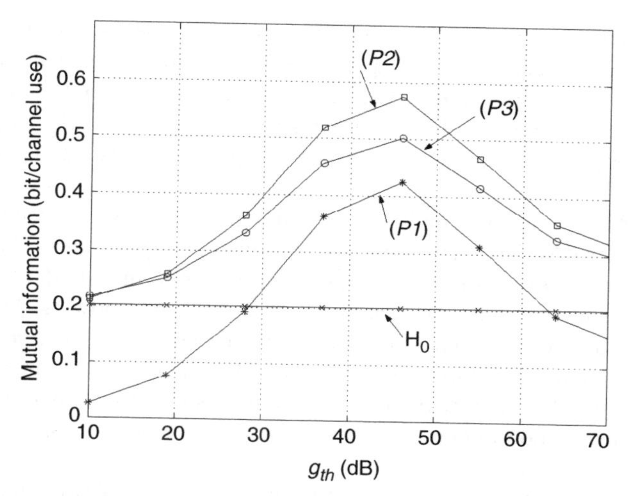

Figure 8.10 *Average mutual information vs. threshold gain g_{th} in dB, $\{N, M, R, P, \sigma_r^2, \sigma^2\} = \{3, 3, 80, 1, 10^{-5}, 10^{-5}\}$.*

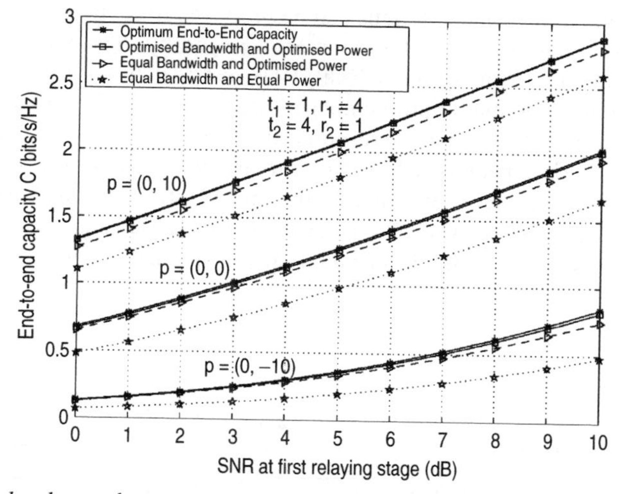

Figure 8.11 *Achieved end-to-end capacity of various fractional resource allocation strategies for a two-stage network.*

and three different pairs of link gains. The figure also underlines that near-optimal solution approaches optimal resource allocation solution.

The second approach to capacity optimisation, which considers the throughput of a data protocol as objective function, is investigated in [MiTZ02], [MaMT03] and [MaVT04] for different MIMO scenarios. Here, the Go-back-N scheme is used as an idealised model of a protocol which includes retransmission and window mechanisms. In [MiTZ02] the application scenario is a single MIMO link with channel information available at the receiver and a limited feedback to the transmitter that allows transmit antenna selection and power allocation. The receiver is based on linear detection and ordered successive interference cancellation. The paper proposes an optimal and a sub-optimal solution for the optimisation problem; the first can be evaluated with a combinatorial/numerical algorithm, the second with a combinatorial/analytical algorithm. It shows that the most effective resource allocation mechanism in this scenario is the antenna selection which provides substantial improvements at small signal-to-noise ratios.

The work in [MaMT03] refers to a multi-user communication to a single access point using Space Division Multiplexing (SDM)/Time Division Multiplexing (TDM) of signals. The space of variables in the optimisation problem includes the channel coordinates in the SDM/TDM frame, the status of the antennas for the antenna selection, and the vector of transmit powers. The proposed solution (sub-optimal) is composed of two steps: the first is a greedy channel allocation (different classes of algorithms are proposed and compared), the second is the antenna selection and power allocation of [MiTZ02]. The results show that with this solution the system throughput increases linearly with the number of active users and saturates to a maximum value which is the single-user capacity times the number of receive antennas.

The multi-user MIMO scheme has been further extended in [MaVT04] to a multicell scenario which introduces the new issue of the optimisation of interfering MIMO schemes. In this new scenario the space of variables in the optimisation becomes very large and the optimal solution, if we are able to find it, should be managed by a centralised controller which also requires a significant amount of signalling. However, sub-optimal solutions can be proposed for network with distributed control as also indicated in Subsection 5.3.1 and a key role is played by those techniques that are able to control intercell interference. Among them [MaVT04] consider the use of channel reuse, stream control, static allocation of the maximum number of spatially multiplexed signals, and investigate the benefits of interference estimation in the resource allocation algorithms. As shown in Figure 8.12, the capacity of the network is heavily affected by the presence of interference; however, the joint use of an estimation of interference covariance matrix and adaptive stream-limited antenna selection prevents the capacity degradation when the offered load increases and introduces a considerable performance improvement over the traditional approach of frequency reuse.

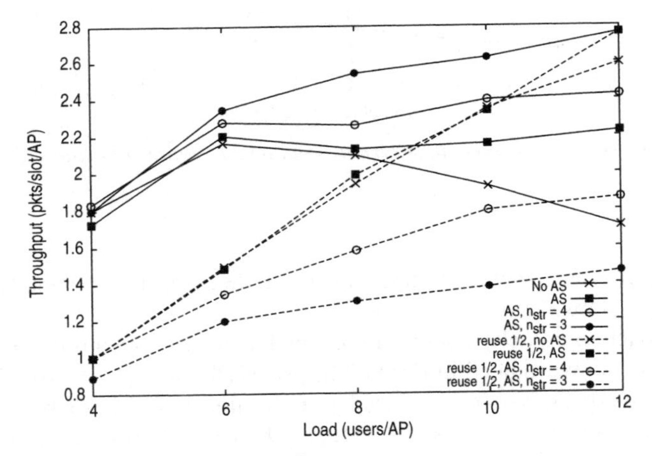

Figure 8.12 *Antenna selection and stream control.*

Finally, it is worth mentioning that some of the concepts related to MIMO communication described in this section, such as multipath transmission and cooperation among networks nodes, can be also applied at the network layer to improve reliability and throughput for the communication across a wireless network. An interesting example is provided in [MiBG03] which proposes the concept of coded route diversity that joins network layer coding and transmission over multiple routes.

8.4 Methodologies for performance evaluation of radio networks

This section begins by extending theoretical connectivity results concerning distance separating mutually audible nodes and numbers of such nodes presented to COST 259 and applying to a situation of log-normal shadowing and inverse power law attenuation to a more general situation which includes Ricean (and Rayleigh) fading and the so-called Suzuki distribution as special cases. There follow applications considering the effect of a hot spot, the number of mobiles using a BS, and the number of BSs in an active set window. A further subsection shows how a Markov chain representation of a protocol may be developed. Finally results concerning radio resource management and mobility models are reported.

8.4.1 Theoretical connectivity results

In this section mobiles and BSs are both assumed to be uniformly and randomly distributed on the (infinite) plane, but with different densities, ρ_M and ρ_B: more precisely, the probability that there is one node in an area of size δA is $\rho\delta A$ (ρ being equal to ρ_M or ρ_B as appropriate) to first order in δA, and the probability of more than one is of higher order than δA. The power loss in decibels at (random) distance R from a mobile, in the most general case to be considered [OrBa02b], [OrBa03b], is

$$L = k_0 + k_1 \ln R + S + T, \tag{8.1}$$

where k_0 and k_1 are constants, S is a shadowing effect, normally distributed with mean zero, variance σ^2, and $T = k_2 \ln(c + Z_1)^2 + Z_2^2/c^2 + 2$, with Z_1 and Z_2 normally and independently distributed with mean zero and variance 1, and $c \geq 0$. (In fact $k_2 = 10/\ln 10$ in all cases, but the analysis is valid for a general value.) The constant c is introduced to simplify the notation, and the Ricean K-factor is equal to $10\log_{10}(c/2)$. Rayleigh fading corresponds to $K = -\infty$ ($c = 0$); when $K = \infty$ ($c = \infty$) there is no fading at all. Finite values of K (finite positive values of c) give Ricean fading. Log-normal shadowing corresponds to $\sigma > 0$, and there is no shadowing when $\sigma = 0$. The special case of no shadowing ($\sigma = 0$) and no fading ($K = \infty$) is referred to as deterministic, and that of no shadowing ($\sigma = 0$) and Rayleigh fading ($K = -\infty$) is referred to as Rayleigh. The special case of no shadowing ($\sigma = 0$) and Ricean fading (K finite) is referred to as Ricean. The combination of shadowing ($\sigma > 0$) and Rayleigh fading ($K = -\infty$) is referred to as the Suzuki distribution. The combination of shadowing ($\sigma > 0$) and Ricean fading ($-\infty < K < \infty$) is the general case, of which all the others are special cases.

By suitable choice of k_1 it is possible to accommodate an inverse square law relationship between power and distance, or an inverse fourth power law, or any other power law of interest, while one which switches from one power to another at some specified distance is handled by means of different values of k_1 within and beyond that distance. Note that, to simplify the mathematical analysis, the formulation uses natural logarithms rather than logarithms to base 10.

The special case $T = 0$ (or $c = \infty$) is considered in a number of papers presented to COST 259.

Distance between mutually audible nodes

The condition for a mobile M and a BS B to be in communication range is that the loss from M to B does not exceed some specified value l_1, i.e. $L \le l_1$. From this it follows that the joint density of R, S, Z_1, and Z_2, is

$$f_{RSZ_1Z_2}(r, s, z_1, z_2) = 2re^{-2(l_1-k_0-s-t)/k_1} \frac{1}{\sigma\sqrt{2\pi}} e^{-s^2/(2\sigma^2)} \frac{1}{\sqrt{2\pi}} e^{-z_1^2/2} \frac{1}{\sqrt{2\pi}} e^{-z_2^2/2}$$

for

$$-\infty < s, z_1, z_2 < \infty, \qquad 0 \le r \le e^{(l_1-k_0-s-t)/k_1}$$

or, equivalently,

$$0 \le r < \infty, \quad -\infty < z_1, z_2 < \infty, \quad -\infty < s \le l_1 - k_0 - k_1 \ln r - t$$

The procedure used to derive this is the same as that already used in the special case, namely to obtain the conditional distribution of R given the other variables (uniform within a finite circle) and then deduce the joint density from the known densities of the other variables.

This leads, after some algebra, to the density of R,

$$f_R(r) = 2K_0 r \int_{-\infty}^{\infty}\int_{-\infty}^{\infty} \frac{1}{\sqrt{2\pi}} e^{-z_1^2/2} \frac{1}{\sqrt{2\pi}} e^{-z_2^2/2} \Phi(a_0 - b_1 \ln r - t/\sigma) dz_1 dz_2 \tag{8.2}$$

for $0 \le r < \infty$, with $K_0 = e^{-(2/k_1)[l_1-k_0-\sigma^2/k_1]}$, $\Phi(x) = \int_{-\infty}^{x} (1/\sqrt{2\pi}) e^{-x^2/2} dx$ and $a_0 = [l_1 - k_0 - 2\sigma^2/k_1]/\sigma$, $\quad b_1 = k_1/\sigma$, with $t = k_2 \ln \frac{(c+z_1)^2+z_2^2}{c^2+2}$

This is the density of the distance between a BS and a mobile which can hear each other, with shadowing and fading effects separated out. There are various special cases.

- The Ricean case arises when there is no log-normal shadowing, so $S = 0$. Then the density of R is

$$f_R(r) = 2K_0 r \iint_{-\infty < t \le l_1 - k_0 - k_1 \ln r} e^{2t/k_1} \frac{1}{\sqrt{2\pi}} e^{-z_1^2/2} \frac{1}{\sqrt{2\pi}} e^{-z_2^2/2} dz_1 dz_2 \tag{8.3}$$

- The Suzuki case corresponds to $c = 0$, and the integrals can then be simplified by defining $U = (Z_1^2 + Z_2^2)/2$, when

$$f_R(r) = 2K_0 r \int_0^{\infty} u^{2k_2/k_1} e^{-u} \Phi(a_0 - b_1 \ln r - b_2 \ln u) du \tag{8.4}$$

  for $0 \le r < \infty$, with $b_2 = k_2/\sigma$, a result earlier obtained directly in [OrBa02b].
- In the special case $\sigma = 0$, which is the Rayleigh case, this becomes

$$f_R(r) = 2K_0 r \int_0^{(rK_0^{\frac{1}{2}})^{-k_1/k_2}} u^{2k_2/k_1} e^{-u} du \tag{8.5}$$

  for $0 \le r < \infty$.

Number of audible nodes

The probability distribution of the number of BSs which can hear a given mobile is obtained, again as in the COST 259 papers, by defining a generating function for the probability that there are exactly n stations within distance r of the mobile and with loss $\leq l_1$, and solving the resulting partial differential equation. Define

$$\mathcal{F}(a, b_1, b_2, c; r) = \int_{-\infty}^{\infty}\int_{-\infty}^{\infty} \frac{1}{2\pi} e^{-(z_1^2+z_2^2)/2} \Phi(a - b_1 \ln r - t/\sigma) dz_1 dz_2 \tag{8.6}$$

which is $P(S + T \leq l_1 - k_0 - k_1 \ln r)$; and then

$$\int_{-\infty}^{\infty}\int_{-\infty}^{\infty} \frac{1}{2\pi} e^{-(z_1^2+z_2^2)/2} \left[r^2 \Phi\left(a - b_1 \ln r - \frac{t}{\sigma}\right) \right.$$
$$\left. - e^{\frac{2(a-t/\sigma)}{b_1} + \frac{2}{b_1^2}} \Phi\left(a - b_1 \ln r - \frac{t}{\sigma} + \frac{2}{b_1}\right) \right] dz_1 dz_2 \tag{8.7}$$

which will be defined as $\Psi(a, b_1, b_2, c; r)$, is an indefinite integral of $2r\mathcal{F}(a, b_1, b_2, c; r)$. Consequently, the number of audible BSs satisfying the above requirement has a Poisson distribution with mean $\pi\rho[\Psi(a, b_1, b_2, c; r) - \Psi(a, b_1, b_2, c; 0)]$.

The limit of $\Psi(a, b_1, b_2, c; r)$ as $r \to \infty$ is zero, so that the total number of stations with loss not exceeding l_1 has a Poisson distribution with mean

$$-\pi\rho\Psi(a, b_1, b_2, c; 0)$$

In the Suzuki case ($c = 0$) and provided $2k_2/k_1 < 1$, this is equal to

$$\pi\rho e^{2(\sigma^2/k_1 - k_0)/k_1} e^{2l_1/k_1} \Gamma(1 - 2k_2/k_1) \tag{8.8}$$

It is worth pointing out that in the case $k_2 = 0$, when only log-normal fading is present, this result agrees with that already obtained for that special case. Also the result does not apply for the inverse square law, for then $2k_2/k_1 = 1$, and the required inequality just fails.

8.4.2 Some applications

The number of audible BSs in a hot spot

The analysis above is used in [OrBa03a] to show that the distribution of the number of audible BSs within an annulus between distances r and $r + \delta r$ and subtending an angle $2\theta(r)$ at the mobile has a Poisson distribution with mean

$$\mu_{r,r+\delta r;\theta} = \theta(r)\rho\left[\Psi(a, b_1, b_2, c; r + \delta r) - \Psi(a, b_1, b_2, c; r)\right] \qquad 0 \leq \theta \leq \pi \tag{8.9}$$

which can be written

$$\mu_{r,r+\delta r;\theta} = \theta(r)\rho \frac{\delta\Psi(a, b_1, b_2, c; r)}{\delta r} \delta r \qquad 0 \leq \theta \leq \pi \tag{8.10}$$

This result makes it possible to consider a specified area over which the density of BSs, while remaining constant, differs from that over the rest of the plane. The former density may be the greater (resulting from the operator's response to an increased intensity of mobile traffic in the same area), and the area may then be called a hot spot. Consequently, if a hot spot lies between radii r_1 and r_2

with its boundary at distance r defined by $\theta(r)$ (which, for a convex region, merely requires that the boundary can be expressed in terms of polar coordinates so that $2\theta(r)$ is the angular distance between the two boundary points at radius r), then the number of audible BSs within the hot spot has a Poisson distribution with mean

$$\mu_{r_1,r_2;\theta(r)} = \int_{r_1}^{r_2} \theta(r)\rho \frac{d\Psi(a, b_1, b_2, c; r)}{dr} dr \tag{8.11}$$

and since $\Psi(a, b_1, b_2, c; r)$ is an indefinite integral of $2r\mathcal{F}(a, b_1, b_2, c; r)$,

$$\mu_{r_1,r_2;\theta(r)} = \int_{r_1}^{r_2} 2\theta(r)\rho r\mathcal{F}(a, b_1, b_2, c; r)dr \tag{8.12}$$

These methods can be applied to the case of the two-slope model (in which the power of the attenuation law changes at some specified distance).

Number of mobiles attached to a BS

Mobiles and BSs are both assumed to be uniformly and randomly distributed on the (infinite) plane, but with different densities, ρ_m and ρ_b. The number of BSs which can hear a given mobile has a Poisson distribution with mean μ_b and equivalently the number of mobiles which can hear a given BS has a Poisson distribution with mean μ_m. These may also be found from any loss model previously described, including two-slope models and models with hot spots. It is shown in [OrBa02a], [VOZB02] that the probability that a given BS is the BS receiving the strongest signal from a given mobile is $1-e^{-\mu_b}/\mu_b$. The mobile is then defined as being attached to the BS. Defining the last probability as p, it is then shown that the number of mobiles which are both audible to a given BS and for which that BS is the BS receiving the strongest signal has a Poisson distribution with mean $\mu_m p$. Since μ_m is the mean number of mobiles audible to a base station and p is the probability that a BS is the base station receiving the strongest signal from a mobile audible to it, this is not an unexpected result.

This result is applied in [OrZB02], [VOZB02] to evaluate the blocking probability of a user served by a network of base stations characterised by a given hard capacity, in other words a maximum to the number of users which can be served by a base station. Access to one BS is modelled as an Engsett loss system, and the Engsett formula then leads to the blocking probability. Numerical results and graphs are provided for a Bluetooth Scatternet.

Number of BSs in an active set window

The active set of a mobile is here defined as the set of BSs which can hear that mobile with the additional requirement that the signal received has a strength within a specified distance (u dB, say) of the largest one received. This range will be described as an active set window of size u. Using the single-slope model with log-normal shadowing, the mean number of audible BSs is $\mu_1 = ke^{2l_1/k_1}$, with $k = \pi\rho e^{2(\sigma^2/k_1 - k_0)/k_1}$. Let Q_m be the probability of at least m BSs in an active set window of size u. This leads to the recurrence relation,

$$Q_{m+1} = (1-p)Q_m - (1-p)^m \mu_1^m e^{-\mu_1}/(m)!, \quad m \geq 1 \tag{8.13}$$

with $Q_1 = 1 - e^{-\mu_1}$, and defining $p = e^{-2u/k_1}$.

Then the probability of exactly m BSs in the window is

$$P_m = Q_m - Q_{m+1} \tag{8.14}$$

and this is equal to

$$pQ_m + (1 - p)^m \mu_1^m e^{-\mu_1} \tag{8.15}$$

for $m \geq 1$.

Thus the probability depends only on p and μ_1.

In particular, when $m = 1$,

$$P_1 = p(1 - e^{-\mu_1}) + (1 - p)\mu_1 e^{-\mu_1} \tag{8.16}$$

When μ_1 is large the result depends only on p, and this is likely to cover situations of practical importance, since μ_1 is the mean number of audible BSs.

The results are developed in [MOBB04], where this statistically based analysis is compared with simulated results based on the deterministic model, and considerable agreement is found.

8.4.3 Markov chain models for protocols

Three papers introduce a Markov chain representation of the various transmitting and receiving states which can be occupied by a node operating under an ALOHA [OrBa04a] or a Carrier Sense Multiple Access with Collision Avoidance (CSMA/CA) [OrBa04b] or a Request to Send (RTS)/Clear to Send (CTS) [OrBa05] protocol. The Markov chain representation gives in particular a steady state probability for occupation of any state and hence offers a means of investigating how the throughput of the protocol relates to the level of traffic. This analysis is based on the statistical model described above. Before this it is necessary to devise a set of states which permit a Markov representation. The ALOHA case is the simplest (or at any rate the smallest) and will illustrate the procedure.

In order to set up a Markov chain time is partitioned into discrete time slots whose length is the length of the acknowledgement packet. Variable length data packets are handled by setting a fixed probability of ending in any given time slot: this is equivalent to an exponential distribution for packet length or a geometric one for length in time slots. A node will be assumed to occupy the state in which it begins a slot for the length of that slot, and may transit to a different state at the end of the slot. In the ALOHA protocol a group of five states, labelled IDLE, Listen for DATA, Transmit DATA, Listen for ACK, and Transmit ACK, is self-explanatory, noting that DATA is an abbreviation for a part of a data packet transmitted in a single time slot, while ACK refers to a transmitted acknowledgement, occupying one time slot.

The next group introduces the concept of the doomed state, which does demand some explanation. Consider, for example, a node which completes transmission of a data packet and therefore expects, or at least hopes for, an acknowledgement. There are two reasons why this may not arrive: either it is not sent, because the receiving node failed to receive some or all of the data packet, or it is sent, but suffers interference. At first sight, therefore, calculation of the probability of non-receipt involves retrospective consideration of what may have happened at some undetermined point in the data transmission, which undermines the Markovian requirement. The introduction of the doomed state concept eliminates this uncertainty: for example, if in any slot the data reception is interfered with by an acknowledgement, then the sender moves from the 'Transmit DATA' state to the 'Transmit DATA doomed' state and the receiver from the 'Listen for DATA' state to the 'Listen for DATA doomed' state. Upon completion of this transmission the sender moves either from the 'Transmit DATA' state to the 'Listen for ACK' state (and in this situation the acknowledgement is sent but may suffer interference), or from the 'Transmit DATA doomed' state to the 'Listen for ACK doomed' state (and here no acknowledgement is sent). In neither case is any retrospective investigation necessary in order to calculate transition probabilities. If the interference is caused by a data packet, then the doomed state is called 'Listen for DATA/INT doomed', the reason for the distinction being that now the interference continues into the next slot and on until the interfering packet is completed, at which point the listening node enters the 'Listen for DATA doomed' state, where the message is compromised although no interference is actually occurring. The distinction is necessary because the presence of an interfering transmission can affect the transition probabilities. There are thus four doomed states, Listen for DATA doomed, Listen for DATA/INT doomed, Transmit DATA doomed, Listen for ACK doomed.

Finally, there is a Backoff state, which is entered after an expected acknowledgement fails to materialise. Like the packet length, its length is allotted an exponential distribution, with consequent fixed probability of ending in any given time slot.

There remains the possibility of backlogged states and, if necessary, doubly or even triply backlogged ones.

Transition probabilities between the various states may be regarded as depending on two major components. The first comprises the values of the various parameters defining the system – arrival rates for data and acknowledgements, mean packet length, node density – and the second the so-called configuration probabilities, which represent the probabilities that the nodes involved – transmitter, receiver, and any potential interferer – are geographically situated in relation to each other in such a way that the corresponding transition will occur. The statistical analysis described in Subsection 8.4.1 makes it possible to give the probability that for example a receiver is within range of a transmitter and simultaneously is far enough away from a potential interferer to preclude interference. Such a probability is given as a multivariate integral over normal variables representing the log-normal shadowings involved and also over the distances separating the nodes.

It is therefore possible formally to give transition probabilities between states. Transition probabilities from a backlogged state are very closely related to those from the corresponding unbacklogged state, so the only complication introduced by backlog is an increase in the dimensions of the matrix of transition probabilities, but fortunately this remains manageably small. On the other hand, unfortunately, the integrals involved in the configuration probabilities have so far resisted numerical evaluation, and the results remain for the moment of theoretical rather than practical interest.

The extension of the method to Carrier Sense Multiple Access (CSMA) and to RTS/CTS protocols requires an increase in the number of states, but no new principles.

8.4.4 Radio Resource Management and mobility models

The impact of economic parameters on Radio Resource Management is examined in [BLZZ03]. A model of users' satisfaction in terms of utility and cost is presented and allows investigation of the relationship between Radio Resource Allocation and the revenue of the provider. Established economic models are included in the Radio Resource Allocation scenario, and numerical results obtained from this analysis confirm that the network management is significantly affected by the economic scenario.

Finally, [DiBQ04] and [DiOe04] extend an existing mobility model for vehicle-borne terminals in urban, suburban, and rural environments to introduce a method of allowing for the case where a car stops at crossroads with or without traffic lights. Simulation analysis produced realistic and plausible values for dwell time and handover rate with, in particular, visible discrepancy between rural and non-rural values.

8.5 Services and traffic modelling

Service characterisation and traffic modelling constitute important steps towards understanding users' demand of mobile and wireless communication systems. They constitute an essential input for planning tools, as well as static and dynamic simulators, in order to investigate ways to efficiently explore resources in order to meet demand. These items are addressed in this section.

8.5.1 Mobile services

Wireless communication systems have grown and evolved enormously. Users claim the availability of new and more services that efficiently satisfy their needs of communication, neglecting the underlined systems in use. Side by side with this idea, a large variety of systems is already available nowadays, and many others, under development, are foreseen in the near future. They appear as suitable solutions for specific issues: WLANs offering high data rates in hot-spot areas (e.g. IEEE 802.11x), cellular systems providing large coverage for speech and some multimedia services (e.g. GSM and UMTS), satellite systems supplying planetary coverage (e.g. Globalstar), and broadcast ones affording high downlink data rates (e.g. DVB-*x*). Their common goal is to provide a diversified set of services to mobile users.

In the 4G concept, this goal is even more emphasised. Both legacy and new systems will constitute possible access interfaces to a common IP-based core network. Services will be offered via the core IP network, not being coupled with any particular system. Services will be available to end users via multisystem terminals, guaranteeing to the user in a transparent way that, for a given service, the best system (in terms of, e.g. price or performance) will be chosen.

Most people now agree that there will not be a single service, the *killer* service, which will make the wireless communications market, but rather a number of services. In [FeCo03], a so-called *killer cocktail* of eight services is proposed and characterised in detail, as illustrated in Figure 8.13. These services are: Speech-telephony, Video-telephony, Streaming MultiMedia (MM), World Wide Web (WWW), Location Based service, File Download, E-Mail and Multimedia Messaging Service (MMS). This set of services is heterogeneous enough, both to meet the foreseen diversified demand of mobile users, and to translate in simulations the different traffic patterns systems should bear.

Beyond 3G, systems will be able to support new applications, with different capacity and requirements. An example is the overview of services supported by Enhanced-UMTS (E-UMTS), an UMTS technological evolution step that provides bit rates higher than 2 Mbps, presented in [FeVe04]. Nearly 30 services were considered, grouped into Sound, High Interactive Multimedia, Narrow-, Wide- and Broadband categories. They are an example of a mixture of applications that may exist in E-UMTS. A selection of the most relevant applications of each group is brought to the foreground, decreasing the burden on simulation work: Voice, Video-telephony, Multimedia Web Browsing, Instant Messaging for Multimedia, Assistance in Travel, High Definition (HD) Video telephony, WLAN Interconnection. It can be seen that this set is very similar to the one presented in Figure 8.13.

In order to better analyse and characterise the panoply of services, various perspectives on services' classification have been proposed by different bodies: ITU-T [ITU-93], ETSI [ETSI98a],

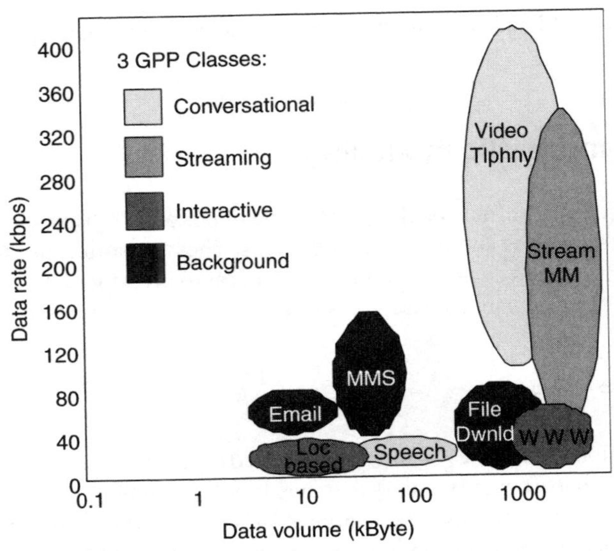

Figure 8.13 *The killer cocktail of services. 3GPP classification, and two differentiating characteristics: bit rate range and downlink session volume [FeCo03].*

UMTS Forum [Foru98] and 3GPP [3GPP99]. Currently, the most popular is the 3GPP one, grouping services according to their characteristics and performance requirements in the following classes:

- Conversational: symmetric and real-time conversational pattern services.
- Streaming: real-time and almost unidirectional data flow with low delay variation, which can be processed as a steady continuous stream.
- Interactive: request-response pattern services, highly asymmetric.
- Background: non-real-time asymmetric services, where destination is insensitive to delivery time.

These classes are represented in Figure 8.13. By analysing the two sets of services proposed in [FeCo03] and [FeVe04], one can see that the four 3GPP service classes are well represented in both cases, evidencing the diversity of service characteristics and traffic patterns of the sets.

8.5.2 Traffic source modelling

In the study and simulation of mobile communication systems, the proper simulation of services behaviour is essential. The central idea in service traffic source modelling lies in constructing models that capture the important statistical properties of the services' behaviour, also taking into consideration the purposes for which they are modelled. As exemplified in Subsection 8.5.1, two types of simulation approaches exist: static and dynamic ones. In static simulations (used, e.g. for cellular planning studies), the characterisation of the service data volume may be sufficient for the construction of a snapshot of users. On the other hand, when performing dynamic simulations (in order, e.g. to dimension RRM functions or assess detailed QoS parameters), detailed traffic source models are required to simulate the dynamic behaviour of each service: while one service may generate a continuous stream of traffic, another generates sporadic bursts of packets.

A plethora of source models may be found in the literature for a variety of services. Nevertheless, statistical features of a service are better synthesised by certain models than others. In fact, many existing models are based on limiting constraints, others are inappropriate for the systems or technologies under simulation. In [SeCo03], [AgCo04], [VePa04] and [MGFC04], overviews of important traffic source models for specific services are presented. In particular, [MGFC04] indicates the traffic source models adopted within the COST 273 MORANS initiative, presented in Subsection 8.5.1. A selection of the most important and useful traffic source models, for typical services, is presented and shortly described below. The purpose of the current section is to give an overview on the most important existing models; references pointing to a complete and exhaustive description of each model are included. It must be noted that some services have more than one associated source model, each of them appropriate to specificities of certain systems or technologies.

The Speech-telephony service may be described by several traffic source models, depending on the technology under consideration:

- Considering a system that establishes a Circuit Switch (CS) connection for a speech call, the two-state active-inactive (ON-OFF) Markov model [Yaco93], with periods modelled by exponentially distributed variables, is the most appropriate and simple model, largely used in simulations.
- In a CDMA system, the effects of the Adaptive Multi-Rate (AMR) speech encoder, the compression device and the air interface characteristics are well captured by the four-state Markov measurement-based model [VaRF99]. Periods are modelled by a Weibull distributed variable. During each state, a packet of specific size is generated each 10 ms. A new state is then selected with a specific probability.
- In the case of a Voice over IP (VoIP) connection, the two-state ON-OFF Markov model with periods modelled by exponentially distributed variables [Mand01] is the most appropriate. The packet size depends on the considered speech codec and packet rate.

For the Video-telephony service, a good reference is 3GPP [3GPP01], which has specified the use of the ITU-T H.261 or H.263 video codecs for the generation of video signals, supporting efficient compression that enables low-bit rate video. 3GPP has also specified that, for associated audio signals, the terminals shall support (mandatory) AMR speech codec [3GPP01] (properly modelled, as already described, by the four-state Markov model). This results in two possible models, depending on using a Variable Bit Rate (VBR) or Constant Bit Rate (CBR) communication:

- The Gamma Beta Auto-Regressive (GBAR) source model [Heym97] is based on two statistical features that can be observed in H.261 and H.263 VBR traffic [LáGD00]: the marginal distribution follows a Gamma distribution, and the autocorrelation function is geometric. The model suggests a first-order autoregressive process that relies on these statistical features.
- In [NyJO01], a model that mimics an H.263 codec with constant frame rate and specified target bit rate is proposed. The considered target bit rates are assumed to be suitable for mobile access (e.g. 32 kbps) in, e.g. the GSM/EDGE Radio Access Network (GERAN) and UTRAN. The proposed model is based on a simple linear function with just a few parameters. The model generates video frames at a certain specified rate. The size of the video frame after coding depends on the target bit rate of the data connection. The parameters are state dependent, and the states are controlled by a Markov chain.

Concerning the Video-streaming service, 3GPP has specified the use of MPEG-4 and H.263 codecs in UMTS [3GPP02b]. For the associated audio signals, 3GPP has specified the mandatory use of

AMR narrowband speech codecs [3GPP02b]. Two types of models are identified:

- In [FrNg00], an MPEG video source model with the features of the GBAR one is presented, explicitly accounting for the presence of Group Of Pictures (GOP) cyclicity, based on real MPEG-4 trace statistics.
- A novel approach, recurring to wavelets, is presented in [LáGD01]. In fact, any kind of video source, i.e. MPEG-4, can be analysed at multiple levels of resolution, by capturing the statistical features of the data being analysed through wavelet transform coefficients. A video sequence can then be obtained by inverse wavelet transforming from created coefficients.

For the WWW service, the following models are available:

- The model adopted by 3GPP [ETSI98a] considers a sequence of packet calls during a session. The user initiates a packet call when requesting an information entity. During a packet call, several packets may be generated, constituting a bursty sequence of packets. The number of packet calls per session, the reading time between packet calls, and the number of packets within a packet call, are geometrically distributed random variables. The interarrival time between packets (within a packet call) is inverse Gaussian distributed. The size of a packet is Pareto distributed.
- Several studies from local area networks [LTWW94] and wide area networks [PaFl95] have shown that WWW traffic data should not be described as a Poisson process, due to the self-similar properties observed in empirical traffic traces. Specifically, these studies have demonstrated that augmenting the traffic load increases burstiness of WWW traffic, whereas exponential traffic, as proposed in [ETSI98a], smoothes. Furthermore, the statistical features of these empirical traces show a long-range dependence, not considered by the exponential model. The model proposed in [BaCr98] exhibits self-similar properties. A session starts by a submission of a Uniform Resource Locator (URL) request by the user. When all the requests related to that URL are completed, the user will take some time to read this information before beginning other requests. The sending of this information request corresponds to the ON-active period, while the users' 'think time' corresponds to the OFF-inactive period. One URL request consists of several files that are transferred in different TCP connections (Hyper Text Transfer Protocol (HTTP) 1.0). The time elapsed between closing a TCP connection and opening the next one to transfer the next object of the same page is called OFF-active time.

For e-mail and file download, a simple traffic source model is available:

- E-mail service is modelled by a two-state ON-OFF Markov model [KlLL01], with periods modelled by exponentially distributed variables; the packet call interarrival time is Pareto distributed. For file download, there is a single packet call within a session. Within each ON period, the packet arrival process is completely captured by the packet interarrival times (log-normal distributed variable) and the corresponding packet sizes (within four sizes, each with a specific probability).

A mid-term approach (in terms of simplicity and assumptions) to model services behaviour is synthesised in [FeVe04]. It consists of modelling the activity within a call/session of any service by a two-state ON-OFF Markov model, defining an average duration of each period together with an adequate statistical distribution and the corresponding file sizes of activity periods, as shown in Table 8.11. It may be noted that Video-telephony service does not exhibit inactive states, therefore the ON state is equal to the call duration and the OFF state duration is zero.

Table 8.11 *Services activity parameters (extracted from [FeVe04]).*

Services		Active state (ON)			Inactive state (OFF)	
	Av. dur. [min]	Av. dur. [s]	Size [kB]	Distr.	Av. dur. [s]	Distr.
Speech	3	1.4	2.1	Exp.	1.7	Exp.
Video-telephony	3	180.0	–	–	0.0	–
Multimedia Web Browsing	15	5.0	240.0	Pareto	13.0	Pareto
Instant Messaging for MM	15	5.0	640.0	Weibull	90.0	Pareto
Assistance in Travel	20	60.0	11520.0	Weibull	14.0	Pareto
HD Video-telephony	30	1800.0	–	–	0.0	–
WLAN Interconnection	60	5.0	7988.0	Weibull	1.0	Pareto

8.5.3 *Spatial traffic demand estimation*

The estimation of users' service demand is a strategic issue, since it is directly related to the capacity to be supported by the network. Nevertheless, this is a non-trivial issue, since, for emerging services, there is no real data to build traffic demands. The intensity of usage of each service may be characterised by the service session arrival rate, modelled by a Poisson process [Yaco93]. The average value characterising this distribution is typically the Busy Hour Call Attempt (BHCA). One can have service BHCA values per user or for a certain area under study.

In [FeVe04], BHCA values per cell are estimated for eight different E-UMTS deployment scenarios: Business City Centre (BCC), urban residential, primary roads, trains, commercial zones, offices, industry, and home. For each deployment scenario, service penetration and usage values are proposed, as formulated in Table 8.12 for a subset of scenarios.

Table 8.12 *Services usage in each simulation deployment scenario (extracted from [FeVe04]).*

Services usage [%]	Office	BCC	Vehicular
Speech	25.0	27.0	42.0
Video-telephony	15.0	16.0	16.0
Multimedia Web Browsing	20.0	26.0	18.5
Instant Messaging for Multimedia	25.0	0.0	0.0
Assistance in Travel	0.0	0.0	23.5
HD Video-telephony	0.0	31.0	0.0
WLAN Interconnection	15.0	0.0	0.0

Population densities are also suggested for the office, BCC and vehicular scenarios (0.150, 0.031 and 0.012 users/m^2, respectively). The estimation of these values results from extrapolations and adaptations of several studies performed for narrow-, wide- and broadband applications in the residential market of fixed networks [OZSI96], [StMu95], as well as for wireless [ETSI98a], and mobile communications, such as UMTS Forum [Foru98] and RACE-Mobile Broadband Systems – RACE Project (MBS) [RoSc94]. For each service, BHCA values are obtained, representing the total number of call attempts by all users covered by a cell. BHCA values depend on the number of potential users and on the session arrival rate per user that characterises each service and deployment scenarios.

Another approach is described in [FeCo03], where the usage of a set of services on a geographical area is estimated using a Geographic Information System (GIS) tool. It combines geographic, demographic, and market segmentation concepts, as well as service usage profiles. The complete processing is illustrated in Figure 8.14. First, the usage of the *killer cocktail* of services, presented in Figure 8.13, is characterised for business and consumer customer segments. This usage is based on the monthly average number of calls and their estimated concentration in the busy hour, adapted from [Foru03]. Average service duration and data rate values are also customer segment dependent. Second, realistic spatial distributions of users of each segment are computed. For each unit of geographical area (a pixel of, e.g. $20 \times 20\,m^2$), the number of users is computed taking into consideration: the number of estimated persons within that unit of area; the penetration of subscribers, per segment, and the operator's market share; the operational environment (e.g. rural or business area) and the associated customer segment share (percentage of business and consumer subscribers in that specific operational environment).

By combining these two components, traffic demand maps are computed for each service and user type. For each pixel, they provide not only the BHCA, and the average number of active users, but also the UL and DL offered traffic, when combined with the average service traffic volume per link.

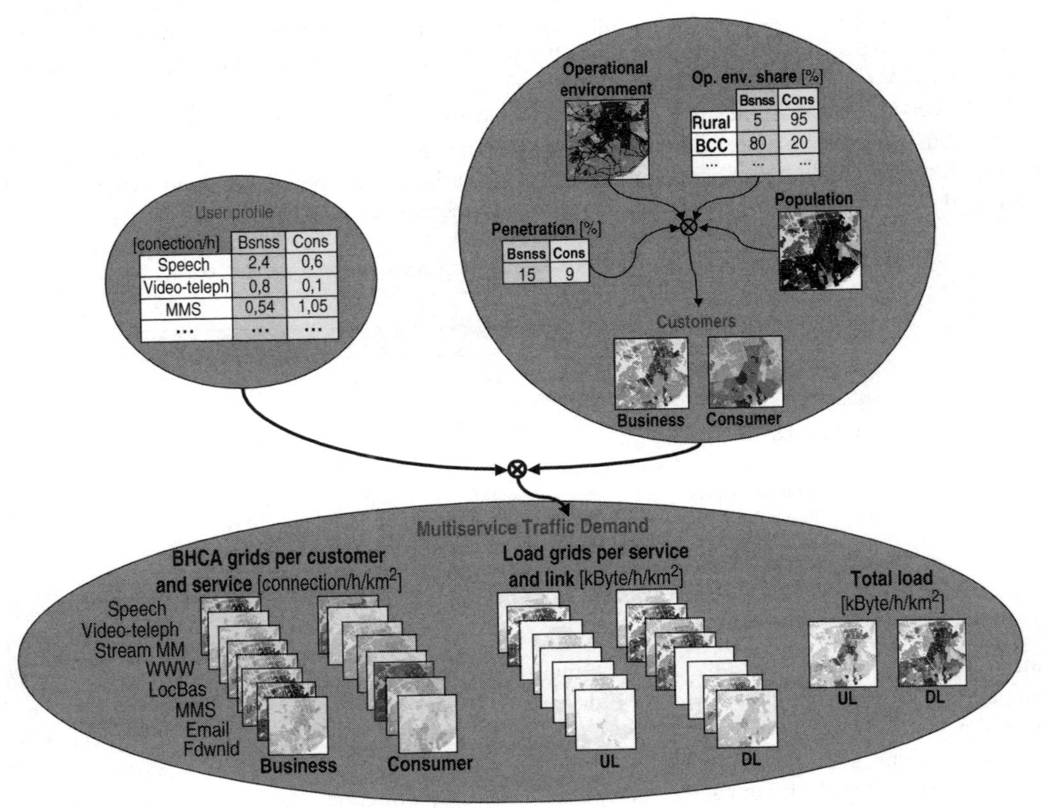

Figure 8.14 *General procedure for the construction of a multiservice traffic demand scenario (example of the city centre of Lisbon) [FCXV03].*

This work [FCXV03] was developed within the IST-MOMENTUM project [MOME03]. Traffic scenarios were built for several European cities; in particular, scenarios for Lisbon, Berlin and The Hague are public and available in XML format at the MOMENTUM site. These three scenarios were included as Reference Scenarios of the MORANS initiative [MuRV04], [MORA02], presented in Subsection 8.5.1. Planning tools, as well as static and dynamic simulators can easily use these traffic scenarios to investigate ways to efficiently explore resources in order to meet demand.

In [VaRF99], a traffic estimation model is developed, where other population characterisation parameters, such as salary, age and education, are also considered. These parameters are multiplied by their respective penetrations in order to generate a non-uniform traffic distribution for users and services in each zone. For these traffic distributions an application gives an estimate of the cell radius in order to take the traffic. It was concluded that, in the zones more densely populated, the cell's radius is smaller due to maximum levels of interference, although this application considers a static approach.

8.5.4 Temporal traffic demand prediction

The performance of a mobile communications network is very sensitive to the incoming traffic; therefore, knowledge on the expected traffic evolution is beneficial. If one observes the traffic load of a cell, periodicity in traffic is noticeable, Figure 8.15, allowing its prediction by examining the autocorrelation coefficients of the traffic sequence [ShBr88]. The autocorrelation function indicates the existence of a linear relation between each traffic sample and its neighbouring ones. In particular, for time shifts equal to one week autocorrelation reaches fairly high values. Two algorithms are proposed, providing an in-advance estimate of the incoming traffic. A linear algorithm, [CFAP01], estimates present unknown traffic load as a linear combination of previous known traffic loads $t(n)$,

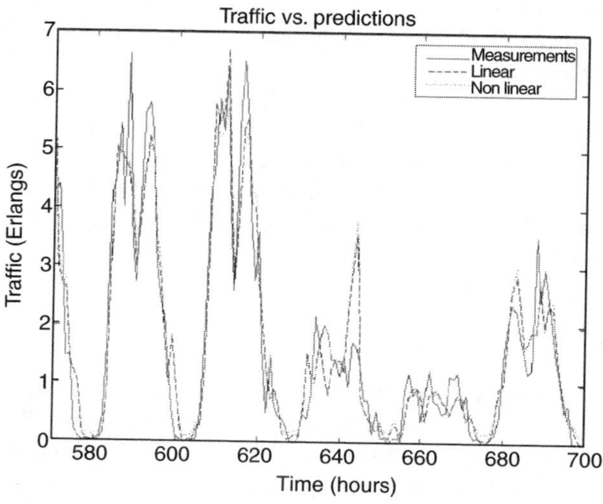

Figure 8.15 *Traffic evolution along several days in one cell, together with the traffic prediction obtained one hour in advance [MFVC02].*

taking into account the high autocorrelation for time shifts of one week (168 hours):

$$\hat{t}(n) = \sum_{i=1}^{l} a_i \cdot t(n-i) + \sum_{i=-p}^{p} b_i \cdot t(n-168-i) \quad (8.17)$$

where n is the present sample, l is the number of considered precedent samples, and p is the half-length of the interval around the same time for the previous week. Optimum a_i and b_i values are estimated recurring to known traffic loads $t(m)$ for times $m < n$ and solving a linear system. A non-linear algorithm, [MFVC02], improves these predictions by incorporating a non-linear function to the output linear predictor. In Figure 8.15, predictions are presented for one hour in advance for an interval of traffic measurements of one cell.

Both algorithms produce good results for short time advances, proving the success of the traffic predictor in a real network. In particular, the non-linear procedure gives lower prediction errors, and this improvement rises when predictions are made more in advance. The procedure is conceptually simple and easy to implement. Significant enhancements of network performance are achievable by adapting the system parameters to the expected future network conditions.

In [FMFC05], the authors have checked that the number of GPRS users that demand data connection services is also predictable, when statistics about their behaviour are available. The predictor is an artificial-neural-network-based algorithm, a feed-forward network called Finite Impulse Response (FIR)-MLP (Finite Impulse Response – Multilayer Perception) is used. This is formed by some layers of computational nodes (neurons), whose synaptic coefficients have been substituted by FIR filters. The filters add memory to the network and allow for a more complex non-linear modelling. The network learning algorithm is called Temporal Back Propagation, and consists of an extension of the well-known Back Propagation that allows the introduction of the FIR filters.

The proposed configuration consists of one neuron with two inputs, one corresponding to information of the previous hour and the other to the previous week. The chosen level of complexity for the neural network is two neurons. A dynamic adaptation of the learning parameters is achieved by using an algorithm called Extended Delta-Bar-Delta. As a result, the velocity of neural network readjusts quickly in each iteration, and the network memory fits in each iteration. These two improvements allow the network to correctly fit all the user distributions at the cost of a higher computational load for the network learning algorithm.

Figure 8.16 illustrates the performance of the neural predictor through comparison of real traffic (number of users) evolution and prediction for one cell. Two distinct phases can be distinguished along time. At first, it takes about 200 hours for the predictor to adapt itself to the specific set of data and, afterwards, it is able to adequately follow traffic variations. Further work shall concentrate on improving the prediction tool in order to adapt it better to other aspects of traffic, such as UL or DL traffic loads in terms of kilobits.

8.6 Wireless LANs

A number of results on radio network aspects in WLANs have been presented during the COST 273 Action. A major block of interconnected results are on LA and Transmit Power Control (TPC) in HIPERLAN Type 2 (H2) networks [Radi01], [Radi02], [RaJo03], [Radi04], [RaJo05].

Other papers with their main focus on radio aspects and, hence, with their main description in this section are [ChEs03] and [FiBa04]. The papers [NaCB03] and [WSWH04] have their main focus in other areas and, therefore, this section contains only short descriptions of their radio network aspects.

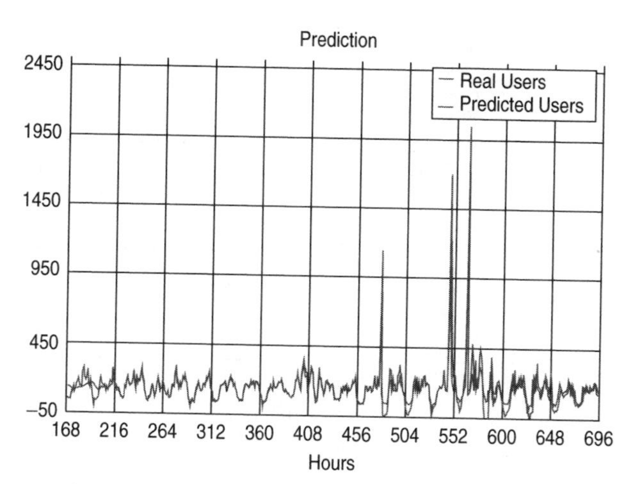

Figure 8.16 *Neural network output (the moment of coupling is between the hour 200 and 220) (extracted from [MFVC02]).*

Link adaptation and transmit power control in HIPERLAN type 2

The engineering of wireless data networks is often based on a soft transition between good and bad state, as opposed to voice networks with a more abrupt transition [RaJo05]. This smooth transition function represents the usefulness or, as we will call it in the sequel, utility of the radio connection for the user. This subsection deals with best effort services for which it is well known that the user has to accept what he gets, so a single performance criterion does not obviously exist.

The first utility considered is throughput and [Radi01], [Radi02], [RaJo05] are devoted to investigate the throughput behaviour of H2. The second measure is energy efficiency and a corresponding optimisation scheme is described in [RaJo03]. Additional investigation results of both schemes are presented in [Radi04].

The example system H2 is a centralised system where an AP is responsible for administering all resources. Communication in the centralised mode happens solely between AP and MT. Transmit opportunities in the UL are requested per connection by the MTs based on their buffer states. The AP allocates transmit opportunities in the DL based on its own buffer states. The H2 standard supports LA, i.e. the ability to use different raw transmission rates on the physical layer (also called Phy Modes, *PM*), where high rates are suitable for good and low rates for bad channel conditions. It also supports TPC, where the AP broadcasts its transmit power $P_{t,\mathrm{AP}}$ and its expected receive power $P_{e,\mathrm{AP}}$. The MTs can calculate the path loss *PLOS* from the received signal and adapt its transmission power such that the received power at the AP, $P_{r,\mathrm{AP}}$, is equal to $P_{e,\mathrm{AP}}$. Further information about H2 and its LA and TPC schemes as well as the scheduling algorithm used in all investigations can be found in [Radi04], [RaJo05], [KMST00].

All investigations are based on PER curves from literature [KMST00]. With an ARQ protocol and the raw data rate $\mathcal{R}_{PM}$ per *PM*, (1-PER) can be interpreted as the frame success rate and the throughput after ARQ is $\mathcal{T}_{PM} = \mathcal{R}_{PM} \cdot$ (1-PER). The throughput curves with $\mathcal{R}_{PM}$ as parameter are shown in Figure 8.17. Note that there is always a Phy mode with maximum throughput for a given SINR value. This has been exploited in the throughput-oriented investigations in the sequel.

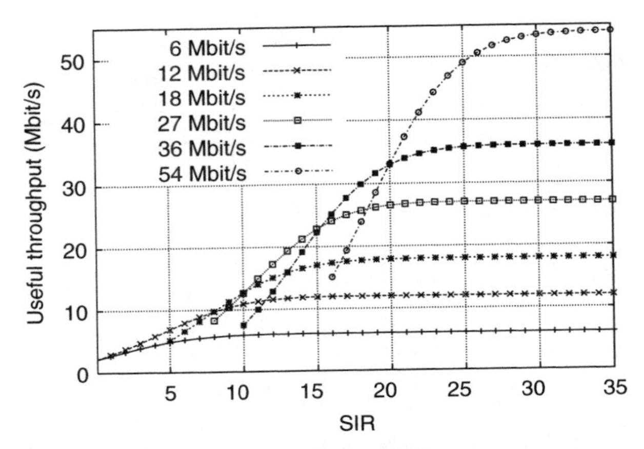

Figure 8.17 *Theoretically achievable useful throughput with $\mathcal{R}_{PM}$ as parameter.*

Intercell interference is inevitable in a cellular network. Given the rules for LA and TPC, an H2 AP can increase the data throughput in its cell by increasing the transmit power, which enables the use of higher data rates. This, however, raises the intercell interference of other cells which, in turn, may increase their transmit power to improve their carrier-to-interference ratio (C/I). Some basic properties of the H2 radio interface with regard to different kinds of interference are analysed in [Radi01]. Based on the course of PER curves for H2 from literature, it investigates whether it is better to transmit the same amount of data during short periods with high power or rather slowly with the lowest transmit power possible. The results indicate that longer transmissions with lower transmit power are better with regard to data throughput after ARQ in interfered cells.

This initial result is used in [Radi02]. H2 is based on MAC frames of equal duration. So if the transmission rates are higher than required, there will be spare resources, indicated by an unused part in the MAC frame which exceeds a certain size. According to the results from [Radi01], it is desirable to decrease transmit power to the minimum. The throughput in Figure 8.17 increases monotonously with the SINR and, assuming constant interference, the SINR is proportional to the transmit power. So in order to decrease transmit powers to the minimum, the algorithm proposed in [Radi02] analyses the unused part in MAC frames and gradually decreases transmit power, until finally the unused part becomes small enough. If the unused part is very small and the buffer states increase, transmit powers are increased again, until the buffers are short enough and the unused part is still small.

This scheme has been investigated and has been shown to operate satisfactorily, unless there is high load in the system. However, it does not consider intercell interference, since the assumption that throughput increases monotonously with transmit powers according to Figure 8.17 is only valid for constant interference. If adjacent radio cells adjust their interference as well, the assumption does not hold any more. All subsequent investigations, therefore, have been conducted for a simple cellular network scenario.

An interesting question in this context is whether the maximum network throughput is achieved at maximum transmission powers. The models in [RaJo05] have been developed to answer this question. Two models are developed for cellular H2 networks for this purpose:

1. A teletraffic model to compute the throughput in a regular cellular network based on H2. The model considers the proposed scheduling algorithm and uses fixed values of the TPC parameter

$P_{t,\mathrm{AP}}$ and $P_{e,\mathrm{AP}}$, i.e. it only considers LA which depends on the SINR at the receiver. The model is based on the computation of the PDF of the SINR at the receiver from the geometry and uses the throughput curves in Figure 8.17 to compute the expected value of the throughput. Models for a single radio cell, for two interfering cells and a regular network are proposed.

2. A simulation model for H2 with models of the physical, the MAC and the ARQ protocol as well as the convergence layer and an implementation of the scheduling algorithm identical to the one used in the analytical model. MTs move in circular areas equally distributed around APs, data packets are generated and transmitted in both directions between AP and MTs.

The analytical model has been developed for high load cases, the parameters of the simulation model have been adjusted accordingly. The results from both models agree quite well and show that, given the scheduling algorithm, the maximum network throughput is not achieved at maximum transmission power. This means that it may be useful from the point of view of throughput to decrease transmit power. For the considered cases, the influence of $P_{t,\mathrm{AP}}$ is much more significant than the one of $P_{e,\mathrm{AP}}$. Additional results from [Radi04] suggest that the ideal value of $P_{t,\mathrm{AP}}$ for maximum mean throughput $\overline{\mathcal{T}}_{\max}$ is constant for constant ratio of distance D between radio cells and the cell radius R. Selected results of the optimum values $P_{t,\mathrm{AP,opt}}$ and $P_{e,\mathrm{AP,opt}}$ to reach $\overline{\mathcal{T}}_{\max}$ with both models for $R = 50\,\mathrm{m}$ and different values of D are listed in Table 8.13.

Table 8.13 *Combination of $P_{t,\mathrm{AP,opt}}$ and $P_{e,\mathrm{AP,opt}}$ with $\overline{\mathcal{T}}_{\max}$ for both models.*

	Theory			Simulation		
D [m]	$P_{t,\mathrm{AP}}$ [dBm]	$P_{e,\mathrm{AP}}$ [dBm]	$\overline{\mathcal{T}}_{\max}$ [Mbit/s]	$P_{t,\mathrm{AP}}$ [dBm]	$P_{e,\mathrm{AP}}$ [dBm]	$\overline{\mathcal{T}}_{\max}$ [Mbit/s]
150	24	−59	17.49	24	−51	18.57
245	27	−63	30.42	24	−55	28.96
335	27	−63	36.76	27	−55	33.72

Another method has been developed which is able to control both LA and TPC. It relies on the energy efficiency as utility, i.e. the number of bits that can be transmitted per unit of energy. This is mainly important for the UL transmission of battery driven MTs. An algorithm to increase energy efficiency $\mathcal{E}$ based on a game theoretic model for LA and TPC is presented in [RaJo03]. The utility for a single transmitter-receiver pair i with Phy mode PM is defined as:

$$n_i^{PM}(p_i, \gamma_i) = \frac{\mathcal{T}_i^{PM}(\gamma_i)}{k_p \cdot p_i + \Delta p} \tag{8.18}$$

It depends mainly on the transmission power p_i, the SINR γ at the receiver γ_i and $\mathcal{T}_i^{PM}(\gamma_i)$ from Figure 8.17. The parameter Δp avoids a degenerate solution, which would result in infinite utility for $p_i \to 0$. Some flexibility is introduced with k_p. Note that, due to the parameters k_p and Δp, the utility function is not equal to the physical definition of $\mathcal{E}$ but rather a measure of $\mathcal{E}$. Example utility functions n_i^{PM} for different PM are shown in Figure 8.18.

Unfortunately, the game cannot be analysed analytically, because its utility function is not continuously differentiable. The analysis of a similar virtual game as well as the convergence behaviour of the game's implementation show that an equilibrium exists. Quantitative results of the game are computed with the same simulation model as the throughput results above and presented in [RaJo03]. Results on throughput and $\mathcal{E}$ are compared with results for maximum throughput, see Table 8.13. The main conclusion is that an increase of $\mathcal{E}$ always means a decrease in throughput, so with the help

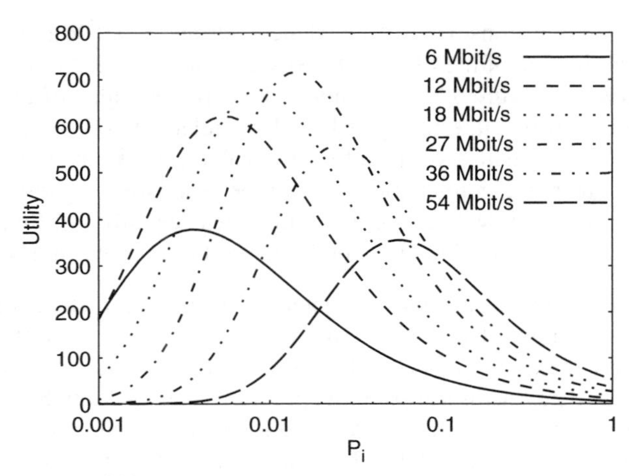

Figure 8.18 *Utility functions* n_i^{PM}.

of parameters in the utility function of the game, the game theoretic method allows to trade off $\mathcal{E}$ and throughput.

Optimum combiner for a smart antenna in an indoor infrastructure WLAN

The subject of investigation in [NaCB03] is an indoor infrastructure WLAN system following IEEE 802.11a or HIPERLAN/2 standards. Multiple users are simultaneously communicating with an access point through SDMA targeting high data rate applications. This is accomplished using a smart antenna array at the access point. Overloaded arrays would not guarantee the required SINR for all the users simultaneously. The number of array elements is hence chosen to be equal to the number of users to ensure an adequate QoS for all users.

The network is modelled using a spatial ray-tracing tool dealing with multiple users arbitrarily located in an indoor environment. A MMSE uplink pattern optimisation algorithm is implemented to mitigate co-channel interference and noise taking into account the multipath behaviour. It is shown that an N element array can mitigate $N - 1$ interfering users rather than directions. The regions of operation of the optimum combiner are identified depending on the SNR per antenna element. These are: (1) zero forcing when interference is dominant (null steering in the LOS case) for high SNR (>40 dB), (2) spatial matched filtering when noise is dominant (beamsteering in the LOS case) for low SNR (<0 dB) and (3) a transition region where a best compromise between zero forcing and spatial matched filtering is reached to maximise the SINR value for each user.

Automatic optimisation algorithms for the planning of WLANs

The planning of WLAN infrastructures that supply large areas requires the consideration of many aspects and therefore needs very sophisticated planning methods [HeMa02], [UnKa03]. Due to the low number of available non-overlapping frequency channels, the problem of co-channel interference has a major impact on the performance of the network. This results in the need to consider the carrier assignment in the planning process depending on the standard used for the network installation.

The proposed approach is based on predictions of the received power to account for the propagation conditions that have a major impact on the performance of WLANs. Therefore, the usage of accurate propagation models is crucial [HoWW03], [CCPS97]. The optimisation is applied to a set of possible locations where access points can be installed. Out of this set, a minimum selection of locations is made to meet the given requirements. These requirements consist of the determination of areas with different priorities and the definition of further parameters.

The optimisation is done in three steps. First, the predicted coverage of every potentially installed transmitter is assessed regarding the priority and the received power. Second, the density of the transmitter locations is reduced depending on the priority settings. In the last, more costly, step, the remaining APs are sequentially added to a solution set that is assessed. The discussed approach may not find the global optimum in all cases, but it results in a proposal based on the locations defined by the network planner. Due to the very short computation time of the discussed approach, different configurations can be analysed very quickly. The resulting selection will be a trade-off between best received power and minimum interference which may be weighted by the user.

The defined test scenario (see Figure 8.19) consists of an office building with an area of approx. 10 000 square metres. Sixty-two possible AP locations were defined. An IEEE 802.11b system with three available carriers was assumed. The finally selected APs provide a good coverage which depends on the entered priorities. Areas marked as unimportant (priority 0) are much more likely to be supplied with less or insufficient received power. Figure 8.20 shows the co-channel interference situation in the entire building. The rooms with the highest priority are not affected by interference problems. Some interference can be tolerated in lower priority areas. The worst values (dark grey area) occur at places with the lowest priority.

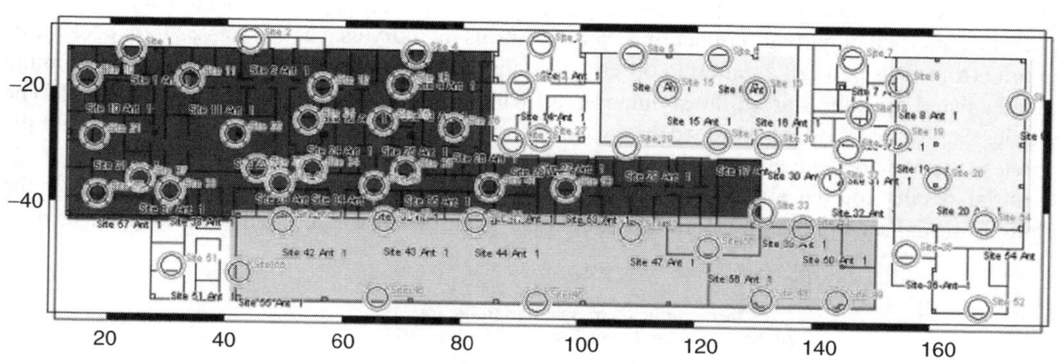

Figure 8.19 *Priority map including the possible access point locations.*

Throughput simulation modelling

Since the 1970s the throughput analysis of wireless networks has been hampered by not being able to derive the saturated throughput for wireless networks except by making severe simplifying assumptions. Even the most complex analytical and simulation models have been unable to take proper account of hidden nodes and other physical layer effects. The contribution in [FiBa04] presents a new simulation methodology using a statistical physical layer propagation model and throughput simulation results showing first results for the benchmark case 'Pure Aloha'.

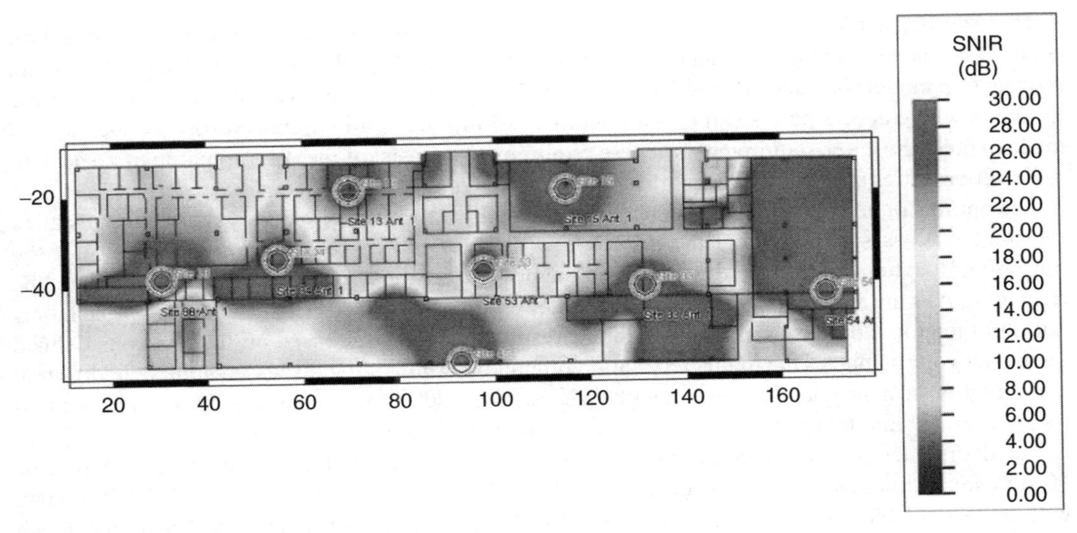

Figure 8.20 *Resulting co-channel interference.*

The methodology adopted is designed to rapidly test as many network topologies as possible. The results are therefore a summed statistical expectation for the behaviour of 'all' possible network topologies. Cycling through multiple topologies rapidly is enabled by creating and destroying wireless nodes whose lifetime is limited to the exchange of a single data packet plus any accompanying control packets. An efficient garbage collection scheme controls memory usage. Wirelesses look backwards in time from their point of creation in order to initialise their state so the wireless behaves as though it had existed from the start of the simulation but without the overhead of actually being there. The wireless's physical world is wrapped around so each wireless perceives a complete world even if the wireless is positioned adjacent to one edge of the world.

Initial results for the simplest historical 'Aloha' case are presented. Figure 8.21 shows the plotted normalised throughput obtained for Aloha using both Magic Genie ACKs (sent using a

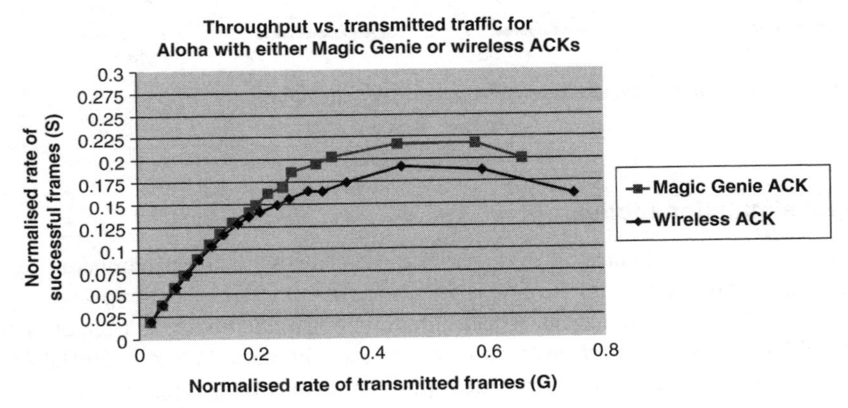

Figure 8.21 *Results for the Aloha case.*

perfect channel) and real ACKs. The x axis shows transmitted frames per vulnerable area per frame length (G). The y axis shows successful frames per vulnerable area per frame length (S). The results show that the maximum throughput is achieved at a normalised packet transmission rate per vulnerable area of approximately 0.45. The Magic Genie ACK version peaks at a normalised success rate per vulnerable area of 0.22 whereas the wireless ACK version peaks slightly lower at 0.19.

Comparison of these results with Abramson's analytic approach shows that they are broadly in agreement.

Voice over IEEE 802.11

With the proposals for a new standard, IEEE 802.11e, there is increasing interest in quality of service issues for real-time voice and multimedia communication over WLANs. Using the pure packet-based MAC protocol for voice communication has proven to be inefficient, which has led to the development of techniques that allow a better integration of voice into the WLAN communication set-up. A proposal from literature called '5-UP' exploits the frequency domain properties of the OFDM-based physical layer of the latest WLAN standard versions and reserves distinct OFDM subcarrier frequencies for voice communication only. This, however, poses serious constraints on accuracy of modulation and transmit power control which are hard to achieve.

The work in [ChEs03] uses the proposed scheme 'Voice communication Over Wireless Local Area Network (VOWAL)' for voice communication over an IEEE 802.11-based wireless Local Area Network (LAN) and compares it with traditional VoIP approaches and 5-UP. It reserves periodic time slots for voice communication and then allocates a subcarrier frequency for each ongoing voice connection. Voice connections use single carrier modulation and need not even be able to participate in the OFDM-based data communication. Multiple voice connections can be multiplexed in each voice communication time slot. Another advantage is the possibility to perform frequency hopping in order to increase robustness. Vowal integrates nicely with the existing IEEE 802.11 standard.

The technical document gives some insight into quantitative measures regarding the capacity share of voice connections. Using, e.g. voice sampling as in DECT systems, approx. 13% of the available channel capacity is connections. Using, e.g. voice sampling as in DECT systems, approx. 13% of the available channel capacity is used for the support of 48 simultaneous voice connections.

8.7 Personal area networks

In the last decade progress in microelectronics and very large-scale integration technology has made possible the realisation of radio transceivers small enough to be utilised in portable, handheld, devices; this opportunity opens new perspectives to wireless networks which are no longer constituted only by personal computers and their peripherals but integrate also Personal Digital Assistants (PDAs), mobile phones, headsets, etc. and pave the way for new and completely different devices and applications.

The most recent development in this direction is constituted by the Bluetooth [IEEE02] wireless technology, which allows users to make effortless, wireless, instant and low-cost connections between various communication devices.

The Bluetooth [Haar00a], [Haar00b], [HaMa00] radio technology, also known as IEEE 802.15.1, was developed as a replacement of cables between electronic devices, but its ability to form small networks, called piconets, opens up a whole new arena for applications where information may be exchanged seamlessly among the devices in the piconet. Typically, such a network, referred to as a

PAN (Personal Area Network), may consist of a mobile phone, laptop, palmtop, headset, and other electronic devices that a person carries around in his everyday life. The PAN may, from time to time, also include devices that are not carried along with the user, e.g. an access point for Internet access or sensors located in a room.

A Bluetooth piconet is constituted by two or more units sharing the same radio channel, defined by a piconet-unique frequency-hop sequence, where one unit acts as a master, controlling the whole traffic in the piconet, while the other units act as slaves.

The master implements a centralised control: only communications between the master and one or more slaves are possible and there is a strict alternation between master and slave transmissions; to this aim the time axis is divided into 625 μs intervals, called time slots, and forward link transmissions (from master to slave) can only start at the beginning of even slots, while reverse link transmissions start at the beginning of odd slots immediately after a forward link transmission.

In [PaCV02] M. Chiani, G. Pasolini and R. Verdone investigate an indoor environment where the users of a Bluetooth-based PAN require non-real-time services. Here, the performance experienced within the PAN is investigated and assessed in terms of link throughput and mean packet delivery time.

The novelty of this work lies in the integrated approach which jointly takes protocol aspects as well as propagation considerations into account. The investigation has been carried out, in fact, considering the ARQ (Automatic Repeat reQuest) and the MAC strategies along with the coding, frequency-hopping and modulation techniques adopted by Bluetooth devices and channel impairments such as thermal noise and Rayleigh flat fading.

In [PaCV02], in particular, a simple polling protocol is proposed as an MAC level solution for Bluetooth-based PANs: it is assumed that the master unit polls with a cyclic order each other unit (the slaves) by sending one packet (a data packet, if any, or an explicit poll packet) and afterward releases the channel, giving its actual counterpart (the same slave unit) the possibility to transmit in the subsequent slot.

This proposal, denoted as roll-call polling or pure round robin [Chun00], is fully compliant with Bluetooth specifications and is a natural choice as far as the MAC protocol is concerned.

The MAC level performance within a piconet is assessed by means of both analysis and simulations, focusing the attention, in particular, on the throughput achievable with the data packet types defined by the specifications, namely DM1, DM3, DM5, DH1, DH3, DH5, that differ for the payload size and the adopted error correcting coding scheme.

As an example result, the curves reported in Figure 8.22 show the MAC level maximum throughput $S_{max}^{(i)}$ of the generic ith Bluetooth link for the different packet types, as a function of the average signal-to-noise ratio in the case of $N_s = 7$ slaves in the piconet; to derive the curves of Figure 8.22 the authors assumed that all terminals in the piconet adopt the same packet type.

The symbols reported in Figure 8.22 refer to analytical results, whereas the lines provide the simulation results: as can be observed the agreement is quite good.

It is worth noting that according to Bluetooth specifications, each master-slave link could use a different packet type; nonetheless specifications give no indication about how to select the proper packet type [IEEE02]. It is obvious, however, that since every unit experiences, in general, a different Signal-to-Noise Ratio (SNR), the adoption of the same packet type for all communications within the piconet does not seem a good choice. LA is a well-known technique providing vertical integration within the protocol stack: in this case, it may consist in selecting the proper packet type (i.e. the payload size and the channel coding scheme) as a function of the mean value of the received power level (let us have in mind that the received power level is measured and reported in the Receiver Signal Strength Indicator (RSSI) by each unit) [IEEE02].

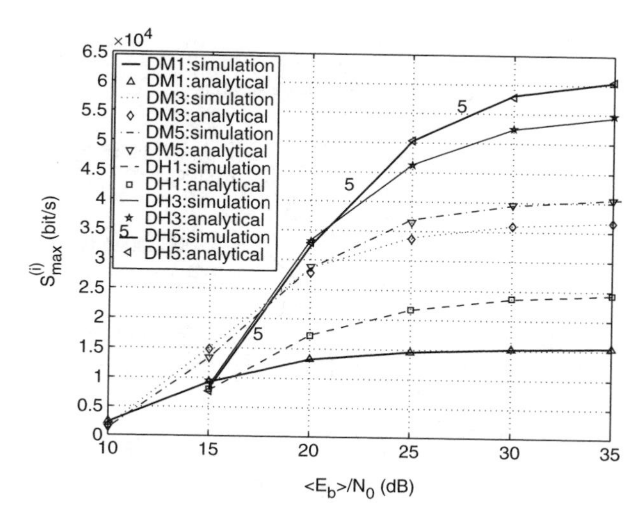

Figure 8.22 *Simulation and analytical results: maximum link throughput, $N_s = 7$ slaves.*

In [PaTV03], G. Pasolini and R. Verdone show that a proper choice of the packet type can provide advantages in terms of overall achieved throughput. The performance is evaluated through a simple analytical model in a reference scenario (analytically characterised), which is also integrated with some results obtained via a ray-tracing tool, to characterise the propagation channel in a realistic scenario (an office environment of about 600 square metres). The results obtained from the analytical model allow suitable comparison between the cases of LA and no LA. They show that a significant improvement can by obtained by properly choosing, on the basis of the knowledge of the channel status experienced, the packet type to be adopted in each link.

In Figure 8.23, for instance, the throughput experienced by a generic slave of the piconet is reported in the case of $N_s = 2$ slaves (dashed curve). The same figure also shows the throughput experienced with the three optimal (in the respective intervals of SNR) packet types in the case of no LA.

It can be immediately observed that, as expected, the LA strategy provides a large performance improvement.

A further investigation on Bluetooth performance is reported in [MiZP04], authored by D. Miorandi and A. Zanella, which is aimed at providing a mathematical framework to evaluate the performance achievable by using multislot packet formats. Here, in particular, the packet-delay statistics for one-hop transmissions is provided, assuming that each node can use different probabilities for the packet type being used.

Furthermore, a characterisation of the channel utilisation parameter and an in-depth investigation of the stability regions achievable under a given SAR policy are provided, showing that the use of multislot packets provides an enlargement of the achievable rate region. Numerical simulations show that the proposed analysis leads to a reasonably accurate prediction of the one-hop delay behaviour, as confirmed in Figure 8.24, and that shows the average packet delay as a function of the offered traffic for a full piconet ($N_s = 7$ slave units) with only downlink (master to slave) traffic and balanced SAR policy, i.e. with equally distributed packet lengths (one or three or five time slots). In the figure, the dotted line is obtained by the analytical equations, while the markers are used for representing simulation results.

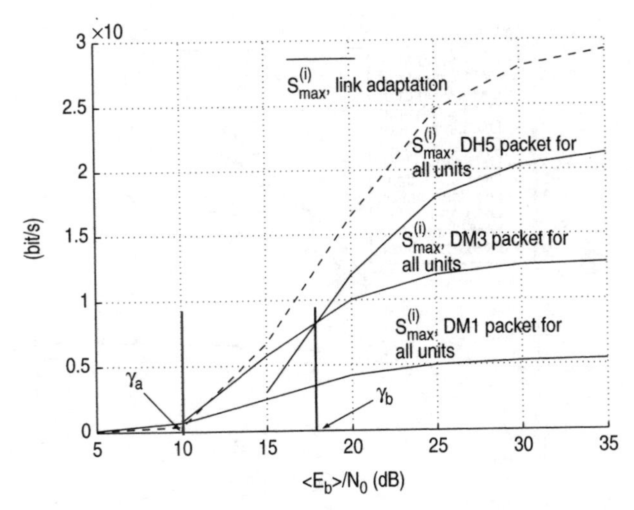

Figure 8.23 *Throughput comparison with link adaptation (in the reference scenario) and without link adaptation, $N_S = 2$.*

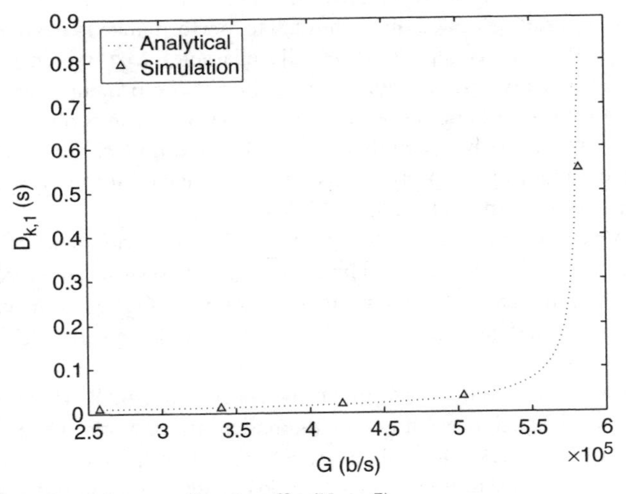

Figure 8.24 *Average packet delay vs. offered traffic ($N_S = 7$).*

Apart from the above-mentioned achievable performance, an important aspect to be considered when dealing with Bluetooth devices, which are usually powered with batteries, is the energy efficiency. As previously recalled, Bluetooth provides six data packet types, which differ for time duration, error protection and data capacity. Unprotected and long packet types show high payload capacity but are sensitive to payload errors. On the contrary, short and protected types are less subject to payload errors to the detriment of capacity. It is straightforward to understand that the choice of the packet type has also an impact on the energy efficiency, expressed in terms of average amount of successfully delivered data bit per unit of energy, since in case of erroneous reception a retransmission, hence a wastage of energy, is required.

This aspect is analytically faced by A. Zanella and S. Pupolin in [ZaMP03], along with the impact on system performance of an important design parameter, namely the receiver-correlator margin *S*. Loosely speaking, this parameter determines the *selectivity* of the receiver with respect to packets containing errors. Low margin values imply strong selectivity, with the risk of dropping packets that could be successfully recovered. On the contrary, high values imply weak selectivity, with the risk of receiving an entire packet before realising that it contains unrecoverable errors. Therefore, the receiver-correlator margin *S* determines a trade-off between throughput and energy consumption that deserves particular attention.

Both aspects are investigated by means of a simple mathematical model for the Bluetooth point-to-point connection. The dynamics of the system is captured by means of a Finite-State Markov Chain (FSMC) model. Hence, following the approach suggested in [ChZo98], the authors resort to the renewal reward analysis to compute the average throughput and energy performance achieved by the system. The analysis is carried out in both AWGN and Rician fading radio channel.

The study has confirmed the presence of a trade-off between average traffic rate and energy efficiency achieved by different packet types.

Furthermore, it is shown that in case of asymmetric data transfer, better performance is achieved by configuring as slave the unit that hosts the server and as master the unit that hosts the client application, respectively. This configuration yields performance improvement in terms of both goodput and energy efficiency, since the server never retransmits packets that were already received by the client.

Finally, the choice of *S* has shown to be critical, since it may significantly impact on performance achieved by short and protected packet types, although long and unprotected packet types show less dependence on this parameter.

Another aspect which deserves great attention when dealing with PAN is networking: issues such as routing and handover are particularly critical in such a scenario.

The networking capabilities of Bluetooth can be further enhanced by interconnecting piconets to form scatternets. However, as opposed to other radio standards, physical proximity does not automatically imply the connection of two Bluetooth radio nodes. As recalled, Bluetooth uses a Frequency Hopping (FH) spread spectrum modulation mechanism to limit interference with other devices operating in the same frequency band. Each Bluetooth piconet, in particular, is associated to a different frequency hopping channel, which is determined on the basis of the Bluetooth physical address and the native clock of the unit that acts as the master in the piconet. All the other units (slaves) in the piconet are synchronised to the same FH channel. Interpiconet communication requires, hence, that some units be synchronised with more than one master.

These units, called interpiconet units or *gateways*, need to share their presence among the piconets of a scatternet on a time division basis. In [ZaTP02], A. Zanella, S. Pupolin and L. Tomba investigate the issue of routing protocols over Bluetooth scatternets. They focus on two well-known routing algorithms, namely the Fisheye State Routing (FSR) [PeGC00b], [PeGC00a] and the Ad-hoc On Demand Distance Vector (AODV) [PeRD01], [PeRo99], proposed for routing on wireless ad-hoc networks. These algorithms are chosen as representative of two wide families of ad-hoc routing algorithms, i.e. *table driven* and *on demand.*

In the table-driven routing protocols, each node computes the routing path for every other node in the network and periodically refreshes it by means of update control messages. The routing information is maintained at each node by using one or more routing-related tables, and its consistency is achieved by propagating updates throughout the network. Opposite to table driven, on-demand routing protocols do not attempt to maintain a valid path to every node in the network, but only to those nodes for which there is an actual need. The route creation procedure in this type of protocols is usually initiated by the source node, by means of a route discovery query.

The analysis reported in [ZaTP02] has been carried out via computer simulations by means of the OpNet tool, which models the Bluetooth baseband, as well as the link manager, the Link layer Control and the Adaptation Protocol (L2CAP) layers, also allowing the creation of piconets and scatternets.

The study has revealed that IP datagram size and the sniff period, which is the time-period that a gateway unit spends in a piconet before switching to the following one, may have a considerable impact on the end-to-end delay experimented by IP datagrams and on the path discovery time needed to the AODV algorithm for finding a new route. Moreover, experiments have shown that the FSR algorithm may lead to a consistent capacity wastage if the refresh interval is not carefully chosen.

The issue of handover, i.e. of the switching of a slave connection between two masters, is finally addressed by R. Corvaja in [Corv02]. Problems related to the handover in an IP over Bluetooth network are considered in [AFMW99], [BFGK00], where, however, the authors assume the presence of a wired connection between the masters. Here, instead, a pure ad-hoc network is considered: devices can communicate only by means of Bluetooth radio connections.

In [Corv02], two handover algorithms are described and analysed. The first is a table-based solution and requires the periodic exchange of handover information among the devices in the scatternet. It is assumed that the devices maintain paging tables with information on other devices, in order to perform an accurate paging. In particular, each master must be aware of the other masters in the scatternet and, possibly, of their availability to receive a new slave. The basic idea is that the neighbour masters perform the paging at precise time instants, when the slave is listening (page-scan state) for the master's paging.

The second algorithm is on demand and, hence, the parameters required by the handover procedure are exchanged only when the handover needs to take place.

In Figure 8.25 the cumulative distribution function of the handover time (expressed in time slots) is shown for both the table-based and on-demand solutions, considering a scatternet with three masters, three or five active devices per piconet, a sniff-time set to $N = 10$ slots, and, for the table-based procedure, two table repetitions and 12 slots dedicated to each master in the table. The traffic corresponds to the use of DH1 packets only with arrival rate 0.9 packets/slot (154.8 kbit/s), which represents almost the full load. These results have been obtained by means of a semi-analytical method, i.e. by numerically generating the random variables that represent the delays in the procedure, according to the system parameters chosen.

As can be seen, the on-demand procedure introduces lower handover delays, due to the absence of the paging table.

A further investigation reported in [Corv02] is related to the dependence of the handover delay on the sniff-time: in particular, it is shown that the mean handover time increases as 1.5 times the increase in the sniff-time.

8.8 Wireless ad-hoc networks

In this section we consider the performance of wireless ad-hoc networks. Here, we distinguish them into *distributed wireless networks* in which a given number of terminals can arbitrarily move in a given scenario and can exchange data without the need for a supporting infrastructure, and *wireless sensor networks*, in which low-cost devices are put in an area with the purpose of collecting information

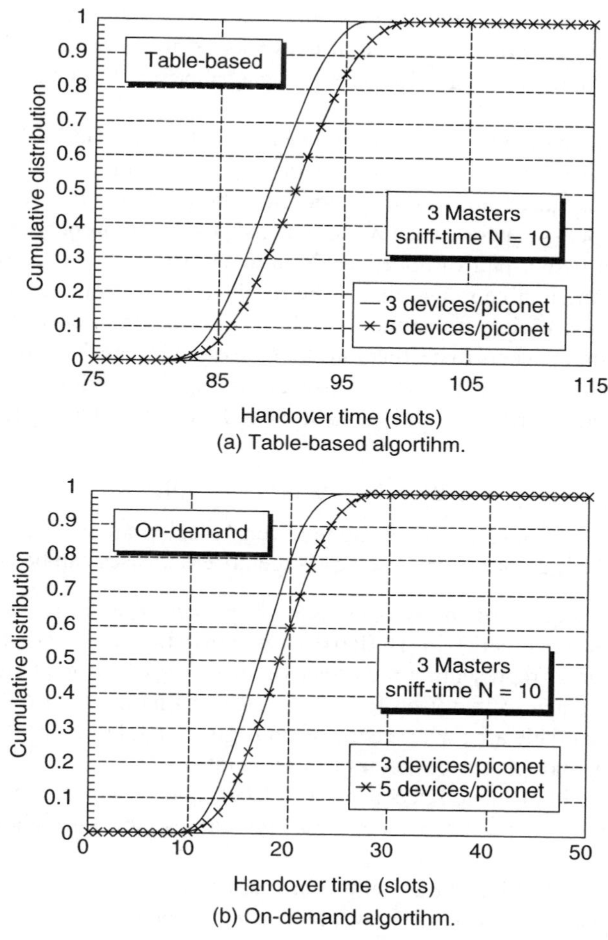

Figure 8.25 *Cumulative distribution function of the handover time in a scatternet with three masters, sniff-time N = 10 and three or five active devices per piconet.*

from a specific environment. These two scenarios do not require infrastructure but present some differences in terms of mobility, cost of terminals, and bit rate of exchanged packets. For these reasons they are commonly analysed by taking different performance figures into account.

8.8.1 Distributed wireless networks

The scenarios considered in this subsection are concerned with distributed wireless networks operating in the ISM band. In this scenario many different networks can operate and coexist in the same band. Here, a network cannot expect to have exclusive access to a dedicated medium and different networks can be operating simultaneously in the same radio frequency band, for example in public areas (conference halls, airports, hotels, etc.). While individual networks may use specific multiple access techniques, in case of coexistence of networks using different modulation schemes and MAC protocols the transmission from other networks may appear as terminals were using the simple unslotted Aloha protocol.

The evaluation of this particular scenario has been carried out in [HaBa01b], [HaBa02] through the use of an analytical model based on the assumption that terminals are randomly distributed in a bidimensional area. Packet arrivals are supposed to be randomly distributed in time, with the probability of a packet being generated in any incremental period, δt, proportional to δt. Every packet is associated with a particular pair of nodes, referred to as its source and destination. These nodes are assumed to be within audible range, that is the received power is larger than a given threshold depending on receiver sensitivity, of one another. This is a key assumption as the model assumes that no source tries to transmit packets to destinations that are out of range. The MAC protocol is similar to the original unslotted Aloha, that is when a packet, having a fixed duration T, is generated, it is immediately transmitted regardless of the status of the radio channel. Owing to shadowing, the path loss a is a random variable

$$a(d) = k_1 + k_2 \ln(d) + Y[\text{dB}] \tag{8.19}$$

where d is the distance between transmitter and receiver, k_1 and k_2 are two deterministic propagation coefficients and Y is the log-normal shadowing. Three different cases can be considered:

1. The probability of success of the weakest audible packet is assumed to apply to all packets (worst case, used as benchmark). In [HaBa01b] it is shown that the expression for the throughput for the vulnerable area (representing the effective area over which the terminal is vulnerable to interference), S, is equal to that used for the classical Aloha protocol Ge^{-2G}, with $G = gA_v$, where A_v is the vulnerable area whose expression is given in [HaBa01b], G is the offered traffic per vulnerable area, and g is the packet transmission rate.
2. The received power is evaluated according to (8.19), and terminals are supposed to transmit using a fixed amount of power. Under these hypotheses, the throughput can be written as $2G^2/(\exp(2G) - 1)$.
3. Power control is considered, and transmitters are assumed to adjust their power in order to guarantee a minimum acceptable received power level which depends on the receiver sensitivity. Now the throughput S can be written as Ge^{-G}. It is worth noting that this expression is formally identical to the throughput achievable in the conventional slotted Aloha case. However, both throughput and offered traffic are here normalised quantities. A comparison between the three different cases is plotted in Figure 8.26 which shows the obtainable throughput as a function of G.

In the previous analysis it is assumed that a packet is successfully received if the received power at the destination terminal is larger than that of the strongest interferer: that is, if P_{r0} represents the received desired power and P_{r_i} are the received interfering powers, a packet is assumed to be successfully received when $P_{r0} > \gamma_T + \max\{P_{r_1}, \ldots, P_{r_N}\}$, where γ_T is a suitable threshold, N is the number of interfering terminals and all terms are expressed either in dB or dBW.

Owing to its simplicity, this model is commonly used in the analysis of multiple access schemes in wireless networks. The major drawback of this model is that it neglects contributions of multiple interferers that do not have individually enough power to prevent capture effect. To overcome this limitation, the model is extended in [HaBa02], now also the contribution of weak interferers is considered. The results show that in case (1) the presence of weak interferers has the effect of increasing the minimum audible power level (in dBW) of a factor $10\log_{10}(1 + 2G/(K_2 - 2))$,

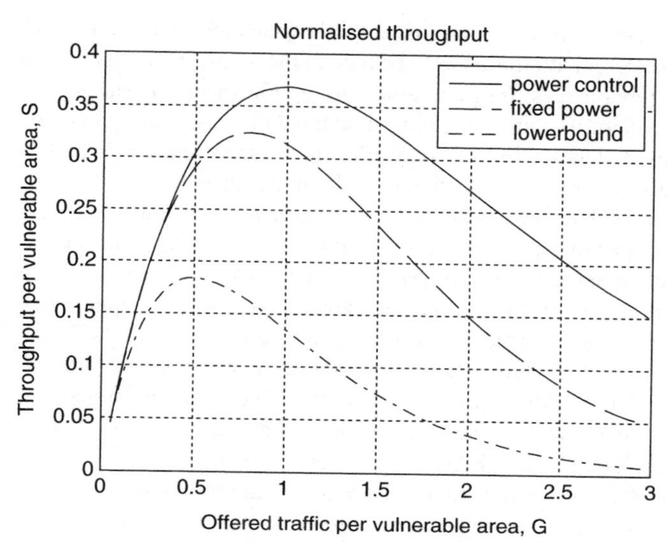

Figure 8.26 *Throughput as a function of offered load for the three cases considered.*

where $K_2 = k_2 \ln 10/10$. In case of power controlled network (3), the analysis carried out in [HaBa02] shows that the presence of weak interferers reduces the receiver sensitivity by a factor $10 \log_{10}(1 + G/(K_2 - 2))$.

The previous analysis considered that packets can transmit regardless of the activities of other terminals, this assumption was made as different wireless technologies were assumed to be active in the same area. In case a unique network is present, a significant performance increase can be obtained using CSMA schemes in which a terminal is required to listen to the common channel before transmission. Furthermore, to reduce the problem of 'hidden' terminals, typical of wireless environments, RTS and CTS protocols can be applied. The impact of these collision avoidance protocols has been considered in [HaBa01a], in which the effect of varying the transmission range of CTS packets is also analysed. The analytical model derived in [HaBa01a] provides an expression for the throughput (normalised for unit area) as a function of the offered traffic and the average rate of audible (that is the received power level is larger than a given threshold) control packets (i.e. RTS and CTS). Results show that the use of adaptive strategies for the choice of the transmission range of CTS packets allows a significant performance improvement in terms of throughput.

8.8.2 Wireless sensor networks

In the past few years, increasing attention has been devoted to the analysis of wireless sensor networks. In the scenario considered in [Verd04], uniformly distributed (in an infinite plane) sensors collect information about environment (temperature, pressure, etc.) and send data to supervisors when polled by them. The supervisors have the task of collecting sensor data from a specific region around them. They are supposed to be uniformly distributed too, and their density is obviously smaller than that of sensors. The region monitored by each supervisor is quite small so that direct communication between sensors and supervisor can take place (multihop between sensors is not considered). As sensors can be put into forests, desert zones or other inaccessible areas, the replacement of batteries is generally impossible and so particular attention has to be paid to the duration of sensors' batteries. Furthermore, as sensors are low-cost devices, transceiver technology has to be as simple as possible; this means that no localisation capabilities are usually implemented in the sensors. This kind of system is subject to the constraints on the latency in receiving information from sensors, which should be minimised, and on the life duration of sensors, which should be maximised.

The goals of [Verd04] are the proposal, the optimisation of an energy-efficient protocol, and the evaluation of its performance using an analytical model which accounts for the propagation environment (modelled using (8.19)), physical as well as MAC layers and application scenario. The main objective of the analysis is to optimise the transmitted power by supervisors during the polling of neighbouring sensors, in order to maximise the average number of packets, N_{ss}, successfully received by the supervisor and sent by those sensors located in the circular area around it. The optimisation of P_{sup}, the power of the polling packet sent by the supervisor, is subject to the constraints on a minimum value for the average life duration of sensors and the maximum latency time.

An example of results is given in Figure 8.27 where N_{ss} is plotted as a function of P_{sup} for different values of the radius, R, of the circular area that defines the monitored region of each supervisor. To plot this figure, the following propagation parameters have been considered: $k_1 = 25$ dB, $k_2 = 30/\ln(10)$, and the standard deviation of shadowing Y, σ, has been fixed equal to 10. The figure clearly shows that the values of P_{sup} maximising N_{ss} are in the range between 15 and 20 dBW regardless of the value of R.

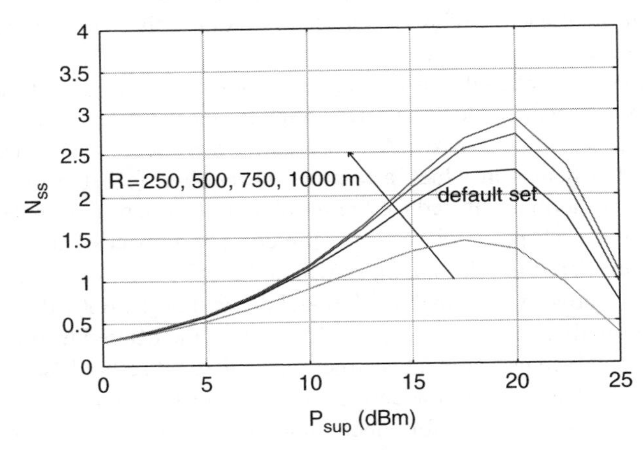

Figure 8.27 *N_{ss} as a function of P_{sup} for different values of the radius of the circular area around the supervisors.*

8.9 Terminal location determination

8.9.1 Introduction

So-called Location-Based Services (LBS) form a new class of services in mobile communication networks. These location-based services supplement existing services in a way that individual information can be provided depending on the user's position. The basis for introducing such services is to determine the location of the MT.

In principle two different approaches can be distinguished. It is either possible to use satellite navigation systems or exploit the data being available in the mobile communication network itself. Satellite navigation systems like GPS provide a very high accuracy in the range of several metres in case there is LoS to at least four satellites. In rural areas this will usually be fulfilled but in narrow street canyons of an urban scenario the LoS is generally obstructed. Hence, satellite navigation fails in indoor environments as well.

As it appears to be desirable to offer LBS at any location with network coverage other positioning methods are needed at least as alternative solutions. A GPS receiver which is integrated into, e.g. a mobile phone, usually reduces the operation time of the telephone due to the increased power consumption. So from this point it is also advantageous to have a location method working with communications network parameters only. This section consequently addresses methods and algorithms that are capable without measurements or references of external systems.

Subsection 8.9.2 employs the knowledge about radio channels to deal with time-based measurements without addressing a specific mobile communications standard. The other subsections assume a cellular mobile communications network like GSM or UMTS.

In cellular systems the basic approach is always to identify the serving cell and deduce the terminal location from the BS site. For many LBSs the achievable accuracy is not sufficient and has to be enhanced by more sophisticated location methods mostly considering wave propagation.

The environments which are covered by mobile communication networks are very different in terms of wave propagation aspects. Multipath propagation with reflected and diffracted waves dominates in dense urban scenarios while in flat rural areas often line-of-sight conditions prevail. Wave propagation models as used for network planning can be utilised to enhance the accuracy of common location methods like Time of Arrival (ToA), TDoA or AoA especially in urban scenarios. Subsection 8.9.3 treats various approaches.

Location methods differ heavily in terms of complexity and accuracy. Comparisons are difficult as the performance also depends on the environment in which a location method is used. Subsection 8.9.4 therefore deals with assessment and validation aspects.

In Subsection 8.9.5 results from field tests in real GSM networks in various environments in Norway and from the city centre of Stuttgart (Germany) are presented.

8.9.2 *ToA techniques employing channel modelling*

Common geometrical location techniques like ToA, TDoA or AoA (also known as DoA) implicitly assume LoS conditions between the antennas of transmitter and receiver. If this direct path is blocked by an obstacle the measured parameters lead to false location estimates. By applying statistical channel knowledge the delay measurements can be improved or with known Channel Impulse Response (ChIR) even a map of obstacles in the environment can be reconstructed.

Besides the identification of an NLoS situation from measurements the subsequent reconstruction of the LoS signal component is the main issue. In doing so, an appropriate statistical model for the NLoS error is inevitable for its mitigation.

There are different theoretical scattering models that describe the distribution of the scatterers in the vicinity of the mobile terminal and the BS. In a macrocellular environment local scatterers are considered most relevant for positioning. Distant scatterers and those located in the vicinity of the base station which are above rooftops in a macrocellular environment are neglected.

From the GSCM according to Fuhl *et al.* [FuMB98] only the contributions created by a single scattering process are considered in [DiOe04]. As only the shortest path is relevant for delay determination this is supposed to suffice. In the model scatterers are arranged uniformly distributed in the angular domain at distances with one-sided Gaussian distribution. That is to say the probability of the first significant multipath component, which accounts for the range estimation, decreases with the distance from the scatterer to the MT.

The error caused by NLoS is the excess path length due to scattering as illustrated in Figure 8.28.

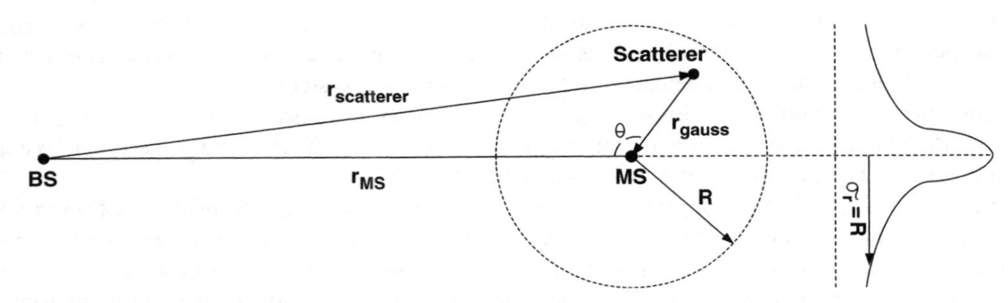

Figure 8.28 *Distance error caused by excess delay and distribution of scatterers in a Gaussian scattering model [DiOe04].*

Statistical properties of the NLoS error obtained with the GSCM are well approximated by a gamma distribution in terms of PDF and CDF.

A completely different technique may be applicable if ChIRs are communicated around in an ad-hoc network where no fixed references like BSs are available. With similarities to hyperbolic navigation and bi-static radar a map of the surrounding environment and the radios in it is generated [GuFB04].

In indoor multipath measurements the ChIR consists of a small number of dominant echoes from large flat surfaces like walls interspersed with noise-like scattering from smaller objects across the entire range of delays. Using suitable thresholding only dominant echoes are extracted and used to map major features of the environment.

With an extended technique a floor plan with walls in an indoor scenario could be created only from unknown positions of mobile terminals. However, already simple scenarios bring about potential ambiguities. In order to solve these ambiguities knowledge of the user motion is required.

8.9.3 Application of propagation models

Especially in urban environments multipath propagation leads to complex scenarios without LoS between the MT and the different BSs. In these situations a location technique based on simple geometric algorithms is not accurate enough. Combined with predicted parameters to compare to, the measurements become significantly more useful for location purposes. Predictions based on wave propagation models, as utilised for the planning of mobile radio networks, improve location accuracy avoiding costly and time-consuming measurements.

The first approach is a simple cell-based method which considers the predicted coverage area of a cell which normally extends over a predefined area. The estimated position of the MT is assumed as the centre of gravity of this area. In the following, several methods are surveyed to define the area of a cell and its centre of gravity.

The area covered by a cell may be assumed to be the area in which the corresponding BS is best server. However, the hierarchical structure of mobile communication networks causes problems. In urban scenarios often macrocells with antennas on raised locations are used as so-called umbrella cells. Hence these cells are only predicted to be best server in gaps between smaller cells and not for a cohesive area. Therefore the resulting centre of gravity of these locations is sometimes located far away from the best server areas.

The consideration of only a single cell for the determination of its cell area avoids the problem of fragmentation. For this purpose a certain power level is defined as a threshold. All locations with predicted values above this threshold are assigned to the area covered by the contemplated cell. The approach also regards the serving cell as not necessarily the strongest one at a given location but may be any other which is strong enough.

An improvement to the use of the centre of gravity of the defined cell area is to weight each predicted point of a coverage area according to its predicted receive power. By this, the weight of a position is increased with higher predicted power moving the centre of gravity towards best covered locations.

After all, the possibility remains to manually appoint the most likely location of residence inside the cell by reviewing the prediction results. The manual effort pays off because one can identify locations with increased residence probability of mobile terminals such as busy squares or other 'hot spots'.

As a more sophisticated solution a method often referred to as the database correlation method yields more precise results in urban environments by applying advanced correlation techniques to predicted power levels of a Look-Up Table (LUT). The general principle is depicted in Figure 8.29. In contrast to all other location methods the database correlation approach offers extensive possibilities for improving location accuracy by optimised processing of the available data. Details about the construction of the LUT and the matching algorithms can be found in [WHZL02] and [ZBLL04].

Propagation models can also be employed to overlay the coverage areas of the received cells [PELW02]. The method defines a coverage area by finding the area with a predicted power level at least equal to the measured signal level minus an error margin. The predictions stem from a network planning tool while the measurements are the actual RxLev values of the GSM network. Finding this area for the serving cell and repeating the procedure for the strongest neighbour cells one after the other the estimated area diminishes. If the delay of the serving cell is applied in the first step the initial area already defines either a complete or part of a ring as illustrated in Figure 8.30. The initial circular area is first reduced by the serving cell coverage plus delay information (a) and then the strongest neighbour coverage (b) is overlaid.

8.9.4 Assessment and validation

Several location methods of different complexity have been suggested in the literature. Even though similar accuracies are achieved, the network scenarios in which the individual authors have tested their algorithms differ tremendously. In some cases only a simple cell model has been used for

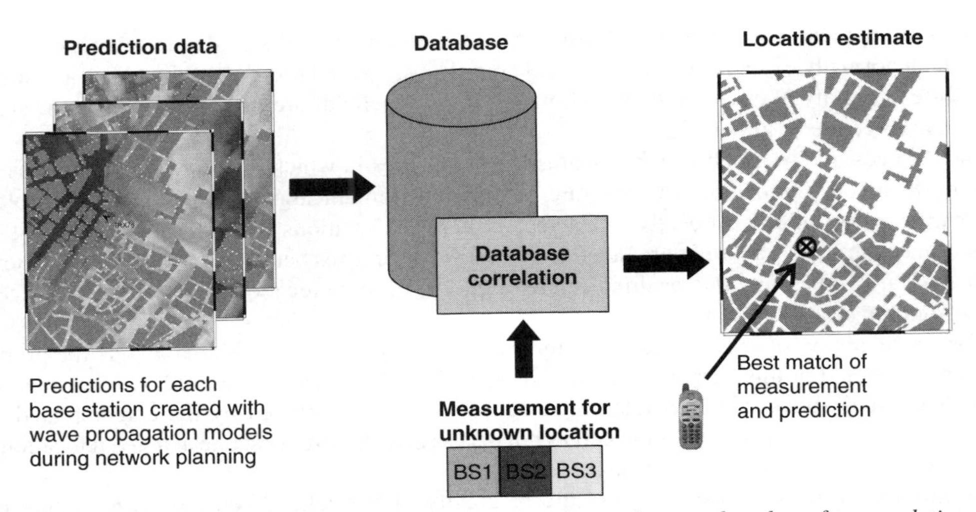

Figure 8.29 *Database correlation method using predicted signal strength values for correlation with measured values to determine the location of an MT.*

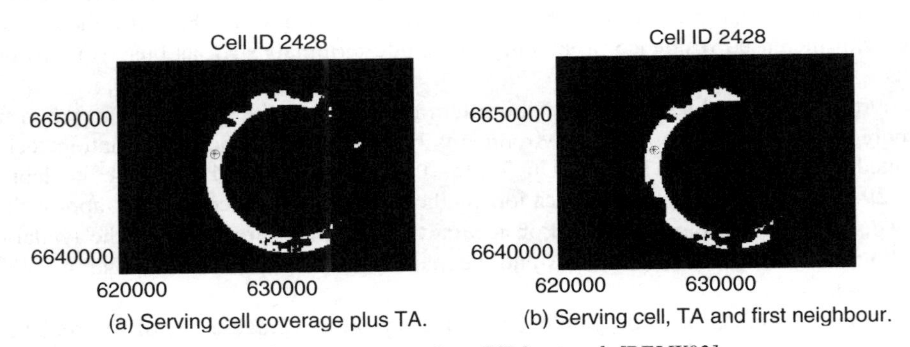

Figure 8.30 *Overlay of estimated coverage areas in a GSM network [PELW02].*

simulations while other methods have already been applied in field tests. These differences hamper an objective capability comparison of the variable location methods in terms of accuracy, cost and system impact.

Additionally, the accuracy of a location method varies depending on BS density, their configuration and propagation conditions. Therefore, the capability even of a single method can hardly be evaluated. So each location method must be assessed under consideration of the network structure. This can be achieved by using a reference method to evaluate the accuracy of various algorithms in the same environment. By this, it becomes possible to compare location methods presented by various authors for different scenarios, too.

An urban area which is usually covered by a dense network will enable different location accuracies compared to a wide meshed network in rural environments. Absolute values of achievable accuracies cannot be used as a significant capability measure of a location technique. The use of basic location methods as a reference is suggested to allow an objective assessment of more sophisticated approaches in terms of additional accuracy and effort. In [ZWBL04] the following simple location methods for GSM are proposed.

The position of the serving antenna is taken as the location estimate. This method simply interprets the cell identification and gets the appropriate site from the location server.

The locations of all BSs appearing in the network monitor are determined and the arithmetic mean of their positions (in an orthogonal coordinate system) is computed. Sites of which more than one cell is received are contributing several-fold.

Both the active site and the GSM parameter Timing Advance (TA) are evaluated. TA is available in dedicated mode only representing a rough measure for the distance between BS and MT. The location result of this method is an area (compare third method in Subsection 8.9.3).

These location methods can easily be applied to any synthetic or real-world scenarios as they require very low computational effort and employ only limited data. The interpretation of the cell identifier may be the most obvious method in any cellular system. The other methods already specialise in GSM as they operate on the data available according to standards. In UMTS such simple approaches could apply the Round Trip Times (RTT) whereas in GSM received power levels (RxLev) are available in an easier way.

If no real data is available to validate the performance of a location method the measurements are acquired from simulations. A verification platform capable of creating up to eight independent channels simultaneously is presented in [KoJä02]. Radio links to several cells can be simulated with a semi-deterministic channel model with an MT moving along a predefined route on a map. Delay and LoS Doppler are modelled geometrically while fading is implemented in different stochastic schemes. Directional effects are only involved when scatterers are defined. The verification platform is useful for testing and evaluating location methods in any mobile communications standard.

8.9.5 GSM results

The presented methods from Subsections 8.9.3 and 8.9.4 were tested in real GSM networks. Besides the GSM parameters Cell Identification (CI), RxLev and TA for the serving cell the test equipment provided a location reference from GPS. This reference is considered to be the exact measurement location as the occurring GPS errors are expected to be small compared to the location estimates obtained from the available GSM data.

The location estimate can either be a single spot or a whole area. In the case of a single spot the location error is defined as the horizontal distance between the estimated and the actual location. For an area, position uncertainty and miss probability are given. These two parameters most often are competing [PELW02]. All the parameters are statistical values obtained from a large number of measurements.

The location error distribution of the estimated area for all measurements of an area type yields the position uncertainty. Miss probability gives the probability of the actual position being outside the predicted area. It is obviously desirable that all the defined parameters are as low as possible.

A GSM network in the city centre of Stuttgart (Germany) was chosen as an example for an urban area with single spot location estimates in [ZBLL04]. The considered area is approximately 10 km^2 large and consists of more than 100 cells with both omnidirectional and sector antennas. Cell sizes are approximately between 100 m and 1 km in diameter. Figure 8.31 shows the cumulated distribution of location error achieved with various methods on a 2 km route consisting of 365 measurements.

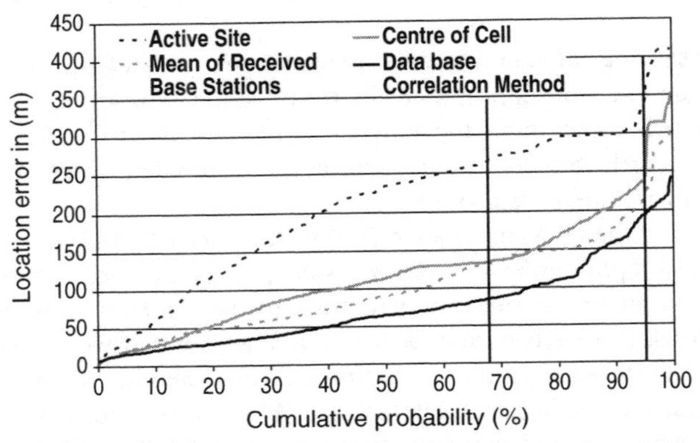

Figure 8.31 *Comparison of the cumulative distribution functions of location errors on the route through the city centre of Stuttgart (Germany) [ZBLL04].*

Analysing the measurements it turned out that TA values are not only very coarse in quantisation but also often incorrect. The deviation between expected delay and measured TA is considerably larger than one would expect from excess delays caused by reflections and diffractions. At least in urban environments the parameter TA cannot be utilised without an empirical correction as introduced in [PELW02].

References

[3GPP99] 3GPP. QoS concept. 3G TR 23.907 V1.2.0, TSG SSA, May 1999.

[3GPP01] 3GPP. Packet switched conversational multimedia applications; default codecs (release 5). TS 26.235, Ver. 5.0.0, TSG SSA, June 2001.

[3GPP02a] 3GPP. RF system scenarios. TR 25.942 - Release 6, TSG SSA, Mar. 2002.

[3GPP02b] 3GPP. Transparent end-to-end packet switched streaming service (PSS); protocols and codecs (release 4). TS. 26.234, Ver. 4.3.0, TSG SSA, Mar. 2002.

[AFMW99] M. Albrecht, M. Frank, P. Martini, A. Wenzel, M. Schetelig, and A. Vilavaara. IP services over bluetooth: Leading the way to a new mobility. In *Proc. LCN 1999 – 24th IEEE Conf. on Local Computer Networks*, Boston, MA, USA, Oct. 1999.

[AgCo04] J. Aguiar and L. M. Correia. Traffic source models for the simulation of next generation mobile networks. In *Proc. WPMC 2004 – Wireless Pers. Multimedia Commun.*, Venice, Italy, Sep. 2004.

[ARRO04] ARROWS, http://www.arrows-ist.upc.es. 2004.

[BaCr98] P. Badford and M. M. Crovella. Generating representative web workloads for network and server performance evaluation. In *Proc. ACM SIGMETRICS 1998 – International Conference on Measurement and Modeling of Computer Systems*, Madison, WI, USA, July 1998.

[BaMV03] C. Balzanelli, A. Munna, and R. Verdone. WCDMA downlink capacity part 1. In *Proc. PIMRC 2003 – IEEE 14th Int. Symp. on Pers., Indoor and Mobile Radio Commun.*, Beijing, China, Sep. 2003.

[BFGK00] S. Baatz, M. Frank, R. Gopffarth, D. Kassaktine, P. Martini, and M. Schetelig. Handoff support for mobility with IP over bluetooth. In *Proc. LCN 2000 – 25th IEEE Conf. on Local Computer Networks*, Tampa, FL, USA., Nov. 2000.

[BLZZ03] L. Badia, M. Lindström, J. Zander, and M. Zorzi. A utility- and price-based approach for the radio resource management in multimedia communication networks. TD(03)147, COST 273, Paris, France, May 2003.

[BüNB02] H. Bühler, T. Neubauer, and E. Bonek. RSSUS – reference system scenario for UMTS simulations based on a real world environment. TD(02)162, COST 273, Lisbon, Portugal, Sep. 2002.

[CaLi01] Y. Cao and V. O. K. Li. Scheduling algorithms in broadband wireless networks. In *Proc. IEEE INFOCOM 2001*, Anchorage, AK, USA, Jan. 2001.

[CCPS97] C. Carciofi, A. Cortina, C. Passerini, and S. Salvietti. Fast field prediction techniques for indoor communication systems. In *Proc. EPMCC 1997 – 2nd European Personal and Mobile Communications Conference*, Bonn, Germany, Nov. 1997.

[CFAP01] N. Cardona, R. Fraile, A. Arregui, and J. Pons. Traffic prediction for adaptive resource allocation: Feasibility analysis. TD(01)032, COST 273, Bologna, Italy, Oct. 2001.

[ChEs03] B. M. G. Cheetham and A. I. Eshhubi. Voice over IEEE 802.11 wireless LANs. TD(03)146, COST 273, Brussels, Belgium, May 2003.

[Chun00] Wah Chun Chan. *Performance Analysis of Telecommunications And Local Area Networks*. Kluwer Academic Publishers, London, UK, 2000.

[ChZo98] A. Chockalingam and M. Zorzi. Energy efficiency of media access protocols for mobile data networks. *IEEE Trans. Commun.*, 46:1418–1421, Nov. 1998.

[Corv02] R. Corvaja. Time analysis of the handover procedure in a bluetooth network. In *Proc. PIMRC 2002 – IEEE 13th Int. Symp. on Pers., Indoor and Mobile Radio Commun.*, Lisbon, Portugal, Sep. 2002. [Also available as TD(02)146].

[CuMB01] F. Cuomo, C. Martello, and A. Baiocchi. Packet scheduling in multiaccess systems jointly adaptive to traffic and transmission quality. TD(01)045, COST 273, Bologna, Italy, Jan. 2001.

[DiBQ04] P. Dintchev, E. Bonek, and B. P. Quiles. An improved mobility model for 2G and 3G cellular systems. In *Proc. MCT 2004 – 5th IEE Int. Conf. on 3G Mobile Commun. Tech.*, London, Great Britain, Oct. 2004. [Also available as TD(04)104].

[DiOe04] P. Dintchev and H. Oezcelik. A stochastic model for the non-line of sight error in mobile location estimation. TD(04)040, COST 273, Athens, Greece, Jan. 2004.

[DoAg04] M. Dohler and H. Aghvami. Outage capacity of distributed STBCs over nakagami fading channels. *IEEE Trans. Commun.*, 8(7):437–439, July 2004. [Also available as TD(04)030].

[EGTK04] A. Eisenblätter, H.-F. Geerdes, U. Türke, and T. Koch. Momentum data scenarios for radio network planning and simulation (extended abstract). In *Proc. WiOpt 2004*, Cambridge, UK, Mar. 2004.

[ETSI98a] ETSI. Evaluation report for ETSI UMTS terrestrial radio access (UTRA), report for ITU-R RTT candidate. Valbonne, France, Sep. 1998.

[ETSI98b] ETSI. Selection procedures for the choice of radio transmission technologies of the UMTS – UMTS 30.03 (version 3.2.0). 1998.

[FCXV03] L. Ferreira, L. M. Correia, D. Xavier, A. Vasconcelos, and E. Fledderus. Final report on traffic estimation and services characterisation. D1.4, IST MOMENTUM Project, IST-TUL, Lisbon, Portugal, May 2003.

[FeCo03] L. Ferreira and L. M. Correia. Generation of traffic demand scenarios for UMTS. In *Proc. 12th IST Summit on Mobile and Wireless Commun.*, Aveiro, Portugal, June 2003. [Also available as TD(03)5].

[FeCo04] L. Ferreira and L. M. Correia. IST-TUL proposal for reference parameters for WCDMA analysis. TD(04)069, COST 273, Bologna, Italy, Apr. 2004.

[FeVe04] J. Ferreira and F. J. Velez. Deployment scenarios and applications characterisation for enhanced UMTS simulation. In *Proc. MCT 2004 - 5th IEE Int. Conf. on 3G Mobile Commun. Tech.*, London, UK, Oct. 2004. [Also available as TD(04)94].

[FGLM04] R. Fraile, J. Gozálvez, O. Lázaro, J. Monserrat, and N. Cardona. Effect of a two dimensional shadowing model on system level performance evaluation. In *Proc. WPMC 2004 – Wireless Pers. Multimedia Commun.*, Duisburg, Germany, Sep. 2004. [Also available as TD(04)190].

[FiBa04] N. Filer and S. K. Barton. Wireless network throughput simulation modelling the physical layer with hidden nodes. TD(04)170, COST 273, Duisburg, Germany, Sep. 2004.

[FMCR04a] R. Fraile, J. Monserrat, C. Cardona, and L. Rubio. MORANS layer 1: Geographic data. synthetic scenario. TD(04)070, COST 273, Gothenburg, Sweden, June 2004.

[FMCR04b] R. Fraile, J. Monserrat, C. Cardona, and L. Rubio. MORANS layer 6: Propagation data. synthetic scenario. TD(04)073, COST 273, Gothenburg, Sweden, June 2004.

[FMFC05] L. Ferre, R. Manzanares, R. Fraile, and N. Cardona. Prediction of GPRS traffic by artificial neural networks. TD(05)037, COST 273, Bologna, Italy, Jan. 2005.

[FMGL04] R. Fraile, J. Monserrat, L. M. González, and O. Lázaro. MORANS layer 7: Node B parameters. synthetic scenario. TD(04)074, COST 273, Gothenburg, Sweden, June 2004.

[FoGa98] G. J. Foschini and M. J. Gans. On limits of wireless communications in a fading environment when using multiple antennas. *Wireless Personal Communications*, 6(3):919–920, Mar. 1998.

[Foru98] UMTS Forum. UMTS/IMT-2000 spectrum. No. 6, London, UK, Dec. 1998.

[Foru03] UMTS Forum. 3G offered traffic characteristics report. No. 33, San Ramon, CA, USA, Nov. 2003.

[FrNg00] M. Frey and S. Nguyen-Quang. A gamma-based framework for modeling variable-rate MPEG video sources: The GOP GBAR model. *IEEE/ACM Trans. Networking*, 8(6):710-719, Dec. 2000.

[FuMB98] J. Fuhl, A. F. Molisch, and E. Bonek. A unified channel model for mobile radio systems with smart antennas. *IEE Proc. Radar, Sonar, Navigation*, 145(1):32-41, Feb. 1998.

[GMFL04] L. González, J. Monserrat, R. Fraile, O. Lázaro, N. Cardona, and S. Ruiz. WCDMA downlink simulator for capacity evaluation. In *Proc. of Telecom I+D 2004*, Madrid, Spain, Nov. 2004. [Also available as TD(04)186].

[GuFB04] W. Guo, N. P. Filer, and S. K. Barton. A novel wireless mapping and positioning technique for impulse radio networks. In *Proc. URSI 2004 – 18th Triennial Intl. Symp. On Electromagnetic Theory*, Pisa, Italy, May 2004. [Also available as TD(04)147].

[Haar00a] J. C. Haartsen. The Bluetooth radio system. *IEEE Personal Commun. Mag.*, 7(1):28–36, Feb. 2000.

[Haar00b] J. C. Haartsen. Bluetooth towards ubiquitous wireless connectivity. *Revue HF, Belgian J. of Electronics and Communications*, pages 8–16, 2000.

[HaBa01a] K. A. Hamdi and S. K. Barton. Capture and spatial reuse in ad hoc wireless networks employing collision avoidance protocols. In *Proc. EPMCC 2001 – 4nd European*

Personal and Mobile Communications Conference, Vienna, Austria, Feb. 2001. [Also available as TD(01)004].

[HaBa01b] K. A. Hamdi and S. K. Barton. On the spatial capacity of randomly distributed wirelss networks. TD(01)018, COST 273, Bologna, Italy, Oct. 2001.

[HaBa02] K. A. Hamdi and S. K. Barton. On the spatial capacity of randomly distributed wirelss networks (part 2). TD(02)013, COST 273, Espoo, Finland, May 2002.

[HaMa00] J. C. Haartsen and S. Mattisson. Bluetooth – a new low-power radio interface providing short-range connectivity. *IEEE Proc. of the IEEE*, 88(10):1651–1661, Oct. 2000.

[HeMa02] M. Hein and B. Maciejewski. *Wireless LAN, Funknetze in der Praxis*. Franzis' Verlag, Poing, Germany, 2002.

[Heym97] D. Heyman. The GBAR source model for VBR videoconferences.*IEEE/ACM Trans. Networking*, 5(4):554-560, Aug. 1997.

[HoWW03] R. Hoppe, G. Wölfle, and P. Wertz. Advanced ray-optical wave propagation modelling for urban and indoor scenarios. *European Transactions on Telecommunica- tions*, 14(1):61–69, 2003. [Also available as TD(02)051].

[IEEE02] Wireless medium access control (MAC) and physical layer (PHY) specifications for wireless personal area networks (WPANs). Standard 802.15.1-2002., IEEE, 2002.

[ITU-93] ITU-T. B-ISDN services aspects, recommendations and reports of the ITU-T, Recommendation I.211. International Telecommunication Union, Geneva, Switzerland, May 1993.

[KlLL01] A. Klemm, C. Lindemann, and M. Lohmann. Traffic modeling and characterization for UMTS networks. In *Proc. Globecom 2001 – IEEE Global Telecommunications Conf.*, San Antonio, TX, USA, Nov. 2001.

[KMST00] J. Khun-Jush, G. Malmgren, P. Schramm, and J. Torsner. HiperLAN type 2 for broadband wireless communication. Ericsson Review, 2000.

[KoJä02] J. Kolu and T. Jämsä. A verification platform for cellular geolocation systems. TD(02)053, COST 273, Espoo, Finland, May 2002.

[LáGD00] O. Lázaro, D. Girma, and J. Dunlop. Statistical analysis and evaluation of modelling techniques for self-similar video source traffic. In *Proc. PIMRC 2000 - IEEE 11th Int. Symp. on Pers., Indoor and Mobile Radio Commun.*, London, UK, Sep. 2000.

[LáGD01] O. Lázaro, D. Girma, and J. Dunlop. A wavelet-based video traffic model for realtime generation of self-similar traffic. In *Proc. EPMCC 2001 – 4nd European Personal and Mobile Communications Conference*, Vienna, Austria, Feb. 2001.

[LTWW94] W. E. Leland, M. S. Taqqu, W. Willinger, and D. V. Wilson. On the self-similar nature of the ethernet traffic. *IEEE/ACM Trans. Networking*, 2(1):1–15, Nov. 1994.

[MaMT03] N. Marchetti, S. Mistrello, and V. Tralli. Resource allocation techniques for wireless packet networks based on V-BLAST architecture. In *Proc. WPMC 2003 – Wireless Pers. Multimedia Commun.*, Yokosuka, Japan, Oct. 2003. [Also available as TD(03)159].

[Mand01] D. Mandato. Concepts for service adaptation, scalability and QoS handling on mobility enabled networks. D1.2, IST-BRAIN Project, Mar. 2001.

[MaVT04] N. Marchetti, R. Veronesi, and V. Tralli. On the impact of interference on data protocol performance in multicellular wireless packet networks with MIMO links. In *Proc. VTC 2004 Fall – IEEE 60th Vehicular Technology Conf.*, Los Angeles, CA, USA, Sep. 2004. [Also available as TD(04)200].

[MFVC02] R. Manzanares, R. Fraile, L. Vergara, and N. Cardona. Non-linear traffic prediction algorithm for cellular networks. TD(02)142, COST 273, Lisbon, Portugal, Sep. 2002.

[MGFC04] J. Monserrat, L. M. González, R. Fraile, and C. Cardona. MORANS layer 2: Traffic information. synthetic scenario. TD(04)071, COST 273, Gothenburg, Sweden, June 2004.

[MGFL04] J. Monserrat, L. M. González, R. Fraile, and O. Lázaro. MORANS layer 5: Site locations. synthetic scenario. TD(04)072, COST 273, Gothenburg, Sweden, June 2004.

[MiBG03] P. D. Mitchell, A. G. Burr, and D. Grace. Performance of two-branch route diversity over a highly correlated rayleigh fading channel. In *Proc. WPMC 2003 – Wireless Pers. Multimedia Commun.*, Yokosuka, Japan, Oct. 2003. [Also available as TD(03)032].

[MiTZ02] A. Milani, V. Tralli, and M. Zorzi. On the use of per-antenna rate and power adaptation in V-BLAST systems for protocol performance improvement. In *Proc. VTC 2002 Fall – IEEE 56th Vehicular Technology Conf.*, Vancouver, Canada, Sep. 2002. [Also available as TD(03)141].

[MiZP04] D. Miorandi, A. Zanella, and S. Pupolin. Achievable rate regions for bluetooth piconets in fading channels. In *Proc. VTC 2004 Spring – IEEE 59th Vehicular Technology Conf.*, Milan, Italy, May 2004. [Also available as TD(03)097].

[MOBB04] A. Munna, J. Orriss, C. Balzanelli, S. K. Barton, and R. Verdone. The probability distribution of the number of base stations within an active set window: Comparison with simulation results. In *Proc. VTC 2004 Fall – IEEE 60th Vehicular Technology Conf.*, Los Angeles, CA, USA, Sep. 2004. [Also available as TD(03)011 and TD(03)133 and TD(03)194].

[MOME03] MOMENTUM (MOdels and SiMulations for Network PlaNning and ConTrol of UMTS) research project, under the EU IST framework, 2003.

[MORA02] Mobile radio access network reference scenarios (MORANS), initiative of the COST273, working group 3 – radio networks aspects, 2002.

[MORA03] MORANS. MObile radio access network reference scenarios CD, R01 beta. Aug. 2003.

[MuRV04] A. Munna, S. Ruiz, and R. Verdone. MORANS – MObile radio access network reference scenarios. In *Proc. WPMC 2004 – Wireless Pers. Multimedia Commun.*, Abano Terme, Italy, Sep. 2004.

[MuVe04] A. Munna and R. Verdone. Downlink WCDMA capacity: usage of a MORANS scenario. TD(04)063, COST 273, Athens, Greece, Jan. 2004.

[NaCB03] K. M. Nasr, F. Costen, and S. K. Barton. An optimum combiner for a smart antenna in an indoor infrastructure WLAN. In *Proc. VTC 2003 Fall – IEEE 58th Vehicular Technology Conf.*, Orlando, FL, USA, Oct. 2003. [Also available as TD(03)088].

[NyJO01] H. Nyberg, C. Johansson, and B. Olin. A streaming video traffic model for the mobile access network. In *Proc. VTC 2001 Fall – IEEE 54th Vehicular Technology Conf.*, Atlantic City, NJ, USA, Oct. 2001.

[OlRG03] J. Olmos, S. Ruiz, and M. García. Description of UTRA FDD link level simulator. description of UTRA FDD link level lookup tables (MORANS). TD(03)095, COST 273, Paris, France, May 2003.

[OrBa02a] J. Orriss and S. K. Barton. Probability distributions for the number of mobiles attached to a base station. TD(02)005, COST 273, Guilford, UK, Jan. 2002.

[OrBa02b] J. Orriss and S. K. Barton. A statistical model for connectivity between mobiles and base stations: the extension to suzuki. TD(02)121, COST 273, Lisbon, Portugal, Sep. 2002.

[OrBa03a] J. Orriss and S. K. Barton. Probability distributions for the number of radio transceivers which can communicate with one another. *IEEE Trans. Commun.*, 51(4):676–681, Apr. 2003. [Also available as TD(01)015].

[OrBa03b] J. Orriss and S. K. Barton. A statistical model for connectivity between mobiles and base stations: from suzuki to rice and beyond. TD(03)089, COST 273, Paris, France, May 2003.

[OrBa04a] J. Orriss and S. K. Barton. A markov chain model for a protocol. part 1: ALOHA with variable packet length. TD(04)081, COST 273, Gothenburg, Sweden, June 2004.

[OrBa04b] J. Orriss and S. K. Barton. A markov chain model for wireless ad-hoc network protocols. part 2: CSMA/CA. TD(04)177, COST 273, Duisburg, Germany, Sep. 2004.

[OrBa05] J. Orriss and S. K. Barton. A markov chain model for wireless ad-hoc network protocols. part 3: RTS/CTS. TD(05)007, COST 273, Bologna, Italy, Jan. 2005.

[OrZB02] R. Verdone, J. Orriss, A. Zanella, and S. Barton. Evaluation of the blocking probability in a cellular environment with hard capacity: A statistical approach. In *Proc. PIMRC 2002 – IEEE 13th Int. Symp. on Pers., Indoor and Mobile Radio Commun.*, Lisbon, Portugal, Sep. 2002. [Also available as TD(02)009].

[OZSI96] B. T. Olsen, A. Zaganiaris, K. Stordahl, L. A. Ims, D. Myhre, T. Overli, M. Tahkokorpi, I.Welling, M. Drieskens, J. Kraushaar, J. Mononen, M. Lahteenoja, S. Markatos, M. De Bortoli, U. Ferrero, M. Ravera, S. Balzaretti, F. Fleuren, N. Gieschen, M. De Oliveira Duarte, and E. de Castro. Techno-economic evaluation of narrowband and broadband access network alternatives and evolution scenario assessment. *IEEE J. Select. Areas Commun.*, 14(6):1184–1203, Aug. 1996.

[PaCV02] G. Pasolini, M. Chiani, and R. Verdone. Performance evaluation of a bluetooth based WLAN adopting a polling MAC protocol under realistic channel conditions. *Int. J. on Wireless Information Networks*, 9(3):141–153, Apr. 2002. [Also available as TD(01)046].

[PaFl95] W. Paxon and S. Floyd. Wide-area traffic: The failure of poisson modeling. *IEEE/ACM Trans. Networking*, 3(3):226–244, June 1995.

[PaTV03] G. Pasolini, M. De Troia, and R. Verdone. Throughput evaluation for a bluetooth piconet with link adaptation. In *Proc. PIMRC 2003 – IEEE 14th Int. Symp. on Pers., Indoor and Mobile Radio Commun.*, Beijing, China, Sep. 2003. [Also available as TD(02)127].

[PBCE05] R. Patelli, H. Buehler, N. Cardona, A. Eisenblätter, M. Feher, H. F. Geerdes, P. Grazzioso, A. Munna, and J. F. Monserrat. Report of the short term mission on MORANS. TD(05)080, COST 273, Leuven, Belgium, June 2005.

[PeGC00a] G. Pei, M. Gerla, and T. W. Chen. Fisheye state routing: A routing scheme for ad hoc wireless networks. In *Proc. ICC 2000 – IEEE Int. Conf. Commun.*, New Orleans, LA, USA, June 2000.

[PeGC00b] G. Pei, M. Gerla, and T.W. Chen. Fisheye state routing in mobile ad-hoc networks. In *Proc. ICDCS – 22th Int. Conf. Distributed Computing Systems*, Taipei, Taiwan, Apr. 2000.

[PELW02] M. Pettersen, R. Eckhoff, P. H. Lehne, T. A. Worren, and E. Melby. An experimental evaluation of network-based methods for mobile station positioning. In *Proc. PIMRC 2002 – IEEE 13th Int. Symp. on Pers., Indoor and Mobile Radio Commun.*, Lisbon, Portugal, Sep. 2002. [Also available as TD(02)067].

[PeRD01] C. E. Perkins, Elizabeth M. Royer, and Samir Das. Ad-hoc on demand distance vector (AODV) routing. draft-ietf-manet-aodv-09.txt, IETF Internet draft, Nov. 2001.

[PeRo99] C. E. Perkins and E. M. Royer. Ad-hoc on demand distance vector routing. In *Proc. of IEEE Workshop on Mobile Computing Systems and Applications*, New Orleans, LA, USA, Feb. 1999.

[Radi01] M. Radimirsch. Analysis of the radio link properties of HiperLAN/2. TD(01)030, COST 273, Bologna, Italy, Oct. 2001.

[Radi02] M. Radimirsch. An algorithm to combine link adaptation and transmit power control in HiperLAN type 2. In *Proc. PIMRC 2002 – IEEE 13th Int. Symp. on Pers., Indoor and Mobile Radio Commun.*, Lisbon, Portugal, Sep. 2002. [Also available as TD(02)082].

[Radi04] M. Radimirsch. *Optimisation of Throughput and Energy Efficiency in HIPERLAN Type 2 Networks.* PhD thesis, Institut für Allgemeine Nachrichtentechnik, University of Hannover (Germany), 2004. published at Shaker Verlag, ISBN 3-8322-3174- 9 [partly available as TD(03)181].

[RaJo03] M. Radimirsch and K. Jobmann. Energy efficiency versus throughput in wireless data networks – a game theoretic model. In *Proc. 2nd Workshop COST 273,* Paris, France, May 2003.

[RaJo05] M. Radimirsch and K. Jobmann. Throughput calculation in a HiperLAN type 2 network considering power control and link adaptation. *IEEE Trans. Wireless Commun.*, 4(4):1798–1807, 2005. Accepted for Publication [Also available as TD(03)004]).

[RaLT98] F. Rashid-Farrokhi, K. J. R. Liu, and L. Tassiulas. Transmit beamforming and power control for cellular wireless systems. *IEEE J. Select. Areas Commun.*, 16(8):1437-1450, Oct. 1998.

[RaWi04] B. Rankov and A. Wittneben. On the capacity of relay-assisted wireless MIMO channels. In *Proc. SPAWC 2004 – Sig. Proc. Advances in Wireless Commun.*, Lisbon, Portugal, July 2004. [Also available as TD(03)124].

[RCFG04] S. Ruiz, L. M. Correia, L. Ferreira, H.-F. Geerdes, A. Grazioso, P. Munna, R. Olmos, J.-J. Patelli, and A. Zanella. Reference parameters for W-CDMA analysis. TD(04)111, COST 273, Gothenburg, Sweden, June 2004.

[RGGH04] S. Ruiz, M. Garcia, G. Guridi, R. Higuero, D. Millas, I. Ramos, and D. Ruiz. RRM parameters obtained through the use of MORANS realistic scenarios and UPC UMTS simulation tool (first simulations). TD(04)181, COST 273, Duisburg, Germany, Sep. 2004.

[RoSc94] C. H. Rokitansky and M. Scheibenborgen. Updated version of SDD. Deliverable R2067/UA/WP 2.1.5/DS/P/68.b, RACE-MBS Project, Brussels, Belgium, 1994.

[SaGV05] D. Sangiorgi, A. Giorgetti, and R. Verdone. Channel based scheduling at the MAC layer in a MC-CDMA system. TD(05)056, COST 273, Bologna, Italy, Jan. 2005.

[SeCo03] A. Serrador and L. M. Correia. Multi-service cell load estimation on UMTS-FDD. In *Proc. WPMC 2003 – Wireless Pers. Multimedia Commun.*, Yokosuka, Japan, Oct. 2003. [Also available as TD(03)118].

[ShBr88] K. S. Shanmugan and A. M. Breipohl. *Random Signals. Detection, Estimation and Data Analysis.* John Wiley & Sons Ltd., New York, NY, USA, 1988.

[StMu95] K. Stordahl and E. Murphy. Forecasting long-term demand for services in the residential market. *IEEE Commun. Mag.*, 33(2):44–49, Feb. 1995.

[Tela99] E. Telatar. Capacity of multi-antenna gaussian channels. *European Transactions on Telecommunications*, 10(6):585–595, Nov. 1999.

[TrVZ02] V. Tralli, R. Veronesi, and M. Zorzi. Resource allocation with power-shaping in TDMA-based mobile radio systems. In *Proc. PIMRC 2002 – IEEE 13th Int. Symp.*

on Pers., Indoor and Mobile Radio Commun., Lisbon, Portugal, Sep. 2002. [Also available as TD(02)075].

[UnKa03] M. Unbehaun and M. Kamenetsky. The evolution of wireless LANs and PANs – on the deployment of picocellular wireless infrastructure. *IEEE Wireless Commun. Mag.*, 10(6):70–80, 2003.

[VaRF99] A. G. Valko, A. Racz, and G. Fodor. Voice qoS in third generation mobile systems. *IEEE J. Select. Areas Commun.*, 17(1):109–123, Jan. 1999.

[VePa04] F. J. Velez and R. R. Paulo. High capacity wideband traffic in enhanced UMTS: a step towards 4G. In *Proc. MCT 2004 – 5th IEE Int. Conf. on 3G Mobile Commun. Tech.*, London, UK, Oct. 2004. [Also available as TD(04)031].

[Verd04] R. Verdone. An energy-efficient communication protocol for a network of uniformly distributed sensors polled by a wireless transceiver. In *Proc. ICC 2004 – IEEE Int. Conf. Commun.*, Paris, France, June 2004. [Also available as TD(04)061].

[VeTr03] R. Veronesi and V. Tralli. DCA with power-shaping (PS-DCA) in TDMA and TDCDMA cellular systems with centralized and distributed control. In *Proc. PIMRC 2003 – IEEE 14th Int. Symp. on Pers., Indoor and Mobile Radio Commun.*, Beijing, China, Sep. 2003. [Also available as TD(02)156].

[VOZB02] R. Verdone, J. Orriss, A. Zanella, and S. K. Barton. Evaluation of the blocking probability in a cellular environment with hard capacity: A statistical approach. In *Proc. PIMRC 2002 – IEEE 13th Int. Symp. on Pers., Indoor and Mobile Radio Commun.*, Lisbon, Portugal, Sep. 2002.

[VTZZ04] R. Veronesi, V. Tralli, J. Zander, and M. Zorzi. Distributed dynamic resource allocation with power shaping for multicell SDMA packet access networks. In *Proc. WCNC 2004 – IEEE Wireless Commun. and Networking Conf.*, Atlanta, GA, USA, Mar. 2004. [Also available as TD(04)052].

[WCFM01] T. Winter, L. M. Correia, E. R. Fledderus, E. Meijerink, R. Perera, A. Serrador, U. Türke, and M. J. G. Uitert. Identification of relevant parameters for traffic modelling and interference estimation. Deliverable D2.1, IST-2000-28088 MOMENTUM, Berlin, Germany, Nov. 2001.

[WHZL02] G. Wölfle, R. Hoppe, D. Zimmermann, and F. M. Landstorfer. Enhanced localization technique within urban and indoor environments based on accurate and fast propagation models. In *Proc. of European Wireless 2002*, Firenze, Italy, Feb. 2002. [Also available as TD(02)033].

[WiRa03] A. Wittneben and B. Rankov. Impact of cooperative relays on the capacity of rank-deficient MIMO channels. In *Proc. 12th IST Summit on Mobile and Wireless Commun.*, Aveiro, Portugal, June 2003. [Also available as TD(03)124].

[WSWH04] P. Wertz, M. Sauter, G. Wölfle, R. Hoppe, and F. M. Landstorfer. Automatic optimization algorithms for the planning of wireless local area networks. In *Proc. VTC 2004 Fall – IEEE 60th Vehicular Technology Conf.*, Los Angeles, CA, USA, Oct. 2004. [Also available as TD(04)130].

[Yaco93] M. D. Yacoub. *Foundations of Mobile Radio Engineering.* CRC Press, Boca Raton, FL, USA, 1993.

[ZaKi01] J. Zander and S.-L. Kim. *Radio resource management for wireless networks.* Artech House Publishers, London, UK, 2001.

[ZaMP03] A. Zanella, D. Miorandi, and S. Pupolin. Mathematical analysis of bluetooth energy efficiency. In *Proc. WPMC 2003 – Wireless Pers. Multimedia Commun.*, Yokosuka, Japan, Oct. 2003. [Also available as TD(03)028].

[ZaTP02] A. Zanella, L. Tomba, and S. Pupolin. On the performance of AODV and FSR routing algorithms on bluetooth scatternets: preliminary results. TD(02)062, COST 273, Espoo, Finland, May 2002.

[ZBLL04] D. Zimmermann, J. Baumann, M. Layh, F. M. Landstorfer, R. Hoppe, and G. Wölfle. Database correlation for positioning of mobile terminals in cellular networks using wave propagation models. In *Proc. VTC 2004 Fall – IEEE 60th Vehicular Technology Conf.*, Los Angeles, CA, USA, Sep. 2004. [Also available as TD(04)195].

[ZWBL04] D. Zimmermann, P. Wertz, J. Bauknecht, and F. M. Landstorfer. Performance of location methods in an urban mobile communication network. TD(04)053, COST 273, Athens, Greece, Jan. 2004.

9 UMTS radio networks

Chapter editors: *Narcis Cardona, Roberto Verdone*
Contributors: *Thomas Kürner, Paolo Grazioso, Jose F. Monserrat, Riccardo Patelli, and Alberto Zanella*

9.1 Introduction

The mobile telecommunication industry throughout the world has shifted its focus to the 3G UMTS technology deployment, investing during the last decade in the design and manufacture of advanced mobile multimedia wireless networks. Since the introduction of 3G technologies in radio access, network planners faced a lot of new challenges when moving from the well-known concepts of 2G to the new and not so predictable 3G networks behaviour. Many of these challenges have been related to the design, planning, tuning and optimisation of operative multiservice radio networks.

Optimisation is a continuous process that is a significant part of the operating costs of the network. It consists not only on fine tuning the BS parameters like antenna tilting, azimuth or pilot power, but also relays on the optimisation process ought to improve the performance of the RRM procedures, which are responsible for ensuring reasonable operation of the network, by providing scenario-specific parameter sets to control the capacity-coverage-cost trade-off. Fast feedback loops in radio access network elements can be considered as adaptive RRM.

Many contributions to UMTS radio network planning and optimisation have been provided from COST 273, as well as some automatic planning algorithms, as explained in Section 9.2 of this chapter. Real-time tuning, which is related to RRM, is the subject of Section 9.3. The fourth section is devoted to present the results on UTRAN performance evaluation by means of simulation tools. Among others, static and dynamic system simulators are compared, and link level effect on system performance is also evaluated. Finally, Section 9.5 deals with the current UMTS evolution and illustrates the additional technologies that are being incorporated into the latest releases of UMTS standards.

9.2 UMTS radio network planning

This section summarises the contributions to UMTS radio network planning. All major topics relevant for the UMTS planning process are covered ranging from propagation and traffic models (Subsection 9.2.1) to the aspects of parameter planning (Subsection 9.2.2) and automatic planning algorithms (Subsection 9.2.3). Finally, in Subsection 9.2.4 alternative infrastructure solutions like site sharing and hybrid networks are covered as well.

9.2.1 Planning fundamentals

UMTS radio networks are very sensitive to variations of the radio environment and the traffic conditions. Theoretical analysis and commercial implementation of CDMA cellular networks trend to show that radio network planning is simplified because of the lack of frequency planning. On the other hand, another complexity is introduced by the fact that interference is not only determined by propagation, but also by the traffic load. This yields the so-called cell-breathing effect, which has to be avoided by proper radio planning. Therefore the fundamentals for any automatic approach for UMTS planning are reliable propagation models and detailed traffic predictions. The more accurate the coverage and traffic predictions are, the closer the expected performance will be to the measured performance. Within COST 273 both traffic and path-loss prediction models have been addressed.

Radio propagation modelling

The importance of accurate radio propagation predictions on urban UMTS planning has been stressed in studies like the one performed by Coinchon *et al.* [CoSW01] for a UMTS network in a 5 km^2 area of Paris. This study concludes that a conventional propagation model could lead to erroneous planning with less than expected quality of service, unacceptable interference and even more BSs than necessary. On the other hand, an accurate ray-tracing model integrated in a UMTS-capable planning tool allows the radio network planner to reach optimal numbers for the BS deployment and configuration while meeting the expected service level requirements. One of the exemplary quantitative results of this case study is that the ratio of rejected calls determined by a ray-tracing model is 14 times higher compared with the ratio derived from predictions using a simple COST 231-Hata model. This means that with ray tracing many more problem areas can be identified in the planning stage.

An analysis of commonly used propagation models applied to UMTS Ultra High Sites (UHS) in urban areas is presented in [HeKü05]. In this context a BS location with antennas mounted at a height of more than 100 m is called a UHS. This principle is used by the German mobile network operator E-Plus for providing coverage in the initial roll-out of its UMTS network. The analysis has shown the Maciel-Xia-Bertoni model is not valid for typical antenna heights occurring with UHS as long as the elevation angle at the BS is larger than 0.15 rad. On the other hand, it is shown that the COST 231-Walfisch-Ikegami model performs well also for BS antenna heights of up to 273 m at distances of up to 2 km. Based on these findings it is possible to define an alternative rules set for the hybrid propagation model in [KüFW96].

UMTS networks and especially the upcoming hybrid networks also consisting of WLAN and broadcast parts, e.g. (see also Subsection 9.2.4) include a huge variety of scenarios ranging from pure indoor systems to large umbrella cell deployments. Hence a prediction model used for automatic planning of such networks has to cover all possible deployment scenarios and operational environments. Therefore adaptive propagation models are required, which enable automatic selection of the propagation model components. In [KEGJ03] a general framework for a fully automatic and adaptive selection of propagation models has been developed taking into account different deployment scenarios (macro-, micro- and indoor cells) and provides transition models for predictions in areas where digital terrain databases of different resolutions are available. Based on this framework a specific implementation uses propagation model components from [KüMe02]. A key issue for such multi-environment radio predictions is the proper representation and handling of digital terrain data. This problem is addressed in two contributions. The approach presented in [KEGJ03] deals mainly with the aspects of transition between areas where the digital terrain models have a different resolution and granularity. This is important for a complete interference calculation between cells located in different areas A_i, see Figure 9.1. Typically data in A_1 consists of digital terrain height and land use data available in a resolution of 50–200 m. Usually such data is available for a whole country

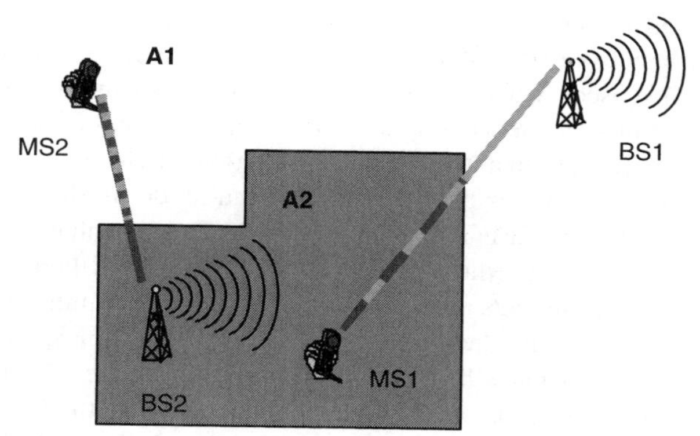

Figure 9.1 *Digital terrain models with different resolution and granularity; terrain profiles between BS and MT.*

or network. Data in A_2 consists of detailed building data with a resolution of a few metres. The two directions of transitions are basically done by transforming the terrain profile vectors. In case of transition from A_2 to A_1 generic buildings are set at those positions of the profile vector where the land use class in A_1 is of urban type.

For the other direction of transition the concept of statistical database is used. Such a statistical database consists of a set of raster layers in A_2, available in the same resolution as the data in A_1, containing mean values of building heights, street width, etc. derived from the detailed building data. In [CoLA03] the importance of perfectly adjusted outdoor map data and 3D building models in a common coordinate system is emphasised. Outdoor map data consists of closed 3D polygons describing the contour of the buildings and of raster layers containing the terrain height. The resolution of this data is a few metres and the accuracy of the height is in the order of 1 m. The 3D building model consists of more detailed data of single buildings derived from architect plans, which are available either on paper or in CAD files. In this data the location and width of all partitions are generally given with high accuracy in the order of a few centimetres. A procedure on how to adjust these two data sets in order to use them in a single multi-environment prediction is given in [CoLA03] allowing a joint analysis of distinct radio systems in outdoor and indoor environments. As an example, results of a joint planning for WLAN transmitters distributed inside a four-floor building and UMTS BSs located in macrocellular sites in the vicinity of this building are shown.

In order to allow such an analysis, prediction models have to incorporate radio propagation into buildings. Martijin and Herben [MaHe03] present results from signal strength measurements in four office buildings at The Hague illuminated by a GSM1800 outdoor BS with antenna above rooftop. From these measurements the main characteristics concerning signal attenuation and variations at different floors of the building are derived from these experiments. While on the lower floors the large-scale fluctuations follow a log-normal distribution, it is shown that large fluctuations occur between average signal levels in line-of-sight and non-line-of-sight areas of the higher floors of multifloor buildings. This yields to a non-linear relationship between the floor height and extra gain with respect to the ground floor level. Similar findings have been made by Kürner and Meier [KüMe02]. Based on these findings they propose a building penetration method, which can be used as an extension to outdoor propagation models using high-resolution building data. In this method the received signal level at the ground floor is derived from the outdoor signal level at those pixels in the vicinity of the building. The signal level at the higher floors is computed based on a simple empirical height gain model, derived from a large measurement campaign, in case no line-of-sight between the BS and the building exists. If line-of-sight between the BS and at least parts of the building exists, a more advanced semi-empirical approach is used. A first verification with measurements shows promising results.

As part of the optimisation of a site's configuration (see Subsection 9.2.4), various sector configurations may be evaluated, differing in antenna type, installation height, azimuth and tilt. The explicit generation of high precision path-loss predictions for a multitude of alternative configurations is neither practical nor necessary. In [EFFG03] an approach to generate antenna-specific path-loss predictions from unmasked path-loss predictions is presented. Such a method is especially important for applications in urban environments with relevant contributions coming from multipath signals using time-consuming ray-tracing prediction models. Among other things, an interpolation scheme for antenna height variations and a new heuristic method to estimate the 3D antenna diagram from a horizontal and vertical antenna diagram are presented. Although the method delivers only an approximation of the path-loss predictions the accuracy is good enough for its application within automatic RF optimisation algorithms.

More details about propagation models developed in COST 273 can be found in Section 4.3.

Traffic modelling

The initial traffic dimensioning in UMTS planning is of crucial importance as has been demonstrated by Ruiz and Olmos [RuOl01]. Traffic maps are derived by assuming uniform traffic densities for four different services in three different environments (urban cells, rural cells and cells covering important roads or highways). Simulations have been done for the uplink using a static multiservice simulator. One of the main results is the observation of the reduction of the coverage area when multiservice traffic is considered. It is concluded that a good estimation of the traffic density for the different services is of extreme importance. Furthermore a more realistic pixel-based approach for traffic maps is recommended. In [KüHe05] Kürner and Hecker compare different models to derive pixel-based traffic and mobility maps with Operation and Maintenance Centre (OMC) measurements from a real network. The traffic models are based on detailed geographic data (population data, street networks, land use, etc.). Although the focus has been set on those relevant parameters for location area planning, this work is considered the first step towards the development of more advanced traffic and mobility models. Inter alia, a strong correlation between the number of handovers in the connected mode and the number of location updates in the idle mode has been observed. Furthermore it was shown that more complex modelling approaches like the MOMENTUM-Traffic model are able to improve the accuracy of traffic predictions. New ways of facing the problem of mobile multimedia source traffic modelling are discussed in [ReVe03] and [VePa04]. Based on the deployment scenario definition from the IST-(Simulation of Enhanced UMTS Access and Core Networks SEACORN) project results in a teletraffic system capacity applicable to cellular planning purposes have been achieved using a macroscopic model. This allows a comparison between capacities of different systems, like GSM/GPRS, UMTS, enhanced UMTS and mobile broadband systems in terms of supported data rate per square km. From this comparison a law for the evolution of data rates per square km similar to Moore's law has been derived.

9.2.2 Radio parameter planning

In the case of UMTS networks, capacity and coverage planning cannot be separated. As a consequence, aspects of radio parameter planning already have to be considered in coverage planning steps.

Means to plan for coverage and capacity in the pre-operational phase of UMTS networks are introduced in [LaWJ02]. The planning process starts with the estimation of the initial site density taking into account both uplink and downlink budgets. At this stage the capacity-coverage trade-off in 3G networks is considered with the interference margin term in the Radio Link Budget (RLB). The coverage limited network performance can be enhanced by improving any item in the RLB. Typically this can be achieved by introducing higher transmit powers, antennas with higher gains, Mast Head Amplifiers (MHA) or by improving diversity solutions. The initial RLB-based coverage planning has to be fine-tuned using a capacity analysis based on information of traffic growth during network evolution. In this analysis step interference control is essential. A typical means to achieve this with radio network planning is the selection of proper antenna parameters. This includes the selection of the antenna type, the main direction of the antenna, the type of sectorisation and especially the antenna tilting. Figure 9.2 shows the impact of antenna tilt on the interference. Furthermore, it is important to understand which link is the limiting one, and whether it is capacity or coverage being limited in order to choose the correct performance enhancement methods. For example, increasing the BS transmit power has only limited potential in the case of a capacity limited scenario.

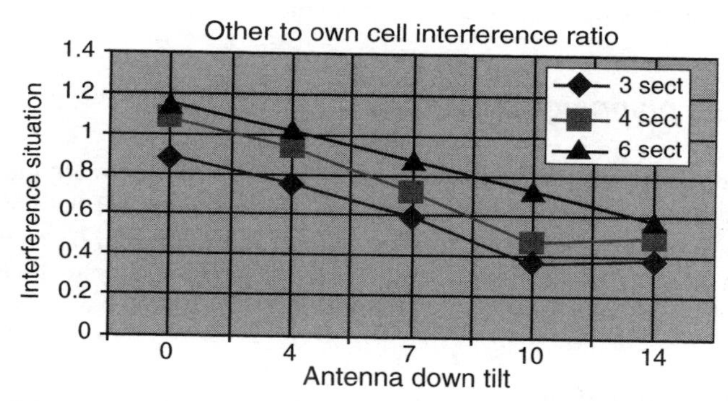

Figure 9.2 *Impact of antenna tilt on interference situation. Interference-wise, all configurations have optimum antenna beamwidth.*

An alternative way of generating input data for the UMTS simulation is proposed by Ruiz *et al.* [RSAG03] based on the assumption that GSM/UMTS co-siting will be used in the 3G network deployment. This means that real data extracted from the GSM network can be used to optimise 3G planning and RRM issues. Therefore a 3G planning simulation tool has been developed which, instead of using theoretical models, benefits from the availability of realistic propagation and mobility as well as spatial and temporal traffic distributions. In [RSAG03] the tool has been used to study the effects of Pilot Power and AS parameters on both UL and DL capacity in a real urban scenario. It is shown that for the UL there is always an absolute capacity increase when increasing the AS size and window with a lower relative increase for AS size higher than 3 and a window size higher than 3 dB. The absolute gain in this case is optimum, when all BSs in the scenario transmit with the same CPICH (Common Pilot Channel) power. The DL is very sensitive to CPICH values and a fast decrease in capacity is obtained when increasing the pilot power and AS size. It was found that the optimum AS configuration is an AS size of 3 for both UL and DL.

Once a UMTS network is operational it has to be continuously monitored and optimised. Laiho [Laih02] explains the control hierarchy in the optimisation loops in a cellular network and stresses its importance for the coverage-capacity trade-off management. Therefore a statistical optimisation loop running in the Network Management System (NMS) is introduced. This feature improves the non-optimum setting of configuration parameter values on a per-cell or cell cluster basis. In the admission control example this feature increases the network capacity, especially when the operator has to set the target values cautiously to a low level in order to ensure the required quality. Other potential areas for the implementation of the control hierarchy are soft handover optimisation and inter-/intrasystem traffic balancing. The main conclusions drawn from the results shown in [Laih02] are that the proposed statistical control layer improves the system performance in comparison to the default settings and manual optimisation.

An approach to incorporate propagation, coverage and network performance measurements from real environments into radio network planning is introduced by Ortiz [Orti02] using a portable UMTS radio beacon. This UMTS radio beacon simulates a basic Node B by radiating a UMTS reference signal, which is generated as a combination of the pilot channel, the synchronisation channel and the broadcast channel. It enables coverage measurements to be made before the deployment of the network and also allows pilot level comparison, interference simulation and the testing of the user equipment synchronisation process. Such measurements are extremely helpful in urban and indoor environments, where propagation and interference characteristics are very difficult to simulate.

9.2.3 Automatic planning algorithms

This subsection reports on contributions to automatic UMTS radio network planning. The primary focus here is on the pre-operational planning of Node B locations and the configuration of Node B antennas (assuming mostly a fixed number of sectors). In addition, one contribution deals with operational network tuning. More details about UTRAN performance evaluation are included in Section 9.4.

Uplink interference estimation

Uplink interferences in UMTS networks are commonly characterised by the so-called 'frequency reuse factor' F. This factor is defined as the ratio of the interference generated with a cell and the total interference (intra- plus intercell interference). A good estimate of F, taking the traffic distribution into account, is of interest, because it may replace the traditional assumption in cell dimensioning that intercell interference is simply a fraction of the intracell interference [HoTo02]. Heideck, Draegert, and Kürner [HeDK02] derive two formulas for estimating uplink interference coupling among a (omnidirectional) cell and its first two neighbouring tiers in a hexagonal network since higher tiers are reported to yield no significant interference contribution to the central cell. The formulas are obtained from Monte Carlo simulations under varying traffic profiles. The first formula is obtained assuming that the traffic is homogeneously distributed within each tier, but may vary between the tiers:

$$F = \frac{1}{1 - 0.0005 \cdot T_0 + \frac{0.530 \cdot T_1 + 0.155 \cdot T_2 - 0.409}{T_0 + 2}} \tag{9.1}$$

The parameters T_0, T_1, and T_2 express the traffic in the centre cell (T_0), each cell (T_1) in the first tier (T_2), and each cell in the second tier, respectively. Extensive simulations suggest that the maximum estimation error is below 5%. The second formula reflects the insight that interference is mostly caused by mobiles close to the cell border of the central cell, through the introduction of coefficient functions μ, ν which generate the segment-based coupling fractions of first and second tiers respectively:

$$F = \frac{1}{1 - 0.0005 \cdot T_0 + \frac{0.530 \cdot \mu \cdot T_1 + 0.155 \cdot \nu \cdot T_2 - 0.409}{T_0 + 2}} \tag{9.2}$$

The prediction accuracy for the second formula is stated to be 6%.

Comparison of single snapshot models for uplink

Amaldi, Capone, and Malucelli [AmCM01] propose discrete optimisation models for the optimised selection of Node B locations. The models consider the SIR as quality measure and differ in the level of detail they account for the specific power control mechanism. These models improve on previous work of the authors, where the effect of power control is modelled assuming a fixed interference level. The most simple model imposes that all users are assigned to a Node B. The number of users a Node B can serve is limited explicitly, since each user has to be received with a predefined reception power, and the ratio between intra- and intercell interference is fixed. The model aims at minimising the cost of installing Node B and the distance between users and their serving Node B. In the second model, the ratio between intra- and intercell interference is no longer assumed constant, but varies with user positions. In the third model, the actual loading of a cell is monitored based on SIR constraints. The required reception power at the Node B depends on the amount of interference. The authors describe greedy and Taboo Search (TS) algorithms for the optimised selection of Node B locations. The different models and algorithms are compared with synthetic 'realistic' planning instances. The results of the TS are comparable to those obtained by the greedy method followed by a short TS, but the TS alone requires less running time. The first optimisation model turns out to be too simplistic. The third (SIR-based) model yields the cheapest solutions, but requires much more computational effort than the other models.

Single snapshot model with up- and downlink

Amaldi *et al.* [ACMS03] describe a mathematical programming model for locating and configuring Node Bs that aims at maximising coverage and minimising cost. All Node Bs are assumed to serve three 120° sectors that, taken together, have an omnidirectional radiation pattern. The sector directions are subject to optimisation. Varying antenna tilts are not considered. The optimisation model takes up- and downlink dedicated traffic links as well as pilot channels into account. The non-homogeneous traffic distribution is modelled via test points (for example, one traffic snapshot) with varying service requirements. A Greedy Randomised Adaptive Search Procedure (GRASP) and a TS algorithm for solving the optimisation problem are proposed. Computational results are reported for synthetic 'realistic' scenarios with multiple services. The authors recommend to apply the TS in two phases to drastically reduce the overall computation time (a factor of 50 is observed). In the first phase, a simplified uplink model is used to determine a starting solution for the second phase. The latter phase is then based on the full model.

Multiple snapshot model with up- and downlink

Eisenblätter *et al.* [EKMA02] introduce a model for the optimisation of the location and configuration of Node Bs, including sectorisation, antenna installation height, type, azimuth, electrical and mechanical tilt. The paper contains a detailed discussion of technical background, and explains competing planning goals and common approaches to network performance simulation. The major part of the work is spent on presenting a detailed mixed integer programming model that allows to capture network coverage and capacity in the presence of heterogeneous multiservice traffic. The model accounts for dedicated traffic links in up- and downlink, the pilot channel, limited downlink code budget, limited maximum interference rise in uplink, reduced SIR requirements in the case of soft(er) handover, Node B and user equipment capabilities, etc. Multiple snapshots are used to test a network design for achieving coverage and capacity goals. No specific objective function is defined, but alternative objectives and applications of the model are described. The model as presented is hardly solvable using state-of-the-art mathematical programming techniques. It is positioned as a reference point during development of simplified models and specialised solution techniques.

Planning realistic networks with mathematical programming.

Based on the model sketched in the previous paragraph, Eisenblätter *et al.* [EFGJ04] report on two developments. Two cases are described in which mathematical programming is used to solve the comprehensive mathematical model for small and/or simplified planning tasks. The solutions for the small tasks are then combined to present an overall optimisation result. In another line of research, a local search procedure is used to obtain heuristic solutions to the comprehensive model (details are given in [EFFG03]). The application of mathematical programming as well as the heuristic procedure are applied to realistic planning scenarios for the cities of The Hague [EGKT04] and downtown Berlin. The solutions obtained from solving the mixed integer program require more sites and cells but produce significantly better network coverage and quality.

Tuning realistic networks based on average traffic load

Network optimisation based on snapshots has a major drawback. Even for reliable statistics on the network performance under heterogeneous multiservice traffic a large number of Monte Carlo simulations have to be performed. The number of snapshots required to reliably optimise the network design is even larger. The available solution methods for the above-mentioned comprehensive optimisation model allow only for, say, up to 10 snapshots. Eisenblätter and Geerdes [EiGe04], [EiGe05] report on a novel optimisation model that is not based on traffic snapshots. The recent characterisation of the up- and downlink cell load in a network through a linear equation system is extended from snapshots to the average traffic distribution. The average-based performance evaluation can be used as a surrogate for multisnapshot analysis of a network, significantly speeding up local search methods. Moreover, a mathematical optimisation model (called matrix design) is sketched to design networks such that a favourable cell coupling matrix is obtained (which is the core of the linear equation system). Computational results are given based on the three public MOMENTUM scenarios, The Hague, Berlin, and Lisbon, see URL http://momentum.zib.de. The task is to optimise the electrical and mechanical tilts of all antennas in a given network. The results obtained from local search and from solving the matrix design model are comparable in solution quality. These results clearly outperform those obtained from the snapshot-based optimisation approach.

Automatic reconfiguration of an operational network

Sharma and Nix [ShNi02] describe a 'situation awareness' functionality and explain how incorporating intelligence in node Bs can complement radio network planning. Such a functionality allows to sustain an initial network design over time by automating the replanning process. Radio resources can be employed more efficiently with an optimised network design. To this end, variations in the propagation environment or the traffic intensity trigger coverage reconfigurations. The monitoring is based on Node B and mobile measurements. A practical implementation of this idea is proposed that adapts antenna beam directions and pilot powers via a (centralised) genetic algorithm. Several examples illustrate the applicability of the approach, e.g. reconfiguration after a sector failure (self-healing) or the deployment of a new Node B. The effects of reconfiguration are discussed with respect to up- and downlink performance.

9.2.4 Alternative infrastructure solutions

UMTS radio networks enable network operators to provide a multitude of different services and data rates. A couple of problems have been identified, which are hard to solve using conventional ways of building network architectures. This situation has fostered the development or suggestions of a large number of different alternative approaches to build a network or to combine different air interface technologies. Some of these approaches have also been presented in COST 273.

One of these alternatives is also to use the TDD mode of the UMTS standard. The currently deployed UMTS networks are using the FDD (Frequency Division Duplex) mode of the UMTS standard. Although the corresponding spectrum has been allocated to many network operators, TDD has not been used yet. In one contribution to COST 273, Butler [Butl02] has highlighted and addressed some of the most critical issues in planning UMTS-TDD networks arising in such systems operated in an interference limited indoor environment.

An alternative roll-out strategy from UMTS networks is to share sites among different operators. This has advantages in terms of costs and mitigates the problem of finding enough suitable sites. Apart from these rather non-technical benefits, site-sharing contributes to the reduction of mutual interference problems. Especially in dense urban areas a large number of new sites are required. In conjunction with the low number of different carriers (typically two to three) per operator it is likely that different operators transmit in adjacent carriers, causing mutual interference problems in high traffic density areas. Grazioso and Varini [GrVa01] have investigated this problem using a dynamic simulator showing that benefits can be achieved in terms of mutual interference if the operators share sites. Apart from this more general result, some more interesting aspects are highlighted by analysis of the details. By considering two operators loaded with uniform and equal traffic density, sharing sites allows significant improvements in terms of several quality criteria: blocking and dropping probabilities, unsatisfied users and noise rise. These advantages are even clearer when cells from two operators are considered loaded with unequal traffic levels. The largest benefit is observed in the DL and for the operator with the higher load. Furthermore the simulation revealed that site-sharing configurations are more resilient to malfunctioning or failures that affect the transmitter or receiver filter, causing a reduction in Adjacent Channel Leakage Power Ratio (ACLR) and Adjacent Channel Selectivity (ACS).

Another drawback of a pure cellular network can be observed, when many users located in the same cell ask for the same content at the same time. In this case a combination of a cellular network (point-to-point) with a broadcast network (point-to-multipoint) is beneficial. In 2004 the DVB-H (Digital Video Broadcast-Handheld) was introduced enabling the reception of digital television on mobile phones. [UnKü04] and [Fusc04] describe different aspects of DVB-H. Fuschini [Fusc04] studies the use of urban and suburban gap fillers for DVB-H. It is shown that due to the constructive combination of more than one useful signal in single frequency networks a satisfactory coverage for portable reception can be achieved by adding an adequate number of transmitters to the conventional planned network. Further advantages of such a solution are a meaningful reduction of power emissions, a more rational management of resources (additional transmitters are used where they are mainly needed) and an easier convergence with 3G cellular networks due to the more widespread structure of the broadcasting network. This convergence is considered by Unger and Kürner [UnKü04] investigating scenarios for radio planning of DVB-H/UMTS hybrid networks. In this investigation a gain parameter for hybrid networks is defined based on data rates and volume of traffic. This gain is used to describe the benefit of a hybrid network and to determine DVB-H cell positions and sizes. Four traffic scenarios from the Berlin MOMENTUM public scenarios including typical unicast and broadcast services have been used in the simulations. For scenarios having scenario-wide distributed DVB-H users, large global cells should be used, whereas for scenarios with locally limited DVB-H usage areas small cells covering these areas should be used. The resulting unloading of UMTS cells for one of the scenarios is depicted in Figure 9.3.

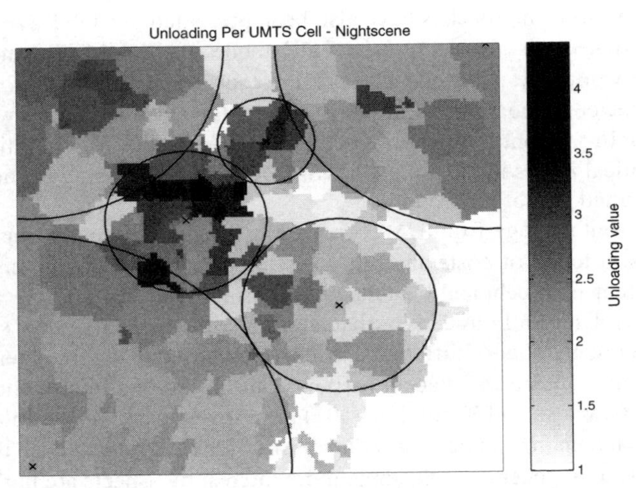

Figure 9.3 *Unloading of UMTS cells when using a hybrid DVB-H/UMTS network.*

9.3 UTRAN optimisation and radio resource management

9.3.1 Soft handover and site selection diversity transmission

UMTS allows two alternative ways to implement macrodiversity: the standard Soft Handover (SHO) when multiple traffic links are contemporarily active between the MT and all the base stations in the AS, and SSDT, where only the best server transmits traffic data to the terminal, while all other BSs within the AS keep only the DPCCH alive, while no power is used to transmit data to the terminal [3GPP04].

The performance of both techniques is affected by a number of parameters, which include:

- AS size, which is obviously equal to one for SSDT and larger than one for SHO;
- the thresholds, in dB, that control the inclusion of BSs into the AS and their removal from it;
- the measurement windows, in s, that are used to determine the inclusion of BSs into the AS and their removal from it.

Furthermore all the above parameters are not characterised by absolute optimum values that can be used in every situation: rather, their values shall be adapted to the main parameters that characterise the cellular layout (cell radius), user mobility (e.g. fraction of vehicular/pedestrian users over the total), environment (clutter, building height, street width, etc.).

The above topics raised interest within COST 273, where several temporary documents were devoted to these techniques, to evaluate their performance as well as their interoperation with other optimisation techniques. For these reasons, they are dealt with also in other subsections of this chapter.

The work presented in [CRBR02] focuses on SSDT, applying it to a hexagonal cell layout with 19 cells, i.e. a central cell and two complete surrounding tiers. The authors derive analytically the probabilities of each BS to be inserted in the AS of a mobile situated in a given location. SIR-based power control is implemented.

The study is performed in the DL, by evaluating both inter- and intracell interferences, which allows to compute E_b/N_0 for each pixel and eventually other outputs, i.e. the minimum required link power, the number of transmitted codes (i.e. the cell capacity) and the number of required codes (i.e. resources).

These outputs are computed as functions of the SHO margins and of other variables, such as the traffic mix, cell radius, user spatial distribution, path-loss exponent, shadowing standard deviation and correlation. As a consequence, the optimum value of the SHO margin can be computed. Figure 9.4 shows the system capacity as a function of the cell radius: we may observe that the optimum value for M_{SH} is generally around 5 to 6 dB.

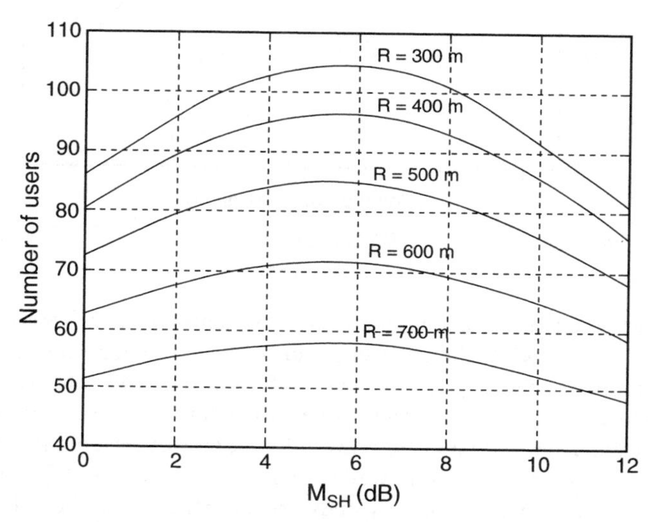

Figure 9.4 *System capacity vs. M_{SH} for different cell radii.*

These and the other results presented in the paper show how SSDT allows an increase in capacity, but optimum soft handover margin and number of channel elements are higher than those obtained for the standard SHO case.

A similar scenario, consisting of 19 hexagonal cells, was adopted by the authors of [BaMV03] (an extension of the model to consider sectorised cells can be found in the paper). The paper deals with the impact of SHO on DL capacity, extending the model and the results already presented in [PiVe02].

The main parameters optimised in this study are the AS size and the SHO margin, M_{SH}. Users are generated according to a uniform distribution, and for each user the E_b/I_0 is evaluated, taking into account also the effect of SHO. The DL transmitted power, set by the BS for each link, is such that the same E_b/I_0 contribution is provided by all BSs in the AS.

The cell capacity is evaluated by computing the total DL transmitted power and maintaining that it remains below its maximum. Users, whose positions are generated with a Monte Carlo method, are added one by one until the power limit is reached. Several such snapshots of the system are taken in order to obtain reliable results. The number of served users which have the observed base station as best server determines the capacity of its cell.

Figure 9.5 shows the number of users served by the reference cell are given as functions of the SHO margin M_{SH} for various values of the maximum AS size.

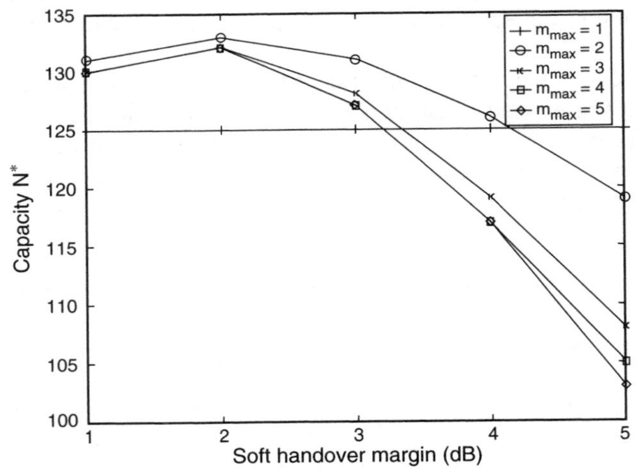

Figure 9.5 *Number of the users N* served by BS_1 vs. M_{SH} for m_{max} = 1, 2, 3, 4, 5. Omnidirectional system.*

From the figure we note that the maximum capacity is achieved with AS size = 2, with a slight degradation for AS size > 2, while the optimum value of the SHO margin is 2 dB. If the margin is increased to 4 dB or more, the capacity is lower even of the case of AS size = 1 (i.e. no macrodiversity at all) which is represented by the horizontal line in the figure. This confirms what was stated previously, i.e. the fact that the optimum M_{SH} is lower for the case of SHO than when SSDT is adopted.

The effect of AS size on system capacity was evaluated also in [RuGO02], which again considers only the DL. The parameters analysed in this study are the AS size and the SHO window size. The maximum AS size is allowed to range from 1 to 3, while the SHO window can be either 3 or 6 dB.

The decision about the inclusion of a given BS in the AS is made on the perceived E_c/I_0. A fast power control algorithm is implemented in the simulator.

The system capacity is defined as the offered traffic when one of the cells has more than 5% of degraded UEs (i.e. UEs which require more DL power than the maximum allowed for each link).

The mobiles are uniformly distributed over the service area, and the authors consider different scenarios: rural, urban with vehicular users, and urban with 60% of indoor users.

For each run a large number of snapshots are taken to ensure statistical reliability of the results. The tool allows to study several outputs, namely: capacity, DL load factor, UL UE distribution according to the macrodiversity status (AS size 1, 2 or 3), percentage of users in SHO, BS and UE transmitted power distribution.

Figure 9.6 shows the system capacity for the urban and rural scenario. We note that there is a significant increase when using AS = 2 instead of 1. Further increasing the AS size (from 2 to 3) as well as increasing the SHO window (from 3 dB to 6 dB) gives significant improvement only in the urban environment.

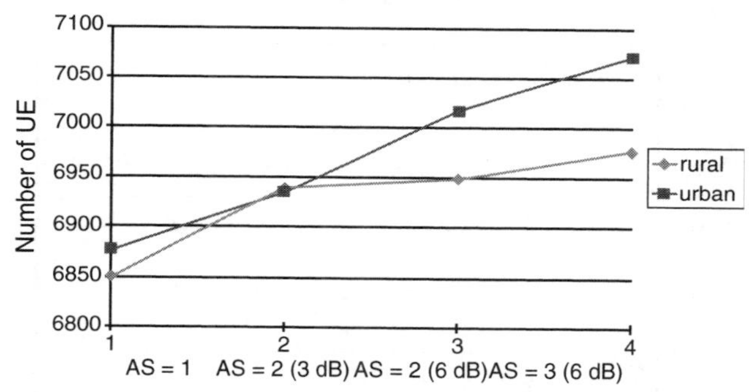

Figure 9.6 *Total scenario capacity (number of simultaneous voice UEs for different AS parameters).*

The main conclusion of the authors is that, by a suitable choice of the AS parameters, a capacity increase around 4% is achievable in the best case. Furthermore, increasing the AS size to more than 2 does not give significant advantages.

Another study of the AS parameters is reported in [GrBa03]. The work is performed with both a static and a dynamic UMTS network simulator. The authors evaluate the impact of the parameters controlling the ADD/DROP thresholds for inclusion and removal of a BS into/from the AS of a given mobile: thresholds in both received power values (named ADD and DROP thresholds) and in time domain (ADD_DELAY and DROP_DELAY) are considered.

Simulations are performed on a regular hexagonal layout with 12 BSs equipped with omnidirectional antennas, the cell radius is 500 m and wrap-around of the service area is implemented to avoid border effects. Users are uniformly distributed over the service area and the considered service is streaming data at 144 kb/s.

The analysis of power thresholds yields the best results for ADD threshold of 3 dB; the DROP threshold has only minor impact. Figure 9.7 shows the dropping probability as a function of the user speed for various values of these thresholds, obtained with the dynamic simulator.

The analysis of time domain thresholds, which can be found in the paper, showed that for UL the best results are achieved with small ADD_DELAY and large DROP_DELAY, while in the DL the situation changes because too many users in SHO imply a waste of resources. A trade-off should be sought.

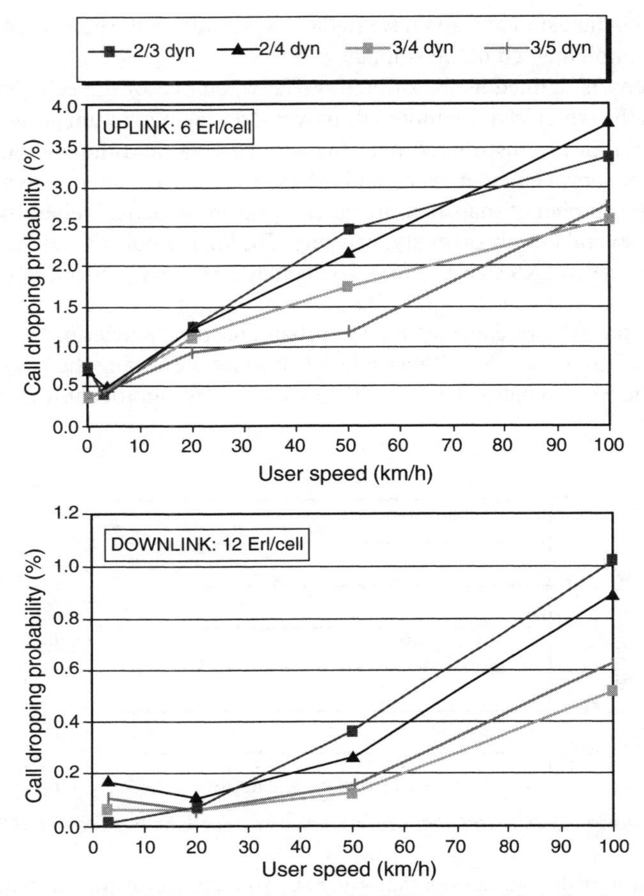

Figure 9.7 *Call dropping probability: effect of the ADD/DROP thresholds (all values in dB).*

The paper shows that the thresholds influence the fraction of users in SHO, which should be controlled. If too few users are in SHO a user may lose all the BSs in his AS before adding a new one, which results in a dropped connection for the user. On the other hand, having too many SHO links implies that some resources are wasted and are not usable to serve other calls in unfavourable conditions. This causes unsatisfied users.

9.3.2 BS parameter adjustment

With the introduction of 3G networks, radio planning has gained in complexity. Coverage and capacity have become tightly coupled and must be treated as a whole. Moreover, the use of new and more complicated RRM algorithms and the WCDMA system nature imply that radio planning will have to deal with many more variables.

Some of the most important ones are BS parameters, which show a significant impact on the network capacity. Examples are: physical antenna settings (height, tilt, azimuth and antenna pattern), number of sectors per site, antenna type (conventional or array), assigned powers to common control channels and SHO policies. These parameters are configurable and contribute to control the operation of the system. Unfortunately, they are highly interdependent and influence the system in a non-linear way. As a consequence, their optimal adjustment is not an easy task. Since efficient network planning appears of utmost importance, different approaches are proposed to achieve an optimal or quasi-optimal network performance.

The authors in [GRHO04] address the problem of adjusting antenna downtilt in the framework of a UTRA-FDD system. Its impact on network capacity and RRM parameters is evaluated by means of a static system level simulator.

Thanks to a downtilt increase, both UL load factor (η_{UL}) and UEs transmitted power (P_{TX-UL}) can be reduced until an absolute minimum. In the case of η_{UL}, the downward trend is explained by a better isolation of intercell interference. The required power also decreases but its minima position depends on the combination of two effects: users' position with respect to the direction of the main radiation lobe and the η_{UL} value itself. Thus, the optimum downtilt angle for individual users is not the same as the optimum for the whole cell. Minima position will just coincide in highly loaded cells.

If minimum P_{TX-UL} is imposed, users will be at a temporary optimal situation, an unstable point. Slight changes in the system will imply sharp changes in the required power and the optimum situation will be left. The closer the angle to the η_{UL} minimum is, the less the changes in P_{TX-UL} are. The system shows a more stable behaviour but, on the other hand, UEs must transmit more power, Figure 9.8. Therefore, fixed downtilt angles would lead the system to poor performance rapidly. Adaptive and remote electrical downtilting arises as an interesting option. This would allow an increase in capacity or the use of services requiring higher QoS.

Regarding pilot powers adjustment, the work presented in [GaRO03] proposes a technique based on simulated annealing to globally optimise the network performance. Pilot signals are modified so that the traffic is balanced. UEs are forced to transmit to the cell that requires less power from them and thus capacity can be increased.

In order to tackle the problem, it has been defined as a resource allocation one. This means that a set of possible pilot powers has to be assigned to BSs so that a cost (or fitness) function is optimised. As the η_{UL} of all BSs is aimed to be jointly reduced, the authors propose the minimisation of the sum of all UL powers as an appropriate cost function with very low computational requirements. Different constraints regarding the range of possible pilot powers and the maximum percentage of users not reaching the E_b/N_0 target are also imposed. The algorithm assesses the network performance for different combinations of pilot powers and accepts a new solution whenever it improves the current cost value. Unlike simple local search, the algorithm explores the solutions space without being directly trapped in local minima, since movements towards worse solutions are allowed with a certain probability.

The simulations are carried out in a realistic UTRA-FDD system level simulation, based on real attenuation measurements from a GSM network. Results show that the algorithm is able to find pilot power combinations that imply a reduction in the load factor (both mean and standard deviation) of all the cells in the system. On the other hand, execution time arises as its main drawback. Nevertheless, on the basis that traffic patterns are repetitive during a day and in different days of the week, the system could be trained just for those particularly differentiated periods of time.

The works in the previous paragraphs focused on pilot powers and downtilt angles separately. On the other hand, [GeJN03], [GJCT03], [GJCT04] propose and compare algorithms which adjust both parameters.

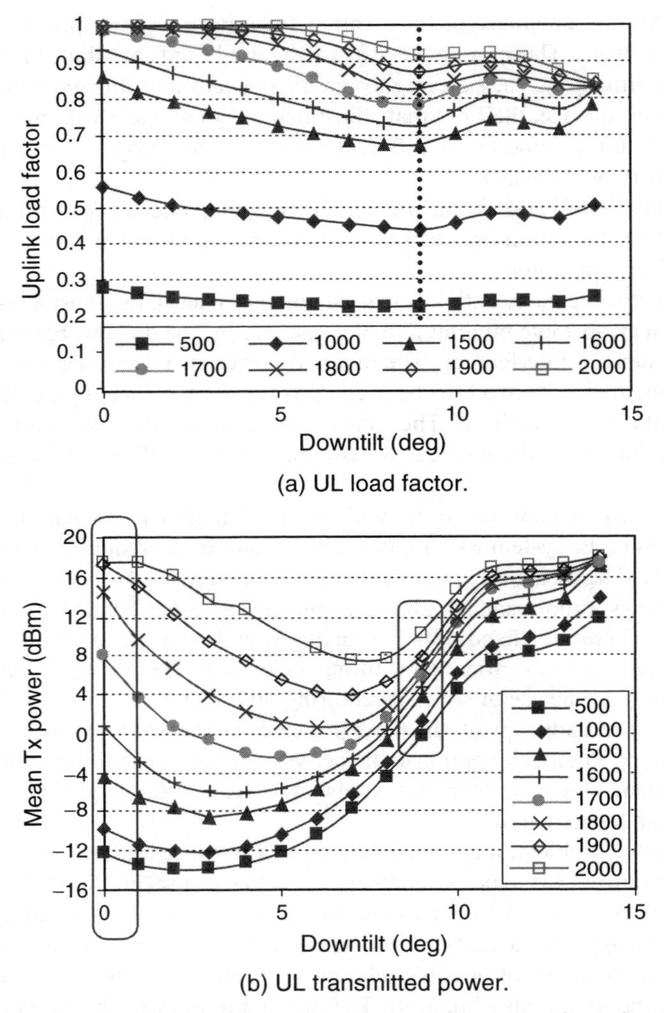

Figure 9.8 *Evolution of UL load factor and mean UL transmitted power vs. downtilt angle.*

Again, a fitness function is defined to represent the optimisation goal, which in this case is the number of served users. Moreover, a certain target value for the Grade of Service (GoS) is introduced and new users are added to the system whenever it is clearly exceeded. The GoS is defined as the quotient between the number of served users and the number of existing users.

Initially, in [GeJN03], a rule-based mechanism is developed. Different rules are defined before running the optimisation itself, some are devoted to pilot powers reduction and the others increase downtilt angles. These modifications are justified because they are aimed at minimising pilot pollution and intercell interference. Lower limits must also be defined so that coverage is guaranteed.

Once this has been done, the optimisation process is run starting from a certain initial network configuration. All the cells with unserved mobiles are sequentially evaluated and their pilot powers and tilt angles are alternatively modified. Each rule is repeatedly run and worse results can be accepted at any time. However, when advancing to the next rule, the best result is always adopted. Subsequently, this approach is extended by modifications derived from simulated annealing. Specifically, the new algorithm allows worse solutions to be accepted with a certain probability.

Results are obtained for a European city scenario with 25 sites (three sectors per site) with nine of them to be optimised. Both approaches imply a capacity gain; nonetheless, figures show the convenience of introducing simulated annealing modifications. In this case the improvement reaches 56.3% in front of 52.2% achieved with the pure rule-based method. Conversely, the second approach is about 20% more time consuming.

The third proposal [GJCT03], [GJCT04] is also a further development of the rule-based approach. One of the main differences is that both pilot powers and downtilt angles are jointly modified by the optimisation process. That means one rule will consist of one instruction for the pilot power and one instruction for the antenna tilt simultaneously.

Moreover, the algorithm checks whether a certain cell is heavily loaded or not and reacts suitably. This is done by means of performance indicators which reflect UL and DL cell loading conditions. Thus, a highly loaded cell will decrease the pilot power and will also downtilt its antenna. On the other hand, the approach also allows power increase and antenna uptilting in those cells with a low user density. So, this implies one more difference with respect to the previous algorithms since load balancing is now actively searched.

Results show a higher gain than that with the previous approaches. The mean number of served users in the area of interest was equal to 511 before optimisation and 831 after optimising 50 snapshots. This leads to a capacity gain of 62.6%. On the other hand, the optimised network is less stable with respect to user distributions. The standard deviation of the number of served users has been increased from 1.48% to 3.7%. Further work concerning these results can be found in [Gerd04].

Up to this point, the described approaches cope with the problem by means of optimisation algorithms which require a significant computational time. The authors in [JGKT04], [GJKT04], however, give more importance to simulation time and present fast strategies to obtain reasonable initial settings. Then, once a network has been pre-optimised, it should be decided whether finer adjustments are needed. If so, the network could be tuned by means of one of the previous algorithms in those areas that need further improvement.

The first mechanisms that are proposed deal with antenna azimuth adjustments. They profit by the knowledge of optimal antenna positioning in regular hexagonal layouts. Two stages are defined, the first one tunes a subset of BS so that critical spots are covered. These spots are defined as the centre of an area with an above-average distance from neighbouring BSs. The second one rotates the remaining sites (all three sectors simultaneously) in order to minimise interference in the network. This is done by analysing their position with respect to already adjusted ones. A modified version of this stage rotates each sector independently. Moreover, the criterion is also modified and a model based on forces of attraction is introduced. Specifically, if a certain area is already covered by a sector, it will generate a repulsive force so that it is minimised that other sectors point to it as well.

Regarding antenna downtilt and pilot powers, mechanisms with low computational complexity are also provided. After a best server analysis in the area under optimisation, mean elevation angles between the covered pixels and the antennas are calculated. This mean value is appropriately scaled according to the antenna type and the resulting cipher is the downtilt angle to apply. Finally, pilot powers are increased one by one so that coverage is guaranteed. These strategies can work either in a pixel manner or directly on user distributions.

The simulator platform described in [GJCT03] is used again and results are obtained for a European city scenario and a planning tool reference scenario. Table 9.1 shows capacity gains obtained by the algorithms. Despite the fact they do not achieve the optimum settings, reasonable default settings are obtained quickly.

Table 9.1 *Capacity gains obtained by the algorithms proposed in [GJCT03].*

	Capacity gain (%)	
Algorithm	European city	Reference
Azimuth (critical spots)	13.5	4.2
Azimuth (site rotation)	5.1	–
Azimuth (sector rotation)	6.8	–
Azimuth (critical spots + site rotation)	–	4.5
Tilt	8.2	10.2
Pilot	1.6	–
Azimuth (c. spots + site rotation) + pilot	9.8	–
Azimuth (c. spots + site rotation) + tilt	23.8	12.1

Along this subsection the power devoted to the pilot channels has been exploited by different algorithms and it has been shown to be a key parameter. However, the strategies used for power management of DL traffic channels will also have an important impact on the network operation. Different authors have tackled this problem usually along with other topics concerning power control. Proposals and their respective performance are discussed next.

Regarding DL power management for SHO users in WCDMA, [Pate03a] compares two algorithms available from 3GPP:

1. Conventional Transmit Power Control (CTPC), which aims at exploiting the macrodiversity effect by providing the service with more than one active link. Specifically in this work, the same level of power is assigned to all BSs within the AS.
2. SSDT. As it aims at minimising the DL interference only the best server link is active per time.

The comparison is carried out by means of dynamic simulations in a realistic microcellular scenario. Simulations for conversational and streaming users have been run separately, considering fixed E_b/N_0 equal to 7.2 dB and 6.9 dB respectively for the DL. Concerning SHO, the simulator takes the decision of including a sector in the AS of a UE according to the received power of the pilot signal. A maximum of four BSs can be included and margins are set to 3 dB and 4 dB to add and release a sector respectively. Time triggers are set to 0 because the corner effect shows a particular relevance in the simulated environment.

From Figure 9.9 it can be seen that DL quality degrades with users' speed in both cases. This effect is caused because power control commands are not instantaneous, aggravated by the sudden degradation of the QoS perceived by users turning a corner. CTPC can clearly react always better than SSDT to this effect. Differences are even more remarkable in the conversational service case. This is because interference peaks are amplified by discontinuous transmission, not present in the streaming case. Moreover, the considered admission control algorithm led the system to a more loaded situation in the conversational case; therefore contributing to a worse performance.

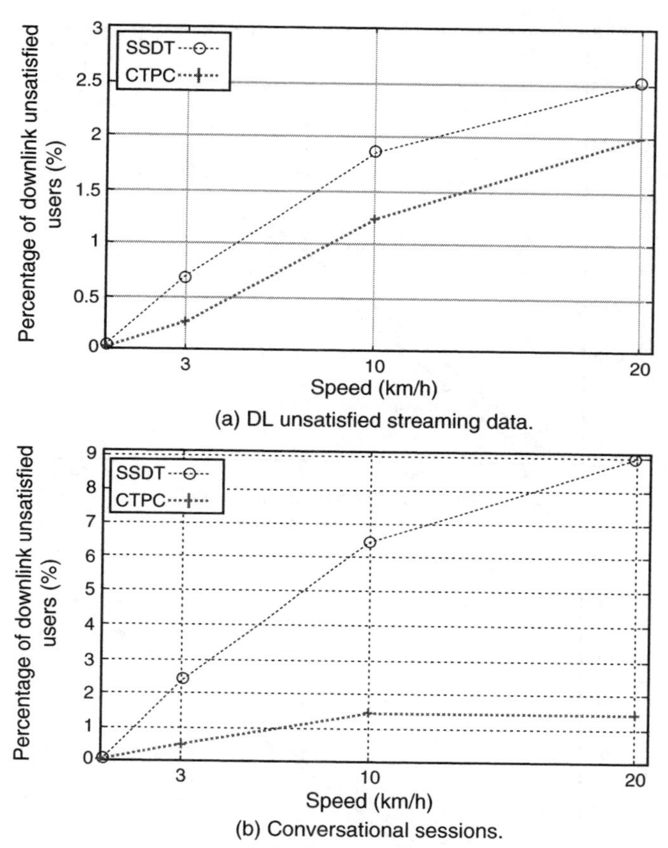

Figure 9.9 *Average percentage of DL unsatisfied streaming data and conversational sessions vs. users' speed.*

In the same line of research, [PiVe02] investigates the impact on capacity of another DL power allocation strategy. In this case the maximum fraction of transmit power per link is limited. This approach is introduced in a semi-analytical model which shows cell capacity as a function of the cell radius and the BS included in the AS. Different types of services as well as sectorised cells are considered. Subsequently, equations are introduced in a computer program and the maximum capacity per cell (sector) is found.

Some of the main assumptions that have been done to obtain numerical results are: users are assigned to cells and SHO areas according to a geometric criterion, all UEs in SHO are always connected to m BSs, a SHO gain of 3 dB has been considered, and finally only voice users are taken into account.

Figure 9.10 shows the fraction of mobiles in outage vs. the total number of users in a cell (sector) when varying the maximum allowed power. An AS size (m) of 2 has been chosen because it was found to be the optimum value under the simulation conditions. From the results, it can be concluded that reducing resources to users requiring more power implies that a larger portion is available. As a consequence, the number of served users, i.e. capacity, can be increased.

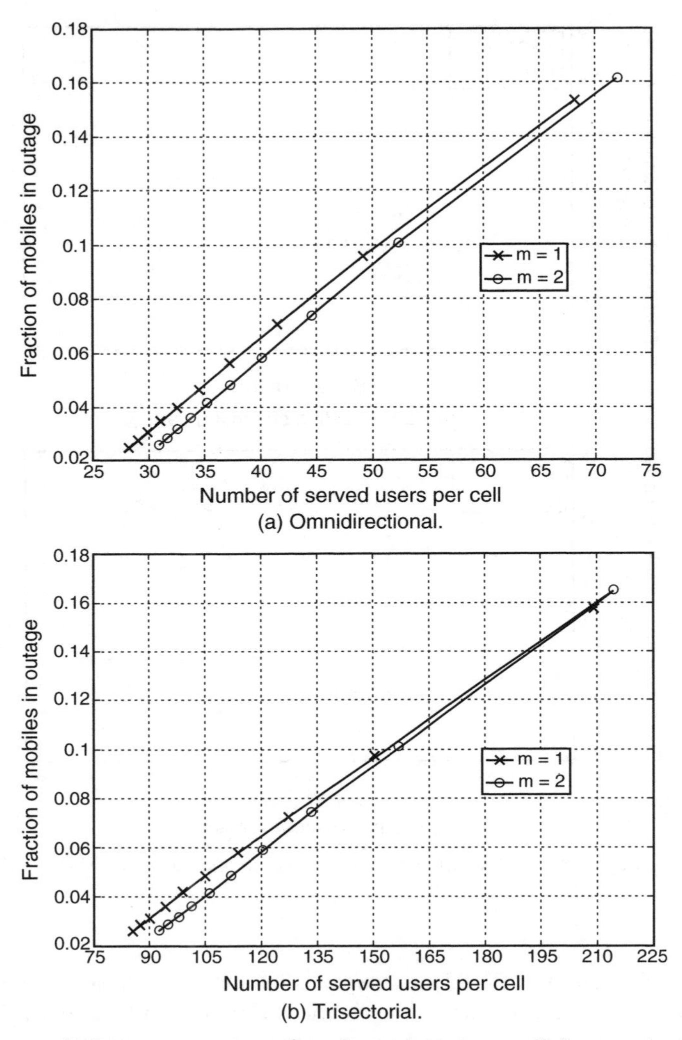

(a) Omnidirectional.

(b) Trisectorial.

Figure 9.10 *Fraction of UEs in outage vs. number of served users per cell, for $m = 1$ and 2 (AS size).*

Another power allocation algorithm for the DL but also for the UL is proposed in [RuGO02]. This paper describes a complete formulation to solve UTRA-FDD power allocation. The model makes use of a macroscopic approach, which only needs to solve a linear equation system with as many unknown variables and equations as BSs in the system. This approach is extended to include SHO users by means of an addition of extra virtual links. Moreover, as this implies a false increase in the system load, a fine adjust must be also introduced once the virtual connections have been eliminated. The resulting procedure can be equally solved in a frame-by-frame basis.

Unfortunately, in overloaded cells, when adding the extra links, the interference could be so high that the macroscopic algorithm could be unable to solve the equation system. Besides, the formulation gives no indications about which are the conflictive UEs or cells. To solve this problem a combination of macroscopic and iterative algorithms is proposed. Resolution speed is guaranteed one more time.

The model returns a set of parameters in a natural way. So, it allows calculating very quickly the load and interference factors for both UL and DL. As a consequence the algorithm could feed different admission and congestion control protocols.

Finally, some numerical results have been derived by means of simulations. UL and DL load factors, interference factors and transmitted power have been obtained for different scenarios and number of UEs. Figures highlight the importance of including realistic indoor traffic when analysing RRM algorithms and UMTS capacity.

9.3.3 Call Admission Control

In general, MAC sublayer determines which access method will be used for communication. In other words, it decides how to communicate through the radio link, i.e. which kind of duplex and which multiple access will be used. MAC tasks include quality supervision of both existing and new calls.

Call Admission Control (CAC) policies protect networks from overloading by determining whether incoming connection requests should be accepted or rejected. In third generation mobile systems, the capacity is not fixed and varies due to changes in the interference, so the design of CAC in that system has two main associated design problems:

1. to set an effective CAC threshold in order to guarantee the QoS for an integration of various service types, and
2. to achieve the maximum efficiency in the resources utilisation.

The CAC really operates at two different levels: the first one characterised by the 'packet level' constraints such as packet loss, delay jitter or average delay and the second one that allows the system capacity to be shared among the various traffic types and/or to protect the handover connections by reserving them a fraction of resources. Several researchers have addressed these issues by means of static and dynamic prioritisation and sharing scheme, associated with static or adaptive reserve of resources.

Although a CAC algorithm has to define an efficient admission policy, a variety of parameters has to be considered in defining this admission policy. Furthermore, the interference level, residual capacity, total transmitted power threshold value, the overload probability and some other criteria are also used for CAC. However, all of these schemes are restricted to specific wireless systems and limited admission policies.

Paper [HeVa02] presented a capacity analysis which allows to set CAC thresholds. Analysis has been associated to the reverse link of the WCDMA, where packet transmission is considered for both real-time and non-real-time services. A centralised demand assignment algorithm has been implemented in the packet level in order to provide QoS, in terms of lost packet and delay constraints. Performance in time variant wireless channel conditions has been evaluated considering channel-stated scheduling algorithms. Results obtained from packet level provide useful information in order to induce capacity bound. On the other hand, delay constraints associated to individual traffic classes have been included in the presented analysis in order to set a more accurate threshold for CAC.

In [AKKC02] the performance of an SIR-based CAC algorithm was investigated. The simulation model considered three different types of services (speech, video, WWW). MTs' mobility and handover functionality were also modelled. The authors assumed perfect power control conditions. An extensive analysis via computer simulations allowed to investigate the impact of several parameters on system performance, measured by blocking probability and call drop probability for each service and handover failure. These considered input parameters were: the users' speed, the offered load per service, and the SIR threshold for each service used in the CAC decision.

The paper [SAGR03] presented an approach for 3G RRM strategies evaluation, with particular applicability to admission control. The intensive GSM/UMTS co-sitting allows extract real propagation losses, mobility patterns and users' distribution from real GSM networks and applying this valuable information on 3G simulations. The presented results compare measurements and statistical-based admission control algorithms under different conditions, assessing the influence of measurement errors, discontinuous traffic sources and multirate transmission.

From the intelligence techniques used for solving the CAC problem there are four basic methods:

1. Markov decision process. The Markov chain is described by a system of probability equations. The Poisson process is a homogeneous Markov process which can be solved using a system of Kolmogor differential equations. Its solution represents a steady-state probability vector, which enables to derive probability of new call blocking and/or probability of handoff call drop-out. As was shown in [DoGo02], [DDMW03] this method is useful for solving threshold-oriented CAC. Threshold optimisation for mobile networks with one CBR service class and with a finite number of time slots can be modelled with a simple one-dimensional (1D) Markov chain model. The model complexity grows up with the number of time slots available and, in addition, depends on time slots requirements for individual service classes. The model dimensionality grows with the number of CBR classes. As was shown, the model can be simply extended to an arbitrary number of time slots and an arbitrary number of CBR classes.
2. Neural networks. In most cases some learning algorithms are applied for determination of the neurons' parameters in the literature. The learning algorithms have some disadvantages. They are either fast but leading only to local optimum (e.g. gradient methods) or they are accurate but very slow (e.g. simulated annealing). In our case we present a direct way for parameter set-up to avoid the above-mentioned drawbacks.
3. Genetic algorithms. Artificial intelligence techniques are promising for their lifelike ability to self-replicate as well as the adaptive ability to learn and control their environment. Among these techniques, Genetic Algorithms (GAs) have been used in a wide variety of optimisation tasks, including numerical optimisation and combinatorial optimisation problems as well as call admission control in wireless systems. Their ability for parallel searching and fast evaluation distinguishes themselves from other decision and optimisation algorithms. In order to solve the call admission problem, several GA-based approaches have been proposed for some specific wireless network architectures. Each local policy represents an organism in GA terminology and is a collection of bits corresponding to admit and reject decisions for the new call set-up and handoff requests at each local state of the system. A group of local policies (a community) is chosen at random initially. Each policy in the community is evaluated using evolution strategy with proper fitness function. In [Dobo02] a simple cellular network with 15 channels per cell was modelled which used a typical two parent-two offspring genetic algorithm in which each local policy represents an organism in GA terminology and is a collection of bits corresponding to admit and reject decisions for the new call set-up and handoff requests at each local state of the system. A group of local policies (a community) is chosen at random initially. Each policy in the community is

evaluated using specific strategy. Four CAC algorithms were compared for evaluating, presented in a genetic CAC algorithm:

CAC I. An algorithm without threshold, in which local and handoff call has been accepted if there was a free channel.
CAC II. An algorithm with fixed threshold, which made only handover call favourable if all channels up to threshold were occupied.
CAC III. A simple genetic CAC algorithm with fixed value of fitness function.
CAC IV. A simple genetic CAC algorithm in which all states were evaluated with respect to overall call blocking probability.

The results are summarised in Table 9.2.

Table 9.2 *The blocking probability of genetic CAC algorithms.*

Blocking probability	CAC I	CAC II	CAC III	CAC IV
Local calls	0.404221	0.746753	0.784091	0.418848
Handoff calls	0.441687	0.047146	0.079404	0.000000
Overall	0.419038	0.470069	0.505394	0.254777

4. Fuzzy logic. This type of algorithm uses a fuzzy controller to adjust the call preblocking load value with the changing traffic parameters. The fuzzy controller makes use of Fuzzy Associative Memory (FAM) to maintain the required QoS. Providing the required level of QoS, the algorithm also increases the channel utilisation.

9.3.4 Genetic algorithms and evolution strategies

The performance and the achievable capacity of the UTRAN ultimately depend on the settings of certain parameters at each BS, such as antenna type, pattern and height, antenna azimuth and downtilt, as well as the transmitted broadcast power of various channels, especially the CPICH and other common channels. Due to the special characteristics of CDMA networks [HoTo02], [LaWN02], these parameters are strongly interdependent, and their influence on the network is highly non-linear. The task of UMTS optimisation can therefore be classified as a non-linear programming problem with a very large search space. One class of algorithm especially suitable for solving such problems are evolutionary algorithms. They create an artificial environment, which models processes of natural evolution.

In literature, several approaches of evolutionary algorithms are known. In COST 273 Action, the two most prominent representatives, genetic algorithms and evolution strategies, have been adapted to the problem of UMTS optimisation. Genetic Algorithms (GAs) [Gold89] differ from other conventional techniques by operating on a group of trial solutions, called a population, in parallel. These solutions are modelled as individuals, described by chromosomes. A chromosome is the coded form of an individual (UMTS network configuration). Stochastic operators (selection, crossover and mutation) are used to explore the solution domain in search of an optimal solution. From generation to generation, according to their fitness, some of the individuals survive, while others become extinct. The fitness of an individual represents the achievable UMTS network capacity with the corresponding network configuration. Crossover and mutation operators are applied to produce new individuals for consecutive generations. With the crossover operator, a new child is produced from two parents' chromosomes. The mutation operator is carried out chromosome by chromosome to introduce new or lost genetic material. Mutation helps to get away from local optima. In [GeJT04] and [Gerd04], a genetic algorithm with improved genetic operators is presented and adapted for the UMTS capacity optimisation problem by taking into account the quality of the network. Additionally, this GA incorporates a local optimisation routine to improve the performance of the algorithm.

Especially for continuous parameter optimisation problems vectors, Evolution Strategies (ES) [Schw81] show good performance. In contrast to GAs, their native parameter encoding is described by vectors of real-valued numbers. Two different population sizes are used: a smaller basic population and a larger working population, which is randomly selected from the former, and onto which the operators are applied. The primary operator is the so-called Gaussian mutation, which perturbs the parameter vectors based on a Gaussian distribution. Thus, small modifications are more probable than large modifications; a principle that can often be found in nature as well. The problem-specific evolution strategies for UMTS optimisation are presented in [Jakl04]. In addition to Gaussian mutation, also a so-called wild mutation operator is investigated, which applies big changes to randomly selected parameters. Various recombination operators supplement the evolutionary changes performed on the population. In contrast to the literature, where most evolution strategies use simple greedy selection schemes, the algorithms presented in [Jakl04] have been improved with advanced selection operators taken from genetic algorithms. An important matter in the design of problem-specific evolutionary algorithms is the issue of finding optimum settings for the algorithm's meta-parameters, which strongly influence their performance. In this approach, self-adaptation mechanisms for several crucial parameters such as mutation step sizes and operator precision are used. This way, not only the initial algorithm configuration is facilitated, but the algorithm can also continuously adapt itself to the current state of the optimisation process. A special control mechanism monitors the algorithm's progress and automatically terminates optimisation, as soon as no further significant gain can be expected.

The only weakness of these evolutionary algorithms (GA and ES) is the considerable execution time: depending on the size of the scenario, a single optimisation may run for up to several days. However, this disadvantage is alleviated not only by the ever-increasing available processor speed, but also by efficient parallelisation approaches. In terms of optimisation gain, the referred algorithms clearly outperform various simple heuristics, such as rule-based algorithms [GJCT04], [GeJN03] and analytical strategies [JGKT04]. On typical scenarios, a capacity gain of more than 50% is shown. Detailed optimisation results on different network scenarios can be found in [Gerd04] and [Jakl04].

9.4 UTRAN performance evaluation

9.4.1 Static vs. dynamic

Estimating system performance for a given network deployment is unfeasible by means of an analytic approach due to the inherent complexity of the WCDMA system. In fact, system capacity depends on the environment, the network deployment, the offered load and performance of RRM algorithms. For this reason great efforts have been devoted by researchers in the development of system simulators, which are able to calculate the effects of plenty of these factors.

System simulators can evaluate system performance according to two different methods: static and dynamic simulations. The former produces snapshots of network performance when resource allocation reaches an equilibrium state. The latter reproduces system behaviours for a time interval of network working. The majority of the contributions to the COST 273 Action dealing with radio network aspects reports results which have been achieved by means of simulation tools. Detailed descriptions of the operation performed by such tools and of the results that can be obtained are reported in [HBWL01] and [Pate03b].

Static tools apply the well-known Monte Carlo method: users are spread all over the simulation area and radio resources are allocated to each of them. Power iterations are applied to the admitted users to maintain the ongoing sessions with the minimum resources at the target quality. When the resources' usage has converged to a stable state, a snapshot of the system performance is taken and results are added to the Monte Carlo simulation results. This process is repeated for a certain number, typically hundreds, of times to increase the confidence of the results then the Monte Carlo simulation results are processed for providing average performance of the simulated network. Usually this kind of tool provides fast and reliable results implementing simplified versions of the RRM algorithms. Static tools have been used for implementing and verifying the efficiency of a network deployment in the planning process since the implementation of second generation networks.

Dynamic tools are time driven: sessions are born and die for the whole duration of the simulated time interval of the network operation. Users are able to move in the simulation area. Resources allocation is continually updated according to the changing condition of the radio channel, so the resource usage varies from a transient state to another. The simulation duration, typically hundreds of seconds, depends on the required confidence of the results. Usually this kind of tool provides reliable results implementing detailed RRM algorithms. Those characteristics make those tools useful for individuating possible causes of system instability, for estimating the QoS provided to packet services in terms of packet delay, for evaluating performance of RRM algorithms and, moreover, for measuring the impact of dynamic effects (e.g. user mobility and system delays) on system performance. The influence of dynamic effects on system coverage and capacity has been investigated since the CDMA access technique was considered for providing commercial services. For this reason dynamic tools have been developed relatively recently.

The characteristics of such tools are sometimes combined in order to obtain semi-static (or semi-dynamic) simulators. For instance, snapshots in a Monte Carlo simulation are obtained from independent traffic realisations of the same process, while in a semi-static simulation users' position in subsequent snapshots could depend on each other in order to take into account the effect of user mobility.

In general, system evaluations could be carried out considering different statistic approaches, system effects and degrees of accuracy in the implementation of system functionalities, and results are produced at different computational costs. For these reasons, in the process of network planning, both static and dynamic tools can be applied to obtain an efficient network exploiting the swiftness of static tools and the accuracy of dynamic simulators.

In [BaGP04] the network planning is carried out by means of analytical methods, static and dynamic simulations in a synthetic macrocellular scenario. The study focuses on the comparison between the network performance estimated by a static and a dynamic tool. This evaluation has been separately carried out for the speech and streaming data services and results show that the static simulator tends to overestimate the system capacity with respect of the dynamic tool. The difference is more evident in the speech service case where the alternation of silence and activity characterising this service superimposes onto the effect of user mobility. The average number of active sessions vs. the average downlink power is reported for the streaming data service considering about 5, 7.5 and 10 offered sessions per cell in Figure 9.11(a) and for the speech service considering about 24, 36 and 48 offered sessions per cell in Figure 9.11(b).

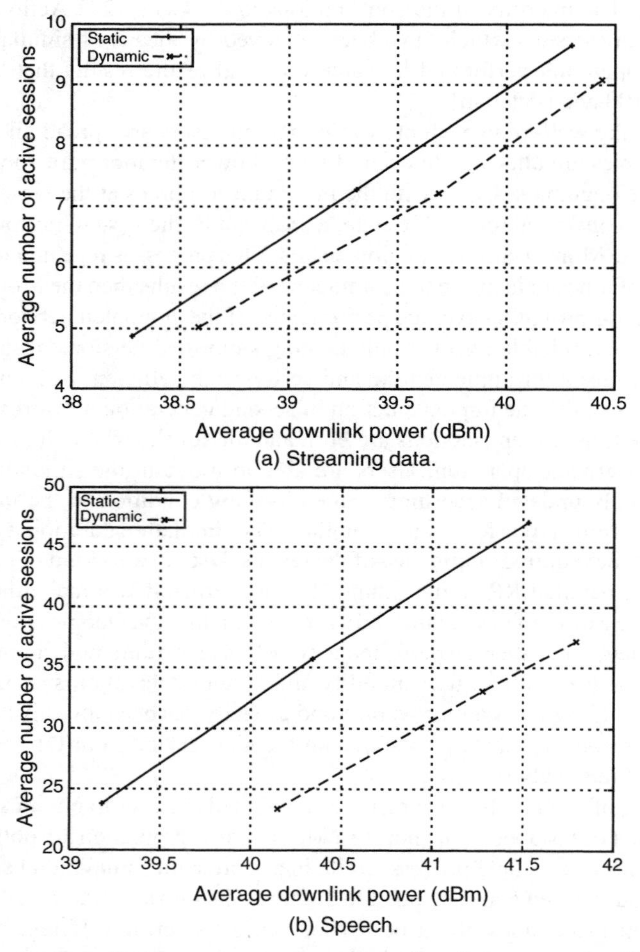

Figure 9.11 *Average number of active sessions vs the average downlink power per cell for static and dynamic simulations.*

The dynamic simulator estimates that about 30% additional downlink power is needed in order to maintain the same number of speech sessions. Conversely, for a given downlink power, the dynamic simulator estimates about 30% reduction of system capacity.

9.4.2 Semi-analytical

Analytical models were successfully applied in the network dimensioning of second generation systems. Given the system requirements in terms of coverage percentage, the characterisation of network devices and MTs and applying a propagation model suitable for the description of the considered environment, the application of a link budget resulted in a reliable evaluation of the system capacity.

Coming to third generation networks the complexity of the planning process has been considerably increased due to plenty of services that should be provided and network behaviours that should be managed. In fact, the performance of third generation systems depends on the mutual interaction between coverage and capacity that results in the soft capacity. The analytical models take into consideration such an effect by means of the interference margin that depends, among others, on the ratio between intercell and intracell interference. This parameter and the path loss, the shadowing effect and soft handover margin strongly influence the network performance and depend on the position of both users and BSs. For this reason a statistical characterisation of such parameters is unfeasible and they are evaluated by means of software simulations. The interaction between analytical models and software simulations produces semi-analytical models.

In [MuVe04] the results of a downlink capacity evaluation performed by means of a semi-analytical model are reported. This model exploits the accuracy of snapshot simulations for the calculation of the position of mobile users and realisations of the shadowing process. The model is applied both in a synthetic scenario and in a real-world scenario and evaluates the system capacity and the performance of soft handover by extending the evaluation carried out in a cell. This real-world scenario is a modification of the Turin scenario developed within the MORANS initiative (see Section 8.2), the propagation is modelled by fitting over the path-loss values of each mobile station towards the selected BS. Results have been adopted for the purpose of further developments of the MORANS Reference Scenario [MORA04] developed within the COST 273 Action.

In [EiGe05] an optimisation model exploiting both an analytical approximation of network performance based on average traffic and an efficient local search algorithm is presented. In this approach static simulations are replaced by the average-based evaluation and network behaviours are described by coupling matrices. In Subsection 9.2.3 detailed descriptions of such methods are reported. System evaluations have been carried out in the real-world scenario of Berlin from the MOMENTUM project [MOME04]. Starting from this reference scenario the local search algorithm is applied to reduce the number of deployed sites and vary tilts and azimuths. Then the matrix design approach is applied to the new network configuration. Figure 9.12 shows the blocking rate obtained by means of snapshot simulations for the three network configurations.

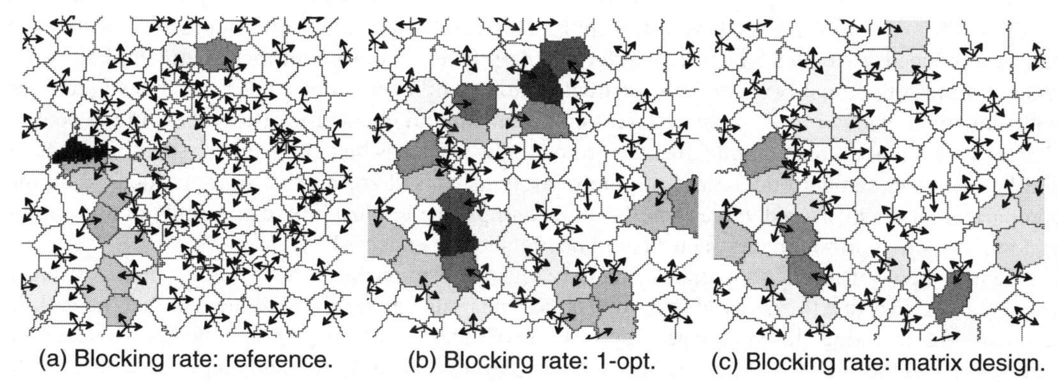

Figure 9.12 *Comparison of average evaluation and snapshot simulation.*

The overall percentage of blocked users is roughly the same for both the reference network and the optimisation result. Finally, the load reduction achieved by the matrix design optimisation significantly reduces blocking probabilities.

9.4.3 Effect of link level on performance

Plenty of parameters are usually included in link budgets as well as in system simulations in order to evaluate the network coverage and capacity. Some of these parameters are meant for considering the performance of the physical and data link layer procedures in typical mobile communication scenarios. Thus the efficiency of code, scrambling, spreading and modulation are held in consideration in system performance evaluations by means of link level parameters (e.g. spreading factor and target E_b/N_0) which describe the bearer configuration under investigation. In order to dimension such parameters, an accurate description of protocols and operations has to be modelled so these evaluations are often carried out by means of link level simulations. The characteristics of MTs and network devices (e.g. noise figure and link loss) act as input parameters for coverage and capacity evaluation too and the accuracy of the results obtained at the system level cannot be achieved, leaving them out of consideration.

The system capacity of a WCDMA network can be increased with the adoption of smart antennas. If these antennas are used with a specific beamforming, the channel estimation is performed on the dedicated pilot bits. This procedure is less accurate than the channel estimation performed on the pilot channel. Thus the use of such antennas could result in a higher quality requirement for the dedicated channel and the capacity enhancement provided by them could be lowered. The accuracy of the channel estimation performed on the dedicated pilot bits depends on the power of these bits. In [BaFW03] the optimum power ratio between the dedicated pilot bits and the information bits that minimises the required E_b/N_0 is investigated by means of a link level simulator.

Figure 9.13 shows the required E_b/N_0 for a BLER of 1% if the receiver uses the dedicated pilot bits for channel estimation, with a ratio of intracell and intercell interference (G) of 3, 6 and 10 dB and various power ratios between dedicated pilot bits and information bits.

The optimum power ratio between dedicated pilot bits and information bits is 4 to 7 dB. Using these power ratios reduces the required E_b/N_0 for a BLER of 1% by more than 1.5 dB compared to a system using equal power for all bits in the dedicated channel.

The effect of multipath propagation is detrimental to the separation of multiple users provided by the OVSF codes in the WCDMA downlink. This results in the so-called intracell interference and in a reduced system capacity. This effect is often taken into account in system simulations by means of the orthogonality factor. The characterisation of this parameter has been investigated for both statistical channel models for different environment and channel measurements in the central Oslo area. Simulation results show that the orthogonality factor is higher for higher spreading factors and has a higher standard deviation in more densely urbanised areas. This study is focused on the comparison of the measured values and the models implemented in a link level simulator and shows that the measured orthogonality factor could be higher than expected.

The efficiency of terminal antennas is often considered in terms of body loss, which is usually defined as the loss due to energy absorption and antenna mismatching due to the fact that the mobile terminals are operating close to the head or the body of the user. The impact of body loss on coverage and capacity of the WCDMA network has been assessed in [Glaz03]. The study highlights that effects of the body loss on system performance are more significant at lower network loads.

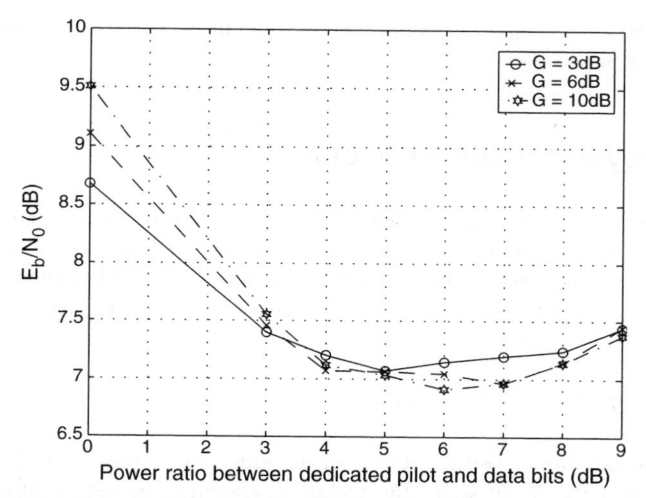

Figure 9.13 *Required E_b/N_0 using different power ratios between dedicated pilot bits and information bits if the dedicated pilot bits are used for channel estimation.*

The effects of codes on the system capacity for packet data services on the Downlink Shared Channel (DSCH) have been investigated in [BCCF01]. The trade-off between the protection of information that results in a lower BLER, and thus in a reduced number of retransmissions, and the code redundancy that involves a lower throughput for each transferred packet is addressed.

In [KaSR05] measurements of downlink dedicated channel BLER in different mobility scenarios are presented. The study illustrates a model for the characterisation of error statistics on the dedicated channel and compares the results with the live network measurements. In Figure 9.14 are reported the measurements for six different mobility scenarios in Vienna.

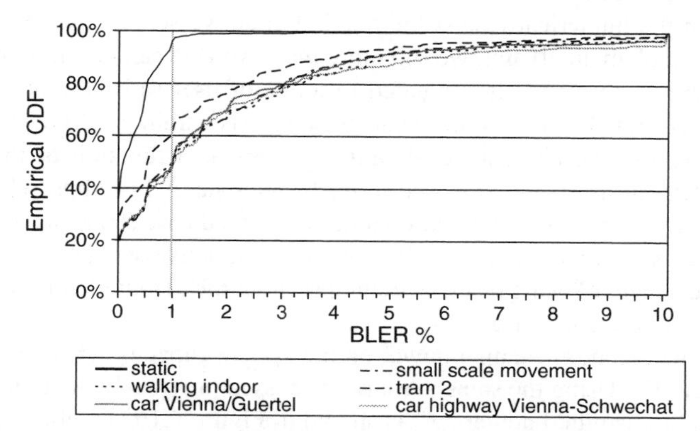

Figure 9.14 *Empirical cumulative distribution function of the block error rate for different mobility scenarios.*

When moving and tilting the mobile at a small scale by hand while still sitting at a table, for example, the power control is not capable of compensating the channel variations in order to reach the quality target of 1% in 30% to 50% of the calculation intervals. In particular, statistics of BLER are almost identical in the case of small-scale movements and in those statistics which can be measured with all other types of movement.

9.5 3G and beyond: network evolution

This section deals with UMTS evolution and in particular describes some additional features that are expected to be implemented in the next few years: beam switching, smart antennas and HSDPA. Furthermore, some new concepts such as Common Radio Resource Management (CRRM) and the Always Best Connected (ABC) paradigm will be discussed and investigated.

9.5.1 Beam switching techniques

Among the possible additional features that can be added to a conventional UMTS network for increasing system capacity, beam switching techniques seem to be very promising as they do not require additional hardware at the terminal user site. The beam switching scheme described here has been proposed and analysed in [BaNB01] and [BaBo02]. In the proposed scheme, which uses ULA, each beam acts like an ordinary BS: that is, it has its own scrambling code, common pilot, synchronisation and paging channels. The main characteristic of this proposal is that only one additional transceiver chain per sector is necessary. The performance improvement obtainable with this beamforming scheme has been evaluated through the use of a snapshot-based system level simulation whose details are discussed in [BaNB01]. The simulation area is composed of 19 sectorised hexagonal cells (three sectors per cell), user terminals are supposed to be uniformly distributed in the whole area. To avoid border effects, only the central cell is considered in the numerical results. Both intercell and intracell interference are considered.

To emphasise the potential benefits of using this kind of beamforming scheme, off-the-shelf sector antennas with 65 degree 3 dB beamwidth have been considered in the numerical results. Soft handover is considered too, in the numerical results the maximum active set size has been set equal to one (no soft handover is implemented) and two (no more than two BSs can be simultaneously connected with an MT). System performance is evaluated in terms of user satisfaction; a user is said to be satisfied if the received signal-to-interference and noise ratio is within 0.5 dB of its predefined quality target. Figure 9.15 shows the percentage of satisfied users as a function of the number of users in the evaluation area; all users are supposed to exchange data at 144 kbps. The figure shows the performance improvement over the conventional three sectored case (taken as reference) when four and six elements are used in the ULA for different values of active set size. If we consider 95% of user satisfaction, the benefit of using beam switching is about 69% (four elements per ULA) and 82% (six elements) over the reference case.

The influence of the common channel power on the system capacity when using beam switching is addressed in [BaBo02]. Using the same scenario discussed previously the impact of the choice of the common-channel transmitted power, p^c, is shown in Figure 9.16. To obtain this figure, a value of 95% of user satisfaction has been considered, with p^c as a parameter ranging from 1 to 3 W. Figure 9.16 shows that the influence of the common-channel power increases when the number of beams increases. As a matter of fact, a decrement of p^c from 2 to 1 W does not bring about a substantial change in the number of users in the reference case but becomes significant when the number of beams per sector increases.

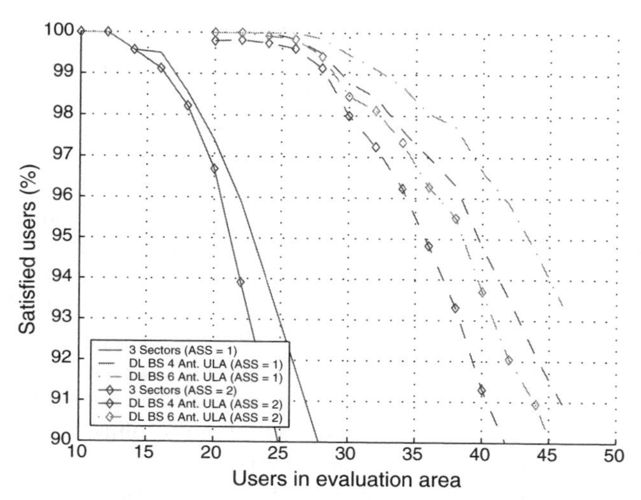

Figure 9.15 *Percentage of satisfied users vs. the number of users in the evaluation area.*

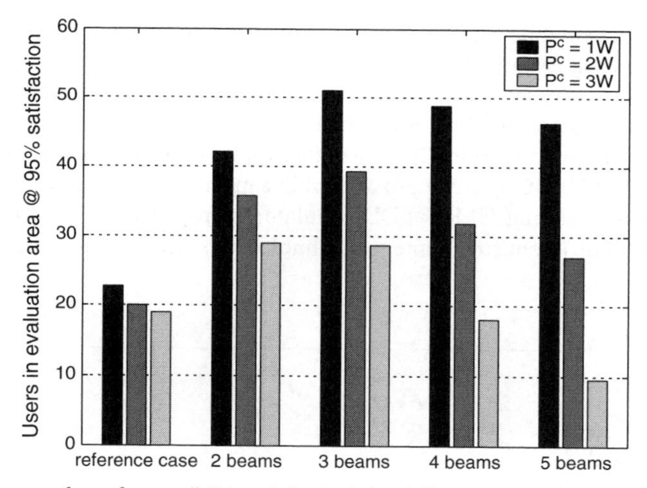

Figure 9.16 *Average number of users (95% satisfaction) for different BS configurations.*

9.5.2 Smart antennas

In recent years, Smart Antenna Systems (SAS) were introduced in order to improve the quality and performance of the wireless networks. An SAS employs a spatio-temporal processing technique such as beamforming to reduce the intercell as well as intracell interference in cellular wireless systems thereby increasing the system capacity significantly. As a result, SAS can be used in interference limited systems like UMTS for capacity improvement. However, much attention has been focused on the link level benefit of the SAS. The system implications have received less attention and are not immediately obvious owing to the complex nature of the adaptive antenna array processing technique and its interaction with the system features such as multiple access technology, power control, and receiver architecture. The only realistic approach for evaluating the system performance of smart antenna techniques for UMTS is through means of simulation. There are a number of commercial UMTS system simulators available, but none of them models the spatial channel response required to evaluate smart antenna techniques.

A dynamic system level simulator is developed in [HaCA03] for investigating the capacity improvement in a UMTS-FDD system employing smart antennas at the BSs. In this work, downlink beamforming with power control is carried out for dedicated channels like the DPDCH while broadcast channels like the CPICH are used for mobile station to BS assignment and soft handover. Dynamic system level simulations are carried out according to 3GPP specifications using simple mobility and traffic models in order to determine overall gain in downlink capacity. The downlink capacity is measured by the number of mobile stations that can be supported in downlink while maintaining the minimum QoS in terms of E_b/N_0 at all mobile stations. The downlink beamforming is carried out based on transformation of the spatial covariance matrix from uplink to downlink frequency. The number of mobile stations (also called active users [HaCA03]) that can be supported in downlink with a data rate of 384 kbits/s for different numbers of antenna elements ($M = 1, 2, 4,$ and 8) using beamforming is shown in Figure 9.17. A UCA is employed at each BS. After a transient the number of mobile stations that can be supported in the downlink remains more or less constant. This indicates that even if new mobile stations are generating calls in each snapshot, the number of mobile stations that can be served is limited. Thus the downlink capacity of the system reaches saturation. The fluctuations in the curves are because of a limited number of simulation runs that are averaged. It can be observed that as the number of antenna elements increases, the downlink capacity also increases. Although the omnidirectionally transmitted CPICH channel causes severe interference to directionally transmitted DPDCH, the improvement in capacity due to the higher number of antenna elements is significant. A similar UMTS-FDD simulator is proposed in [BrSP03] for comparing the relative performance of different smart antenna technologies such as switched beams, phased arrays,

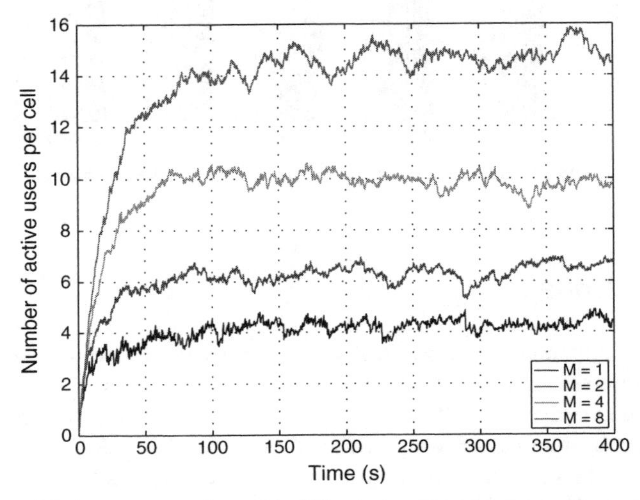

Figure 9.17 *Number of active users vs. time with different number of antenna elements.*

and adaptive arrays. Although not all of the real-time UMTS parameters are included, the simulator is flexible and allows different parameter settings such as asymmetric traffic in uplink and downlink. The main task of this simulator, given the network layout with mobiles, BSs and channel propagation conditions, is to iteratively find the maximum number of users that can be supported with a given service within the system limitations. The simulation scenario consisted of seven sites with 21 base stations. The simulated network capacity [BrSP03] for specific scenarios is displayed in Table 9.3 and the capacity per BS is given in Table 9.4. For the first scenario where an adaptive antenna was only deployed at the central location, the average network capacity increased by 10% on the downlink relative to that of the sector antenna. Smart antennas at all the BSs result in more than a doubling of capacity compared to sector antennas for the given example network. Most of this increase is obtained by using the least complex form of smart antenna, the switched beam. The phased array solution is slightly inferior to adaptive arrays in the downlink. The performance of an adaptive beamformer degrades significantly if channel state information is not known exactly at the base station. The effect of errors in spatial covariance matrix estimates on uplink beamforming of a UMTS FDD system is investigated in [ChCz04]. The assumed statistical distribution of the error is used to minimise the outage probability of each mobile station in the uplink. The design leads to a convex optimisation problem that can be efficiently computed using known algorithms. The overall objective is to find a beamforming solution that is robust to errors in the covariance matrix estimate. The number of active users per main area (which consists of seven cells [ChCz04]), vs. time that can be supported in the uplink with a data rate of 384 kbits/s with the proposed robust beamforming algorithm is compared with the non-robust max-SINR method as shown in Figure 9.18 for $M = 4$. Note that for illustrative

Table 9.3 *Average network capacity.*

Scenario	Capacity (Mbits/s) Up	Capacity (Mbits/s) Down	Relative cap. (%) Up	Relative cap. (%) Down
All sector	22.76	22.25	100	100
Centre switched lobe	27.09	24.40	119	110
Centre phased array	27.85	24.73	122	111
Centre adaptive array	29.59	25.11	130	113
All switched lobe	49.82	44.16	219	199
All phased array	51.89	45.15	228	201
All adaptive array	63.13	52.17	277	235

Table 9.4 *Average capacity per BS.*

Scenario	BS capacity (Mbits/s) Up	BS capacity (Mbits/s) Down
All sector	1.08	1.06
Centre switched lobe	1.29	1.16
Centre phased array	1.33	1.18
Centre adaptive array	1.41	1.20
All switched lobe	2.37	2.10
All phased array	2.47	2.15
All adaptive array	3.01	2.48

convenience, only 10 out of 2000 sampling periods have been shown with the marks joined by straight lines. As expected, the proposed uplink beamforming method outperforms the non-robust method based upon max-SINR by supporting 25% more mobile stations on average than that of max-SINR. The receiver at the BS uses a rake receiver and multi-user detection algorithms are not used in order to avoid huge computational tasks necessary for system level simulations. Power control, soft handover along with mobility and traffic models were used while evaluating the performance of the proposed beamforming algorithm. A similar result can be shown for a higher number of antenna elements.

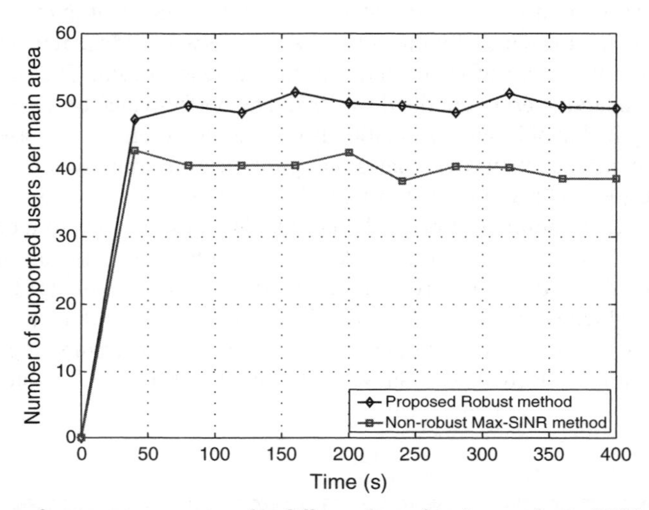

Figure 9.18 *Number of active users vs. time for different beamforming methods (UCA with $M = 4$).*

An assessment of capacity gain and/or coverage extension achieved by the deployment of smart antennas in downlink WCDMA-FDD networks together with efficient UE antennas is carried out in [Glaz04]. This study shows that BSs equipped with smart antennas and user equipment antennas with high mean effective gain together have a significantly positive impact on the system performance. Improvements obtained in hot spots are equivalent to the improvement obtained in an average cell when smart antennas are deployed in each BS of the entire network.

9.5.3 HSDPA channels: impact of multiple antennas

The HSDPA channel, which allows the downlink transmission of packet data for streaming, interactive and background services at high bit rate, represents one of the major innovations of release 5 of UMTS. The key feature of HSDPA is the concept of AM, where the modulation format (QPSK or 16-QAM) and effective code rate are changed by the BS according to system load and channel conditions. A complete description of the transmission aspects of HSDPA is given in Section 3.6; here, we focus on the impact of such a technique to network performance. This impact is evaluated in [Dött03], where several Modulation and Coding Schemes (MCS) are considered and analysed and the effect of an imperfect channel quality indicator on the link and system performance is taken into account too.

To evaluate system level results, a network simulator has been implemented. The scenario is characterised by an hexagonal grid of sectorised cells. Terminal mobility is uniformly distributed in the whole area. The simulation tool is static, that is results are based on a collection of snapshots. The state of the whole system is constant for the duration of a single Transmission Time Interval (TTI) and changes to an independent realisation from one TTI to another. Link level aspects are accounted for separately by using values of signal-to-interference and noise ratios from the system level tool. Fast fading samples are generated at the system level side. Performance of HSDPA in terms of sector throughput as a function of the number of users per sector are shown in Figure 9.19. Here, the following implementations are compared: CC used as reference as it does not require signalling, Bit Mapping Incremental Redundancy (BMIR) and Modulation Limited Bit Mapping Incremental Redundancy (MLBMIR) [Dött03], for different values of the standard deviation of the (supposed to be log-normally distributed) SIR estimation error. The figure shows that with 50 users per sector, the sector throughput is about 6.5 Mbps in the case of ideal channel feedback (standard deviation of SIR error equal to zero); whereas a feedback error of only 2 dB provides throughput degradation of about 12% (from 6.5 to 5.7 Mbps). Note that no restrictions on the available transmitted power at node B for the HSDPA channel are present in Figure 9.19. In case of restriction, the sector throughput tends to saturate and reaches a maximum for a given number of users.

The use of a dynamic simulation tool to evaluate cell throughput in the presence of MIMO is discussed in [Poll03]. The dynamic simulator used to derive the results has a time resolution equal to the UTRA FDD time slot (0.677 ms). The main features of the simulator are: inner loop power control (both uplink and downlink), mobility of terminals, soft and softer handover, propagation model taking both shadowing and fast fading into account, downlink common channels, and several scheduling algorithms.

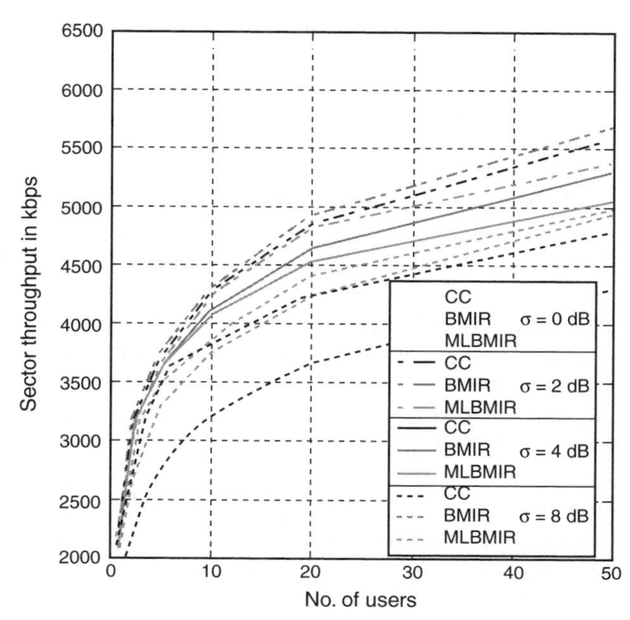

Figure 9.19 *Sector throughput as a function of number of users. System level effects of imperfect channel quality indicator.*

More details about simulation parameters can be found in [Poll03]. Results can be summarised in Figure 9.20 in which the cumulative density function of the cell throughput is shown for different receiver schemes: SISO with Round-Robin (RR) scheduling; dual rake receiver, Space-Time-Transmit-Diversity (STTD) with scheduling based on the received SIR, vertical-Bell Labs space time (V-BLAST) MIMO with Ordered Successive Interference Cancellation (OSIC). The figure shows that the STTD receiver provides a lower throughput than the reference receiver but all the other techniques improve the system performance.

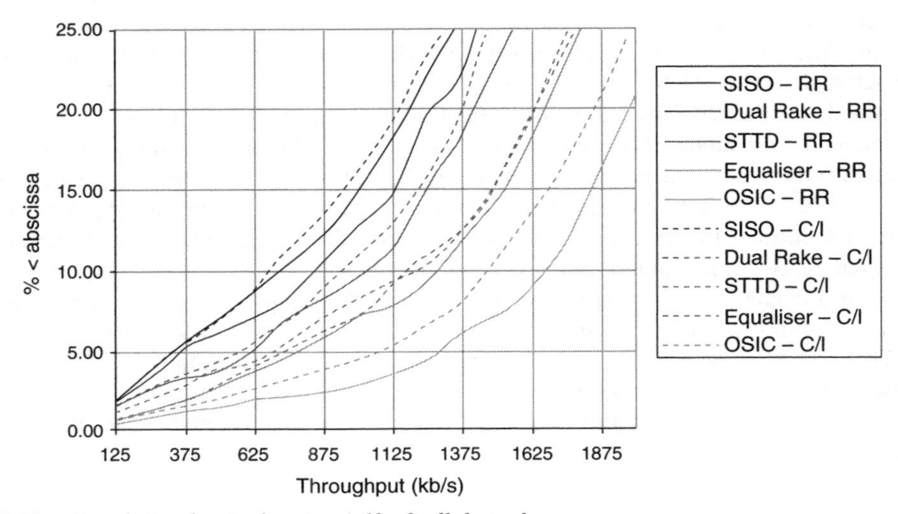

Figure 9.20 *Cumulative density function (cdf) of cell throughput.*

9.5.4 Future network scenarios

In the near future, terminals will be able to access services over a variety of radio access technologies, including 2G and 3G cellular systems and WLANs. This will require integration of networks using different technologies and transmission concepts. Among the possible new paradigms of future networks involving different wireless systems, here we focus on CRRM [Meag02] and the concept of *simultaneous use* in the convergence of wireless networks [FSCS04].

The CRRM model is based on the definition of two types of functional entities: an RRM entity responsible for RRM procedures within a particular system, and a CRRM entity in charge of the coordination between RRM entities [Meag02]. Both entities exchange signalling information, including reporting (on cell capacity and current load) as well as RRM decision support (allowing the CRRM entity to influence handover decisions). CRRM decisions may be binding for the RRM entities or simply used as assistance data. Moreover, the CRRM entity may be directly involved in RRM decisions such as handover (tight coupling) or it may just provide policies for the RRM entities and keep them up to date (loose coupling). In addition, CRRM could be implemented either as a stand-alone server or integrated with already existing BSC/Radio Network Controllers (RNCs), allowing for either centralised or distributed implementations. 3GPP has studied the usage of CRRM limited to UMTS and GSM [Meag02]. In release 5, Nokia proposed the introduction of a CRRM server using tight coupling, whereas Ericsson suggested an integrated solution based on small enhancements of existing protocols. As a result of the discussion, signalling protocols in GSM and UMTS have been enhanced to allow the exchange of load-related information, including indicators of cell capacity and current load value, for both real-time and non-real-time services. Since the release 5 solution lacks the definition of a common CRRM strategy among the different RNCs/BSCs, it could potentially lead to interoperability problems in multivendor scenarios. Siemens has submitted a proposal for release 6 based on the definition of CRRM policies and the standardisation of an open interface between RRM and CRRM entities. This policy-based approach requires the introduction of additional mechanisms for RRM decision support, including the possibility of setting a load target for each cell based on centralised CRRM policies, which can be managed effortlessly by the operator. The CRRM entity acts only as an advisor and is not directly involved in each particular RRM decision, and hence no extra delay is experienced for RRM decisions. Moreover, even in the case of a CRRM failure, the RRM entities would still be able to continue operating autonomously.

In the future, wireless communications users will claim availability of services, aiming to be ABC, i.e. to obtain the best possible performance-price ratio at any given time, neglecting the choice of a particular underlying system. To satisfy this need, integration and interworking are key aspects leading to simultaneous use concepts [FSCS04]. Different scenarios can be represented by means of coordinates in an 'orthogonal space' based on three components, Services (Se), Systems (Sy) and Operators (Op). The most simple scenario is the usage of different services via a single system (nSe, 1Sy, 1Op), for instance in the case of simultaneous transmission of voice and data across a cellular system. As a next step, different services could be served by different systems (nSe, nSy, 1Op), optimally mapping services to systems based on their characteristics, or even by different operators (nSe, 1Sy, nOp). In the latter case, a third-party broker could be in charge of operator selection on behalf of the user. A tighter integration between systems enables new mechanisms such as cooperative transmission of a service across several systems (1Se, nSy, 1Op). On the one hand, cooperative transmission may be redundant, duplicating information and aiming at improving robustness and reliability. On the other hand, it may be based on the transmission of different portions of data across different systems, either on a system basis (using several systems in parallel to increase bandwidth), on a bearer basis (mapping different bearers on different systems) or on a link basis (using different systems for uplink and downlink). With respect to handover mechanisms, it is possible to distinguish between vertical handover, in which a connection is switched to another system that best suits user needs (1Se, nSy, 1Op), and Horizontal Handover (HHO), where the handover takes place between similar systems belonging to different operators in order to keep a specific connection (1Se, 1Sy, nOp). Another concept of simultaneous use is range extension, where access to a cellular system is provided through the use of ad hoc extensions via an intermediate gateway device [FSCS04]. This is a special case (1Se, nSy, 0.5Op), where 0.5 means a combination of ad hoc and cellular. In the absence of a service provider or operator, multisystem terminals can directly communicate through the use of generalised ad hoc networks, in which different wireless systems may be simultaneously used (1Se, nSy, 0Op).

References

[ACMS03] E. Amaldi, A. Capone, F. Malucelli, and F. Signori. Radio planning and optimization of W-CDMA systems. TD(03)078, COST 273, Paris, France, May 2003.

[AKKC02] E. S. Angelou, A. G. Kanatas, N. Koutsokeras, and P. Constantinou. SIR-based uplink terrestrial call admission control scheme with handoff for mixed-traffic WCDMA networks. TD(02)093, COST 273, Espoo, Finland, May 2002.

[AmCM01] E. Amaldi, A. Capone, and F. Malucelli. Improved models and algorithms for UMTS radio planning. *In Proc. VTC 2001 Fall - IEEE 54th Vehicular Technology Conf.,* Atlantic City, NJ, USA, Oct. 2001. [Also available as COST 273 TD-01-042].

[BaBo02] T. Baumgartner and E. Bonek. Influence of the common-channel power on the system capacity of UMTS FDD systems that use beam switching. TD(02)049, COST 273, Espoo, Finland, May 2002.

[BaFW03] T. Baumgartner, A. G. Ferrás, and W. Weichselberger. Optimum power ratio between dedicated pilot bits and information bits if user specific beamforming is used in UMTS FDD. TD(03)019, COST 273, Barcelona, Spain, Jan. 2003.

[BaGP04] M. Barbiroli, P. Grazioso, and R. Patelli. Dimensioning and capacity evaluation of UMTS networks by means of system simulation tools. TD(04)029, COST 273, Athens, Greece, Jan. 2004.

[BaMV03] C. Balzanelli, A. Munna, and R. Verdone. WCDMA downlink capacity -grow part 1. *In Proc. PIMRC 2003 - IEEE 14th Int. Symp. on Pers., Indoor and Mobile Radio Commun.,* Beijing, China, Sep. 2003. [Also available as TD(03)034].

[BaNB01] T. Baumgartner, T. Neubauer, and E. Bonek. Performance of a simple downlink beam switching scheme for UMTS FDD. *In Proc. CIC 2001 - 6th CDMA Int. Conf.,* Seoul, Korea, Oct. 2001. [Also available as TD(01)022].

[BCCF01] F. Borgonovo, A. Capone, M. Cesana, and L. Fratta. Delay-throughput performance of UMTS downlink shared channel (DSCH). TD(01)021, COST 273, Bologna, Italy, Oct. 2001.

[BrSP03] L. E. Braten, A. Spilling, and M. Pettersen. A UMTS FDD simulator for smart antennas; general description and preliminary results. TD(03)052, COST 273, Barcelona, Spain, Jan. 2003.

[Butl02] G. Butler. Some issues in planning indoor TDD UMTS networks (a personal view). TD(02)089, COST 273, Espoo, Finland, May 2002.

[ChCz04] B. K. Chalise and A. Czylwik. Uplink user capacity of UMTS FDD with robust beamforming based upon minimum outage probability. TD(04)179, COST 273, Duisburg, Germany, Sep. 2004.

[CoLA03] Y. Corre, Y. Lostlanen, and S. Aubin. Multi-environment radio predictions involving an in-building WLAN network and outdoor base stations). TD(03)156, COST 273, Prague, Czech Republic, Sep. 2003.

[CoSW01] M. Coinchon, A.-P. Salovaara, and J.-F. Wagen. The impact of radio propagation predictions on urban UMTS planning). TD(01)41, COST 273, Bologna, Italy, Oct. 2001.

[CRBR02] N. Cardona, J. Reig, L. F. Bueno, D. Romero, J. C. del Rio, J. A. Pons, and P. Guerediaga. Capacity analysis in downlink WCDMA systems using soft handover techniques with SIR-based power control and site selection diversity transmission. TD(02)058, COST 273, Espoo, Finland, May 2002.

[DDMW03] L. Doboš, J. Dúha, S. Marchevský, and V. Wieser. *Mobile Radio Networks. Mobilné rádiové siete.* EDIS Press, ´ Zilina, Slovakia, 2003. [Also available as TD(03)077].

[Dobo02] L. Doboš. CAC algorithms for wireless ATM networks. (CAC algoritmus pre bezdrôtové ATM siete). *In Proc. COFAX-Telekomunikácie 2002,* Bratislava, Slovakia, Apr. 2002. [Also available as TD(04)085].

[DoGo02] L. Doboš and J. Goril. Call admission control in mobile wireless. *Radioengineering*, 11(4):17–23, Dec. 2002. ISSN 1210-2512. [Also available as TD(03)077].

[Dött03] M. Döttling. Impact of imperfect channel quality feedback and user terminal capabilities on throughput of HSDPA. TD(03)009, COST 273, Barcelona, Spain, Jan. 2003.

[EFFG03] A. Eisenblätter, E. R. Fledderus, A. Fügenschuh, H.-F. Geerdes, B. Heideck, D. Junglas, T. Koch, T. Kürner, and A. Martin. Mathematical methods for automatic optimisation of UMTS radio networks. D4.3, MOMENTUM IST-2000-28088, 2003. [Partly also available as TD-04-20].

[EFGJ04] A. Eisenblätter, A. Fügenschuh, H.-F. Geerdes, D. Junglas, T. Koch, and A. Martin. Integer programming methods for UMTS radio network planning. *In Proc. WiOpt 2004*, Cambridge, UK, Mar. 2004. [Also available as TD-04-35].

[EGKT04] A. Eisenblätter, H.-F. Geerdes, T. Koch, and U. Türke. MOMENTUM data scenarios for radio network planning and simulation (extended abstract). *In Proc. WiOpt 2004*, Cambridge, UK, Mar. 2004.

[EiGe04] A. Eisenblätter and H.-F. Geerdes. UMTS radio network tuning. TD(04)135, COST 273, Duisburg, Germany, June 2004.

[EiGe05] A. Eisenblätter and H.-F. Geerdes. A novel view on cell coverage and coupling for UMTS radio network evaluation and design. *In Proc. of INOC'05*, Lisbon, Portugal, Mar. 2005. [Also available as COST 273 TD-05-64].

[EKMA02] A. Eisenblätter, T. Koch, A. Martin, T. Achterberg, A. Fügenschuh, A. Koster, O. Wegel, and R. Wessäly. *Modelling Feasible Network Configurations for UMTS*, In G. Anandalingam and S. Raghavan, editors, *Telecommunications Network Design and Management*, pages 1–24. Kluwer Academic Publishers, London, UK, 2002. [Also available as TD(02)122].

[FSCS04] L. Ferreira, A. Serrador, L. M. Correia, and S. Svaet. Concepts of simultaneous use in the convergence of wireless systems. *In Proc. 13th IST Summit on Mobile and Wireless Commun.*, Lyon, France, June 2004. [Also available as TD(04)102].

[Fusc04] F. Fuschini. A study on urban and suburban gap fillers for DVB-H system. TD(04)201, COST 273, Duisburg, Germany, Sep. 2004.

[GaRO03] M. García-Lozano, S. Ruiz, and J. J. Olmos. CPICH powers optimization by means of simulated annealing in an UTRA-FDD environment. *Elect. Lett.*, 39(23):1676–1677, Nov. 2003. [Also available as TD(03)103].

[GeJN03] A. Gerdenitsch, M. Jakl, S., Toeltsch, and T. Neubauer. Intelligent algorithms for system capacity optimization of UMTS FDD networks. *In Proc. MCT 2003 - 4th IEE Int. Conf. on 3G Mobile Commun. Tech.*, London, UK, June 2003. [Also available as TD(03)101].

[GeJT04] A. Gerdenitsch, S. Jakl, and M. Toeltsch. The use of genetic algorithms for capacity optimization in UMTS FDD networks. *In Proc. 13th IST Summit on Mobile and Wireless Commun.*, Gosier, Guadeloupe, French Caribbean, June 2004.

[Gerd04] A. Gerdenitsch. *System Capacity Optimization of UMTS FDD Networks*. PhD thesis, Technische Universität Wien, Vienna, Austria, July 2004.

[GJCT03] A. Gerdenitsch, S. Jakl, Y. Y. Chong, and M. Toeltsch. An adaptive algorithm for CPICH and antenna tilt optimization in UMTS FDD networks. *In Proc. CIC 2003 - 8th Int. Conf. on Cellular and Intelligent Communications,* Seoul, Korea, Oct. 2003. [Also available as TD(03)162].

[GJCT04] A. Gerdenitsch, S. Jakl, Y. Y. Chong, and M. Toeltsch. A rule-based algorithm for CPICH power and antenna tilt optimization in UMTS FDD networks. *ETRI Electronics and Telecommunication Research Institute Journal*, 26(5):437–442, Oct. 2004. [Also available as TD(03)162].

[GJKT04] A. Gerdenitsch, S. Jakl, W. Karner, and M. Toeltsch. Influence of antenna azimuth in non-regular UMTS networks. *In Proc. WWC 2004 - 5th World Wireless Congress*, San Francisco, CA, USA, Jan. 2004. [Also available as TD(04)039].

[Glaz03] A. Alayon Glazunov. UE antenna efficiency impact on UMTS system coverage/ capacity. R4-030546, 3GPPTSG-RAN Working Group 4 (Radio) meeting #27, Paris, France, May 2003. [Also available as TD(03)186].

[Glaz04] A. Alayon Glazunov. Joint impact of the mean effective gain and base station smart antennas on WCDMA-FDD systems performance. TD(04)158, COST 273, Duisburg, Germany, Sep. 2004.

[Gold89] D. E. Goldberg. *Genetic Algorithm in Search, Optimization and Machine Learning.* Addison-Wesley, Boston, MA, USA, 1989.

[GrBa03] P. Grazioso and M. Barbiroli. Sensitivity analysis of UMTS performance with respect to 'add' and 'drop' parameters. TD(03)002, COST 273, Barcelona, Spain, Jan. 2003.

[GRHO04] M. García-Lozano, S. Ruiz, R. Higuero, and J. J. Olmos. UMTS network optimisation: Impact of downtilted antennas. TD(04)033, COST 273, Athens, Greece, Jan. 2004.

[GrVa01] P. Grazioso and A. Varini. The advantages of site sharing in third generation systems: a simulation-based approach. TD(01)038, COST 273, Bologna, Italy, Oct. 2001.

[HaCA03] L. Haring, B. K. Chalise, and Czylwik A. Dynamic system level simulations of downlink beamforming for UMTS FDD. *In Proc. Globecom 2003 - IEEE Global Telecommunications Conf.,* San Francisco, CA, USA, Dec. 2003. [Also available as TD(03)018].

[HBWL01] R. Hoppe, H. Buddendick, G. Wölfle, and F. M. Landstorfer. Dynamic simulator for studying WCDMA radio network performance. *In Proc. VTC 2001 Spring - IEEE 53rd Vehicular Technology Conf.,* Rhodos, Greece, May 2001. [Also available as TD(02)148].

[HeDK02] B. Heideck, A. Draegert, and T. Kürner. Heuristics for the reduction of complexity in UMTS radio network quality assessment. *In Proc. PIMRC 2002 - IEEE 13th Int. Symp. on Pers., Indoor and Mobile Radio Commun.,* Lisbon Portugal, Sep. 2002. [Also available as TD(02)110].

[HeKü05] A. Hecker and T. Kürner. Analysis of propagation models for UMTS Ultra High Sites in urban areas. TD(05)033, COST 273, Bologna, Italy, Jan. 2005.

[HeVa02] A. Hernández-Solana and A. Valdovinos-Bardají. Uplink admission control for multimedia packet transmission in WCDMA. TD(02)119, COST 273, Lisbon, Portugal, Sep. 2002.

[HoTo02] H. Holma and A. Toskala, editors. *WCDMA for UMTS - Radio Access For Third Generation Mobile Communications.* JohnWiley & Sons Ltd., New York, NY, USA, 2nd edition, 2002.

[Jakl04] S. Jakl. *Evolutionary Algorithms for UMTS Network Optimization.* PhD thesis, Technische Universität Wien, Vienna, Austria, Nov. 2004.

[JGKT04] S. Jakl, A. Gerdenitsch, W. Karner, and M. Toeltsch. An approach for the initial adjustment of antenna azimuth and other parameters in UMTS networks. *In Proc. 13th IST Summit on Mobile and Wireless Commun.,* Lyon, France, June 2004. [Also available as TD(04)039].

[KaSR05] W. Karner, P. Svoboda, and M. Rupp. A UMTS DL DCH error model based on measurements in live networks. *In Proc. ICT 2005 - 12th Int. Conf. on Telecommunications,* Capetown, South Africa, May 2005. [Also available as TD(05)016].

[KEGJ03] T. Kürner, A. Eisenblätter, H.-F. Geerdes, D. Junglas, T. Koch, and A. Martin. Final report on automatic planning and optimisation. D4.7, IST-2000-28088 MOMENTUM, 2003.

[KüFW96] T. Kürner, R. Fauß, and A. Wäsch. A hybrid propagation modeling approach for DCS1800 macro cells. *In Proc. VTC 1996 - IEEE 46th Vehicular Technology Conf.,* Atlanta, GA, USA, Apr. 1996.

[KüHe05] T. Kürner and A. Hecker. Performance of traffic and mobility models for Location Area Code planning. *In Proc. VTC 2005 Spring - IEEE 61st Vehicular Technology Conf.,* Stockholm, Sweden, May 2005. [Also available as TD-04-148].

[KüMe02] T. Kürner and A. Meier. Prediction of outdoor and outdoor-to-indoor coverage in urban areas at 1.8 Ghz. *IEEE J. Select. Areas Commun.,* 20(3):496-506, Apr. 2002. [Partly also available as TD-01-13].

[Laih02] J. Laiho. Control hierarchy in utran cellular networks. TD(02)057, COST 273, Espoo, Finland, May 2002.

[LaWJ02] J. Laiho, A. Wacker, and C. Johnson. Radio network planning for 3G. TD(02)061, COST 273, Espoo, Finland, May 2002.

[LaWN02] J. Laiho, A.Wacker, and T. Novosad. *Radio Network Planning and Optimisation for UMTS.* John Wiley & Sons Ltd., New York, NY, USA, 2nd edition, 2002.

[MaHe03] E. F. T. Martijn and M. H. A. J. Herben. Radio propagation into buildings at 1.8 Ghz - empirical characterisation and its importance to UMTS radio planning. TD(03)191, COST 273, Prague, Czech Republic, Sep. 2003.

[Meag02] F. Meago. Common radio resource management. TD(02)046, COST 273, Barcelona, Spain, Jan. 2002.

[MOME04] http://momentum.zib.de. 2004.

[MORA04] http://morans.cost273.org. 2004.

[MuVe04] A. Munna and R. Verdone. Downlink capacity: usage of a MORANS scenario. TD(04)063, COST 273, Athens, Greece, Jan. 2004.

[Orti02] M. Ortiz. UMTS radio beacon for network planning. TD(02)103, COST 273, Lisbon, Portugal, Sep. 2002.

[Pate03a] R. Patelli. Analysis of downlink power management strategies for soft handover in WCDMA. TD(03)037, COST 273, Barcelona, Spain, Jan. 2003.

[Pate03b] R. Patelli. A fully dynamic simulation tool for UMTS planning and optimization. TD(03)195, COST 273, Prague, Czech, Sep. 2003.

[PiVe02] W.-U. Pistelli and R. Verdone. Power allocation strategies for the downlink in a W-CDMA system with soft and softer handover: the impact on capacity. *In Proc. PIMRC 2002 - IEEE 13th Int. Symp. on Pers., Indoor and Mobile Radio Commun.,* Lisbon, Portugal, Sep. 2002. [Also available as TD(02)081].

[Poll03] A. Pollard. Prelimnary system level results on HSDPA with multiple antenna methods. TD(03)180, COST 273, Prague, Czech Republic, Sep. 2003.

[ReVe03] E. Reguera and F. J. Velez. Tele-traffic engineering for enahnced UMTS multi-rate applications. *In Proc. EPMCC 2003 - 5th European Personal and Mobile Communications Conference,* Glasgow, UK, Apr. 2003. [Also available as TD(02)140].

[RSAG03] S. Ruiz, O. Sallent, R. Augusti, M. García-Lozano, F. Adelanto, M. A. Diaz-Guerra, J. Montero, E. Gago, and J. L. Miranda. 3G planning using 2G measurements. TD(03)073, COST 273, Barcelona, Spain, Jan. 2003.

[RuGO02] S. Ruiz, M. García-Lozano, and J. Olmos. Influence of active set configuration on UMTS capacity. TD(02)160, COST 273, Lisbon, Portugal, Sep. 2002.

[RuOl01] S. Ruiz and J. J. Olmos. The importance of initial traffic dimensioning in UMTS planning (by the use of simulation tools). TD(01)010, COST 273, Brussels, Belgium, May 2001.

[SAGR03] O. Sallent, F. Adelantado, M. García-Lozano, S. Ruiz, R. Agustí, M. A. Díaz-Guerra, J. Montero, E. Gago, and J. L. Miranda. Analysing UMTS asmission control strategies by using GSM measurements. TD(03)096, COST 273, Paris, France, May 2003.

[Schw81] H. P. Schwefel. *Numerical Optimization of Computer Models.* John Wiley & Sons Ltd., New York, NY, USA, 1981.

[ShNi02] S. Sharma and A. R. Nix. Automatic replanning for W-CDMA situation aware networks. TD(02)096, COST 273, Espoo, Finland, May 2002.

[3GPP04] 3GPP. Radio resource management strategies. TR 25.922, V6.0.1, TSG SSA, Apr. 2004.

[UnKü04] P. Unger and T. Kürner. Scenarios for radio planning of DVB-H/UMTS hybrid networks. TD(04)151, COST 273, Duisburg, Germany, Sep. 2004.

[VePa04] F. J. Velez and R. R. Paulo. High capacity wideband traffic in enhanced UMTS: a step towards 4G. *In Proc. MCT 2004 - 5th IEE Int. Conf. on 3G Mobile Commun. Tech.,* London, UK, Oct. 2004. [Also available as TD(04)031].

Appendix A
About COST 273

Founded in 1971, European Cooperation in the Field of Scientific and Technical Research (COST) [COST05a] is an intergovernmental framework for European Cooperation in the Field of Scientific and Technical Research, allowing the coordination of nationally funded research on a European level. COST projects, designated by Actions, cover basic and pre-competitive research as well as the activities of public utility. The goal of COST is to ensure that Europe holds a strong position in the field of scientific and technical research for peaceful purposes, by increasing European cooperation and interaction within this field. COST has clearly shown its strength in non-competitive research, in pre-normative cooperation and in solving environmental and cross-border problems and problems of public utility. It has been successfully used to maximise European synergy, and has added value in research cooperation and is a useful tool to further European integration. Ease of access for institutions from non-member countries also makes COST a very interesting and successful tool for tackling topics of a truly global nature.

To emphasise that the initiative came from scientists and technical experts themselves, and from those with a direct interest in furthering international collaboration, the founding fathers of COST opted for a flexible and pragmatic approach. COST activities have in the past paved the way for community activities and its flexibility allows COST Actions to be used as a testing and exploratory field for emerging topics. Member countries participate in an 'à la carte' principle, and activities are launched on a 'bottom-up' approach. One of its main features is its built-in flexibility. This concept clearly meets a growing demand and, in addition, it complements European Union programmes. COST has a geographical scope beyond the European Union, and it also welcomes the participation of interested institutions from non-COST member states without any geographical restriction.

COST has developed into one of the largest frameworks for research cooperation in Europe and is a valuable mechanism for coordinating national research activities in Europe. Today, there are more than 220 Actions running in 13 scientific domains (ranging from agriculture and biotechnology and medicine and health to transportation and social sciences, among others), involving scientists from 34 European member countries and many institutions from 17 non-member countries. In particular, COST TIST (Telecommunications and Information Science and Technology) [COST05b] involves scientists from network operators, research institutes, universities and manufacturers, dealing with: optical communications; user requirements, including special needs; speech technology; multimedia communications; broadband networking; space and satellite networks; antennas, radiowaves propagation and system aspects; mobile and wireless communications; telecommunication software and user interfaces; and electromagnetic impact.

In the telecommunications world, there is no doubt today that mobile and wireless communications have an increasing importance, and this trend will continue in the next few years. Moreover, Europe wants to continue to play a leading role in this area, as was the case with Global System for Mobile Communications (GSM). This is also expected to be the situation with Universal Mobile Telecommunications System (UMTS), due to its importance to European industry. As a consequence,

R&D continues to be a key factor, and issues related to the next generation of mobile and wireless systems, dealing with broadband multimedia communications (with bandwidths and data rates much larger than the 3rd generation ones), are already being addressed by a large number of people in the European R&D community. It has also been recognised for many years now that better and faster results are achieved by joint efforts at the European level, rather than countries conducting their national programmes individually. Besides COST, the RACE, ACTS [ACTS05] and IST [IST05] frameworks are the result of this recognition, and many projects are being developed in the area of mobile and wireless communications within these frameworks.

The telecommunications area of COST has already in its curriculum very successful Actions, which deal with mobile and wireless communications, and have contributed to the development and standardisation of commercial systems:

- COST 207, 'Digital Land Mobile Radio Communications', Mar. 1984–Sep. 1988, which contributed to the development of GSM;
- COST 231, 'Evolution of Land Mobile Radio (Including Personal) Communications', Apr. 1989–Apr. 1996, [C23105], which contributed to the deployment of GSM1800 and to the development of DECT, HIPERLAN 1 and UMTS;
- COST 259, 'Wireless Flexible Personalised Communications', Dec. 1996–Apr. 2000, [C25905], which contributed to the deployment of DECT and to the development of UMTS and HIPERLAN/2, as well as initial inputs to the next generations of HIPERLAN and 4th generation systems.

Each of these projects has published a Final Report, which constitutes the summary of the main results achieved in it [Fail89], [DaCo99], [Corr01].

The main objective of COST 273 – 'Towards Mobile Broadband Multimedia Networks', May 2001–May 2005 [C27305], was to increase the knowledge on the radio aspects of mobile and wireless broadband multimedia networks, by exploring and developing new methods, models, techniques, strategies and tools towards the implementation of 4th generation mobile and wireless communication systems. As a secondary objective, it was intended that it should continue to play a supporting role similar to the one played by the previous Actions in the mobile and wireless communications area, that is, besides giving inputs to the development of systems beyond the 3rd generation, it was also expected that it would contribute to the deployment of systems that are more or less standardised, like UMTS and WLANs.

The Action was structured into Working Groups (WGs), three in total, within which the technical work was carried out:

- WG 1 – Radio System Aspects,
- WG 2 – Propagation and Antennas,
- WG 3 – Radio Network Aspects.

Sub-Working Groups (SWGs) have also been created, devoted to more specific topics, at the moment being as follows:

- SWG 2.1 – MIMO channel model,
- SWG 2.2 – Antenna performance of small Mobile Terminals (MTs),
- SWG 2.3 – Channel measurements,
- SWG 3.1 – Mobile radio networks reference scenarios.

The description of their activities is presented in what follows, Figure A.1 showing their interrelation.

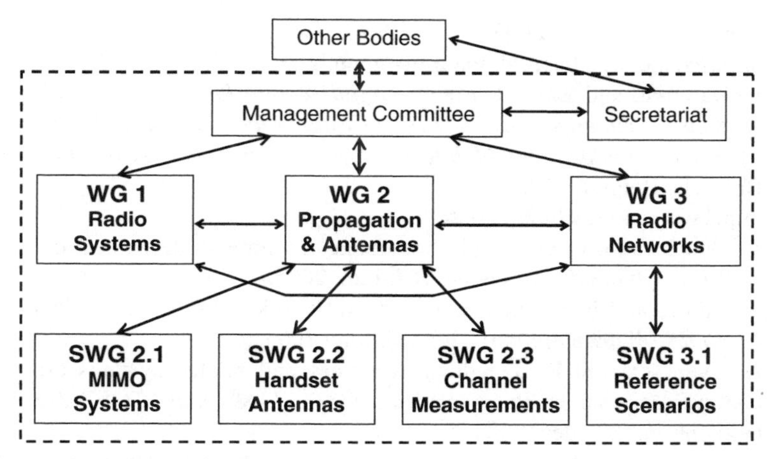

Figure A.1 *Organisation of COST 273.*

The Management Committee, together with WGs and SWGs, met three times per year, meetings circulating among the participating European countries. Besides administrative matters, which occupied a small fraction thereof, essentially these meetings were used to present and discuss Temporary Documents (TDs), which consisted of the technical contributions of each participating institution. A full list of TDs, with abstracts, is available in [C27305]. Although the documents themselves are only available inside the project, authors can be addressed directly to release their own TDs. Usually, meetings began with one or two half day tutorials, given by participants, on topics dealt within the project. At the end of each project year, a workshop was held, sometimes in conjunction with other Actions or other bodies, hence, putting together people with interest in a given research area [Vain02], [SiDe03], [SvKi04], [Verd05], [Lil05]. Each chapter of this book corresponds to some of the work developed in each WG, sometimes a section being the result of the work produced in an SWG.

More than 350 people have participated, belonging to 137 institutions from 29 countries (including four outside Europe); these numbers show the large effort involved, as well as the interest of the participating institutions in the project. Some of the participants have volunteered to summarise the results into the sections of this book. A good balance was obtained between academia and industry (roughly 50:50) which shows the interest of both in this type of close cooperation. There was definitely much to gain by participating in COST 259, since it enabled an exchange of information and cooperation among its participants that would not have been possible any other way.

References

[ACTS05] http://www.cordis.lu/acts/home.html, 2005.

[C23105] http://www.lx.it.pt/cost231, 2005.

[C25905] http://www.lx.it.pt/cost259, 2005.

[C27305] http://www.lx.it.pt/cost273, 2005.

[Corr01] L. M. Correia, editor. *Wireless Flexible Personalised Communications, COST 259: European Co-operation in Mobile Radio Research.* John Wiley & Sons, Chichester, UK, 2001.

[COST05a] http://cost.cordis.lu, 2005.

[COST05b] http://cost.cordis.lu/src/domain/tist/main.html, 2005.

[DaCo99] E. Damosso and L. M. Correia. *Digital Mobile Radio Towards Future Generation Systems Communications - COST 231 Final Report.* Brussels, Belgium, 1999.

[Fail89] M. Failli. *Digital Land Mobile Radio Communications - COST 207 Final Report.* Brussels, Belgium, 1989.

[IST05] http://www.cordis.lu/ist, 2005.

[Lil05] E. Van Lil. Towards mobile broadband multimedia networks. In *Proc. of the 5th COST 273 Workshop*, Leuven, Belgium, 2005.

[SiDe03] A. Sibille and P. Degauque. Broadband wireless local access. In *Proc. of the 2nd COST 273 Workshop*, Paris, France, May 2003.

[SvKi04] A. Svensson and P. S. Kildal. Antennas and related systems aspects in wireless communications. In *Proc. of 3rd COST 273 & Joint COST 273/284 Workshop*, Gothenburg, Sweden, 2004.

[Vain02] P. Vainikainen. Opportunities of the multidimensional propagation channel. In *Proc. of the 1st COST 273 Workshop*, Helsinki, Finland, May 2002.

[Verd05] R. Verdone. Wireless communications. In *Proc. of the 4th COST 273 & Joint COST 273 / IST-NEWCOM Workshop*, Bologna, Italy, Jan. 2005.

Appendix B
List of participating institutions

Austria

- ARC Seibersdorf Research
- Graz University of Technology
- DI Dr Hermann Bühler
- Vienna University of Technology
- Telecommunications Research Centre Vienna

Belgium

- K. U. Leuven
- Université Catholique de Louvain
- Université Libre de Bruxelles

Bulgaria

- Institute of Electronics – BAS

Canada

- Communications Research Centre

Croatia

- University of Zagreb

Cyprus

- University of Cyprus

Czech Republic

- Czech Technical University of Prague

Denmark

- Aalborg University
- Bang & Olufsen Telecom
- Maxon Telecom
- Siemens Mobile Phones

Finland

- Elektrobit
- Filtronic
- Helsinki University of Technology
- Nokia
- University of Oulu

France

- Alcatel
- DéCom – Lab. of Decision & Communication Systems
- DiBcom
- Ecole Nationale Supérieure de Techniques Avancées
- Ecole Nouvelle de Ingénieurs en Communication
- France Télécom R&D
- Institut National des Télécommunications
- Mitsubishi Electric ITE
- SAGEM
- SATIMO
- SIRADEL
- SUPELEC
- TéléDiffusion de France
- Thales Communications
- University of Lille
- University of Paris-Sud

Germany

- AWE Communications
- Dresden University of Technology
- Ilmenau Technical University
- IMST
- Konrad Zuse Zentrum fuer Informationstechnik Berlin
- Medav
- Ruhr-Universität Bochum
- Siemens
- Technical University of Braunschweig
- University of Duisburg-Essen
- University of Hannover
- University of Kaiserslautern
- University of Karlsruhe
- University of Kassel
- University of Stuttgart
- University of Ulm
- University of Wuppertal

Greece

- National Technical University of Athens
- University of Piraeus

Ireland

- Trinity College Dublin

Italy

- Fondazione Ugo Bordoni
- Politecnico di Milano
- Siemens Information and Communication Networks
- Telecom Italia Lab
- University of Siena
- University of Ferrara
- University of Padova
- University of Bologna
- University of Florence
- University of Modena and Reggio Emilia
- Wireless Future

Japan

- National Institute of Information and Communications Technology
- Tokyo Institute of Technology

The Netherlands

- Eindhoven Institute of Technology

Norway

- Telenor R&D
- The Norwegian University of Science and Technology

Poland

- National Institute of Telecommunications

Portugal

- Instituto Superior Técnico/Technical University of Lisbon
- Instituto de Telecomunicações – Aveiro
- Universidade da Beira Interior

Slovak Republic

- Technical University of Košice

Slovenia

- Institut Jozef Stefan
- Mobitel

Spain

- Telefonica Investigacion y Desarrollo
- University of Cantabria
- University of Oviedo
- University of Zaragoza
- Politecnical University of Cartagena
- Politecnical University of Valencia
- Politecnical University of Catalunya

Sweden

- AMC Centurion
- Bluetest
- Chalmers University of Technology
- Ericsson
- Flextronics
- Lund Technical University
- Perlos
- Sony Ericsson Mobile Communication
- Swedish Defence Research Agency
- TeliaSonera

Switzerland

- Elektrobit
- Swiss Federal Institute of Technology
- University of Applied Sciences in Fribourg
- Wavecall

Taiwan

- National Chiao Tung University

United Kingdom

- Centre for Telecommunications Research
- Comsearch an Andrew Company
- Durham University
- Ofcom
- Philips Research Laboratories
- Toshiba Telecommunications Research Laboratory
- UMIST
- University of Bristol
- University of Edinburgh

- University of Essex
- University of Leeds
- University of Manchester
- University of York
- Vodafone Group

USA

- Lucent Technologies
- Motorola

Yugoslavia

- University of Novi Sad

Index